Neosporosis in Animals

Neosporosis in Animals

Neosporosis in Animals

J. P. Dubey

Animal Parasitic Diseases Laboratory
Beltsville Agricultural Research Center
Agricultural Research Service
U.S. Department of Agriculture
Beltsville, Maryland 20705

A. Hemphill

Institute of Parasitology,
Department of Infectious Diseases and Pathobiology
Vetsuisse Faculty
University of Bern
Länggass-Strasse 122
CH-3012 Bern, Switzerland

R. Calero-Bernal

Animal Parasitic Diseases Laboratory
Beltsville Agricultural Research Center
Agricultural Research Service
U.S. Department of Agriculture
Beltsville, Maryland 20705

G. Schares

Institute of Epidemiology
Friedrich-Loeffler-Institut
Bundesforschungsinstitut für Tiergesundheit
Federal Research Institute for Animal Health
Südufer 10,
17493 Greifswald—Insel Riems, Germany

CRC Press
Taylor & Francis Group
Boca Raton London New York

CRC Press is an imprint of the
Taylor & Francis Group, an **informa** business

CRC Press
Taylor & Francis Group
6000 Broken Sound Parkway NW, Suite 300
Boca Raton, FL 33487-2742

First issued in paperback 2020

© 2017 by Taylor & Francis Group, LLC
CRC Press is an imprint of Taylor & Francis Group, an Informa business

No claim to original U.S. Government works

ISBN 13: 978-0-367-57370-6 (pbk)
ISBN 13: 978-1-4987-5254-1 (hbk)

Library of Congress Cataloging-in-Publication Data

Names: Dubey, J. P., author. | Hemphill, A., author. | Schares, G., author. |
Calero-Bernal, R. (Rafael), 1983- author.
Title: Neosporosis in animals / J.P. Dubey, A. Hemphill, G. Schares, and R.
Calero-Bernal.
Description: Boca Raton : Taylor & Francis, 2017. | Includes bibliographical
references.
Identifiers: LCCN 2016042634| ISBN 9781498752541 (hardback : alk. paper) |
ISBN 9781498752565 (ebook)
Subjects: LCSH: Veterinary parasitology. | MESH: Parasitic Diseases, Animal
Classification: LCC SF810.A3 D79 2017 | NLM SF 810.A3 | DDC 636.089/696--dc23
LC record available at https://lccn.loc.gov/2016042634

**Visit the Taylor & Francis Web site at
http://www.taylorandfrancis.com**

**and the CRC Press Web site at
http://www.crcpress.com**

Contents

Chapter 18
N. hughesi and Neosporosis in Horses and Other Equids 397

Preface

In the 1980s a neuromuscular syndrome of dogs simulating toxoplasmosis was recognized. In 1988, a new genus, *Neospora*, and the type species, *Neospora caninum*, were named, cultivated *in vitro*, and differentiated from *Toxoplasma gondii*. A year later, *N. caninum* was identified as an etiological agent for bovine abortions. Considerable progress in understanding the biology of neosporosis has been made in the last 30 years, resulting in more than 2000 scientific publications. The economic importance of abortion in cattle, and the availability of knowledge, reagents, and technology used to study toxoplasmosis have contributed to the rapid progress in understanding the biology of neosporosis. Whole genome sequencing of *N. caninum* confirmed close similarities between *N. caninum* and *T. gondii*. However, these 2 protozoans are biologically different: *N. caninum* causes a major disease in cattle, and canids are its definitive hosts, whereas toxoplasmosis is a major public health problem and felids are its definitive host. Both parasites have a wide host range.

Here we summarize information on the biology of neosporosis, starting with Chapter 1 on the historical background. Subsequent chapters deal with general aspects of the biology of *N. caninum* (Chapter 2), techniques (Chapter 3), and the disease caused by this parasite in cattle (Chapter 4), dogs (Chapter 5), and all other animals including primates and humans (Chapters 6 through 18).

Abortion is a worldwide problem in the livestock industry accounting for annual economic losses of billions of dollars, and *N. caninum* is a major cause of this. Neosporosis causes abortion in both dairy and beef cattle. Abortions not only occur in cattle that have been exposed recently but also in chronically infected cattle, which poses a major challenge for vaccine development. There is no effective vaccine or therapy to eliminate *N. caninum* in cattle, but progress is being made.

In this book, we provide an up-to-date account of structure, biology, clinical disease, diagnosis, epidemiology, treatment, attempts at immunoprophylaxis, and control in all hosts. There are 175 illustrations on the life cycle, structure of parasitic stages, and of lesions. More than 2100 references are cited.

It is hoped that this book will be useful to biologists, veterinarians, and researchers.

We would like to acknowledge those who made this book possible; we feel we cannot possibly list all. Camila K. Cerqueira-Cézar was a big help in compiling bibliography, and coordinating efforts among the 4 of us; we are truly grateful to her for this. Many scientists contributed illustrations, unpublished information, and help with this book; chief among them being S. Almería, I. Bjerkås, J. F. Edwards, A. L. Hattel, D. S. Lindsay, M. M. McAllister, L. M. Ortega-Mora, C. A. Speer, and W. Wouda. Many others contributed to the making of this book including D. Alves, M. Anderson, B. C. Barr, W. Basso, C. Björkman, D. Buxton, J. L. Carpenter, O. Cabezón, P. A. Conrad, F. J. Conraths, B. Daft, T. Dijkstra, J. T. Ellis, E. A. Innes, L. C. Gasbarre, C. Genchi, S. M. Gennari, B. Gottstein, C. E. Green, M. E. Grigg, W. J. Hartley, D. K. Howe, M. C. Jenkins, A. Khan, O. C. H. Kwok, M. R. Lappin, A. E. Marsh, D. P. Moore, J. A. Morales, D. O'Toole, K. Peperkamp, M. P. Reichel, J. R. Šlapeta, C. Sreekumar, J. P. Thilsted, M. J. Topper, W. Tuo, A. J. Trees, A. Uggla, C. Venturini, and I. Villena.

J. P. Dubey
A. Hemphill
R. Calero-Bernal
G. Schares

Authors

J. P. Dubey, MVSc, PhD, was born in India. He earned his veterinary degree in 1960, and master in veterinary parasitology in 1963, from India. He earned his PhD in medical microbiology in 1966 from the University of Sheffield, England. Dr. Dubey received postdoctoral training from 1968 to 1973 with Dr. J. K. Frenkel, Department of Pathology and Oncology, University of Kansas Medical Center, Kansas City. From 1973 to 1978, he was an associate professor of veterinary parasitology, Department of Pathobiology, Ohio State University, Columbus and a professor of veterinary parasitology, Department of Veterinary Science, Montana State University, Bozeman, from 1978 to 1982. He is currently a senior scientist, Animal Parasitic Diseases Laboratory, Beltsville Agricultural Research Institute, Agricultural Research Service, U.S. Department of Agriculture, Beltsville, Maryland.

Dr. Dubey has spent more than 50 years researching protozoa, including *Toxoplasma, Neospora, Sarcocystis*, and related cyst-forming coccidian parasites of humans and animals. He has published more than 1400 research papers in international journals, more than 250 of which are on neosporosis. In 1985, he was chosen to be the first recipient of the "Distinguished Veterinary Parasitologist Award" by the American Association of Veterinary Parasitologists. Dr. Dubey is recipient of the 1995 WAAVP Pfizer Award for outstanding contributions to research in veterinary parasitology. He also received the 2005 Eminent Parasitologists Award by the American Society of Parasitologists. The Thomas/Institute for Scientific Information identified him as one of the world's most cited authors in plant and animal sciences for the last decade. In 2003, he was selected for the newly created Senior Science and Technology Service (SSTS), and is one of the few scientists and executives within the USDA's Agricultural Research Service; selection for this position is by invitation only, on approval by the Secretary of Agriculture. In 2010, Dr. Dubey was elected to the U.S. National Academy of Sciences, Washington, DC, and inducted in the USDA-ARS Hall of Fame. He has made seminal contributions to the biology of neosporosis, including naming of the parasite, *Neospora caninum.*

Andrew Hemphill, PhD, was born in Lucerne, Switzerland. He grew up in central Switzerland, studied microbiology with an emphasis on cell and molecular biology at the University of Bern, and completed his PhD on cytoskeletal elements of African trypanosomes (*Trypanosoma brucei*) in 1991, at the University of Bern. He did postdoctoral training at the University College London and at the London School of Hygiene and Tropical Medicine and continued research on trypanosomiasis. In 1994, Dr. Hemphill was appointed as a researcher at the Institute of Parasitology of the University of Bern. From 2002 to 2007, he was the president of the Swiss Society of Tropical Medicine and Parasitology. He became an associate professor in 2008 and is teaching parasitology at the Vetsuisse Faculty, University of Bern. Dr. Hemphill has been an external advisor for Swissmedic since 2013. He has been acting as peer reviewer and editorial board member of several journals dealing with infectious diseases, guest-edited several special issues, and is currently an editor of *Parasitology*, an international peer-reviewed journal in the field. His main research interests relate to the development of options for the prevention and treatment of protozoan and helminth infections, especially *Neospora caninum* and *Echinococcus multilocularis*, the causative agent of alveolar echinococcosis in humans. Echinococcosis is an important public health problem and its treatment options are limited. He established a parasite culture and efficacy assessment system for drug screening activities. He pioneered studies on cell biology of *N. caninum*, including the mechanism of parasite interaction with host cells, identification, and characterization of parasite proteins for vaccine development, studies on drug targets in both parasite and host cells, and immunoprophylaxis. He is an author and coauthor of more than 200 peer-reviewed research publications, and contributed to several reviews and book chapters.

Rafael Calero-Bernal, DVM, MSc, PhD, was born in Badajoz, Spain. He earned his degree in veterinary medicine at the University of Extremadura, Spain in 2006. One year later he attended the Official Master in Meat Science and Technology at the same institution. In 2011, Dr. Calero-Bernal earned PhD in European framework in Veterinary Medicine. From 2008 to 2015 he was a professor of the Animal Health Department at the University of Extremadura, developing teaching periods at the Faculty of Veterinary Medicine of the University of Lisbon (Portugal). He has been a researcher in the Spanish National Microbiology Centre and the Tropical Medicine National Centre, both belonging to the Instituto de Salud Carlos III (Spain). Dr. Calero-Bernal has developed several research protocols at Istituto Superiore di Sanità (Italy), Instituto Nacional de Saúde (Portugal), Fundação Oswaldo Cruz (Brazil), and Centro de Referencia para el Control de Endemias (Equatorial Guinea). Currently he is a postdoctoral researcher in the Agricultural Research Service, United States Department of Agriculture. He has authored more than 50 articles and 4 books related to veterinary sciences, the latest being *Sarcocystosis of Animals and Humans*, CRC Press. His research interests are wildlife parasites and zoonoses, especially meat-borne pathogens and tissue cysts forming coccidia.

Gereon Schares, DVM, was born in Bitburg, Germany. He is a veterinary parasitologist with primary interests in the diagnosis and epidemiology of various parasitic diseases of animals, including those with zoonotic importance such as toxoplasmosis. He earned a degree in veterinary medicine in 1987 at Justus-Liebig-University, Giessen, Germany and a doctorate in veterinary medicine at the Institute of Parasitology, Justus-Liebig-University in 1992. As a postdoctoral trainee at the Institute of Parasitology and Tropical Veterinary Medicine, Free University, Berlin, Germany, Dr. Schares worked on the diagnosis and typing of African trypanosomes. In 1995, he joined the Institute of Epidemiological Diagnostics and the Institute of Epidemiology, Federal Research Centre for Virus Diseases of Animals, Wusterhausen, Germany as an independent scientist. There he initiated research on tissue cyst forming coccidia, including *Toxoplasma*, *Neospora*, *Sarcocystis*, and *Besnoitia*. He is a senior researcher at the Institute of Epidemiology, Friedrich-Loeffler-Institute, Federal Research Institute for Animal Health, Greifswald-Insel Riems, Germany, since 2013, and the head of the National Reference Laboratories for Toxoplasmosis and Dourine in Germany. He has authored and coauthored more than 150 peer-reviewed research publications. He has made seminal contributions to the diagnosis of neosporosis and besnoitiosis.

Abbreviations Commonly Used

GENERAL

CNS	central nervous system
CSF	cerebrospinal fluid
DPI	day post inoculation
ELISA	enzyme-linked immunosorbent assay
h	hour
HE	hematoxylin and eosin
IFAT	indirect fluorescent antibody test
IFN-γ	interferon-gamma
IG	Intragastric
IgG	immunoglobulin G
IgM	immunoglobulin M
IHC	immunohistochemical staining
IMB	immunoblot
IM	intramuscular
IP	intraperitoneal
ITS	internal transcriber space
IV	intravenous
KO mice	interferon gamma gene knockout mice
LM	light microscopy
min	minutes
NAT	*Neospora* direct agglutination test
PAS	periodic acid Schiff reaction
PBS	phosphate buffered saline
PCR	polymerase chain reaction
PI	post inoculation
PV	parasitophorous vacuole
PVM	parasitophorous vacuolar membrane
RAPD	random amplified polymorphic DNA
RBC	red blood cells
RFLP	restriction fragment length polymorphism
SC	subcutaneous
SCID	severe combined immune deficiency syndrome
SEM	scanning electron microscopy
TEM	transmission electron microscopy
TNF	tumor necrosis factor

FOR ELECTRON MICROSCOPY

Amylopectin granules	am
Conoid	co
Golgi body	go
Ground substance layer	gs
Inner membrane	im

Micronemes	mn
Micropore	mp
Microtubules	mt
Mitochondrion	mc
Nucleus	nu
Outer membrane	om
Plasmalemma membrane	pm
Rhoptries	rh
Subpellicular microtubules	st

CHAPTER 1

History of *Neospora* and Neosporosis

1.1 HISTORY

This section is based on the personal experience of one of us, JPD.

1.1.1 Discovery of the Organism in Dogs

Three Norwegian veterinarians, Inge Bjerkås (an anatomic pathologist), Svein Fredrik Mohn (serologist and diagnostician), and John Presthus (a neurologist, now deceased) reported in a short communication the finding of an unidentified protozoan in a litter of 6 congenitally infected pups born to a Boxer dog in Norway.[205] The pups appeared to be healthy until 2 months old. Five of these pups had neurological signs for several months. All 6 pups were examined at necropsy and were diagnosed with encephalitis and myositis with protozoa in lesions. There were numerous tachyzoites and a few tissue cysts in the brain. Ultrastructurally, tachyzoites were like *Toxoplasma gondii* but with more rhoptries (up to 11) than seen in *T. gondii*. Tissue cysts were not examined ultrastructurally. Antibodies to *T. gondii* were not found in the sera of 5 dogs by the dye test but the dilution of the serum tested was not stated. Several attempts to culture the parasite failed.

The report of Bjerkås et al.[205] remained vague until one of us, JPD, requested a tissue slide of the brain of the affected dog for inclusion in the book on *Toxoplasmosis of Animals and Man* that was being prepared.[527,563] Because of no reply from Dr. Bjerkås, JPD examined tissue sections and case histories from all dogs and cats that had died of toxoplasmosis-like illness from 1952 to 1987 and were archived at the Angell Memorial Animal Hospital (AMAH), Boston, Massachusetts, USA. The AMAH is the largest hospital for dogs and cats in USA and meticulously keeps records of pathology cases. Dr. James Carpenter, one of the pathologists at AMAH, and JPD examined thousands of slides from dogs, and also cats, and reached a conclusion that the syndrome recognized by Bjerkås et al. was not toxoplasmosis. In addition to neuromuscular clinical signs, the dogs suffered severe disease involving the heart, lungs, liver, and the skin. The parasite in dogs did not react in immunohistochemical tests to *T. gondii* antibodies; sera were not available for antibody determination. The presence of thick walled (up to 4 μm thick) tissue cysts was considered an identifying feature and inspired naming the organism as a new genus and species *Neospora caninum* in collaboration with Drs. Carpenter, Speer, Uggla, and Topper.[529] The focus of this paper "Newly recognized fatal protozoan disease of dogs" was to draw attention to the new clinical syndrome[529]; and the name *N. caninum* helped to achieve that. Thanks are due to Dr. A. J. Koltveit, then the editor of the *Journal of the American Veterinary Medical Association*, for making an exception to publish our article with new taxa and numerous histopathology illustrations in a clinical journal.

He also advised JPD to separately publish findings of neosporosis and toxoplasmosis in dogs and cats. Clinically, neosporosis was found to be a primary disease of dogs versus canine toxoplasmosis often associated with tumors or canine distemper virus infection.[536] Additionally, neosporosis was not found to be a disease of cats.[553]

The name *N. caninum* aroused considerable scientific controversy because the description was based on parasites in tissue sections, and many scientists felt that it should have been a species of *Toxoplasma*. In collaboration with many scientists around the world the parasite was redescribed and specimens were deposited in museums.[589]

1.1.2 *In Vitro* Cultivation of Viable *N. caninum*

Luck, opportunity, perseverance, and confidence in one's own findings are an integral part of discovery; all of these were a factor in the discovery of *N. caninum*. JPD contacted practicing veterinarians and pathologists in USA for help to send tissues from paralyzed dogs for isolation of *N. caninum*. Such an opportunity arose when Dr. E. J. Stanley, a veterinarian in Pennsylvania, telephoned JPD that one of his clients had a litter of dogs with hind limb paralysis. He also indicated that the previous litter from the same bitch had died of toxoplasmosis-like illness.[404] JPD became very interested in this case because there were no previous confirmed cases of congenital toxoplasmosis in dogs in sequential pregnancies. JPD contacted Dr. Arthur Hattel, a veterinary pathologist at the Animal Diagnostic Laboratory, University Park, Pennsylvania who made arrangements for the donation of the affected litter, necropsy, and sending of fresh, unfixed tissues to JPD. By luck, Dr. David Lindsay (DL) who was a postdoctoral scientist in JPD's laboratory had expertise in cell culture. DL succeeded in growing the parasite on first attempt. For several months DL and JPD could not decide if the parasite was *N. caninum* or *T. gondii*, until bioassay results in mice became available. Twenty-five outbred Swiss Webster mice had been inoculated with homogenized tissues of the 4 affected dogs. The mice remained seronegative for *T. gondii*. Three thick-walled tissue cysts[528] were found in unstained brain smears of the brains of 25 mice; entire brains of all 25 mice had been examined microscopically. This is an example of perseverance; it was subsequently discovered that outbred mice are not susceptible to *N. caninum* unless immunosuppressed. *In vitro* cultivation of *N. caninum* made it possible to develop diagnostic tests to fulfill Koch's postulates, and induce clinical disease in several hosts within 3 years of the discovery of the parasite (Table 1.1). None of this would have been possible without the flexibility in the Agricultural Research Service, U.S. Department of Agriculture system to pursue new areas of research and the support of JPD's supervisors who allowed him to perform this boot-legged research while performing his assigned research on toxoplasmosis.

1.1.3 Linking *N. caninum* to Abortions in Cattle

Although there were isolated reports of protozoa-associated encephalitis in calves (Table 1.1), protozoa were not known to be a major bovine abortifacient. In 1987, Dr. John Thilsted, a veterinarian pathologist from New Mexico, USA contacted JPD concerning abortions in a 240 dairy cow herd where 29 (12%) cows had aborted; and 4 to 8 cows aborted in mid gestation during 5 months. Tests for bacterial, fungal, and viral causes were negative. Histologically, focal necrotic encephalitis and nonsuppurative myocarditis were the main lesions in 7 of 9 fetuses examined histologically. When an immunohistochemical test became available[1157] the slides were stained with *N. caninum* antibodies. *Neospora* parasites were found in the brain of 2 fetuses and in the kidney of 1 fetus.[1954] This was the first report of an epidemic type of abortion in cattle associated with protozoa. Subsequently, Barr et al.,[130,131] and Anderson et al.[61,62] documented that *N. caninum* is a major cause of abortion in cattle, accounting for 18% of all bovine abortions in California, so much so that some dairies went out of business because of these abortions (Table 1.2).

Table 1.1   Earliest Reports of Encephalitis or Placentitis in Animals, Now Considered Neosporosis

Host	Country	Year	Main Finding	Original Diagnosis	Original Reference	Reevaluation and Confirmation of Neosporosis
Sheep	England	1974(?)	Congenital defect, spinal cord atrophy	Suspected *Toxoplasma* congenital encephalomyelitis	854	545
Cattle	Australia	1974(?)	Congenital defect, spinal cord atrophy	Encephalomyelitis with free and intracytoplasmic *Toxoplasma*-like cysts in neurons	854	542
Cattle	USA	1985	Myelitis in 4 calves from 4 herds, born ill, recumbent	Myelitis in spinal cord with *Toxoplasma*-related protozoan cysts	1560	535
Cattle	England	1986	Born ill, recumbent	Protozoan myeloencephalitis caused by *Sarcocystis* or *Toxoplasma* species	1495	530
Cattle	USA	1982	5-month gestational age aborted fetus. Placenta and kidneys studied	Necrosis in cotyledonary villi caused by a parasite resembling *N. caninum*	1849	JPD-unpublished

(?) Uncertain, not stated in papers.

1.1.4 Retrospective Studies

1.1.4.1 Dogs in USA

After the discovery of *N. caninum* in 1988, one of the questions asked was whether neosporosis is a new disease. In late 1970s while a faculty member at the Ohio State University, JPD recalled reading Richard Piper's PhD thesis (1960, The Ohio State University, http://library.ohio-state.edu/record=b2781251~S7) where Piper mentioned finding small thick-walled tissue cysts in the retina of dogs. In an outbreak starting in 1957, 4 litters of German Shorthaired Pointers from 1 owner developed clinical illness, initially diagnosed as toxoplasmosis; 29 of the 39 dogs had pelvic limb paralysis. Six pups from 2 litters were necropsied and their tissues had been studied by several pathologists at the Ohio State University, including Koestner and Cole[1084] and Piper et al.[1620] However, Piper did not mention these thick-walled tissue cysts in a formal publication of his findings in a refereed journal.[1620] In collaboration with Drs. Piper and Koestner, JPD reevaluated the case histories and original histological sections of these 6 dogs. Paraffin sections were stained with *N. caninum* antibodies. The presence of *N. caninum,* and not *T. gondii,* was confirmed in all 6 dogs, including thick-walled tissue cysts in the retina.[541] This finding in 1957 is the earliest record of neosporosis in any host, worldwide.

1.1.4.2 Dogs in Norway

Bjerkås published details of the cases he reported initially in 1984[207–209] allowing the following conclusions to be drawn. (i) The 6 dogs were born to a single Boxer bitch from 3 successive litters, starting in 1982. (ii) The dogs developed ataxia starting at 2–5.5 months of age. (iii) Lesions were confined to the central nervous system (CNS) and muscles, and protozoa were found in the histological sections of tissues of all 6 dogs, including the thick-walled tissue cysts demonstrated by electron microscopy. (iv) There was another Saluki dog in Norway with similar disease that had died in 1967. (v) Comparison of the parasite in dogs in Norway and USA revealed that both organisms were identical. JPD invited Dr. Bjerkås to collaborate with scientists at USDA; his studies in the laboratory of Dr. Mark Jenkins resulted in the first characterization of antigens of *N. caninum* using the culture derived tachyzoites.[211]

If this case[205] had not been reported it is likely that the recognition of neosporosis and *N. caninum* would have been delayed several years.[205]

Table 1.2   History of *N. caninum* and Neosporosis

	Contribution	Reference
1	Disease first recognized in dogs in Norway, but not named	205
2	New genus, *Neospora* and the type species *N. caninum* proposed for the protozoan from dogs in USA	529
3	*N. caninum* isolated in cell culture and mice	528, 1158
4	Indirect fluorescent antibody test developed for serologic diagnosis of neosporosis	528
5	Immunohistochemical test developed to identify *Neospora* organisms in tissues	1157
6	Neosporosis identified as cause of abortion in dairy cattle	1954
7	Transplacental transmission of *N. caninum* induced in dogs, cats, sheep, and cattle	532, 533, 538, 552
8	Experimental models for neosporosis developed in mice and rats	1159, 1161
9	Drugs screened for chemotherapy of neosporosis	1160, 1162
10	The Norwegian dog parasite identified as *N. caninum*	209
11	Neosporosis recognized as a major cause of bovine abortion in California drylot dairies	61, 131
12	*Neospora* isolated from bovine aborted fetuses and disease induced in cattle with bovine isolate	136, 373
13	*N. caninum* shown to be a common asymptomatic infection in adult dairy cattle	1553
14	ELISA developed for diagnosis of neosporosis in dogs and cattle	213, 568, 1555
15	First recombinant *N. caninum* proteins produced for diagnosis	1113
16	Dog found as the definitive host of *N. caninum*	1180, 1309
17	*Neospora hughesi* proposed for the species causing neosporosis in horses	1284
18	Prevention of vertical transmission in mice by immunization of dams with killed tachyzoites	1154
19	Oral infection of cattle with *N. caninum* oocysts	445
20	First report of *N. caninum* oocysts in a naturally infected dog	164
21	*N. caninum* redescribed, specimens deposited in museums	589
22	Multilocus microsatellites typing revealed genetic diversity	1673
23	Coyote (*Canis latrans*), Australian dingo (*Canis lupus dingo*), and wolf (*Canis lupus*) reported additional definitive hosts of *N. canInum*	615, 781, 1067
24	First *Neospora* genome annotated	1698

Source: Modified from Dubey, J. P., Lindsay, D. S. 1996. A review of *Neospora caninum* and neosporosis. *Vet. Parasitol.* 67, 1–59.

1.1.5 Cattle and Other Hosts

Encephalitis in sheep and cattle associated with undiagnosed protozoa had been recorded as early as 1974 by Hartley and Bridge[854] (Table 1.1). In 1976, Bill Hartley (now deceased) from New Zealand sent paraffin blocks of his sporozoan-associated encephalitis cases in sheep and cattle for further diagnosis to JPD. Some of these turned out to be *Sarcocystis*.[622] One case in sheep and one in cattle were confirmed neosporosis (Table 1.1). After the discovery of neosporosis in cattle in 1989, JPD contacted veterinary pathology laboratories in several countries seeking tissues from aborted bovine fetuses for retrospective studies; none was available earlier than 1980. The reason given was that brains of aborted fetuses were not examined routinely. Thus, 1974 remains as the earliest case reported.

1.2 LANDMARKS IN THE BIOLOGY OF *NEOSPORA*

The information is summarized in Table 1.2.

BIBLIOGRAPHY

61, 62, 130, 131, 136, 164, 205, 207, 208, 209, 211, 213, 373, 404, 445, 527, 528, 529, 530, 532, 533, 535, 538, 541, 542, 545, 552, 553, 563, 564, 568, 589, 615, 622, 781, 854, 1067, 1084, 1113, 1154, 1157, 1158, 1159, 1160, 1161, 1162, 1180, 1284, 1309, 1495, 1553, 1555, 1560, 1620, 1673, 1698, 1849, 1954.

CHAPTER 2

General Biology

2.1 INTRODUCTION

N. caninum is a recently recognized protozoan parasite. Until 1988, it was misdiagnosed as *T. gondii*. It is structurally, antigenically, and molecularly related to *T. gondii* but these organisms are biologically distinct. A lot of information gained from the biology of *T. gondii* has been applied to *N. caninum*.

2.2 TAXONOMIC CLASSIFICATION

N. caninum is a coccidian parasite with canids as the definitive hosts, and warm-blooded animals as intermediate hosts. It belongs to

Phylum: Apicomplexa; Levine, 1970
Class: Sporozoasida; Leukart, 1879
Subclass: Coccidiasina; Leukart, 1879
Order: Eimeriorina; Leger, 1911
Family: Toxoplasmatidae, Biocca, 1956
Genus: *Neospora* Dubey, Carpenter, Speer, Topper and Uggla, 1988
Genus definition: Tissue cysts in several cell types, primarily in the neural tissues. Tissue cyst wall up to 4-μm thick, numerous bradyzoites, not separated by septa. Tachyzoites with numerous electron dense rhoptries, some posterior to nucleus. Canids (dog, coyote, wolf) as definitive hosts, and many intermediate hosts. Tachyzoites and tissue cysts in both intermediate and definitive hosts. Oocyst excreted unsporulated. Transmission by carnivorism, transplacental, and fecal. Tachyzoites, tissue cysts, and oocysts infectious to both intermediate and definitive hosts
Type species: *N. caninum* Dubey, Carpenter, Speer, Topper and Uggla, 1988

N. caninum is a coccidian parasite. Coccidia are among the most important parasites of animals, and they were the first protozoa discovered. The oocyst is the key stage of all coccidians, and their classification was based on the structure of the oocyst. Oocysts with 4 sporocysts, each with 2 sporozoites (total 8 sporozoites), are classified as *Eimeria*. Oocysts containing 2 sporocysts, each with 4 sporozoites were historically classified as *Isospora*. Coccidiosis due to *Eimeria* species is one of the most economically important diseases of poultry, cattle, sheep, goats, and many other herbivores; it is difficult to raise livestock coccidia free.

Before the discovery of the life cycle of *T. gondii,* coccidians were considered to be host-specific with a simple 1-host life cycle. Infection was confined to the intestines and usually to enterocytes.

With few exceptions, eimerians still follow this life cycle. The host becomes infected by ingesting sporulated oocysts of *Eimeria*. After excystation, the sporozoites penetrate intestinal epithelial cells and multiply asexually before forming male and female gamonts. Oocysts are produced after fertilization, and are passed in feces in an unsporulated stage. Sporulation occurs outside the host. Unlike *Eimeria* for which the life cycle has been known for many years, little was known of the complete life cycle of most *Isospora* species until 1970, when the life cycle of *T. gondii* was discovered. Until then, *Isospora* species were considered parasites of carnivores (dogs, cats) and birds and were not thought to be host specific. In 1970, *T. gondii*, a parasite previously known to parasitize extraintestinal tissues of virtually all warm-blooded hosts, was found to be an intestinal coccidium of cats and to have an isosporan-like oocyst. This finding was a major breakthrough in medical and veterinary sciences and eventually led to the recognition of several new taxa of economically important *Toxoplasma*-like parasites (e.g., *Neospora, Sarcocystis*) and discovery of their life cycles.

Historically, *T. gondii* originated probably as a coccidian parasite of felids with a fecal–oral cycle. Later on, probably driven by domestication of cats and farm animals, it's adaptation and transmission became possible by several other modes, including transmission by fecal–oral cycle, by carnivorism, and transplacentally.

From the evolutionary point of view, *Neospora* and *Toxoplasma* probably were the same fecally transmitted coccidian parasite of canids and felids and they diverged from their common ancestors around 28–57 million years ago.[794,1502a,1698] *Toxoplasma* became adapted to a domestic cycle with felids as the definitive hosts and humans and sheep as the main intermediate hosts, whereas *Neospora* diversed as a coccidian parasite of canids as the definitive host and cattle as the primary intermediate host. Both of these organisms have retained restricted definitive hosts (only canids for *N. caninum,* and only felids for *T. gondii*). Their adaptation by congenital transmission became medically important. Although <1% of *T. gondii* infections in humans are transmitted congenitally, they can cause devastating disease in congenitally infected children. On the other hand, *N. caninum* is a major cause of abortion in cattle, and it is one of the most commonly congenitally transmitted organisms of microbial infections (a large proportion of calves from *N. caninum*-infected cows are born congenitally infected). Genetically, *N. caninum* and *T. gondii* are remarkably similar.[1698]

2.3 HOSTS

2.3.1 Definitive Hosts

N. caninum has a restricted definitive host range. Dogs (*Canis familiaris*, domestic and wild including dingo), coyote (*Canis latrans*), and gray wolf (*Canis lupus*) are definitive hosts.

2.3.2 Intermediate Hosts

Many species are intermediate hosts (Tables 2.1 and 2.2). These hosts are grouped as proven hosts with isolation of viable *Neospora*, hosts from whom only parasite DNA (Table 2.2) was demonstrated, hosts in which only antibodies were found, and hosts with clinical neosporosis. In marked contrast to *T. gondii,* neosporosis is mainly a clinical disease of cattle and dogs.

2.4 LIFE CYCLE STAGES

There are 3 infectious stages of *N. caninum*: the tachyzoites, the bradyzoites, and the oocysts. These stages are linked in a complex life cycle (Figure 2.1).

Table 2.1 Natural Host Range of *N. caninum*

Viable Parasite Isolated (for Details See Chapters)	Only Parasite DNA Found (for Details See Table 2.2)	Only Antibody Found	Clinical Disease Confirmed
Axis deer (14)	Badger	Cat	Alpaca
Cattle (4)	Brown bear	Humans–Primates	Antelope
Dog (5)	Capybaras	Pig	Axis deer
European bison (14)	Ferret	Several avian species	Black-tailed deer
Gray wolf (9)	Hoary fox		Cattle
Sheep (6)	Mink		Dog
Water buffalo (8)	Mouse		Eld's deer
White-tailed deer (14)	Otter		Equids
	Pine marten		Goat
	Rabbit		Llama
	Rat		Parma wallaby
	Red fox		Pine marten
	Shrew		Red fox
	Squirrel		Rhinoceros
	Stoat		Sheep
	Voles		
	Several avian species		

2.4.1 Tachyzoite

Tachyzoites (tachos = speed in Greek) represent the rapidly multiplying stage. Their shape and size can vary (Figures 2.2a–c and 2.3). Following division, they are crescent-shaped, approximately 2×6–7.5 µm (Figure 2.2c), with a pointed anterior (conoidal) end and a rounded posterior end. In histological sections, tachyzoites are often round with a central nucleus (Figure 2.2d,f).

Tachyzoites can proliferate in almost all cell types in the body, including neural cells, endothelial cells, dermal cells, retinal cells, macrophages, hepatocytes, and fibroblasts (Figures 2.4 and 2.5). Intracellular tachyzoites are located in the host cell cytoplasm within a parasitophorous vacuole (PV), surrounded by a parasitophorous vacuolar membrane (PVM). Often the PVM is so closely butted against the host cytoplasm that it appears that the PVM is absent. Within the PV, tachyzoites are surrounded by a matrix consisting of tubules, some appearing as whorls, and granules.[132,208,529,562,573,598,1339,1599,1884] Ultrastructurally, the tachyzoite is comprised of various organelles and inclusion bodies including a pellicle (outer plasma membrane and 2 inner membranes), cytoskeletal elements (such as subpellicular microtubules, apical and polar rings, and the conoid), secretory organelles (rhoptries, micronemes, dense granules), a mitochondrion, lipid body, a Golgi complex, ribosomes, rough and smooth endoplasmic reticula, micropore, nucleus, amylopectin granules, and an apicoplast (Figures 2.6 through 2.10). The pellicle consists of 3 membranes, a plasmalemma, and 2 closely applied membranes that form an inner membrane complex (IMC). The IMC is formed from a patchwork of flattened vesicles. The inner membrane is discontinuous at the anterior tip above the polar rings, at the micropore, and at the basal complex posterior pore at the extreme posterior tip of the zoite.

There are 2 apical and 2 polar rings (Figure 2.7). The apical rings are located at the anterior tip of the parasite and consist of electron-dense material (Figure 2.7). The apical ring 1 encircles the top of the resting conoid. The polar ring 1 is an electron-dense thickening of the IMC at the anterior end of the tachyzoite. The polar ring 2 anchors the subpellicular microtubules (Figure 2.10). The conoid is a truncated hollow cone and consists of tubulin structures wound like compressed springs (Figures 2.7 and 2.8). Twenty-two subpellicular microtubules originate from the inner polar ring

Table 2.2 Detection of *N. caninum* DNA in Wild Animals

Host[a]	Country	No. Tested	No. Positive (%)	PCR Method	Remarks	Reference
Badger (*Meles meles*)	Ireland	50	0	Nested ITS1 PCR[259]	200 mg of brain tested	1918
	Great Britain	64	7 (10.9)	Nested ITS1 PCR[259]	DNA extraction from 1 g of tissue (brain, muscle, heart, lung, liver, kidney, spleen, spinal cord, blood, lymph node)	152
Bank vole (*Myodes glareolus*)	Germany	2	0	pNc5-based Taqman-real-time PCR	Brain	380
	Germany	2	0	pNc5-based Taqman-real-time PCR	Brain	380
Black rat (*Rattus rattus*)	Italy	78	0	pNc5-based Taqman-real-time PCR	Brain, skeletal muscle	2167
Brown bear (*Ursus arctos*)	Slovakia	45	11 (24.4)	Conventional pNc5 PCR[2141] Np-6/Np21, Nested ITS1 PCR[259]	Muscle, liver, or spleen; 25 mg based on kit protocol	354
Capybaras (*Hydrochaeris hydrochaeris*)	Brazil	26	3 (11.5)	Nested pNc5 PCR[2141] Np6/Np21, NP21/Np4, Np7/Np4 ITS1 sequencing	Lymph node 25 mg, liver, 25 mg, heart, 25 mg, blood	1997
Common vole (*Microtus arvalis*)	Austria	268	4 (1.5)	Conventional pNc5-PCR[1426]	Brain; 25 mg based on kit protocol	711
	The Netherlands	24	1 (4.2)	pNc5-based Taqman-real-time PCR	Brain (half brain)	1338
	Germany	198	0	pNc5-based Taqman-real-time PCR	Brain	380
Common shrew (*Sorex araneus*)	The Netherlands	9	3 (33.3)	pNc5-based Taqman-real-time PCR	Brain (half brain)	1338
	Germany	12	0	pNc5-based Taqman-real-time PCR	Brain	380
Ferret (*Mustela furo*)	Great Britain	99	10 (10.1)	Nested ITS1 PCR[259]	DNA extraction from 1 g of tissue (brain)	152
Field mouse (*Apodemus sylvaticus*)	Italy	55	2 (3.6)	Conventional pNc5 PCR[1426] Np6plus/Np21plus	Brain, kidney, muscle (25 mg)	680
Harvest mouse (*Micromys minutus*)	The Netherlands	6	1 (16.7)	pNc5-based Taqman-real-time PCR	Brain (half brain)	1338
Hoary fox (*Pseudalopex vetulus*)	Brazil	49	6 (12.2)	Conventional pNc5 PCR[2141] Np6/Np21	25 mg fox brain based on kit protocol	1450

(Continued)

Table 2.2 (Continued) Detection of *N. caninum* DNA in Wild Animals

Host[a]	Country	No. Tested	No. Positive (%)	PCR Method	Remarks	Reference
Mink (*Neovison vison*)	Great Britain	65	3 (4.6)	Nested ITS1 PCR[259]	DNA extraction from 1 g of tissue (brain, muscle)	152
House mouse (*Mus musculus*)	Ireland	197	6 (3.1)	Nested ITS1 PCR[259]	200 mg of brain tested	1918
	UK	100	3 (3.0)	Nested pNc5 PCR[174,1152,2141] Np21plus/Np6plus, Np6/Np7	Brain	953
	Australia	104	28 (27.8)	Nested PCR	Brain, heart, or liver	138
	USA	105	9 (8.6)	Nested pNc5 PCR[1426,2141] Np7/Np10, Nested ITS1 PCR[259]	Brain	1008
	Mexico	13	10 (77.0)	Single-tube nested PCR[643]	Brain, liver, heart, spinal cord (tissue section of paraffin-embedded material)	1336
	Italy	75	9 (13.8)	Conventional pNc5 PCR[1426]	Brain, kidney, muscle (25 mg)	680
	Brazil	2	0	Nested pNc5 PCR[174,2141] Np4/Np7, Np-6/Np7, Nested ITS1 PCR, and PCR-RFLP	Brain, heart (up to 250 mg)	1438
	The Netherlands	78	12 (15.4)	Nc5-based Taqman-real-time PCR	Brain (half brain)	1338
Otter (*Lutra lutra*)	Ireland	24	0	Nested ITS1 PCR[259]	200 mg of brain tested	1918
Pine marten (*Martes martes*)	Ireland	8	0	Nested ITS1 PCR[259]	200 mg of brain tested	1918
Polecat (*Mustela putorius*)	Great Britain	70	13 (18.5)	Nested ITS1 PCR[259]	DNA extraction from 1 g of tissue (brain)	152
Rabbit (*Oryctolagus cuniculus*)	UK	57	6 (10.6)	Nested pNc5 PCR[174,1152,2141] Np21plus/Np6plus, Np6/Np7	Brain, heart, liver, tongue (about 1 g of tissue sampled; amount extracted not mentioned)	954
Norway rat (*Rattus norvegicus*)	UK	45	2 (4.4)	Nested pNc5 PCR[174,1152,2141] Np21plus/Np6plus, Np6/Np7	Brain	953
	Taiwan	55	9 (16.4)	Conventional pNc5 PCR[2141] Np6, Np21	Brain	949
	Grenada, West Indies	242	96 (39.7)	Nested pNc5 PCR[1426,2141] Np7/Np10, Nested ITS1 PCR[259]	Brain	1008

(Continued)

Table 2.2 (Continued)   Detection of *N. caninum* DNA in Wild Animals

Host[a]	Country	No. Tested	No. Positive (%)	PCR Method	Remarks	Reference
	Mexico	14	10 (71.0)	Single tube nested PCR[643]	Brain, liver, heart, spinal cord (tissue section of paraffin-embedded material)	1336
	Italy	103	14 (13.6)	Conventional pNc5 PCR[1426]	Brain, kidney, muscle	680
	Brazil	112	0	ITS1 Nested pNc5 PCR[174,2141] Np4/Np7, Np-6/Np7, Nested ITS1 PCR, and PCR-RFLP	Brain, heart (up to 250 mg)	1438
Red fox (*Vulpes vulpes*)	Great Britain	83	4 (4.8)	Nested ITS1 PCR[259]	DNA extraction from 1 g of tissue (brain)	152
	Spain	122	13 (10.7)	Conventional pNc5 PCR[1152]	DNA extraction from 0.5 g of brain tissue	36
	Ireland	33	0	Conventional pNc5 PCR[2141] Np6, Np21	DNA from frozen fox brain with histological evidence for parasitic encephalitis	1439
	Czech Republic	152	7 (4.6)	Conventional pNc5 PCR[1426]	40 mg of frozen brain	958
	Romania	182	0	Conventional pNc5 PCR[2141] Np6, Np21	40 mg of brain sampled	1925
	Ireland	151	9 (6.0)	Nested ITS1 PCR[259]	200 mg of brain tested	1918
	Slovakia	137	22 (16.1)	Conventional pNc5 PCR[2141] Np-6, Np21	Brain, skeletal muscle, blood	1703a
Squirrel (*Spermophilus variegatus*)	Mexico	6	3 (59.0)	Single tube nested PCR[643]	Brain, liver, heart, spinal cord (tissue section of paraffin-embedded material)	1336
Stoat (*Mustela erminea*)	Great Britain	9	0	Nested ITS1 PCR[259]	DNA extraction from 1 g of tissue (brain, tongue, muscle, heart, lung, liver, kidney, spleen, spinal cord, blood, lymph node)	152
	Ireland	33	0	Nested ITS1 PCR[259]	200 mg of brain tested	1918
Water vole (*Arvicola terrestris*)	Austria	86	2 (2.3)	Conventional pNc5-PCR[1426]	Brain (half brain)	711
	Germany	11	0	pNc5-based Taqman-real-time PCR	Brain	380
White-toothed shrew (*Crocidura russula*)	The Netherlands	102	11 (10.8)	pNc5-based Taqman-real-time PCR	Brain (half brain)	1338
Wood mouse (*A. sylvaticus*)	The Netherlands	17	3 (17.6)	pNc5-based Taqman-real-time PCR	Brain (half brain)	1338
Yellow-necked mouse (*Apodemus flavicollis*)	Germany	1	0	pNc5-based Taqman-real-time PCR	Brain	380

[a] In addition, *N. caninum* DNA has been found in tissues from several avian species; such information has been summarized in Table 12.1 from Chapter 12 (avian).

Table 2.3 Prevalence of *N. caninum*-Like Oocysts in Feces of Dogs

Country	No. of Dogs	Type	No. Positive	Microscopic (n)	NC-PCR (n)	Bioassay	Average No. Ooocysts/g of Feces	*N. caninum* Strain Designation	Reference
Argentina	1	Pet	1	Yes	Not done	Yes	NR	NC-6 Argentina	164
Australia	132	Wild, stray	2	2	2	ND	Low	–	1070
	160	Pet	0	2	0	ND		–	92a
	36	Puppies from 5 litters	1	Yes	1	ND			1548
Costa Rica	34	Dairy farms	3	0	3	Negative	–	–	
China	212	Farms	74	Not reported	74 (34.9%)	ND	NR	–	1142
	315 fecal samples	Pet	5	ND	5 (1.59)	ND	–	–	1147
	78	Dairy farms	2	Yes	7	–	400	Nc-LY1	1649
Czech Republic	2240 (3135 fecal samples)	Various sources	2 (1 more in puppies)	Yes	Negative	ND	NR	–	1865
	1	Pet	1	Yes	1	ND	106 oocysts recovered	Cz-4	1866
	470	Army	0	Yes	No	No	NR	–	2017
Ethiopia	383	Pet 218, farm 265	37	yes	7 (4.4%)	No	NR		96
Germany	24,089	Owned (pet and others)	47	Yes, 47	Yes, 4	7	1.9×10^1–1.14×10^5	NC-GER 1–6	1805
	3	Pet	3	3	3	3	6.5×10^3–1.43×10^5	NC-GER 7–9	168
	8438	Pet, clinics	59	Yes	–	–	NR	–	163
Iran	174	89 farm, 85 house-hold	4	4	2	ND	5–10	–	1668
	428	Dairy farms	Not clear	Yes	9	–	NR	–	416
Italy	230	160 farms, 70 kennel	0	0	ND	In calves	–	–	1552
Portugal	1	Stray	1	1	1	1	NR	NC-P1	167
Romania	386	Various sources	19	19	ND	ND	2–24 oocysts/slide	–	1367
Spain	285	Farms	1	1	ND	ND	1–2 (5 oocysts/slide)	–	1676
Switzerland	249 (3289 feces)	Pet	25	Yes	Negative	Negative	Low (<20 oocysts/slide)	–	1760
UK	10 + 5 (140 fecal samples)	Kennel	2	Yes	1	ND	84–400	–	1324

Note: NR = not reported; ND = not done.

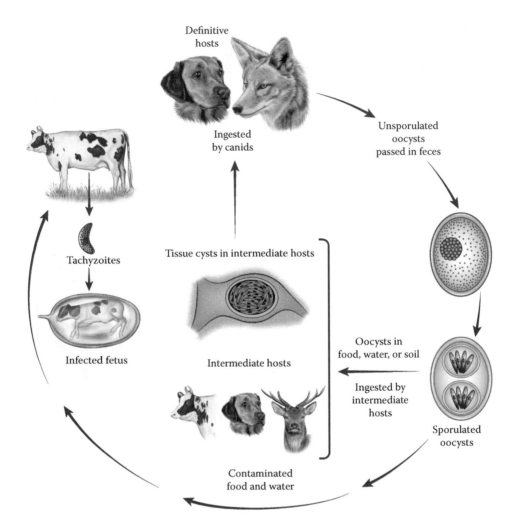

Figure 2.1 Life cycle of *N. caninum*.

and run longitudinally almost the entire length of the cell, closely apposed to the IMC. They are evenly spaced, and their distal ends are not capped. In addition, there are two 400 nm long intraconoidal tightly bound microtubules. The subpellicular microtubules form a rib cage and are arranged in a gentle counterclockwise spiral. Individual microtubules have prominent transverse striations.

Between the anterior tip and the nucleus there are up to 16 secretory organelles called rhoptries (Figures 2.8 and 2.9). A rhoptry consists of an anterior narrow neck that extends into the interior of the conoid, and a sac-like posterior end. Up to 6 rhoptries may extend posterior to the nucleus. A few rhoptries may be looped back toward the conoid. The contents of the rhoptries are homogenously electron dense. Tachyzoites also contain numerous micronemes, which are electron-dense rod-like structures occurring mostly at the anterior end of the parasite but a few are found posterior to the nucleus (Figures 2.8 and 2.9). Some of the micronemes at the conoidal end are oriented perpendicular to the pellicle.[1883] The number of micronemes is highly variable (Figure 2.9). Tachyzoites also contain numerous dense granules, which are round electron-dense organelles scattered throughout the tachyzoite, but predominantly found at the posterior end (Figure 2.8).

The nucleus is usually situated toward the central area of the cell. It consists of a nuclear envelope with pores, clumps of chromatin, and a centrally located nucleolus. Tachyzoites contain a

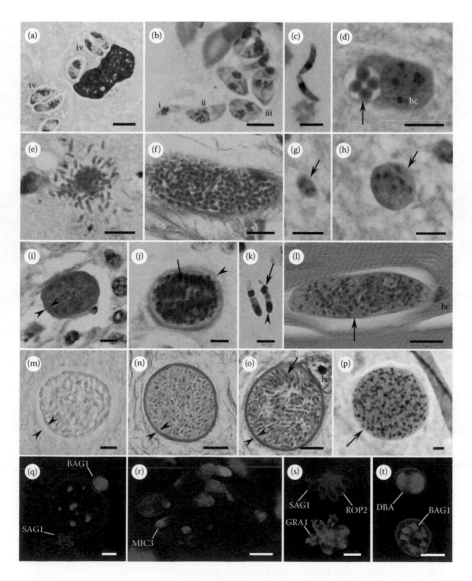

Figure 2.2 Tachyzoites and tissue cysts of *N. caninum*. a, b, c, k, p = smears, Giemsa stain. d–i, l, o = histologic sections, HE stain. j = histologic section, PAS-counter stained with hematoxylin. m = smear, brain, unstained. n = histologic section, Wilder's ammoniacal silver stain. q–t = infected cell cultures, fluorescent staining with different reagents. a–e, g–i, m = mouse brain. f, l = muscle dog. k = brain dog. j, n, o = cattle CNS. (a–c) Tachyzoites of different shapes and sizes, i—uninucleate slender, crescenteric, ii—plump uninucleate, iii—binucleated oval, iv—paired organisms. (d) Cross-section of 4 intracellular tachyzoites (arrow). (e) Group of tachyzoites/bradyzoites. (f) Large group of intracellular tachyzoites. (g) Tissue cyst with 4 bradyzoites. (h) Thin-walled tissue cyst with 7 bradyzoites. (i) Thick-walled (arrowheads) tissue cyst. (j) Thick-walled tissue cyst with bradyzoites (arrow) stained red with PAS; cyst wall is PAS-negative (arrowhead). (k) Two bradyzoites with apical staining (arrowhead) and subterminal nucleus (arrow). (l) Elongated intramuscular tissue cyst with thin cyst wall (arrow). (m) Thick-walled (arrowheads) tissue cyst. (n) Tissue cyst with silver-positive (arrowheads) cyst wall. (o) Intraneuronal tissue cyst with thick wall (arrowheads) and bradyzoites with terminal nuclei (arrow). (p) Thin-walled (arrow) tissue cyst. (q) SAG1-positive (red) tachyzoites, and BAG1-positive (green) bradyzoites. (r) Micronemes stained with anti-MIC3 antibodies (green) in tachyzoites stained with anti-SAG1 (red). (s) Double-stained tachyzoites with anti-ROP2 antiserum (green) and anti-SAG1 (red); dense granules labeled with anti-GRA1 antibodies (green). (t) Two tissue cysts, stained with anti-BAG1 (green) and DBA (red). hc = host cell. Bars in a–d, g–i, j, k, q–t = 5 μm, in e, f, l–p = 10 μm.

Figure 2.3 SEM images of *N. caninum* tachyzoites from cell culture. Note different shapes and sizes of tachyzoites, a conoid (co), micropore (mp), and a degenerated tachyzoite (arrow).

single convoluted mitochondrion, up to 6 μm in length (Figures 2.8 and 2.9f). Considerable biological curiosity has been focused on a nonphotosynthetic plastid organelle called the apicoplast (Figure 2.7). The apicoplast is a membrane bound, algae-derived, obligatory endosymbiont structure which is no longer photosynthetic, but has retained its own genome.[771,1322,1742]

Although tachyzoites can move by gliding, flexing, undulating, and rotating, they have no visible means of locomotion such as cilia, flagella, or pseudopodia. Instead, their motility is powered by the actin–myosin motor complex called glideosome, which is anchored to the IMC.[308,505] The outermost IMC membrane is studded with myosin complexes, similar to a conveyor–belt system.

Functions of the conoid and the secretory components of rhoptries, micronemes, and dense granules are not fully known, but they were shown to be associated with host cell penetration and creating an intracellular environment suitable for parasite growth and development, and exit from the host cell. The conoid can rotate, tilt, extend, and retract as the parasite probes the host cell plasmalemma immediately before penetration. During host cell invasion, rhoptries and micronemes secrete their contents through the plasmalemma just above the conoid to the exterior.

The mechanical events involved in zoite attachment and penetration of a host cell (Figures 2.1 and 2.12) include the following steps: (1) gliding motility of the zoite; (2) probing of the host cell with the zoite's conoidal tip; (3) indenting the host cell plasmalemma; (4) forming a moving junction that moves posteriorly along the zoite as it penetrates into the host cell; and (5) sequential exocytosis of micronemes, rhoptries, and dense granules. *N. caninum* can penetrate a variety of

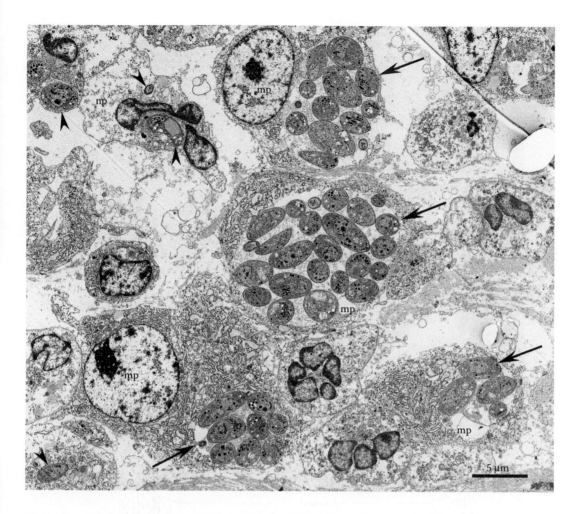

Figure 2.4 TEM of infected skin of a dog. Note several groups (arrows) and individual tachyzoites (arrowheads) in macrophages (mp) and neutrophils (np). (Adapted from Dubey, J. P. et al. 1995. *Vet. Dermatol.* 6, 37–43.)

cell types from a wide range of hosts, indicating that the biochemical receptors involved in attachment and penetration are probably common to most animal cells. The actual invasion process takes only a few seconds. Before entry the parasite mobilizes apical invasion proteins on to the parasite surface to mediate apical attachment. Invasion involves 2 Ca^{2+}-dependent events of protrusion of the conoid and secretion of microneme and rhoptry contents, which form a junction between the parasite and host cell surface membrane (Figure 2.11), which then moves toward the posterior end of the tachyzoite as it invades the host cell.[197,1913,1979] During this process, the tachyzoite is temporarily deformed (Figure 2.12c). Following invasion, the intracellular tachyzoite is surrounded by the PVM that is evidently derived from the host cell plasmalemma, but lacks host proteins. The PVM is not a homogenous structure of uniform thickness. A number of parasite molecules, secreted from dense granules and rhoptries, associate and modify the PVM, and a tubular membranous network (TMN) within the PV matrix is formed within 1 h following invasion.[1852] The TMN components are antigenically and structurally different from, but are physically connected to, the PVM. Rhoptry proteins (ROPs) were found to be located on the host cell cytoplasmic side of the PVM, suggesting a role in host/parasite biochemical communication. The PVM acquires pore structures that

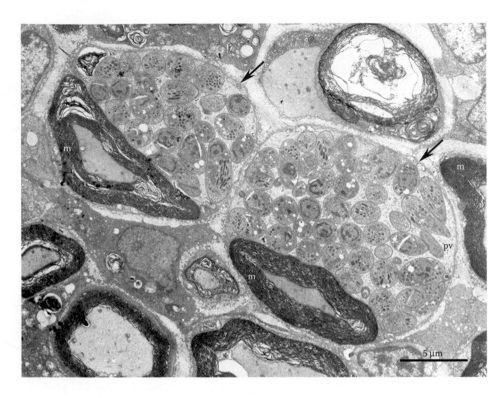

Figure 2.5 TEM of spinal nerve roots of a congenitally infected pup reported in Reference **404**. Two myelin-ated Schwann cells are distended with numerous tachyzoites (arrows). (Courtesy of late Dr. John Cummings.) Also note parasitophorous vacuoles (pv) and myelin (m).

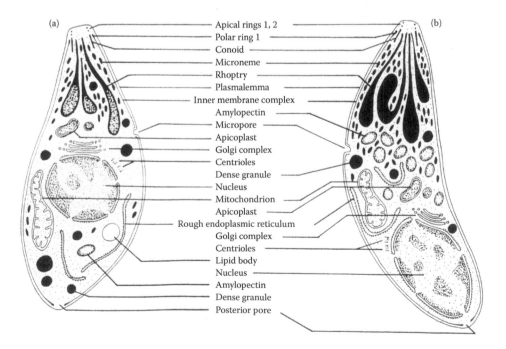

Figure 2.6 Schematic drawings of *N. caninum* tachyzoite (a) and bradyzoite (b).

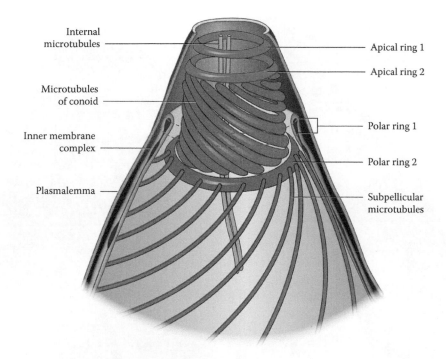

Internal microtubules

Apical ring 1

Apical ring 2

Microtubules of conoid

Inner membrane complex

Polar ring 1

Polar ring 2

Plasmalemma

Subpellicular microtubules

Figure 2.7 Schematic drawings of apical complex of *N. caninum*.

allow bidirectional diffusion of charged molecules of certain size between the PV and host cell cytoplasm.[469,1258] Collectively, these modifications establish a parasite friendly environment within the host cell cytoplasm conducive to parasite replication. The volume of the PV may increase many fold within 24 h of its formation. Similar to host cell entry, egress from host cells is also regulated by Ca^{2+}.[184]

One interesting and elusive phenomenon is the association of the host cell mitochondria and endoplasmic reticulum (ER) to the PV, but the intensity may be variable depending on the host cell type (Figure 2.8) and the strain of the parasite.[469,882,1487,1884] Within minutes of tachyzoite invasion, host mitochondria and ER are found in close vicinity, and often physically linked, to the PVM, and the extent of this juxtaposed association does not change with enlargement of the PV. A close physical association of host cell mitochondria is also found in cells infected with *T. gondii* tachyzoites, although only in type I and III, but not in type II strains, and it was also shown that the presence of host cell mitochondrial association had an impact on host cell cytokine transcription and cytokine expression during infection in mice.[1594] The *N. caninum* vacuole does not fuse with host lysosomes.

Tachyzoites multiply asexually within the PV by repeated endodyogeny (Figures 2.13 through 2.16), a specialized form of reproduction in which 2 progeny form within the parent parasite (Figure 2.16d), consuming it.[1986] At the onset of endodyogeny, the Golgi complex and the apicoplast divide, at the anterior end of the nucleus, and 2 centrosomes are formed within the nucleus. During division, the nuclear membrane remains intact and microtubules form an intranuclear spindle. Next, the conoid, the anterior portions of IMC, and the subpellicular microtubules of the progeny cells appear as 2 dome-shaped structures. The parasite nucleus becomes bilobed and the ends of the nucleus move into the dome-shaped anterior portions of the developing progeny. The IMC and subpellicular microtubules continue to extend posteriorly and surround one half of the nucleus, which eventually pinches into 2. Partitioning of the ER and the mitochondrion continues. Rhoptries and micronemes are synthesized de novo in the progeny. The progeny continue to grow until they reach the surface of the parent. The inner membrane complex of the parent disappears, and its outer membrane becomes

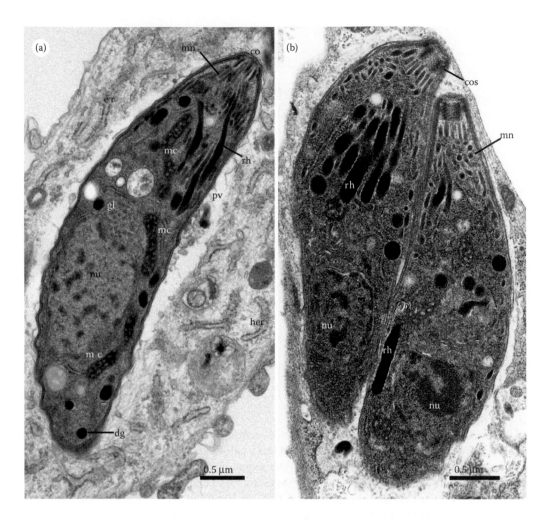

Figure 2.8   TEM of intracellular *N. caninum* tachyzoites in cell cultures. Note conoid (co), conoidal spring (cos), micronemes (mn), rhoptries (rh), a long convoluted mitochondrion (mc), dense granules (dg), Golgi body (go), parasitophorous vacuole (pv), and nucleus (nu). (a) Note accumulations of host endoplasmic reticulum (her) around pv. (b) Paired tachyzoites. Note a posteriorly located rh. (Adapted from Dubey, J. P., Lindsay, D. S. 1993. *Parasitol. Today* 9, 452–458.)

the plasmalemma of the progeny cells. *In vivo*, most groups of tachyzoites are arranged randomly due to asynchronous cycles of endodyogeny. However, occasionally rosettes are formed due to synchronous division. The host cell ruptures when it can no longer support growth of tachyzoites. As many as 100 tachyzoites can be present in 1 group of intracellular tachyzoites.

The rate of invasion and growth can vary depending on the strain of *N. caninum* and the type of host cells. After entry of tachyzoites into a host cell, there is a variable lag phase of 8 h or more before the parasite divides, and this lag phase is partly parasite dependent. The doubling time of 8–14 h or more can also vary depending on the strain of *N. caninum*.[1136,1677,1921]

2.4.2 Bradyzoites and Tissue Cysts

The bradyzoite is the encysted stage of the parasite in tissues. Bradyzoites are also called cystozoites. Tissue cysts (Figures 2.2g–p and 2.17) grow and remain intracellular as the bradyzoites

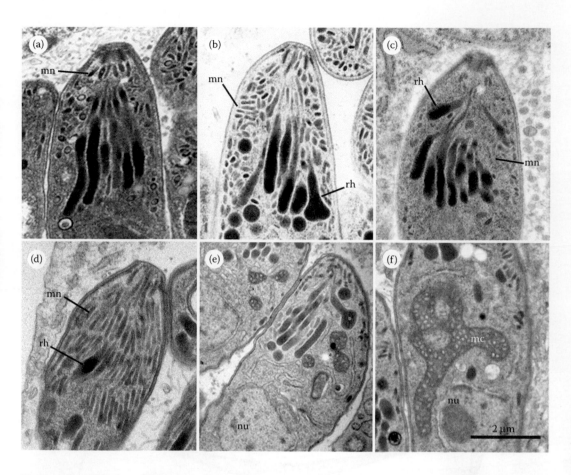

Figure 2.9   TEM of conoidal ends of tachyzoites. Note variable number, shapes, and sizes of micronemes (mn) and rhoptries (rh). (a) (Courtesy of Dr. David Lindsay). (b) Note a bell-shaped blind end of a rhoptry. (Courtesy of Dr. C. A. Speer.) (c) Nine rhoptries in 1 plane of section. (Adapted from Dubey, J. P. et al. 2002. *Int. J. Parasitol.* 32, 929–946.) (d) Numerous micronemes. (e) Expanded blind end of a rhoptry (*). (f) Sections of a convoluted mitochondrion (mc). nu = nucleus.

divide by endodyogeny. Tissue cysts vary in size, depending on the host, and the type of cell parasitized. Young tissue cysts may be as small as 5 μm in diameter and contain only few brady-zoites (Figure 2.2g), while older cysts may contain more than 200 organisms. Most tissue cysts studied were from the brain and the spinal cord, and mostly from naturally infected dogs. Tissue cysts in brain are round and were found to be up to 107 μm in diameter. The cyst wall is 0.5–4.0 μm-thick, both in live, unstained preparations, and in histological sections (Figure 2.2g–p,t). The intramuscular cysts are elongated and may be 100 μm long, and have a thin cyst wall (Figure 2.2l).[598,1599] Histologically confirmed tissue cysts are rare in other organs of naturally infected animals.

The tissue cyst wall is elastic, argyrophilic, has a wavy contour, and is periodic acid Schiff reaction (PAS) negative (Figure 2.2i–p). The outer limiting membrane is lined with granular material. The granular layer contains tubules, and electron-dense vesicles of different densities[132,207,1883,18...] (Figures 2.16 through 2.18). The ground substance in the interior of cyst also contains vesicles, tubules. The bradyzoites are loosely arranged inside the cyst. Bradyzoites and tissue cysts are fluorescent in UV light.[1137]

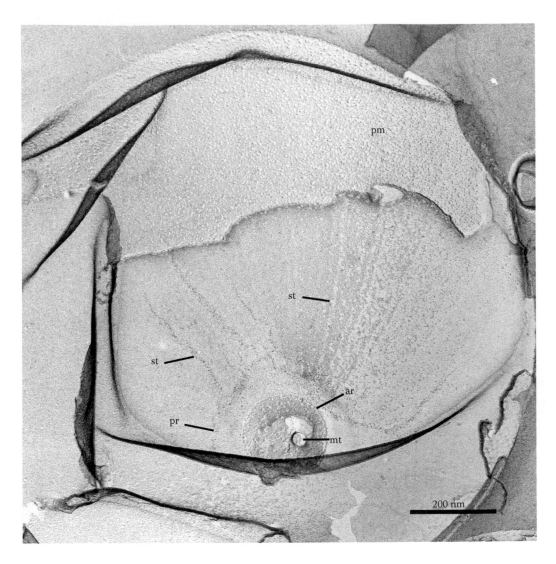

Figure 2.10   Freeze fracture SEM of conoidal end of a *N. caninum* tachyzoite. Note rough plasma membrane (pm), subpellicular microtubules (st) arising from the polar ring (pr), conoidal internal microtubules (mt) within anular ring (ar).

The size of bradyzoites appears to be variable based on techniques used for measurement. In Giemsa-stained smears, bradyzoites (Figure 2.2k) were curved with the nucleus often located at the posterior (non-conoidal) end.[598] Bradyzoites (n = 100) released from 2 tissue cysts from the brain of a naturally infected dog were 4.8–8.0 × 1.0–1.9 and 5.6–7.1 × 1.3–2.3 μm in size with an average of 6.5 × 1.5 μm.[598] Few longitudinally cut bradyzoites in transmission electron microscopy (TEM) sections were 8 × 2 μm. Bradyzoites are slender and contain several amylopectin granules, often concentrated in the middle of the zoite which stains red with PAS reagent (Figures 2.2j and 2.19). A membrane body has been described, located toward the conoidal end[132,209,998] (Figures 2.19 and 2.20). There are 6–12 rhoptries located anterior to the nucleus; rhoptries reaching beyond the nucleus toward the posterior part are rare.[1884] Bradyzoites contain fewer rhoptries compared to tachyzoites, while they contain more micronemes, also posterior to the nucleus. Micropores are more common in bradyzoites versus tachyzoites.[1884]

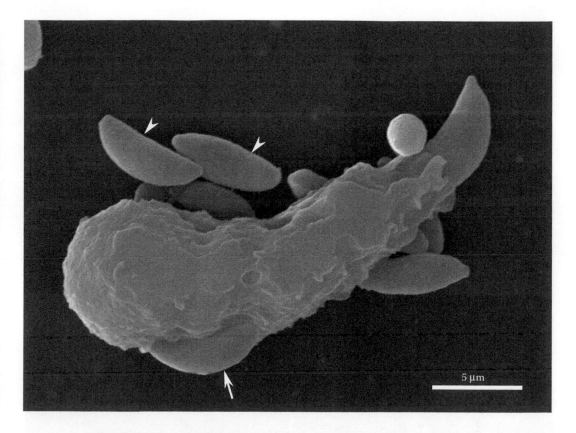

Figure 2.11 SEM of *N. caninum* tachyzoites attaching to murine dendritic cells. Note that multiple tachyzoites adhere to the surface of the DC. Note a tachyzoite attached along its whole cell surface (arrow). Also note spiral arrangement of subpellicular microtubules (arrowheads).

2.4.2.1 Conversion of Tachyzoites to Bradyzoites

Tissue cysts represent an integral part of the life cycle of *N. caninum*. Little is known concerning the genesis of *N. caninum* tissue cysts in naturally infected animals. Most information is from congenitally infected dogs or cattle. In a naturally infected full-term stillborn bovine fetus,[546] individual bradyzoites, and small 8–10 μm tissue cysts were scattered in and around encephalitic lesions (see Chapter 4, Figure 4.9d).

The knowledge gained from *in vitro* and *in vivo* formation of *T. gondii* tissue cysts has been useful for studying *in vitro* formation of *N. caninum* tissue cysts. Several stress factors, including higher incubation temperature, high pH, sodium arsenite treatment, and mitochondrial inhibitors stimulated *in vitro* conversion of *T. gondii* tachyzoites to bradyzoites (reviewed in Reference **570**) and most of these factors were either not suitable for *in vitro* conversion of *N. caninum* tachyzoites to bradyzoites, or only with low efficiency.[1998,2080] Instead, treatment of cell cultures for 9 days with 70 μM sodium nitroprusside was most successful using a variety of cell lines[34,631,888,1713,2052,2053] and different *N. caninum* isolates. Using Vero cells and the Nc-Liv isolate, the first evidence of bradyzoite stage conversion occurred after 3–4 days.[2053] Several markers such as bradyzoite antigen 1 (BAG1) that stains bradyzoites, monoclonal antibody CC2 (mAbCC2) and the lectin Dolichos biflorus agglutinin (DBA) that stains the cyst wall, NcSAG1 (surface antigen 1; specific for tachyzoites), and TEM were used to study stage conversion (Figure 2.2q,t). The bradyzoite formation was asynchronous, and doubly stained parasites (with NcSAG1 and BAG1) within the same PV were

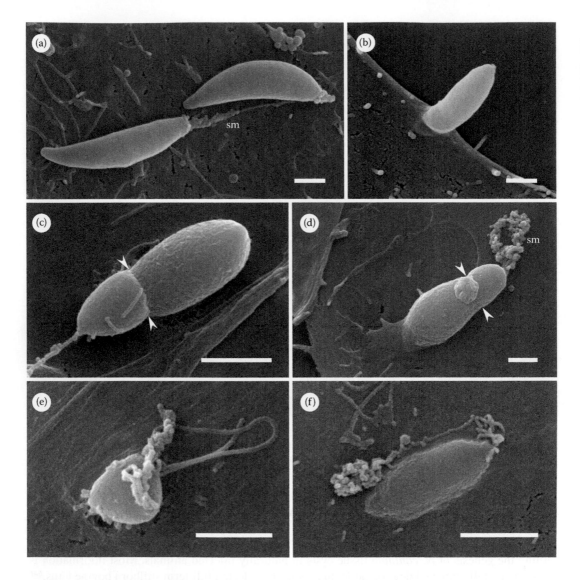

Figure 2.12   SEM of the sequence of events leading to invasion of human foreskin fibroblasts (HFF) by *N. caninum* tachyzoites. (a) Two tachyzoites adhering to the HFF surface move along by gliding motility, leaving tracks of shed surface material (sm) behind. (b) Orientation of a tachyzoite perpendicular to the host cell surface and initiation of the host cell invasion process. (c–e) Tachyzoites at different stages moving into the HFF cytoplasm. Note the constrictions (arrowheads) forming at the contacts of the host and parasite membrane at entry site, and shed material left behind by the tachyzoites on the host cell surface. (f) Freshly invaded tachyzoite, the outlines of which are barely visible. Bars = 2 μm.

seen (Figure 2.2q). Immunofluorescence showed that dense granule proteins such as GRA1, GRA2, and GRA7 were incorporated into the tissue cyst wall during differentiation[2053] and differences in protein profiles and transcription of tachyzoites and bradyzoites were useful in differentiating tachyzoites and bradyzoites.[16,17,1030,1031,1289] By TEM, 1–5 organisms (bradyzoites) were enclosed in a cyst wall within the PVM. The cyst wall was up to 1 μm thick and consisted of granular material with few vesicles; the same type of material was inside the cyst.[2051,2053] Interestingly, monoclonal antibodies recognizing the sporocyst wall of *T. gondii* are also able to recognize material both in the tissue cyst walls of *N. caninum* and *T. gondii*.[774] The bradyzoites contained amylopectin granules.

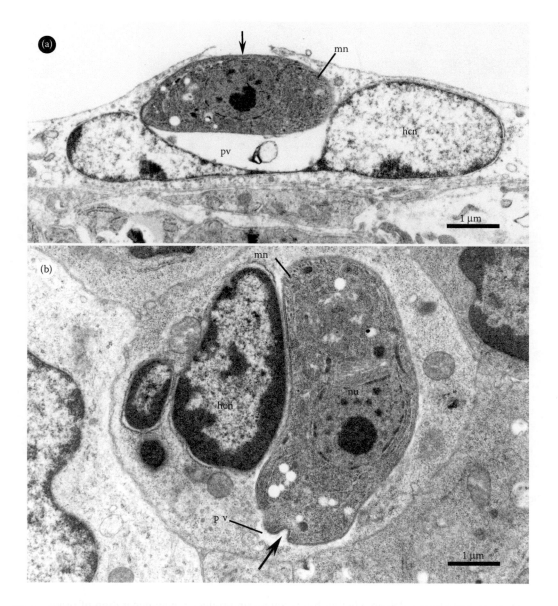

Figure 2.13   TEM of intracellular tachyzoites in the liver of a KO mouse infected with NC1 strain and perfused with glutaraldehyde. (a) Tachyzoite (arrow) within the parasitophorous vacuole (pv), probably soon after entry into host cell. Note host cell nucleus (hcn), and micronemes (mn). (b) Pre-divisional stage with few organelles. Note a mouth-like opening at the non-conoidal end (arrow), hcn, mn, nu, and pv. (Courtesy of Dr. Bjerkås.)

Bradyzoites could be separated from the cells by passing the infected cultures through a 25-gauge needle and centrifugation; their identity was confirmed by BAG1 staining. Unlike tachyzoites, the infectivity of bradyzoites was low for initiating new infections in Vero cells. Removal of sialic acid residues from the Vero cell surface impaired bradyzoite adhesion, but not tachyzoite host cell binding.[2053]

Whether the *in vitro* cultured bradyzoite-like stages are biologically bradyzoites is not known. In limited tests, dogs fed cultured bradyzoites did not excrete oocysts.[888,1998] Oral infectivity of *in vitro* cultured bradyzoites to mice and their resistance to acid-pepsin digestion were not tested.

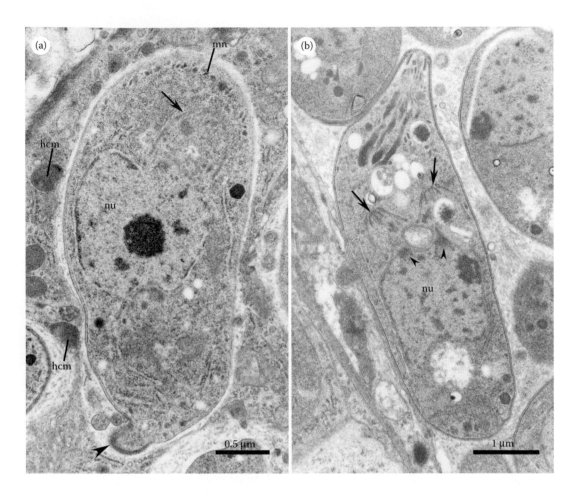

Figure 2.14 TEM of tachyzoites undergoing division by endodyogeny in the liver of a KO mouse infected with NC1 strain and perfused with glutaraldehyde. The host mitochondria (hmc) have accumulated at the periphery of the parasitophorous vacuolar membrane. (a) Ovoid mother cell without visible rhoptries, tail-shaped posterior end (arrowhead), and formation of daughter anlagen (arrow) at the pole of the nucleus. (b) Elongated mother cell with few rhoptries. The conoidal ends of both daughter cells (arrows) have formed. The mother nucleus (nu) is in very early stage of division and 2 cones (arrowheads) have formed. (Courtesy of Dr. Inge Bjerkås.)

There is considerable debate whether bradyzoites can produce a new generation of bradyzoites without first converting to tachyzoites.[599] In mice inoculated with *T. gondii* sporozoites (oocysts) and bradyzoites (tissue cysts), tissue cyst formation is delayed by 2–3 days following sporozoite and bradyzoite inoculation compared to mice that are inoculated with tachyzoites; these results suggest that sporozoite and bradyzoites first convert to tachyzoites, and then to bradyzoites.[523] *In vitro* experiments with *N. caninum* using NcSAG1/BAG1 double immunofluorescence labeling suggested that BAG1-positive bradyzoites may develop only if the host cell is invaded by a parasite that already began tachyzoite-to-bradyzoite stage conversion prior to host cell entry.[1998]

2.4.3 Oocyst

Oocysts represent the only unambiguosly known sexual stage of *N. caninum*. Oocysts are excreted unsporulated in canid feces (Figure 2.21). Unsporulated oocysts are 10–11 μm in diameter. Sporulated oocysts are slightly bigger in size than the unsporulated oocysts, namely 11.7 × 11.3 μm

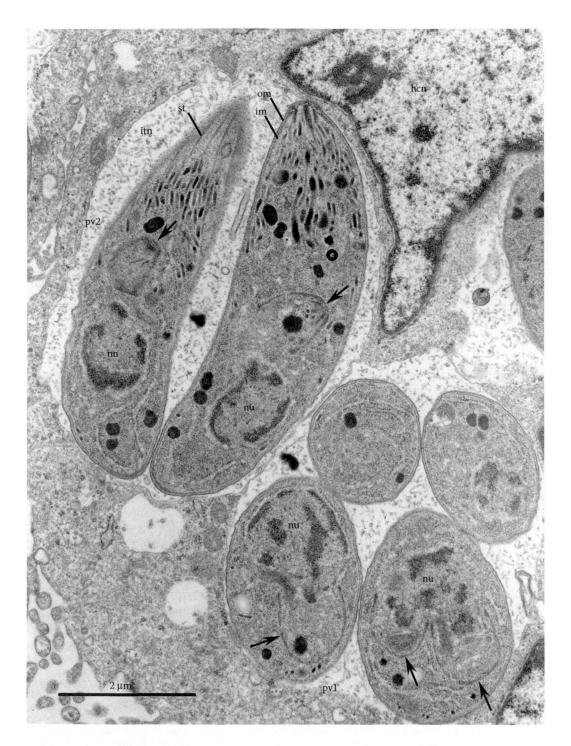

Figure 2.15 Tachyzoites undergoing endodyogeny in 2 parasitophorous vacuoles (pv1, pv2) of a host cell *in vitro*. In pv1, 2 of the 4 tachyzoites are dividing and the mother nucleus is being incorporated in to daughter cells. Note 2 elongated tachyzoites with formation of apical end of 2 daughter cells in pv2. Note numerous micronemes (mn) in mother cells and profuse intravacuolar tubular network (itn), subpellicular microtubules (st), outer membrane of pellicle (om), inner membrane (im) of pellicle, parasite nucleus (nu), and host cell nucleus (hcn). (Courtesy of Dr. C. A. Speer.)

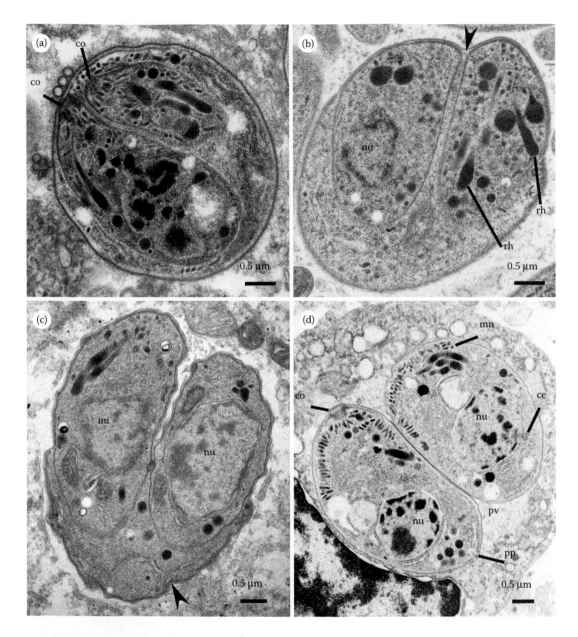

Figure 2.16 *In vivo* divisional process in *N. caninum* tachyzoites. a and d are from the skin of the naturally infected 13-year-old dog. (Adapted from Case no. 10 in Dubey, J. P. et al. 1988. *J. Am. Vet. Med. Assoc.* 192, 1269–1285.) The skin biopsy was fixed in glutaraldehyde. b and c are from liver of a KO mouse infected with NC1 strain and specimens perfused with glutaraldehyde (courtsey of Dr. I. Bjerkås). (a) Almost nearly formed daughter tachyzoites are still within the globular mother cell. (b) Two daughter tachyzoites within an overstretched mother cell. Parasite nucleus (nu). (c) Two daughter tachyzoites being released from mother cell but still attached (arrow) by their posterior (non-coinoidal) ends. Parasite nucleus (nu). (d) Two tachyzoites in a macrophage soon after division. Note only a few micronemes (mn) and rhoptries (rh) in all 4 images. Also note conoid (co), centriole (ce), and a break in pellicle inner membrane (pp), parasite nucleus (nu), and parasitophorous vacuole (pv).

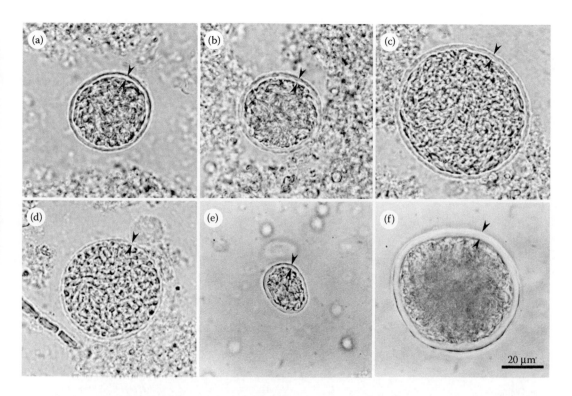

Figure 2.17 *N. caninum* tissue cysts in homogenates of brains of 2 congenitally infected dogs. Unstained. Note different thickness of the cyst walls (arrowheads). (a–c from Dubey et al. 1998. *Int. J. Parasitol.* 28, 1293–1304; d–f from Dubey et al. 2004. *Int. J. Parasitol.* 34, 1157–1167.)

(10.6–12.4 × 10.6–12.0) with a length-to-width ratio of 1.04. The oocyst wall is colorless, and 0.6–0.8 μm-thick.

In concentrated sucrose, oocysts appeared to be smaller than in water. The largest 75th percentile for length was 10.7 μm.[1805] The length-to-width ratio reported for concentrated sucrose (1.06) was similar as reported for water.[1805]

Sporulated oocysts contain 2 sporocysts and a residual body that may be compact or dispersed. Sporocysts measure 8.4 × 6.1 μm with a length-to-width ratio of 1.37. There are no polar granules, oocyst residuum, or Stieda bodies.[1181] Sporozoites are elongated and measure 6.5 × 2.0 (5.8–7.0 × 1.8–2.2) μm in size. They have a centrally located nucleus, amylopectin granules but no round bodies. The ultrastructure of sporozoites has not been described, so it is uncertain if, as with *T. gondii* sporozoites, a crystalloid body is absent.[613]

The developmental stages preceding the formation of *N. caninum* oocysts during schizogony and gametogony are unknown. In 1 report schizonts, gamonts, and oocysts were found in histological sections of small intestine of a 1.5-month-old dog that had systemic neosporosis.[1105] The main evidence is based on immunoreactivity to *N. caninum* to monoclonal antibody (210/70, VMRD). There were no measurements of these stages. In Figure 2.2 in this reference,[1105] the illustrated oocysts appear to be more than 20 μm long, based on the scale bar provided; thus they are not *N. caninum* oocysts. One of us, JPD, reviewed histological slides of small intestine from a naturally infected dog from South Africa, kindly provided by June Williams. In that dog coccidian gamonts reacted brilliantly with polyclonal rabbit *N. caninum* antibodies but the oocysts were *Cystoisospora ohioensis*-like with a diameter of about 20 μm.

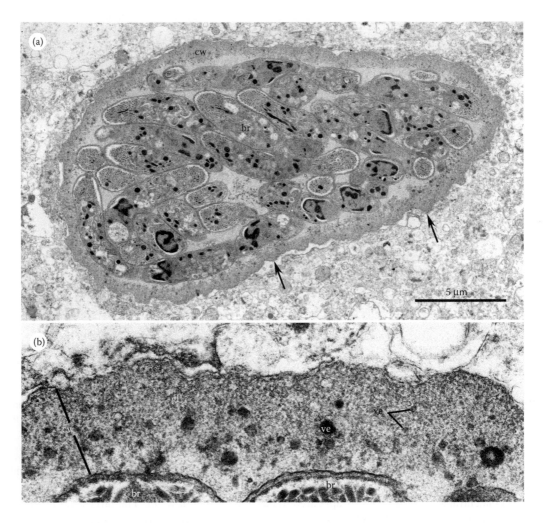

Figure 2.18 Tissue cyst of *N. caninum* in the brain of a naturally infected dog. (From Dubey et al. 1998. *Int. J. Parasitol*. 28, 1293–1304). Note tissue cyst wall (cw) with a wavy outline, granular layer (gl) containing vesicles (ve) of different densities, and bradyzoites (br). (a) Low magnification. (b) Higher magnification of the tissue cyst wall. (Adapted from Speer, C. A., Dubey, J. P. 1989. *J. Protozool*. 36, 458–463.)

2.5 TRANSMISSION AND EPIDEMIOLOGY

N. caninum can be transmitted postnatally (horizontally, laterally) by ingestion of tissues infected with tachyzoites or tissue cysts, by ingestion of food or drinking water contaminated by sporulated oocysts, or it can be transmitted transplacentally (vertically, congenitally), from an infected dam to her fetus during pregnancy. The 2 terms "exogenous transplacental transmission" and "endogenous transplacental transmission" have been proposed to describe more precisely the origin of the transplacental infection of the fetus.[1996] Exogenous transplacental transmission occurs after a primary, oocyst-derived infection of a pregnant dam, while endogenous transplacental transmission occurs in a persistently infected dam after reactivation (recrudescence) of the infection during pregnancy.

Transplacental transmission has been documented in naturally infected cattle, sheep, goats, dogs, various species of deer, and induced experimentally in cattle, dogs, sheep, goats, monkeys, pigs, cats, and mice (discussed in Chapters 4 through 18). Transplacental transmission occurs when

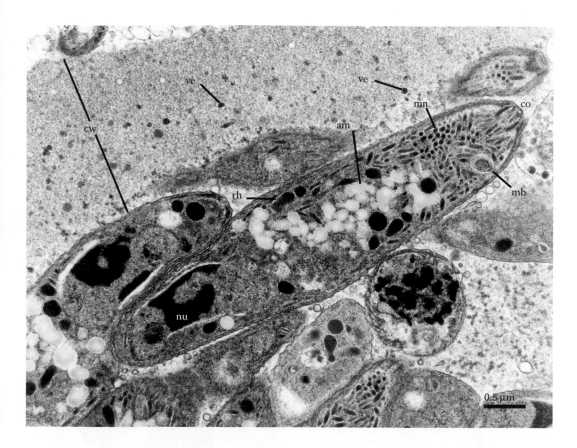

Figure 2.19   Tissue cyst of *N. caninum* in brain of a naturally infected dog. Note thick cyst wall (cw), with vesicles (ve) of different densities, bradyzoites butted against the cyst wall. Note conoid (co), a membrane bound body (mb), micronemes (mn), amylopectin granules (am), rhoptries (rh), and nucleus (nu). (Adapted from Dubey, J. P. et al. 2004. *Int. J. Parasitol.* 34, 1157–1167.)

tachyzoites from the dam cross the placenta, during the second week after infection. *Neospora* tachyzoites initially multiply in the maternal part of the placenta and then spread to the fetal part, and eventually to fetal tissues. In Figure 2.28, numerous tachyzoites are present in the maternal placenta. Venereal transmission is a remote possibility because *N. caninum* has been detected in testis and semen but very high doses of tachyzoites were needed to infect dams by intra-vaginal inoculation.[268,285,1293a,1830,1831]

Mice were infected successfully by oral inoculation of tachyzoites or bradyzoites.[1163] These results are of interest because tachyzoites treated with acidic pepsin were rendered noninfective for cell cultures whereas bradyzoites survived the acidic pepsin.[1163] Tissue cysts and bradyzoites can survive up to 2 weeks at refrigeration temperature (4°C) but were killed by freezing.[598,1167]

The ingestion of oocysts is the only demonstrated mode for postnatal (horizontal) transmission in herbivores. Oocysts are the key in the epidemiology of neosporosis. Although data concerning the environmental resistance of *N. caninum* oocysts are limited, it is likely to be similar to other coccidian oocysts.[1458] In 1 study, oocysts were not infective after storage for 46 months at 4°C.[2014]

N. caninum oocysts have been identified in only a few dogs worldwide (Table 2.3). The number of oocysts shed by dogs is usually low. Because *N. caninum* oocysts structurally resemble another coccidium in dog feces, *Hammondia heydorni*, it is epidemiologically important to properly identify *N. caninum* oocysts.[1805,1865,1866,1871] In this respect, the reported detection of *N. caninum* DNA in 34.9% of 212 fecal samples of dogs from China[1142] is unexpected and needs confirmation.

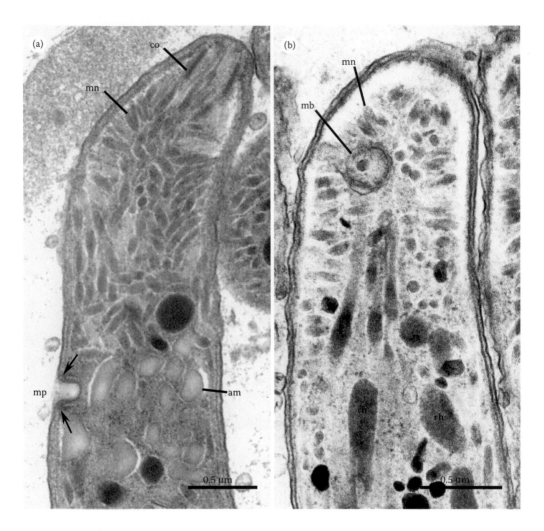

Figure 2.20   TEM of conoidal ends of 2 bradyzoites (a,b) of *N. caninum* from tissue cysts in brain of naturally infected dogs. Note conoid (co in (a)), haphazardly arranged micronemes (mn in (a) and (b)), a micropore (mp) with a rim (arrowheads in (a)), amylopectin granules (am), rhoptries (rh) and a membrane bound body in (b) (mb). (Courtesy of Dr. Inge Bjerkås.)

Although oocysts were detected in the feces of experimentally infected coyotes and dingoes, until now, the presence of viable *N. caninum* oocysts has not been confirmed in the feces of naturally infected coyotes or dingoes (for details see Chapter 16).

Because of the epidemiological importance, the transmission of *N. caninum* in dogs and cattle is discussed separately in Chapters 4 and 5, respectively.

2.6 HOST–PARASITE RELATIONSHIP

N. caninum usually parasitizes the host, both definitive and intermediate, without producing clinical signs. Only rarely does it cause severe clinical manifestations. Natural infections are probably acquired by ingestion of tissue cysts in infected meat or oocysts from food and water contaminated with dog feces. After ingestion of oocysts, sporozoites excyst in the intestine and spread locally to mesenteric lymph nodes (MLN) and to distant organs via lymph and blood[583] (Figures 2.22 and 2.23).

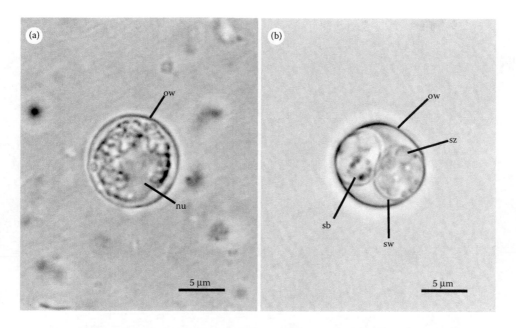

Figure 2.21 Oocysts of *N. caninum*. Unstained. Note oocyst wall (ow), sporocyst wall (sw), sporozoites (sz), sporocyst residual body (sb), and nucleus (nu). (a) Unsporulated. (b) Sporulated.

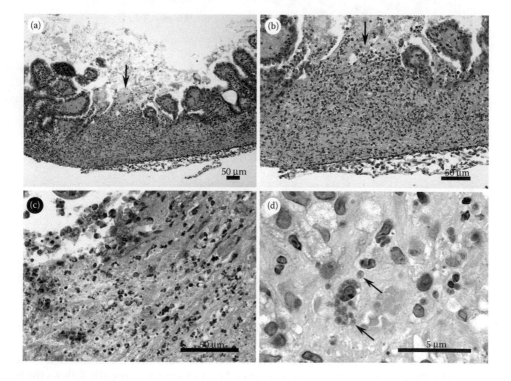

Figure 2.22 Enteritis in the small intestine of a gerbil, day 9 after feeding *N. caninum* oocysts. HE stain. (a, b) Focal ulceration (arrows) with denudation of epithelium and necrosis of subepithelial tissue. (c) Necrosis of tunica muscularis and the lamina propria. Numerous tachyzoites are present but not visible at this magnification. (d) Intact tachyzoites (arrows) among necrotic host tissue. (See Dubey, J. P., Lindsay, D. S. 2000. *Parasitol. Res.* 86, 165–168.)

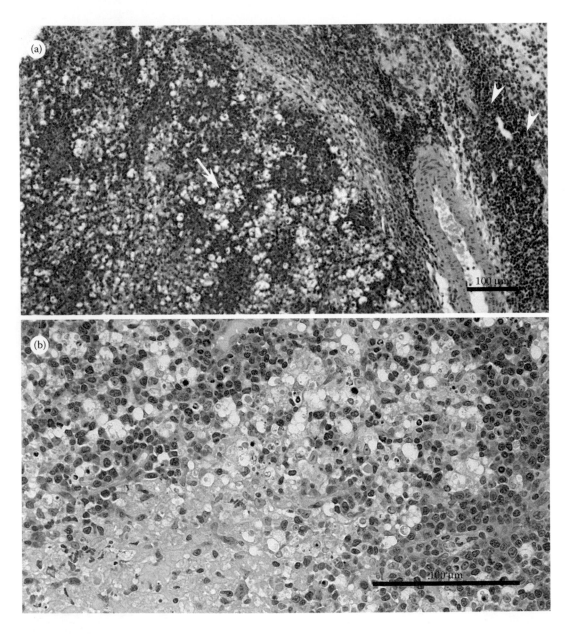

Figure 2.23 Section of mesenteric lymph node of the gerbil in Figure 2.22. HE stain. (a) Note epitheloid cell proliferation (arrow) and peritonitis around the lymph node (arrowheads). (b) Necrosis and depletion of lymphoid cells.

Focal areas of necrosis may develop in many organs (Figures 2.24 and 2.25). Necrosis is caused by the intracellular growth of tachyzoites; a toxin has not been demonstrated.

The host may die of acute neosporosis, but more often recovers with the acquisition of immunity coincident with the appearance of humoral antibodies. Inflammation usually follows the initial necrosis (Figure 2.26). By about the 3rd week after infection, tachyzoites begin to disappear from visceral tissues and may localize as tissue cysts in neural and muscular tissues.

Latent neosporosis may be reactivated by rupture of tissue cysts. When and why tissue cysts rupture is not known. Latent neosporosis has been reactivated in experimentally infected dogs by

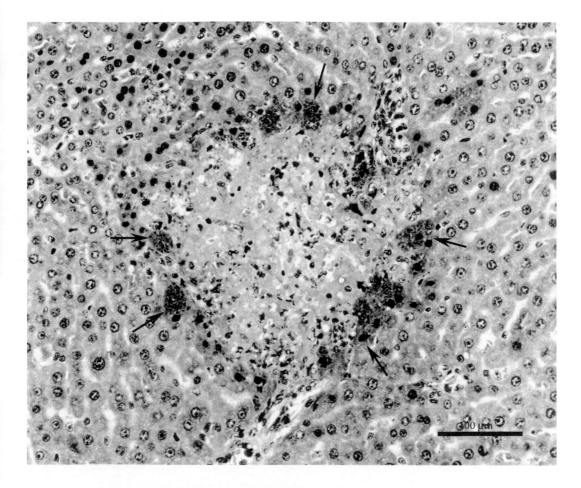

Figure 2.24 Focal hepatic necrosis in an experimentally infected rat. Note hepatocytes filled with numerous tachyzoites (arrows) at the periphery of the lesion. HE stain. (Adapted from Lindsay, D. S., Dubey, J. P. 1990. *Can. J. Zool.* 68, 1595–1599.)

administration of high doses of corticosteroids.[539] Granulomatous inflammation follows after rupture of tissue cysts (Figure 2.27).

Placental invasion (Figure 2.28) during acute or chronic infection can lead to infection of the fetus. The transport of tachyzoites within dendritic cells (DCs) may facilitate to cross blood and brain or dam-fetal placental barriers.[370]

2.7 MOLECULAR AND CELL BIOLOGY

2.7.1 Genome and Transcriptome Analysis of *N. caninum*

Neospora and *Toxoplasma* diverged from each other from their common ancestor around 28 million years ago and their genomes are highly similar with respect to gene content and synteny. In spite of such similarities, they differ markedly in their host range and their etiology.

Whole genome sequencing of the *N. caninum* Liverpool isolate (in the following refered to as Nc-Liv) was achieved using Sanger sequencing to ~8-fold depth, and respective results and

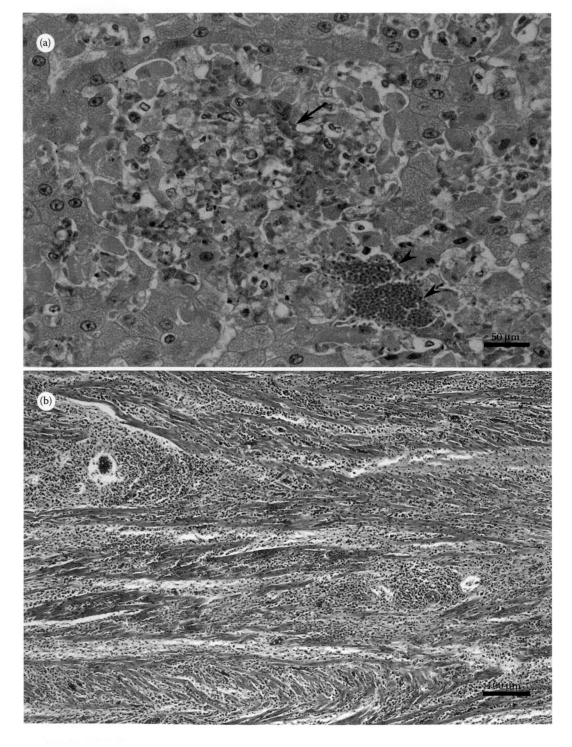

Figure 2.25 Lesions of neosporosis. HE stain. (a) Hemorrhage and necrosis (arrow) in liver of a naturally infected dog. Note 2 large groups of tachyzoites (arrowheads). (b) Neosporosis-assoclated severe fatal myocarditis in a naturally infected rhinoceros. Many tachyzoites are present but not visible at this magnification. (Courtesy of Dr. June Williams, Reference **2102**.)

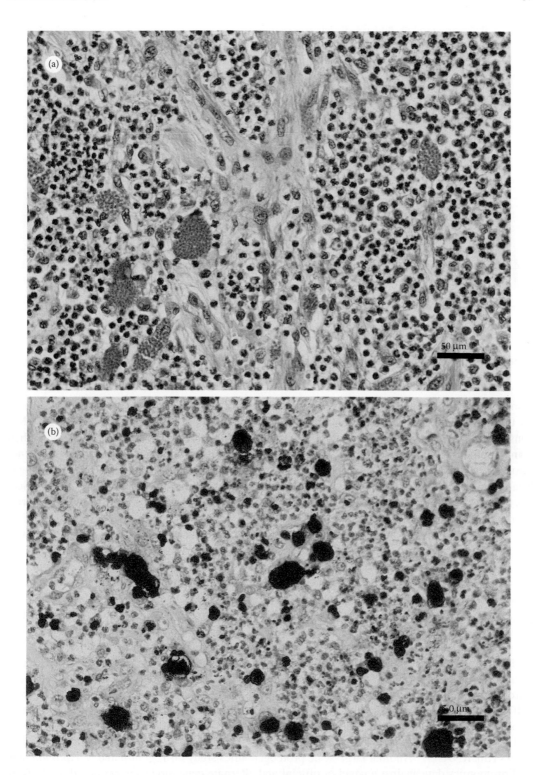

Figure 2.26 Pyogranulomatous dermatitis in a 13-year-old dog. (Adapted from Case no. 10 of Dubey, J. P. et al. 1988. *J. Am. Vet. Med. Assoc.* 192, 1269–1285). (a) Numerous groups of tachyzoites among the mixed leukocyte cell infiltration. HE stain. (b) Section stained with *N. caninum* antibodies. Many more tachyzoites are visible than seen in the HE-stained section.

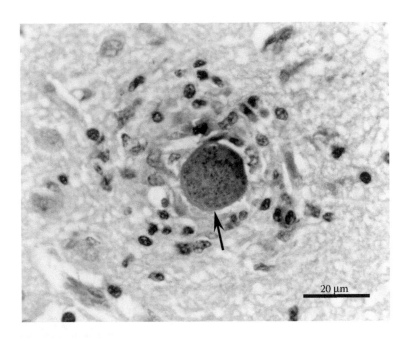

Figure 2.27 Inflammatory response around a degenerating tissue cyst (arrow) in the brain of an aborted goat fetus. Arrow points to irregular outline of the tissue cyst wall. HE stain.

comparison with *Toxoplasma* was first published in 2012.[1698] The genome was constructed based on the available *T. gondii* Me49 genome (www.toxodb.org) to 14 chromosomes totaling 61 MB consisting of 7121 genes.[1698] In addition, the parasite mitochondrion and the apicoplast have their own genomes; these have not yet been sequenced for *N. caninum*.[794] The entire genome sequence information on *N. caninum* (Nc-Liv isolate) is available on http://www.genedb.org/Homepage/Ncaninum. Recently, high-throughput second-generation transcriptome sequencing (mRNA-Seq) has been used to improve annotation of the genome sequence and also to study differential expression in the tachyzoite life cycle stage.

As expected, the 2 genomes (*Neospora* and *Toxoplasma*) show a high degree of synteny with very few chromosomal rearrangements. Although there is very little gain or loss of genetic content, there are significant differences between *N. caninum* and *T. gondii*.[1666,1698] In *N. caninum* an unexpected expansion of surface antigen gene families and the divergence of secreted virulence factors, including rhoptry kinases were noted.[1698] A genomic sequence coverage analysis to identify *T. gondii* and *N. caninum* loci that have undergone duplication and expansion (Els) showed that *N. caninum* has more (63) Els compared to *T. gondii* (53), and the *N. caninum* Els were enriched for surface-associated glycoprotein (SAG)-related surface antigen (SRS) families.[7] A subsequent study, compared the genomes, transcriptomes, and proteomes of Nc-Liv and *T. gondii* VEG-strain tachyzoites and improved the annotation of genes.[1666] Corrected predicted structures of over one third of the previously annotated gene models and annotated untranslated regions (UTRs) in over half of the predicted protein-coding genes were presented. Distinctly long UTRs in both the organisms, almost 4 times longer than other model eukaryotes were observed. The authors also identified a putative set of cis-natural antisense transcripts (cis-NATs) and long intergenic noncoding RNAs (lincRNAs).[1666] Another large-scale proteogenomics study on *T. gondii* and *N. caninum* queried proteomics data against a panel of official and alternate gene models generated directly from RNASeq data, using several newly generated and some previously published MS datasets for this meta-analysis. This resulted in the identification of loci apparently absent from the official annotation (release 10 from EuPathDB) of these species. The new data were integrated into EuPathDB.[1101]

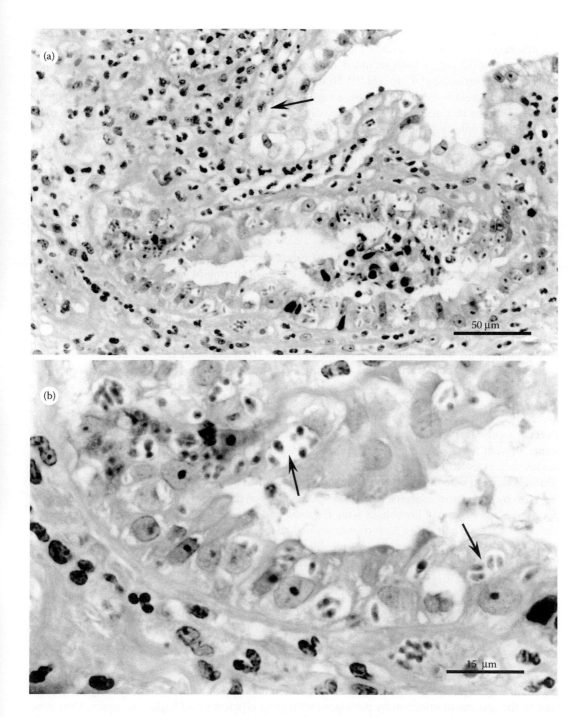

Figure 2.28 Uterus of a queen, 3 days after giving birth to an infected kitten. The pregnant cat was inoculated SC with NC1 tachyzoites, 20 days previously. HE stain. (a) Metritis with numerous tachyzoites (arrow) in the epithelium. (b) Higher magnification to show tachyzoites (arrows). (Adapted from Dubey, J. P. and Lindsay, D. S. 1989. *J. Parasitol.* 75, 765–771.)

Previously, information on gene expression in apicomplexan parasites was obtained through construction of expressed sequence tag (EST) libraries, which were available in a corresponding database (ApiEST-DB), which included *Eimeria tenella, N. caninum, Plasmodium falciparum, Sarcocystis neurona,* and *T. gondii*.[1144] To determine the specific differences in gene expression profiles of different *N. caninum* strains, a high density oligo DNA microarray has been developed using 63,000 distinct oligonucleotides consisting of 5692 unique *N. caninum* sequences including 1980 tentative consensus sequences and 3712 singleton ESTs from the TIGR *N. caninum* Gene Index (NCGI, release 5.0). Comparative gene expression profiling of *N. caninum* wild type and a temperature sensitive clone Ncts-8[1183] was performed using this high-density microarray and identified 111 genes that were repressed in Ncts-8, of which 58 were hypothetical protein products and 53 were annotated genes.[1145]

2.7.2 Modulation of Host Gene Expression upon Infection

Historically, the first study to describe host gene expression changes upon *N. caninum* infection employed microarray analysis and examined the events occurring in spleens of naive nonpregnant Quackenbush (Qs) and BALB/c mice at 6 h PI with Nc-NOWRA.[638] Such infections are known to induce protective host responses that will prevent transplacental transmission of a challenge given during pregnancy. Important differences between Qs and BALB/c mice were identified with regard to their transcriptional responses in the spleen to infection, and detected alterations in Jak-STAT signaling pathway (as well as interferon regulatory factors [Irfs] and other IFN-γ-regulated molecules such as GTPases) were detected. Gene ontology analyses also assigned some of the transcriptional changes to well-known cellular pathways associated with cancer, Parkinson's and Alzheimer's diseases. In a follow-up study,[639] gene set (enrichment) analyses of microarray data were carried out and additionally included also samples taken at 10 DPI. The analyses show that the major signal in the core mouse response to infection occurs early in infection and can be defined by gene ontology terms Protein Kinase Activity, Cell Proliferation, and Transcription Initiation. In addition, processes such as signaling, morphogenesis, response, and fat metabolism are also affected. At 10 DPI, genes associated with fatty acid metabolism were identified as upregulated in expression.

Also by microarray analysis, a genome-wide expression profiling of host transcriptional changes upon infection of human fibroblasts with *T. gondii* and *N. caninum* tachyzoites was conducted.[185] Unlike eukaryotic pathogens, host transcriptional changes revealed that *Neospora* is a potent activator of the type I (alpha/beta) interferon pathways typically associated with antiviral responses, and confirmed the activation of IFN-γ, expression in *T. gondii* infected cells.[186,1065] Co-infection studies reveal that *T. gondii* actively suppresses the production of type I interferon.[185] Another study showed that while *Toxoplasma* tachyzoites of all canonical types I–III mediate rapid and sustained induction of a pivotal regulator of host cell transcription, c-Myc, this induction is not seen in *N. caninum*.[697] This indicates that *Toxoplasma* and *Neospora* alter host cell functions differently, and these findings revealed the parasite specific interaction with host cells, which will have different outcome in disease progression.

More recently, high-throughput RNA sequencing (RNA-seq) technology has demonstrated a powerful way to profile the transciptome to understand the systemic alteration of host gene expression in the process of infection by pathogens with great efficiency and higher accuracy. Host gene transcriptional changes in the CNS upon *N. caninum* infection were analyzed by RNAseq in acutely ill BALB/c mice infected with *N. caninum* NC1.[1483] In infected mice, the expression of 772 mouse brain genes involved in immune responses was upregulated. Genes whose expression correlated positively and negatively with parasite numbers in the CNS were involved in the host immune response, and neuronal morphogenesis and lipid metabolic processes, respectively, suggesting that these processes contribute to the pathogenesis in *N. caninum*-infected animals. In another study, C57BL76 mice were infected with *N. caninum* NC1 to establish a chronic infection.[971] RNA-seq revealed that expression levels of IFN-γ and tumor necrosis factor (TNF)-α were highly upregulated, and the levels of neurotransmitters glutamate, glycine, gamma-aminobutyric acid, dopamine, and 5-hydroxytryptamine were altered in infected mice compared with those of uninfected mice. The expression levels of immediately early genes, c-Fos and Arc, in the brain of infected mice were lower than those of in uninfected mice. Thus, changes in the host gene expression profile associated with neuronal functions as well as immune responses lead to the better understanding of progression of the pathogenesis upon *Neospora* infections.

N. caninum protein sequences are mostly derived from gene predictions, and annotated inaccuracies can lead to erroneously predicted vaccine candidates by bioinformatics programs.[796] An RNASeq experiment was performed and the current annotation was validated. Potential discrepancies originating from a questionable start codon context and exon boundaries were identified in 1943 protein coding sequences (28% of the predicted genes). To provide a more accurate annotation of predicted proteins in *N. caninum* is a requirement for the development of an integrative bioinformatic tool that would allow us to predict the usefulness of any given protein as a vaccine candidate *in silico*.[795]

2.7.3 Proteomics Approaches

Mass spectrometry (MALDI-TOF) and 2D gel electrophoresis have been applied to study *N. caninum* proteomics.[1128,1129] Immunoproteomics was then employed to identify antigenic proteins from *N. caninum* recognized by bovine immunoglobulins M, E, A, and G,[869,1846,1847] and antigens stimulating bovine CD4[+ve] T-cell responses were identified through immunopotency screening and proteomic approaches.[1721] Protein profiles of 2 isolates of *N. caninum* (KBA-2 and JPA1) and the *T. gondii* RH strain were investigated by proteomics,[1130] demonstrating high similarities between the 2 *Neospora* isolates and clear differences to *T. gondii*. In addition, the proteome and antigenic proteome of 2 *N. caninum* isolates from geographically distinct areas (KBA-2 and VMDL-1) were comparatively assessed, and were also found to exhibit a high degree of similarities.[1848]

Comparative proteomics also identified cross-reactive and species-specific antigens between *N. caninum* and *T. gondii* tachyzoites by 2D gel electrophoresis and immunoblotting.[2175] *N. caninum* proteins were identified with differentially regulated expression during sodium nitropusside induced tachyzoite-to-bradyzoite differentiation *in vitro*.[1289] The proteome expression changes among virulent (Nc-Liv and Nc-Spain7) and attenuated *N. caninum* isolates were studied,[1678] followed by a study on the immunome, which revealed differences among 3 biologically different isolates.[1681] Another interesting study[1632] focused exclusively on the secretome of *N. caninum* by 2 approaches: The first approach was identifying the proteins present in the tachyzoite-secreted fraction (ESA) by MALDI-TOF. The second approach was determining the relative quantification through peptide stable isotope labeling of the tachyzoites submitted to an ethanol secretion stimulus (discharged tachyzoite). An interaction network was built by computational prediction involving the up- and down-regulated proteins.

2.7.4 Genetic Manipulation of *N. caninum*

The development of molecular genetic tools for *N. caninum* has opened the door to address the functional relevance of *Neospora* antigens and their role in the host–parasite interactions. Essentially the same vectors carrying *Toxoplasma* promotors and similar protocols for transfection and heterologous protein expression previously developed for *T. gondii* could be applied for *N. caninum*.[939] Several *Toxoplasma* proteins, including SAG1, GRA2, NTPase3, and ROP2 were expressed and correctly targeted in *N. caninum*.[182,940] Heterologous expression of a previously unknown ORF from *T. gondii* in *N. caninum* led to the identification of the TgROP8 gene.[182] In addition, an

N. caninum strain that exhibits stable expression of bacterial beta-glactosidase was generated,[940] and has become an important tool for the first-line screening of anti-parasitic compounds.[1421] The bradyzoite antigen NcSAG4 was overexpressed in *N. caninum* tachyzoites,[1290] and these transgenic parasites were evaluated as live vaccine candidates.[1291] In addition, a transgenic *N. caninum* strain expressing the major *T. gondii* surface antigen TgSAG1 was generated, and its protective potential as a live vaccine against *T. gondii* infection was demonstrated in mice.[2171] Another *N. caninum* strain genetically modified to express the *T. gondii* rhoptry protein ROP18 had increased virulence compared to the parental strain.[1135]

The first vector for stable expression of proteins based on chloramphenicol resistance in *N. caninum* was developed in 2014,[1586] for which expression is controlled by *Neospora* promotors. In addition, a method for the stable insertion of genes based on resistance to pyrimethamine was developed. For that, the coding sequence of NcDHFR-TS (dihydrofolate reductase-thymidylate synthase) was point mutated in 2 amino acids, generating DHFRM2M3. The DHFRM2M3 flanked by the promoter and 3′UTR of Ncdhfr-ts (Ncdhfr-DHFRM2M3) conferred resistance to pyrimethamine after transfection.[1588]

While the above-mentioned approaches aimed to express heterologous proteins in *N. caninum*, *Neospora* proteins were also expressed in other apicomplexans. An example is the expression of the *N. caninum* Kazal-type inhibitor, originally identified in *T. gondii* tachyzoites and *Plasmodium berghei* ookinetes,[1408] which triggered developmental defects in these 2 species.[1934]

The recent completion of whole genome sequencing of *Neospora* and comparative analysis with other closely related apicomplexan parasites will provide an unprecedented amount of data, which will lead us to develop multiple genetic manipulation approaches including CRISPR-Cas in the post genomic era.

2.7.5 *Neospora*–Host Cell Interactions and Invasion

The first steps that *N. caninum* tachyzoites have to achieve to ensure host cell entry are the establishment of a low-affinity contact between tachyzoite and host cell surface membranes, followed by a more stable association between tachyzoites and the host cell surface. In order to initiate host cell entry, tachyzoites then reorientate themselves perpendicularly to the host cell surface membrane, and enter the host cell cytoplasm, by advancing conoidal end first (Figures 2.11 and 2.12), until they are located inside a parasitophorous vacuole (PV).[878] Host cell invasion is an active process requiring metabolic energy solely on part of the parasite, but not on part of the host cell. However, by far not all *N. caninum* tachyzoites adhering to the host cell surface also achieve successful host cell entry.[1445] Thus, specific signals and/or receptor–ligand interactions are required that enable tachyzoites to exploit their invasive capacity.[1444] These interactions involve proteins constitutively expressed on the parasite surface, as well as secretory components,[356,651] and motility that allows host cell entry is mediated by myosin as the main motor protein.[873] Once inside the host cell, *N. caninum* resides within a PV, surrounded by a PVM. There are many similarities with *T. gondii*-infected cells, where the PV resists acidification and phagolysosomal maturation.[887] Following invasion, the lumen of the PV as well as its membrane is extensively modified through secretory products, most likely originating from rhoptry and dense granule organelles.[819,881,2053]

Initial molecular characterizations of antigens revealed 4 immunodominant antigens of 17, 29, 30, and 37 kDA.[211] Immuno-electron microscopical examination revealed that an antigen migrating at 17 kDa was associated with rhoptries (Figure 2.29) while the 29 and 30 kDa antigens were associated with the dense granules (Figure 2.30), network, and the PVM. Subsequently further antigens, mainly surface and dense granule antigens were characterized (reviewed in Reference **883**). Information on subsequent characterization and GenBank information of different antigens/ proteins of *N. caninum* is summarized in Table 2.4.

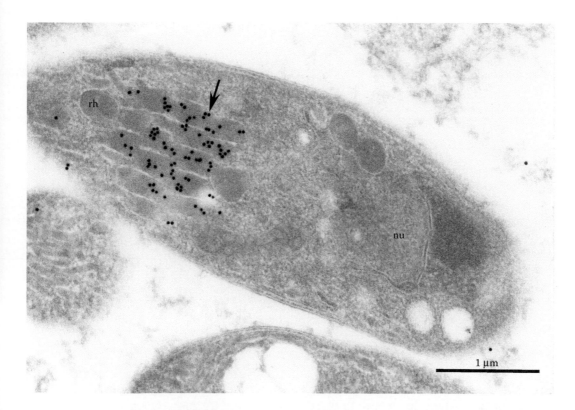

Figure 2.29 TEM of tachyzoites incubated with antibodies against a 17-kDa *N. caninum* protein. Note exclusive labeling of rhoptries (rh) (arrow). (Adapted from Bjerkås, I. et al. 1994. *Clin. Diagn. Lab. Immunol.* 1, 214–221.)

2.7.5.1 Initial Host Cell Contact is Mediated by Parasite Surface Antigens (SAGs)

The initial low affinity host–parasite contact is mediated, at least in part, through NcSAG1 and NcSRS2, the 2 major immunodominant surface antigens (SAGs) of *N. caninum* tachyzoites, both of which are inserted into the plasma membrane by a GPI-anchor.[709,941,1791,1793,1878] SAG1 and SRS2 are members of the SAG-related sequences (SRS) family of surface proteins, which plays a key role in modulation of host immunity and regulation of parasite virulence. Interestingly, comparative genomics and transcriptomics analysis of *N. caninum* and *T. gondii*[1698] has demonstrated that in *T. gondii* this family is composed of approximately 100 functional genes and 35 pseudogenes, while in *N. caninum* (Nc-Liv), the SRS gene family is expanded to 223 SRS genes and 52 pseudogenes. However, a greater number of genes is expressed in *T. gondii*. The same accounts for another gene family, SUSA (SAG-unrelated surface antigen), postulated to interact with the immune system.

The surface of *N. caninum* tachyzoites has considerable differences to *T. gondii* also with regard to surface carbohydrate content. A monoclonal antibody directed against a periodate sensitive carbohydrate epitope on a 65 kDa surface protein is available,[172] which is useful for serological diagnosis of *N. caninum* infection by competitive ELISA.[176] The presence of surface carbohydrates on *N. caninum* tachyzoites, and respective absence in *T. gondii*, was demonstrated by ruthenium red labeling and staining with the lectin Concanavalin A[710,885] and a monoclonal antibody directed against *N. caninum* carbohydrate epitopes reacted specifically with apical complex sialylated beta tubulin.[1896]

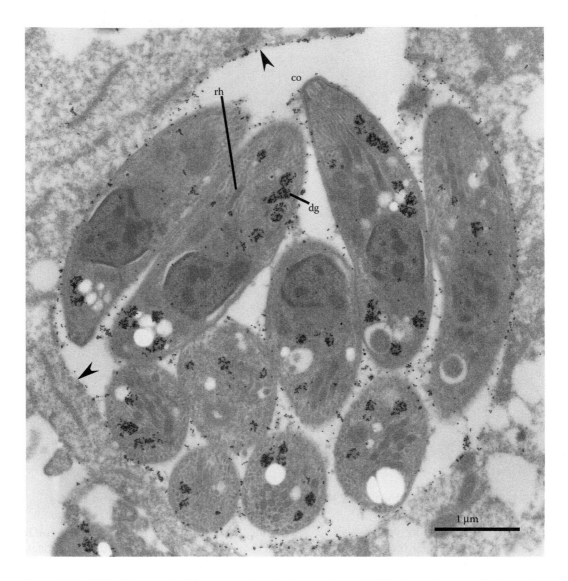

Figure 2.30   TEM of tachyzoites incubated with antibodies against a 30-kDa *N. caninum* protein. Note labeling of dense granules (dg) and the PV membrane (arrowheads). Note conoid (co) and unlabelled
rhoptries (rh). (Adapted from Bjerkås, I. et al. 1994. *Clin. Diagn. Lab. Immunol.* 1, 214–221.)

Polyclonal and monoclonal antibodies directed against SAG1 and SRS2 inhibited host cell invasion.[832,876,1464] Monoclonal antibodies directed against a 73 kDa *Neospora* surface antigen have also
been shown to inhibit host cell invasion.[2005] Pellicle and plasmalemma fractions of *N. caninum*
tachyzoites have been isolated using biochemical subcellular fractionation and have identified a
number of proteins using specific antibodies.[1138]

On the host cell side, both *N. caninum* and *T. gondii* bind to the host cell surface via binding to
sulphated host cell surface glycosaminoglycans (GAGs). However, despite the obvious similarities
between the 2 species, there are distinct differences with regard to the actual host cell surface receptors. While *N. caninum* tachyzoites preferentially bind to chondroitin sulfate GAGs and bradyzoites
appear to bind to terminal sialic acid residues on the host cell surface,[2053] it was shown that *T. gondii*
tachyzoites preferentially interact with heparan sulfate residues.[1444]

Table 2.4 Compilation of the Major Features of the Currently Characterized *N. caninum* Antigens

Name	Genbank Accession No.[a] (No. of Additional Entries)	Approx. M_r in kDa[b]	Tachyzoite Expression/ Localization	Bradyzoite/ Tissue Cyst Expression/ Localization	Proposed Function	Major Applications	Selected References
NcSAG1 (surface antigen 1)	AAD25091.1 (8)	36 (red), 29 (non-red)	Major immuno-dominant surface antigen	No detectable expression	Mediates tachyzoite–host cell adhesion and invasion	RecSAG1 for immunodiagnosis of infection in cattle and dogs by ELISA, latex agglutination or printing immunoassay test RecSAG1, vaccinia virus expressed SAG1 or DNA vaccine as experimental vaccine in mice and cattle	288, 867, 941, 942, 943, 1020, 1400, 1471, 1878, 2093, 2158
NcSRS2 (SAG related sequence 2)	CEL67525.1 (12)	43 (red), 35 (non-red)	Major immunodominant surface antigen	No detectable expression	Mediates tachyzoite–host cell adhesion and invasion	Immunodiagnosis by ELISA using bacterially expressed, baculovirus and *Pichia pastoris* expressed SRS2. Experimental vaccination in mice and cattle using native SRS2, vaccinia virus-expressed and bacterially expressed SRS2, ISCOMs, SRS2-derived lipo-peptides, recSRS2 and DNA vaccines	177, 288, 330, 334, 755, 831, 832, 876, 877, 883, 941, 942, 1465, 1466, 1471, 1472, 1476, 1608, 1616, 1664, 1863, 1877, 1901, 2002, 2162, 2178, 2179
NcIMP1 (immune-mapped protein1)	AFB35814.1 (4)	42.9 (non-red)	On the surface membrane	Not known	Homologous to a highly antigenic *Eimeria* IMP1, function is unknown, likely involved in host cell invasion	Putative vaccine candidate	403
NcSOD (superoxide dismutase)	CEL70197.1 (3)	25	Surface	Surface	High homology to other protozoan SOD. Binds Fe, possibly important for intracellular survival	Putative vaccine candidate	343
P0 phospho-protein	CEL70501.1 (4)	34	Surface and cytoplasm	Not known	Cytoplasmic P0 is possibly involved in translation, surface localized P0 function is unknown	Antigenic crossreactive with *Toxoplasma*, identified through immunoscreening of a *Neospora* cDNA expression library with anti-*Toxoplasma* antiserum	2173

(Continued)

Table 2.4 (Continued) Compilation of the Major Features of the Currently Characterized *N. caninum* Antigens

Name	Genbank Accession No.[a] (No. of Additional Entries)	Approx. M_r in kDa[b]	Tachyzoite Expression/ Localization	Bradyzoite/ Tissue Cyst Expression/ Localization	Proposed Function	Major Applications	Selected References
NcBAG1 (bradyzoite antigen 1)	BAI44436.1 (4)	25	Not detected	Cytoplasmic, not in the nucleus	Heat shock protein. Represents a stress marker and is therefore higher expressed in parasites undergoing tachyzoite-to-bradyzoite conversion	Lower cerebral parasite load in nonpregnant mice vaccinated with recombinant BAG1	**1082, 2003**
NcSAG4 (surface antigen 4)	AAW88532.1 (0)	18	Not detected	Surface antigen	Not clear, antibody responses in chronically infected mice correlated with virulence	Antibodies against recSAG4 as a marker for chronic infection. Live vaccination of mice with transgenic *Neospora* expressing SAG4 confers partial protection against vertical transmission in mice, recombinant formulations do not	**14, 15, 16, 668, 1015, 1018, 1291**
NcSRS9 (SAG-related sequence)	ABQ43784.1 (0)	40	Not detected	Surface antigen	Antibody responses in chronically infected mice correlated with virulence	Low protection in mice upon vaccination with recombinant nanoparticle encapsulated SRS9	**1015, 1715**
NcBSR4	ABQ43783.1 (0)	51	Not detected	Surface antigen	Antibody responses in chronically infected mice correlated with virulence	Low protection in mice upon vaccination with recombinant nanoparticle encapsulated SRS9	**1015, 1714**
NcCyP (cyclophilin)	CEL67263.1 (0)	19, 21, and 24	Cytoplasmic localization. Secretory protein. Caused the CCR5-dependent migration of murine and bovine cells	Not reported	Could contribute to host cell migration	Induces high IFN-γ responses by PBMC and CD4+ antigen specific T cells. Vaccination with recombinant NcCyp results in decreased cerebral parasite load in mice	**1025, 1362, 2000, 2002**

(Continued)

Table 2.4 (Continued) Compilation of the Major Features of the Currently Characterized *N. caninum* Antigens

Name	Genbank Accession No.[a] (No. of Additional Entries)	Approx. M_r in kDa[b]	Tachyzoite Expression/ Localization	Bradyzoite/ Tissue Cyst Expression/ Localization	Proposed Function	Major Applications	Selected References
Nc56	AAR02100.1	56, 43, 39	Apical end, secreted	Apical end	Unknown	Immunogenic and recognized by sera from infected cattle	1002
NcAMA1 (apical membrane antigen)	BAF45372.1 (1)	69	Micronemes	Not reported	Antibodies inhibit tachyzoite invasion into host cells	Antigenic crossreactivity with *Toxoplasma*. Vaccination with AMA1 encapsulated in oligomannose liposomes reduced offspring mortality in mice	2172, 2174
NcSUB1 (subtilisin-like serine protease1)	CEL64642.1 (0)	65 (red), 55 (non-red)	Micronemes	Not known	Secreted subtilisin-like serine protease, possibly involved in protein processing	Serodiagnosis (ELISA) using highly antigenic NcSUB1 domain	1229, 1230, 2152
NcMIC1 (microneme antigen 1)	CEL68574.1 (8)	60 (red), 50 (non-red)	Micronemes	No expression	Mediates host cell adhesion/invasion by interacting with sulfated glycosaminoglycans	Protective experimental vaccine in nonpregnant mice, protective in pregnant mice when expressed in *Brucella abortus* vaccine strain	31, 1046, 1664
NcMIC3 (microneme antigen 3)	CEL65204.1 (3)	38 (red)	Micronemes	Not known	Involved in adhesion and invasion of host cells, binds to chondroitin sulfate proteoglycans	Protection upon experimental vaccination in nonpregnant mice. Protective in pregnant mice when expressed in *B. abortus* vaccine strain. Antibodies are crossreactive with *T. gondii* MIC3	287, 1444, 1664, 1879, 2147
NcMIC4 (microneme antigen 4)	CEL64395.1 (3)	70 (red), 55 (non-red)	Micronemes	No expression	Exhibits unique lactose-binding properties and is involved in host cell invasion	Immunodominant antigen, but vaccination increases cerebral infection and clinical signs in infected mice	1047, 1447, 1897
NcMIC10 (microneme antigen 10)	CEL70962.1 (3)	18	Micronemes	Not done	Not known	Partial protection using recombinant NcMIC10 in the pregnant mouse model; Development of a circulating *Neospora* antigen-based diagnostic assay	637, 2157

(Continued)

Table 2.4 (Continued) Compilation of the Major Features of the Currently Characterized *N. caninum* Antigens

Name	Genbank Accession No.[a] (No. of Additional Entries)	Approx. *Mr* in kDa[b]	Tachyzoite Expression/ Localization	Bradyzoite/ Tissue Cyst Expression/ Localization	Proposed Function	Major Applications	Selected References
NcPDI (protein disulfide isomerase)	CEL65286.1 (5)	52	Nuclear periphery, tachyzoite surface and micronemes, found in excretory fraction	No	Chaperone activity for glycine-rich proteins. Represents one of the drug targets for the nitrothiazolide nitazoxanide	Involved in host cell adhesion. Confers protection in nonpregnant mice upon intranasal vaccination	478, 1150, 1419, 1447
NcGRA1 (dense granule antigen 1)	CEL67853.1 (7)	24	Dense granules, PV matrix	Associated with the cyst wall	Modification of the PV	Immunogenic, but no protection in pregnant mouse model	637, 2053
NcGRA2 (dense granule antigen 2)	CEL68832.1 (8)	28	Dense granules	Associated with the cyst wall	Modification of the PV	RecGRA2 as immunodiagnostic marker for acute *Neospora* infection in dogs. GRA2 as a vaccine protective only when expressed in *B. abortus* vaccine strain	637, 646, 1020, 1664
NcGRA6 (dense granule antigen 6)	CEL69583.1 (5)	37	Dense granules	Not known	Modification of the PV	Partial protection in mice when applied in *B. abortus* vaccine strain	1151
NcGRA7 (dense granule antigen 7)	CEL66347.1 (8)	17 and 33	Dense granules, PV matrix and PVM	Associated with cyst wall	Antibodies inhibit host cell invasion, but not intracellular proliferation. Proposed to be an important virulence factor since more efficiently recognized in mice infected with virulent isolates	Bacterially expressed recombinant protein or fragments for immunodiagnosis (ELISA). mAb for IHC. Good protection in mice when applied as DNA vaccine with CpG adjuvant. Low efficacy in mice as nanoparticle formulated recombinant antigen. Good protection against vertical transmission in mice and in nonpregnant cattle when applied as oligomannose-coated liposomes	14, 17, 55, 102, 1001, 1045, 1112, 1155, 1238, 1482, 1681, 2015

(Continued)

Table 2.4 (Continued) Compilation of the Major Features of the Currently Characterized *N. caninum* Antigens

Name	Genbank Accession No.[a] (No. of Additional Entries)	Approx. *Mr* in kDa[b]	Tachyzoite Expression/ Localization	Bradyzoite/ Tissue Cyst Expression/ Localization	Proposed Function	Major Applications	Selected References
NcNTPase (nucleoside tri-phosphate hydrolase)	CEL71174.1 (3)	64	Dense granules and PVM, higher expression levels in tachyzoites of virulent *Neospora* isolates	Not shown	Potentially involved in transport within the PV or in mediating targeted insertion of GRA proteins into the PV membrane and the intravacuolar network. Regulated expression, secretion, and phosphorylation suggest a potential role of different isoforms during the lytic cycle	Immunodominant protein, associated with virulence due to higher expression in isolates of high virulence	92, 1566, 1678, 1846
NcMAG1 (matrix-associated antigen 1)	CEL67854.1 (4)	67	Dense granules and PV matrix	Highly upregulated expression, dense granules, PV matrix, and tissue cyst wall	Not known	Bacterially expressed recMAG1 applied as vaccine conferred 50% protection against cerebral infection in nonpregnant mice	473, 819
NcROP2-Fam1 (ROP2-family member 1)	ADM48813.1 (4)	58, 48, 40, 36, 34, and 26	Rhoptries, PVM, surface of adherent and invaded tachyzoites, vacuoles in invaded cells	Accumulation in the cyst wall	Antibodies inhibit host cell invasion. Secreted into the host cell during invasion, could play a role in the establishment of the PV. Higher expression during egress and invasion	Protection of mice against cerebral infection by vaccination with recROP2. Only limited efficacy in pregnant mice. Combinations with MIC1 and MIC3, or ROP40, result in increased inhibition of vertical transmission and pup survival	32, 471, 473, 1564
NcROP40 (Rhoptry kinase family protein)	ALT04898.1 (3)	53, 44, 38, 32, and 28	Rhoptry bulbs, higher expression levels in tachyzoites of virulent *Neospora* isolates	Rhoptry bulbs, no association with the cyst wall detected	Unknown. Secretion into the host cell could not be detected by immunofluorescence	Vaccination of mice using a ROP40-ROP2-Fam1 combination vaccine increased pup survival	1564, 1678

Note: PBMC, peripheral blood mononuclear cell.

[a] Only 1 GenBank number of multiple entries provided, number of other identical entries are given in parentheses.

[b] Refers to relative molecular mass (*Mr*) as determined by SDS-PAGE, proteins separated under reducing conditions (red), or nonreducing conditions (non-red). Some antigens display multiple bands.

2.7.5.2 Secretory Organelle Discharge Governs Host Cell Invasion

Once the contact to the host cell surface is established, *N. caninum* tachyzoites discharge secretory organelles named micronemes, rhoptries, and dense granules in a sequential manner, either prior, during, or following host cell invasion. In this context, proteases have been shown to be critical for assembly and trafficking of microneme and ROPs in *T. gondii*.[308,516] In *N. caninum*, the serine protease NcSUB1 (a subtilisin-like serin protease), formerly known as NC-p65[1229,1230] was identified and characterized. Pepstatin, an inhibitor of aspartyl proteases, has a profound impact on *N. caninum* invasion,[1445] but the corresponding protease has not been identified. A fetuin-binding fraction of *N. caninum* tachyzoites was also shown to contain 2 proteins of 96 and 140 kDa exhibiting metalloprotease activity.[2054]

2.7.5.3 Microneme Proteins

Microneme proteins (MICs) (Figure 2.2r) that have been characterized include NcMIC1,[1046] NcMIC2,[1231] NcMIC2-like,[1585] NcMIC3,[1879] NcMIC4,[1047] NcMIC6,[1146] NcMIC10,[2157] NcSPATR (sporozoite protein with an altered thrombospondin repeat),[962] NcPDI (protein disulfide isomerase),[1447] and NcSUB1,[1229] which was shown to contain major proteolytic activity by zymography.[1230] For most MICs, secretion is initiated *in vitro* as soon as parasites initiate egress from the host cells.

As for other apicomplexans, *Neospora* MICs are deployed, and function, as protein complexes.[1019] Several adhesive domains that could interact with receptors on the surface of target cells have been identified, similar to related domains found in vertebrate extracellular matrix proteins. These include thrombospondin-(TSP-) like domains and sialic acid binding adhesive repeat-(MAR) domains in NcMIC1,[701,1046] integrin- and TSP-type I-like domains in NcMIC2[1231] and NcSPATR,[962] epidermal growth factor (EGF)-like domains in NcMIC3[1443] and NcMIC6,[1146] and apple-domains in NcMIC4.[1047]

2.7.5.4 ROPs and Rhoptry Neck Proteins (RONs)

Invasion of host cells by *Toxoplasma* takes place through the formation of a moving junction, which selectively excludes host cell plasma transmembrane proteins on the basis of their membrane anchoring.[197] Parasite actin is required for efficient host cell invasion[518] as determined by the use of specific inhibitors and parasite mutants. This mechanism also applies to *N. caninum*. Simultaneously, or right after the formation of the moving junction complex, other ROPs are injected into the host cell cytoplasm at the invasion site. Some ROPs remain associated with small vesicles that fuse with the developing PVM, whereas others remain soluble and are targeted to other sites within the host cell.

The most prominent group of ROPs is the ROP2 family, which has recently been cataloged in *N. caninum*.[1933] NcROP2-Fam1, and NcROP40, 2 members of the ROP2 family in *N. caninum*, represent the best characterized.[32,1565] Both are promising vaccine candidates and exerted a synergistic effect in terms of protection against vertical transmission in mouse models,[472,1564] which suggests that they may be relevant for parasite pathogenicity. Transcription of both ROPs is highly increased during tachyzoite egress and invasion. NcROP2-Fam1 is secreted during invasion, and antibodies against NcROP2 peptides inhibited host cell entry.[471] NcROP2-Fam1 was also found to be associated with the nascent PVM and vacuoles surrounding the host cell nucleus and, in some instances, the surface of intracellular parasites. In contrast, NcROP40 secretion could not be detected by antibody labeling techniques.[32,1565]

The *Neospora* rhoptry proteome was studied by isolation of subcellular rhoptry fractions, 2D SDS-PAGE and LC/MS–MS and resulted in the identification of 8 potentially novel rhoptry components (NcROP1, 5, 8, 30 and NcRON2, 3, 4, 8), several kinases, proteases, and phosphatases with a high homology to those previously found in *T. gondii*.[1292,1874] Comparative genomics and transcriptomics analyses of *T. gondii* and *N. caninum* rhoptry genes have shown that several *T. gondii* rhoptry genes are missing from *N. caninum*, including the locus that encodes ROP2A, ROP2B, and ROP8.[1698]

2.7.5.5 Dense Granule Proteins

Another set of secretory proteins, dense granule (GRA) proteins (Figure 2.2s), are released into the PV once the invasion process is achieved. Some GRA proteins have been identified as components of the PV membranous network that constitutes the matrix of the PV and provides the metabolically active compartment that favors parasite replication. Other GRA proteins are involved in the modification of the PVM, and contribute to the membrane resistance to fusion with host lysosomes.[1345] More than 20 GRA proteins have been reported for *T. gondii*, and 15 have been identified in *N. caninum*, either as proteins or on the transcriptional level.[1698] In *T. gondii*, a number of GRA proteins, including TgGRA7, TgGRA15, TgGRA16, TgGRA24, TgGRA25, and TgNTPase I, have been implicated in virulence mechanisms, and TgNTPase and the TgGRA7 are essential for the replication of *T. gondii* tachyzoites.

In contrast, GRA proteins in *N. caninum* have not been as extensively characterized. NcGRA7 and NcGRA6, identified as immunodominant antigens through cDNA expression library screening using sera from *N. caninum* infected cattle, were originally named NcDG1[1112] and NcDG2,[1151] respectively. NcGRA7 was shown to be secreted into the PV and localized to the PV matrix and the PVM.[881] Immunohistochemistry employing monoclonal and affinity purified polyclonal antibodies showed that NcGRA7 was expressed in both tachyzoites and bradyzoites.[709,1791] It was shown that antibodies directed against NcGRA7 inhibited the entry of *N. caninum* tachyzoites into baby hamster kidney cell cultures, while addition of antibodies after host cell entry did not affect the intracellular proliferation of tachyzoites.[102] A similar observation was made using a monoclonal antibody against NcGRA7 and MARC-145 cells.[17] A 17–19 kDa antigen protein fraction that is recognized by sera of *N. caninum*-infected cattle and used for immunodiagnosis was found to be encoded by the NcGRA7 gene.[55] As evidenced by a comparative immunomics study, NcGRA7 is efficiently recognized by sera of mice infected with virulent isolates of *N. caninum* such as Nc-Liv or Nc-Spain7, but much less by sera of mice infected with a non-virulent isolate, Nc-SpainH1, and it was suggested that this protein could represent an important virulence factor.[1681]

Similar to its counterpart in *T. gondii*, *Neospora* NcGRA3 contains an N-terminal secretory signal sequence and a transmembrane domain, which is consistent with its insertion into the PVM as shown by monoclonal antibody labeling.[894] Other GRA proteins such as NcGRA1, NcGRA2, and NcMAG1 (matrix antigen 1), also homologs to respective *T. gondii* GRA antigens, are abundantly expressed within the tachyzoite dense granules and are also found within the matrix of the PV, and during tachyzoite-to-bradyzoite differentiation *in vitro*, their localization shifts toward the periphery of the tissue cyst.[819,2053] In analogy to *T. gondii*, *N. caninum* expresses a potent type-I nucleoside triphosphate hydrolase (NTPase), which is also localized within dense granules.[92] Comparative proteomics showed an upregulated expression of NcNTPase in virulent versus non-virulent isolates.[1678] Further investigations demonstrated the existence of up to four NcNTPase alleles in Nc-Liv, and that transcription, NTPase phosphorylation and secretion coincides with tachyzoite egress.[1566]

An unusual single-domain Kazal serine protease inhibitor (serpin) that is localized in the dense granules and discharged into the PV was described and named NcPI-S.[1408] Recombinant NcPI-S efficiently inhibited subtilisin activity, while little or no activity against elastase or chymotrypsin was found.[250] Serpins are believed to protect parasites from proteinase-rich environments and to modulate the host inflammatory response, and the action of proteinases secreted by the parasites themselves.[250]

In *T. gondii*, a set of dense granule proteins (also named MAF1 proteins, mitochondria association factors) are secreted by the parasites and associate with the PVM and mediate the association between PV and host cell mitochondria for type I and type III strains.[7,1594] Host cell mitochondrial association has also been described in *N. caninum*-infected cells, but this can vary between different strains, as it does for *T. gondii*. A recent study[8] has shown that the *MAF1* gene locus in *T. gondii* type I and III strains has duplicated, while no such gene duplication has been identified in *N. caninum* NC1. *T. gondii* type I and III strains evolved a *MAF1* paralog which is much more efficient in recruiting host cell mitochondria to the PV, and expression of the *T. gondii MAF1* paralog in *N. caninum* NC1 also resulted in a much more efficient mitochondrial association with the PV. This is especially interesting since it was shown that mitochondrial association was linked to cytokine responses and virulence of *T. gondii* strains *in vivo*, and the same is most likely the case for *N. caninum*.

2.7.5.6 *Intracellular Host Cell Modulation and Parasite–Host Cell Crosstalk*

Earlier studies focused on the relationship between host cell apoptosis and *N. caninum* infection. It was demonstrated that infection of IFN-γ-treated BALB/3T3 clone A31 fibroblasts with *N. caninum* tachyzoites *in vitro* causes apoptosis as evidenced by increased DNA-fragmentation, and increased caspase-3 and caspase-8 activity.[1468] The administration of respective inhibitors such as anti-mouse FasL monoclonal antibody inhibited cell death.[1475] Inhibition of host cell apoptosis is a critical issue, especially for the chronic phase of infection, during which *N. caninum* bradyzoites form tissue cysts that can persist within infected tissue for a long time. These *N. caninum* bradyzoites, as well as their host cells, are predestined for long-term survival. It is conceivable, that host cells are modulated accordingly, by either utilizing the PVM as a signaling platform, or by actively secreting bioactive parasite-derived factors into the cytoplasm. *T. gondii* inhibits host cell apoptosis by inducing the activation of the transcription factor Nf-κB, which in turn regulates the expression of inhibitors of apoptosis in the host cell.[226] The activation of Nf-κB pathway by *T. gondii* correlates with the localization of phosphorylated I-κB at the PVM. In an interesting study[895] it was shown that, in contrast to *T. gondii*, *N. caninum* inhibits host cell apoptosis in the absence of discernible Nf-κB activation. Thus, receptor-mediated apoptosis is repressed and the executioner caspase 3 does not become activated in the infected cells. Although a putative I-κB kinase activity was detected in *N. caninum* extracts, thereby implying that this parasite is capable of modulating Nf-κB translocation into the host cell nucleus, a profound and sustained activation of the Nf-κB pathway is not central to the ability of *N. caninum* to prevent host cell apoptosis.

In *T. gondii*, some ROPs injected into the host cell cytoplasm have been shown to be involved in manipulation of host cell gene expression through kinases, and thus represent crucial virulence factors involved in the crosstalk between parasite and host cells.[486] TgROP16 and TgROP18 act as hypervariable protein kinases that are responsible for the high virulence of certain strains of *T. gondii*. TgROP18 phosphorylates host immunity-related GTPases (IRGs) and activating transcription factor-6 beta (ATF-6 beta), and TgROP16 is translocated to the host cell nucleus, subverts STAT3/6 (signal transducer and activator of transcription 3/6) signaling, and as a consequence interleukin 12 (IL-12) production in infected host cells. In this respect, there are clear differences between *T. gondii* and *N. caninum*, as shown by genomic and transcriptomic studies.[1698] In *N. caninum* Nc-Liv, ROP18 is a pseudogene due to several interrupting stop codons in the sequence syntenic with the *Toxoplasma* genes, and *N. caninum* is not able to phosphorylate IRGs without a functional copy of ROP18. However, phenotype and virulence assays showed that the expression of TgROP18 in *N. caninum* did not affect the motility and cell invasion, but resulted in a significant increase in intracellular parasite proliferation and virulence in mice. IRG-phosphorylation assay showed that a transgenic parasite NC1-TgROP18 was able to phosphorylate IRGs as *T. gondii* did.[1135] In addition, while NcROP16 possesses the key active-site leucine residue for STAT3 activation, the gene is not expressed in Nc-Liv tachyzoites cultured in HFF, thus it is likely that *N. caninum* infection does not activate STAT3 due to the lack of NcROP16 expression.[1698]

Infection by apicomplexan parasites modulates the host cell cytoskeleton dynamics.[305,306] Microtubules start to surround the parasite at the first minute of invasion by forming a cone-shape microtubule network, and a microtubule ring on the host cell is observed around the parasite entrance site. In a recent study that compared microtubule interactions in *N. caninum* and *T. gondii* infected cells, microtubules were also closely apposed to *N. caninum* containing PVs, but at a statistically significant lower density compared to what was observed in *T. gondii* infected cells.[1487] Both parasites also affect the positioning of the centrosome, which is closely associated with the nucleus in uninfected cells, but during *N. caninum* and *T. gondii* infection the host cell centrosome is recruited away from the nuclear membrane and positioned near to the PVM. *N. caninum* infection also modulated the spatial organization of host cell actin microfilaments and intermediate filaments, as visualized in astrocytes of organotypic rat brain slice cultures.[2051] F-actin bundles and glial fibrillary acid protein filaments were found in close juxtaposition to the cytoplasmic side of the PVM.

These pronounced changes in the cytoskeletal organization also reflect on an altered migration of the host cell: infection of DCs with *N. caninum* tachyzoites was shown to confer a hypermigratory phenotype.[370] In addition, an *N. caninum* macrophage migration inhibitory factor (NcMIF) was described.[1651] Immunoelectron microscopy localized NcMIF in the micronemes, rhoptries, dense granules, and nuclei. NcMIF was abundant in the tachyzoite lysate and present in excretory and secretory antigen preparations. Total and secretory NcMIF was more abundant in a nonpathologic temperature-sensitive *N. caninum* mutant Ncts-8, than in the wild-type NC1 isolate. However, the functional and immunoregulatory properties of NcMIF were not demonstrated.

2.7.5.7 Monoclonal Antibodies

Numerous monoclonal antibodies have been generated and some of them cross-react with *T. gondii* (Table 2.5).

2.8 RODENT MODELS OF NEOSPOROSIS

2.8.1 Mice

2.8.1.1 Outbred Mice

During the first attempt to culture *N. caninum* in 1988, it became clear that the outbred mice are resistant to this parasite.[528] Administration of immunosppressive drugs can overcome part of this resistance. Cortisone, methylprednisolone acetate (MPA), and dexamethasone have been used for this purpose. Cortisone acetate (2.5 mg injected SC) is relatively inexpensive but needs to be given twice weekly. MPA (2–4 mg MPA injected IM) is more expensive but its efficacy lasts for 3 weeks after a single dose. Dexamethasone oral (10 μg/mL) can be given in drinking water. Although results are better if these drugs are given a week before the start of the experiment, immunosuppressive effects are achieved even when administered on the day of the *N. caninum* inoculation. Other immunosuppressive drugs (cyclophosphamide 15–20 mg daily inoculated IP, vinblastine 1.2 mg, daily IP) have also been used. In our experience MPA is most convenient and efficacious. These drugs can be administered until mice become ill. To overcome the effect of concurrent bacterial infections, mice can be administered penicillin and streptomycin that have no effect on *N. caninum* infection. Immunosuppressed outbred mice have been used to study pathogenesis,[365,1159] tissue cyst formation,[1159,1326,1327] parasite propagation,[1733] chemotherapy,[1162] and for bioassays.[603]

Swiss Webster (SW) 25–30 g mice inoculated with 2×10^5 NC1 tachyzoites and given 4 mg MPA died of hepatitis, pneumonia, pancreatitis, and encephalitis by 30 DPI, while those given 2 mg MPA developed milder illness, predominantly encephalitis. Tissue cysts were seen in histological

Table 2.5 Monoclonal Antibodies against *Neospora*

mAb Name(s)	Immunoglobulin Isotype	Reactivity under Nonreducing Conditions (*Mr*)	Reactivity under Reducing Conditions (*Mr*)	Localization[a]; Identity of Recognized Antigen	Remarks	Reference
6G7	IgG2a	ND	97.4, 90, 80, 70, 43, 38.5, 34, 31	Surface, PV, DG, MIC, ROP	Staining for *N. caninum* by IHC. React with a common epitope on many proteins, cross-reactivity with a 107 kDa antigen from *T. gondii*	**360, 361**
4A4-2	IgM	ND	65	Surface	*m*-Periodate-sensitive epitope, mAb used for serodiagnosis	**172**
D9	IgG2a	ND	42	DG	IFAT localization, IHC	**1920**
Ncmab-4	IgG2a	30, 32	30, 32 (NcSAG1)	Surface, DG; NcSAG1	IB, IFAT, immunogold on-section labeling	**217**
Ncmab-7	IgG2b	18	18	Surface	IB, *m*-periodate-sensitive epitope	**217**
Ncmab-10/-17	IgG2a/IgG1	41	41 (NcSRS2)	Surface; NcSRS2	Pre-embedding immuno-EM	**217**
Ncmab-13/-24	IgG2a	Not reactive	65	Apical part	Only binds to permeabilized tachyzoites	**217**
5H5/4H7	ND	35	ND	Surface; NcSRS2	Generated against tachyzoite membrane fractions; IB	**941**
6C11	ND	29	ND	Surface; NcSAG1	Generated against tachyzoite membrane fractions; IB; does not recognize *N. hughesi*[1285]	**941**
6D12-G10	ND	21	21	Apical part	Reacts with a conoid epitope common to 7 *Eimeria* species and *Toxoplasma*; IFAT	**1781**
1.11.1	IgG1	40 (38, 36)[a]	Not reactive	Surface	IB, *m*-periodate-sensitive epitope; IFAT	**1791**
5.2.15	IgG1	38 (36, 33)	Not reactive	Surface; NcSRS2	IB; localization by IFAT and immunogold TEM	**1791**
4.11.5	IgG2	33 (28, 24, 22)	33 (28, 24, 22)	DG, PV; NcGRA7	IB; IFAT localization and immunogold TEM	**1791**
4.7.12	IgG2	19	19	Surface	IB; IFAT localization	**1791**
1E3/1F3	IgM	Not reactive	70	Surface	IB; IFAT localization; Invasion inhibition	**1464**

(Continued)

Table 2.5 (Continued) Monoclonal Antibodies against Neospora

mAb Name(s)	Immunoglobulin Isotype	Reactivity under Nonreducing Conditions (Mr)	Reactivity under Reducing Conditions (Mr)	Localization[a]; Identity of Recognized Antigen	Remarks	Reference
1B8/2C8/2G2	IgG1	43	43	Surface; NcSRS2	IB; IFAT; invasion inhibition, react with NcSRS2 expressed in vaccinia virus	1464
1A5	IgG2a	42	42	Apical end	IB; IFAT; invasion inhibition	1464
1C7/2C11	IgM/IgG1	36	Not reactive/36	Surface; NcSAG1	IB; IFAT localization; invasion inhibition, react with NcSAG1 expressed in vaccinia virus	1464
3NCT/1-3NCT/12	11xIgG1/1xIgM	ND	73	Surface	IB; inhibition of invasion	2005
mAbCC2	IgG2b	ND	40 (T. gondii tachyzoites); 115 (bradyzoites)	DG, PV. Tissue cyst wall	IB, Toxoplasma cross-reactive epitope, ideal for monitoring stage conversion	2052, 2053
9D12/10C7/11C4	IgG2a/IgG2a/IgG1	ND	28–76	Surface	Cross-reactive with Toxoplasma; growth inhibition	1149
9E8	IgG2b	ND	50	Interior; protein disulfide isomerase (NcPDI)	Cross-reactive with Toxoplasma; growth inhibition	1149
10F7	IgG1	ND	35, 14	Interior; heat-shock protein 70 (HSP70)	Cross-reactive with Toxoplasma; growth inhibition	1149
10A10/11A10/11D5	IgG1	ND	64	Interior; ribosomal protein 1 (RP1)	Cross-reactive with Toxoplasma	1149
10G6	IgG1	Not reactive	Not reactive	Surface	Cross-reactive with Toxoplasma; growth inhibition	1149
10F10	IgM	Not reactive	Not reactive	Apical end	Cross-reactive with Toxoplasma	1149
100.2.4/119.4.9.10	IgG2a/IgG1	–	42	Surface; NcSRS2	73/78% inhibition of trophoblasts and Vero cells	832
119.14.11.14/119.16.13. 37/119.20.14.22	IgE	–	42	Surface; NcSRS2	No reduction of invasion	832
119/3.13.1	IgG1	–	–	Polyspecific carbohydrate epitope on several proteins	m-Periodate sensitive epitope	832

(Continued)

Table 2.5 (Continued) Monoclonal Antibodies against Neospora

mAb Name(s)	Immunoglobulin Isotype	Reactivity under Nonreducing Conditions (Mr)	Reactivity under Reducing Conditions (Mr)	Localization[a]; Identity of Recognized Antigen	Remarks	Reference
6D12/6C6	IgG1/IgM	48–52/180–190	48–52/180–190	Interior, apical two-thirds; glycosylated form of beta-tubulin	8 mAb against m-periodate sensitive epitopes. MS/MS identification of the 52 kDa protein. Sialidase treatment impairs 6D12 binding	1896
46 different mAbs	ND	5 reactive proteins	41 reactive proteins: surface 16, 61 inner membrane complex IMC; 2 dense granules, rhoptries	2 surface: 16, 61; 3 inner membrane complex IMC: 43, 36 >250; 5 mitochondria: 33–60; 2 apicoplast: 60; 10 ROP: 34–80; 3 ROP neck: 22–145; 6 PV: 15–50; 2 DG: 20, 40; 11 MIC: 32–72; 2 internal spots: 32, 70	IFAT localization of all mAbs, selected proteins defined by MS/MS Providing a toolbox for further studies on Neospora proteins	1874
A10/H3	mAbs against rSAG1 generated by phage display technology	Binding to recombinant NcSAG1 by ELISA	–	Surface: NcSAG1	Binding of A10 and H3 to recombinant SAG1 is inhibited by cattle sera; IFAT on tachyzoites for localization	510
4.15.15/4.11.5/1/24–12	IgG2 IgG2a/IgG2b	ND[1791]	ND[1791]	Surface; NcSRS2 (4.15.15) DG; NcGRA7 (4.11.5, 1/24–12)	–	1791, 2015
A6/E1/H3	mAbs against rSRS2 generated by phage display technology	Binding to recombinant SRS2 by ELISA	ND	Surface; NcSRS2	–	511
K8/15-15	IgG	ND	80–350	Tissue cysts of N. caninum	Crossreactive with T. gondii sporozoites. Tissue cyst and sporocyst walls of H. hammondi and H. heydorni, and sporocyst walls of Cystoisospora felis	774

Note: ND = no data available; PV = parasitophorous vacuole; DG = dense granules; MIC = micronemes; ROP = rhoptries.

a Localization assessed either by immunofluorescence or immunogold TEM.

b Parentheses indicate Mr of weakly reactive bands on IBs.

brain sections 21 DPI.[1159] Similar results were obtained in other investigations. Additionally, parasite DNA and lesions were quantified in liver, lungs, heart, and brain of ICR mice given immunosuppressive drugs.[365]

SW mice orally inoculated with bradyzoites became infected indicating the bradyzoites are infectious by the oral route, and clinical disease was more severe in mice inoculated with NC1 versus NC3 strain.[1163]

Outbred Qs mice, are rather resistant against *N. caninum*, show a Th1-type immune response characterized by high IFN-γ and low IL-4 production as well as dominant IgG2a antibody responses.[1352,1653]

A recent study[1303] demonstrated that outbred CD1 mice infected with a low virulence *N. caninum* isolate (NcIs491) transmitted the parasite to their offspring during repeated pregnancies, thus this represents an interesting model to study congenital infection.

Transplacental and lactogenic transmission of *N. caninum* has been demonstrated in mice.[362] Information is summarized in Table 2.6.

2.8.1.2 Inbred Mice

BALB/c, C57BL/6, B10.D3, and other inbred mouse strains have been used to study pathogenesis, assess immunity parameters, and congenital transmission of neosporosis.

2.8.1.2.1 Pathogenesis

The outcome of infection varies with the strain of the mouse, parasite strain, and the dose. BALB/c mice infected with 10^5 NC1 tachyzoites developed neurological signs with tachyzoites-associated meningoencephalitis.[1172] Lesions were confined to the CNS. Depending on the dose, DNA was detected in blood 1–7 DPI. During the acute phase of infection *N. caninum* was detected in visceral tissues, and during the chronic phase the infection was confined to the brain.[31,366]

Of *N. caninum* NC1-induced infections in 3 strains of inbred mice (BALB/c, C57BL/6, B10.D3) compared, there was no difference in susceptibility of BALB/c and C57BL/6 mice and in general the number of lesions increased with dose; B10.D3 mice were least susceptible.[1207] The predominant lesion was encephalitis associated with intralesional tachyzoites demonstrable by immunohistochemical (IHC) staining. These results were confirmed by another group of researchers using detection of parasite DNA by quantitative polymerase chain reaction (PCR).[365] No clinical signs of neosporosis were observed throughout an observation period of 21 DPI with 5×10^6 NC1 tachyzoites,[1663] but also it was shown that these mice were highly susceptible to an infection with 2×10^7 tachyzoites. Other studies failed to observe clinical signs in C57BL/6 mice after a challenge with 10^5 NC1 tachyzoites as long as 44 days, but detected severe lesion in μMT (B-cell deficient) mice.[652]

Nc-Liv-induced infections were compared in BALB/c, C57BL/6, and CBA/Ca mice.[1372] Groups of 5 mice received 10^6, 5×10^6, or 25×10^6 tachyzoites by IP injection. The mortality rates were the highest in C57BL76 mice, in BALB/c mice mortality was 0% at the lowest and 100% at the highest infection dose. CBA/Ca mice were the most resistant, with no animal succumbing to infection at a dose of 10^6 and 5×10^6, but 100% mortality after infection with 25×10^6 tachyzoites.

BALB/c mice have been useful in defining virulence of different *Neospora* isolates[98,366,730]; certain strains (Nc-Spain5H, Nc-Spain7 and 9; and Nc-SweB1) were found more pathogenic to BALB/c mice than others.

2.8.1.2.2 Congenital Model

In 1996 the suitability of BALB/c mice as a model for reproductive loss due to *N. caninum* infection was reported.[1206] In mice infected prior to pregnancy, no difference was present in resorptions

Table 2.6 Experimental Congenital Neosporosis in Mice

Mouse Strain	Parasite Isolate	Infection Dose, Route	Infection Day	Experimental Remarks	Parasite Detection in Pups	Fate of Pups	Reference
				Outbred			
SW	NC1	$5.8-8.0 \times 10^6$, SC	Days 5–18	Pups delivered by Caesarean section or naturally	Bioassay in cell culture of pup tissues on day of delivery. Grossly visible foci of necrosis on fetal brains	Not tested	362
Qs	Nc-Liv, Nc-SweB1	10^4, 10^6, or 10^7, SC or IP	Day 5 or day 8 of gestation	ELISA for analysis of maternal immune responses	Offspring were collected at day 7 PP, PCR on brain tissue	Vertical transmission in mice given 10^6 Nc-Liv at day 5 of gestation. Maternal antibody response observed in mice infected with Nc-Liv or Nc-SweB1 at days 5 and 8 of gestation	1653
CD1	NcIs491	2×10^7, IP	30 days pre-pregnancy to establish chronic infection, then mating	The same dams were mated to achieve 4 consecutive pregnancies	PCR analyses of brain tissue, serology for measuring IgG levels by IFAT	No clinical signs in adult female mice, but high antibody titers by IFAT. No impact on fertility, but decreasing mortality of pups with successive pregnancies (from 17%, 6.2% to 2.5%), and decreasing transmission rate (74%–40%)	1303
				Inbred			
BALB/c	NC1	2×10^6, SC	4 weeks pre-pregnancy, day 5 or day 10 of gestation	Mice euthanized 11–19 day of gestation	Histopathology and IHC. *Neospora* was detected histologically in placenta, fetal brain, and muscle when dams were infected at 10 day of gestation	Pre-pregnancy infection: no increased resorptions, increased resorption when infected at days 5 and 10. No reproductive loss in the group infected at day 10 of gestation	1206

(Continued)

Table 2.6 (Continued) Experimental Congenital Neosporosis in Mice

Mouse Strain	Parasite Isolate	Infection Dose, Route	Infection Day	Experimental Remarks	Parasite Detection in Pups	Fate of Pups	Reference
BALB/c	NC1	10^5, SC	Days 8–15 of gestation	Pups analyzed at days 2–23 PP	PCR in brain and lung tissue	Upon infection at days 13–15 of gestation, transmission was only partial, while between days 8 and 12 of gestation, infection resulted in 100% transmission to offspring. In pups of 2–4 days of age, parasites were detected in lung and brain, in 7–23 days old pups, only the brain was PCR positive	1153
BALB/c	Ovine isolate	2×10^6, IP	Group 1: 28–100 days pre-pregnancy; group 2: days 7–19 of gestation; group 3: offspring from group 2 rebred at days 28 after delivery	Dams and pups euthanized at birth	PCR in brain, lymph nodes, and spleen	50% congenital transmission in group 1; 76% in mice from group 2, 86% in the pups from infected and rebred offspring from group 3. Dams and pups euthanized at birth	1511
BALB/c	NC1	2×10^6, SC	Infection at day 0, 7, or 14 of gestation	Dams killed during gestation or PP. Pups killed on days 1 and 7 PP	In different tissues in dams by nested PCR	Infection at day 7 of gestation produces highest level of vertical transmission, increased fetal mortality and stillbirth and a decrease in litter size	1221
BALB/c	NC1	2×10^6, SC	Infection at day 0, 7, or 14 of gestation	Focus on postnatal development of offspring for 60 days PP. Some litters sacrificed on day 30 PP	Nested PCR in brain and lung, histopathology in 3 neonates. Parasite DNA was detected in placenta on day 3 and in fetal tissues on day 7	Infection during pregnancy on day 0 results in 69% mortality of pups within 60 days of postnatal development, on day 7 100%, and day 14 46%	1222, 1223

(Continued)

Table 2.6 (Continued) Experimental Congenital Neosporosis in Mice

Mouse Strain	Parasite Isolate	Infection Dose, Route	Infection Day	Experimental Remarks	Parasite Detection in Pups	Fate of Pups	Reference
BALB/c	Nc-Spain7, Nc-Spain 3H	10^6, SC	Infection 90 days pre-pregnancy, then mating	Morbidity, mortality, vertical transmission, and humoral immune responses recorded for 2 consecutive generations. Dams were sacrificed 30 days PP, male pups at 50 days PP, seropositive female pups were re-mated	PCR in brain and lung tissue	First generation pups with 40%–50% transmission and high mortality rates in Nc-Spain7 mice and lower in Nc-Spain3H infected mice. second-generation pups: 7.7% and 17.1% transmission, respectively, and even lower (0% and 8.7%) in the third generation	**1016**
BALB/c	Nc-Spain3H, Nc-Spain8	10^6, SC	Infection 90 days pre-pregnancy, then mating	Recording of morbidity, mortality, and vertical transmission as above. NcGRA7 serology to detect reactivation during pregnancy. Focus on endogenous transplacental transmission	PCR in brain and lung tissue	Reactivation of infection in several congenitally infected nonpregnant females (second generation) as shown by clinical signs after mating and increased antibody levels (IgG1, IgG2a, and specific antibodies against rNcGRA7). Low level of congenital transmission	**1017**
BALB/c	NC-GER2, NC-GER3, NC-GER6, NC-6Argentina NC-Bahia Nc-Liv	2×10^6, SC	Infection on day 7 of gestation	Assessment of morbidity, mortality, and antibody responses in dams and offspring, and parasite transmission to the progeny. Postnatal follow-up for dams until day 30 PP, for pups until day 50 PP	PCR on brain and lung tissue	The mortality rates ranged from 19% to 100%, and rates of vertical transmission were between 9% and 100%, depending on the isolate. Highest IgG responses in mice infected with Nc-Liv and NC-Bahia, which were the most virulent isolates	**482**

(Continued)

Table 2.6 (Continued) Experimental Congenital Neosporosis in Mice

Mouse Strain	Parasite Isolate	Infection Dose, Route	Infection Day	Experimental Remarks	Parasite Detection in Pups	Fate of Pups	Reference
BALB/c	Nc-Spain 2H, 3H, 4H, 5H, Nc-Spain 6, 7, 8, 9, 10, Nc-Liv	2×10^6, SC	Infection on day 7 of gestation	Vertical transmission rates and neonatal mortality assessed	PCR on brain tissue	Vertical transmission rates varied from 52.6% to 100%. Postnatal follow-up revealed large differences in survival time and neonatal mortality between isolates	**1675**
BALB/c	Nc-Spain7	10^2 to 2×10^6, SC	Day 7 of gestation	Assessment of clinical outcome, vertical transmission, parasite burden, and antibody responses in dams and pups after 30 days PP. Follow-up period	PCR on brain tissue of dams and pups	Dams from all infected groups had neurological signs. Only 24% survival of pups was found in the group infected with 100 tachyzoites. Infection with 10^5 tachyzoites resulted in seropositivity in dams, cerebral parasite burden in dams and 100% mortality rate in pups, similar to the group infected with 2×10^6 tachyzoites	**86**
C57BL/6	NC1	5×10^6, IP	14 days pre-pregnancy or 12–14 days of gestation	Assessment of histopathological lesions in brain, morbidity, mortality, and vertical transmission. Fetuses were collected in utero between days 18 and 20 of pregnancy	PCR on whole macerated fetal tissue	The rate of vertical transmission was 100% and 90.5% for mice infected during pregnancy and mice infected before mating, respectively. No postnatal follow-up	**1663**
CBA/Ca	NC1	5×10^6, SC	8–10 days gestation for acute infection. Subsequent 3 litters tested from chronically infected dams		PCR on pooled tissues of each fetus	18 (90%) of 20 mice infected when dams inoculated 8–10 days gestation. In chronically infected dams transmission decreased to 14.6% (6/41), 3.7% (2/54), and 4.4% (2/45) in subsequent litters	**1707**

Note: PP = post partum.

between infected and control mice, although litter size was decreased in the infected mice. In mice infected at day 5 of gestation, resorptions were increased. In mice infected at day 10 of gestation, *N. caninum* tachyzoites were identified in placenta and fetal muscle and neural tissue. In the placenta, there was multifocal necrosis and hemorrhage with intralesional tachyzoites.

Not only *N. caninum* can be transmitted congenitally during acute infection in BALB/c, it is also transmissible congenitally from chronically infected mice.[1511] In BALB/c mice inoculated with *N. caninum* NC1 tachyzoites, 76% of the neonates were infected when dams were inoculated during pregnancy and 50% of neonates born to chronically infected dams were infected.[1511] The offspring of the group infected during pregnancy were mated again, and 86% of their resulting offspring mice were found to be infected.

Some studies have focused on the postnatal development of BALB/c offspring mice born from dams infected with *N. caninum* tachyzoites during pregnancy.[1222] Infection with *N. caninum* on day 7 of pregnancy resulted in 100% mortality, while mortality was lower when infection took place at days 0 and 14 of gestation. In general, infection provoked a delay in the development of neonates, and caused clinical signs and severe histopathological lesions.

Attempts to establish a BALB/c model to study endogenous transplacental transmission failed. While mice that had been congenitally infected with a low-to-moderate virulence *N. caninum* isolate exhibited clinical reactivation during the mating period, transmission to the next generation did not occur.[1016,1017]

Various studies on murine congenital neosporosis are summarized in Table 2.6. However, results are not comparable because they lack standardization: different groups have worked with different mouse strains, different parasite isolates, they have employed different culture techniques and varying routes of inoculation. The C57BL/6 mouse model was optimized for its use in studies on congenital *N. caninum* infections.[1663] In addition, conditions for BALB/c mouse model for exogenous transplacental *N. caninum* transmission were standardized employing the virulent Nc-Spain7 isolate.[86]

2.8.1.2.3 Immune Mediation

2.8.1.2.3.1 Immunity in Nonpregnant Mice — Studies concerning immunity in nonpregnant mice are summarized in Table 2.7.

Innate Immune Response: Exposure of bone marrow derived dendritic cells (DCs) to *N. caninum* tachyzoites resulted in enhanced expression of IL-12p40, IL-10, and tumor necrosis factor α (TNF-α), whereas IL-4 RNA expression was not detected.[1916] Similarly, exposure of DCs and spleen cells to whole *Neospora* tachyzoites (live, freeze-killed) or tachyzoite lysates (whole or insoluble antigen) stimulated high levels of IL-12, IFN-γ, and TNF-α, while heat-killed tachyzoites and soluble antigens stimulated low levels of these cytokines. Whole *N. caninum* tachyzoites were more effective in inducing IL-12, IFN-γ, and TNF-α than the lysate antigen preparations.[664a]

It was shown that *N. caninum* invaded and activated DC, as well as macrophages, which in turn activated T-cells.[502] The production of IFN-γ by T-cells was IL-12 dependent when they were activated by DC, but not when they were activated by macrophages. Macrophages from *N. caninum*-infected mice showed increased activation, and enhanced IL-6, IL-12p40, and IFN-γ production.[3] However, live *N. caninum* tachyzoites as well as *N. caninum* tachyzoite extracts were also shown to induce the rapid phosphorylation and thus activation of the p38 mitogen activated protein kinase (p38MAPK) in macrophages, as a consequence reduced B7/MHC expression and reduced antigen presentation were observed. Inhibition of p38 activation using a specific inhibitor in experimentally infected mice resulted in diminished parasite burden and increased survival, demonstrating that *N. caninum*-mediated phosphorylation of p38MAPK is induced to modulate the immune response.[1415a] Macrophage-depleted mice were more susceptible to *N. caninum* infection. Increased mortality and neurological impairment were observed in the *N. caninum*-infected mice lacking the chemokine receptor CCR5. Poor migration of DCs and natural killer T cells to the site of infection was

Table 2.7 Studies on Immunity against *N. caninum* Infection in Nonpregnant Mice

Mouse Strain	Isolate	Infection Dose/Route	Experimental Remarks	Parasite Detection	Results	Reference
AJ, BALB/c, C57BL6 (inbred), CD1 (outbred)	NC1	1×10^6/SC	Only in AJ mice: Splenocyte proliferation assays and cytokine measurements to assess recall responses, treatment of infected mice with antibodies against IFN-γ and IL-12	Histology	In all strains no clinical and little histologic evidence of infection. T-cell hyporesponsiveness to parasite antigen and mitogen at day 7 PI. *In vivo* depletion of IFN-γ and IL-12 renders mice more susceptible	1051
ICR	BT3	1×10^6/SC	Three out of 8 mice were immunosuppressed with prednisolone acetate prior to infection. Sera were collected in days 15, 20, 25, 48, 77, and 125 PI, and probed by WB on tachyzoite lysates	Not done	At days 20–25 PI, recognition of a 36–38 kDa, from day 49 PI onward, 43 kDa antigens were consistently recognized until day 125 PI	1974
BALB/c	NC1	2×10^7/IP	Treatment concomitant to infection with (1) anti-IFN-γ; (2) recombinant murine interleukin-12; or (3) recombinant murine interleukin-12 plus anti-IFN-γ. Measurement of cytokines *in vivo* and IgG isotypes	Histology and IHC-based quantification	Neutralization of IFN-γ and addition of IL-12 increased susceptibility to infection. Effects of IL-12 (decreased brain lesions and increased IFN-γ) were abolished by concomitant administration of both. Both cytokines drive IgG2a > IgG1 production resulting in a protective Th1-type immune response	173
SCID	KBA-1; KBA-2, NC1	2×10^5–2×10^7/ IP, SC, and PO	Pathogenicity was followed by progression of clinical signs, histopathology, and IHC	IHC	KBA-2 with higher pathogenicity than KBA-1, but less than NC-1. Pathology similar upon IP or SC inoculation. No infection by PO route	109
Athymic nude, BALB/c (wt^a)	JPA1	2×10^5/IP	Histopathology and IHC to visualize tissue damage and parasite localization. IFN-γ and IL-6 measurements in sera	IHC	Nude mice died latest after 28 days, no clinical signs in BALB/c mice. Parasites in uterus, pancreas etc., within the epithelium of the venules and capillaries. Nude mice with high IFN-γ and IL-6 at later stages of infection	1844
B-cell deficient μM T mice, C57BL/6 (wt)	NC1	10^5/IP	Euthanasia at 10, 24, and 44 DPI. Measurement of antibody responses. *In vitro* splenocyte-proliferation, and recall responses were measured for IL-2, IFN-γ, IL-12, IL-4, and IL-10	PCR of 1 brain hemisphere, IHC with the other	μMT mice died after 29 days onward, with multifocal lesions in the brain, but no clinical signs nor brain lesions in wt C57BL/6. Depression of lymphocyte proliferation after 10 days in both strains, which remained in μMT, but was only transient in wt mice. μMT mice secreted less IFN-γ and IL-10	652

(Continued)

Table 2.7 (Continued)   Studies on Immunity against *N. caninum* Infection in Nonpregnant Mice

Mouse Strain	Isolate	Infection Dose/Route	Experimental Remarks	Parasite Detection	Results	Reference
BALB/c, C57BL/6, B10.D2	NC1	2×10^6, SC	Euthanasia at 6 weeks PI. Serum Ig isotype analysis and quantification of brain lesions, antigen-stimulated splenocyte-proliferation, and IFN-γ and IL-4 measurements in supernatants	Quantification of brain lesions by histopathology	BALB/c and C57BL/6 much more susceptible than B10.D2 mice. Resistance in B10.D2 associated with high IFN-γ in antigen-stimulated splenocytes. *Neospora*-specific IgG2a in serum from B10.D2 mice and high IgG1 in serum of the other strains	1207
BALB/c	NC1	10^6, IP	Infection and treatment with anti-CD4, anti-CD8, and anti-IFN-γ antibodies	PCR-based detection in the brain	Inoculation of anti-CD4 and combined anti-CD4/CD8 antibodies leads to death within 30 DPI, with low IgG levels in treated mice. Anti-CD8 less effective. Anti-IFN-γ antibodies lead to high IgG levels, but death within 18 DPI	1936
KO mice, BALB/c (wt)	NC1	2.5×10^3/IP	IL-4 and IFN-γ measurements. Treatment of mice with IFN-γ or antiCD4+ antibodies. Measurements of MHC class II and T-cell proliferation	—	Resistance in wt mice, high susceptibility of KO mice within 9 days. Increased survival by adding exogenuous IFN-γ. KO macrophages do not express MHC class II, and no T-cell proliferation occurs in KO mice. Increased morbidity in wt mice treated with anti-CD4 antibody, but no effect with anti-CD8 antibody. High levels of IFN-γ and IL-4 in sera of wt mice	1469
KO, TNFR2 KO, IL-10 KO, beta2M KO, iNOS2 KO, C57BL/6 (wt)	NC1, NC2, NCts-8 (attenuated)	1×10^6	Infection and monitoring of clinical signs and disease. Histopathology in different organs. Adoptive transfer of immune cells from NCts-8 infected mice to KO mice	Histopathology lesions graded based on HE-stained sections of liver, spleen, lung, brain	NC1 100% lethal in KO mice within 10–13 days. TNFR2 and beta2M KO mice are partially susceptible to NC-1, but not to NCts-8. Adoptive transfer of immune cells prolonged survival time, but did not protect against mortality. Tissue tropism in KO mice to the liver and spleen	1717
BALB/c A	NcBT-3	1.5×10^6/SC	Infection of prednisolone-treated and ovariectomized mice were treated with physiological concentrations of the steroid hormones for 1 or 2 weeks. Histopathology, IHC on tissues and serology	PCR of brain, liver, lung, and pancreas	No mortality in any of the mice, and no significant difference in the parasite distribution and histopathological changes between the hormone-injected and control groups	1081
KO mice, BALB/c (wt)	NC1	2.5×10^3/IP	Measurements of cytokines levels by ELISA	—	In the acute infection of IFN-γ-deficient mice that were sensitive to the *N. caninum* infection high levels of IL-10 production were detected, whereas significant levels of IFN-γ and IL-4 production were observed in resistant wt mice	1477

(Continued)

Table 2.7 (Continued) Studies on Immunity against *N. caninum* Infection in Nonpregnant Mice

Mouse Strain	Isolate	Infection Dose/Route	Experimental Remarks	Parasite Detection	Results	Reference
Swiss white (outbred), CBA/Ca (inbred)	NC1	2×10^5/SC, 1×10^6/SC, 5×10^6/SC	Infection + and – with MPA treatments. Lymphoproliferative and humoral responses and cytokine production were evaluated 8 weeks PI, and tissue cyst production and histopathology assessed 4, 6, and 10 weeks PI in immunosuppressed mice	IHC	Tissue cysts were observed 10 weeks after infection only in CBA/Ca mice receiving the two highest infection doses. CBA/Ca mice showed the highest specific lymphoproliferative response. A mixed cytokine response with elevated IFN-γ and fairly low IL-4 and IL-10 secretion was recorded. In both strains, no lesions were observed in the tissues of infected mice	1706
BALB/c	NC1	2×10^6/SC for initial priming of CD4+ and CD8+ cells. 2×10^5/SC for challenge infections	Infection of mice and isolation of immune CD4+ and CD8+ splenocytes at day 28 PI and adoptive transfer into tail vein of recipient mice 60 h prior to challenge with NC1	—	Mice receiving immune-enriched CD8+ cells had severe neurological signs by 19 days PI. Mice receiving immune enriched CD4+ cells had mild neurological signs on day 22 PI. Control-infected mice did not show any clinical signs	1889
BALB/c	NC (sheep isolate[1097])	2×10^6/IP	The footpads of BALB/c mice infected with *N. caninum* and those of noninfected mice were injected with either tachyzoite extract, or para-formaldehyde-fixed tachyzoites at different timepoints PI.	Not done	Injections resulted in footpad swelling in infected, but not in uninfected mice. In mice given anti-CD4+ mAbs, swelling decreased at 24-h post injection of the extract	1514
BALB/c	NC1	1×10^7/IP	The maturation and activation of splenic conventional dendritic cells (cDCs) and plasmacytoid dendritic cells (pDCs) upon *N. caninum* infection were studied. Experimental depletion of pDCs was done prior to infection by anti-GR1 antibodies	Real-time PCR of lungs and brain	The number of cDCs decreased in the spleen of infected mice 12 h and 2 days PI, and increased at day 5 PI. After infection it was significantly above that of mock-infected controls. In contrast, the number of pDCs did not change upon infection. Both DC subtypes displayed an activated phenotype with upregulation of co-stimulatory and MHC class II molecules, especially at 12 h PI, with increased frequency of cDCs and pDCs producing IL-12	1946

(Continued)

Table 2.7 (Continued)   Studies on Immunity against *N. caninum* Infection in Nonpregnant Mice

Mouse Strain	Isolate	Infection Dose/Route	Experimental Remarks	Parasite Detection	Results	Reference
BALB/c	NC1	5×10^6/IP or IG	IP and intragastric infection comparatively assessed. Flow cytometry or ELISA for analysis of the B- and T-cell-mediated immune responses. Different organs for histopathology and IHC, spleen and Peyer's patches for immunological analysis, blood for antibody detection	Real-time PCR on brain tissues	Expansion of splenic B- and activated and regulatory T-cells, and increased levels of IFN-γ and IL-10 mRNA in IP and IG infected groups. In Peyer's patches only IFN-γ upregulation. Parasite-specific IgG1, IgG2a, and IgA antibody levels elevated in the sera of all infected mice	1945
C57BL/10 ScCr (lack TLR4 and functional IL-12R); C57BL/10 ScSn (immunocompetent)	NC1	5×10^5/IP	Collection of spleen for cytokine mRNA expression, and various organs fixed for IHC	IHC in pancreas, liver, lung, intestine, heart, and brain	Death of C57BL/10 ScCr at day 8 PI, and spread of parasites in all organs, while not for C57BL710ScSn. Susceptibility due to low IFN-γ and high IL-4 mRNA expression	233
BALB/c	NC1	5×10^6/IP	NC1 virulent = passage 34, NC1 attenuated = passage 78. Euthanasia at day 14, 28, or 48 PI. Spleen, brain, and serum collected for splenocyte proliferation assays, PCR, and histopathology	IHC and real time PCR	Milder lesions, lower parasite load and less IFN-γ in mice infected with NC1 attenuated. Mice infected with attenuated strain were protected against challenge with virulent NC1	150
C57BL/6	NC1	2×10^6/IP	Two groups plus minus progesterone treatment. Measurement of IFN-γ and IL-4, and number of IL-4 producing cells, at different timepoints after infection	Not done	IFN-γ production in the progesterone group was significantly lower than in the control group on day 40 PI. IL-4 producing cell population in the progesterone group was larger than that in the intact group	1034
BALB/c (males and females)	NC1	8×10^6 and 5×10^5	Euthanasia at different timepoints, collection of different organs	IHC and PCR	PCR detected parasite DNA in lung and heart 12 h PI, and in the brain at 8 DPI. Males appeared more susceptible than females	485
BALB/c	NC1	2×10^4/IP	Anti-CD122 mAb was used for the depletion of NK and NKT cells and polyclonal anti-asialoGM1 Ab was used for the depletion of NK cells, 4 days prior and 4 days after infection. The depletion of cell populations were confirmed by flow cytometry using PBMC prepared 2 days after the first injection of antibodies	Brains were collected at days 3 and 6 after infection and analyzed by PCR	Higher parasite burden in the brain of mice treated with anti-CD122 mAb, but no difference to controls in mice treated with anti-asialoGM1 antibody. Activation of CD4+ T cells was suppressed in the mice treated with anti-CD122 mAb. Depletion of CD122+ cells or NK cells did not affect the number of activated CD8+ T cells, DCs, and B cells following *N. caninum* infection	1480

(Continued)

Table 2.7 (Continued) Studies on Immunity against *N. caninum* Infection in Nonpregnant Mice

Mouse Strain	Isolate	Infection Dose/Route	Experimental Remarks	Parasite Detection	Results	Reference
C57/BL6 IL-12 P40$^{-/-}$	NC1	5×10^7/PO	Mice were infected orally and sacrificed at 6 h, 12 h, 18 h, 48 h, and 4, 7, and 21 days after challenge. T-cell-mediated immunity evaluated in the intestinal epithelium and MLN and spleen	IHC and real-time PCR	Increased TCRαβ, CD8, IFN-γ, lymphocytes detected, splenic and MLN CD4+, CD25+ T cells sorted from infected mice presented a suppressive activity on *in vitro* T-cell proliferation and cytokine production	386
C57/BL6	NC1	10^6/IP	0, 5, and 10 DPI. Euthanasia, isolation of peritoneal exudate cells, staining with anti-CD11b and flow cytometry. Macrophage depletion by clodronate treatment. Cytokine analysis of macrophage culture supernatants	—	Macrophages in *N. caninum* infected mice showed greater activation and increased IL-6, IL-12p40, and IFN-γ levels, and migrated to the site of infection. Macrophage-depleted mice exhibited increased sensitivity to *N. caninum* infection	3
C57/BL6 IL-12 P40$^{-/-}$	NC1	1×10^7/IP or 1×10^7/PO	6 h, 7 days, 21 days, and 2 months PI adipose tissue from different locations was isolated, fixed for IHC and stored for DNA and/or RNA extraction. Stromal vascular fraction (SVF) cells and macrophages from gonadal adipose tissue were isolated and analyzed by flow cytometry, and serum leptin, adiponectin, and insulin were measured	Real-time PCR of brain, lung, liver, and adipose tissue	Early PI, high numbers of parasites detected in adipose tissue. By day 7 PI, macrophages, Treg cells, and T-bet+ cells are increased in gonadal, mesenteric, omental, and subcutaneous adipose tissue. High IFN-γ in gonadal adipose tissue. 2 months PI, parasites are cleared, but Th1 cells still increased, and Th1/Treg cell ratio higher than that of controls in the mesenteric and subcutaneous adipose tissue. Chronically infected mice presented a marked increase of serum leptin	1947
WT B6 IL-12 P40$^{-/-}$	NC1	10^7/IP	24 h, 7 days, 21 days, and 12 months PI adipose tissue from different locations was isolated, SVF cells were isolated, and analyzed by flow cytometry or cultured and supernatants collected for cytokine analysis	Real-time PCR of brain, lung, liver, and adipose tissue	IFN-γ is increased in adipose tissues of infected mice. In B6 mice, NK, NK T, and TCRγδ+ cells, CD4+ and CD8+ TCRβ+ lymphocytes contributed to IFN-γ production early on, with upregulated expression of genes encoding interferon-inducible GTPases and nitric oxide synthase but this was not found in IL-12 P40$^{-/-}$ mice	1948

a Wild type.

observed in CCR5$^{-/-}$ mice, and higher levels of IFN-γ and CCL5 expression, which are associated with brain tissue damage, were observed in the brain tissue of CCR5$^{-/-}$ mice during the acute phase of the infection.[4] A recent study showed that *N. caninum* also interacts with canine polymorphonuclear neutrophils by inducing the formation of extracellular traps (NETS). NETS have been recognized as novel effector mechanism in many immune processes.[2709a]

N. caninum triggers a type I IFN-α/β response in infected host cells.[185] Host responses to *Neospora* are dependent on the toll-like receptor TLR3 and the adapter protein Trif, and RNA from *Neospora* elicits TLR3-dependent type I interferon responses when targeted to the host endo-lyso-somal system.[185] In contrast, *Toxoplasma* actually inhibits the production of type I interferon, and promotes the synthesis of IFN-γ,[186] and this marks an important difference between the 2 species with regard to their innate immune response. C57BL/10ScCr mice that lack toll-like receptor 4 and a functional IL-12 receptor is highly susceptible to neosporosis.[233] Others have demonstrated that initial *N. caninum* recognition occurs by engaging TLR2 and triggers the activation of the adaptor molecules Myd88 and its respective pathways.[1359,1361] Another important component governing the innate immune response is nucleotide-binding oligomerization domain (NOD)-like receptors NOD1 and NOD2, which detect intracellular pathogens. In previous studies it was proposed that engagement of NOD2 is necessary for generating protective Th1 immunity against *T. gondii*. Their role in nonpregnant mice was investigated,[435] and it was shown that NOD2 is involved in macrophage responsiveness against *N. caninum* infection *in vitro* and *in vivo*, that NOD2-triggered responses are required but not essential for *N. caninum* growth restriction during sublethal infection, and that NOD2 induces aggravated inflammation and lethality during lethal *N. caninum* infection in nonpregnant mice.

Acquired Immunity: Both, cellular and humoral immune responses are important to control the infection in mice.[887] The immune response in inbred strains may differ among the strains, due to their specific relative haplotype.[1207,1372] It appears however that the principal mechanism of protection against *N. caninum* infection involves mainly IFN-γ and IL-12.[173,1040,1051,1469,1717] Indeed, more resistant mouse strains such as B10.D2[1207] or CBA/Ca[1706] exhibited high IFN-γ/IL-4 ratios. However, other cytokines such as TNF-α, IL-10, or transforming growth factor β (TGF-β) have been shown to be involved in the control of parasite proliferation, at least *in vitro*.[441] Interestingly, it was shown that CD4+ T-cells are crucial for protection against *N. caninum* infection, while CD8+ T-cells are not.[1936] One has to keep in mind, however, that different *Neospora* isolates are likely to induce different cytokine responses, similar to what has been previously observed for different *T. gondii* strains.[82] The importance of indoleamine 2,3-dioxygenase in the defense against *N. caninum* infection in human and bovine cells was identified.[1885] More recently mesenchymal stromal cells were reported as putative effector cells against *N. caninum*, since IFN-γ activation restricted parasite proliferation through involvement of guanylate-binding proteins (GBPs) and immunity-related IRGs.[1886] The importance of the humoral immune response was also demonstrated. μMT-antibody KO mice were more susceptible to *Neospora* infection than wild-type C57BL/6 mice.[652] BALB/c mice infected with *N. caninum* tachyzoites underwent B-cell expansion and exhibited a dominant IgG2a production.[1944] Although these results speak in favor of a Th1-type immune response for protection, other studies have documented that an appropriate Th1/Th2 balance is required.[150,1477]

2.8.1.2.3.2 Immunity in Pregnant Mice — Similar to the situation in cattle,[44] an immunomodulation toward a Th2-type response associated with high IL-4 production is usually observed during pregnancy, thus favoring the proliferation of the parasite and the vertical transmission. Respective studies in the pregnant mouse model are summarized in Table 2.8. Neutralization of IL-4 levels by application of anti-IL-4 antibodies in pregnant mice modulates the immune response and increases fetal loss,[1208] and this was confirmed by other studies.[1033,1655] Immunological parameters responsible for reactivation of the parasitemia in mice are not well understood. However, vertical transmission in mice infected before mating, as well as recrudescence in the first generation, were observed. Interestingly, different patterns leading to the same outcome were seen between different strains of

Table 2.8 Studies on Immunity against *N. caninum* Infection in Pregnant Mice

Mouse Strain	Isolate	Infection Dose, Route	Infection Day (Gestation)	Experimental Remarks	Parasite Detection	Results	Reference
BALB/c	NC1	(a) 10^4, IP; (b) 2×10^6, IP	(a) For *in vivo* priming: 8 weeks prior to pregnancy. (b) For challenge infections either day 5 or day 7–9 of gestation	Treatment with anti-IL-4 concomitant to *in vivo* priming, and 8 weeks prior to mating, and on day 5 of gestation. IFN-γ and IL-4 mRNA by real-time PCR of splenocytes. Cytokines by ELISA from antigen stimulated lymphocytes, antibody measurements	PCR on 1 day neonatal mice	In mice that were naive before pregnancy, neutralization of IL-4 during gestational challenge had no effect. In mice that were primed and modulated before pregnancy, congenital transmission was significantly decreased and associated with lower maternal splenocyte IL-4 secretion, lower IL-4 mRNA levels, higher IFN-γ secretion and decreased *Neospora*-specific IgG1	1208
CBA/Ca	NC1	5×10^6, SC	4–6 months prior to pregnancy under MPA treatments to establish chronic infection, then mating	4 mice sacrificed on days 0, 18, and after delivery of their first pregnancy. The others were mated again and monitoring of 2 additional pregnancies. Maternal cytokine and antibody analyses	PCR of fetal organs or fetuses as a whole	Vertical transmission decreased with successive pregnancies. During the first pregnancy, modification of the immune response occurred, but after 2 successive pregnancies, the specific cellular response decreased, and Th2 cytokine mRNA expression and increased IgG1 levels were noted At the third delivery, a partial restoration of the proliferative response was observed	1707
Qs (outbred). Comparative study in pregnant versus nonpregnant mice	Nc-Liv, Nc-SweB1	3×10^6, SC	Days 6–7 of gestation	All mice euthanized 6 DPI (day 12 of pregnancy in pregnant animals). Spleen cell cultures and proliferation assays, cytokine detection in supernatants and IgG isotypes in sera		Spleen cell production of IFN-γ, IL-12, and TNF-α were lower in infected/pregnant mice than in nonpregnant mice. IL-4 was only produced in infected/pregnant mice	1654
BALB/c	Ovine isolate	2×10^6, IP	4 weeks prior to infection (chronic infection) or between days 7 and 9 of gestation	Blood taken at days 1, 3, 5, 7 and 10 PI. The offspring in all groups were killed by cervical dislocation on the day of birth. Adults were re-infected and immune responses (antibodies, IFN-γ, IL-4) measured	The brain, mesenteric lymph nodes and spleen of each neonate and dam were removed and then examined by PCR	Mice infected during pregnancy may show an enhanced-type 2 immune response (lower IFN-γ, higher IL-4) in the recrudescence of the infection	1033

mice. It was shown that in chronically infected CBA/Ca mice that the vertical transmission in successive pregnancies was first associated with a high IgG2a antibody response during the first two pregnancies, and only during the third pregnancy an increase in IgG1 antibodies was noted.[1707] This indicated that immunomodulation was not responsible for the reactivation of chronic infection. On the other hand, others showed an association of vertical transmission with downregulation of the Th1 response.[1033,1511] It was suggested that the overall reduction of specific anti-*Neospora* IgG in successful pregnancies is most likely a consequence of the reduction in parasite numbers,[1016] and thus does not give a clue to which type of immune response lead to the clearance of the parasites. However, these studies are hardly comparable with the former ones because of the different experimental parameters and different mouse strains employed.

2.8.1.3 Nude Mice and IFN-γ-KO Mice

Nude mice that lack cells for humoral and cellular immunity are susceptible to *N. caninum* infection.[1561,1784,1844,2141] *N. caninum* multiplied in many organs, including intestines.[1561] Nude mice inoculated IP with homogenate of infected bovine brain and spinal cord developed neurological signs 2–4 months PI.[1784] However nude mice are very expensive and offer no additional advantage over KO mice.[564]

When in 1998, KO mice became available commercially; they became widely used for bioassays for *N. caninum* (see Section 9 in Chapter 3). Both BALB/c-derived and C57BL/6 derived-mice are equally susceptible; we use C57BL/6 because they are more robust. The KO mice that die during the acute stage of the infection develop hepatitis and pneumonia.

2.8.2 Gerbils

2.8.2.1 Common Pet Gerbil (Meriones unguiculatus)

Gerbils have been very useful to study pathogenesis and for the isolation of viable *N. caninum* from feces and tissues of animals (Table 2.9).

Little is known of the fate of sporozoites in animals after ingestion of *N. caninum* sporulated oocysts. In one experiment, 10-fold dilutions of oocysts were inoculated orally into 18 gerbils and their tissues were examined by IHC.[583] Undiluted suspension estimated to contain 1000 oocysts were fed to 8 gerbils. Four of these gerbils were euthanized daily at 1–4 DPI; these gerbils remained asymptomatic. No lesions or parasites were detected in their tissues. However, bioassays in KO mice and cell culture inoculated with MLN of gerbils killed on days 1 and 2 PI revealed the presence of viable organisms. These results indicated that the sporozoites had excysted and invaded the MLN. Tachyzoites and lesions were first detected in sections of intestines of gerbils killed on days 3 and 4 (Figure 2.22). The gerbils became ill at 6 DPI and tachyzoites and lesions were demonstrable in several tissues. Parasites and encephalitis were detected in the brain of the gerbil at 9 DPI. Although not demonstrated, similar scenarios might occur in other animals after ingesting *N. caninum* oocysts. In conclusion, enteritis, MLN necrosis (Figure 2.23), and hepatitis may preceed encephalitis.

In other experiments, gerbils were very useful in determining the viability of oocysts fed to cattle.[1994] Not all *N. caninum* strains appear to be pathogenic to gerbils (Table 2.9).

Gerbils have been also useful in isolating the parasite from oocysts and tissues of animals infected with *N. caninum*. All gerbils inoculated with bradyzoites liberated from tissues of naturally infected dogs or cattle remained asymptomatic, suggesting that this stage may be less pathogenic than tachyzoites or sporozoites. In the initial experiments, gerbils inoculated IP with *N. caninum*-infected tissues developed ascites with tachyzoites in the peritoneal exudates.[402] This provided false hope of the procedure to maintain this parasite in the laboratory by serial passages in gerbils. However, tachyzoites were no longer demonstrable in gerbils after a few passages.

Table 2.9 Experimental Neosporosis in Gerbils (*M. unguiculatus*)

Stage	Strain	Dose	Route	No.	Remarks	Reference
					Gerbils	
	N. caninum					
Oocysts	Nc-Liv	1000 (estimated)	Oral	8	7 eu. at day 1, 2, 3, 4, 7, 9, 9, and 1 died at day 11 PI. All gerbils became infected. Disseminated neosporosis by 7 days	583
		100 (estimated)		2	Eu. 13, 18 DPI disseminated neosporosis	
		10 (estimated)		2	Eu. 21, 25 DPI. Parasites seen day 21 PI *Neospora* not detected day 25	
		1 (estimated)[a]		1	Eu. 16 DPI. Disseminated neosporosis	
	NC-6 Argentina	Unknown	Oral, gerbils treated with MPA	3[b]	Remained healthy. Necropsied 63 and 75 DPI. Two tissue cysts in the brain of 1 seropositive gerbil; all gerbils infected as evidenced by bioassay in KO mice	164
	No designation	Unknown	Oral	6	Remained healthy but developed *N. caninum* antibodies as shown by immunoblot	493
	NC-GER2, 3, 4, 5	1×10^3 (NC-GER2), dose unknown (NC-GER3), 2.5×10^2 (NC-GER4), dose unknown or 1×10^2 (NC-GER5)	Oral	4	Remained asymptomatic, necropsied 200, 246, 277, 332 DPI. Became infected as revealed by bioassay in KO mice	1805
	Isolate from sheep	1000	Oral	2	Remained healthy. Necropsied 60 DPI. Tissue cysts in brains of both gerbils	1579
	Nc-Liv	1, 10, 100	Oral	6	2 gerbils per each dose. All became infected. 1 gerbil fed on 10 oocysts died 14 DPI	1994
	NCBrBuf-1 (buffaloes, buffalo no. 403, buffalo no. 412)	Unknown (NCBrBuf-1), Unknown or at least 1000 (NCBrBuf-4)	Oral	2	Remained asymptomatic but seroconverted (NAT). No parasites detected histologically, 70 DPI	1722
Naturally infected dog tissues	NC3	Unknown	SC, IP	3	Gerbil became sick 10 DPI. Numerous tachyzoites found in peritoneal fluid[c]	402
	NC6	Unknown	SC	5	Died of acute neosporosis, 9–11 DPI	598
	NC7	Unknown	SC	2	Remained healthy but seroconverted (NAT). No parasites seen when necropsied 60 DPI	598
	NC8	Unknown	SC	3	Remained healthy. Tissue cysts found in brain when necropsied 73 DPI	598
	NC9	Unknown	SC	4	Remained healthy but seroconverted (NAT). No parasites seen when necropsied 149 DPI	608

(Continued)

Table 2.9 (Continued) Experimental Neosporosis in Gerbils (*M. unguiculatus*)

Stage	N. caninum Strain	Dose	Route	Gerbils No.	Remarks	Reference
Naturally infected cow tissues	NC-SKB1	Unknown	IP	2	Gerbils remained asymptomatic but became infected as revealed by bioassay in cell culture	1702
Tachyzoites	NCJPA-1	2×10^5	IP	4	Died 6–8 DPI. Peritonitis. Tachyzoites in peritoneal fluid[d]	778
	NC-Kr2	$5 \times 10^1, 5 \times 10^5, 5 \times 10^6, 4 \times 10^7$	IP	32 (8 gerbils each dose)	5×10 not infective, lethal dose 5×10^5, parasite DNA in several tissues including blood of gerbils infected with 5×10^6 or 4×10^7 tachyzoites	1032
	NC1	10^3	SC	10	Remained asymptomatic. No cross protection with *B. besnoiti* but slight serological cross-reactivity	1850
	NC1	$1–5 \times 10^6$	IP	20	All became ill. Disseminated neosporosis in all examined 6–20 DPI	1661
	NC1	2×10^6	IP	6	Became ill 5 DPI, tachyzoites in peritoneal exudates	234
	NC1	(a) 2.5×10^6; (b) 5×10^4	IP	5 (a); 10 (b)	Gerbils necropsied 7, 15, 30 DPI. Gerbils developed ascites with tachyzoites. Biomarkers of oxidative stress and cholinesterases and purine activities determined	1975, 1976, 1977

Note: Eu. = euthanized.
[a] Based on 10-fold serial dilutions of inocula, the next dilution was not infective to gerbils.
[b] Administered 8 mg (1 gerbil), or 4 mg (2 gerbils) of methylprednisolone acetate intramuscularly.
[c] *Neospora* could not be propagated in gerbils by serial passage of tachyzoites inoculated IP.
[d] Passage made every 3–4 days in gerbils inoculated with peritoneal exudates.

The lethal dose of *N. caninum* in gerbils is around 10^6 tachyzoites inoculated IP for testing the efficacy of vaccines and determining pathogenic factors (Table 2.9). Four to five day old gerbils were successfully infected by oral inoculation of tachyzoites.[672a] All 17 gerbils died of neosporosis between 8 and 17 DPI (1 died on each of days 8, 9, 15, 16, and 17, 4 died day 12, and 5 died on day 15—personal communication to JPD on October 24, 2016) with 4×10^5 NC1 tachyzoites. *N. caninum* DNA was found in several tissues of gerbils.

2.8.2.2 Tristram's Jird (Meriones tristrami)

Severe neosporosis was induced in *M. tristrami*.[1619] Groups of 5 gerbils were inoculated SC or IP (total 10 per dose) with 10-fold dilutions of 1 to 10^7 NC1 tachyzoites. Of the 70 jirds, 53 (75%) developed clinical signs, ranging from acute death to development of neuromuscular signs and survival. Sixteen jirds had paralysis of the rear limbs, simulating clinical signs in dogs. Parasitemia was demonstrable by bioassays of blood in all 8 jirds tested. Tachyzoites were found both in pleural and peritoneal cavity. The virulence of tachyzoites decreased after third subpassages, and tachyzoites eventually disappeared from peritoneal exudates with subpassages. Gerbils that survived infection developed antibodies to *N. caninum*.[1619] Results of this investigation were confirmed in another report.[959] Three adult jirds were inoculated IP with 10^6 NC1 tachyzoites. One jird died on day 11 PI and had peritoneal exudate. The second jird became emaciated and was euthanized on day 11 PI. The third jird developed paresis of the rear limbs and was euthanized on day 19 PI. *Neospora* DNA was detected in the brains of all 3 jirds.

2.8.2.3 Wagner Gerbil (Gerbillus dasyurus)

Three gerbils were inoculated IP with 10^6 NC1 tachyzoites.[959] All became ill and were euthanized 7, 8, and 9 DPI because of paralysis of the rear limbs (1 gerbil), severe apathy (1 gerbil), and hepatomegaly (1 gerbil). *Neospora* DNA was detected in the brains of all 3 gerbils.

2.8.3 Multimammate Rat (Mastomys natalensis)

Six adult multimammate rats were inoculated IP with 10^6 NC1 tachyzoites.[959] All 6 rats became ill and died or had to be euthanized on days 10 (4 rats), 11 (1 rat), and 19 (1 rat). Lesions and tachyzoites were seen in several tissues, including accessory sex glands.

2.8.4 Sand Rat (Psammomys obesus)

Using the same experimental design as that used to infect the jirds, 40 (10 per each dose) laboratory raised sand rats were inoculated SC or IP with 10, 100, 1000, and 10,000 NC1 tachyzoites.[1619] Of all the hosts so far tested, sand rats were the most susceptible to neosporosis. Eight of the 10 rats inoculated with 10 tachyzoites died or were euthanized when ill, irrespective of the route of inoculation; the 2 survivors were not infected. Parasitemia was demonstrable by bioassay of peripheral blood in rats inoculated with both routes. Death was delayed by 3–4 days when infection took place through the SC route. *N. caninum* was passaged successfully by IP inoculation of peritoneal exudates up to 7 times.

2.8.5 Norwegian Rat (Rattus norvegicus)

Rats, similar to outbred mice, are resistant to *N. caninum* infection.[1161] A total of 54 rats were divided into 3 groups and inoculated each with 5×10^5 NC1 tachyzoites and observed up to 65 DPI. Complete necropsies were performed on all rats and their tissues examined histologically. Group A (n = 14 rats) were inoculated but not medicated; *N. caninum* was not found in tissues of any of these examined 3–55 DPI. Group B rats (n = 20) were given 4 mg MPA by IM route on –7 and 0 day of inoculation. These rats developed disseminated neosporosis with severe hepatitis (Figure 2.24), and they died or were killed by 12 DPI. Group C rats (n = 20) were medicated with 2 mg MPA –7 and 0 DPI; these rats were examined 3–55 DPI. Group C rats developed milder disease with lesions in several organs up to the last day of examination (day 45). *N. caninum* tachyzoites were demonstrable in tissues 6–17 DPI. A single small (12 × 14 μm) tissue cyst was found in the brain of a rat 17 DPI.

2.8.6 Djungarian Hamster (*Phodopus sungorus*)

This hamster species was successfully infected with *N. caninum*.[2004] Hamsters (n = 29) were inoculated IP with 5 × 10⁶ JPA1 strain tachyzoites; 3 of them became ill. One of the hamsters exhibiting clinical signs died 9 DPI, and was not examined, the other two were killed 16 DPI; many tissue cysts were found in their cerebrum, and 1 tissue cyst was detected in the *tunica muscularis* of the gastric mucosa. Tissue cysts were also found in 26 of the remaining hamsters.

2.9 GENERAL DIAGNOSIS

Diagnosis of neosporosis can be achieved in many different ways.[600,603,607,1316,1525] Most important are (i) cytological, histological, immunohistochemical, and electron microscopical examinations, (ii) detection of parasitic DNA by PCR, (iii) detection of specific antibodies in individual animals, and (iv) epidemiologic observations. Only general aspects are discussed here, and technical details are provided in Chapter 3. Diagnosis of neosporosis in cattle and dogs is discussed in Chapters 4 and 5, focussed to infection in these animals.

2.9.1 Cytology

A rapid diagnosis maybe made by microscopic examination of cytospin smears or impression smears after Diff-Quick rapid staining or conventional Giemsa stain. The tachyzoite stage is the most likely to be present. Intracellular organisms often appear smaller in size than extracellular, and dividing tachyzoites are oval to round. By light microscopy tachyzoites of *N. caninum* and *T. gondii* tachyzoites appear similar.

2.9.2 Conventional Histopathology

Examination of tissue sections stained with HE is very useful in recognition of lesions and arriving at provisional diagnosis. However, it is difficult to identify *Neospora* unless thick-walled tissue cysts are present. In HE sections the tachyzoites are often round to slightly elongated and it is important to look for the vesicular nucleus to distinguish them from degenerating host cells (Figure 2.2d). Rarely the tachyzoites are cut longitudinally to be seen as crescentric. In HE-stained sections (Figure 2.2e), tachyzoites of *Neospora* and *Toxoplasma* are similar. Although tissue cysts of *Neospora* have a thicker wall (up to 4 μm-thick) compared to tissue cysts of *Toxoplasma* (<0.5 μm-thick), in some instances tissue cysts are indistinguishable.[573] *Sarcocystis* schizonts can be confused with *Neospora* tachyzoites but the presence of immature stages in *Sarcocystis* is diagnostic. *Sarcocystis* schizonts divide by endopolygony, where the nucleus becomes highly lobulated and merozoites are formed at the periphery (see Chapter 4), whereas *Neospora* tachyzoites divide into 2 by endodyogeny.

2.9.3 Transmission Electron Microscopy

The structure of rhoptries is helpful in differenting tachyzoites of *Neospora*, *Toxoplasma*, and merozoites of *Sarcocystis*. Rhoptries in *Neospora* are electron dense with an amorphous matrix, and

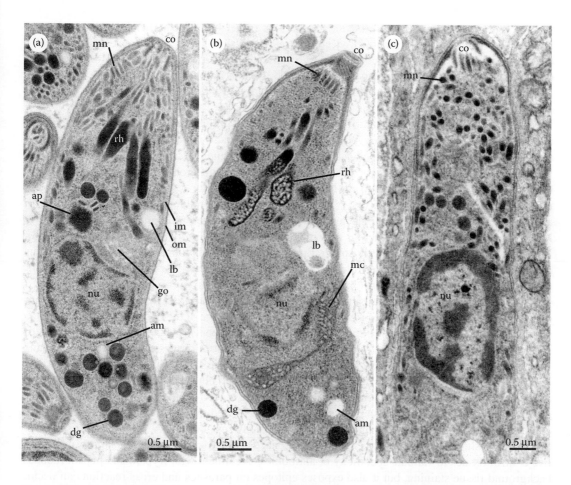

Figure 2.31 Tachyzoites of *Neospora*, and *Toxoplasma* compared with a merozoite of *Sarcocystis*. Note conoid (co), micronemes (mn), mitochondrion (mc), nucleus (nu), Golgi body (go), apicoplast (ap), dense granules (dg), amylopectin granules (am), rhoptries (rh), lipid body (lb), inner membrane complex (im), outer membrane complex (om). (a) *N. caninum*. Note electron dense rhoptries. (Adapted from Speer, C. A. et al. 1999. *Int. J. Parasitol.* 29, 1509–1519.) (b) *T. gondii*. Note electron lucent rhoptries. (Adapted from Dubey, J. P., Lindsay, D. S. 1993. *Parasitol. Today* 9, 452–458.) (c) *S. cruzi*. Note the absence of rhoptries. (Courtesy of Dr. C. A. Speer.)

few extend posterior to the nucleus whereas *T. gondii* tachyzoite rhoptries are electron lucent with a structure interior and located anterior to nucleus. *Sarcocystis* merozoites lack rhoptries (Figure 2.31).

2.9.4 Immunohistochemical Staining (IHC)

IHC is necessary to confirm histologic diagnosis. The method is both sensitive and convenient and is used with fixed tissues, including archived tissues. Tissues with high concentration of peroxidase, especially placenta, should be treated with trypsin or pepsin prior to analysis to avoid background in immunoperoxidase staining. The diagnosis should not be made unless parasite outlines are visible because diffuse staining may be nonspecific. Groups of tachyzoites often cannot be distinguished from tissue cysts based on IHC unless bradyzoite-specific antibodies (either polyclonal or monoclonal), for example against BAG1 (bradyzoite antigen 1) are used.[1307,1599]

Formalin-fixed paraffin-embedded tissues have been commonly used for IHC for specimens submitted to diagnostic laboratories. Although *Neospora* antigens can be detected even in tissues preserved in formalin for 20 years, fixation for short periods (24 h) is recommended. Sections should be thin (3–5 μm thick), and mounted on grease-free clean slides with an adhesive or "electrically charged" slides. The avidin biotin complex (ABC) indirect immunoperoxidase method and the peroxidase-antiperoxidase (PAP) technique are equally good. Various immunogen substrate kits used in IHC are commercially available. The AEC (3-amino-9-ethylcarbazole) method is detailed in the Chapter 3.

Both polyclonal and monoclonal antibodies specific to *N. caninum* can be used[361,986,1157,1307, 1599,1704,2015] and both polyclonal and also monoclonal *N. caninum* antibodies are commercially available. For diagnostic purposes, polyclonal antibodies made in rabbits seem to be more reliable than the mouse-derived monoclonal antibodies (Schares and Dubey, unpublished observation). However, a combination of 2 monoclonal antibodies targeting different antigens, that is, a surface (NcSRS2) and a dense granule antigen (NcGRA7) revealed similar results by IHC as a polyclonal rabbit antibody.[2015] Cross-reactivity of *N. caninum* antibodies to related apicomplexans, *T. gondii*, and *Sarcocysti*s sp. may occur in polyclonal antibodies but are often mild or even absent.[1127,1157,1503] Monoclonal antibodies, cross-reacting with related species like *T. gondii* are reported (Table 2.5). However, *T. gondii*, and *Sarcocystis* sp. are rarely associated with abortion in cattle[62,280,1536]; thus cross-reactivities play only a minor role for diagnosis in cattle. It must be remembered that none of the IHC reagents are 100% uniform, which may result in minor differences in specificity observed in various laboratories. Additionally, there can be considerable variation in reactivity of antiserum depending on the source of rabbits used, the type of antigens, stage of the parasite used to immunize rabbits, and fixation of the parasite antigens. Other technical aspects such as incubation times, trypsin or pepsin treatment of tissues, or lack of enzyme treatment of tissues can influence the results of these IHC tests. Occasionally, some blocks of tissues from the same animal and fixed identically will react differently in IHC tests. Therefore, diagnosis should not be based solely on the results of a single IHC examination. An interlaboratory comparison of IHC protocols revealed false-positive *N. caninum* reactions in tissue sections of a *T. gondii* infected animal[2022] and in another study a commercial polyclonal antibody against *T. gondii* strongly reacted with *N. caninum* in tissue section.[1920] This shows that a potential cross-reactivity needs to be excluded by using appropriate negative control tissues to validate specificity of antibodies. Treatment with trypsin and pepsin reduces background tissue staining, but it also exposes epitopes on parasites and cross-reaction can occur, especially in tissue cysts of related parasites.

2.9.5 Isolation of Viable *N. caninum* by Bioassay

Isolation has little practical importance for routine diagnosis but is essential to unambiguously confirm that a particular animal species represents a host of *N. caninum*. In addition, isolation is essential to study the population structure of *N. caninum*[1679] because often, only by isolation sufficient parasitic DNA can be generated; in addition *in vitro* isolation provides the option to generate cloned populations. A large variety of isolates with differences in virulence (either *in vitro* or *in vivo*) represent the base for virulence and immunological studies.[730,1678,1681,1848] Parasite isolates are also essential to study differences in expression of various proteins between strains of *Neospora* to explain differences in virulence and immunogenicity. Several proteins (e.g., NcMIC1, NcNTPase, NcROP40, aspartyl tRNA synthetase, and G6PD) have been identified as being more abundant in highly virulent isolates.[1678]

Viable *Neospora* has been isolated from tissues of cattle, sheep, water buffalo, horses, white-tailed deer, and axis deer (see respective chapters) by bioassay of infected tissues in cell culture, in immunosuppressed mice, or by feeding to dogs. The number of viable *Neospora* in naturally infected tissues is usually low, especially in asymptomatic animals. Additionally, only a small volume of tissues can be inoculated on to cell cultures or injected into mice, thus limiting chances of isolation. Bioassay in dogs can overcome this difficulty because larger volumes (up to 1 kg) of tissues can be fed to dogs. Dogs may excrete *Neospora* oocysts in feces after ingesting infected tissues.

Bioassay in dogs has been useful to isolate *Neospora* from tissues of naturally infected buffaloes and sheep in Brazil.[116,1579] However, this method is expensive, needs careful ethical considerations and oocyst excretion by dogs is erratic.

Details of bioassay in cell culture, mice, and dogs are provided in Chapter 3.

2.9.6 Serological Methods

There are several serological tests used for neosporosis (Table 2.10). Only general aspects of these tests are stated here. The serological methods for the diagnosis of neosporosis were reviewed in detail previously.[219a] Protocols for these tests are described in the Chapter 3.

2.9.6.1 Antibody Detection

Antibody detection tests have the advantage that they can be applied antemortem and may provide information on the stage of infection. In addition, the level of the specific antibody response (e.g., expressed as a titer) may provide information on the risk of an animal to develop disease or to be diseased (i.e., abortion in ruminants, likelihood of transplacental transmission, likelihood of canine neosporosis). The maturity of the antibody response can be assessed using so called avidity ELISAs (for details see serological tests in cattle in Chapter 4) which could provide valuable information on the stage of infection (acute versus chronic). Also the differentiation between specific IgM and IgG may also help to identify cases of recent infection. The recognition of specific epitopes (as e.g., determined by competitive or blocking ELISAs), or specific antigens (as e.g., established by using recombinant antigens or affinity purified antigens) may provide further insights into the chronology or phase of infection, on the extent of the infection or even the route of infection.

Prenatally infected animals mount specific antibodies, which can aid diagnosis, especially in ruminants because there is no passive transfer of antibodies across the placenta. When testing newborn or young animals sampling is optimally done prior to the ingestion of colostrum. Colostral antibodies might be detectable for several months in noninfected animals, depending on the analytical sensitivity of the test used.

In addition to serum or plasma, also other body fluids (as e.g., pleural or peritoneal fluids, or exudates) may contain enough specific antibodies for detection. However, serum or plasma are optimal, because other fluids only partially reflect the antibody response measurable in serum or plasma, and levels are often influenced by other parameters than the *N. caninum* infection, for example, in case of milk, the stage of lactation may affect the outcome of analyses.

Most of the serological tests developed for one animal species can be easily adapted for the examination of samples in another species. In case of ELISA, indirect fluorescent antibody test (IFAT) or immunoblot, just the conjugate (i.e., the secondary antibody used to visualize the reactivity of sample antibodies with specific antigen) needs to be changed. For several domestic or wildlife species no specific conjugates exist and heterologous conjugates, that is, conjugates generated against a phylogenetically related species, might be helpful. For example, anti-dog IgG conjugates can be used to detect antibodies in various canid species.[1796,1857] Others replaced species-specific conjugates by Protein G conjugates, for example, to detect IgG in new world camelids.[358] Anti-bovine conjugates also worked in new world camelids[2108] and goat anti-llama IgG conjugates are available commercially.[330,331] Protein G is known to bind IgG of many but not all animal species at neutral pH.[1100] In addition, Protein A or Protein L conjugates might be helpful for other species in which IgG does not bind to Protein G.[1442]

Validation of a given test in a given species is time consuming, and data may not be valid for other related host species. Additionally, validation may not be possible for all species. In general, such serological results should be interpreted with caution and, for example, claims on the presence of *N. caninum* in particular intermediate and definitive host species only based on serological

Table 2.10　Serological Assays for Detection of *N. caninum* Antibodies

Type of Assay	Type of Antigen, Name of Assay	Antigen Characteristics (other antigen names; T, Tachyzoite, or B, Bradyzoite Specific)	Host Species	Tested for Cross-Reaction	Main References
NAT	Whole fixed tachyzoite	Cell-culture-derived	Cattle, dogs, and various other species	*T. gondii, H. hammondi, S. cruzi, S. neurona, S. falcatula, Tritrichomonas foetus,* infectious bronchitis virus	1545
		Mouse-derived	Various species	*T. gondii, H. hammondi, S. cruzi, S. neurona, S. muris*	1733
LAT	Recombinant protein	Recombinant NcGRA6 (T, B)	Cattle, dogs	Not tested	759
	Recombinant protein	Recombinant NcSAG1 (T)	Cattle	*T. gondii*	1400
IFAT	Whole fixed tachyzoites	Air-dried	Various species	Tested in Reference 1545 for *T. gondii, H. hammondi, S. cruzi, S. neurona, S. falcatula, T. foetus,* infectious bronchitis virus	374, 1990, 2049
		No fixation reported	Various species	*T. gondii*	528
		Formaldehyde fixed, air-dried	Various species	Not tested	255
		Air-dried, methanol fixed	Various species	Not tested	1556
		Air-dried, acetone fixed	Various species	Not tested	1789
Indirect ELISA (for serum)	Whole tachyzoite lysate	Extracted with PBS, Kinetic ELISA, commercially available	Cattle	*T. gondii, C. muris, Cryptosporidium* sp., *Sarcocystis* sp., *Eimeria* sp.	1555
		Extracted with PBS	Cattle, dog, horse	*T. gondii, H. hammondi, S. cruzi, S. hominis, S. hirsuta, Eimeria bovis, Cryptosporidium parvum*	172, 1228, 1546, 1858
		Extracted with distilled water	Cattle, sheep, goats, IgG	*T. gondii, S. cruzi, S. tenella, Babesia divergens, Babesia bigemina, Babesia bovis*	1527, 2049
		Extracted with PBS (*Neospora*-SA-ELISA)	Cattle, goats, IgG	Not tested	797, 1446, 2049
		Extracted with PBS, 1% Triton X-100 (AHS-ELISA)	Cattle, total Ig	*T. gondii, S. cruzi, C. parvum, B. bigemina, B. divergens, B. bovis,* or *Eimeria alabamensis*	2049, 2124
		Extracted with PBS, 6 mM Nonidet P-40 in Tris-buffer	Dog, IgG and IgE	Not tested	1012
		Extracted with PBS, based on Reference 344	Cattle, IgG	Not tested	108

(Continued)

Table 2.10 (*Continued*) Serological Assays for Detection of *N. caninum* Antibodies

Type of Assay	Type of Antigen, Name of Assay	Antigen Characteristics (other antigen names; T, Tachyzoite, or B, Bradyzoite Specific)	Host Species	Tested for Cross-Reaction	Main References
Whole tachyzoite		Formaldehyde fixed, commercially available	Cattle, IgG	*T. gondii, C. parvum, B. divergens, S. cruzi, E. alabamensis, E. bovis*	2094
ISCOM antigen		Based on Reference 213, antigens further characterized in Reference 217, commercially available	Cattle, dogs, IgG	*T. gondii, S. cruzi, E. alabamensis, B. divergens*	216, 218, 225, 703, 2049
Recombinant protein		NCGRA1	Cattle, IgG	Not tested	101
		MBP-NcGRA2	Cattle, IgG	Not tested	509
		NcGRA2t (truncated recombinant NcGRA2)	Dog, mouse, IgG	*T. gondii* (mouse serum)	1020
		rNcGRA6; HPLC purified (NcGRA6d, NcGRA6s; T, B)	Cattle, IgG	Tested in Reference 1113	1007
		rNcGRA6 (Nc4.1, NcDG1; T, B)[1112]	Cattle, IgG	*T. gondii, S. cruzi, S. hominis, S. hirsuta*	1113
		rNcGRA7 (Nc14.1, NCDG2; T, B)[1045,1151]	Cattle, IgG	*T. gondii, S. cruzi, S. hominis, S. hirsuta*	1003, 1113
		rNcGRA7 (N57; T, B)	Cattle, horse, IgG	*T. foetus*	1228, 1546
		rNcGRA7t (truncated recombinant NcGRA7; T, B)	Cattle, dog, IgG or cattle IgG, IgG1, IgG2, IgA, IgM[950]	*T. gondii, Babesia gibsoni, Babesia canis canis, Babesia canis rossi, Babesia Canis vogeli, Leishmania infantum.*	850, 950, 1904
		rNcGRA7 (T, B)	Cattle, mouse, IgG1, IgG2	*S. cruzi, B. besnoiti,* or *T. gondii*	14, 17, 1018
		MBPNcGRA7 (T, B)	Cattle, water buffalo, IgG	*T. gondii*	842
		Truncated rNcSAG1 (rNcp29; T)	Cattle, IgG (via Protein G)	*T. gondii, S. cruzi*	943
		Truncated rNcSAG1 (NcSAG1t; T)	Cattle, IgG; also used in humans, cattle, and rabbits[967]	*T. gondii*	320, 950, 2152
		Recombinant NcSAG1 expressed in silkworms (T)	Cattle, IgG	Not tested	509
		rNhSAG1 (cross-reacting with *N. caninum*; T)	Horse, IgG	*T. gondii, S. neurona*	914
		rNcSRS2 (Ba/SRS2p44) expressed in baculovirus (T)	Cattle, dog, IgG	*H. heydorni, T. gondii*	1472

(*Continued*)

Table 2.10 (Continued) Serological Assays for Detection of N. caninum Antibodies

Type of Assay	Type of Antigen, Name of Assay	Antigen Characteristics (other antigen names; T, Tachyzoite, or B, Bradyzoite Specific)	Host Species	Tested for Cross-Reaction	Main References
		Truncated rNcSRS2 (Ncp43P; T)	Cattle, sheep, IgG	Not tested	21, 73, 1156
		Recombinant NcSRS2 in expressed *Escherichia coli* (T)[232,1474]	Cattle, dog, mouse IgG	*H. heydorni*, *T. gondii*	1474
		Truncated rNcSRS2 (NcSRS2t) expressed in *E. coli* (T)	Cattle, IgG or cattle IgG+A+M[950]	*T. gondii*	737, 950
		Truncated rNcSRS2 (tNCSRS2) based on Reference **737** expressed in *E. coli* (T)	Cattle, IgG	*T. gondii*	1193
		Recombinant NcSRS2 expressed in *P. pastoris* (T)	Cattle, sheep, dog, IgG	Not tested	1607, 1608
		Recombinant NcSRS2 expressed in silkworms (T)	Cattle, IgG	Not tested	1534
		MBP-NcSRS2 (T)	Cattle	Not tested	509
		rNcSAG4 (Early B)	Cattle, sheep, IgG, IgG1, IgG2	*S. cruzi*, *B. besnoiti*, or *T. gondii*	14, 17, 947, 1018
		rNcBSR4 (Late B)	Mouse, IgG	Not tested	1018
		rNcSRS9 (Late B)	Mouse, IgG	Not tested	1018
		Recombinant Ncp40 (surface antigen, related to NcSAG1, NcSRS2)	Cattle, mouse, IgG	*T. gondii*	864
		NcMIC10 (NcMiC10M, NMIC10N, NcMIC10C), NCP20	Cattle, sheep, goat, IgG	*T. gondii*	101, 2157
		NcPF (recombinant *Neospora* profilin)	Dog, mouse, IgG1, IgG2	Not tested	905
		Recombinant subtilisin-like serine protease (N54, NcSUB1, NcSUB1t, 5 tandem repeats, NcSUB1tr, 5 tandem repeats)	Cattle, mouse, IgG, IgG1, IgG2	*T. gondii*,[2152] *T. foetus*[1228]	1228, 1229, 2152
		NcCyp (recombinant cyclophilin)	Cattle, mouse, IgG (no specific antibodies detected)	Not tested	1025
	Affinity purified antigen	Native NcSRS2 (p38), commercially available (T)	Cattle, dog, IgG	*T. gondii*,[936,1794] *S. cruzi*, *S. hominis*, *S. hirsuta*, *B. divergens*, *C. parvum*, *E. bovis*, *L. infantum*[936]	936, 1794, 2049

(Continued)

Table 2.10 (*Continued*) Serological Assays for Detection of *N. caninum* Antibodies

Type of Assay	Type of Antigen, Name of Assay	Antigen Characteristics (other antigen names; T, Tachyzoite, or B, Bradyzoite Specific)	Host Species	Tested for Cross-Reaction	Main References
	Antibody captured antigen	Polyclonal anti-*N. caninum* antiserum for antigen capture	Cattle, IgG	*T. gondii, S. cruzi, S. hominis, S. hirsuta, B. divergens, C. parvum, E. bovis*	1792
	Antibody captured antigen	Monoclonal antibody captured NcSRS2, commercially available (T)	Cattle, dog, IgG	Not tested	755
Competitive indirect ELISA (for serum)	Antibody-captured whole tachyzoite antigen	Monoclonal antibodies for antigen capture (mAb 5B6-25) and inhibition (mAb 4A4-2), commercially available	Cattle, dog, and various other species	mAb 5B6-25 tested by dot-blot for cross-reactivity with *T. gondii* and *S. cruzi*. mAb 4A4-2 was tested for cross-reactivity in Reference **172**. Infectious bovine rhinotracheitis herpesvirus, bovine virus diarrhea virus, *Leptospira* sp., and *B. abortus*	176, 569
	Whole tachyzoite lysate	Based on a monoclonal antibody (mAb 4A4-2)		*T. gondii, S. cruzi, S. hominis, or S. hirsuta*	172
	Fixed whole tachyzoites	Based on a polyclonal antibody and an ELISA using fixed tachyzoites	Cattle, sheep, dog, human	*B. divergens, C. parvum, S. cruzi, E. bovis, T. gondii*	1318, 1323
	Recombinant antigen	rNcSRS2 (recombinant NcSRS2) and polyclonal antibody against rNcSRS2 (anti-rNcp-43; T)[232,462]	Cattle	*T. gondii*[462]	462, 1863
		Recombinant NcSRS2 and monoclonal antibody against native NcSRS2, commercially available (T)	Cattle	Not tested	Unpublished, applied in References 972, 1096
Avidity ELISA	ISCOM incorporated antigen	Based on Reference **213**, antigens further characterized in Reference **217**	Cattle, IgG	Tested in Reference **216**	219
	Whole tachyzoite lysate	Extracted with distilled water	Cattle, IgG	Tested in Reference **1527**	1263
		Extracted with PBS	Cattle, IgG	Tested in Reference **797**	13, 1758
		Antigen extracted with PBS, 6 mM Nonidet P-40 in Tris-buffer	Dog, IgG	Not tested	1012
	Affinity purified antigen	Native NcSRS2 (p38; T)	Cattle, IgG	Tested in Reference **1794**	1800

(Continued)

Table 2.10 (Continued)   Serological Assays for Detection of N. caninum Antibodies

Type of Assay	Type of Antigen, Name of Assay	Antigen Characteristics (other antigen names; T, Tachyzoite, or B, Bradyzoite Specific)	Host Species	Tested for Cross-Reaction	Main References
ELISA (milk and bulk milk)	ISCOM antigen	ISCOM incorporated antigen	Cattle, IgG	Tested with serum for *T. gondii*, *S. cruzi*	216, 321, 322, 706, 2033
	Affinity purified antigen	NcSRS2 (T)	Cattle, dog, IgG	Tested with serum for *T. gondii*, *S. cruzi*, *S. hominis*, *S. hirsuta*, *B. divergens*, *C. parvum*, *E. bovis*, *L. infantum*[936,1794]	936, 1793
	Whole tachyzoite lysate antigen	Extracted with PBS, 1% Triton X-100 (AHS-ELISA)	Cattle, total Ig	Tested with serum for *T. gondii*, *S. cruzi*, *C. parvum*, *B. bigemina*, *B. divergens*, *B. bovis* or *E. alabamensis* in Reference **2124**	143
Immunoblot (see also Table 2.12)	Reduced whole tachyzoite antigen (major bands)	17, 31, 34, 37, 40.5, 47, 50, 55, 65 kDa[140]	Rabbit, IgG	Not tested	140
		36.5–38, 45.5–48.5, 52–53.5, 58, 58.5, 59.5, 60.5, 62, 63.5, 64, 66.5, 67, 67.5, 68.5, 69.5 kDa[172]	Cattle, sheep, and goat, IgG	Tested with serum of cattle, sheep, and goat experimentally infected with *T. gondii*	172
		14, 20, 31, 35, 42, 54, 65, 76, 116 kDa[852]	Cattle, IgG	Tested with sera of cattle naturally or experimentally infected with *N. caninum*, *T. gondii*, *S. cruzi*, *S. hirsuta*, *S. hominis*	852
		17–18, 34–35, 37, 60–62 kDa[53]	Cattle, IgG	Not tested	53
		18, 25, 33, 35–36, 45, 46, 47, 60–62 kDa[1505]	Cattle, IgG	Not tested	1505
		Reduced p17 antigenic fraction of tachyzoites	Cattle, IgG	Not tested	52
	Nonreduced whole tachyzoite antigen (major bands)	16, 29, 31, 32, 37, 46, 51, 56, 79 kDa[140]	Rabbit, IgG	Not tested	140
		17, 29, 30, 37 kDa[211]	Cow, dog, sheep, goat, rabbit, IgG	*T. gondii*,[211,213,1858] *H. hammondi*;[211] *S. cruzi*[211,213]	211, 213, 1555, 1858
		17–19, 29, 30, 33, 37 kDa[224]	Cattle and other species (i.e., foxes, pig, alpaca, antelopes), IgG	*T. gondii*[438,2108]	224, 278, 438, 1600, 1796, 2108

(Continued)

Table 2.10 (Continued) Serological Assays for Detection of N. caninum Antibodies

Type of Assay	Type of Antigen, Name of Assay	Antigen Characteristics (other antigen names; T, Tachyzoite, or B, Bradyzoite Specific)	Host Species	Tested for Cross-Reaction	Main References
		16–17, 29/30, 30–36, 37, 45 kDa[100]	Cattle, IgG	Not tested	100
		14, 29, 30–33, 35, 40, 84, 97, 170 kDa[1012]	Dog, IgG, IgE	Not tested	1012
		29, 36 kDa[1902]	Cattle, IgG	T. gondii[1902]	1902
Avidity IB	Reduced whole tachyzoite antigen	17, 34–35, 36–37, 60–62 kDa[13]	Cattle, IgG	Not tested	13
Other assays	Dot ELISA	No details on extraction after sonication	Dog, IgG	T. gondii	1609
	Dot ELISA	2500 whole tachyzoites/µL fixed to nitrocellulose	Cattle, IgG	Not tested	839
	Dot ELISA		Cattle, comparison with NAT		840
	RIT (rapid immuno-chromato-graphic test)	Truncated rNcSAG1 (T), based on[320]	Dog, cattle, mouse	T. gondii	1148
	APIA (antigen print immuno-assay)	Truncated NcSAG1 (T), based on[320]	Cattle, IgG	T. gondii[320]	2093
	Disperse dye immunoassay (DDIA)	Dot blot with tachyzoite lysate antigen; secondary antibody conjugated to colloidal dye particles	Cattle	T. gondii	1823

findings are not justified unless the viable parasite has been isolated from this species (Table 2.1). And even after this, the lack of knowledge on the diagnostic parameters of a serological test makes it impossible to make reliable estimates on the true prevalence of infection.

Once estimates on sensitivity and specificity are available, true prevalence estimates can be calculated. Most prevalence values reported in the literature do not necessarily reflect the true prevalence, because no estimates on sensitivity and specificity of the tests were available or used during data analysis. Serological tests based on some of the methods (IFAT, NAT, various ELISAs) were validated and many reports are found in the literature (Tables 2.10 and 2.11).

2.9.6.1.1 Antigens Used for Antibody Detection

Most antibody detection assays make use of tachyzoite antigens obtained by *in vitro* cultivation and purification. Some assays apply crude tachyzoite antigens, others antigen preparations in which specific antigens have been enriched in various ways. A number of assays employ recombinant parasite antigens.

2.9.6.1.2 Indirect Fluorescent Antibody Test

The IFATs are based on intact tachyzoites which can be fixed to the wells of a glass slide in different ways.[255,374,528,1556,1789,1990,2049] There are no comparative studies available to clarify how the way of fixation influences sensitivity or specificity of *Neospora* IFATs. IFATs are regarded among others as reference serological tests ("gold standard tests"). To avoid misinterpretation only a bright, continuous peripheral fluorescence of the tachyzoites should be regarded as positive. A partial, only polar, or cap fluorescence has to be interpreted as negative. Because the IFAT outcome is influenced by the person who reads the test and by the equipment (e.g., fluorescence microscope), the protocol, and the reagents (e.g., quality of specific conjugates) used, it is difficult to propose general cut-offs for IFAT. Establishing a diagnostic cut-off titer is a major concern, because of nonspecific reactivity with low dilutions of serum. In a comparative study for tests to detect infected cattle, a cut-off of 1:200 turned out as sensitive and specific[2049] but other applications may require different, eventually less stringent or more stringent cut-offs to ensure optimal performance of the IFAT (see species-specific chapters). When cell-culture-derived parasites are used to generate tachyzoite antigen, fetal calf serum should be avoided (at least 24 h prior to harvest of tachyzoites) to prevent that fetal antibodies against *N. caninum*, often reported from various batches of fetal calf serum, influence test results. A protocol for IFAT is detailed in Chapter 3.

2.9.6.1.3 Direct Neospora Agglutination Test

A simple direct agglutination test was originally developed for the serological diagnosis of toxoplasmosis in humans and other animals.[488,525] In this test, no special equipment or conjugates are needed. Whole formalin-killed tachyzoites are used as antigen and sera are treated with 2-mercaptoethanol to remove nonspecific IgM or IgM-like substances. Therefore, this assay detects only IgG antibodies. The *Toxoplasma* test was termed modified agglutination test (MAT). *T. gondii* antigen was substituted with *Neospora* antigen and called *Neospora* agglutination test (NAT).[1545,1733] The NAT was validated using a larger number of sera from several species. A protocol for this assay is discussed in Chapter 3.

2.9.6.1.4 Latex Agglutination Test

Two latex agglutination tests (LATs) have been published, both based on recombinant antigens, a dense granule antigen NcGRA6[759] and a surface antigen NCSAG1.[1400] Similar to tachyzoite-based

Table 2.11 Commercial ELISA Tests Kits for *N. caninum* Antibodies

Trademark (Abbreviation)	Availability, Link to Supplier	Antigen	Type of ELISA	Target Species, According to Supplier	Analyte, According to Supplier	Main References of Validation	Off Label Use, Often without Validation
CIVTEST Bovis Neospora, HIPRA (CIVTEST Neospora)	https://www.hipra.com	Sonicate lysate of tachyzoites	iELISA	Bovine	Serum, individual milk, and tank milk	54,56,791,1214	NA
IDVET ID Screen *N. caninum* Indirect Multi-species (IDVET Neospora)	http://www.id-vet.com	Sonicate lysate of tachyzoites	iELISA	Multispecies including cattle and dogs	Serum, plasma, milk	56	Small ruminants[97,491]
LSIVet™ Bovine Neosporosis Advanced Serum ELISA Kit, Thermo Fisher (LSIVet Neospora)	https://www.thermofisher.com	Sonicate lysate of tachyzoites	iELISA	Bovine	Serum	ND	ND
LSIVet *N. caninum* milk, Laboratoire Service International	No longer marketed	Not reported	iELISA	Bovine	Bovine milk	Bulk milk[142]	ND
LSIVet Ruminant, Laboratoire Service International	No longer marketed	Sonicate lysate of tachyzoites	iELISA	Bovine	Serum, plasma	56	ND
N. caninum ELISA antibody KIT BIO K 192, Bio-X Diagnostics (BIO K 192 Neospora)	http://www.biox.com	NcSRS2 purified protein from cultured parasites captured by Mabs	iELISA	Bovine	Serum, milk	56,755	Small ruminants[5]; dog[755]
N. caninum ELISA antibody KIT BIO K 218, Bio-X Diagnostics (BIO K 218 Neospora)	http://www.biox.com	Recombinant protein	cELISA	Bovine	Serum	ND	ND
N. caninum Antibody Test Kit, cELISA, VMRD Inc. (VMRD cELISA Neospora)	https://www.vmrd.com	Whole tachyzoite/ monoclonal antibody based	cELISA	Bovine	Serum	172,176,2074, dog[299,1070], roe deer (*Capreolus capreolus*)[1639]	Small ruminants[156,203,1453,1550,] buffalo[1337,1451,1824,1660]; roe deer (*C. capreolus*)[1550,1639,] horse[157,161]; pig[161]; dog[90,265,1070,1107,1259,1456,1841,] cat[90,1821]; fox[2165]

(Continued)

Table 2.11 (Continued) Commercial ELISA Tests Kits for *N. caninum* Antibodies

Trademark (Abbreviation)	Availability, Link to Supplier	Antigen	Type of ELISA	Target Species, According to Supplier	Analyte, According to Supplier	Main References of Validation	Off Label Use, Often without Validation
IDEXX Neospora X2 (IDEXX Bov)	https://www.idexx.com	Sonicate lysate of tachyzoites	iELISA	Bovine	Serum	142 (bulk milk), [145] (bulk milk), [262] (milk), [650,2049] (milk, bulk milk), [1214,1555, 1686,1802] (milk), [2067] (milk), [2074] (bulk milk), [2131]	Small ruminants[1197], zebu[1280], dog (anti-canine IgG conjugate)[504,1770]
IDEXX Chekit Neospora Ab, IDEXX (IDEXX Rum)	https://www.idexx.com	Detergent lysate of tachyzoites	iELISA	Bovine, ovine and caprine	Serum, plasma	1692,2049 (sheep)	Small ruminants[26,408,946,985,1692], buffalo[1459]
SVANOVIR Neospora-Ab ELISA (Boehringer Ingelheim)	http://www.svanova.com	Tachyzoite proteins incorporated into ISCOMs	iELISA	Bovine, ovine and caprine	Serum, plasma, bovine milk	56,216,2033 (milk)	ND
BOVICHEK Neospora (BIOVET)	http://www.biovet-alquermes.com	Sonicate lysate of tachyzoites	iELISA	Bovine	Serum, skimmed milk	56,1555,2062,2074,2131	ND
Neospora-ELISA Cattle, (AFOSA—Animal Welfare & Food Safety GmbH (p38 ELISA AFOSA)	http://www.afosa.de	Affinity purified NcSRS2 from cultured tachyzoites	iELISA	Bovine	Serum, plasma	2049	ND
Neospora-ELISA, Cypress Diagnostics C.V.	No longer marketed	Detergent lysate of tachyzoites	iELISA	Bovine	Serum, plasma	2049	ND
Mastazyme™ Neospora, MAST Diagnostics	No longer marketed	Whole tachyzoite	iELISA	Bovine	Serum, plasma	2049,2094	Dog[1552] (blocking ELISA)
N. caninum monocupule screening (P00510/02), Pourquier	No longer marketed	Whole tachyzoite/ monoclonal antibody based	cELISA	Bovine	Serum	39,62,834,835 (validation for milk)	Red deer (Cervus elaphus), Barbary sheep (Ammotragus lervia), roe deer (C. capreolus), wild boar (Sus Scrofa), Spanish ibex (Capra pyrenaica hispanica), wild rabbit (O. cuniculus), hare (Lepus granatensis), fallow deer (Dama dama), mouflon (Ovis ammon), chamois (Rupicapra pyrenaica)[39]
N. caninum blocking ELISA, Pourquier	No longer marketed	ND	cELISA	Bovine	Serum	834,835 (validation for milk)	ND

Note: ND = no data; NA = not applicable; iELISA = indirect ELISA; cELISA = competitive ELISA.

agglutination tests an examination of a large variety of animal species is possible. A comparative validation of several ELISA tests, an agglutination test and immunoblot with cattle and dog reference sera revealed that the immunoblot and the ELISA were superior to agglutination tests in terms of sensitivity.[759]

2.9.6.1.5 Enzyme-Linked Immunosorbent Assays

In most ELISA tests, soluble antigen is absorbed or specifically bound on a plastic surface (e.g., in microtiter plates), the antigen–antibody reaction is visualized by the addition of a secondary enzyme-linked antibody–antigen system, and the reaction can be assessed objectively by quantification of the color that develops. ELISA can be automated so that a large number of sera can be examined rapidly in a central laboratory. It does, however, require an ELISA reader to quantify the color reaction.

Numerous modifications of ELISA have been published, and some of them are commercially available. Most are indirect ELISAs (Tables 2.10 and 2.11), all based on lysate antigen, ISCOM incorporated antigen or affinity purified NcSRS2; some of these tests showed excellent diagnostic parameters when validated with a set of reference sera.[56,219a,2049] Some of the ELISAs are designed as competitive or blocking ELISAs (cELISAs). These have the option be applied also to other host species than initially validated for. Competitive ELISAs also showed excellent or acceptable diagnostic parameters when validated for the examination of cattle[176,569,1323,1863] or dog sera.[299,1323] However, cELISAs are not always excellent for diagnosis. Standardization of antigen preparations, test protocols, and subsequent validation studies are essential to confirm and sustain an excellent diagnostic performance.

Recombinant antigens have been used in some ELISAs. In addition, several further *Neospora* recombinant proteins have been established, but have not been used for serological but for other purposes (e.g., as vaccines), yet. Antigens used or validated for serological purposes are based on different parasitic compartments (Table 2.10), including surface antigens (NcSAG1, NhSAG1, NcSRS2, NcSAG4, rNcSRS9), dense granule antigens (NcGRA2, NcGRA6, NcGRA7) microneme antigens (NcSUB1, NcMIC19), and other antigens (e.g., *Neospora* profilin or cyclophilin). By using single antigens in ELISA, it is possible to examine the antibody response simultaneously against different antigens to which a host is differentially exposed depending on the stage of infection (i.e., the acute or chronic stage, see host species-specific chapters).

Most of the commercialized assays are ELISAs (indirect ELISAs [iELISAs] or competitive ELISAs [cELISAs]). Most of the commercialized ELISAs to detect *N. caninum*-specific antibodies are based on total tachyzoite lysate antigen (Table 2.11). In addition, there was a commercialized test using fixed *N. caninum* tachyzoites.[2095] ELISAs using ISCOM incorporated tachyzoite antigens, native antigens captured by monoclonal antibodies against a 65 kDa *N. caninum* tachyzoite antigen[176] or native NcSRS2 captured by a monoclonal antibody against this immunodominant tachyzoite surface antigen[755] are still marketed. One commercial iELISA uses native affinity-purified NcSRS2 as an antigen absorbed to the ELISA plate.[2049] One commercialized cELISA uses a recombinant antigen (NcSRS2, *E. coli*) in combination with a specific monoclonal antibody labeled with peroxidase (Table 2.11).

2.9.6.1.6 Immunoblots

Gel electrophoresis (SDS-PAGE) is applied to separate native proteins by size of their 3D structure or denatured proteins by length of the polypeptide. The proteins are then transferred to a membrane (typically nitrocellulose or polyvinylidene fluoride) by IB, where they are stained with antibodies specific to the target protein. Immunoblots are important to detect antibody reactions in sera to particular antigens, and are used to generate well-characterized reference samples for the validation of new tests. In addition, immunoblots are important to confirm inconclusive findings. Several ways to treat *N. caninum*-antigen and several banding patterns have been described (Table 2.12).

Table 2.12 Immunodominant Bands Recognized by *N. caninum* Infected or Immunized Animals

Major Antigens, Molecular Weights in kDa (Antigen)	Source of Antibodies	Parasite Species Used to Examine Specificity, Details, and Comments	Reference
	Nonreduced antigen		
16, 29, 31, 32, 37, 46, 51, 56, 79 (total lysate antigen)	Hyperimmune serum from a rabbit (*N. caninum*)	No cross-reactivity examined	140
17, 29, 30, 37 (total lysate antigen, detergent soluble supernatant after sonication)	Sera from several animal species, either naturally or experimentally infected with *N. caninum*; hyperimmune sera (*T. gondii, S. cruzi, H. hammondi*)	*T. gondii*: clear cross-reactivities with bands at 46, 88, 97 kDa, faint and inconsistent reactions with the 30 kDa band; *N. caninum*: cross-reactivities observed when *N. caninum* seropositive sera were probed with *T. gondii* antigen; *S. cruzi, H. hammondi*: no cross-reactivities reported	211
17–19, 30–45, 60 kDa (either total lysate antigen, water soluble fraction or antigens incorporated into ISCOMs)	Serum from experimentally infected rabbits (*N. caninum, T. gondii, S cruzi*) and from naturally infected dogs (*N. caninum, T. gondii*)	*T. gondii, S. cruzi*: no cross-reactivity reported	213
28–35, 45–52, 64–78 kDa (antigens incorporated into ISCOMs)	Sera and milk from naturally infected bovines, serum from experimentally infected calf (*T. gondii*)	*T. gondii*: no cross-reactivities reported	216
16, 17, 29/30, 30–36, 37, 45 (total lysate antigen)	Sera from experimentally and naturally infected bovines	No cross-reaction reported with *T. gondii* serum of an experimentally infected calf	100
17, 29, 30, 33, 37, 40 (total lysate antigen)	Sera from naturally infected bovine fetuses and their dams	No cross-reactivity examined	1880
29, 36 (total lysate antigen)	Sera from naturally infected bovines	*T. gondii*: faint cross-reactivities observed in 3 of 17 sera when *T. gondii* seropositive sera were probed with *N. caninum* antigen (double infection with *T. gondii* and *N. caninum* suspected)	1902
14, 29, 30–33, 35, 40, 55, 60, 77, 84, 97, 170 (total lysate antigen)	Sera from experimentally infected dogs (infected by feeding tissue cysts infection, subsequently shedding oocysts) and naturally infected dogs	No cross-reactivity reported	1012
17, 29–32, 35–37, 45–50, 60, 82, 88, 97, 120 (total lysate antigen)	Sera from naturally infected dogs, experimentally infected mice	*T. gondii*: *T. gondii* seropositive dogs and mice recognized *N. caninum* antigens of 45–50 kDa and higher but not the antigens at 17, 29–32 kDa	1858

(Continued)

Table 2.12 (Continued) Immunodominant Bands Recognized by *N. caninum* Infected or Immunized Animals

Major Antigens, Molecular Weights in kDa (Antigen)	Source of Antibodies	Parasite Species Used to Examine Specificity, Details, and Comments	Reference
	Reduced antigen		
17, 31, 34, 37, 40.5, 47, 50, 55, 65 (total lysate antigen)	Hyperimmune serum from a rabbit (*N. caninum*)	No cross-reactivity examined	140
17, 29, 30, 37 (total lysate antigen, detergent soluble supernatant after sonication)	Sera from several animal species, either naturally or experimentally infected with *N. caninum*; hyperimmune sera (*T. gondii, S. cruzi, H. hammondi*)	*T. gondii*: cross-reactivities existed but were not described in detail; *S. cruzi; H. hammondi*; no cross-reactivity reported	211
17–18, 34–35, 37, 60–62 (total lysate antigen, supernatant after sonication)	Sera from naturally infected pregnant cattle and aborted foetuses	No cross-reactivity examined	53
18, 25, 33, 35–36, 45–46, 47, 60–62 (total lysate antigen, supernatant after sonication)	Sera from naturally infected bovines	No cross-reactivity examined	1505
14, 20, 31, 35, 42, 54, 65, 76, 116 (total lysate antigen, supernatant after sonication)	Sera from naturally (*N. caninum*) and experimentally infected cattle (*T. gondii, S. cruzi, S. hirsuta*)	*T. gondii*: a *T. gondii* serum recognized major bands of 14, 31, and 42 kDa and in addition minor bands of 37, 55, and 80 kDa; *S. cruzi, S. hirsuta, S. hominis, S. hirsuta*: a *S. hirsuta* serum recognized bands of 11, 25, 34, 37 kDa	172
36.5–38, 45.5–48.5, 52–53.5, 58, 58.5, 59.5, 60.5, 62, 63.5, 64, 66.5, 67, 67.5, 68.5, 69.5 (total lysate antigen)	Sera from *N. caninum* or *T. gondii* experimentally infected sheep, goats, cattle	*T. gondii*: *T. gondii* infected animals recognized major bands of 63.5, 64, 66.5, 67, 67.5, 68.5, 69.5 kDa	852

All immunoblots used for diagnostic purposes employ cell culture-derived tachyzoites as an antigen source. These antigens are separated either under reducing or nonreducing conditions (i.e., reduce disulphide bridges or not, respectively) and the focus is more or less on antigens with a molecular mass between 10 and 70 kDa. Both reduced and nonreduced antigens can be used to diagnose *N. caninum* infections.[140,211] Different immunodominant *N. caninum*-specific antigens were identified. Stronger reactions are observed against nonreduced antigens, suggesting that conformational epitopes are predominantly involved in the *N. caninum*-specific antibody response (Table 2.12). Under nonreduced conditions antigens around 17–19, 29–30, and 35–37 kDa relative molecular weight are among the most often reported (Table 2.12). Laboratories that use nonreduced antigens report fewer cross-reactions between sera from animals infected with *N. caninum* and those infected with *T. gondii*, *Hammondia hammondi*, or *Sarcocystis* sp.[140,213,216,1902] compared to those employing reduced antigens (Table 2.12). A potential reason for this is that in *N. caninum* conformational epitopes might have a higher species-specificity than linear epitopes. Patterns of recognized antigens change during the course of infection.[13,99,1974,1996,2108]

2.9.6.1.7 Avidity Tests

The binding (avidity) of parasite antigen to specific antibodies can change during the course of infection. During the early (acute) stage of infection, avidity values are low and increase with duration of infection. In this test, the reactivities of sera are assessed with or without urea treatment (or any other protein denaturing agents), and the difference in titers can be used to determine whether an infection has occurred recently or has been established some time ago. The test can be used with IgG, IgA, and IgE antibodies using different serological procedures, most often ELISA. In avidity ELISAs, antibodies are allowed to bind as usual in the avidity test wells and in control wells, both coated with antigen. In avidity test wells, low affinity antibodies are then eluted by incubation in 6–8 M urea, while high avidity antibodies remain bound to the antigen. Antibody titers obtained with and without urea treatment are then used to calculate the IgG avidity indices.

Avidity tests have been used to diagnose neosporosis.[13,219,221,1758,1800] A comparison of the avidity tests used in several laboratories has been performed.[223]

A protocol for a *Neospora* avidity ELISA is provided in Chapter 3.

2.9.6.2 Antigen Detection

An assay to detect *N. caninum* parasitic antigen by ELISA was established based on polyclonal antibodies developed against recombinant microneme antigens.[2157] The motivation to develop this assay was that NcMIC10 is discharged during invasion and that this antigen may circulate in blood or body fluids during active infection. Polyclonal antibodies to 2 recombinant antigens, covering different areas of the protein (i.e., rNcMIC10N and rNcMIC10M) were used to establish a capture ELISA. Antibodies against rNcMIC10M seem to be species-specific since *T. gondii* excretory/secretory antigens did not react. It was reported that in two groups of goats infected by 10^4 and 10^6 tachyzoites the capture ELISA was able to detect antigen up to 5 weeks post infection in most of the animals.[2157] Further studies are needed to confirm these findings and to determine the practical relevance of antigen detection.

2.9.7 Detection of DNA (PCR)

2.9.7.1 General Aspects of PCR

Neospora DNA can be detected by several methods using several gene targets. PCR is a molecular genetic technique that permits the detection of any short sequence of DNA (or RNA) in samples containing even minute quantities of DNA or RNA. PCR is used to reproduce selected sections of nucleic acids for analysis. Most PCRs reported are conventional endpoint PCRs, including 1-step PCRs, 1-step nested PCRs, or 2-step nested PCRs (Table 2.13). Some PCRs are able to amplify the same target in *N. caninum* and in other related parasitic species (e.g., *T. gondii*). In these cases species diagnosis is achieved by analysing the amplicon for restriction fragment length polymorphism using restriction enzymes (PCR-RFLP),[1253,1378,1805] after sequencing[515,1347,2032] or after hybridization.[911]

PCR protocols were developed that not only detect, but also quantify, *N. caninum* DNA in samples. Quantitative PCR has become one of the key methodologies to examine the pathogenesis of neosporosis and to assess the activity of vaccines and therapeutic or prophylactic drugs[287,288,365,367,653,1584,1617,1618] or for diagnostic purposes.[380,439,753,1338,1508,1704] Quantitative PCR was also used in epidemiological studies to estimate parasitic load in *N. caninum* positive bull semen[268,671,672,1524] or semen of rams[1929] or the parasitic burden in tissues of aborted fetuses, calves, lambs, or kids.[87,1481,1635,1680]

Different target DNAs were chosen to establish *N. caninum*-specific primer pairs. A repetitive character of some of the target DNAs is an advantage since PCRs amplifying repetitive elements usually have a higher analytical sensitivity compared to PCRs amplifying fragments of single copy genes. Because of their repetitive character the genes coding for rRNA and the pNc5 gene have become important targets for diagnostic and quantitative PCRs. The suitability of particular targets, as for example, the pNc5 gene, need to be further validated because false positive reactions have been reported if used to examine rodent tissues.[1438,1735]

2.9.7.2 Target Genes for Diagnostic N. caninum PCRs

Using several *N. caninum* isolates, including NC1, NC2, NC3, Nc-Liv, BPA-1, Nc-SweB1, several PCRs were developed to specifically amplify *N. caninum* DNA (Table 2.13). Since the genomic sequences of DNA coding for ribosomal RNA (rDNA) can be used for phylogenetic studies among related apicomplexan species[307,636,641,642,644,646a,922,1418] (e.g., *T. gondii*, *N. caninum*, *H. hammondi*, *H. heydorni*, and *Besnoitia besnoiti*) rDNA sequences (18S rDNA, 28S rDNA, ITS1) are promising targets for the development of species-specific PCRs. The most important is the pNc5 gene,[2141] a multi-copy gene.[1427] Other diagnostic targets are single-copy genes, 14-3-3, and the HSP70 gene.[1378]

The genome of the *N. caninum* Liverpool strain (Nc-Liv) has been sequenced[1698] and corresponding data are available at http://www.genedb.org/Homepage/Ncaninum. Based on these sequence data it is possible to develop further targets for diagnostic PCRs and further sequence databases are available to have access to specific sequences.[794]

2.9.7.2.1 28S rDNA

The D2 domain of the 28S rDNA was often used for phylogenetic studies. Based on the species-specific sequences an *N. caninum*-specific primer pair was identified which was able to distinguish *N. caninum* from *T. gondii*, *Hammondia* sp., or *B. besnoiti*.[642] No data on the analytical sensitivity of this PCR were provided.

2.9.7.2.2 18S rDNA

Only minor differences have been found between the 18S rDNA genes of *T. gondii* and *N. caninum* suitable for the development of species-specific primers.[307,1281] Therefore, universal primers had to be used to amplify the 18S rDNA.[640,911,1253] In 1 study, the application of species-specific chemiluminescent DNA hybridization probes allowed differentiation of *N. caninum* and *T. gondii* 18S rDNA amplicons.[911] This protocol was applied in a subsequent study on the distribution of *N. caninum* DNA in bovine tissues,[912] but also in a study to characterize a bovine *N. caninum*

Table 2.13 Analytical Sensitivity and Specificity of PCR for the Detection *N. caninum* DNA

Target DNA	Primer Names	Type of PCR	Analytical Sensitivity According to the Original Description	Parasites Used to Test Analytical Specificity	Reference
28S rDNA	GA1/NF6	1-step PCR	ND	*T. gondii, H. hammondi, B. besnoiti*	642
18S rDNA	COC-1/COC-2	1-step PCR + hybridization	1 tachyzoite in medium, 5 tachyzoites in blood or amniotic fluid	*T. gondii, S. cruzi, S. tenella, S. capracanis, C. parvum, E. bovis*	911
	AP1/D, SP4/A	2-step nested PCR	ND	*T. gondii, S. cruzi*	640
	COC-1/COC-2	1-step PCR, RFLP (restriction enzyme BsaJI)	ND	*T. gondii*	1253
ITS1	NS1/SR1	1-step PCR	ND	*T. gondii, S. cruzi*	1575
	PN1/PN2	1-step PCR	5 tachyzoites heated for 2 min at 100°C in distilled water	*T. gondii, S. cruzi, S. fusiformis, S. gigantea, S. tenella*	924
	NN1/NN2, NP1/NP2	2-step nested PCR	ND	ND	259
	TIM3/TIM11, NS1/SR1	2-step nested PCR	ND	See Reference **1575**	640
	F6/5.8B, PN3/PN4	2-step nested PCR	ND[a]	ND	2009
	NF1/SR1, NS2/NR1	1-step nested PCR	10^{-1} fg DNA (0.1–0.01 tachyzoites)	*T. gondii, S. cruzi*	643
	CR3/CR4	ND	ND	ND	1353
	NS3/SR1	1-step PCR	ND	*T. gondii, Toxocara canis, Toxascaris leonina, Trichuris vulpis, Sarcocystis sp., Isospora ohioensis-group, Isospora canis, C. parvum*	1866
ITS1 and 18S	JB1/JB2, SF1/SF2	2-step nested PCR	10^{-2} fg/µL in 500 ng/µL mouse DNA (0.01–0.001 tachyzoites)	*T. gondii, H. heydorni, B. besnoiti*	138
14-3-3 gene	Nc13F3/ Nc13R2 Nc13F1/ Nc13R4	2-step nested PCR	25 tachyzoites in 5 mg brain tissue	*T. gondii, S. muris, S. tenella, S. cruzi*	1114
HSP70 gene	Hsp400F/Hsp400R	PCR, RFLP (restriction enzymes MunI, StuI, and KpnI)	100 oocysts/mL	*T. gondii, H. hammondi, H. heydorni*	1378
pNc5	Np1/Np 2	1-step PCR	100 pg genomic tachyzoite DNA	*T. gondii, S. cruzi, S. ovifelis, S. capracanis, S. moulei, S. miescheriana*	1043
pNc5 gene	Np6/Np21 (and other primer pairs Np21/Np4, Np7/Np4)	1-step PCR	1 tachyzoite in 1 mg brain tissue (Np6/Np21)	*T. gondii, H. hammondi, S. cruzi, S. tenella, S. capracanis, S. moulei, S. miescheriana, H. heydorni[b], Toxocara canis[b]*	2141
	Np6plus/Np21plus	1-step PCR	DNA equivalent to 1–10 tachyzoite genomes	*T. gondii, H. hammondi, S. cruzi, S. tenella, S. capracanis, S. moulei, S. miescheriana*	1426

(Continued)

Table 2.13 (Continued) Analytical Sensitivity and Specificity of PCR for the Detection N. caninum DNA

Target DNA	Primer Names	Type of PCR	Analytical Sensitivity According to the Original Description	Parasites Used to Test Analytical Specificity	Reference
	Np6plus/Np21plus	1-step PCR + hybridization ELISA	DNA equivalent to 1 tachyzoite genomes	T. gondii, H. hammondi, S. cruzi, S. tenella, S. capracanis, S. moulei, S. miescheriana	1426
	Np4/Np7	1-step PCR	1–2 tachyzoite equivalents per DNA sample (150 ng brain tissue DNA)	See Reference 2141; note that primers were found to amplify house mouse[1735] or rat DNA[1438]	174
	Np4/Np7, Np6/Np7	2-step seminested PCR	ND; sensitivity of seminested PCR was not superior to 1-step Np6/Np7 PCR	See Reference 2141	174
	Np6plus/Np21plus (sequence of Np21plus modified according to Genbank X84238), Nc5 competitor	1-step quantitative PCR	9 fg DNA per 250 ng of mouse DNA	H. heydorni, C. parvum, S. cruzi, S. tenella, and I. ohioensis, H. hammondi, T. gondii, E. tenella, Strongyloides stercoralis, Toxocara canis, Dipylidium caninum, Ancylostoma caninum,[910] see also Reference 1426	910, 1152
	Np4B/Np21B; identical with Np4/Np21[1735]	1-step PCR	ND	See Reference 2141	196
	Np6plus, Np21plus	LightCycler Real-time PCR (including detection and anchor probes)	DNA equivalent to 1 tachyzoite	See Reference 1426	1428
	Nc5fwd/ Nc5rev	Real-time PCR (SYBR Green)	DNA equivalent to 0.1 tachyzoite genomes (10 fg) in 100 mg mouse brain DNA;	T. gondii	364
	Np6/Np21	Multiplex PCR for N. caninum and T. gondii	ND	See Reference 2141	1460, 1810
	"Outer primer pair" (based on Np6/Np21) and "inner primer pair"	In-situ PCR	ND	ND	1226
	Np6plus/Np21plus	1-step PCR	See Reference 1752	See Reference 1752	1506
	NeoF/NeoR[c]	Real-time PCR (SYBRGreen)	ND	ND	1508
	Np6plus/Np21plus and Np6/Np7	2-step nested PCR	Reported to be 4-fold more sensitive than nested PCR published[174]	T. gondii, and see Reference 1752	953
	Np6plus/Np21plus followed by Np9/Np10[d]	2-step nested PCR	ND	ND	1330
	Np6/Np21 followed by Np7/Np10	2-step nested PCR	ND	H. heydorni	2072

(Continued)

Table 2.13 (Continued) Analytical Sensitivity and Specificity of PCR for the Detection N. caninum DNA

Target DNA	Primer Names	Type of PCR	Analytical Sensitivity According to the Original Description	Parasites Used to Test Analytical Specificity	Reference
	External for/rev; Internal for/rev	2-step nested PCR	ND	T. gondii, B. besnoiti	692, 1299
	Np6plus/Np21plus	Real-time PCR (SYBRGreen)	0.008 pg	See Reference 1426	1617
	"Forward primer/Reverse primer", "Probe"	Real-time PCR (Taqman)	ND	ND	1704
	561U20/806L20 (based on Np7/Np4,[T4]	Real-time PCR (SYBR green)	DNA equivalent to 1 tachyzoite genome; DNA equivalent to 10 genomes in chicken liver DNA (100 µg/mL)	ND	753
	NC5PLEX/NC5FRIEND	Real-time PCR (Plexor)	DNA equivalent to 1 tachyzoite genome; DNA equivalent to 10 genomes in chicken liver DNA (100 µg/mL)	ND	753
	NC5-550/NC5-596, NC5 probe	Real-time PCR (Taqman)	DNA equivalent to 1 tachyzoite genome; DNA equivalent to 1 genome in chicken liver DNA (100 µg/mL)	T. gondii	753, 756
	NC5-F1/ NC5-R1	Real-time PCR (SYBRGreen)	ND	ND	715[e]
	Np6/Np21, Np6/Np7	2-step nested PCR	ND	ND	2165
	Sense, Antisense, Probe	Real-time PCR (Taqman)	ND	ND	1458
	NeoF/NeoR[c], NeoProbe	Real-time PCR (Taqman)	ND	ND	380
	NF1/NR1, NP	Real-time PCR (Taqman)	ND	T. gondii	439
	Np6plus/Np21plus and Np6plus/Np7	2-step nested PCR	ND	ND	986
	"Forward"/"Reverse", "Probe"	Real-time PCR (Taqman)	ND	ND	1584
	OF/OR (based on Np6/Np21), IF/IR	2-step nested PCR	ND	ND	1142

a Some information of analytical sensitivity can be found in Reference 825.
b Analyzed in Reference 910.
c Note: NeoF/NeoR sequences published in Reference 1508 or 380 are different.
d Note: Np10 sequences published in Reference 1330 or 2072 are different.
e First published in Furuta PI. 2008. Infecção experimental em cães com ovos embrionados de galinha (Gallus gallus domesticus) infectados com taquizoítas de N. caninum. Thesis. Jaboticabal, Brazil: Universidade Estadual Paulista, Faculdade de Ciências Agrárias e Veterinárias.

isolate.[1059] It was used further to demonstrate vertical transmission in rhesus macaques[913] and up to now used to diagnose equine protozoal myeloencephalitis caused by *N. hughesi*.[1283,1645] Another protocol also applied universal primers to amplify 18S rDNA in the first round PCR.[640] In a second round, an *N. caninum*-specific primer (containing two mismatches) was used to differentiate *N. caninum* 18S rDNA from that of *T. gondii*, *Sarcocystis cruzi* and host cell DNA.[640]

A third protocol applied a published primer pair[911] after modification and differentiated *N. caninum*, *T. gondii*, and *Sarcocystis* sp. 18S rDNA by differences in fragmentation using the restriction enzyme BsaJI.[1253] This PCR was used to confirm neosporosis in an aborted water buffalo fetus,[103] in a goat fetus[635] and was also validated in an interlaboratory comparison.[2022] Also *Hammondia* 18S rDNA was amplified by these universal primers and species-specific fragments are obtained after using SecI, an isoenzyme of BsaJI.[1805] The non-species-specific, universal primers have the advantage that the DNA of different species is amplified simultaneously in a single reaction, and the DNA of different species is amplified with the same sensitivity.[1805]

2.9.7.2.3 Internal Transcribed Spacer 1 Region

Among the internal transcribed spacer 1 (ITS1) regions of the rRNA gene of *T. gondii*, *N. caninum*, *H. heydorni*, *H. hammondi*, and *B. besnoiti* there are several sequence differences that allow the establishment of species-specific PCRs.[794,926] The ITS1 region represents as a gene array, which has about 110 copies in *T. gondii*.[767] Many PCR protocols using the ITS1 region as the target have been published. In addition to 1-step conventional PCRs,[924] 2-step nested PCRs were developed[259,2009] and applied in studies on the epidemiology and pathogenesis of bovine neosporosis[96,189,260,268,364,368,671,825,976,1264,1324,1524,1590,1680,1827,1904,1994,2022,2097] and canine neosporosis[1936a] or to test tissues of new world camelids,[1828] of capybaras,[1997] wild living carnivores,[152,1918] and rodents.[1008] One study compared the performance of an ITS1-based[924] and a pNc5-based PCR (Np4/Np7)[174] and observed a higher sensitivity in the ITS1-based PCR.[1771] In a large-scale Canadian study on aborted fetuses, ITS1 primers targeting at a ribosomal SSU rDNA gene array[767] were used and the amplified fragments were subsequently sequenced.[2104] A comparison with existing and validated methods was not provided and sensitivity and specificity of this protocol is not known.

Although 2-step nested PCRs are usually superior in sensitivity, they have the disadvantage of having a higher risk of carry-over contamination. As an alternative, 1-step (single-tube) nested PCRs combine the higher sensitivity of a nested PCR with the lower risk of carry-over contamination in 1-step PCRs. An ITS1-based one-tube nested PCR for *N. caninum* has been developed[643] and an analytical sensitivity of 1–10 fg genomic DNA of *N. caninum* tachyzoites was reported. 1–10 fg DNA is supposed to be equivalent to the genomic DNA of 0.1–0.01 tachyzoites.[643] This PCR was used in to examine paraffin-embedded material from aborted bovine fetuses,[1335] brain tissues from aborted bovine fetuses[2106] or blood of experimentally aborted lambs and their ewes,[2086] ovine and caprine abortion material,[1293] or rodents from dairy farms.[1336] Other studies also employed the primers reported[643] but in a 2-step nested PCR.[434,1324,2097] Whether variations in the ITS1 sequences between and within individual strains[782] affect analytical or diagnostic sensitivity of ITS1 is not known.

A further 2-step nested PCR targeting 18S RNA gene and ITS1 gene region was established to test rodents from Australia for *N. caninum* infection.[138] An analytical sensitivity of 10 fg/µL (equivalent to the DNA of 0.1 tachyzoite) was reported. This PCR was subsequently used for the examination of aborted bovine fetuses,[266] brains from adult cattle with clinical signs of encephalitis[1262] or a liver biopsy of a dog.[708]

2.9.7.2.4 HSP70 Gene

The HSP70 genes code for a set of chaperones that are involved in a large variety of protein folding processes. HSP70 proteins possess a highly conserved NH_2-terminal ATPase domain and a

COOH-terminal region, including a conserved substrate binding domain.[252,1377] HSP70 proteins are highly conserved across prokaryotes and eukaryotes. In studies aiming to characterize the HSP70 genes of *N. caninum*, *H. heydorni*, *H. hammondi*, and *T. gondii*, a PCR was established that amplified a ~400 bp large fragment of a HSP70 gene of *N. caninum* and *H. heydorni*, but not from *T. gondii* and *H. hammondi*.[1378] A subsequent fragmentation using restriction enzymes MunI, StuI, and KpnI,[1378] or StuI and KpnI[116] enabled the differentiation between *N. caninum* and *H. hammondi*. This PCR-RFLP method was used to differentiate the morphologically similar oocysts of *N. caninum* and *H. heydorni* in dog feces.[116,1378,1458]

2.9.7.2.5 The 14-3-3 Gene

A 2-step nested PCR based on the 14-3-3 gene was developed.[1114] Although this gene is evolutionarily conserved among eukaryotic taxa it was possible to identify primers that proofed to be *N. caninum*-specific using DNA from *T. gondii*, *S. cruzi*, *Sarcocystis tenella*, and *Sarcocystis muris*. This PCR was shown to be able to detect 25 *N. caninum* tachyzoites in 5 mg brain tissue. This PCR was applied to examine tissues of naturally infected dogs suffering from canine neosporosis.[573]

2.9.7.2.6 pNc5 Gene

Soon after the detection of the pNc5 gene[1043,2141] it became obvious that there are multiple copies of this gene in the genome of *N. caninum*. Obviously the genome of a single organism may contain several variants of this gene.[1427] The pNc5 gene has not been found in related parasite species such as *T. gondii*, *S. cruzi*, or *H. hammondi* although cross-reactivity with *S. cruzi* was reported for one of the evaluated primer pairs (see below Np5, Np6).[2141] The pNc5 gene has not been shown or sequenced in *N. hughesi*.[1887] Initially, several primer pairs were developed (Np1 to Np8, and Np21). In addition to the primer pair Np1/Np2[1043] the pair Np6/Np21 appeared to be the most suitable one.[2141] A conventional 1-step PCR using these primers was able to detect 1 tachyzoite in 2 mg brain tissue. This PCR has been used in numerous studies in bovine fetuses, embryos, or placenta,[196,633,948,1413,2104,2176] in adult cattle,[311] colostrum/milk of cattle,[1411,1412] in water buffaloes,[103,1722] in sheep,[91,1083,1097] in dog tissues,[417,1346,1810] on feces or oocysts from dogs or other (potential) definitive hosts[164,781,910,1722,2072] and tissues of other (experimental) intermediate hosts, or cultivated parasites.[164,354,517,779,949,1054,1140,1306,1309,1328,1433,1439,1450,1722,1786,1810,1887,1893,1894,1925] Primers Np6 and Np21 were also used in a modified version to establish an *in situ* PCR to detect *N. caninum* DNA in histological sections.[1226]

Some researchers used the primers Np4, Np6, and NP7 in a semi-nested PCR to detect the pNc5 target.[174] It has to be noted that primers Np4 and Np7 amplified host DNA in *Mus musculus* but not in *Rattus ratus* or *Apodemus sylvaticus*.[1735] In a study from Brazil this nested PCR was also used to test rodents,[1437,1438] which confirmed these findings, and in addition recognized false positive reactions in rats (*R. ratus*, *R. novegicus*).[1438] This questioned the suitability of the pNc5 gene target for the examination of rodents.[1438] This semi-nested PCR was also applied for the analysis of tissues of bovine and water buffalo fetuses including also paraffin-embedded tissues.[760,1572,1707,1771,1875] One of these studies compared the performance of a PCR based on Np4 and Np7[174] with a ITS1-based PCR[924] and observed a higher sensitivity in the ITS1-based PCR.[1771] Another study reported that it is essential to use a semi-nested protocol based on the primers Np7/Np6 and Np4/Np7[174] to reach maximum sensitivity.[1572]

Others modified the originally reported primers to amplify fragments of the pNc5 gene,[1152,1426] or developed new primers for the Np5 gene, which were used in conventional or nested PCR,[36,266,692,986,1299] or quantitative PCR.[364,380,439,715,753,756,1508,1584,1704] Available sets of primers were combined to establish nested PCRs[2,81a,953,1008,2165] or at least one additional primer was established for application in nested PCR.[1142,1330,2072]

Using the primers Np6plus and Np21plus (modified versions of Np6 and Np21) a 1-step PCR was developed and the amplicons were detected using a hybridization ELISA.[1426] This PCR method was able to detect 1 tachyzoite in 1 mg tissue. Without hybridization, but using the modified primers, the PCR had a sensitivity of 1–10 tachyzoites in 1 mg tissue. The primer pair Np6plus and Np21plus or their variants have been used by the majority of studies were an *N. caninum* PCR was applied.

PCRs using Np6plus and Np21plus primers or their variants were used to test bovine fetuses or placenta,[83,493,797,798,860,893,1757,1788,1880,1924] bovine calves,[1600] blood of adult cows,[1506] small ruminants,[861] horses,[1624] dogs,[573,598,1133,1598,1684] or fecal samples and oocysts,[1067,1070,1324,1427,1805,1865] wildlife including birds,[36,426,680,958] experimental mice[58,799,1152,1153,1154] or other animal species,[958,1069] and cell culture-derived parasites.[2022] Previously published primers, Np6plus and Np21plus[1426] were included into a multiplex PCR to identify a number of infectious agents from aborted bovine clinical samples.[1985] This multiplex PCR was able to identify, in addition to *N. caninum*, also *H. heydorni* and *T. gondii*.[1985] Other nested PCRs, employing the primers pairs Np6/Np21 and Np9/Np10 were applied to demonstrate the presence of *N. caninum* stages in the intestine of a dog.[1105]

2.9.7.3 Quantitative PCR

Quantitative PCRs are important tools to study the pathogenesis of neosporosis and to assess the activity of vaccines and therapeutic or prophylactic drugs. Conventional 1- or 2-step endpoint PCRs are qualitative, but not quantitative, detection methods. All quantitative PCRs, published as yet, are based on the pNc5 gene. A first quantitative PCR was established as a so called quantitative-competitive PCR (QC-PCR).[1152,2022] An *N. caninum*-specific DNA is amplified in the presence of an artificial competitor and examined in ethidium bromide gels. Competitor titration allows the estimation of copies of pNc5 gene per sample. This method is labor-intensive and there are better methods available, now.

PCR approaches based on SYBR Green incorporation are less laborious.[364,715,753,1508,1617,2022] In 1 SYBR Green assay, results were normalized based on 28S real-time PCR results amplifying host cell 28S rDNA (i.e., bovine or murine DNA).[364] This methodology was used to estimate parasite multiplication *in vitro*,[481] parasite load in murine tissues,[16,1269] or the infection intensity in various tissues of bovine fetuses aborted at different stages of gestation.[368] It was also used to determine parasite burden in experimental ovine abortion.[87] By using a SYBR Green I-based real-time PCR the number of tachyzoites in brain samples of aborted bovine fetuses was estimated 2.9–26.6 per mg of brain. The disadvantage of SYBR Green assays is that unspecific amplification products are labeled, and amplicons need to be carefully checked for specificity.

A number of real-time PCRs employ Taqman probes. Probes increase the specificity of the reaction[380,753,1584,1704] and were employed in a number of epidemiological studies,[380,756] experimental studies in cattle[1584] or mice,[2003] and for diagnostic purposes in cattle[1596] and in dogs.[1133] Taqman assays can be multiplexed, enabling the simultaneous examination of a variety of pathogens; a multiplex quantitative real-time PCR panel for detecting neurologic pathogens in dogs with meningo-encephalitis was reported[847] which was based on a real-time PCR previously published.[753] Another multiplex assay was reported for the simultaneous examination of abortion material from ruminants for *Chlamydia* sp., *Coxiella burnetii*, and *N. caninum*.[1699] This multiplex PCR employed a primer pair and a previously reported probe.[380] A further multiplex real-time PCR was established to examine animals from the wild employing primers and probes to detect DNA from *T. gondii*, *N. caninum* and host DNA at the same time.[439] This PCR was also employed for *N. caninum* testing in rodents, but with a Zen-modified probe.[1338]

One real-time PCR, based on detection of PCR products by specific fluorescent probes,[1428] was used to demonstrate the development of infection in an organotypic slice culture system for

N. caninum,[2051] for assessing vaccination or drug trials,[287,288,471,472,473,478,653,1424,1504,1815,1914] and to study the cell biology of *N. caninum*.[1445]

2.9.7.4 Typing Different Neospora sp. Strains

Strain typing of *Neospora* sp. has been achieved by using PCR amplification (by applying random or species-specific primers) and subsequently characterizing the amplicons in terms of DNA sequence or size (in agarose gel electrophoresis or gel chromatography).

2.9.7.4.1 Random Amplified Polymorphic DNA

The random amplified polymorphic DNA (RAPD) PCR technique was able to differentiate between *N. caninum*, several *T. gondii* strains, and *Sarcocystis* species,[821] or between *N. caninum* and *H. heydorni*,[1893] and early attempts were also made to type *N. caninum* from *N. hughesi*[432,1284,1813,1887] (see Chapter 18). With other random primers it was also possible to detect differences among *N. caninum* isolates[432,1813] but it was not possible to show clusters related to hosts of isolation or geographic origin.[1813] With random primers it was possible to show genetic differences between NC1, an American isolate, and 2 European isolates (Nc-SweB1, Nc-Liv)[1887] or between European isolates (Nc-SweB1, Nc-Liv).[98] Although relatively easy to perform, one of the challenges of RAPD is to standardize the outcome, due to differences in RAPD patterns observed after other PCR reagents, thermocyclers or DNA concentration are used for experimentation. In addition, RAPD cannot be performed on clinical material because it requires purified *N. caninum* DNA.

2.9.7.4.2 Sequencing

Using ITS1 sequencing it was possible to find genetic differences among *Neospora* isolates.[782,794,1284] It was possible to separate the isolates Nc-Nowra[1353] and Nc-NZ from the remaining, but the bootstrap-value observed was low (i.e., 0.77).[794] One study reported variations in the ITS1 sequences of individual strains (Nc-Liv, NC2, NC-Bahia, Nc-beef, Nc-Illinois).[782] These variations occurred mainly in the last 70 nucleotides (i.e., the 3′ end) of the ITS1 sequence and it was suspected that these variations at the sequence ends are sequencing errors.[794] However, since ITS1 rDNA sequence exists in several copies in the genome (i.e., most likely in about 100 copies) the sequence variations observed could indeed represent true sequence variations in a single organism. In addition, 2 variants of ITS1 amplicons were visualized by polyacrylamide gel electrophoresis for *N. caninum* but not for *T. gondii*, which is in accordance with the hypothesis that the second variant observed in *N. caninum* is not an artifact.[782] Further studies are necessary to clarify this point. The only significant intra-strain difference was reported for the NC-Bahia strain ITS1 sequence which differed in 12 bp from those of North American and European strains.[782] No major differences were identified in sequence among other *N. caninum* isolates.[119,585,642,924,1284,1353,1575,1865] Also regions of genes coding for alpha-tubulin, beta-tubulin, and heat shock protein 70 (HSP70) seem to be very similar among *N. caninum* isolates.[1329,1864]

In some studies pNc5 gene sequences are used to characterize isolates or *N. caninum* DNA.[2,81a,200,390,615,680,953,954,1427,1702,2072,2176] Five different sequence variants were observed in the pNc-5 gene when examining a single *N. caninum* isolate.[1427] This suggests that there are several variants of this multicopy gene in a single organism. This questions the validity of sequence differences reported among isolates based on the pNc5 gene because direct sequencing of amplicons most likely represent a mixture of several individual sequences and cloned amplicons may represent only a small proportion of sequence variants existing in a single organism.

2.9.7.4.3 Mini- and Microsatellite Markers

Minisatellites and microsatellites are repetitive DNA sequences (tandem repeats of a DNA motif) in the genomes of eukaryotic organisms. They are highly polymorphic in sequence and length.

Microsatellite markers represent simple sequence repeats (SSRs) which consist of tandemly repeated units of 2–6 base pair (bp) length, present in the genome of eukaryotic and prokaryotic organisms. They can be found both in protein-coding and noncoding regions.[30,1983] Microsatellite loci are highly variable, have a higher mutation rate than other loci, and the polymorphisms in these sequences are caused by gain or loss of single repeat units. This results in repeat length variations between different *N. caninum* isolates which can be used for strain typing[189,352,1673] (Table 2.14).

In contrast to microsatellites, minisatellite tandem repeats have longer repeat units (8–100 base pairs)[836,1941] and minisatellite units have been characterized in *N. caninum* with the potential to type *N. caninum* isolates[28,29] (Table 2.14).

The length polymorphisms of individual mini- or microsatellite loci can be analyzed after PCR amplification with primers that anneal to their flanking regions, using PAGE, high-resolution agarose gel electrophoresis, capillary gel electrophoresis, or DNA sequencing.[1499,1814]

In relative high concentrations of DNA extracted from purified parasites, single-step PCR amplification is sufficient to amplify these marker regions. However, to amplify microsatellite DNA fragments from *N. caninum* parasites that are only available in clinical material or fecal or environmental samples, nested PCR protocols are necessary. A number of nested PCR protocols are published[168,169,189,1578] which allow typing in material containing only a low concentration of *N. caninum* DNA. However, often not all marker regions are amplified when using clinical samples[169,247,1767] A multiplex assay by amplifying several microsatellite markers was simultaneously developed.[29]

Microsatellite markers are used to characterize new *N. caninum* isolates and prove that these new isolates do not represent laboratory contaminations[167,168,729,1510a,1674,1729] and confirm the identity of strains applied in experimental animal infections.[28,106,189,1091,1680] Microsatellite marker regions seem to be relatively stable over time as no differences were observed between microsatellite patterns in isolated oocysts and cell-culture isolates generated from these oocysts even after passages through mice.[168]

In addition, microsatellite typing was used to clarify epidemiological situations. In clinical material only identical or almost identical microsatellite patterns were observed in DNA collected from several fetuses aborted in individual herds during bovine abortion epidemics. This shows that a single source of infection for all aborting cattle on an individual farm is responsible for this type of *N. caninum*-associated abortion.[169] In a Spanish sheep flock, in 13 of 14 aborted fetuses, stillborn or newborn sheep lambs with neurological disorders *N. caninum* DNA was observed and microsatellite analysis revealed an almost identical pattern in *N. caninum* DNA isolated from all the material (fetuses, lambs) which suggested that all ewes were infected by *N. caninum* of a common source.[793] In a dog with neosporosis a mixed infection of more than one *N. caninum* strain was shown by using microsatellite typing.[1636]

2.9.7.4.4 Genetic Diversity among Neospora Isolates

To elucidate details on the population structure of *N. caninum*, attempts were made to characterize the worldwide population structure of *N. caninum* using 96 isolates, specifically focusing on isolates from 4 countries, that is, Spain, Argentina, Scotland, and Germany and using multilocus microsatellite markers.[1679] The results suggested that there are different clusters of multilocus genotypes among *N. caninum* isolates and that these clusters are related to the geographic origin of these isolates.[1679] Furthermore, the results showed a close genetic relationship between those isolates from Spain and Argentina which might have been a result of the introduction of Iberian cattle into South America by Spanish and Portuguese colonizers at the end of the fifteenth century.[1679] Iberian breeds

Table 2.14 Chromosomal Location of Mini- and Microsatellite Markers Suitable for Typing *N. caninum* Strains

Marker	Mini- and Microsatellite Sequences[a]	Chromosomal Location (Representative GenBank Accession No. for Nc-Liv, NC1, Nh-Oregon)	Representative References
Tand-13	TG-$(CGTCGCCTCCCGCCGACAGTG)_n$-CG	Chr1a (FJ824983, FJ824987, FJ883956)	28
MRI42	ATAT-$(ATAG)_x$-ATTTG-$(TA)_y$-GTGC	Chr1b (NA, NA, NA)	189
MS2	TT-$(AT)_n$-CC	Chr2 (AY935167, EU872351, NA)	1673
MS8	TGAC-$(AT)_n$-GG	Chr2 (AY935173, NA, NA)	1673
MS1A	AG-$(TA)_n$-tGA	Chr7a (AY935165, NA, NA)	1673
MS1B	AC-$(AT)_n$-GC	Chr7a (AY935166, EU872336, NA)	1673
MS7	AT-TA-$(TA)_n$-GG or AT-AA-$(TA)_n$-GG	Chr7a (AY935172, NA, NA)	1673
MS21	TG-$(TACA)_3$-TACC-$(TACA)_n$-TT	Chr7a (AY935176, NA, NA)	1673
MS10 or Tand-3	AGT-$(ACT)_x$-$(AGA)_y$-$(TGA)_z$-CAA	Chr8 (AY935174 or FJ824929, EU872411 or FJ824932, FJ883954)	28, 29, 1673
Cont-14	AG-$(GAA)_n$-CT	Chr8 (FJ883928, FJ883929, FJ883958)	29
Cont-16	AG-$(GAAGAGGAAGGGGAAGAAGAAGAGGAA)_n$-GG	Chr8 (FJ883944, FJ883952, FJ883959)	29
MS4	GC-$(AT)_n$-ACATTT-$(AT)_2$-AC	Chr9 (AY935169, EU872381, NA)	1673
MS5	CG-$(TA)_n$-TG-TA-GG	Chr9 (AY935170, EU872396, NA)	1673
MS6A	GC-$(TA)_n$-AC	Chr10 (AY935171, NA, NA)	1673
MS6B	CC-$(AT)_n$-GT	Chr10 (AY935171, NA, NA)	1673
Tand-12	TA-$(AGTTTTGCCGTTTTGCTAACGTGAA)_n$-AG	Chr11 (FJ824958, FJ824963, FJ883955)	28
Cont-6	TC-$(TGCTTGATCTCCCTCATATCCTGCTTCGTCTCC GTCACATCCTTCTTGGTTCCCCGCATCTCA)_n$-TGCTT-AT	Chr11 (FJ883900, FJ883912, FJ883957)	29
MS3	GC-$(AT)_n$-AA	Chr12 (AY935168, EU872366.2, NA)	1673
MS12	GC-$(GT)_n$-GC	Chr12 (AY935175, NA, NA)	1673

a Repeat region in italic types.

were the founders of the adapted Creole breeds, unique South American bovines for more than 300 years until the introduction of new selected European and Zebu breeds.[1679]

2.10 TREATMENT

Many aspects related to treatment options and respective molecular targets for intervention have been reviewed previously.[884,886,892,1376a] Sulfonamides and pyrimethamine, 2 drugs widely used for therapy of toxoplasmosis, have been used to treat neosporosis in dogs (see Chapter 5). These 2 drugs act synergistically by blocking the metabolic pathway involving p-aminobenzoic acid and the folic–folinic acid cycle, respectively. They are usually well tolerated, but sometimes thrombocytopenia and/or leukopenia may develop. These effects can be overcome by giving patients folinic acid and yeast without interfering with the treatment because the vertebrate host can utilize presynthesized folinic acid while *T. gondii* cannot. The commonly used sulfonamides, sulfadiazine, sulfamethazine, and sulfamerazine are all considered effective against neosporosis. Generally, any sulfonamide that diffuses across the host cell membrane is useful in anti-*N. caninum* therapy.[810] While these drugs have beneficial action when given in the acute stage of the disease process when there is active multiplication of the parasite, they will not eradicate infection. Sulfonamide compounds are excreted within a few hours of administration; therefore, treatment has to be administered in daily divided doses usually for several weeks or months. Because pyrimethamine is toxic and tablets smaller than 25 mg are not available, a combination of trimethoprim and sulfamethoxazole as possible alternatives to pyrimethamine and sulfadiazine has been used. Trimethoprim, like pyrimethamine, is a folic acid antagonist, and it has synergistic effect in combination with sulfonamides in *in vitro* tests.[1174] Experimentally, sulfadiazine given in drinking water (1 mg/mL) was effective in controlling tachyzoite proliferation[1162] but had no effect on tissue cyst formation.[1327] The antibiotic clindamycin has been used successfully to treat neosporosis in dogs. It is available as pill for oral administration, as injection, and as cream for topical application.

Certain other anticoccidials (toltrazuril, ponazuril) were found efficacious against neosporosis in mice.[58,423,799,1915] These drugs, when added to cell cultures infected with *N. caninum*, damaged the parasite apicoplast and mitochondrion.[424] Amprolium, known to be an anticoccidial for poultry and bovine coccidiosis, was not effective against neosporosis in mice.[1162] Tunicamycins, a class of nucleoside antibiotics similar to corynetoxins, did not diminish, but rather increased, the susceptibility of mice to experimental *N. caninum* infection.[297]

Most of the drug screening for neosporosis has been performed in infected cell cultures or in mice (Table 2.15). *In vitro* screening is usually done by light microscopical assessment of parasite proliferation, quantitative real-time PCR,[187] or by employing transgenic parasites that express an easily detectable marker such as beta-galactosidase, green fluorescent protein (GFP), or any other marker.[1420,1421,1422,1587] Alternatively, a simple protocol for parasite quantification using prestained *N. caninum* tachyzoites and fluorescent probes based on ester compounds has been developed.[1415] Toltrazuril, a triazinone derivative effective against various coccidians including *Eimeria* and commercialized under the proprietary name Baycox™ was investigated intensively *in vitro*,[423,1914] and in nonpregnant and pregnant mice toltrazuril exhibited promising efficacy. However, it remains unclear whether toltrazuril is a suitable drug against neosporosis because overall the efficacy results in cattle do not support this conclusion.[407,828,1102] Thiazolides such as nitazoxanide and a range of derivatives targeting protein disulfide isomerase activity in *N. caninum*[1419] exhibited promising effects against *N. caninum in vitro*,[654,655] but nitazoxanide failed to be active in mice when applied orally or was even toxic when applied intraperitoneally.[474]

Compounds initially developed against *Plasmodium* (such as mefloquine and artemisinin-derivatives[1209,1422,1423,2030a]) were highly active against *N. caninum in vitro*, but did not exhibit protection in the mouse model.[1424,1425] Other inhibitors of *P. falciparium* Atg8-Atg3 protein–protein

Table 2.15 Summary of Chemotherapeutics against Neosporosis[a]

Compound	Class	Result	Reference
Compound screen	9 anti-coccidials	Activities of lasalocid, monensin, piritrexim, pyrimethamine, trimethoprim, and decoquinate against tachyzoites *in vitro*	1160, 1177
Sulfadiazine, amprolium	Sulfonamide, thiamine analog	Amprolium ineffective. Sulfadiazine (1 mg/mL in drinking water for 14 days) protected 90% of mice	1162
Compound screen	43 from various classes	Few selected compounds with *in vitro* activities inhibit proliferation of tachyzoites	1171
Toltrazuril, ponazuril (*in vivo*)	Triazinone	Protection against cerebral infection by daily application (20–100 mg/kg) in drinking water	423, 799
Ponazuril (*in vivo*, calves)	Triazinone	Protection against symptoms, lower parasite burden in organs	1102
Artemisinin	Sesquiterpene lactone	1 µg/mL for 14 days has parasiticidal activity	1064
Apicidin, depudecin	Histone deacetylase inhibitors	Depudecin at 500 ng/mL inhibits tachyzoite proliferation similar to apicidin at 15 ng/mL	425, 1108
Alcoholic herb extracts	Various plants	98%–99% inhibition with *Torilis japonica* and *Sophora flavescens* extracts at 156 ng/mL	2159
HPLC fractions	*T. japonica*, *S. flavescens*	Identification of one (Tj) and four (Sf) with activities of 97%–99% at 2.85 ng/mL	2160
Active components in HPLC fractions	*T. japonica*, *S. flavescens*	Active components identified by GC/MS	1825
Toltrazuril	Triazinone	Protection only in immunocompetent mice	58
52 dihydroxyiso-flavone and trihydroxydeoxy-benzoin-derivatives	Isoflavones targeting EGF receptor tyrosine kinases	Several isoflavones with good efficacy against *N. caninum* (IC$_{50}$ at 5–10 µg/mL) and low cytotoxicity in BM and HCT-8 cell cultures	734
Toltrazuril (*in vivo*)	Triazinone	Reduction of placental transmission	800
Toltrazuril (*in vivo*, calves)	Triazinone	Treatment of congenitally infected calves does not affect seropositivity	828
Toltrazuril	Triazinone	Treatment with 30 mg/L during 14 days is parasiticidal	424, 1914
Toltrazuril	Triazinone	Reduction of placental transmission (3 treatment with 30 mg/kg each)	1915
Nitro- and bromo-thiazolides	Thiazolides	Inhibition of proliferation is independent of nitro group	653
Nitro- and bromo-thiazolides	Thiazolides	Induction of egress of tachyzoites from infected cells	654, 655
Nitazoxanide	Thiazolide	150 mg/kg/day during 6 days PO has no effect, IP kills the mice after 2 days	474
DB750	Dicationic arylimidamide	IC$_{50}$ 230 nM. 2 mg/kg/day, during 6 days IP is well supported, better survival, reduction of cerebral parasite burden	474, 1132
			(*Continued*)

Table 2.15 (Continued) Summary of Chemotherapeutics against Neosporosis[a]

Compound	Class	Result	Reference
DB745	Dicationic arylimidamide	IC_{50} 80 nM. Reduction of cerebral burden in mice after 14 daily applications of 2 mg/kg/day	1815
Mefloquine	Trifluoromethylhinolin	IC_{50} 0.5 µM, EC_{50} for HFF 3 µM	1420
Monensin (in vivo, cattle)	Polyether antibiotic	Treatment of experimentally challenged cattle with slow-release monensin bolus for 100 days had no noticeable effect	2030
Miltefosine	Alkylphosphocholine	IC_{50} 5.2 µM. Treatment with 25 µM for 20 h parasiticidal. Reduction of symptoms and of cerebral burden	476
Artemisone (in vivo, gerbil)	Sesquiterpene lactone	Inhibition of infection by 15 mg/L. Partial clearance of pre-infected cells by 50 mg/L. Reduction of symptoms and of cerebral burden	1300
Artemisone	Sesquiterpene lactone	IC_{50} around 3 nM. Parasiticidal only by long-term treatment with 5 µM. No effect	1423, 1425
Flavonoids	Polyphenolic compounds	Inhibition of PV formation in glial cells at 50 µM by 3′,4′-dihydroxyflavone, 3′,4′,5,7-tetrahydroxyflavone (luteolin), and 3,3′,4′,5,6-pentahydroxyflavone (quercetin)	446
Ruthenium	Heavy metal	IC_{50} around 10 nm. Parasiticidal activity only by long-term treatment with 100 nM.	129
Bumped kinase inhibitors (in vivo)	Substituted pyrazolopyrimidines	Good structure-activity correlation. IC_{50} is around 100 nM. Inhibition of infection. Reduction of symptoms and cerebral parasite burden in nonpregnant mice	1504
Bumped kinase inhibitors	Substituted pyrazolopyrimidines	Protection against vertical transmission	2105
Buparvaquone	Naphtoquinone	IC_{50} of approx. 5 nM. Parasiticidal activity after 6 days treatment with 1 µM. Prevention of acute neosporosis and promising efficacy against vertical transmission	1424, 1425
Compound screen	Various anti-parasitic compounds	Curcumin, artemether, atrazine, toltrazuril, and ponazuril with IC_{50}s of 1–33 µg/mL. Curcumin not active	1648
Fluridone	Abscisic acid (ABA)–mediated signaling inhibitor	IC_{50} 143 µM, inhibition of egress. 15 days treatment ameliorated acute disease symptoms	2153

Note: IP = intraperitoneal; PO = per os (oral application).

[a] Unless indicated otherwise tests were conducted in *in vitro* in HFF: human foreskin fibroblasts. *In vivo* studies were performed with mice.

interactions were identified through virtual screening and predicted to be active, but this still lacks experimental validation.[829] In contrast, dicationic pentamidine derivatives as well as miltefosine, both originally developed for use against *Leishmania* infections, showed promising effects *in vitro* and *in vivo*,[474,476,1815,1873] but with a lot of room for improvement. This highlights the need to develop measures to increase delivery of drugs into the cells. One potential solution are protein transduction domains (PTDs), which have the ability to traverse lipid bilayers and can act as peptide-mediated delivery vehicles in parasites. For instance, the HIV-derived TAT PTD fused with GFP (TAT-GFP) was shown to efficiently penetrate the *N. caninum* membrane and could be an option for enhanced drug delivery.[341]

There is a link between anticancer chemotherapeutics and antiparasitic activities, since many anticancer drugs, which target mechanisms that lead to increased cellular proliferation, also affect the proliferative stages of parasites. For instance, artemisinin and derivatives with activity against *N. caninum* and *T. gondii*[1423] also impact the proliferation and viability of many cancer cells. Another example are the organometallic ruthenium compounds originally developed for the treatment of cancer, and also active against *N. caninum* and *T. gondii* tachyzoites *in vitro* in the low nanomolar range.[129]

Recently, calcium dependent kinase 1 (CDPK1), which is essential for microneme secretion, host cell invasion, and egress of *T. gondii*, has been extensively investigated. A particular class of inhibitors, bumped kinase inhibitors (BKIs), has bulky C3 aryl moieties entering a hydrophobic pocket in the ATP binding site. BKIs selectively inhibit CDPK1 from apicomplexans in a good structure–activity relationship,[1049] but do not inhibit mammalian kinases because they have larger amino acid residues adjacent to the hydrophobic pocket, thereby blocking the entry of the bulky C3 aryl group. Some BKIs, especially BKI-1294,[1504] have a good efficacy against *N. caninum in vitro* and *in vivo* (Table 2.15). *In vitro* studies, however, indicate that the BKI 1294 is not directly parasiticidal. Only upon long-term *in vitro* treatment of infected HFF monolayers, a complete clearance of viable tachyzoites can be observed.[1504] For different *Neospora* isolates, but also for *T. gondii* strain RH and ME49, the clearance of intracellular parasites is preceded by the formation of large, multinucleated complexes with deregulated gene expression as evidenced by the expression of bradyzoite as well as tachyzoite antigens. In the pregnant mouse model, BKI-1294 achieved a good protection against vertical transmission of *N. caninum*.[2105]

2.11 VACCINES

2.11.1 Vaccination Studies in Small Laboratory Animals

Most vaccine studies in small laboratory animals were done in murine models (information is summarized in Table 2.16). One study was performed in nonpregnant *M. unguiculatus* vaccinated with recombinant surface and dense granule antigens, and only assessed protection against clinical signs of neosporosis and survival.[342] Partial protection was found against clinical signs with all antigens, most promising was the efficacy when the NcSRS2-NcGRA7 combination was applied.[342]

2.11.1.1 Live Vaccines

Live vaccines composed of either isolates of low virulence or experimentally induced attenuated parasites have clearly shown superior efficacy in experimental trials compared to any subunit vaccine formulation (Table 2.16). While live vaccines show promising efficacy, the setbacks are high production costs, short shelf life of the product, maintenance of a cool-chain prior to application of the vaccine, and the potential risk of reversion to virulence. In addition, as live vaccines might result

Table 2.16 Overview of Selected Vaccine Studies on Neosporosis in Mice

Vaccine	Mouse Strain	Set-up	Results	Reference
Live Vaccines				
NC1 tachyzoites	A/J, BALB/c, C57BL/6 nonpregnant	Mice vaccinated with live NC1 tachyzoites; challenge after various days with *T. gondii* tachyzoites	Complete protection in mice vaccinated with *N. caninum* against acute infection by *T. gondii*. Early stimulation of CD8+ T-cells	1040
NC1, NC3 tachyzoites	SW, nonpregnant	Mice vaccinated with live NC1 or NC3 tachyzoites; challenge after 3 weeks with *T. gondii* tachyzoites	No protection as judged by day of death after *T. gondii* challenge	1164
NC1 tachyzoites (sublethal dose, 2 injections SC)	SW, nonpregnant	Mice vaccinated with live NC1 tachyzoites; challenge with sporulated VEG strain *T. gondii* oocysts or *T. gondii* RH tachyzoites	Mice vaccinated with NC1 survived significantly longer after VEG oocyst challenge, but not after challenge with *T. gondii* RH tachyzoites	1178
NC1 tachyzoites (sublethal dose)	BALB/c nonpregnant	Mice vaccinated with live NC-1 tachyzoites; challenge after 56 days with *N. caninum* tachyzoites	Vaccinated mice showed only mild symptoms, low lesion scores in the brain, and high levels of IFN-γ in splenocytes.	1235
NC-Nowra tachyzoites (live, crude extracts)	Qs (outbred) pregnant	Outbred Qs, vaccinated with tachyzoites or crude extracts before pregnancy, mated and challenged with Nc Liv (10^5) on day 5 of gestation	Reduction of transmission to pups by live vaccine and to a lower extent by crude extracts	637, 1352
NC1 tachyzoites (γ-irradiated)	C57BL/6 nonpregnant	Vaccination with irradiated tachyzoites (2×). Challenge 6 weeks after last boost (lethal; 10^7; sublethal, 10^6), euthanasia 25 days after challenge	All lethally challenged control mice died within 1 week, all vaccinated mice survived. Protection associated with mixed Th1/Th2 response	1662
NC1 (parental) and NCts8 (temperature-sensitive strain induced by chemical mutagenesis)	BALB/c,	Vaccination with live or frozen NCts8, then challenge with NC1	Significant protection in mice vaccinated with NCts8 tachyzoites prior to NC1 challenge	1183
Transgenic Nc expressing TgSAG1 and GFP	BALB/c nonpregnant	BALB/c, vaccinated 2× with 10^5 NC1 expressing TgSAG1 or GFP. Challenge with 500 *T. gondii* tachyzoites at 4 weeks after last boost	Moderate protection by Nc/GFP, good protection by Nc/TgSAG1. Immune response Th1-dominant	2171
Transgenic Nc expressing NcSAG4	BALB/c pregnant/ nonpregnant	Vaccination twice with NC1 expressing SAG4, some were mated, challenge at day 7 of gestation	Protection against vertical transmission by NC1 wild type and NC1 expressing SAG4, not associated with constant Th1- or Th2-type-immune response	1291
Mic1-mic3 ko T. gondii tachyzoites	Swiss OF1 (outbred)	Vaccination with 1×10^4 (PO) or 20 (IP) mic1-mic3 ko *T. gondii*, challenge with 5×10^7 NC1 IP. Monitoring daily for 30 days	Mice immunized with mic1-mic3 KO tachyzoites developed a strong cellular Th1 response and displayed significant protection with survival rates of 70% and 80% compared to 30% in controls	1580

(Continued)

Table 2.16 (Continued) Overview of Selected Vaccine Studies on Neosporosis in Mice

Vaccine	Mouse Strain	Set-up	Results	Reference
Naturally low-virulence NcSpainH-1	BALB/c pregnant/nonpregnant	Immunization with live Nc SpainH-1 tachyzoites (SC 2× at 3-week intervals). For pregnant model, mating, challenge at mid-gestation with Nc-Liv	Reduction of neonatal mortality, reduction of vertical transmission, lower cerebral parasite load in nonpregnant mice	1731
NcSpainH-1 tachyzoite extracts at different dosage in different adjuvants		Immunization twice with 3-week intervals. Adjuvants were water-in-oil emulsion, aluminum hydroxide with CpG oligodeoxynucleotides, and aluminum hydroxide with ginseng extract. Challenge IP with 10^6 NC1 tachyzoites	Efficacy can significantly vary depending on the adjuvant, the dose of antigen and the phase of *N. caninum* infection in which the vaccine is tested	1730
Attenuated NC1 tachyzoites after long-term culture	BALB/c nonpregnant	Mice inoculated with 10^6 live attenuated NC1 tachyzoites or virulent NC1. Infection with 5×10^6 tachyzoites 28 days later. Analysis latest 28 days post-challenge	Protection against a lethal challenge and reduction of pathology by vaccination with low dose of live attenuated tachyzoites	149
Subunit Vaccines				
Nc tachyzoite extract	BALB/c pregnant	Inoculation once with Nc-extract, mating and challenge infection with NC1 at days 10–12 of gestation. PCR evaluation of offspring shortly 7 days after birth	Much higher survival of offspring from vaccinated mice, complete protection against infection in offspring	1154
Nc tachyzoite extract	BALB/c nonpregnant	Immunization 2 times SC with soluble tachyzoite antigen entrapped in nonionic surfactant vesicles or administered with Freund's complete adjuvant. Challenge 3 weeks later SC with 2×10^5 NC1 tachyzoites	Immunization with Nc-extract in these two adjuvants results in Th2-type immune response and increased cerebral lesions	175
SAG1, SRS2 (vaccinia virus [vv] expression)	BALB/c pregnant	Vaccination with 5×10^6 PFU of vv-SRS2, vvSAG1, or vv GFP. Mating after 3 weeks, and challenge infection with NC1 (10^5 tachyzoites) after 4 weeks. Analysis 28 days PP	Much higher survival of offspring from mice vaccinated with vv-SRS2, associated with cellular and humoral immune response. Less with vv-SAG1.	1470, 1471
SRS2 (vaccinia virus [vv] expression)	BALB/c pregnant	Vaccination with 5×10^6 PFU of vv-SRS2, vvSAG1, offspring of survivors with high anti-SRS2 antibody levels mated and challenged at mid-gestation	Higher levels of anti-NcSRS2 antibody in surviving offspring from vaccinated dams than controls. After mating, and challenge in mid-gestation, transferring antibodies disappeared during pregnancy upon parasite infection. No significant difference on the parasite burden in dams and the survival rates of their offspring	1478

(Continued)

Table 2.16 (Continued) Overview of Selected Vaccine Studies on Neosporosis in Mice

Vaccine	Mouse Strain	Set-up	Results	Reference
Neospora ISCOMs or lysates	BALB/c nonpregnant	Mice vaccinated with *N. caninum* ISCOMs, NC1 lysates mixed with Quil A, or lysate alone. Challenge after 56 days with *N. caninum* tachyzoites	Groups immunized with ISCOMs or lysate and Quil A had reduced lesion scores compared to the group vaccinated with lysate only. Protection correlated with high IFN-γ in splenocyte culture supernatants	1235, 1615
SAG1, SRS2 (rE, DNA) alone or combined; crude somatic antigen	C57BL/6 nonpregnant	Vaccine + RIBI (2×), challenge 28 days after the first injection (proteins). pcDNA vector with genes IM, challenge 69 days after first injection. Euthanasia after 21 days	Protection with crude antigen. No protection with recombinant antigens as compared to adjuvant control. Protection with pcDNA in combination with recombinant antigens	288
GRA7, sHSP33 (DNA)	BALB/c pregnant	Vaccination of BALB/c mice IM with pCMVi-GRA7 or HSP33, booster after 21 days. After mating, challenge at days 10–12 of gestation. Pups euthanized at day 7 PP	100% of control pups infected, but only 46% of pCMVi-GRA7 and 55% of pCMVi-HSP33 pups	1155
MIC3 (rE)	C57BL/6 nonpregnant	Vaccination with MIC3+RIBI (3× IP), challenge 7 days after last boost, euthanasia after 21 days	Reduced cerebral infection in MIC3 vaccinated mice as compared to adjuvant control. Th2-type humoral response associated with protection	287
MIC1 (rE and DNA)	C57BL/6 nonpregnant	Vaccination with recMIC1 (3× IP), pcDNA-MIC1 (3× IM) alone or in combination. Challenge with Nc-Liv, euthanasia after 21 days	No clinical signs in vaccinated mice. Cerebral infection reduced in mice vaccinated with recombinant protein, enhanced in group with combined vaccination	31
SRS2 (native purified antigen)	BALB/c pregnant	Vaccination with native protein 3× SC mating, challenge with 10⁷ NC1 tachyzoites on day 10–12 of gestation. Pups and dams analyzed at day 7 PP	Decreased frequency of transmission in vaccinated dams, associated with Th2-type immune response	831
MIC1, MIC3, GRA2, GRA6, SRS2 (expressed in *Brucella abortus* RB51 vaccine strain)	C57BL/6 pregnant/ nonpregnant	Vaccination with live *B. abortus* expressing the antigens, lethal challenge 14 days after last boost. Euthanasia after 28 days	All control mice died within 8 days. Complete protection by MIC1 and GRA6	1665
MIC1, MIC3, GRA2, GRA6, SRS2 (in *Brucella abortus* RB51 vaccine strain)	C57BL/6 pregnant/ nonpregnant	Vaccine application twice with live *B. abortus* expressing the antigens, mating (2×), sublethal challenge. Euthanasia of pups after 21 days	Protection against vertical transmission by *B. abortus* expressing antigens	1664
GRA7, SRS2 (in *Brucella abortus* RB51 vaccine strain)	BALB/c nonpregnant	Vaccination with live *B. abortus* expressing the antigens, lethal challenge 7 weeks later. Euthanasia after 6 weeks	Protection against cerebral infection in SRS2-vaccinated mice, but not in mice vaccinated with GRA7	2038

(Continued)

Table 2.16 (Continued) Overview of Selected Vaccine Studies on Neosporosis in Mice

Vaccine	Mouse Strain	Set-up	Results	Reference
MIC4 (native, rE, DNA)	C57BL/6 nonpregnant	Vaccination 3× in 4-week intervals, sublethal challenge, euthanasia after 21 days	Mice of all vaccinated groups showed neosporosis symptoms and had an increased mortality as compared to the control group	**1897**
ROP2 (rE)	C57BL/6 nonpregnant	Vaccine application 3× in 2-week-intervals either with Freund's incomplete adjuvant or saponin, challenge, euthanasia after 35 days	No symptoms in vaccinated mice, reduced parasite burden in brains of vaccinated mice. Th1- or Th2-type humoral response depending on the adjuvant	471
GRA1, GRA2, MIC10, p24B (rE)	Qs (outbred) pregnant/ nonpregnant	Vaccination SC 2× with antigens formulated in VSA-3 adjuvant. Mating after 8–11 days, infection 5 days post-mating SC with 10⁶ Nc-Liv. Analysis 7 days PP	Single recombinant antigens were ineffective, partial protection against transplacental transmission with MIC10-p24B mixture	637
ROP2+MIC1+MIC3 (rE)	BALB/c pregnant/ nonpregnant	BALB/c, single vaccines or in combination (applied 3×), mating, challenge at day 7 post-mating, euthanasia of dams, offspring and nonpregnant mice after 1 month PP	Reduced vertical transmission by ROP2 alone or in combination. Humoral and cytokine responses associated with a Th2 immune response	472
PDI, ROP2, MAG1 (rE)	C57BL/6 nonpregnant	Vaccination with saponin as adjuvant (IP) or intranasally (IN) with cholera toxin as adjuvant (3×, 15 days intervals). Challenge 2 weeks after last boost, euthanasia after 28 days	Reduced cerebral loads with ROP1 (IP, IN) and PDI (IN only). Protection against clinical symptoms only by IN PDI vaccine	473
PDI	BALB/c	Vaccination of PDI incorporated or not into chitosan-based nanogels (IN), compared to vaccination with saponin (IP) as adjuvant (3×, 15 days intervals). Challenge 2 weeks after last boost. Euthanasia after 28 days	Reduced cerebral loads with PDI applied IN, but not IP Nanogel encapsulation altered the cytokine mRNA patterns	475
Nc-extract	BALB/c	Vaccination of Balb/c mice IP with tachyzoite extract, either in saponin, or by intracisternal inoculation of tachzyoite extract.	Decreased parasite burden in the CNS of mice vaccinated intracisternally compared to IP vaccinate mice	477
GRA7, AMA1 (rE)	BALB/c pregnant/ nonpregnant	BALB/c vaccinated SC with GRA7 or AMA1 formulated in oligomannose-coated liposomes, 3× with 1 week intervals. Mating 1 week later and infection IP with 1 × 10⁵ NC1 tachyzoites on days 6–9 of gestation. Analysis until 30 days PP	Decreased parasite burden in the CNS of vaccinated dams and increased survival rate in respective offsprings, associated with a Th1 immune response. No protection with non-entrapped NcGRA7	**1479, 2174**
NCP78 GRA7 (rE)	BALB/c pregnant	Application of antigens SC, 3 times with 2 weeks intervals. After 2 weeks mating, then challenge with NC1 (2 × 10⁶ each)	Improved survival of pups, but not dams, decreased cerebral parasite load in vaccinated groups	1238

(Continued)

Table 2.16 (Continued) Overview of Selected Vaccine Studies on Neosporosis in Mice

Vaccine	Mouse Strain	Set-up	Results	Reference
Cyclophilin, SRS2 (rE)	BALB/c nonpregnant	Application of antigens alone or in combination with adjuvants (SC, 2×, 2-week intervals). Control with irrelevant bacterial antigen. Challenge 3 weeks after last boost Euthanasia 3 weeks after challenge	Humoral response against antigens. Higher protection against cerebral infection when cyclophilin was present. Lower protection with SRS2 alone	2002
MIC1-MIC3-ROP2, (different versions of chimeric antigens, rE)	BALB/c nonpregnant	Female BALB/c, immunized with combinations of antigenic domains from MIC1, MIC3, and ROP2 with saponin as adjuvant (IP 3×, 2-week-intervals). Challenge 2 weeks after last boost. Euthanasia 36 days post-challenge	Complete protection by one combination only (MIC3-1-R), correlated with lower parasite load in brains	1374
MIC3-1-R (chimeric, rE)	BALB/c pregnant/ nonpregnant	Pregnant female BALB/c, vaccinated with MIC3-1-R either in saponin or Freund's incomplete adjuvants, or physically linked to a TLR2-ligand (Oprl)	No protection achieved no matter whether a Th1- or a Th2-immune response was elicited by vaccination. Oprl linkage to MIC3-1-R induced high IFN-γ in dams but did not achieve protection	18, 1375, 1376
GRA7, SAG4, and combination (rE)	BALB/c pregnant/ nonpregnant	Pregnant female BALB/c mice vaccinated with GRA7, SAG4, or both in Titermax Gold adjuvant	No significant protection observed. High IFN-γ and high antibody titers in GRA7 groups	15
GRA7, SAG4, BSR4, SRS9 (rE)	BALB/c pregnant/ nonpregnant	Vaccinated with recombinant proteins encapsulated in poly-epsilon-caprolactone	High morbidity and mortality. No protection against vertical transmission. Low IFN-γ levels	1015
Nc tachyzoite extract	BALB/c nonpregnant	BALB/c, vaccinated with different amounts of extract formulated with various adjuvants (SC 2×, 2-week intervals). Challenge at 38 days post-vaccination with NC1. Euthanasia 21 days post-challenge.	Immune responses depend on formulation, protection achieved independent from the formulation and immune response.	1273
Nc tachyzoite lysate (NLA) and secreted antigens (NcESA)	C57BL/6 nonpregnant	Mice vaccinated with NLA and NcESA with or without CpG adjuvant (SC, 3 times 2-week intervals), then lethal challenge IP with 2×10^7 NC-1 tachyzoites.	CpG-ODN combined with NLA, but not with excreted–secreted antigen, enhances protection against infection in mice.	1709
Nc tachyzoite extract	C57BL/6 nonpregnant	C57BL/6 mice immunized SC 3′ at 2 week intervals with Nc-extract emulsified in D-galactose binding lectin from Synadenium carinatum latex, then challenged IP. 28 days later with 2×10^7 NC1 tachyzoites	Increased adjuvant and immunostimulatory effect of the S. carinatum latex adjuvant formulation resulting in higher protection and lower degree of inflammation compared to antigen without adjuvant or adjuvant alone	304

(Continued)

Table 2.16 (Continued) Overview of Selected Vaccine Studies on Neosporosis in Mice

Vaccine	Mouse Strain	Set-up	Results	Reference
BAG1, BSR4, MAG1, SAG4 (rE)	BALB/c nonpregnant	BALB/c, vaccinated with four bradyzoite antigens with PBS or bitter gourd extract as adjuvants (IM, 2×). Challenge with Nc-Liv at 3 weeks after last boost	Antigen specific IgG1 and IgG2 and IFN-γ responses to all antigens. Protection from acute infection and lower parasite load in mice vaccinated with BAG1, MAG1, and SAG4	2003
PDI (rE)	BALB/c pregnant/ nonpregnant	BALB/c, female, vaccinated with PDI and cholera toxin (subunits A and B or subunit B alone) as adjuvants, IN application (3×, 2-week intervals), mating, challenge at day 7 post-mating	Good protection against cerebral infection by PDI with subunits A and B as compared to cholera toxin alone. No effect with subunit B alone. No protection against vertical transmission	478
NC-1 cell membrane extracts	C57BL/6 nonpregnant	Mice were vaccinated IN with NC1 membrane extract in CpG adjuvant 2 times with 3 weeks interval, and challenged IG with 5 × 10⁷ tachyzoites 3 weeks later. Analysis 1 week or 4 months PI	Immunized mice presented a lower parasite burden in the brain. IgG1: IgG2a ratio <1, but no IFN-γ in spleen or mesenteric lymph nodes	673
SAG1, SRS2, MIC3 (Bombyx mori nucleopolyhedrovirus, rE)	BALB/c nonpregnant	Mice, vaccinated with BmNPV displaying antigens or wild-type BmNPV (IM, 3× at 2-week intervals). Challenge with Nc-Liv after last challenge	IgG2-based humoral response to antigens. Reduced parasite load in brains in all groups vaccinated with NPV as compared to placebo. No effect due to displayed antigens	1042
ROP2+ROP40+GRA7+NTPase (rE)	BALB/c pregnant/ nonpregnant	Mice vaccinated with recombinant antigens in QuilA adjuvant, mating, challenge	Limited degree of protection by ROP2 and ROP2 + ROP40 combination	1564
SAG1, SRS2, MIC3 (rE in Bombyx mori)	BALB/c nonpregnant	BALB/c mice vaccinated IM 3× (14 days intervals) with B. mori expressed antigens in Freund's incomplete adjuvant. Challenge with 4 × 10⁶ Nc-Liv, experiment terminated at 5 weeks PI	Significantly lower parasite load in SAG-vaccinated group, with high IL-4 and IgG1 levels. SRS2 no protection and higher IFN-γ and low IL-4. MIC3 did not elicit a response	2158
Profilin (rE)	BALB/c nonpregnant	BALB/c mice vaccinated SC twice (14-day interval) with profilin emulsified in soy-lecithin-based adjuvant. Challenge IP with 10⁶ NC1 tachyzoites, and termination of experiment at day 21 PI	Profilin vaccination elicited profiling-specific IgM and IgG3, and regulatory T cells, and induced limited protection and reduced cellular immune response to N. caninum antigens	1273

Note: rE: recombinantly expressed in E. coli; DNA = DNA vaccine; SC = subcutaneous; IP = intraperitoneal; IM = intramuscular; IN = intranasal; IG = intragastric; PI = post-infection; PP = post partum.

in chronic infection of the host there is a risk that the life cycle could ultimately be completed again if tissues from vaccinated animals were fed to canid definitive hosts.[1697]

2.11.1.2 Subunit Vaccines

The concept of using subunit vaccines has largely relied on identifying defined parasite fractions or proteins that play essential roles in host cell invasion and/or tachyzoite-to-bradyzoite stage differentiation.[891] Different approaches have been used, such as the expression of antigens in viral or bacterial strains used for vaccination, and the application of subunit antigens as DNA vaccines or as purified recombinant antigens expressed in *E. coli* or other expression systems, and additional incorporation into ISCOMS.[891,892,2092] Clearly, not only the antigens confer protection (or not), but this is also highly dependent on the formulation, choice of adjuvant, application route, and many additional parameters (Table 2.16).

In both pregnant and nonpregnant mouse models, combinations of antigens as polyvalent vaccine had a higher efficacy compared to single antigens applied as monovalent vaccines, indicating that only a combination of recombinant antigens will induce protective immunological responses. However, in some cases, vaccination rendered mice more susceptible to infection, demonstrating that some antigenic components of the parasite could exhibit immune-modulating properties. More recently, it was shown that intranasal immunization employing hydrophobic proteins of *N. caninum* emulsified in CpG adjuvant conferred long-term (>20 weeks) protection against intragastric (IG) challenge with *N. caninum* tachyzoites, and NcGRA7 and NcMIC1 were identified as immunodominant antigens.[673a] Most subunit vaccines that had promising protective efficacy in nonpregnant mouse models have been found to be largely nonprotective in pregnant mice (reviewed in Reference **892**). Thus, in these cases pregnancy has led to the loss of subunit vaccine-induced protective immunity.

An important difficulty in interpreting the results from studies employing recombinant vaccines comes from the fact that many of these antigens are expressed in *E. coli*. The presence of immunomodulating agents derived from bacterial contaminants such as lipopolysaccharides (LPS) cannot be ruled out, and controls with irrelevant proteins expressed in the same system or with lipopolysaccharide-depleted protein fractions should be included. Surprisingly, only a minority of *Neospora* vaccine studies performed to date have addressed this point. LPS and other bacterial contaminants will greatly influence the way innate immune pathways are activated, and to what extent polarization of the cellular immune response is elicited later during infection.[171] This fact could be exploited, by actually generating LPS-free subunit vaccine candidate formulations, and then mixing them with pathogen-associated molecular patterns (PAMPs) in a controlled way, or even physically linking these antigens with PAMPs.[18]

2.12 PREVENTION, PROPHYLAXIS

Prevention of contamination of feed and water with oocysts, not feeding raw meat to canids, and selective breeding are the 3 broad preventive measures. Risk factor studies provide further information on routes how *N. caninum* infections in animals can be avoided or how disease caused by the infection (i.e., abortion) can be avoided. Specific measures are specially discussed in Chapters 4 and 5.

BIBLIOGRAPHY

General Biology: 34, 132, 184, 197, 207, 208, 209, 267, 308, 318, 370, 386, 469, 505, 523, 529, 539, 546, 562, 570, 573, 583, 598, 599, 613, 631, 771, 774, 794, 882, 888, 998, 1014, 1030, 1031, 1105, 1136, 1137, 1168, 1181, 1258, 1289, 1322, 1339, 1487, 1594, 1599, 1677, 1698, 1713, 1805, 1852, 1883, 1884, 1913, 1921, 1935, 1979, 1986, 1998, 2052, 2053, 2080.

Epidemiology: 36, 152, 163, 164, 167, 168, 354, 380, 416, 598, 680, 711, 949, 953, 954, 958, 960, 1008, 1070, 1142, 1147, 1163, 1167, 1324, 1336, 1338, 1367, 1438, 1439, 1450, 1458, 1548, 1552, 1649, 1668, 1676, 1760, 1805, 1865, 1866, 1871, 1918, 1925, 1996, 1997, 2014, 2017, 2167.

Molecular Biology: 7, 14, 15, 16, 17, 31, 32, 55, 92, 102, 177, 182, 185, 287, 288, 330, 334, 343, 403, 471, 473, 478, 487, 637, 638, 639, 646, 668, 697, 755, 794, 795, 796, 806, 819, 831, 832, 867, 869, 876, 877, 879, 880, 883, 939, 940, 941, 942, 943, 971, 1001, 1002, 1009, 1013, 1015, 1018, 1020, 1025, 1045, 1046, 1047, 1065, 1082, 1101, 1112, 1128, 1129, 1130, 1135, 1144, 1145, 1150, 1151, 1155, 1229, 1230, 1238, 1289, 1291, 1362, 1400, 1408, 1419, 1421, 1444, 1447, 1465, 1466, 1467, 1471, 1472, 1476, 1482, 1483, 1489, 1564, 1566, 1586, 1588, 1608, 1616, 1632, 1664, 1666, 1678, 1681, 1698, 1714, 1715, 1721, 1846, 1847, 1848, 1863, 1877, 1878, 1879, 1897, 1901, 1934, 2000, 2002, 2003, 2015, 2053, 2093, 2147, 2152, 2157, 2158, 2162, 2171, 2172, 2173, 2174, 2175, 2178, 2179.

Cell Biology: 7, 8, 17, 18, 32, 55, 92, 102, 172, 176, 197, 211, 217, 226, 250, 305, 306, 308, 356, 360, 361, 370, 471, 472, 486, 510, 511, 516, 518, 651, 681, 701, 709, 710, 774, 819, 832, 873, 876, 878, 881, 883, 885, 887, 894, 895, 941, 962, 1019, 1046, 1047, 1112, 1119a, 1135, 1138, 1146, 1149, 1151, 1229, 1230, 1231, 1292, 1345, 1408, 1443, 1444, 1445, 1447, 1464, 1468, 1475, 1487, 1564, 1565, 1566, 1585, 1594, 1651, 1678, 1681, 1698, 1781, 1791, 1793, 1874, 1878, 1879, 1896, 1920, 1933, 2005, 2015, 2051, 2052, 2053, 2054, 2157.

Rodent Models: 3, 4, 31, 44, 82, 86, 98, 109, 150, 164, 173, 186, 233, 234, 362, 365, 366, 386, 402, 435, 441, 482, 485, 493, 502, 528, 564, 583, 598, 603, 608, 652, 664a, 730, 778, 887, 959, 1016, 1017, 1032, 1033, 1034, 1040, 1051, 1081, 1153, 1159, 1161, 1162, 1163, 1172, 1206, 1207, 1208, 1221, 1222, 1223, 1303, 1326, 1327, 1352, 1359, 1361, 1372, 1469, 1477, 1480, 1511, 1514, 1561, 1579, 1619, 1653, 1654, 1655, 1661, 1663, 1675, 1702, 1706, 1707, 1717, 1722, 1733, 1784, 1805, 1844, 1850, 1885, 1886, 1889, 1916, 1936, 1944, 1945, 1946, 1947, 1948, 1974, 1975, 1976, 1977, 1994, 2004, 2141.

Diagnosis by Histology and Bioassay: 62, 116, 280, 361, 573, 600, 603, 607, 730, 986, 1127, 1157, 1307, 1316, 1503, 1525, 1536, 1579, 1599, 1678, 1679, 1681, 1704, 1848, 1920, 2015, 2022.

Diagnosis by Serological Methods: 1, 5, 13, 14, 17, 21, 26, 39, 52, 53, 54, 56, 62, 73, 90, 97, 99, 100, 101, 108, 140, 142, 143, 145, 156, 157, 161, 172, 176, 203, 211, 213, 216, 218, 219, 221, 223, 224, 225, 255, 262, 265, 278, 299, 320, 321, 322, 330, 331, 358, 374, 378, 408, 438, 462, 488, 491, 504, 509, 525, 528, 569, 650, 703, 706, 737, 755, 759, 791, 797, 834, 835, 839, 840, 842, 850, 852, 864, 904, 905, 914, 936, 943, 946, 947, 950, 972, 985, 1003, 1007, 1012, 1018, 1020, 1025, 1070, 1096, 1100, 1107, 1113, 1148, 1156, 1193, 1197, 1214, 1228, 1229, 1259, 1263, 1280, 1318, 1323, 1337, 1400, 1442, 1446, 1451, 1453, 1456, 1459, 1472, 1474, 1505, 1527, 1534, 1545, 1546, 1550, 1552, 1555, 1556, 1600, 1607, 1608, 1609, 1639, 1660, 1686, 1692, 1733, 1758, 1770, 1789, 1792, 1793, 1794, 1796, 1800, 1802, 1821, 1823, 1824, 1841, 1857, 1858, 1863, 1880, 1902, 1974, 1990, 1996, 2033, 2049, 2062, 2067, 2074, 2093, 2094, 2095, 2108, 2124, 2131, 2152, 2157, 2165.

Diagnosis by Molecular Methods: 2, 16, 28, 29, 36, 44, 58, 81a, 83, 87, 91, 96, 98, 103, 106, 116, 119, 138, 152, 164, 167, 168, 169, 174, 189, 196, 200, 247, 252, 259, 260, 266, 268, 287, 288, 307, 311, 352, 354, 364, 365, 367, 368, 380, 390, 417, 426, 432, 434, 439, 471, 472, 473, 478, 493, 515, 517, 573, 585, 598, 615, 633, 635, 636, 640, 641, 642, 643, 646a, 653, 671, 672, 680, 692, 708, 715, 728, 729, 753, 756, 760, 767, 779, 781, 782, 793, 794, 797, 798, 799, 821, 825, 836, 847, 860, 861, 893, 910, 911, 912, 913, 922, 924, 926, 948, 949, 953, 954, 958, 976, 986, 1008, 1043, 1054, 1059, 1067, 1069, 1070, 1083, 1088, 1091, 1097, 1105, 1114, 1133, 1140, 1142, 1152, 1153, 1154, 1226, 1253, 1262, 1264, 1269, 1281, 1283, 1284, 1293, 1299, 1306, 1309, 1324, 1328, 1329, 1330, 1335, 1336, 1338, 1346, 1347, 1353, 1377, 1378, 1411, 1412, 1413, 1418, 1424, 1426, 1427, 1428, 1433, 1437, 1438, 1439, 1445, 1450, 1458, 1460, 1481, 1499, 1504, 1506, 1508, 1524, 1572, 1575, 1578, 1584, 1590, 1596, 1598, 1600, 1617, 1618, 1624, 1635, 1645, 1673, 1674, 1679, 1680, 1684, 1698, 1699, 1702, 1704, 1707, 1722, 1729, 1735, 1752, 1757, 1767, 1771, 1786, 1788, 1805, 1810, 1813, 1814, 1815, 1827, 1828, 1864, 1865, 1866, 1875, 1880, 1887, 1893, 1894, 1904, 1914, 1918, 1924, 1925, 1929, 1941, 1985, 1994, 1997, 2003, 2009, 2022, 2032, 2051, 2072, 2086, 2097, 2104, 2106, 2141, 2165, 2176.

Treatment: 58, 129, 187, 297, 343, 407, 423, 424, 425, 446, 474, 476, 653, 654, 655, 734, 799, 800, 810, 828, 829, 884, 886, 892, 1049, 1064, 1102, 1108, 1132, 1160, 1162, 1171, 1174, 1177, 1300, 1376a, 1415, 1419, 1420, 1421, 1422, 1423, 1424, 1425, 1504, 1587, 1648, 1815, 1825, 1873, 1914, 1915, 2030, 2105, 2153, 2159, 2160.

Vaccines: **15, 18, 31, 75, 149, 171, 175, 287, 288, 304, 342, 471, 472, 473, 475, 477, 478, 637, 673, 831, 891, 892, 1015, 1040, 1042, 1154, 1155, 1164, 1178, 1183, 1235, 1238, 1273, 1291, 1352, 1374, 1375, 1376, 1470, 1471, 1478, 1479, 1564, 1580, 1615, 1662, 1664, 1665, 1697, 1709, 1730, 1731, 1897, 2002, 2003, 2038, 2158, 2171, 2174, 2092.**

Additional References: **9, 12, 30, 33, 37, 45, 171, 183, 187, 214, 261, 341, 388, 516, 565, 574, 575, 590, 594, 636, 637, 638, 639, 641, 645, 746, 750, 786, 794, 795, 812, 884, 888, 889, 896, 898, 902, 903, 921, 975, 980, 982, 983, 1041, 1094, 1182, 1294, 1304, 1311, 1312, 1313, 1323, 1332, 1473, 1938, 1952, 2035, 2092, 2132, 2160, 2181.**

CHAPTER 3

Techniques

3.1 BIOASSAY OF TISSUES FOR ISOLATION OF *NEOSPORA*

3.1.1 Tissues of Acutely Infected Animals

For the isolation of *N. caninum* from clinical specimens (e.g., cerebrospinal fluid [CSF], fluid from dermal ulcers in dogs, tissue biopsy), the samples are homogenized, centrifuged for 10 min at 2000 rpm ($400 \times g$), the sediment is suspended in saline, mixed with antibiotics (1000 units of penicillin, 100 µg of streptomycin/mL), and injected SC into KO mice.

3.1.2 Tissues of Chronically Infected Animals

The number of *Neospora* in tissues of chronically infected animals is low, and tissue cysts are more likely to be found than tachyzoites. Therefore, it may be necessary to use a digestion method to concentrate *Neospora* in the inocula. The tissue cyst wall in isolated tissue cysts is destroyed immediately by pepsin or trypsin, but released bradyzoites survive for a few hours whereas tachyzoites are killed by pepsin, but not by trypsin. Thus, host tissues can be digested in pepsin or trypsin without affecting the viability of parasites. Acid pepsin is preferred to trypsin because acid pepsin digests muscle tissue faster and more efficiently and kills many bacteria. Moreover, trypsin is more toxic to mice. Therefore, all traces of trypsin should be removed before inoculation into mice, and this process is time consuming. However, trypsin digestion is better for the brain than muscle. One should remember that a proportion of *Neospora* is destroyed by pepsin or trypsin as the host tissue is digested. Therefore, homogenized tissue is digested for 60 min, 37°C, allowing enough time for digestion of host tissue without destroying many organisms. The final concentration of trypsin in the tissue homogenate should not be more than 0.25%. The procedure is as follows:

1. Trim connective tissue, fat, epithelium (e.g., from tongue) from muscular tissues using nonporous, hard plastic cutting boards, scissors, or disposable razors. Cut muscle into small (1–2 cm) pieces (25 g) and store in plastic bags or cups.
2. Grind tissue in a blender for 15 seconds at low speed without saline. Then, add 125 mL of saline and blend at top speed for about 30 s. Rinse blender with 125 mL of saline and add the washings to the muscle homogenate. To save expense and time in cleaning the lid of the blender every time to avoid cross-contamination, line the lid of the blender with a disposable plastic sheet (commercially available 16.5 × 14 cm sandwich bags are convenient to use).
3. Pour the tissue homogenate into a 1000 mL widemouth plastic jar with a disposable plastic liner. Make 2 labels for each jar with the type of tissue, and the animal number using a good adhesive tape and water-resistant marker; transfer of 1 of these labels through the procedure helps to reduce

mislabeling. Homogenates can be left at room temperature for 1–3 h until all specimens have been processed.

4. To the prewarmed (37°C) homogenate, add 250 mL of freshly prepared, prewarmed (37°C) acid pepsin solution (pepsin 5.2 g, NaCl 10.0 g, 25% HCl 14 mL, and distilled water to make 1000 mL, pH 1.10–1.20). Incubate at 37°C in a shaking water bath for 60 min. The source and purity of pepsin used is probably not critical, but porcine stomach pepsin (1:10,000 biological activity, Sigma Chemical Co., St. Louis, Missouri, USA) has been used routinely and successfully.

5. Filter the homogenate through 2 layers of gauze and centrifuge 250 mL of filtered homogenate in a 250 mL widemouth polypropylene centrifuge bottle (Nalgene®) at $1200 \times g$ for 10 min.

6. Pour off the supernatant. Depending on the tissue, fatty scum may stick to the rim of the centrifuge bottle. To prevent this, suspend the sediment in 20 mL of PBS (pH 7.2) using disposable plastic pipettes. Transfer the homogenate in a 50 mL centrifuge tube with a conical bottom. Neutralize the homogenate with 12–15 mL of freshly prepared 1.2% sodium bicarbonate (pH 8.3) with phenol red as a pH indicator until the color changes to orange. After mixing, centrifuge at $1200 \times g$ for 10 min.

7. Pour off the supernatant and add 5–10 mL of saline that contains 1000 IU penicillin and 100 µg of streptomycin/mL.

8. *N. caninum* stages (tachyzoites and bradyzoites) in tissues are killed by water and by heating to 60°C, and so blenders, cutting boards, and other materials can be cleaned with soap and hot water, then rinsed with cold water, and finally with saline before using them for the next specimen.

9. Inoculate 0.5–1 mL of tissue homogenate SC into each of 2–5 KO mice over the back using a 4 cm long 21–23 gauge needle. It is preferable that each mouse is identified with a rodent ear tag (Ear tag size No. 1, National Band and Tag Company, P.O. Box 430, Newport, Kentucky 41092, USA).

10. Examine all inoculated mice for clinical signs of *N. caninum* infection daily.

Use of other rodents for bioassay is discussed in Chapter 2.

3.2 PARASITE CULTURES

3.2.1 Obtaining Tachyzoites

Numerous mammalian cell lines have been used to grow *N. caninum* in cell cultures and the procedures used to isolate viruses in diagnostic laboratories are suitable to grow *N. caninum*. Although most authors used Vero cells (African green monkey cells) to grow *N. caninum* from bovine tissues (Table 4.6 in Chapter 4), *N. caninum* has no cell culture preference.[1136] Usually, the cell cultures are incubated with the tissue homogenate for 1 h and then replaced with cell culture medium. It is important to observe the cell culture flask for 2 months because most *N. caninum* strains grow slowly and tachyzoites may not be visible microscopically for 60 days from the time of seeding with the homogenate.

3.2.1.1 In Vitro Cultivation

3.2.1.1.1 Tachyzoite Inoculum

In vitro culture of *N. caninum* tachyzoites has been established in a variety of primary cells, and also established cell lines.[34,565,882,885,1421] This has enabled researchers to generate and characterize a wide range of *Neospora* isolates, including strains to be considered as live vaccines, and has been the basis for the production of *Neospora* antigens for immunodiagnosis, immunohistochemical, and molecular detection, and genotyping of the parasite by PCR. They have also been instrumental for the identification of potential immunoprotective antigens and targets for intervention. Finally, genetic manipulation of *N. caninum*, and the more recent developments related to genomics, transcriptomics, and proteomics, would not have been possible without the extensive use of *in vitro* culture models.

Essentially, the same techniques for cultivation and cryopreservation can be used as previously described for *T. gondii*.[613] In analogy to *T. gondii*, *N. caninum* cultures contain varying amounts of fetal calf serum (FCS), which is commercially available. However, many batches of FCS contain antibodies against *N. caninum* antigens[1981]; FCS is pooled from many bovine fetuses, for commercial purposes, and some of the fetuses are infected with *N. caninum*. This can lead to agglutination of parasites and rapid death once they are released into the medium. A serum-free cell culture medium (DefCell) was shown to support the growth of *N. caninum in vitro*.[449]

In general, the proliferation rate of *Neospora* tachyzoites *in vitro* varies considerably, depending on the host cells used and the parasite isolates.[1677] Ovine fibroblasts have been employed[973] as host cells and comparatively the growth of 2 isolates (NC1 and Nc-Liv) was quantified by ^{3}H uracil uptake, and it was shown showed that NC1 tachyzoites proliferated more rapidly than Nc-Liv. Addition of IFN-γ markedly reduced tachyzoite proliferation. *In vitro* culture of tachyzoites in human foreskin fibroblasts (HFF) was used to assess growth and competition of *N. caninum* (NC1 isolate) and *T. gondii* (RH strain).[1921] This study showed that initially *N. caninum* invaded fibroblasts more efficiently, but the generation time of *N. caninum* was longer compared to *T. gondii*, resulting in out-competition of *N. caninum* by *T. gondii* in the longer term. Both species could occupy and multiply in the same host cell simultaneously. In contrast, another study compared host cell invasion by these 2 parasites using dog fibroblasts, cat kidney cells, and Vero cells, and showed that *T. gondii* invaded all cell types with greater efficiency.[1136] *N. caninum* tachyzoites were also cultured in bovine angioendothelial cells or human Caco-2 cell cultures, and it was found that besides FCS, human serum also supports the growth of the parasite.[1512]

There are limiting aspects of scaling up the *N. caninum* tachyzoite production, and 1 paper suggested that *Theileria annulata*-infected lymphoblastoid (TIL) suspension cultures could be useful for the propagation of tachyzoites, as these parasites were continuously released into the medium supernatant.[1036] However, *N. caninum* tachyzoites have only a limited time span of few hours to remain infective when maintained extracellularly,[878,1445] and this approach never really made it to wider use.

A further aspect of *in vitro* culture is the adaptation of isolates to culture conditions. This was investigated for 3 *N. caninum* isolates (NC1, Nc-Liv, and Nc-SweB1),[1592] also it was shown that the 3 isolates could adapt to environmental conditions without the help of sexual recombination, which supports the idea that *N. caninum* has the capacity for maintaining clonal propagation in nature. Treatment of *N. caninum*-infected cultures with a commercially available *Mycoplasma* removal agent (MRA) resulted in a mutant population that differed in antigen content.[951] The treatment with MRA is therefore not recommended for *Mycoplasma* eradication in *Neospora* isolates.

N. caninum tachyzoites cultured through serial passages for 8 years have been able to retain their infectivity for mice.[565] However, others have shown that the prolonged passage of tachyzoites in tissue culture could attenuate the virulence of *N. caninum in vivo*: BALB/c mice infected with parasites that had a low passage number were more susceptible to infection, died earlier, and had a higher parasite load compared to those infected with tachyzoites cultured for extended periods of time.[148] Culture of *N. caninum* for 3 months in a murine macrophage cell line (J774) resulted in parasites had an impaired virulence upon infection of broiler chicken embryonated eggs compared to those cultured in Vero cells.[1055]

3.2.1.1.1.1 Brain Cell and Organotypic Cultures as Models for Neospora–CNS Interactions —
The CNS represents the predilection site for *N. caninum* infection. One way to study CNS-relevant infection *in vitro* is by applying CNS-derived cell cultures. For instance, the effects of *N. caninum*

infection have been investigated in primary astroglial cultures obtained from neonatal rats.[1610–1614] Infected cells exhibited hypertrophy, gliofilament reorganization, metabolic changes suggesting hypoxia, strong IL-10 TNF-α, and NO production. *N. caninum* infection did not induce IFN-γ release. Addition of IFN-γ and TNF-α treatments resulted in reduction of tachyzoite loads, while depletion of IL-10 and TGF-β also had a similar effect. In addition, *N. caninum* infection in these cells induced lysosomal activity.[1614]

In 1 study,[2051] organotypic slice cultures of rat central nervous tissue were employed and the infection kinetics were studied by qPCR, and the interactions with actin microfilaments and glial fibrillary acidic proteins by immunofluorescence, and by TEM. The majority of tachyzoites were found within microglia cells.

N. caninum infection was also studied in human brain microvascular endothelial cells. Infection of these cells did not affect mitochondrial integrity in the host cells, but resulted in increased overall oxygen consumption,[647] and metabolic footprinting of extracellular metabolites showed that β hydroxybutyrate, pyruvate, adenosine triphosphate (ATP), total protein, non-esterified fatty acids, and triglycerides are significantly different in infected cells compared to control cells.[648]

3.2.1.1.1.2 In Vitro Culture in Canine Intestinal Cells — To study the sexual development of *N. caninum* in the canine intestine, canine intestinal cells were isolated from neonatal canine duodenum, grown to polarized monolayers, and infected with *N. caninum* tachyzoites or cell culture-derived bradyzoites. The development of the parasites and the host cells was monitored by TEM and immunofluorescence using antibodies against cytokeratins, desmosomes, tight junctions, tachyzoite and bradyzoite antigens, and measurements of alkaline phosphatase activity. No sexual stages were detected within the 2-week culture period.[890]

3.2.1.1.2 Bradyzoite Inoculum

Neospora has been cultivated in cell cultures directly seeded with bradyzoites from the tissues of naturally infected dogs.[598] Tissue cysts were isolated on Percoll from the brain of a naturally infected dog. Bradyzoites were released from the tissue cysts by incubation in acid pepsin solution. Intracellular bradyzoites could be seen at 15 h PI, and endodyogeny stages were seen at 40 h. Two endodyogenous divisions were completed by 64 h. Tachyzoites were first seen 6 days after seeding cultures with bradyzoites.[598]

3.2.1.1.3 Oocyst Inoculum

Coccidian oocysts are normally not known to excyst in cell culture and excystation of sporozoites is necessary to initiate *in vitro* cultivation. Trypsin, bile, and/or sodium taurocholate are used to excyst coccidian oocysts. However, attempts were successful in growing tachyzoites during 3 of 6 attempts after cultures were seeded with sporulated oocysts.[780] Oocysts were treated with acid–pepsin for 5 min at 37°C; after washing, oocysts were seeded on to Vero cells.

3.2.1.1.4 Separation of Tachyzoites from Host Cells

Various methods (differential centrifugation gradients, treatment with lectins, filtration through sintered glass filters, nylon wool, glass wool, and digestion) have been described to separate *T. gondii* tachyzoites from host cells with varying success.[483] PD-10 Sephadex column (Pharmacia Biotech) filtration to remove host cells has also been used.[1545] Filtration through 3–5 μm millipore filtration to remove host cells from tachyzoites has also been reported, but there is a loss of some organisms.

3.2.2 Obtaining Tissue Cysts

Initially, tissue cysts were obtained from outbred mice given corticosteriods (MPA or dexamethasone). Different doses of *N. caninum* tachyzoites and MPA have been used based on the multiplication rate of the parasite and the weight of the mice.[1159,1327] Tissue cysts were first seen 17–21 DPI and were confined to the brain. Tissue cysts were not found in all inoculated mice. In other experiments conducted[1327] the highest number of tissue cysts was in an outbred ICR mouse inoculated with Nc-Liv and given 2.0–2.5 mg MPA on day 7 and the day of infection—the mouse died 39 DPI. Individual and paired BAG1-positive bradyzoites were scattered among well-developed tissue cysts. Even in very small-sized tissue cysts, the bradyzoites were slender, had a subterminal nucleus, and because of compactness individual bradyzoites were not discernible (Figure 2.2g and h, Chapter 2).

Although inbred BALB/c mice inoculated with *N. caninum* not given corticosteroids developed neurological signs, tissue cysts were not detected histologically.[1172] Even in KO mice, the number of tissue cysts were few and confined to the CNS. Tissue cysts were also found in the brains of gerbils but the number was always low.

Unlike other rodents, numerous tissue cysts were found in the shrew-like marsupial, the fat-tailed dunnart (FTD, *Sminthopsis crassicaudata*) from Australia.[1069] The FTD weighed 12–16 g. Tissue cysts were found in the FTD (group 1) inoculated IP with 10^5 tachyzoites but surprisingly not in these inoculated 10^4 *N. caninum* tachyzoites. The group 1 inoculated FTD became severely ill and were euthanized 13–18 DPI. Tissue cysts and bradyzoites were found in virtually all organs, including pancreas, adrenal gland, urinary and skeletal muscle, heart, lungs, lymph nodes, liver, brain, and accessory sex glands.[1069]

Several methods have been described to separate tissue cysts from the brain. Percoll, dextran, and gum Arabic have been used to separate *T. gondii* tissue cysts from brain tissue.[613] A Percoll method was modified to separate *Neospora* tissue cysts from mouse brain.[1326] A continuous density gradient was prepared by centrifuging 30 mL of 35% Percoll in PBS at $27,000 \times g$ for 20 min in a fixed angle head centrifuge. Brain homogenate (2–5 mL) was layered on top of the Percoll gradient and then centrifuged in a fixed angle rotor at $4400 \times g$ for 15 min.[1326] The brain material floated on top of the gradient and was carefully removed with a pipette. The remaining gradient was mixed with 3 volumes of PBS, and centrifuged in a swinging bucket centrifuge. After pouring off the supernatant, tissue cysts were collected from the pellet.

The degree of success in purifying tissue cysts depends on the host tissue and the amount of blood contamination. To minimize red blood cell and tissue contamination, the mice should be bled out before harvesting brains, and the brain homogenate passed through a 90 μm sieve. Tissue cysts of *Neospora* are usually smaller than 90 μm and pass through the sieve.

In another study,[598] *Neospora* tissue cysts were separated from the brain of a naturally infected dog by using 35% isotonic Percoll. To prepare isotonic Percoll, mix 9 parts of Percoll and 1 part of 10× saline.

3.2.3 Obtaining Oocysts

Oocysts can be obtained by feeding infected tissues to dogs. It is preferable to use recently weaned (10–12-week-old) pups. The number of oocysts excreted and the frequency is erratic. The number of oocysts excreted is normally very low. More success has been obtained by feeding dogs naturally infected water buffalo muscle or infected bovine placenta. Better success was obtained by feeding beef experimentally infected with Nc-Illinois strain.[780] The reasons for failure are unknown. Sporulated oocysts are stored in capped plastic bottles with a plastic liner under the cap.

Procedures for optimal excystation of *N. caninum* sporozoites have not been published. Bile salts and trypsin, used for coccidian oocysts, also excyst *N. caninum*. Although intact oocysts can be excysted, the success rate is better with free sporocysts. Sporocysts can be released from oocysts by vigorously shaking (or vortexing) sporulated oocysts with 500-μm glass beads until 80% of sporocysts are released. Incubate sporocysts with 5% ox bile at 37°C for 20 min or longer. Sporozoites can be separated from debris by filtration through 3-μm filters.

3.2.4 Cryopreservation

Neospora tachyzoites, bradyzoites, or sporozoites can be preserved by freezing by using techniques described for other coccidia. Organisms survive freezing better in dimethylsulfoxide (DMSO) than without it. It is advisable to start with a high number (100,000 or more/mL) in the inoculum because not all organisms survive freezing. For preserving *N. caninum*, suspend organisms in tissue culture medium (TCM), mix with DMSO, and slowly freeze in liquid nitrogen (LN). For this, prepare a sterile stock solution of 16% bovine serum albumin (BSA) in TCM (solution A) and a sterile 50% concentration of DMSO in TCM (solution B). Mix equal volumes of solutions A and B to produce solution C. Finally, mix equal volume of solution C with equal volume of extracellular parasites or *Neospora*-infected cells in the TCM (thus, the final concentration of DMSO is 12.5%) and let the mixture incubate for 30 min at room temperature. Freeze samples in a freezing box. We use NALGENE-Cryo 1°C freezing container 18 vials with isopropyl alcohol to achieve a −1°C/min rate of cooling (Nalgene Cat. No. 5100-0001). Store the box at −70°C overnight and then transfer the frozen vials to LN. In case such a freezing box is not available, a practical method of doing this is to immerse the sealed vial of tachyzoites in 95% ethanol, store at −70°C overnight, and the next day, transfer the frozen vial to LN. For experimentation, thaw the tachyzoites quickly (37°C water bath, 1 min), and inoculate into mice or cell culture. We have revived *N. caninum* after up to 20 years of storage in LN.

Intracellular tachyzoites exhibit better survival after storage in LN compared to extracellular tachyzoites, but free bradyzoites and sporozoites survive better than intact tissue cysts and oocysts. We routinely cryopreserve parasites infected tissues on their first (primary) isolation to minimize effect of sub-passage in the laboratory. For cryopreservation of bradyzoites, bradyzoites are liberated from tissue cysts by digestion in pepsin for 5–10 min, depending on the tissue, neutralized with sodium bicarbonate, centrifuged, and mixed with TCM.

3.3 DIAGNOSTIC PROCEDURES

3.3.1 Examination of Canine Feces for Oocysts

The number of *Neospora* oocysts in feces is usually low; therefore, concentration methods are often necessary to detect oocysts in feces.[519] Either sugar or salt (NaCl or $ZnSO_4$) solutions of specific gravity 1.15 or higher can be used to float sporocysts free of fecal debris. Sugar solution is less deleterious to sporocysts than salt solutions. We use the following method:

1. Mix 5–10 g of feces thoroughly with 10–20 mL of water using a tongue depressor to obtain a homogenous suspension. Mix this fecal suspension with 50–100 mL of Sheather's sugar solution (sugar, 500 g; water, 320 mL; and add liquid phenol, 6.5 mL if the solution needs to be stored for more than a few days to prevent fungal growth). To dissolve sugar, heat water to 70°C and stir solution continuously, then cool before use. The specific gravity of Sheather's sugar solution is 1.18 and the specific gravity of *Neospora* oocysts is lower than 1.15; therefore, mixing the watery suspension with sugar solution does not affect the flotation procedure.
2. Filter through cheesecloth or a tea strainer.
3. Centrifuge fecal suspension in 50-mL centrifuge tubes with a cap at about $400 \times g$ for 10 min.
4. With a pipette or with a loop, remove several small drops from the very top, put drops on a glass slide and cover the drops with a coverslip.

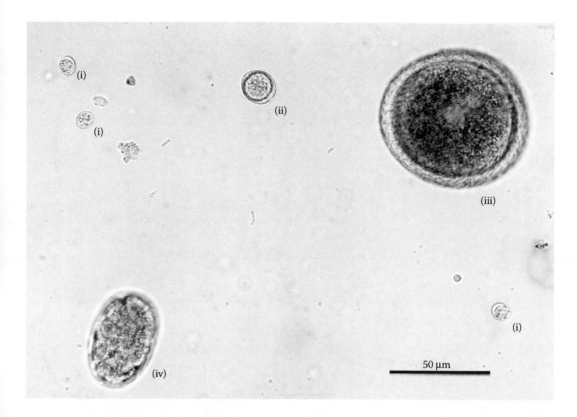

Figure 3.1 *Neospora-Hammondia*-like oocysts and other parasitic stages in sugar solution float of dog feces. Unstained. *Neospora*-like oocysts (i), *Cystoisospora ohioensis*-like oocyst (ii), *Toxocara canis* egg (iii), and hookworm sp. egg (iv). (From Dubey, J. P. 1993. *Toxoplasma, Neospora, Sarcocystis*, and other tissue cyst-forming coccidia of humans and animals. *Parasitic Protozoa*. In Kreir, J. P. (Editor). Academic Press, New York. Volume 6, 1–158.)

5. Let the slide–coverslip lie flat for about 5–10 min so that the fecal particles can settle and the sporocysts can rise to the top before the slide is examined.
6. Examine at magnification of 100× or higher. *Neospora* oocysts are about 10×12 μm in diameter and are almost identical to *Hammondia heydorni* oocysts illustrated in Figures 3.1 and 3.2. They are about half the size of *Cystoisospora ohioensis*-like oocysts, and about one-eighth the size of round worm eggs (Figure 3.1).
7. To collect sporocysts from feces, aspirate the top 5 mL of the solution from the 50-mL centrifuge tube. Mix the aspirate with 45 mL of water, and centrifuge at $400 \times g$ for 10 min. Discard the supernatant fluid and resuspend the sediment in water, centrifuge, and repeat the process. Suspend the final sediment in 2% aqueous sulfuric acid or 2.5% potassium dichromate. For sporulation, shake oocysts at room temperature for 1 week and then store at 4°C. During incubation in 2% sulfuric acid, oocysts lose the spherical shape whereas the oocyst shape is preserved in dichromate. The chromation of the oocyst wall hardens it and the potassium dichromate is difficult to remove from the oocyst wall. Repeated washings are needed to completely remove the dichromate. Potassium dichromate is also an environmental hazard.

3.3.2 Cytology, Histopathologic, and IHC Procedures

Cytology preparations, impression smears of exudates, and tissue impression smears can be stained with conventional Giemsa stain, Leishman stain, or rapid Diff Quick®. Tachyzoites stain well with any of these procedures (Figure 3.3). *Neospora* tachyzoites and tissue cysts stain relatively

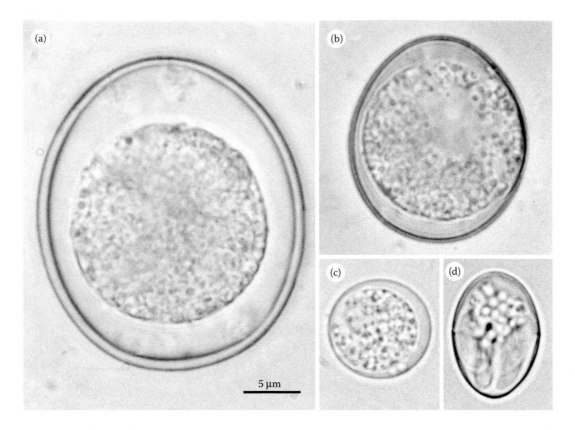

Figure 3.2 Higher magnification of coccidia in dog feces. *Cystoisospora canis* (a), *C. ohioensis*-like (b), *Neospora-Hammondia*-like (c), and *Sarcocystis* sp. sporocyst (d).

faintly with the routine hematoxylin and eosin stain, commonly used for routine histopathology. The strength of the hematoxylin could be doubled for staining tachyzoites. The PAS reaction can be used to stain tissue cysts; the bradyzoites in tissue cysts stain red (Figure 2.2j, Chapter 2).

IHC procedure for AEC 3-amino-9-ethylcarbozole staining is as follows:

1. De-paraffinize slides in xylene I for 30 min. Discard xylene I. Pour xylene II into xylene I container. Soak for 5 min.
2. Pour fresh xylene into xylene II. Soak slides in fresh xylene for 5 min.
3. Soak in 100% ethanol I for 5 min. Discard ethanol I. Pour ethanol II into ethanol I container. Soak for 5 min. Pour fresh 100% ethanol into ethanol II container. Soak for another 5 min.
4. Soak in 95% ethanol for 5 min.
5. Quench endogenous peroxidase in 3% H_2O_2 in methanol for 15 min (20 mL 30% hydrogen peroxide and 180 mL methanol).
6. Soak in 80% ethanol for 5 min.
7. Soak in 70% ethanol for 5 min.
8. Soak in 50% ethanol for 5 min.
9. Soak in saline for 5 min.
10. Soak in warm saline (37°C) for 5 min.
11. Pepsin digestion: Incubate slides in 0.4% pepsin in 0.01 N HCl for 15 min, at 37°C (0.8 g pepsin, 0.333 mL 6 N HCl, 200 mL saline, prewarmed to 37°C for 30 min).
12. Soak in 0.75% biotinylation reagent Brij® 35–PBS for 5 min twice (2.5 mL 30% Brij® 35/L PBS); Brij® 35 is a detergent.
13. Block nonspecific binding with 0.5% Na caseinate in Brij® 35–PBS for 10 min.

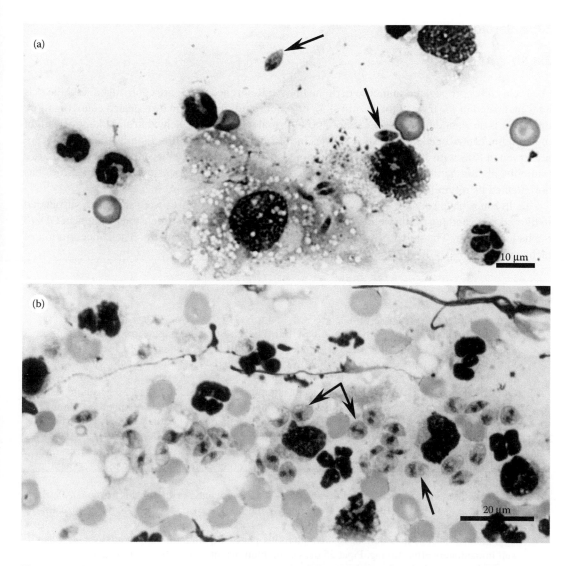

Figure 3.3   *N. caninum* tachyzoites (arrows) compared with neutrophils, macrophage, and red blood cells in impression smears. Giemsa stain. Note most tachyzoites are dividing and are crescentic to globular. (a) Lung aspirate from a dog. (From Greig, B. et al. 1995. *J. Am. Vet. Med. Assoc.* 206, 1000–1001.) (b) Exudate from dermal ulcer from a dog. (Dubey, J. P. et al. 1988. *J. Am. Vet. Med. Assoc.* 192, 1269–1285.)

14. Drain and without rinsing apply the properly diluted primary antibody to the tissues for 30 min, at 37°C.
15. Soak slides in Brij® 35–PBS for 5 min, twice.
16. Treat tissues with Dako EnVision Rabbit peroxidase solution at 37°C for 30 min.
17. Soak slides in Brij® 35–PBS for 5 min, twice.
18. Apply Dako AEC substrate chromogen solution to slides at 37°C for 15 min.
19. Collect chromogen in hazard waste bottle, and soak slides in Brij® 35–PBS for 5 min.
20. Rinse slides in running tap water for 2 min. Rinse in deionized water for 2 min, twice.
21. Counterstain with Mayer's hematoxylin for 1 min.
22. Wash slides in running tap water for 1 min.
23. Soak slides in Scott's tap water substitute (MgSO$_4$ · 7H$_2$O 20 g, NaHCO$_3$ 2 g, water 1 L) for 1 min.
24. Soak in running tap water for 1 min. Soak in deionized water for 2 min, twice.
25. Apply Crystal Mount and incubate slides overnight at room temperature.
26. Apply Permount and cover slip, if desired.

3.3.3 Serologic Procedures

3.3.3.1 NAT

Tachyzoites propagated in mouse peritoneum (see Reference **1733**) are centrifuged, and the pellet is incubated in 1% trypsin for 45 min at 37°C. After removing trypsin by repeated centrifugation, the pellet is suspended in 6% solution of 100% formaldehyde. The tachyzoites are kept overnight in formalin, centrifuged, and washed 3 times in PBS to remove formalin. Finally, the parasites are preserved in PBS with 1% sodium azide and stored at 4°C until use. We have used this antigen after 6 months of storage at 4°C without loss of reactivity. Short-time storage at room temperature has not affected its reliability.

Tachyzoites have been grown in Vero cells.[1545] Infected cell cultures were scrapped, suspended in filtered PBS, and host cells were removed by passing the tachyzoite suspension through an OD-10 Sephadex column. The pellet of tachyzoites was suspended in 6% formalin. The concentration of tachyzoites was $1–2 \times 10^4/\mu L$.[1545,1733]

Protocol:

1. Serum diluting buffer: Dissolve 42.5 g NaCl, 1.54 g NaH_2PO_4 (M.W. 120), and 5.4 g Na_2HPO_4 (M.W. 142) in 900 mL deionized water. Adjust the pH to 7.2. Bring the volume to 1 L with deionized water. Store at 4°C. This is the 5× stock solution. Dilute this stock solution 1:5 to give 0.01 M PBS (1 part stock and 4 parts deionized water). PBS should be filtered just before use through a 0.22 μM membrane.
2. Antigen diluting buffer: Dissolve 7.01 g sodium chloride, 3.09 g boric acid, and 2.0 g sodium azide in 900 mL deionized water. Add 24 mL 1 N NaOH and adjust the pH to 8.95. Bring the volume to 1 L. This is the stock solution and can be stored at room temperature. For the working antigen diluting buffer, dissolve 0.4 g BSA in 100 mL borate buffer. Store at 4°C.
3. Dilute serum samples with serum diluting buffer (no. 1) in small test tubes (1.2 mL in strips of 8 or 12) with a multichannel pipette, starting at 1:25. Microtiter plates may also be used for making serum dilutions.
4. Prepare antigen mixture as follows: For each plate, mix 2.5 mL antigen diluting buffer (no. 2), 35 μL 2-mercaptoethanol, 50 μL Evans blue dye solution (2 mg/mL water), and 0.15 mL antigen (formalin-fixed whole parasites).
5. Agglutination is done in U bottom 96-well microtiter plates. Pipet 25 μL antigen mixture into each well immediately after mixing. Pipet 25 μL serum dilutions into the wells and mix gently with the antigen by repeated pipetting action.
   a. A positive control should be included in each plate. The control should have a titer of 1:200, and 2-fold dilutions from 1:25 to 1:3200 should be used. A negative control ensures the specificity of reaction.
   b. Cover the plates with sealing tape and incubate at 37°C overnight.
   c. Read results using a magnifying mirror. A blue button at the bottom of the well means negative. A clear bottom means positive.

3.3.3.2 IFAT

The protocol presented here has been described previously.[1789]
Tachyzoites from cell culture:
Use tachyzoites grown with FBS as supplement. One day before harvesting the tachyzoites, the FBS-supplemented medium is replaced by a medium not supplemented with FBS. Isolate the parasite by scraping off the host cell monolayer with a rubber policeman and release tachyzoites by

vigorous pipetting, supporting cell rupture. To remove the host cell debris, the preparation should be filtered through 5-μm filters. Wash the parasites 2–3 times with cold PBS (centrifugation, 10 min, $1500 \times g$, 4°C). Dilute tachyzoites to 5–6×10^7 parasites/mL in PBS.

Preparation of slides for immunofluorescence:

Wash printed microscope slides in 96% ethanol; let slides dry. Add 10 μL of a parasite suspension (5–6×10^7 parasites/mL in PBS) to each well and let the slides dry. Slides are kept at −20°C until use.

Buffers:

1. $4 \times$ fluorescence assay (FA) rinse buffer: Na_2CO_3—11.4 g, $NaHCO_3$—33.6 g, NaCl—8.5 g, double-distilled water—make volume to 1000 mL, pH 9.0
2. Anti-fading-buffer: 3.5 g 1,4 diazobicyclo [2,2,2]-octane in 90 mL glycerol and 10 mL PBS

Procedure:

Avoid drying of the slides during the procedure. Avoid cross-contamination of wells. Carefully remove excessive fluids.

1. Fix slides 10 min in acetone (−20°C).
2. Incubate slides for 10 min in PBS.
3. Dilute samples in 2-fold serial dilutions, for example, from 1:25 to 1:3200.
4. Add 10 μL of the sample dilutions to each well.
5. Incubate the slides for 30 min in a moist chamber.
6. Wash the slides gently with FA rinse buffer and incubate for 5 min in FA rinse buffer.
7. Add 10 μL freshly prepared conjugate to each well. We use FITC-labeled species-specific conjugates diluted PBS/0.05% Evans blue (e.g., 1:50, optimal conjugate concentrations need to be established depending on conjugate).
8. Incubate slides in a moist chamber for 30 min.
9. Wash the slides gently with FA rinse buffer and incubate for 5 min in FA rinse buffer.
10. Mount slides with anti-fading-buffer.
11. Read slides using a fluorescent microscope; a serum is considered negative if no fluorescence or only apical fluorescence is detected at a serum dilution of 1:200; a serum is regarded positive if there is a bright, complete, and unbroken peripheral fluorescence. The establishment of a lab specific cut-off is necessary. Depending on application the use of other cut-offs lower than 1:200 might be better.

3.3.3.3 ELISAs

In the following, a protocol for an iELISA assay to detect *N. caninum* antibodies is presented. Parasitic antigen is coated to the wells of an ELISA plate. After incubation with test sera, enzyme-labeled, species-specific anti-immunoglobulin antibodies or antibody-binding proteins conjugated to an enzyme (e.g., to peroxidase or alkaline phosphatase) are applied to report the antigen-specific reactions, that is, the extent to which antibodies have bound to parasitic antigens. In a final step, a substrate is added which is converted by the enzyme into a colored reaction product. The absorbance or optical density (OD) is measured by a spectrophotometer.

A similar protocol as described here was published for the examination of cattle[224,797] but was modified. For this test, no complex purification[1794] or other treatments like incorporation into ISCOMS[216] are needed which may have the disadvantage that unspecific components have not been removed from the antigenic preparation.

Antigen source:

N. caninum tachyzoites from cell cultures

Buffers and reagents:

1. PBS (NaCl—8 g, KH_2PO_4—0.2 g, Na_2HPO_4 12 H_2O—2.9 g, KCl—0.2 g).
2. Coating buffer: 100 mM sodium bicarbonate/carbonate, pH 8.3.
3. Washing buffer: PBS-T: 0.05% Tween 20 in PBS (washing).
4. *Neospora* sp. seronegative horse serum.
5. Blocking buffer: PBS-T, 20% horse serum: 0.05% Tween 20 in PBS, 20% horse serum (blocking, sample dilution).
6. Conjugate buffer: PBS-T, 1% horse serum: 0.05% Tween 20 in PBS, 1% horse serum (conjugate dilution).
7. PBS-T-Urea: PBS-T with 6 M Urea (for avidity testing).
8. Species-specific conjugates or conjugates based on antibody-binding proteins (Protein A, G, L, with confirmed binding to immunoglobulins of species under examination); the substrate protocol (see below) is established for peroxidase as a reporter enzyme.
    a. Substrate and stop solution (appropriate to the enzyme, also commercially available). Here a peroxidase substrate solution is described:
    b. Substrate buffer: 0.2 M sodium acetate/0.2 M citric acid
    c. TMB: 3,3′,5,5′-tetramethylbenzidine solution (50 mg dissolved in DMSO)
    d. H_2O_2, 30%
    e. Prepare fresh: 10 mL substrate buffer, 100 μL TMB, 1.2 μL H_2O_2, 30%, 100 μL/well
    f. Stop solution: 4 N H_2SO_4, 50 μL/well

Process of preparing antigen and coating of plates:

1. Purified *N. caninum* tachyzoites harvested from cell culture by filtration through 5 μm filters.
2. Extensive washing of parasites to remove cell culture proteins and debris: Centrifuge parasite suspension in a 15 mL tube (2000 × *g*, 10 min, 4°C, no brake), discard supernatant and resuspend parasite pellet in ice-cold PBS. Repeat the washing 4–6 times. To glass tubes, parasites may adhere. To avoid this, we siliconize tubes using Sigmacote® (Sigma-Aldrich).
3. Resuspend tachyzoites in PBS to a final concentration of 1×10^8 tachyzoites/mL.
4. Lysate tachyzoites by 3 freezing-thawing cycles (−50°C/+37°C) and ultrasonication (e.g., Branson Sonifier, 90 s, 50% active cycle, output control level 2, on ice).
5. Centrifuge the solution at 10,000 × *g* at 4°C for 30 min.
6. Collect the supernatant, determine protein content (e.g., via a Bradford test) and use it directly or store at −80°C.
7. Coat Polysorp 96-well microtiter plates (e.g., NUNC, Roskilde, Denmark) with *N. caninum* antigen diluted in coating buffer (usually 1:100 to 1:1000; optimal dilutions need to be adjusted; add 100 μL into each well). Note: Optimal dilutions need to be determined by checkerboard titrations (serum, conjugate, and antigen).
8. Incubate for 1 h at 37°C.

Process of testing sample sera:

1. Discard the coating buffer.
2. Wash the plate 3 times with PBS-Tween.
3. Add 300 μL PBS-Tween containing 20% horse serum (blocking buffer) to each well. Incubate for 30 min at 37°C.
4. Remove blocking solution. No washing necessary.
5. Add serum samples diluted in blocking buffer (100 μL/well). Include a standard positive and negative control serum on each plate. Incubate for 30 min at 37°C. Note: Optimal sample dilution needs to be determined by checkerboard titrations (serum, conjugate, and antigen). Usually, a 1:100 dilution is appropriate.
6. Note: At this point of the protocol, additional steps for avidity ELISA can be added.
7. Discard sample dilutions from wells. Wash 3 times with PBS-Tween.

8. Add conjugate appropriately diluted (to be determined in advance by checkerboard titrations) in PBS-Tween with 1% horse serum (100 µL/well).
9. Incubate for 30 min at 37°C.
10. Wash 3 times with PBS-Tween and 2 times with double-distilled water.
11. Add 100 µL substrate solution per well.
12. Incubate for 15 min at 37°C.
13. Stop the reaction by adding 50 µL stop solution per well.
14. Read OD at 450 nm in a microplate spectrophotometer.
15. Calculate ELISA indices (S-Idx = (ODS-ODN)/(ODP-ODN); S-Idx, sample index; ODS, mean sample OD; ODN, mean negative control OD; ODP, mean positive control OD).
16. To determine an appropriate cut-off, a set of known negative and known positive reference sera have to be analyzed (e.g., 100 positive sera, 400 negative sera; for test validation, please refer to OIE manual, Section 1.1.6, http://www.oie.int/international-standard-setting/terrestrial-manual/access-online/).

Procedure for avidity ELISA (insert at step 6 of the process of testing sample sera in ELISA):
For avidity testing, each serum has to be titrated with and without urea incubation (e.g., serial dilutions in 2 of the columns of an ELISA plate; 1 column is later examined with and the other column without urea incubation step). For more accurate results, each serum sample dilution could be tested twice with and without urea.
Follow the ELISA protocol until step 5.

1. Wash ELISA plates with PBS-Tween 20.
2. Incubate wells in odd numbered columns with PBS-T, 6 M urea; incubate wells in even numbered columns with PBS-T; incubate for 5 min at 37°C.

Proceed with the ELISA protocol, step 7.
Appropriate cut-offs to decide that the antibody response of acutely or chronically infected animals need to be evaluated. For validation, use appropriate sera from experimentally infected animals or from animals for which the time of natural infection is known.

3.3.3.4 Immunoblots

The protocol presented has been described previously.[224,1789] Tachyzoite antigens are separated under nonreducing conditions.

Tachyzoites from cell culture:
Use tachyzoites grown with FBS as supplement. One day before harvesting the tachyzoites, the FBS-supplemented medium is replaced by a medium not supplemented with FBS. Isolate the parasite by scraping off the host cell monolayer with a rubber policeman and release tachyzoites by vigorous pipetting, to support cell rupture. To remove the host cell debris, the preparation should be filtered through 5-µm filters. Wash the parasites 5 times with cold PBS (centrifugation, 10 min, $500 \times g$). Prepare a parasite pellet (a pellet of 4×10^7 tachyzoites is sufficient for 1 Minigel).

Reagents for SDS-PAGE:

1. *Sample buffer (5×)*: 3.75 g tris(hydroxymethyl)aminomethane, 10 g sodium dodecyl sulfate (SDS), 35 mL glycerol, and 25 mg bromophenol blue. Adjust to pH 6.8 with HCl and add distilled water up to 100 mL.
2. *Solution A (SDS-PAGE stacking gel buffer)*: 6.06 g tris(hydroxymethyl)aminomethane, 4 mL 10% (w/v) SDS. Adjust to pH 6.8 with HCl and add distilled water up to 100 mL.
3. *Solution B (SDS-PAGE separation gel buffer)*: 18.17 g tris(hydroxymethyl)aminomethane, 4 mL 10% (w/v) SDS. Adjust to pH 8.8 with HCl, add distilled water up to 100 mL.

4. *Solution C (SDS-PAGE acrylamide stock solution)*: 30 g acrylamide, 0.8 g N,N'-methylen-bis(acrylamide), add distilled water up to 100 mL.
5. *Electrophoresis buffer (5×)*: 15.2 g tris(hydroxymethyl)aminomethane, 72.1 g glycine, 5 g SDS, add distilled water up to 1000 mL.
6. *Ammonium persulfate 40% (w/v) in distilled water.*
7. *N,N,N',N'-tetramethylethylenediamine* (TEMED).
8. *Electrophoresis calibration kit* (low molecular weight).

Buffers for antigen transfer to nylon membranes (PVDF membranes, Polyvinylidenfluorid membranes):

1. *Stock transfer buffer (10× stock; 25 mM tris base, 192 mM glycine)*: Tris(hydroxymethyl)aminomethane—30.3 g, glycine—144.1 g, add distilled water up to 1000 mL.
2. *Transfer buffer (1×; with 0.1% [w/v] SDS and 10% [v/v] methanol)*: 100 mL stock transfer buffer (10×), 10 mL 10% (w/v) SDS, 100 mL methanol, add distilled water up to 1000 mL.
3. *India ink staining solution.*[848]
4. *PBS-T-G*: PBS with 0.05% (v/v) Tween 20 and 2% (v/v) fish gelatine liquid (or another appropriate blocking reagent).

Buffers for immunoblot:

1. *PBS-T-G*: PBS with 0.05% (v/v) Tween 20 and 2% (v/v) fish gelatine liquid (or another appropriate blocking reagent).
2. *PBS-Tween*: 0.05% Tween 20 in PBS.
3. *Conjugate*: Species-specific conjugate, for example, anti-bovine IgG[H+L] peroxidase or anti-dog IgG[H+L] peroxidase are used for testing cattle or dog sera, respectively.
4. *Substrate*: 30 mg 4-chloro-1-naphthol, 10 mL methanol, 30 mL PBS, and 40 µL 30% H_2O_2.

Specific equipment:

1. Minigel SDS-PAGE system.
2. Semidry blotting system for antigen transfer.
3. Block heater.

SDS-PAGE procedure:

1. Add 80 µL distilled water and 20 µL sample buffer (5×) to a pellet containing 4×10^7 tachyzoites. This amount is sufficient for 1 Minigel ($60 \times 70 \times 1$ mm).
2. Heat antigen in sample buffer in block heater for 10 min at 94°C.
3. Prepare the separation gel solution (Table 3.1).
4. Cast the separation gel, overlay it with distilled water and let it stand for approximately 30 min (i.e., until the gel has polymerized).
5. Prepare the stacking gel solution (Table 3.1).
6. Pour on the stacking gel solution and insert the comb into the solution to form the slots. Wait at least for 30 min (i.e., until the gel has polymerized) before the comb is removed.
7. Submerge the gel in 1 × electrophoresis buffer.
8. Load the tachyzoite antigen into the preparative slot (about 60 mm length) and the marker protein into the narrow slot (about 2 mm).
9. Run the gel at 120 V until the bromophenol blue front leaves the gel.

Western blotting procedure:

We prefer the electric transfer of SDS-PAGE-separated antigens to PVDF membranes using a continuous buffer system. Using a discontinuous buffer system as described[224] had very similar results.

Table 3.1 Composition of Separation and Stacking Gels; the Volumes are Sufficient for 2 Minigels (60 × 70 × 1 mm)[a]

	5% [w/v] Separation Gel	12.5% [w/v] Separation Gel	Stacking Gel
Solution A	–	–	1.5 mL
Solution B	2.5 mL	2.5 mL	–
Solution C	1.7 mL	4.2 mL	0.7 mL
H_2O	5.8 mL	3.3 mL	3.6 mL
APS (40% [w/v])	25 µL	25 µL	8 µL
TEMED	10 µL	10 µL	16 µL

[a] The 12.5% (w/v) SDS-polyacrylamide separation gel is used for demonstration of *N. caninum*-specific antibodies in sera. To investigate canine sera for antibody reactions toward the 152 kDa antigen, a 5% (w/v) SDS-polyacrylamide separation gel is used.

1. Assemble the transfer unit according to manufacturer. To assemble the transfer unit gel blotting filter paper is needed (e.g., GB 001).
2. Run the transfer unit at a current of 1.5 mA/cm² gel for 90 min and a continuous buffer system.
3. Stain the section of the membrane containing the molecular weight marker and a small part of the antigen section with India ink staining solution.
4. Incubate the remaining part of the membrane with PBS-T-G control for at least 30 min to block all left protein binding sites.
5. The membrane is dried overnight and can then be stored frozen (−20°C) until used.

Antibody detection procedure:

1. Cut the required number of 3–4 mm wide strips from the membrane and label each strip.
2. Incubate the strips for 10–30 min with PBS-T-G.
3. Remove PBS-T-G and incubate each strip with a test serum diluted 1:100 in PBS-T-G (or undiluted fetal fluid, or other dilutions as appropriate), appropriately diluted positive and negative control serum or PBS-T-G for 1 h (500–1000 µL serum dilution are required, depending on the slot size of the incubation tray). A positive and a negative control as well as PBS-T-G control must be included in each test run.
4. Wash the strips 4 times with PBS-Tween.
5. Incubate the strips with conjugate for 1 h (300–1000 µL required per strip, conjugate dilution has to be established; usually 1:2000 is a good starting point).
6. Wash the strips 3× for 3 min with PBS-Tween and twice for 3 min with PBS.
7. Incubate the strips with substrate for 20 min (500–1000 µL required per strip).
8. Wash the strips once for 3 min with PBS and dry them on filter paper.
9. Read results. When bovine and canine sera are tested for *N. caninum*-specific antibodies, reactions against 5 immunodominant antigens (IDAs) with relative molecular masses of 19, 29, 30, 33, and 37 kDa are recorded. Figures with example reaction patterns of various animal species are available in the literature (References **224, 1600, 1796,** and **2108**). The results are interpreted as follows:
    a. >2 IDAs recognized: positive
    b. 1 IDA recognized: inconclusive
    c. No IDA recognized: negative

When canine sera are tested for antibody reactions against the 152 kDa antigen, a serum is judged positive if a band at this relative molecular mass is visible (an example of a positive reaction is displayed in References **1797** and **1805**).

3.3.4 PCR

Most PCRs reported are conventional endpoint PCRs, including 1-step PCRs, 1-step nested PCRs, or 2-step nested PCRs as discussed in Chapter 2. No detailed PCR protocols are provided, because of the large number of available target sequences. Specific protocols need to be established in each lab, adjusted to available reagents and equipment.

3.3.4.1 Sampling, Sample-Treatment and DNA Extraction

For diagnostic purposes and studies on the epidemiology of neosporosis, a large set of different sample types, including tissues of diseased animals (e.g., aborted fetuses), body fluids, placenta, milk, fecal samples, environmental samples, fodder, or water may be examined by PCR for the presence of *N. caninum* DNA. During collection and manipulation, DNA-free equipment should be used and cross-contamination and carry-over needs to be avoided and controlled using negative processing controls.

3.3.4.1.1 Fresh or Fixed Samples

Fresh samples (tissues, body fluids, blood, and milk) are the most suitable for DNA analyses. For short-term storage of samples, 4°C is sufficient. For long-term storage, −20°C and lower are recommended. A large number of in-house and commercialized protocols are available in the literature to extract DNA from these types of samples.

It is well known that formaldehyde significantly alters the quality of DNA. Prolonged fixation with 4% or 10% formaldehyde buffers prior to paraffin embedding may impair a subsequent PCR analysis. However, sometimes no other materials were preserved during necropsy and also retrospective examinations rely on formalin-fixed or already paraffin-embedded material only. Paraffin-embedded material which has been fixed only for a short time, that is, for 2 h or a few days is better suited than long-term fixed material. An alternative to formalin fixation might be Weigners fixative (3% pickling salt, 30% ethanol, and 20% Pluriol®) which was shown to better preserve DNA.[1079]

Nevertheless, successful examinations of sections of formaldehyde-fixed and paraffin-embedded tissues have been described,[174,364,643,1131,1335,1336,1389,1771,1810,1828] although the number of PCR positive findings seems to be much higher in fresh than in formalin-fixed paraffin-embedded samples.[1389] A PCR based on the primer pair PN1/PN2[924] was superior to a PCR with using the Np4/Np7 primers,[174,1771] which suggests that a selection of the most suitable PCR protocol is necessary prior to the examination of formalin-fixed material. In most PCR studies, fresh samples from different tissues were collected and stored frozen (at least −20°C) until used, but commercial DNA extraction kits are marketed along with protocols how to treat formalin-fixed or paraffin-embedded material.

Protocol to extract nucleic acid from fresh or formaldehyde-fixed tissue samples
Reagents:

1. Digestion buffer: 100 mM NaCl, 10 mM Tris-Cl, 25 mM EDTA, 0.5% SDS pH 8, 0.1 mg/mL proteinase K (Note: add fresh)
2. Phenol/chloroform/isoamyl alcohol (25/24/1)
3. Chloroform/isoamyl alcohol (24/1)
4. 3 M sodium acetate (pH 5.2)
5. 96% (v/v) ethanol
6. 70% (v/v) ethanol

DNA extraction from fresh or fixed tissue samples:

1. Collect sample into a reaction tube

    a. Collect and homogenize fresh tissue, cells:
        i. Homogenize tissue using mortar and pestle after snap freezing using LN. Transfer powder to reaction tube.
       ii. Alternatively, a number of cryostat sections can be added to a reaction tube.
      iii. Alternatively, cells isolated from blood, tissue fluids, or cell culture-derived cells are pelleted by centrifugation ($1500 \times g$, 10 min, 4°C) and the supernatant is removed from reaction tube.
    b. Collect from a paraffin-embedded tissue: Tissue sections (~5 μm thick) are directly added to a microfuge tube, without dewaxing.
    c. HE-stained sections: Incubate slide in fresh xylene (Note: up to several days). Lift and remove coverslip with a scalpel blade. Scrape material into a reaction tube.

2. Add digestion buffer (1.2 mL/100 mg tissue) for 1 h, 37°C and agitate (in case of paraffin-embedded or HE-stained sections incubation may have to be extended for several days).
3. Mix with an equal volume phenol/chloroform/isoamyl alcohol (25/24/1) by inverting 50 times. Centrifuge for 7 min at $13,000 \times g$.
4. Mix with an equal volume of chloroform/isoamyl alcohol (24/1) by inverting 50 times. Centrifuge for 7 min at $13,000 \times g$.
5. Transfer the supernatant to a fresh tube and add 0.1 volumes of 3 M sodium acetate (pH 5.2) and 2 volumes of −20°C—cold 96% (v/v) ethanol to precipitate DNA (keep at least 20–30 min at −20°C).
6. Centrifuge for 15 min at $13,000 \times g$. Decant the supernatant.
7. Wash the pellet using 70% (v/v) ethanol and centrifuge for 15 min at $13,000 \times g$.
8. Discard the ethanol solution and air dry the pellet.
9. Resolve DNA in sterile, double-distilled water for at least 12 h at 4°C.
10. Use 2.5–10 μL aliquots for PCR.

3.3.4.2 Fecal, Environmental, and Water Samples

Canine feces or environmental samples contaminated by dog feces may contain *N. caninum* oocysts. DNA extracted from feces or environmental samples may contain PCR inhibitors.[1519] However, only sparse information is available on the best way to enrich and purify *N. caninum* DNA from such samples.[96,164,781,910,1067,1069,1070,1324,1805,1854,1865,2072] Protocols developed or validated for the extraction of oocysts DNA of *Hammondia heydorni*,[644,1866] *H. hammondi*, and *T. gondii*,[413,900,1807] or other parasites even when not successful[413] may help to establish suitable techniques to examine feces or environmental samples for *N. caninum*. Methods used to detect *T. gondii* oocysts in soil or water may also be applicable after modification.[627,628,1139,1836]

In the following section, a protocol that is used at Friedrich-Loeffler-Institute, Federal Research Institute for Animal Health, Germany to extract DNA from isolated oocysts is described. This protocol consists of lysis steps utilizing freezing and thawing cycles, treatment with saline saturated buffers, proteinase K, and cetyl-trimethyl ammonium bromide (CTAB). The method is partially based on the recommendations of others.[2177] Other protocols for the isolation of DNA from oocysts employ grinding with glass beads to disintegrate the oocysts.[910,1752,1865] We used the following protocol to extract DNA from oocysts in dog feces, enriched and isolated by a combined sedimentation–flotation technique.[1805] The protocol is almost identical to that used by us for related parasites, for example, *T. gondii* oocysts in further studies.[899,1807] Details of the following protocol are also provided previously.[613,1806]

Protocol to extract nucleic acid from oocysts or fecal samples
Reagents:

1. PBS: 300 mM NaCl, 2.7 mM KCl, 10 mM Na_2HPO_4, 1.7 mM NaH_2PO_4
2. Sodium hypochlorite, aqueous solution, ≥4% as active chlorine

3. Oocysts-lysis buffer (pH 9.5): 600 mM EDTA, 1.3% (v/v) N-lauroylsarcosine, 2 mg/mL proteinase K
4. CTAB buffer: 2% (w/v) cetyl-trimethyl ammonium bromide, 1.4 M NaCl, 0.2% (v/v) mercapto-ethanol, 20 mM EDTA, 100 mM tris(hydroxymethyl)aminomethane
5. Phenol/chloroform/isoamyl alcohol (25/24/1)
6. 3 M sodium acetate (pH 5.2)
7. 96% (v/v) ethanol
8. 70% (v/v) ethanol

Procedure:

1. Isolate oocysts by sucrose floatation.
2. Use oocysts directly after isolation or store them in 1%–2% $K_2Cr_2O_7$ or in 2% sulfuric acid in a refrigerator at 4–8°C.
3. Wash oocysts 4 times by centrifugation ($1100 \times g$, 7 min, without the use of the brake) in 15 mL PBS in a 15 mL centrifugation tube.
4. Incubate the pellet with oocysts and contaminants in 2 mL 5.75% NaOCl (30 min, 37°C).
5. Add double-distilled H_2O up to 15 mL.
6. Centrifuge supernatant in a 15 mL tube (7 min, $1100 \times g$, without the use of the brake) and resuspend the pellet with PBS. Wash pellet 3 times with PBS (7 min, $1100 \times g$, without brake).
7. After a last centrifugation, resuspend the pellet in 1 mL PBS, transfer into a 1.5 mL—reaction tube and spin down (7 min, without $1100 \times g$, no brake).
8. Carefully remove as much of the supernatant as possible and apply 3 freezing-thawing cycles (freezing: 10 min, −20°C; thawing: 2 min, room temperature) to the pellet.
9. Resuspend the pellet in 100 μL oocyst-lysis buffer (45 min, 65°C).
10. Add 400 μL CTAB buffer (60 min, 60°C).
11. Mix with 500 μL phenol/chloroform/isoamyl alcohol (25/24/1) by inverting 50 times. Centrifuge for 7 min at $13,000 \times g$. Repeat.
12. Transfer the supernatant to a fresh tube and add 0.1 volumes of 3 M sodium acetate (pH 5.2) and add 2 volumes of −20°C—cold 96% (v/v) ethanol to precipitate DNA (keep at least 20–30 min at −20°C).
13. Centrifuge for 15 min at $13,000 \times g$. Decant the supernatant.
14. Wash the pellet using 70% (v/v) ethanol and centrifuge for 15 min at $13,000 \times g$.
15. Discard the ethanol solution and air dry the pellet.
16. Resolve DNA in sterile, double-distilled water for at least 12 h at 4°C.
17. Use 2.5–10 μL aliquots for PCR.

3.3.4.3 Options for Transporting DNA Prior to Analysis

Aqueous extracts of DNA should be kept cold before being analyzed, including during shipping. Drying a DNA pellet, or applying an aqueous DNA solution to FTA® cards (Whatman®) (FTA® cards = fast technology for analysis of nucleic acids cards, impregnated with DNAse inhibitors) allows DNA to be transported at ambient temperatures (at significant cost savings). In any event, templates should not be subjected to x-ray or direct sunlight.

BIBLIOGRAPHY

96, 148, 164, 174, 216, 224, 364, 413, 449, 483, 484, 565, 598, 613, 627, 628, 643, 644, 647, 648, 780, 781, 797, 848, 878, 882, 885, 890, 899, 900, 910, 924, 951, 973, 1055, 1067, 1069, 1070, 1079, 1136, 1139, 1159, 1172, 1324, 1326, 1327, 1335, 1336, 1389, 1421, 1445, 1512, 1519, 1545, 1592, 1600, 1610–1614, 1677, 1733, 1752, 1771, 1789, 1794, 1796, 1797, 1805–1807, 1810, 1828, 1836, 1865, 1866, 1921, 1981, 2051, 2072, 2108, 2177.
Additional references: **519**.

Neosporosis in Cattle

4.1 NATURAL INFECTIONS

4.1.1 Serologic Prevalence

N. caninum antibodies in cattle have been reported from most parts of the world (Table 4.1). Prevalence varied with age, breed, and geography, and other factors (see epidemiology Section 4.4 of this chapter).

4.1.2 Clinical Infections

Reproductive losses are the main clinical outcome of neosporosis in cattle. Assessment of these economic losses is difficult because many costs have to be considered including, for example, the loss of the calf, the diagnostic examination of abortus and the cow, the loss of milk yield, and losses caused by culling and replacement of an affected dam. In the early 1990s, there were many reports of abortion due to neosporosis and the number of these reports has decreased once neosporosis was firmly established as a major abortifacient in cattle. We have attempted to summarize information on bovine neosporosis. Evidence for abortion has been based on finding *Neospora* parasites or its DNA directly in aborted fetal tissues (Table 4.2), in diseased newborns (Table 4.3), and by comparing serological responses to *Neospora* in cows that aborted versus control cows that did not abort (Table 4.4). Data in Table 4.4 were extracted from studies that varied in design and methodology. Our objective in summarizing the data was to indicate an association between serology and abortion.

4.1.2.1 Abortion and Stillbirth

N. caninum causes abortion in both beef and dairy cattle. Fetuses may die *in utero* or be resorbed, mummified, autolyzed, stillborn, born alive with clinical signs, or born clinically normal but being persistently infected (Figure 4.1). Among aborted twins, one or both maybe infected with *N. caninum*.[995,996] Unlike bacterial infections, mummification is common in neosporosis. Worldwide reports are summarized in Table 4.2. Most of these reports are from Europe and the Americas, probably related to the diagnostic facilities available. Cows of any age may abort. Cows may abort repeatedly in consecutive pregnancies or after skipping a lactation.[64,1497] Lesions in abortion vary, depending on several factors. Important factors are as follows.

Table 4.1 Serologic Prevalence of *N. caninum* Antibodies in Cattle

Country	Region	No. of Animals	Type	No. of Herds	No. Positives	Percent of Positive (%)	Test	Titer/ Company/ Type of Test/ Antigen	Remarks	Reference
Algeria	–	102	NS	7	4	3.9	ELISA	iBio-X	–	755
Algeria	North and Northeast	799	NS	87 farms	157	19.6	IFAT	1:200	Population, area, dog presence, season, global farm hygiene, abortion	758
Algeria	–	100	NS		40	40	IFAT	1:200	–	759
Andorra	–	65	Beef	1	6	9.2	ELISA	CIVTEST	1 of 6 seropositive cows had confirmed *Neospora* abortion	83
Argentina	–	1758	Beef	26	130	7.4	ELISA	CIVTEST	Area, age, breed	84
	La Plata	79	Dairy	3	27	34.1	IFAT	1:800	–	2039
	La Plata	189 (abortion history)	Dairy	19	122	64.5	IFAT	1:25	Abortion history	2040
	Northeast	880	Beef	4 ranches	252	28.6	IFAT	1:100	Age	1393
	–	290 (abortion history)	Beef	1	59	20.3	IFAT	1:200	–	1380, 1381
	–	400	Beef	17	18	4.7	IFAT	1:200	Presence of seropositive dogs	1380, 1385
	–	1048		52 farms	173	16.6	IFAT	1:200	Presence of seropositive dogs	1380, 1385
	–	750 (abortion history)	Beef	49	323	43.1	IFAT	1:200	Production system	1380, 1385
	–	216 (abortion history)	Beef	39	41	18.9	IFAT	1:200	(dairy/beef)	1380, 1385
	–	305 (bulls)	Beef	19	15	4.9	IFAT	1:200	–	1383
	–	4190	Dairy/beef	5,594	594	14.2	IFAT	1:200	Type of production, history of abortion	1390
	–	90	Beef	Abattoir	65	73.0	IFAT	1:25	–	1401
	–	173	Dairy	1 farm	140	80.9	IFAT	1:25	–	1403

(Continued)

Table 4.1 (Continued) Serologic Prevalence of *N. caninum* Antibodies in Cattle

Country	Region	No. of Animals	Type	No. of Herds	No. Positives	Percent of Positive (%)	Test	Titer/ Company/ Type of Test/ Antigen	Remarks	Reference
Australia	Northern	192	Beef	–	61	31.8	ELISA	IDEXX	–	1459
	New South Wales	266	Dairy	1	63	24	IFAT	1:160	Abortion history	100
					78	29	IB	–		
	Queensland	1673	Beef	45	249	14.9	IFAT	1:200	–	1911
	Queensland	711	Dairy	3 herds	210	29.5	ELISA	IDEXX	Abortion	1115
	South	943	Dairy/beef	60	25	2.7	ELISA	IDEXX	–	1452
Belgium	–	711	Dairy/beef	52	86	12.2	IFAT	1:200	Breed	448
	–	70	Dairy	NS	20	28.6	IFAT	1:200	–	450
	–	93	Beef	NS	13	13.8	IFAT	1:200	–	450
	–	18,858	Dairy/beef	957	1927	10.2	ELISA	iBio-X	–	2023
Brazil	Bahia	447	Dairy	14	63	14.0	IFAT	1:200	–	776
	Goiás	444	Dairy	11	135	30.4	IFAT	1:250	–	1342
	Goiás	30	Mixed	1	13	43.3	IFAT	1:250	–	1342
	Goiás	456	Beef	9	135	29.6	IFAT	1:250	–	1342
	Maranhão	812	Dairy	27	412	50.7	IFAT	1:200	–	1949a
	Mato Grosso	932	Dairy	24 farms	499	53.5	IFAT	1:200	–	192
	Mato Grosso do Sul	2448	NS	205	449	14.9	IFAT	1:50	Production system (dairy/beef)	1530
	Mato Grosso do Sul	197	NS	6	66	33.5	ELISA		–	320
	Mato Grosso do Sul	275	Beef	2 farms	81	29.5	cELISA	VMRD	–	1576
	Mato Grosso do Sul	392	Dairy/beef	4 farms	43	9.1	IFAT	1:50	–	1341
	Mato Grosso do Sul	23	Dairy	NS	5	21.7	IFAT	1:25	–	1658
	Mato Grosso do Sul	90	Beef (history of abortions)	NS	38	43	ELISA	IDEXX	–	71
	Mato Grosso do Sul	91	NS	Healthy	7	7.7	ELISA	IDEXX	–	71
	Mato Grosso do Sul	60	NS	Heifers	18	30.0	ELISA	IDEXX	–	71
	Mato Grosso do Sul	87	Beef	NS	26	29.9	IFAT	1:25	–	1658

(Continued)

Table 4.1 (Continued)   Serologic Prevalence of *N. caninum* Antibodies in Cattle

Country	Region	No. of Animals	Type	No. of Herds	No. Positives	Percent of Positive (%)	Test	Titer/ Company/ Type of Test/ Antigen	Remarks	Reference
	Mato Grosso do Sul	1098	Beef	1 farm	687	62.5	IFAT	1:50	Reproductive failure, 15% higher in seropositive cows	74
	Minas Gerais	559	Dairy	18	510	91.2	IFAT	1:200	Farm size, number of cows lactating, milk production per day	814
	Minas Gerais	575	Dairy	Abattoir	559	97.2	IFAT	1:200	Farm size, number of cows lactating, milk production per day	814
	Minas Gerais	503	Dairy	Abattoir (fetuses)	64	12.7	IFAT	1:25	Farm size, number of cows lactating, milk production per day	814
	Minas Gerais	584	Dairy	18	109	18.7	ELISA	IDEXX	–	452
	Minas Gerais	576	Dairy	18	106	18.4	ELISA	IDEXX	–	455
	Minas Gerais	126	Dairy	2	43	34.4	IFAT	1:25	–	1658
	Minas Gerais	243	Dairy	2	41	16.8	ELISA	IH-ISCOM	–	1358
	Minas Gerais	1204	Dairy	40 farms	260	21.6	IFAT	1:200	Reproductive failure	249
	Minas Gerais	36	Beef	NS	4	11.1	IFAT	1:25	–	1658
	Pará	40	Dairy	4 farms	7	17.5	IFAT	1:100	–	1364
	Pará	120	Beef	12	23	19.2	IFAT	1:100	–	1364
	Paraná	1263	NS	77 farms	423	33	ELISA	IDEXX	–	1204
	Paraná	165 (abortion history)	NS	1	69	42.1	ELISA	IDEXX	–	1200
	Paraná	159	Beef	Abattoir	24	15.1	ELISA	IDEXX	–	1280
	Paraná	15	Beef	NS	4	26.7	IFAT	1:25	–	1658
	Paraná	76	Beef	4	23	30.3	ELISA	IDEXX	–	504
	Paraná	172	Dairy	1	60	34.8	ELISA	IDEXX	Abortion	1199
	Paraná	623	Dairy	23 farms	89	14.3	IFAT	1:25	Breed, presence of dogs, age, feed	818
	Paraná	75	Dairy	NS	16	21.3	IFAT	1:25	–	1658
	Paraná	385	Dairy	90 farms	45	12	IFAT	1:200	–	1501

(Continued)

Table 4.1 (Continued) Serologic Prevalence of *N. caninum* Antibodies in Cattle

Country	Region	No. of Animals	Type	No. of Herds	No. Positives	Percent of Positive (%)	Test	Titer/ Company/ Type of Test/ Antigen	Remarks	Reference
	Paraná	309	Dairy	15 farms	63	20.4	IFAT	1:100	Feed, wild animal access, artificial insemination	1287
	Pernambuco	469	Dairy	20 farms	163	31.7	IFAT	1:200	Veterinary assistance, nutritional condition, presence of wetlands, manipulation of newborn calves, destination of cows that had aborted, abortion history, abortion period	1859
	Pernambuco	158	Dairy	Several farms	31	19.6	IFAT	1:200		1666a
					57	36.0	ELISA			
	Rio de Janeiro	75	Dairy	NS	17	22.7	IFAT	1:25	–	1658
	Rio de Janeiro	563	Dairy	57 farms	131	23.2	ELISA	IDEXX	Breed	1430, 1431
	Rio de Janeiro	75	Beef	NS	5	6.7	IFAT	1:25	–	1658
	Rio Grande do Sul	223 (abortion history)	Dairy	5	25	11.2	IFAT	1:200	Reproductive failure	383
	Rio Grande do Sul	1549	Dairy	60 farms	276	17.8	IFAT	1:200	Several	385
	Rio Grande do Sul	70	Dairy	NS	13	18.6	IFAT	1:25	–	1658
	Rio Grande do Sul	70	Beef	NS	15	21.4	IFAT	1:25	–	1658
	Rio Grande do Sul	781	Dairy/beef	NS	89	11.4	ELISA	CHEKIT	–	2048
	Rondônia	1011	Dairy	50	114	11.2	IFAT	1:25	Farm size, number of cows	20
	Rondônia	584	Beef	11 farms	56	9.5	IFAT	1:25	–	20
	Rondônia	514	Mixed	25 farms	50	9.7	IFAT	1:25	–	20
	Rondônia	621	Dairy	63 farms	66	10.6	IFAT	1:100	Abortion, birth of weak calves	228
	Santa Catarina	130	Dairy	29 farms	57	43.8	IFAT	1:200	Age, no. of pregnancies	1072

(Continued)

Table 4.1 (Continued) Serologic Prevalence of *N. caninum* Antibodies in Cattle

Country	Region	No. of Animals	Type	No. of Herds	No. Positives	Percent of Positive (%)	Test	Titer/ Company/ Type of Test/ Antigen	Remarks	Reference
	São Paulo and Minas Gerais	600	NS	NS	101	16h.8	IFAT	1:200	Area	389
	São Paulo	777	Beef	8 Farms	121	15.5	IFAT	1:200	–	855
	São Paulo	505	Beef	11 Herds	101	20.0	ELISA	1:100	Production system	1780[a]
	São Paulo	150	NS	NS	41	27.3	IFAT	1:25	–	1658
	São Paulo	521	Dairy	NS	82	15.9	IFAT	1:200	–	1779
	São Paulo	521	Dairy	NS	159	30.5	ELISA	IDEXX	–	1779
	São Paulo	408	Dairy	6 herds	145	35.5	ELISA	1:100	*N. caninum* significatively higher in dairy cattle	1780[a]
	São Paulo	1027	Dairy	3 farms	107	10.4	IFAT	1:100	High degree of association between *N. caninum* serological status of dams and daughter	302
	Tocantins	192	Dairy	10 farms	48	25.0	IFAT	1:200		1289a
Canada	Alberta	2816	Dairy	77	521	18.5	ELISA	IDEXX	Agroecological area	1817
	Alberta	1976 (steers)	Beef	4 feedlots	128	6.5	ELISA	IDEXX	13.5% of additional samples collected between 1984 and 1986 from the same area were positive	2056
	Alberta	840	Dairy	28 ranches	53	6.3	cELISA	VMRD	Purchase new animals, carcass removal	1640
	Alberta	1806	Beef	174	162	9.0	ELISA	IDEXX	–	2060
	Alberta	2996	Beef	100	290	9.7	ELISA	IDEXX	Area	1818
	Manitoba	1204	Dairy	40	100	8.3	ELISA	IDEXX	–	2027
	Manitoba	1425	Beef	49	129	9.1	ELISA	IDEXX	–	2027
	Maritime	2594	Dairy	NS	497	19.2	ELISA	BIOVET	–	1044
	New Brunswick (NB), Nova Scotia (NS), Prince Edward Island (PEI)	2425	Dairy	90	229	20.3	ELISA	BIOVET	–	2024, 2025

(Continued)

Table 4.1 (Continued) Serologic Prevalence of *N. caninum* Antibodies in Cattle

Country	Region	No. of Animals	Type	No. of Herds	No. Positives	Percent of Positive (%)	Test	Titer/ Company/ Type of Test/ Antigen	Remarks	Reference
	Ontario	758	Dairy	25	51	6.7	ELISA	WT-IHCA	Retained placenta, abomasal displaced	625
	Ontario	3412	Dairy	56	359	10.5	ELISA	WT-IHCA	—	394
	Ontario	3702	Dairy	82	448	12.1	ELISA	WT-IHCA	—	917
	Ontario	3162	Dairy	57	332	10.5	ELISA	WT-IHCA	—	917
	Ontario	1704	Dairy	57	190	11.2	ELISA	WT-IHCA	—	917
	Ontario	9723	Dairy	125	104	11.2	ELISA	WT-IHCA	Vertical transmission rate of 40.7%	1549
	Ontario, PEI, NB, NS	3531	Dairy	134	448	12.7	ELISA	BIOVET		1972
	Ontario	3449	Dairy	57	283	8.2	ELISA	WT-IHCA	—	1582
	Québec	437	Dairy	11	43	9.8	ELISA	BIOVET	Embryo transfer as a measure of prevention against vertical transmission	111
	Québec	2037	Dairy	23	447	21.9	ELISA	BIOVET	Vertical transmission rate	194
	Québec	3059	Dairy	46	507	16.6	ELISA	WT-IHCA	Presence and no. of dogs	1559
	Saskatchewan	1530	Dairy	51	85	5.6	ELISA	BIOVET		2026
	Western Provinces	112	Beef	1	75	66.9	ELISA	NS	Age, birth weight, calf sex, age at fall weighing	2061
	—	2484	Beef	66	128	5.2	ELISA	BIOVET	—	2063
Chile	IX Region	198	Dairy	1	31	15.7	IFAT	1:200	Age	1569
	IX Region	173	Dairy	1	52	30.2	IFAT	1:200	Age	1569
China	Beijing	283	NS	1 farm	175	61.8	ELISA	NS	—	845
	Beijing	212	Dairy	NS	92	43.4	ELISA	IH-NcSRS2	0.0% blood positive by PCR	2151
	Beijing	94	Dairy	NS	17	18.1	ELISA	IDEXX	—	2170
	Hebei	55	Dairy	NS	13	23.6	ELISA	IDEXX	—	2170
	Jilin	1091	Not clear	Not clear	189	17.3	IFAT	1:200	Not clear	270

(Continued)

Table 4.1 (Continued) Serologic Prevalence of N. caninum Antibodies in Cattle

Country	Region	No. of Animals	Type	No. of Herds	No. Positives	Percent of Positive (%)	Test	Titer/ Company/ Type of Test/ Antigen	Remarks	Reference
	Jilin	153	Beef	Not clear	18	11.7	ELISA	NcSAG1	–	501
	Qinghai	35	Dairy		4	11.4	ELISA	NcSAG1	–	328
	Qinghai	946 yaks	NS	8 areas	21	2.2	ELISA	IH-tNcSRS2	–	1194
	Shanghai	1239	Dairy	41 farms	63	5.08	ELISA	IDEXX	–	1843
	Southern	370	Dairy	5 farms	70	18.9	ELISA	IDEXX	–	2134
	Tianjin	601	Dairy	NS	34	5.7	ELISA	IH-NcSRS2	1.2% blood positive by PCR	2151
	Five provinces	2147	Dairy	134	389	18.1	ELISA	IDEXX	Source of water, management system, presence of canids and felids	1919
	Five provinces	2340	Beef	134	380	16.2	ELISA	IDEXX	Source of water, management system, presence of canids and felids	1919
	–	262	Dairy	9	45	17.2	ELISA	CIVTEST	Area (herd)	2163
	–	300	Dairy	9 regions	61	20.3	ELISA	IH-NcSRS2	Abortion history, area	1193
	–	10	Beef	1	0	0.0	ELISA	CIVTEST	–	2163
	–	540	Dairy	92 farms	72	13.3	ELISA	IDEXX	Grazing	2068
	–	365	Dairy	6 farms	55	15.0	ELISA	IDEXX	Area	2135
	–	161	Dairy	Not clear	7	4.3	ELISA	IH-NcSAG1t	–	1240
	–	30 (abortion history)	Dairy	Not clear	8	26.7	ELISA	IDEXX	–	1196
	–	10 (healthy)	Dairy	Not clear	0	0.0	ELISA	IDEXX	–	1196
	–	221	Beef	Not clear	35	15.8	ELISA	IH-GST-NcSAG1	–	1241
Colombia	–	262	Dairy	9	45	17.2	ELISA	CIVTEST	Area (herd)	2163
	Cundinamarca	397	Dairy	1	84	21.2	IFAT	1:400	Age	722
	Monteria, Cordoba	162 (28 bulls, 134 cows)	NA	28 farms	121	74.7	ELISA	IDEXX	Tested for Sarcocystis	299a

(Continued)

Table 4.1 (Continued) Serologic Prevalence of *N. caninum* Antibodies in Cattle

Country	Region	No. of Animals	Type	No. of Herds	No. Positives	Percent of Positive (%)	Test	Titer/ Company/ Type of Test/ Antigen	Remarks	Reference
	Boyacá	100	Dairy	1 farm	64	64.0	ELISA	INGEZIM	Abortion	1641a
Costa Rica	—	3002	Dairy	20	1191	39.7	ELISA	WT-IHCA	Age of daughters	1738
	—	2743	Dairy	94 farms	1185	43.3	ELISA	WT-IHCA	—	1740
Croatia		395	Dairy	NS	23	5.8	ELISA	SVANOVIR	Abortion, purchase of animals, age	181
Czech Republic	—	407 (abortion)	Dairy	5	13	3.1	IFAT	1:200	Abortion	2016
	—	546 (healthy)	Dairy	49 farms	3	0.5	cELISA	VMRD	No concurrent *T. gondii* infection	160
	—	463 (abortion)	Dairy	137	18	3.9	ELISA	IDEXX	—	2016
Denmark	—	1561	Dairy	31	343	22	ELISA, IFAT	IH-ISCOM	Age, calving, abortion, no. of gestation	1010
Egypt	—	93	NS	NS	19	20.4	ELISA	IH-tNcSAG1	—	967
Estonia	—	320 (bulk milk)	NS	NS	51	16.0	ELISA	SVANOVA	—	1124
Ethiopia	—	266 (history of abortions)	Dairy	NS	63	23.8	ELISA	IDEXX	Tested for *Brucella*	95
	—	465 (control)	Dairy	NS	59	12.7	ELISA	IDEXX	Tested for *Brucella*	95
	—	2334	Dairy	273 farms	335	13.3	ELISA	IDEXX	Area, age, breed, origin, reproductive disorders	93
	—	134 (abortion history)	Dairy	NS	40	29.8	ELISA	IDEXX	Tested for *Brucella*	94
	—	268 (control)	Dairy	NS	29	10.8	ELISA			
Finland	—	40	Dairy	—	26	65.0	ELISA	Svanovir Neospora Ab	—	1825a
France	Calvados	127 (with history of abortions and without it)	Dairy	13	37	29.1	ELISA	IDEXX	Area	1621

(Continued)

Table 4.1 (Continued) Serologic Prevalence of *N. caninum* Antibodies in Cattle

Country	Region	No. of Animals	Type	No. of Herds	No. Positives	Percent of Positive (%)	Test	Titer/ Company/ Type of Test/ Antigen	Remarks	Reference
	Calvados	895	Dairy	13	237	26.5	ELISA	IDEXX	–	**1621**
	Calvados	2087	Dairy	NS	652	31.2	ELISA	NS	–	**1022**
	Normandy	575	Beef	NS	148	25.7	ELISA	IDEXX	Age, presence of cats and dogs	**1074**
	Saône et Loire	219	Beef	NS	30	13.7	ELISA	IDEXX	–	**1074**
	West	1373	Dairy	13	142	10.4	ELISA	IDEXX	–	**1621**
	West	1170	Dairy	12	130	11.1	ELISA	IDEXX	–	**1622**
	West	2141 (abortion history)	Dairy	NS	354	16.5	ELISA	IDEXX	–	**1622**
	–	1924	Dairy	42	107	5.6	ELISA	IDEXX	Type of housing, presence of birds, cats, and dogs, pregnancy trimester, long calving periods, somatic cell counts, pond water supply	**1075, 1541**
Germany	Hesse	388 (abortion history)	NS	22	16	4.1	IFAT	1:400	–	**375**
	Rhineland-Palatinate	–	Dairy (bulk milk)	3260 herds	258	7.9	ELISA	IH-p38	Risk factors examined by logistic regression model	**1801**
	Rhineland-Palatinate	4343		100	172	4.0	ELISA	IH-p38	Risk factors examined	**2050**
	Rhineland-Palatinate	4261	Dairy	100 herds	134	1.6	ELISA	IH-p38 (milk samples)	Production system	**144**
	Sleswick-Holstein	–	Dairy (bulk milk)	1,950 farms	20	1.0	ELISA	IH-p38	–	**1808**

(Continued)

Table 4.1 (Continued) Serologic Prevalence of *N. caninum* Antibodies in Cattle

Country	Region	No. of Animals	Type	No. of Herds	No. Positives	Percent of Positive (%)	Test	Titer/ Company/ Type of Test/ Antigen	Remarks	Reference
	Bavaria	1357	Dairy	220 farms	92	6.8	ELISA	IDEXX	–	2077
	–	100	Dairy	1	27	27.0	IFAT	1:50	Familial association with congenital infections	1789
	Rhineland-Palatinate	2022	Beef	106 herds	99	4.1	ELISA	IH-p38	Production system	144
Greece	Epirus, Thessaly, Macedonia, Thrace	777	Dairy	10	118	15.2	ELISA	IH-p38 (milk samples)	–	1882
Grenada	–	148	NS	35	10	6.8	ELISA	IDvet	–	1837
Hungary	–	97 (abortion)	Dairy	NS	9	10.0	ELISA	IH-ISCOM	–	929
	Northeast	518	Dairy	39	17	3.3	IFAT	1:100	–	930
	Northeast	545	Beef	49	10	1.8	IFAT	1:100	–	930
India	–	427	Dairy	7 farms	35	8.2	ELISA	VMRD	–	1337
Indonesia	Western Java	991	Beef	21 locations	165	16.6	ELISA	IH-NcSAG1	–	970
Iran	Eastern Islands	438		NS	24	5.5	ELISA	IH-p38	–	419
	Hamedon	492	Dairy	41	63	12.8	ELISA	IDEXX	Presence of dogs	764
	Kerman	285	NS	NS	36	12.6	ELISA	SVANOVA	–	659
	Khorasan	337	Dairy	30	156	46.3	ELISA	IDEXX	Abortion, age	1669
	Kurdistan	368	Beef	Slaughter-houses	29	7.8	ELISA	IDEXX	–	871
	Mashhad	810 (abortion history)	Dairy	4	123	15.1	IFAT	1:200	–	1753
	Tabriz	266	Dairy	NS	28	10.5	ELISA	IDEXX	Age	1457
	–	237	NS	NS	76	32.0	ELISA	IDEXX	Origin	2161
	–	768	Dairy	4	298	38.8	ELISA	IDEXX	Abortion	1765
Ireland	–	324 (abortion)	Dairy	NS	40	12.6	IFAT	1:640	Reproductive failure	1334

(Continued)

Table 4.1 (Continued) Serologic Prevalence of *N. caninum* Antibodies in Cattle

Country	Region	No. of Animals	Type	No. of Herds	No. Positives	Percent of Positive (%)	Test	Titer/ Company/ Type of Test/ Antigen	Remarks	Reference
Israel	Central	1078 pregnant	Dairy	1	382	35.5	IFAT	1:400	–	1301
	–	7,951 (abortion history)	NS	NS	81	18.2	IFAT	1:200	18.2% of 446 thoracic fluids from aborted fetuses	1299
Italy	Italian Apennines	948	–	81	303	32.0	ELISA	IDEXX	BVH-1, age	1712
	Lombardia, Emilia Romagna	10,684	NS	986	2,596	24.3	IFAT	1:640	–	1254, 1257
	Parma	820 (abortion history)	Dairy	85 farms	236	28.7	IFAT	1:160	History of abortion	670
	Potenza, Padua	385	Beef	39	23	6.0	ELISA	CHEKIT	Grazing	1533
	Potenza, Padua	387	Dairy	39	44	11.4	ELISA	CHEKIT	Farm size, grazing	1533
	Southern Italy	350	Dairy	35	65	18.8	ELISA	MASTAZYME	–	1552
	North	5912 (abortion history)	Dairy	1	1442	24.4	IFAT	1:640	–	1253
	–	71 (45 aborting, 26 control)	NS	6 farms	20	28.1	IFAT	1:1280	–	669
	–	880 (healthy)	Dairy	85 farms	123	13.9	IFAT	1:160	History of abortion	670
Japan	Gifu Prefecture	129	NS	3	71	55	ELISA	NS	–	1930
	Nationwide	2420	Dairy	NS	139	5.7	IFAT	1:200	Area	1086, 1087
	–	145 (abortion)	Dairy	NS	29	20	IFAT	1:200	Reproductive failure	1086
	–	65	Beef	NS	1	1.5	IFAT	1:200	Reproductive failure	1086
Jordan	–	671	Dairy	62	236	35.2	ELISA	IDEXX	Presence of workers visiting other farms was considered a risk factor	1931

(Continued)

Table 4.1 (*Continued*) Serologic Prevalence of *N. caninum* Antibodies in Cattle

Country	Region	No. of Animals	Type	No. of Herds	No. Positives	Percent of Positive (%)	Test	Titer/ Company/ Type of Test/ Antigen	Remarks	Reference
Mexico	Aguascalientes	187	Dairy	13	110	59.0	ELISA	IDEXX	–	731
	Aguascalientes	150	Dairy	NS	45	30.0	ELISA	IDEXX	*N. caninum* DNA in water	1854
	Coahuila, Chihuahua, Hidalgo Queretaro, Jalisco	813 (history of abortion)	Dairy	20	341	42.0	ELISA	IDEXX	–	732
	Coahuila	185	Dairy	12	84	45.4	ELISA	IDEXX		1340
	Colima	920	Dual purpose/ beef	90	149	16.2	ELISA	CIVTEST	Commercial concentrate supplementation, metritis, and placental retention history	89
	Hidalgo	390	NS	13	201	51.5	ELISA	IDEXX	Presence of dogs	1770
	Nuevo León	262	Dairy/beef	21	104	39.7	ELISA	IDEXX	–	1340
	Nuevo León	813	NS	48	94	11.6	ELISA	NS	–	1822
	Tamaulipas	144	Dairy	11	23	15.9	ELISA	IDEXX	–	1340
	Veracruz	863	Dairy, beef, and crossbreed	78	224	26.0	ELISA	IDEXX	Age, breed, reproductive status, abortion history	1741
	Veracruz	555	Dual purpose	28	105	18.9	ELISA	Pourquier	Abortion	685
	4 provinces	596	Beef	31	69	11.6	ELISA	IDEXX	–	733
	–	1003	Dairy	50	561	55.9	ELISA	WT-IH	–	1396
The Nether- lands	–	1676	Dairy	21	601	35.8	ELISA	WT-IH		498
	–	1601	Beef	82	208	13.3	ELISA	WT-IH	Production system (dairy or beef)	144
	–	6910	Dairy	108	684	9.9	ELISA	WT-IH	Production system (dairy or beef)	144
	–	6910	Beef	108	684	9.9	ELISA	WT-IH	Risk factors examined	144

(Continued)

Table 4.1 (Continued) Serologic Prevalence of N. caninum Antibodies in Cattle

Country	Region	No. of Animals	Type	No. of Herds	No. Positives	Percent of Positive (%)	Test	Titer/ Company/ Type of Test/ Antigen	Remarks	Reference
New Zealand	–	77	Dairy	1	36	46.7	IFAT	1:200	–	1960
	–	499	Beef	40	14	2.8	ELISA	WT-IH	–	1951
	–	97 (abortion history)	Dairy	1	29	30.7	IFAT	1:200	–	393
	–	800	Dairy	40	60	7.6	ELISA	WT-IH	–	1683
	–	194 (abortion)	Dairy	1	102	53.0	ELISA	WT-IH	Comparison of serological methods	1790
	–	600 (abortion history)	Dairy	1	300	50.0	ELISA	WT-IH	–	1602
	–	1199 (abortion history)	Dairy	3	403	33.6	IFAT	1:200	–	1686, 1688
	–	164 (abortion history)	Diary	1	18	10.9	IFAT	1:200	–	2085
	Taranaki	226	Dairy	1	57	25.2	IFAT	1:400	Age	1568
	–	815	Beef	45	22	2.7	ELISA	IDEXX	Fetal loss	1774
Norway	–	1657 bulk milk	Dairy	1,657	11	0.7	ELISA	SVANOVA	Area	1077
Pakistan	Punjab	240 (141 aborting, 99 at risk)	Dairy	5	105	43.8	cELISA	VMRD	Abortion	1835
	Punjab and Sindh	641	Dairy	12	277	43.2	cELISA	VMRD	Breed	1455
Paraguay	–	297	Dairy	6	107	36.0	ELISA	WT-IH	Abortion	1528
	–	582	Beef	5	155	26.6	ELISA	WT-IH	Abortion	1528
Peru	Junín	347	Dairy	NS	45	12.4	IFAT	NS		1643
	Arequipa	199	Dairy	NS	84	42.2	ELISA	IH-ISCOM	Abortion	1899
Philippines	–	96	Beef	1	16	16.7	ELISA	IH		1089

(Continued)

Table 4.1 (Continued) Serologic Prevalence of *N. caninum* Antibodies in Cattle

Country	Region	No. of Animals	Type	No. of Herds	No. Positives	Pecent of Positive (%)	Test	Titer/ Company/ Type of Test/ Antigen	Remarks	Reference
Poland	North	45 (abortion history)	Dairy	6	7	15.6	ELISA	IDEXX	–	263
	–	416	NS	32	38	9.3	ELISA	IDEXX	–	2091
	–	734	Dairy	97	143	19.5	ELISA	cBio-X	–	1096
Portugal	–	119 (abortion history)	Dairy	1	58	48.7	ELISA	IDEXX	Abortion rate from 14% (1996) to 27% (1999)	1957
	–	114	Dairy	49	13	32.0	NAT	1:40	Reproductive failure	282
	–	1237 (abortion)	Dairy	36	569	46.0	NAT	1:40	Reproductive failure	282
Romania	Cluj, Statu-Mare, Mures, Sibiu, Alba	193	NS	NS	108	55.9	ELISA	IDEXX	–	740
	Northwest and Center	901	Dairy	16 farms	312	34.6	ELISA	IDEXX	Abortion, pregnancy (at least once)	739
	South	258	Dairy	9 farms	104	40.3	ELISA	IDEXX	–	1366
	Southern	44	Dairy	4 farms	25	56.8	ELISA	IDvet	–	650
	Western	376	Dairy	25	104	27.7	ELISA	cBio-X	–	972
	–	60	Dairy	4 farms	27	45.0	ELISA	IDEXX	–	650
Russia	Moskow and Kalugy	391	NS	8	39	9.9	ELISA	MASTAZYME	Abortion in 2 farms	377
Senegal	Dakar	196	Dairy	4	35	17.9	cELISA	VMRD	Age, breed	1028
Serbia	Vojvodina	100	Dairy	NS	7	7.0	cELISA	VMRD	Farm size	1107
	Vojvodina	256	Dairy	NS	48	18.8	IFAT	1:200	Farm size	1107
	Vojvodina	52	Dairy	NS	9	17.3	IFAT	1:200	–	1783
	South Banat	500	Dairy	NS	23	4.6	cELISA	IDEXX	7 of 11 sera from aborted cow were positive	741
Slovakia	East	716	Dairy	NS	144	20.1	cELISA	Pourquier	Abortion, region. 2.3% in 247 non-aborting cattle	1700
	–	105 (abortion history)	Dairy	Not clear	23	22.9	ELISA	IDEXX	–	623

(Continued)

Table 4.1 (Continued) Serologic Prevalence of *N. caninum* Antibodies in Cattle

Country	Region	No. of Animals	Type	No. of Herds	No. Positives	Percent of Positive (%)	Test	Titer/Company/Type of Test/Antigen	Remarks	Reference
South Korea	9 provinces	793	Dairy	168	164	20.7	IFAT	1:200	–	956
	9 provinces	438	Beef	78	18	4.1	IFAT	1:200	–	1061
	–	895 (abortion)	Dairy	30	437	48.7	IFAT	1:200	–	956
	–	492	NS	NS	113	23.0	ELISA	IgG-IH	*T. gondii* in 6 (1.2%)	108
	9 provinces	852	Stock farms	NS	103	12.1	ELISA	IH-Ncp43P	Sera negative for *T. gondii*	21
Spain	Galicia	2407	Beef	372	386	15.8	ELISA	CIVTEST	Age, production system	144
	Galicia	20,206	Beef	1,464	5173	25.6	ELISA	IDEXX	Age	634
	Galicia	2292	Mixed	141	582	25.4	ELISA	IDEXX	Age	634
	Galicia	37,090	Dairy	1147	8345	22.5	ELISA	IDEXX	Age	634
	Galicia	2292	Mixed	141	582	25.4	ELISA	IDEXX	–	634
	–	889	Dairy	43	272	30.6	ELISA	WT-IHCA	Abortion history	1260
	–	1121	Dairy	143	402	35.9	ELISA	WT-IH	Production system	1656
	Galicia	178	Beef	NS	43	24.1	cELISA	VMRD	7.3% seroprevalence for *T. gondii*	1550
	–	13	Breeder bulls	Artificial insemination center	8	61.5	IFAT	1:250	6 out of 8 bulls were positive by PCR in semen	671
	–	237 (abortion history)	Dairy	1	84	35.4	ELISA	IDEXX	Reproductive failure	1215
	–	285	Breeder bulls	Artificial insemination center	32	11.2	IFAT	1:50	–	269
					32	11.2	ELISA	CIVTEST		
					38	13.3	ELISA	IDEXX		
	–	1712	Beef	216	306	17.9	ELISA	WT-IH	Production system, herd size	1656
	–	3360	Dairy	291	554	16.2	ELISA	WT-IH	Area, production system	144

(Continued)

Table 4.1 (*Continued*) Serologic Prevalence of *N. caninum* Antibodies in Cattle

Country	Region	No. of Animals	Type	No. of Herds	No. Positives	Percent of Positive (%)	Test	Titer/ Company/ Type of Test/ Antigen	Remarks	Reference
—	—	1970 (abortion history)	Dairy	3	230	15.1	ELISA	CIVTEST, IDEXX	Abortion	**1214**
	—	1331	Dairy	2	236	17.7	ELISA	CIVTEST	Reproductive failure	**1216**
	—	1331	Dairy	2	357	26.8	ELISA	CIVTEST	Insemination and dry conditions reduced the risk	**1218**
	—	5196	Dairy	276	816	15.7	IFAT	1:50	Reproductive failure	**791**
Sudan	Khartoum, Gazira	276	NS	NS	44	15.9	cELISA	VMRD	Crossbred cows	**961**
		262	NS	NS	28	10.7	cELISA	VMRD	Abortion	**966**
Sweden	—	70 (abortion)	NS	1	44	63.0	ELISA	IH-ISCOM	—	**1908**
	—	4665	Dairy (bulk milk)	124	28 (herds)	22.6 (herds)	ELISA	IH-ISCOM	—	**706**
	—	2754	Beef	2130	77	2.8	ELISA	IH-ISCOM	Area	**1210**
	—	4252	Dairy	112	57	1.3	ELISA	IH-ISCOM	—	**144**
	—	780	Dairy	NS	16	2.0	ELISA	IH-ISCOM	BVDV	**220**
	NS	52–84	Dairy	1	37–73	63.0–87.0	ELISA	IH-ISCOM	Longitudinal study between 1994 and 1999	**1909**
Switzer-land	—	—	Dairy (bulk milk)	2978	247	8.3	ELISA	IH-ISCOM	—	**707**
Taiwan	—	1689	Dairy	113	194	11.5	IFAT	1:40	—	**797, 798**
Thailand	—	613	Dairy	25	275	44.9	IFAT	1:200	—	**1518**
	Khon Kean	424	Dairy	11	32	7.5	ELISA	IH-ISCOM	Herds	**326**
	Northeast	83	Dairy	16		37.5–70	IFAT	1:100	—	**1038**
	Northern	642	Dairy	42	301	46.9	ELISA	IH-NcSAG1	—	**984**
	Northern	642	Dairy	42	220	34.3	IFAT	1:200	—	**984**
	11 provinces	904	Dairy	NS	54	6.0	IFAT	1:200	Abortion	**1922**

(Continued)

Table 4.1 (Continued) Serologic Prevalence of N. caninum Antibodies in Cattle

Country	Region	No. of Animals	Type	No. of Herds	No. Positives	Percent of Positive (%)	Test	Titer/Company/Type of Test/Antigen	Remarks	Reference
		549	Dairy	59	30	5.5	cELISA	VMRD	Herd size	1109
Turkey	Anatolia	164	Dairy	11	24	15.0	ELISA	IH-ISCOM	Area	322
		3287	Dairy	32	459	13.9	ELISA	IDEXX	Area, abortion	2055
	Ankara	60 (12 aborted, 48 not)	NS	NS	6	10.0	Immuno-comb	NS	—	1106
	Elazig	89 (repeat breeder)	Dairy	NS	12	13.4	cELISA	VMRD	Breeder repeated	1862
	Elazig	94 healthy	Dairy	NS	3	3.1	cELISA	VMRD	—	1862
	Gebze	97	Dairy	3	5	5.1	cELISA	VMRD	Herd 1 5 of 27 positive, but all 70 from herd 2 negative	23
	Kars	73	Dairy	3	6	8.2	ELISA	MASTAZYME	Origin/breed	22
	Kars	228	NS (local breeds)	14	0	0.0	ELISA	MASTAZYME	Origin/breed	22
	Nidge	264 (207 male, 57 female)		Local breed	264	26.5	cELISA	VMRD	Males more often positive	1034a
	Nidge	264	Abattoir	Not clear	70	26.5	cELISA	VMRD	Gender, age	1035a
	Sakarya	92	Dairy	Not clear	10	9.2	cELISA	VMRD	—	1516
	Sanliurfa	305	Dairy	NS	23	7.5	cELISA	VMRD	—	1834
	Thrace	274	Dairy	6	22	8.0	ELISA	IDEXX	—	204
		186	NS	Not clear	13	7.0	cELISA	VMRD	Reproductive failure	969
	4 provinces	513	NS	Not clear	36	7.0	cELISA	VMRD	7 of 149 from Bingül, 18 of 120 from Elazig, 4 of 10 from Malatya, and 7 of 144 from Mas	24
		32 (aborting)	NS	Not clear	1	3.1	cELISA	VMRD	Not clear	24
		25 (aborted)	Dairy	1 ranch	15	60.0	cELISA	VMRD	—	1104
		40 (heifers, abortion)	Dairy	1 ranch	16	40.0	cELISA	VMRD	—	1104

(Continued)

Table 4.1 (Continued) Serologic Prevalence of *N. caninum* Antibodies in Cattle

Country	Region	No. of Animals	Type	No. of Herds	No. Positives	Percent of Positive (%)	Test	Titer/ Company/ Type of Test/ Antigen	Remarks	Reference
	—	6 (calves)	Dairy	1 ranch	2	33.3	cELISA	VMRD	—	1104
	—	234 aborted	Dairy	40	16	6.8	cELISA	VMRD	Pregnant cows, age	2155
	6 regions	377	Dairy and crossbred	NS	15	3.9	ELISA	MASTAZYME	—	2180
UK	Northern Ireland	73 (dis-eased)	NS	NS	4	5.5	IFAT	1:320	—	804
	Northern Ireland	165 (healthy)	Dairy	NS	5	3.0	IFAT	1:640	Reproductive failure	1334
	—	40 (abortion history)	Dairy	1	24	60.0	ELISA	MASTAZYME	—	422
	—	4,295	Dairy	14	734	17.1	ELISA	MASTAZYME	Age, herd	429
	—	460	Dairy	18	33	7.2	ELISA	MASTAZYME	Gestational loss	246
	—	15,736	Dairy	11	2,030	12.9	ELISA	IDEXX	—	2111
Uruguay	—	186	Dairy	1	114	61.3	IFAT	1:200	Abortion	1039
	—	4,444	Beef	229	589	13.2	ELISA	WT-IH	No difference in prevalence in cows and heifers	115
USA	California	176	Dairy	1	60	34.0	IFAT	1:640	—	1553
	California	277	Dairy	1	119	43.0	IFAT	1:640	—	1553
	California	405	Dairy	2	153	37.7	ELISA	WT-IHCA	Infection of calf	1557
	California	254	Dairy	1	154	60.6	ELISA	WT-IHCA	Abortion	1558
	Georgia	327	Dairy (milk)	3	105	32.1	IB		Area	1523
	Illinois	60	Beef	1	49	81.6	ELISA	IH-ISCOM	Abortion episode	1315
	Idaho, Montana, Oregon, Washington, Wyoming	2,585	Beef	55	594	23	cELISA	VMRD	Density of cows in herd during winter	1772
	Maryland	1,029	Dairy	1	288	28	IFAT	1:200	Age	630
	Nebraska	208 (abortion history)	Beef	1	164	81.7	ELISA	IH-ISCOM	—	1314

(Continued)

Table 4.1 (Continued) Serologic Prevalence of *N. caninum* Antibodies in Cattle

Country	Region	No. of Animals	Type	No. of Herds	No. Positives	Percent of Positive (%)	Test	Titer/ Company/ Type of Test/ Antigen	Remarks	Reference
North Dakota		212	Beef	7	11	5.2	ELISA	IDEXX	—	1050
Oklahoma		1,000 (no history of abortions)	Dairy	16	147	14.7	ELISA	IDEXX	—	1134
Texas		87	Dairy (milk)	2	9	10.3	IB	—	Area	1523
Texas		1,009	Beef	92	131	12.9	NAT	1:80	Daily gain of weight, live body weight, hot carcass weight, income	126
	—	4,907	Dairy and beef	98	788	16	ELISA	IDEXX	—	1725
	—	900	Dairy	2 pens	150	16.7	Kinetic ELISA	WT-IHCA	Area (pens)	916
Venezuela	—	459	Dual purpose	15 farms	52	11.3	ELISA	IDEXX	Reproductive failure	1191
Vietnam	—	200	Dairy	>30	11	5.5	ELISA	IH-ISCOM	—	955
	—	254	Dairy (milk)	199	76.2	30.0	ELISA	SVANOVA	—	751
	—	215	Dairy	5 state farms, 97 smallhold- ers herds	88	41.0	ELISA	IH-ISCOM	Herds with cow importation, BVDV	629

a Summarized local surveys.

Abbreviation: IFAT = Indirect fluorescent antibody test; ELISA = Enzyme-linked immunosorbent assay; WT = Whole tachyzoite extract; IH = In house; WT-IHCA = Kinetic ELISA-California[1555]; BIOVET = BIOVET-*Neospora caninum*, indirect ELISA, sonicate lysate of tachyzoites, BIOVET Laboratories, Canada; cBio-X = BIO K 218 Monoscreen AbELISA Neospora caninum (rNcSRS2)/blocking, BIO-X Diagnostics, Belgium; iBio-X = BIO K 192 NEOSPORA CANINUM ELISA antibody KIT, BIO-X Diagnostics, Belgium; CHEKIT = CHEKIT *Neospora*, indirect ELISA, detergent lysate of tachyzoites, IDEXX Laboratories, The Netherlands; IDEXX = IDEXX HerdChek *Neospora caninum* antibody, indirect ELISA, sonicate lysate of tachyzoites, IDEXX Laboratories, USA; IDvet = Neospora caninum Indirect Multispecies, ID.vet, France; Ingezim = indirect ELISA, INGEZIM NEOSPORA, Ingenasa, Spain; MASTAZYME = MASTAZYME NEOSPORA, indirect ELISA, formaldehyde-fixed whole tachyzoites, MAST GROUP, UK; VMRD = *Neospora caninum* cELISA Competitive ELISA GP65 surface antigen of tachyzoites VMRD, USA; Pourquier = Indirect ELISA, Institut Pourquier, France; CIVTEST = CIVTEST BOVIS NEOSPORA, indirect ELISA, sonicate lysate of tachyzoites, Laboratorios Hipra S.A., Spain; SVANOVA = SVANOVIR Neospora-Ab, Detergent extracted tachyzoite antigen incorporated in ISCOMs; ISCOM = Detergent extracted tachyzoite antigen incorporated in immune stimulating complex particles; Immunocomb = Immunocomb bovine Neospora antibody test kit, Biogal, Israel; IH-p38 = Native immune-affinity-purified surface antigen NcSRS2; IH-Ncp43P = Recombinant NcSRS2; NhSAG1 = Recombinant NhSAG1; NAT = *Neospora* agglutination test; IB = Immunoblotting; NS = not stated.

Table 4.2 Reports of *N. caninum*-Associated Abortion in Cattle

Country	Area	No. Aborted	No. Positive (%)	Diagnosis				Remarks[a]	Reference
				Histo	IHC	PCR	Fetal Antibodies		
Algeria	–	5	3 (60.0)	Yes	ND	Yes	ND	–	757
Andorra	–	1	1 (100.0)	Yes	Yes	Yes	ND	7 months gestational age fetus from a beef cow	83
Argentina	Buenos Aires	2	2 (100.0)	Yes	Yes	ND	ND	18 of 330 cows aborted in 4 weeks. Fetuses were 5 and 6 months gestational age	274
	Buenos Aires	104 (slaughterhouse)	7 (6.7)	Yes	No	ND	IFAT, 1:80	Lesions in 7 of 8 fetuses. *N. caninum* antibodies in 21 (20 dairy, 1beef) fetuses. The fetuses were 5–8 months gestational age and were from slaughterhouse	2040
	Buenos Aires	188	29 (15.4)	Yes	Yes	ND	IFAT, 1:25	Lesions in 43; 29 of these 43 were IHC positive. 30 fetuses were seropositive. Of 29 IHC positive fetuses, *N. caninum* was found in heart of 24, CNS of 21, liver of 18, lung of 16, striated muscle of 17, kidneys of 17, periorbital muscle of 9, adrenal glands of 10, intestines of 8, spleen of 12, and placentas of 5. Of the 8 mummified fetuses, 3 were positive by IHC and 2 were seropositive	1380
	Buenos Aires	354	26 (7.3%)	Yes	Yes	ND	–	42 fetuses and 1 premature calf were examined by IHC; *N. caninum* was found in 26. Fetal gestational ages were 4–9 months, average 6.5 months. *N. caninum* found in 3 mummified fetuses	275
	Buenos Aires	2	2 (100.0)	Yes	Yes	ND	ND	69 abortions in a beef herd of 857 heifers. IHC confirmed diagnosis in 1 aborted fetus and 1 premature calf	1381
	Buenos Aires	666	70 (10.5)	Yes	Yes	Yes	IFAT, 1:25	In 49 fetuses *N. caninum* was found by IHC, 31 of 55 had *N. caninum* antibodies, and 34 of 80 were positive by PCR	1389
Argentina	Jujuy	2	2 (100.0)	Yes	ND	ND	ND	Fetuses were 5 and 8 month gestational age	1277
Australia	New South Wales	729	152 (21.0%)	Yes	Yes	ND	ND	Encephalitis in all 152 fetuses and myocarditis in 138 of 144 (96%) of fetuses examined	235
	Tasmania	11	11 (100.0)	Yes	Yes	ND	ND	27 of 118 cows aborted in few weeks. Lesions in all 11, 3 were IHC confirmed. Viable *Neospora* isolated from 1 fetus by bioassay in immunosuppressed mice. One cow aborted for 3 in 3 pregnancies with IHC confirmed diagnosis in alternate years	1497
	New South Wales	12	8 (66.6)	Yes	Yes	ND	ND	*Neospora*-like lesions in 8 fetuses	100

(Continued)

Table 4.2 (Continued) Reports of *N. caninum*-Associated Abortion in Cattle

Country	Area	No. Aborted	No. Positive (%)	Histo	IHC	PCR	Fetal Antibodies	Remarks[a]	Reference
Austria	–	1	1 (100.0)	No	ND	Yes	ND	7 month gestational age fetus, autolyzed	633
Belgium	–	224	17 (7.6)	Yes	Yes	–	IFAT, 1:25	6 confirmed by IHC. 10 fetuses had *N. caninum* antibodies	450
Brazil	Bahia	1	1 (100.0)	Yes	Yes	ND	ND	8 month gestational age fetus	775
	Goiás	195 dead fetuses from abattoir	40 (920.5)	No	No	Yes	No	31 brains, 3 hearts, 2 livers positive. Tissues not simultaneously positive	247
	Paraná	34	8 (23.5)	Yes	Yes	ND	ND	6 were IHC positive. **Fetuses were 4–7 months gestational age**	513
	Rio Grande do Sul	30	1 (3.3)	Yes	Yes	ND	ND	Lesions in 6 fetuses, 1 positive by IHC	381
	Rio Grande do Sul	46	22 (47.8)	Yes	Yes	ND	–	Samples from 12 dairy herds. **Fetal gestational ages were 3–7 months, most were 5 months**	383
	Rio Grande do Sul	161	37 (17.4)	Yes	Yes	ND	–	Samples from 149 farms. 34 fetuses were IHC positive. 51.5%. **Most aborted fetuses were of 4.4 months gestational age.** Cows aborting a *Neospora* infected fetuses were 2.4 times likely to have aborted previously	384
	Rio Grande do Sul, Santa Catarina	258	55 (21.3)	Yes	Yes	ND	ND	**Macroscopic (0.5–1.0 mm) foci in lungs of 2. Microscopic lesions in 89, 55 were positive by IHC. Lesions in skeletal muscle of 51, heart of 42, brain of 41, and lungs of 41. Fetal gestational age of 3–8 months**	1595
	São Paulo	105	26 (24.7)	Yes	Yes	Yes	ND	Lesions in the brains of 38; 4 of these were IHC positive. Lesions were seen in hearts of 53 with 2 IHC confirmed, in livers of 58 with 1 IHC positive, in placentas of 11 with 1 IHC positive, and in kidneys of 55 with 1 IHC positive. *N. caninum* DNA was detected in the brain of 16, placentas of 4, heart and liver of 8, and in pooled organs of 7 fetuses	266
Canada	Atlantic	10	5 (20.0)	Yes	Yes	ND	–	Out of 265 cases of abortions examined in a diagnostic center	1053
	British Columbia	1	1 (100.0)	Yes	Yes	ND	–	8-month-old gestational age fetus	1331
		1	1	Yes	ND	ND	ND	10 of 15 cows that aborted had IFAT titers of 1:640, 3 cows that did not abort were seronegative	915a
	Ontario	4	4 (100.0)	Yes	Yes	ND	ND	**15 of 80 cows from 1 herd aborted in 18 days. All fetuses had characteristic lesions in their brains and diagnosis was confirmed by IHC in 1**	626

(Continued)

Table 4.2 (Continued) Reports of *N. caninum*-Associated Abortion in Cattle

Country	Area	No. Aborted	No. Positive (%)	Diagnosis				Remarks[a]	Reference
				Histo	IHC	PCR	Fetal Antibodies		
	Prince Edward Island	1	1 (100.0)	Yes	Yes	ND	ND	5 of 70 cows from 1 herd aborted fetuses at 3–8 gestational age. Tachyzoites seen in encephalitic lesions by IHC	202
	Quebec	26	3 (11.5)	ND	ND	Yes	ND	—	717
	—	15	4 (26.6)	Yes	Yes	ND	ND	62 of 350 beef cows aborted	2058
	—	1019	38 (3.7)	Yes	ND	ND	ND	8 cases (1.6%) diagnosed in 1993–1994 and 30 (5.7%) diagnosed in 1994–1995	57
China	Beijing	4	4 (100.0)	Yes	Yes	Yes	ND	12 aborted fetuses from seropositive cows were tested. Tissue cysts were found in brain of 1 of the 2 fetuses examined by IHC. *N. caninum* DNA was found in brains of 4 of 12 fetuses	2176
	Beijing and Tianjin	26	13 (50.0)	ND	ND	Yes	ND	15 fetuses were PCR positive; in brains of 13 of 25, hearts of 1 of 24, kidneys of 5 of 23, and gluteus muscle of 4 of 22 fetuses	2151
Costa Rica	—	22	2	Yes	Yes	ND	Yes	2 of 6 tested by IHC. *N. caninum* confirmed in lesions in heart, liver, lungs, CNS in 9%–63% of fetuses	1591
Ethiopia	—	4	4 (100.0)	ND	ND	Yes	ND	Late term abortions	96
Denmark	—	2	2 (100.0)	Yes	Yes	ND	ND	Two fetuses from different herds. A 7-month gestational age fetus from a herd with 6 abortions in few weeks. The 5-month gestational age fetus from a herd with 40 abortions in 2 years	10
	—	218	18 (8.2)	Yes	Yes	ND	ND	Causes identified in 65 of 218 submissions from 18 of 186 case herds and 0 of 32 control herds. Complete specimens from 105 cases	11
Finland	—	1	1	ND	ND	Yes	ND	5.5 month gestational age fetus	1825a
Germany	North Rhine Westphalia	4	4 (100.0)	Yes	Yes	Yes	ND	Lesions in brains of 2 fetuses and in liver of 2 fetuses. 3 fetuses were positive by PCR	1788
	North Rhine Westphalia	135	17 (12.6) by IHC	Yes	Yes	ND	(Yes)	Up to 4 locations in the brain (cerebrum, cerebellum, mesencephalon, medulla oblongata) were examined histologically. 12.6% (17/135) of the fetuses showed lesions characteristic for neosporosis or were positive by IHC	1880
	North Rhine Westphalia	139	30 (21.6) by PCR	ND	ND	Yes	(Yes)		

(Continued)

Table 4.2 (Continued) Reports of *N. caninum*-Associated Abortion in Cattle

Country	Area	No. Aborted	No. Positive (%)	Diagnosis Histo	IHC	PCR	Fetal Antibodies	Remarks[a]	Reference
–	–	20	18 (90.0)	ND	ND	Yes	ND	**Point source abortions from 5 herds (45 fetuses aborted) as indicated by low avidity antibodies in dams and microsatellite DNA characteristics. N. caninum DNA found in 18 of 20 fetal brains**	169
	Bavaria	232	23 (10.0)	Yes	Yes	Yes	ND	**IHC in 16, PCR in 16–19, fetal antibody in 14**	1881
Hungary	–	1	1 (100.0)	Yes	ND	Yes	Yes	6-month gestational age fetus	107
Iran	Marshad	100	12 (12.0)	Yes	Yes	ND	ND	Lesions in brains of 13, 3 brains positive by IHC. A tissue cyst-like structure was seen in brain homogenate of 1 fetus	1670
	Marshad	12	3 (25.0)	Yes	Yes	Yes	ND	Lesions in 3 fetuses, tachyzoites in 2 fetuses, 4 fetuses PCR positive	1755
	Marshad	151	22 (14.5)	No	Yes	Yes	ELISA (IDEXX)	6 of 52 brains were positive by IHC, 18 positive by PCR, 15 of 151 were positive by fetal serology	1671
	Tehran	12	12 (100.0)	Yes	ND	Yes	ND	Brains of all 12 fetuses were positive by PCR. An N. caninum-like tissue cyst in 1 brain of 1 fetus	1764
Ireland	–	5	5 (100.0)	Yes	Yes	ND	ND	6 cows of a 70 cow herd aborted in 5 days	371
	–	5	4 (80.0)	Yes	Yes	ND	–	–	1333
Israel	–	20	3 (15.0)	Yes	Yes	ND	ND	–	853
Japan	–	115	3 (2.6)	Yes	Yes	ND	ND	**Three aborted infected fetuses were of 5 months gestational age. 1 of 48 fetuses from slaughterhouse was IHC positive. 1 calf that died on the day of birth was also IHC positive**	1503
	–	15	4 (26.6)	No	No	Yes	–	15 mummified fetuses were tested. Four 100, 113, 123, and 131 days gestational age fetuses were positive. Tests for complex vertebral malformation (CVM) were negative	760

(Continued)

Table 4.2 (Continued) Reports of *N. caninum*-Associated Abortion in Cattle

Country	Area	No. Aborted	No. Positive (%)	Diagnosis				Remarks[a]	Reference
				Histo	IHC	PCR	Fetal Antibodies		
South Korea	–	1	1 (100.0)	Yes	Yes	ND	ND	**Macroscopic 2–3 mm yellowish foci in heart and skeletal muscle**	1057
	–	180	45 (25.0)	Yes	Yes	Yes	–	Lesions were seen in 45 fetuses; 34 of these were positive by PCR. Of the 11 PCR negative fetuses *N. caninum* antibodies were found in 4	1062
	–	2	2 (100.0)	Yes	Yes	–	–	**Confirmed repeat abortion in 2 consecutive years from seropositive cow**	1058
Mexico	–	6	6 (100.0)	Yes	Yes	ND	ND	Misdiagnosed as *Hammondia pardalis*	1, 554
	Aguascalientes	44	35 (79.5)	Yes	No	Yes	ND	**Lesions in 20 fetuses. 35 fetuses were PCR positive. Gestational age 3–9 months, average 5.6 months**	1335
	Aguascalientes	63	27 (42.8)	ND	ND	Yes	ND	Fetuses from 11 herds. DNA detected in 27. Microsatellite typing from 11 samples	1336a
	–	1	1 (100.0)	Yes	Yes	ND	ND	Fetus 5 month gestational age	1395
	–	211	41 (19.4)	Yes	Yes	ND	ND	73 had lesions, 41 were positive by IHC. Encephalitis in 39, myocarditis in 58, hepatitis in 39, and myositis in 19 fetuses	1397
	–	48	29 (60.4)	Yes	Yes	Yes	ND	**Lesions in 29, 21 fetuses positive by IHC, and 15 brains were PCR positive**	1771
The Netherland	–	3	3 (100.0)	Yes	Yes	ND	ND	Initial report of neosporosis abortion from the Netherlands	2115
	–	2053	349 (17.0)	Yes	Yes	ND	Yes	**Comprehensive descriptions of lesions in epidemic and sporadic abortions**	2114, 2118, 2120, 2121, 2122
New Zealand	–	320	88 (28.0)	Yes	Yes	ND	ND	Lesions in brains of 88 fetuses. 4 fetuses were IHC positive	1959
	–	2	2 (100.0)	Yes	–	–	–	–	393
	–	18	7 (38.8)	Yes	?	ND	ND	Retrospective analysis of diagnostic data. Abortive material submitted from 230 cases. Retrospective data analyzed	1955
	–	4	2 (50.0)	Yes	ND	ND	–	41 of 158 cows from 1 herd aborted	1960

(Continued)

Table 4.2 (Continued) Reports of *N. caninum*-Associated Abortion in Cattle

Country	Area	No. Aborted	No. Positive (%)	Diagnosis — Histo	IHC	PCR	Fetal Antibodies	Remarks[a]	Reference
—	—	34	9 (26.5)	Yes	?	ND	ND	**Comparison of abortions in cows vaccinated with killed Neospora vaccine and unvaccinated cows; 3 abortions in vaccinated and 6 abortions in non-vaccinated cows**	2088
Portugal	—	42	15 (35.7)	No	ND	ND	ND	**Viable *N. caninum* isolated by bioassay in mice from 15 of 42 fetuses**	282
	—	15	2 (13.3)	Yes	Yes	ND	—	Lesions in heart and liver of 5 fetuses. 2 cases IHC confirmed	1957
Romania	—	9	3 (33.3)	No	ND	Yes	—	**All fetuses from 1 herd. Infected fetuses were 3, 4, and 7 (mummified) month gestational age.** Lesions could not be detected in histologic sections of any fetus because of autolysis	1924
Slovakia	—	4	4 (100.0)	No	ND	Yes	No	—	1892
South Africa	—	144	2 (1.4)	Yes	Yes	ND	ND	**Infection in twin fetuses, 7 months gestational age**	995, 996
Spain	Asturias, Galicia, Castilla, León, Madrid	80	25 (31.2)	—	—	—	ELISA, IFAT	25 had lesions, 7 were IHC positive, 6 were seropositive, 9 were PCR positive	1590
	Barcelona	1	1 (100.0)	Yes	Yes	ND	—	4–5 months gestational age fetus. Lesions in brain, heart, liver, and kidneys	695
	Barcelona	2	2 (100.0)	Yes	Yes	Yes	ND	—	1215
	—	220	72 (32.7)	?	?	Yes	—	**Fetus gestational age was 2.5–8 months.** *N. caninum* DNA was detected in 95.8% of brains, 49.3% of hearts, 22.5% of livers; in 3 cases brain was negative but heart and livers were positive. *N. caninum* DNA was also found in other organs including kidneys, lungs, diaphragm, spleen, thymus, lymph nodes, and adrenal glands. **In 12 epidemic cases, lesions were seen in the heart and liver of all 12 and brains of 7 of 8 cases examined; however, parasite DNA was detected in the brains of all 14 studied**	367, 368
	Northern	81	8 (9.8)	Yes	Yes	ND	IFAT, 16	Samples from 71 herds. Lesions in brains of 36 fetuses, 8 were IHC confirmed. Fetal *N. caninum* antibodies in 34	790

(Continued)

Table 4.2 (*Continued*)   Reports of *N. caninum*-Associated Abortion in Cattle

Country	Area	No. Aborted	No. Positive (%)	Histo	IHC	PCR	Fetal Antibodies	Remarks[a]	Reference
	Lérida	2	2 (100.0)	Yes	Yes	Yes	ND	One whole herd tested. Overall abortion 23.2% (38 aborted cows in the herd). 76.3% seropositive cows aborted	1215
	Lérida	10	10 (100.0)	Yes	Yes	Yes	ND	Lesions in 4, 6 fetuses in autolytic conditions were PCR positive	1220
	Lérida	6	6 (100.0)	Yes	Yes	Yes	ND	31 pregnant seropositive dams in a herd, 11 aborted. 6 submitted fetuses with 1 positive by PCR	40
	Lérida	28	28 (100.0)	Yes	Yes	Yes	ND	750 seropositive pregnant dams tested, 23.6% aborted, 28 fetuses submitted to the lab for analysis	2150
	Lérida	19	9 (47.3)	Yes	Yes	Yes	ND	Lesions in 9, 4 autolytic fetuses were PCR positive	726
Sweden	—	1	1 (100.0)	Yes	Yes	ND	ND	**Fetus was 4.5 months gestational age**	923
Switzerland	—	242	51 (21.1)	Yes	Yes	Yes	ND	**Lesions in brains of 47 fetuses; 42 of these were PCR positive. 8 of 91 fetuses without neural lesions were PCR positive. Fetuses were 5–7 months gestational age**	691, 1757
	—	223	36 (16.1)	Yes	Yes	Yes	—	**4 were confirmed positive by IHC, 21 were questionable positive by IHC. 36 fetuses were PCR positive. Fetal gestational ages were 4–8.5 months with median of 5.75 months**	1704
UK	—	8	8 (100.0)	ND	ND	Yes	—	**8 of 73 cows from 1 herd aborted within 2 weeks**	1759
	England	6	4 (66.6)	Yes	Yes	ND	—	**26 of 110 cows from 1 herd aborted in 1 year; 13 aborted within 19 days. Infection probably acquired through ingestion of oocysts**	395
	England	7	2 (28.6)	Yes	Yes	ND	ND	34 of 80 cows in 1 herd aborted within 1 year	624
	England and Wales	190	8 (4.2)	Yes	Yes	ND	—	**Lesions in 20 (10.5%), 8 were confirmed by IHC. Fetal gestation ages 4.5–25 months**	1535, 1536, 1537
	Scotland	324	3 (0.9)	Yes	Yes	Yes	ELISA, IFAT	Lesions in 13, 3 were positive by IHC. *N. caninum* antibodies were detected in 34 fetuses by both ELISA and IFAT	1812

(Continued)

Table 4.2 (*Continued*) Reports of *N. caninum*-Associated Abortion in Cattle

Country	Area	No. Aborted	No. Positive (%)	Histo	IHC	PCR	Fetal Antibodies	Remarks[a]	Reference
	Scotland	356	5 (1.4)	Yes	Yes	ND	–	**36 fetuses had antibodies in fetal fluids. Thus, congenital infection rate was 10.1%**	272, 273
USA	California	86 with lesions	66 (76.7)	Yes	Yes	ND	ND	Encephalitis and myocarditis were found in 100%, adrenalitis in 80%, myositis in 72%, nephritis in 66%, hepatitis in 62%, pneumonia in 44%, and placentitis in 53% of 82 fetuses. **Tissue cysts were more common in the brains of 3 months gestational ages fetuses and their number decreased with maturity of the fetus.** The number of fetuses with gestational ages were 3 (3 months), 6 (4 months), 16 (5 months), 30 (6 months), 13 (7 months), 4 (8 months), and 2 (9 months). **Thus, 46 of 74 (62%) of aborted fetuses were 5–6 months gestational age**	130, 131
	California	468	89 (19.0)	Yes	Yes	ND	ND	Causes identified in 213. **63% were 5–6 month gestational age**	62
	California	698	170 (24.3)	Yes	Yes	ND	ND	Samples from 311 dairies submitted to 1 diagnostic lab (Tulare) from 1985 to 1990. Causes identified in 323. **No change in abortion rate in 6 years. Aborted fetuses were 3–8 month gestational age; 78% were 4–6 month gestational age. Higher abortion rate November to February**	59, 1962
	California	266	113 (42.5)	Yes	Yes	ND	ND	Samples from 24 dairies in 1991. In addition to 113 confirmed neosporosis abortions, 17 additional cases had protozoal-like lesions. Of 112 cows with confirmed neosporosis abortion, **4 cows aborted twice with confirmed neosporosis. Mean age of abortion was 5.5 months**	64
	California	665	66 (9.3)	Yes	Yes	ND	Yes	Retrospective cases from 2007 to 2013. Lesions in brain, kidneys, and placenta. **Abortions both in dairy (35.2%) and beef (8.4%) cattle**	353

(*Continued*)

Table 4.2 (Continued) Reports of *N. caninum*-Associated Abortion in Cattle

Country	Area	No. Aborted	No. Positive (%)	Histo	IHC	PCR	Fetal Antibodies	Remarks[a]	Reference
	California	8	7 (87.5)	Yes	Yes	ND	ND	**66 of 360 (18%) cows from 1 herd aborted in 2 months. Mummified fetuses observed on day 44 of the outbreak considered from a point source. Encephalitis, myocarditis, myositis, and hepatitis seen. Lesions seen in all 8 fetuses tested**	1305
	Colorado	20	6 (30.0)	Yes	Yes	ND	ND	**126 of 450 dry lot cows from 1 herd aborted in 3 years. Tissues of 20 fetuses, including a mummified fetus were studied by IHC. Myocarditis was seen in all 20, encephalitis in 10, hepatitis in 7, myositis in 4, and placentitis in 4 fetuses. 6 fetuses were positive by IHC**	1727
	Illinois	3	1	Yes	ND	ND	ND	**Abortion storm** in a 60 cow dairy herd; one-third aborted in 11 weeks or gave birth to premature calves. Viable NC-Illinois strain isolated from a premature calf	1315
	Kentucky	85	11 (12.9)	Yes	Yes	Yes	ELISA	**Abortion storm**. Within 3 weeks 85 abnormal pregnancies and abortions among 154 cows. 11 fetuses tested histologically. **Gestational age of fetuses 7–8 months**	85
	Maryland	1	1 (100.0)	Yes	Yes	ND	ND	**Heart was grossly enlarged and had disseminated myocarditis with many tachyzoites**	546
	Nebraska	14	14 (100)	Yes	Yes	ND	ND	**Abortion storm. 4 fetuses *N. caninum* positive by IHC. Viable NC-Beef isolated from 1 calf**	1314
	New Mexico	9	7 (77.8)	Yes	Yes	ND	ND	**Lesions in several organs, including encephalitis and myocarditis in all 7. 29 of 240 cows from 1 herd aborted in 5 months. Gestational ages 5–9 months. *N. caninum* seen in sections of 2 fetal brains**	1954
	North Carolina	6	3 (50.0)	Yes	Yes	ND	–	28 of 240 cows from 1 herd aborted in 4 months	1004
	Oklahoma	6	6 (100.0)	Yes	Yes	ND	–	**Fetuses were 3–7 months gestational age. Lesions were seen in brains of all 6, heart of 5, liver and kidneys of 1. All 6 fetuses IHC confirmed**	874

(Continued)

Table 4.2 (Continued)   Reports of *N. caninum*-Associated Abortion in Cattle

Country	Area	No. Aborted	No. Positive (%)	Diagnosis				Remarks[a]	Reference
				Histo	IHC	PCR	Fetal Antibodies		
	Pennsylvania	688	34 (4.9)	Yes	Yes	ND	ND	Samples from 13 counties. Of the 34 IHC confirmed **fetuses, lesions were seen in the brains of 27, hearts of 17, placentas of 8, livers of 5, kidneys and skeletal muscles of 2 each. Fetuses 3–8 months gestational age; maximum 5 months gestational age**	862
	Pennsylvania	144	12 (8.3)	Yes	Yes	ND	—	Retrospective study of 144 previously undiagnosed cases. **Tachyzoites were seen in tissues of 5 cases without histologic lesions. *N. caninum* lesions and tachyzoites in the brains of 2, liver of 3, kidneys of 2, heart of 1, placenta of 3, and placenta and lungs of 1 fetus**	230
	Texas	1	1 (100.0)	Yes	Yes	ND	ND	Hydrocephalus in a 7-month gestational age fetus. Only known case with numerous tachyzoites in a bovine case	571
	Washington	1	1 (100.0)	Yes	No	ND	—	5-month gestational age fetus. Only placenta and kidneys were examined. Tachyzoites in placenta	1849
	Arizona, Iowa, Massachusetts, Minnesota, Missouri, Nebraska, South Dakota, Texas, Virginia, Vermont, Wisconsin	2552	68 (2.7)	Yes	Yes	ND	ND	**All 68 abortions were confirmed by IHC. Seasonal trend-peak in April, September, and November. *N. caninum* as the primary cause of abortions. In 1 point source abortion, 11 of 90 cows in a herd in South Dakota aborted in 2 weeks. Fetuses were 4–7 months gestational age; *N. caninum* was found by IHC in 7 of 7 fetuses**	2136
	Arizona, Iowa, Minnesota, South Dakota, Wisconsin, Texas	655	19 (2.9)	Yes	Yes	ND	ND	***N. caninum* in brains of 15, hearts of 15, livers of 15, kidneys of 8, and several other organs**	1462
Zimbabwe	—	6	4 (66.7)	Yes	Yes	ND	ND	—	997

Note: ND = not done or no data.
[a] Data in bold are salient features.

Table 4.3 Reports of Confirmed Clinical Neosporosis in Calves

Country	Cases	Age	Clinical Signs	Histology	IHC	Remarks[a]	Reference
Argentina	2	2 days, 10 days	First case born premature, 8 months gestation, weak, recumbent. Second case ataxia from birth	Yes	Yes	**Two of 324 calves necropsied had neosporosis.** PCR positive	**1349**
	1	20 days	Neuromuscular signs, hyperflexion of hind limb	Yes	No	Nonsuppurative necrotizing encephalitis, gliosis, cuffing, myocarditis, and tissue cysts in cerebellum	**1277**
Australia	1	Born dead, full term	Arthrogryposis	Yes	ND	The caudal cervical and cranial thoracic **segments of the spinal cord asymmetrical with marked unilateral reduction of ventral gray matter and focal cavitation** (Figure 4.2c). Nonsuppurative myelitis	**542, 854**
Canada-Alberta	1	3 days	Hyperextension of limbs, moribund. Calf euthanized	–	–	**Lateral deviation of the vertebral column** between eighth thoracic and third lumbar vertebrae. Spinal cord reduced in diameter. Lesions confined to CNS. Marked meningoencephalomyelitis with tachyzoites and tissue cysts. **Viable N. caninum was isolated in mice inoculated with homogenate of brain stored for 4 months at –52°C**	**251**
Ethiopia	1	2 months	Not described	ND	ND	Pelvic limb arthorgryposis. Brain PCR positive	**96**
Germany	2	2 days	Not reported	Yes	Yes	Few glial scars in metencephalon of first calf. Myositis in left gastrocnemius muscle. Only myositis in tongue, and no other lesions of the second calf. Both calves PCR positive	**1599**
Turkey	1	20 days	Neurological, hyperextension of all limbs	Yes	Yes	Encephalitis and myocarditis	**1104**
UK	1	5 days	Weak, unable to stand, exophthalmia. Euthanized	Yes	Yes	Macroscopic multifocal brown discoloration in cervical spinal cord due to necrotizing myelitis	**530, 1495**
	1	1	Neurological, born 9 days prematurely	Yes	Yes	Unable to stand, hyperextension forelimbs. Internal hydrocephalus. Gliosis and necrosis in brain. Tissue cysts in brain. Spinal cord not examined	**820**
USA	4	1–3 days	Recumbent	Yes	Yes	Microscopic lesions confined to spinal cord. Necrosis, nonsuppurative meningitis. Tachyzoites and tissue cysts found	**535, 1560**

(Continued)

Table 4.3 (Continued)   Reports of Confirmed Clinical Neosporosis in Calves

Country	Cases	Age	Clinical Signs	Histology	IHC	Remarks[a]	Reference
USA	1	3 days	Stunted, recumbent, contracted forelegs tendons, mildly domed skull. Euthanized. Complete necropsy performed	Yes	Yes	Gliosis, perivaculitis, tachyzoites, and tissue cysts. **Slight lateral deviation to the cervical vertebral column.** Tissue cysts up to 49 μm wide and the cyst wall up to 3 μm thick. Gastrointestinal erosions; etiology unknown. *Neospora* confined to brain and spinal cord	132
	2	Calf 1–2 days, calf 2–6 days	Ataxia and neurological deficits in all 4 limbs in calf 1, and pelvic limbs in calf 2. Both calves euthanized and complete necropsy performed	Yes	Yes	**Dams of these cows had aborted fetuses with histologically confirmed neosporosis.** Lesions confined to brain and spinal cord. Encephalomyelitis in both calves with tissue cysts. Tachyzoites in 1 necrotic focus in the spinal cord of calf 2. Both calves had precolostral *N. caninum* antibodies	134
	1	4 weeks	Normal at birth, became ill 2 weeks later. Weak, unable to get up. Euthanized. Brain skeletal muscle, heart, lungs, and kidneys fixed	Yes	Yes	Main lesions in brain and muscle associated with tachyzoites, tissue cysts not found. Areas of necrosis (up to 6 mm) and cavitation with neovascularization. Severe myositis. Interstitial nephritis and pneumonia but *N. caninum* not identified	550
	1	2 weeks	Born normal, except the right forelimb permanently extended. The shoulder, elbow, and carpus could not be flexed	Yes	Yes	No gross lesions in the brain. Asymmetrical tan discolorations were found in transverse sections of formalin preserved spinal cord, most evident in the white matter of the ventral funiculi and ventral part of the lateral funiculi (Figure 4.2b). **Only tachyzoites and no tissue cysts found in lesions of severe meningoencephalitis**	557
	7	Neonatal	4 calves recumbent	Yes	Yes	Out of 25 seropositive cows 7 calves were congenitally infected. All 7 calves had histological evidence of encephalitis. Tissue cysts found in 5 calves	65

Note: ND = no data.

[a] Salient features in bold.

Table 4.4 Seropositivity as Evidence of Abortion in Cattle

| Country | Total Number of Cattle/Farms Examined | Test, Cut-Off | Aborting | | Non-Aborting or Control | | Risk of Abortion Indicated by Odd Ratio (OR), Significant Association (SA), Remarks | Reference |
			No. Tested	No. Seropositive (%)	No. Tested	No. Seropositive (%)		
Algeria	723 dairy cattle, 87 farms	IFAT, 200	54	26 (48.1)	669	(17.0)	OR 4.5	758
Argentina	Beef herd A: 57	IFAT, 200	11 (heifers)	7 (63.6)	46	12 (26.1)	SA in epidemic abortion (herd A, OR 4.96)	271
	Beef herd B: 44	IFAT, 200	14	2 (14.3)	30	3 (10.0)	No SA in endemic abortion (herd B OR 0.69)	271
	290 beef cattle, 1 herd, 58 aborted	IFAT, 200	58	34 (58.6)	214	22 (10.3)	OR 12.0	1381
	5594 dairy and beef heifers, 1404 aborted	IFAT, 200	1404	362 (25.7)	4190	598 (14.2)	OR 2.4	1390
Australia	266 dairy cattle, 1 farm, 21 aborted	IFAT, 160	21	18 (85.7)	164	50 (30.4)	OR 8.0	100
	140 dairy cattle, 1 farm, 16 aborted	cELISA-Pourquier	16	5 (31.3)	124	3 (2.4)	OR 13.0	833
	183 dairy cattle, 1 herd, 39 aborted	ELISA-IDEXX	39	11 (28.2)	183	20 (10.9)	OR 13.0	1655
	711 (3 dairy farms); abortion rate: 12.0%–20.0% in positive, 3.6%–7.0% in negative animals	ELISA-IDEXX	—	NS	—	NS	OR 3.5	1116
Belgium	711 dairy and beef cattle	IFAT, 200	370	67 (18.0)	341	20 (6.0)	OR 3.0	448
Brazil	1256, 41 aborted	IFAT, 200	41	24 (58.5)	1215	199 (16.4)	OR 7.2	872
	621 cattle, 63 farms, 36 farms with abortion	IFAT, 100	—	26/36 farms (72.2)	—	12/27 farms (44.4)	SA	228
	1204 dairy cows from 40 farms	IFAT, 200	NS	NS (31.1)	NS	NS (17.7)	OR 1.98	249
	3428 cattle from 174 herds	IFAT, 100	—	99/108 herds (91.7)	—	11/52 herds (21.1)	SA	339
	223 dairy cows, 5 herds	IFAT, 200	NS	NS (23.3)	NS	NS (8.3)	OR 3.3	383
	2448 cattle from 205 herds, beef, dairy	IFAT, 50	—	55/68 herds (80.9)	—	84/134 herds (62.7)	OR 2.5	1530
	1273 cattle from 6 dairy herds	IFAT, 200	305	122 (40.0)	968	40 (4.1)	SA	1596

(Continued)

Table 4.4 (Continued) Seropositivity as Evidence of Abortion in Cattle

Country	Total Number of Cattle/Farms Examined	Test, Cut-Off	Aborting		Non-Aborting or Control			Reference
			No. Tested	No. Seropositive (%)	No. Tested	No. Seropositive (%)	Risk of Abortion Indicated by Odd Ratio (OR), Significant Association (SA), Remarks	
Canada	347 beef cattle, 1 herd, 282 (43.3%) of seropositive and 7 (10.8%) of seronegative cows were nonpregnant or aborted	cELISA-VMRD	122	–	–	–	OR 6.2	2058
China	245	IH-ELISA	43	35 (81.4)	202	114 (56.4)	Not clear	845
Czech Republic	407 dairy cattle, 137 farms, 44 aborted	IFAT, 200	44	6 (13.6)	363	7 (1.9)	OR 8.0	2016
Ethiopia	2334 dairy and breeding cattle, 273 farms, 350 aborted	ELISA-IDEXX	350	83 (23.7)	1573	214 (13.6)	OR 2.3	93
Germany	222 dairy cattle at risk from 5 herds with epidemic abortion, 46 aborted	IH-p38 ELISA	46	37 (80.4)	176	31	SA	1800
	336 dairy cattle at risk from 5 herds with endemic abortion, 33 aborted	IH-p38 ELISA	33	24 (72.7)	303	68	SA	1800
	396 cattle at risk from herds with epidemic and endemic abortion	IFAT, 50 and IB	63	51 (80.9)	333	81	Higher antibody levels in aborting vs. non-aborting	1794
Iran	492 dairy cattle, 41 herds, 35 with abortion	ELISA-IDEXX	35	20 (57.1)	457	43 (9.4)	SA	764
	337 dairy cattle, 30 herds, 158 with abortion	ELISA-IDEXX	158	85 (53.8)	179	71 (39.6)	OR 1.8	1669
Israel	1078 pregnancies, 1 dairy herd, 240 dams examined	IFAT, 200	63	34 (53.7)	177	35 (19.8)	SA	1301
Italy	820 dairy cattle, 85 foci of abortion	IFAT, 640	340	95 (28)	480	56 (12)	Higher titers in aborting cows	670
Japan	143 dairy cows, 25 aborted	IFAT, 200	88	23 (26.1)	54	2 (3.7)	OR 9.2	1086
	168 selected cows from dairy herds, 8 aborted	IFAT, 200	8	6 (75.0)	160	50 (31.3)	OR 6.1	1085

(Continued)

Table 4.4 (*Continued*) Seropositivity as Evidence of Abortion in Cattle

Country	Total Number of Cattle/Farms Examined	Test, Cut-Off	Aborting		Non-Aborting or Control			Reference
			No. Tested	No. Seropositive (%)	No. Tested	No. Seropositive (%)	Risk of Abortion Indicated by Odd Ratio (OR), Significant Association (SA), Remarks	
Mexico	Of 187 dairy cows, 13 herds, 110 were seropositive	ELISA-IDEXX	123	76 (61.8)	64	34 (53.1)	OR 1.4 (not significant)	731
	813 dairy cows, 20 herds, 106 abortions	ELISA-IDEXX	204	106 (51.9)	609	215 (35.3)	OR 2.0	732
The Netherlands	Dairy cows from 4 herds during abortion storms	IH ELISA	55	40 (72.7)	317	116 (36.6)	All abortions due to neosporosis	1370
	Seropositive (n = 106) and seronegative (n = 108) F1 progeny born after abortion outbreaks	IH ELISA	52 pregnancies ending with abortion or interval >120	40 (76.9)	241 normal pregnancies	111 (46.1)	SA, seropositive F1 progeny after an epidemic has an increased abortion risk	2123
New Zealand	320 dairy cattle, 1 farm, 69 aborted	IFAT, 200	69	58 (84.0)	31	10 (32.2)	Higher titers in aborting cows. Titers decreased 2 months after abortion	393
	812 beef cattle, 45 herds, 379 aborted	IDEXX-Checkit (cut-off ≥30)	NS	NS	NS	NS	OR 3.36	1774
	164 dairy cattle, 1 farm with previous *Neospora* abortions	IFAT, 200	15	11 (73.3)	49	7 (4.7)	SA	2085
	2246 dairy cattle, 5 farms selected for vaccine trial, 148 aborted	IFAT, 200	256	39 (15.2)	1990	72 (3.6)	OR 4.21	2088
Slovakia	716 post abortion dairy cows, 247 control cows	cELISA	716	144 (20.1)	247	6 (2.3)	SA	1700
Spain	889, 111 abortions	ELISA	111	63 (62.5)	200	48 (26.1)	OR 3.3	1260
	5196 dairy cattle, 276 herds, 15.7% seropositive, 70% of abortions occurred in seropositive cows	NAT, 40	—	NS	—	NS	OR 5.3	791
	1134 lactating dairy cows, 38 farms	iELISA-Bio-X	69	24 (34.8)	721	29 (4.0)	OR 9.1	792
	237 dairy cattle, 1 herd, 164 pregnant, 38 aborted	ELISA-IDEXX, ≥0.50	38	29 (76.3)	126	9 (9.2)	OR 12.2	1215

(*Continued*)

Table 4.4 (*Continued*)   Seropositivity as Evidence of Abortion in Cattle

Country	Total Number of Cattle/Farms Examined	Test, Cut-Off	Aborting		Non-Aborting or Control			Reference
			No. Tested	No. Seropositive (%)	No. Tested	No. Seropositive (%)	Risk of Abortion Indicated by Odd Ratio (OR), Significant Association (SA), Remarks	
	2773 dairy cattle, 6 herds, 146 (after 90 days of gestation)	ELISA-IDEXX (≥0.50) and ELISA-CIVTEST, ≥6.0 relative index	146	105 (71.9)	–	–	OR 18.9	1214
	273 pregnancies in seropositive dairy cattle, 2 herds	ELISA-CIVTEST, ≥6.0 relative index	273	77 (28.2)	–	–	OR 1.01 by antibody level. Risk of abortion 2.8 times lower for pregnant cows inseminated with beef bulls instead of Holstein-Friesian	1217
	86 seropositive and 40 seronegative dairy cattle	ELISA-CIVTEST, ≥6.0 relative index	22	21 (95), only 1/21 had detectable IFN-γ	104	65 (62.5), 15/65 had detectable IFN-γ	Risk of abortion 15.6 times higher from seropositive cows without detectable IFN-γ than in seronegative dams	1220
	414 pregnancies in dairy cows from 1 herd tested for 3 consecutive years	ELISA-CIVTEST	53 pregnancies ending in abortion	49 (92.5)	361 normal pregnancies	75 (20.8)	SA, abortions predominantly in seropositive cows	1542
	1115 pregnancies in 5 dairy herds of seropositive cows, 251 (22.6%) aborted; of 7432 seronegative cows, 239 (3.2%) aborted	ELISA-CIVTEST	490 pregnancies ending in abortion	251 (51.2)	8057 normal pregnancies	864 (10.7)	Higher risk of abortion in cows with high *N. caninum* antibody titers; lowest incidence of abortions in Limousin semen inseminated cows with low antibody titers	41[a]
	750 pregnancies of seropositive dairy cows (30.9% aborted), 2709 pregnancies of seronegative dairy cows (2.3% aborted)	ELISA-CIVTEST ≥6.0	238	177 (74.4)	3221	573 (17.8)	Likelihood of abortion 3.2 times lower for parous cows with low antibody titers against *N. caninum*; in heifers this variable had no effect	2150
	19 aborting, 53 non-aborting dairy cattle, 1 herd	ELISA-CIVTEST ≥6.0	19	19 (100.0)	72	53 (35.8)	Likelihood of abortion 7.0 higher in seropositive than in seronegative cows	726

(*Continued*)

Table 4.4 (Continued) Seropositivity as Evidence of Abortion in Cattle

Country	Total Number of Cattle/Farms Examined	Test, Cut-Off	Aborting			Non-Aborting or Control		Risk of Abortion Indicated by Odd Ratio (OR), Significant Association (SA), Remarks	Reference
			No. Tested	No. Seropositive (%)	No. Tested	No. Seropositive (%)			
Switzer-land	113 dairy cattle in 24 case (abortion) and 64 dairy cattle in 24 control herds	IFAT, 160	364	113 (31.0)	251	64 (25.5)	Seroprevalence in case herds 31% and in control herds 25.5%; difference not significant	860
Taiwan	38 dairy cattle examined on 1 farm; 18 aborted	IFAT, 200	18	16 (88.9)	20	13 (65.0)	None of 30 cows was seropositive before abortion storm	948
Turkey	186, 7% seropositive	cELISA-VMRD	9	3 (33.3)	177	10 (5.6)	SA	969
UK	Heifers from 18 dairy farms, 34 had gestational loss during second pregnancy	ELISA-MAST	34	7 (20.6)	323	12 (3.0)	Seropositive heifers were more likely to suffer gestational loss (late embryonic/early fetal loss and abortion) during their second pregnancy (OR 6.0)	246
	95 dairy cattle, 10 presented abortions	IFAT, 640	–	9 (90.0)	–	–	57 of 95 cows in this had IFAT titers of 1:640, 2 weeks after last abortion	421
	1051 dairy cattle, 633 aborting (cases), 418 non-aborting (controls)	ELISA-MAST	633	114 (18.0)	418	25 (5.9)	OR 3.5	431
	35 cows with confirmed *N. caninum* abortion, 100 cows with aborted calves with lesions typical for neosporo-sis, 128 cows with healthy calves	IFAT, 640	35	33 (94.3)	228	11 (4.8)	Fetal fluids from 21 of 25 (84%) had IFAT titers of 1:50 or higher, compared with fetal fluids from 7 of 100 abortions of other causes	1538
	120 aborting cows, 97 non-aborting control cows	IFAT, 200	120	11 (9.2)	97	1 (1.0)	SA, specificity of IFAT assayed	1991

(Continued)

Table 4.4 (Continued) Seropositivity as Evidence of Abortion in Cattle

Country	Total Number of Cattle/Farms Examined	Test, Cut-Off	Aborting		Non-Aborting or Control			Reference
			No. Tested	No. Seropositive (%)	No. Tested	No. Seropositive (%)	Risk of Abortion Indicated by Odd Ratio (OR), Significant Association (SA), Remarks	
USA	460 dairy cows, 110 aborting	Kinetic ELISA	110	40 (36.4)	350	62 (17.7)	OR 2.8	898
	Aborting and non-aborting cows from 20 herds with epidemic from several states (detailed data for 14 herds)	ELISA	221	153 (69.2)	336	116 (34.5)	SA in 6 of 14 herds, OR 1.4–40.0	1966
	Aborting and non-aborting cows from 2 herds with endemic abortion	ELISA	76	42 (55.3)	98	33 (33.7)	SA in 1 of 2 herds association, OR 3.4–7.0	1966
	Dairy herd, epidemic abortion in 28 dams, 43 calved normal	IFAT, 25	27	24 (88.8)	43	33 (76.7)	Differences not significant. Three other assays similar results, avidity ELISA revealed a higher proportion of low avidity responses in aborting cows	1004

Note: NS = not specified, iELISA-Bio-X = BIO K 192 Neospora caninum ELISA antibody Kit, BIO-X, Belgium; cELISA-Pourquier = Indirect ELISA, Institut Pourquier, France; ELISAIH-p38 = Native immune-affinity-purified surface antigen NcSRS2; IDEXX-Checkit = IDEXX Chekit Neospora, indirect ELISA, sonicate lysate of tachyzoites, IDEXX Laboratories, The Netherlands; ELISA-IDEXX = IDEXX HerdChek Neospora caninum antibody, indirect ELISA, sonicate lysate of tachyzoites, IDEXX Laboratories, USA; ELISA-CIVTEST = CIVTEST BOVIS NEOSPORA, indirect ELISA, sonicate lysate of tachyzoites, Laboratorios Hipra S.A., Spain; cELISA-VMRD = Neospora caninum cELISA Competitive ELISA GP65 surface antigen of tachyzoites VMRD, USA; ELISA-MAST = MASTAZYME NEOSPORA, indirect ELISA, formaldehyde-fixed whole tachyzoites, MAST GROUP, UK; Kinetic ELISA = Kinetic ELISA-California.[1555]
a The likelihood of abortion was 7.6 times lower for pregnant heifers inseminated with Limousin bull semen versus Holstein-Friesian bull semen.

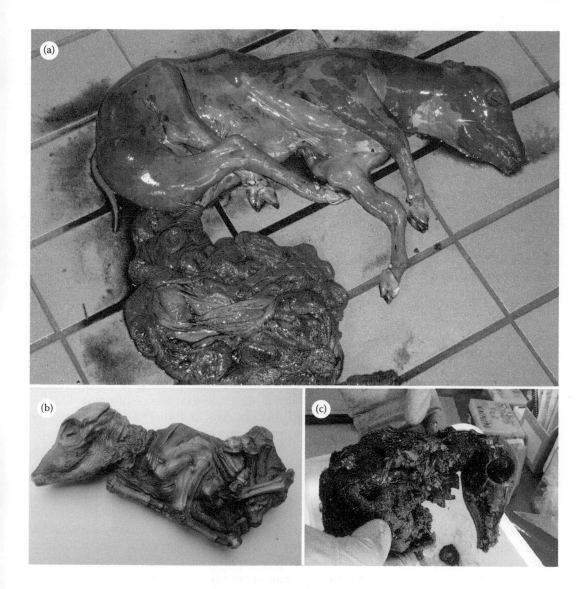

Figure 4.1 Neosporosis in fetuses. (a) Seven-month gestational age autolyzed fetus with mineralization of placenta. (Courtesy of Dr. John Edwards.) (b) Mummified fetus. (Courtesy of Dr. William Wouda.) (c) Mummified fetus, 3.5-month gestational age. (Courtesy of Ghanem et al., see Reference **760**.)

4.1.2.1.1 Gestational Age

Fetuses of any gestational age can be aborted, from 2.5 months to term (Table 4.2). Most neosporosis abortions occur at 5–6 months of gestation.[62,64,130,131,274,367,383,691,862,1335,1536,1595,1704,1757,1954,2136]

It is likely that fetuses aborted earlier than 3 months are not retrieved, since they may be reabsorbed. Fetuses dying before 5 months gestation may be mummified and retained in the uterus for several months, those dying later are usually expelled.

4.1.2.1.2 Epidemic and Endemic (Sporadic) Abortions

Abortions in cattle may have an epidemic or endemic (sporadic) character. Abortions have been considered as epidemic if more than 10% of cows at risk abort within 6–8 weeks.[85,169,395,1305,1370,1497, 1800,1966,2121,2125] In 1 study as many as 37.5% of cows at risk (i.e., of the pregnant herd) experienced abortion within a few weeks.[1800]

4.1.2.1.3 Pathological Changes

4.1.2.1.3.1 Gross Lesions—Aborted fetuses may be in different stages of autolysis (Figure 4.1). There are no pathognomonic lesions of neosporosis. In few fetuses, 1–2 mm foci of discoloration (indicative of necrosis) were found in the heart, liver, lungs, and placenta.[1057,1595] In 1 fetus, there was hydrocephalus,[571] associated with dilated lateral ventricles and hypoplasia of cerebellum and medulla (Figure 4.2a). In 1 stillborn fetus, the heart was grossly enlarged.[546] The entire heart of this fetus was fixed in formalin, and paraffin blocks from this animal were shared with many diagnostic laboratories of the world because it had many intralesional tachyzoites.

4.1.2.1.3.2 Microscopic Lesions—Lesions may differ depending upon the gestational age and the pattern of abortions. Necrosis is more prominent in younger fetuses before immunocompetence develops (5 months or later) (Figure 4.3). Inflammation is more prominent in older fetuses. Although several tissues may be affected, lesions are most common in the brain, heart, liver, and placenta (Figures 4.3 through 4.10). Because of the costs involved in examination of multiple tissues, there are only a few reports on the extent of lesions in different tissues. In 1 study, lesions were seen in the brains and hearts of all 82, livers of 48/77, skeletal muscle of 46/64, kidneys of 52/79, adrenal glands of 53/54, lungs of 35/80, and placentas of 10/19.[130,131] The presence of *N. caninum* in tissues of fetuses aborted due to neosporosis confirmed that lesions and parasites can be present in many fetal tissues (Table 4.5). In another study, not listed in Table 4.5, *N. caninum* was identified by IHC in 27/27 brains, 10/13 hearts, 5/6 placentas, 1/4 livers, 2/2 kidneys, and 1/1 skeletal muscle.[862]

Lesions may be present in any part of the brain, but are more common in gray versus white matter, and more common in the cerebrum.[874] The initial lesion is focal necrosis with or without inflammatory response. In older fetuses, inflammatory responses are evidenced by the presence of microglia, reactive astrocytes, and cells of the monocyte and lymphoid system. A focus of central necrosis with peripheral inflammatory cells is the most characteristic feature of bovine neosporosis (Figure 4.3). Mineral deposits may occur in necrotic foci (Figure 4.4b). Spinal cords have rarely been examined histologically but are expected to have lesions similar to those in brains. There is no difference in neural lesions with respect to epidemic or endemic (sporadic) abortions.[2121]

Hepatic lesions may vary depending on whether abortions are sporadic or epidemic. Hepatitis is more severe in epidemic cases, and consists of periportal hepatitis and multifocal hepatocellular necrosis[2121] (Figure 4.6).

Myositis involving heart, tongue, and diaphragm is common. Lesions are mainly inflammatory with minimal necrosis. Even in autolyzed fetuses, inflammatory foci can be recognized in the epicardium, while in most of fetuses the myocardium had degenerated (Figure 4.7). There are only few tachyzoites, and generalized myocarditis is rare. We are aware of only 1 stillborn fetus that had generalized myocarditis with many tachyzoites (Figure 4.9). In that fetus, tachyzoite groups were up to 150 μm long and contained more than 200 organisms, but no tissue cysts.[546]

Placental lesions consist of focal necrosis and nonsuppurative placentitis (Figures 4.3d and 4.8). The intercotyledonary areas are not affected. Occasionally, miliary mineralized foci are grossly visible (Figure 4.1a). Tachyzoites have been demonstrated in trophoblasts, but rarely.[196,1849]

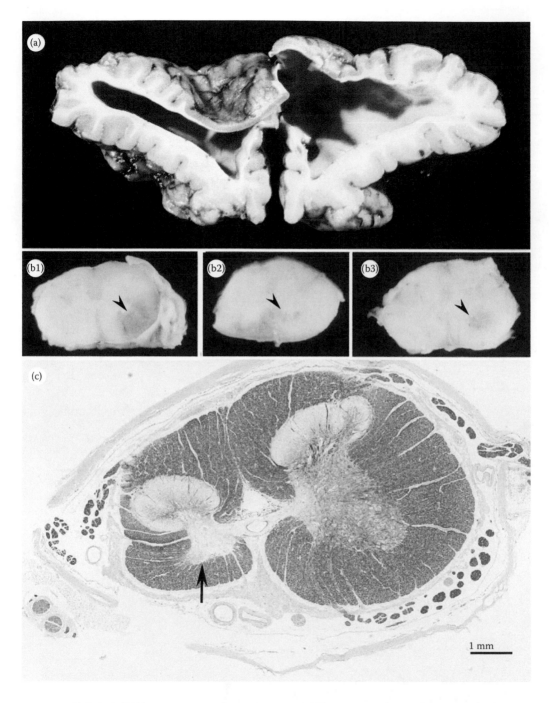

Figure 4.2 Lesions of bovine neosporosis. (a) Hydrocephalus in a 7-month gestational age aborted fetus reported in Reference **571**. The lateral ventricles of the cerebrum are dilated. Unstained. (Courtesy of Dr. John Edwards.) (b) Transverse sections of the cervical (b1), thoracic (b2) and the lumbosacral spinal cord (b3) segments of a 2-week-old calf. The asymmetrical areas of gray discoloration (arrowheads) are due to necrotizing inflammation and were present throughout the spinal cord. Unstained. (From Dubey, J. P., de Lahunta, A. 1993. *Appl. Parasitol.* 34, 229–233.) (c) Cross-section of spinal cord of a congenitally infected calf. The lesion was grossly visible. Note unilateral reduction of ventral gray matter (arrow). PAS reaction. (From Dubey, J. P. et al. 1990. *J. Am. Vet. Med. Assoc.* 197, 1043–1044.)

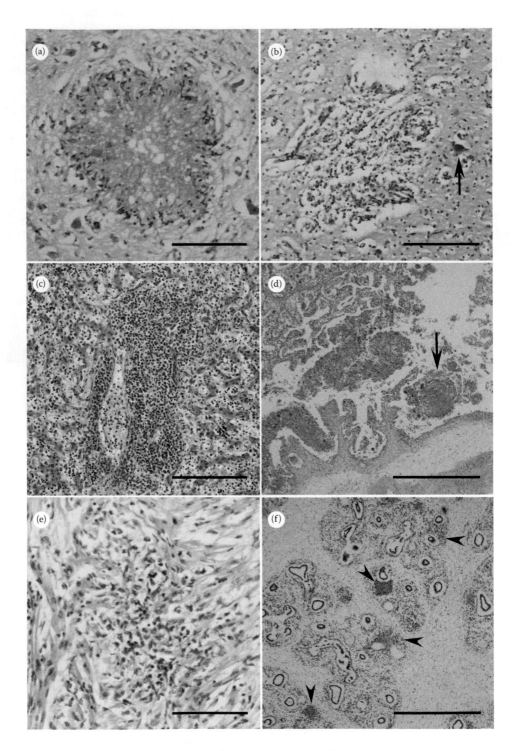

Figure 4.3   Lesions in young (3–5-month gestational age) fetuses naturally aborted due to confirmed neo-sporosis. HE stain. (a) Focal necrosis with scattering of mononuclear cells at the periphery in cerebrum. (b) Focal infiltration of mononuclear cells in the cerebrum. Arrow points to a tissue cyst. (c) Periportal hepatitis. (d) Focal placental necrosis (arrows). (e) Inflammatory focus in the myocardium. (f) Foci of necrosis and inflammation in lung (arrowheads). (b–f, Courtesy of Dr. William Wouda; see Reference **605**). Bars in a–c, and e = 250 μm, in d and f = 100 μm.

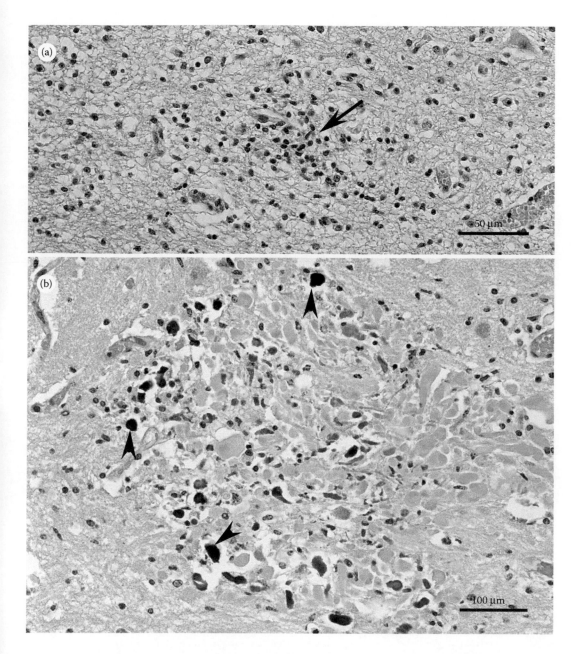

Figure 4.4 Encephalitis in aborted fetuses. HE stain. (a) Small inflammatory focus (arrow). (b) A large focus of necrosis and mineralization (arrowheads).

4.1.2.2 Congenitally Infected Calves

A small percentage of congenitally infected calves are born ill or die within 2 weeks of birth and some of these isolated cases are summarized in Table 4.3. In addition, *N. caninum* DNA was found in 5 of 7 placentas from full-term calves born to seropositive cows in a herd in Iran.[1764] Thick-walled *N. caninum* tissue cysts were identified by IHC in the brain of a calf born to a seropositive cow.[1110]

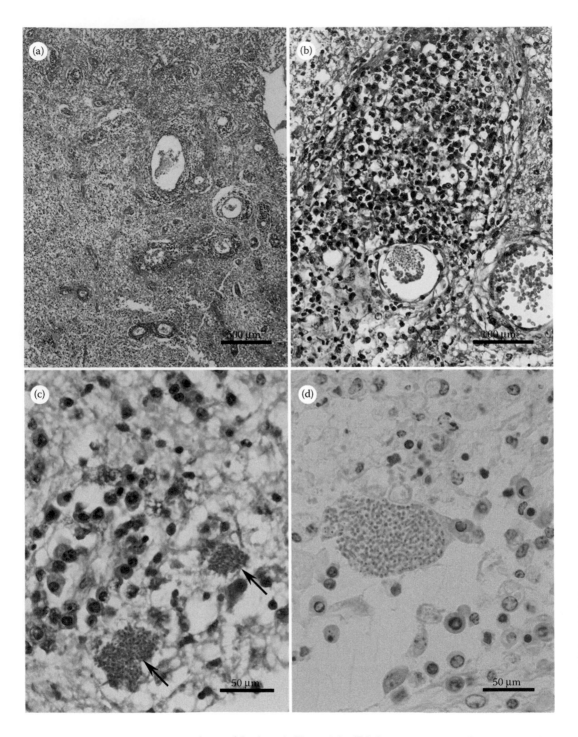

Figure 4.5   Encephalitis in the cerebrum of the fetus in Figure 4.2a. This is a rare example of severe encephalitis with numerous tachyzoites. HE stain. (a) Necrosis and severe vasculitis. (b) Mixed leukocyte
cell infiltration in the tunica media and adventitia of a blood vessel. (c) Plasmacytic cell infiltration
of a blood vessel. Two groups of tachyzoites are present (arrows). (d) A large group of tachyzoites among necrotic tissue and infiltration of plasma cells. (Adapted from Dubey, J. P. et al. 1988.
J. Comp. Pathol 118, 169–173.)

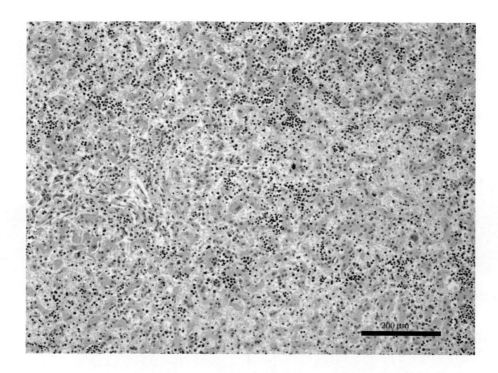

Figure 4.6 Severe generalized hepatitis, common in fetuses aborted during a storm of abortion, presumably acquired postnatally. HE stain. (Specimen courtesy of Dr. William Wouda.)

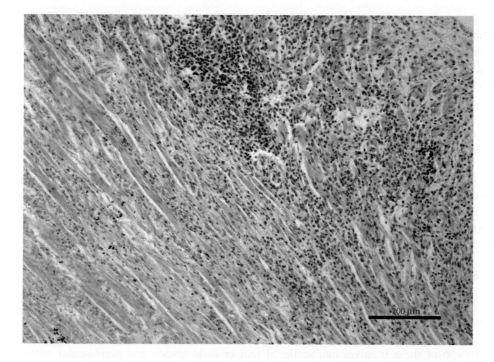

Figure 4.7 Epicardial myocarditis in an aborted fetus, the deeper myocardium was unaffected. HE stain. (Specimen courtesy of Dr. William Wouda.)

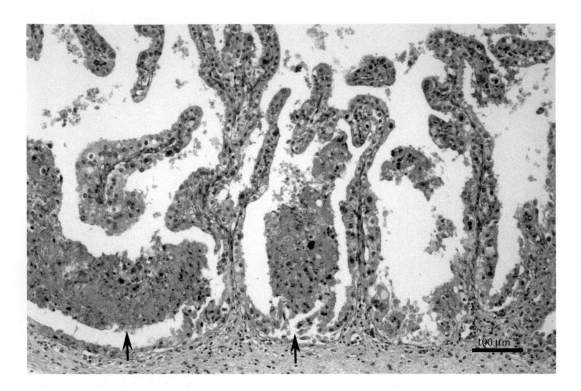

Figure 4.8 Focal placental necrosis (arrows). HE stain. Note that adjoining cotyledonary villi are unaffected. (Specimen courtesy of Dr. William Wouda.)

The predominant clinical signs are ataxia and weakness of, sometimes all, limbs, and there may be permanent hyperextension (Figure 4.11). These calves were identified because of epidemiological studies on farms where cows had aborted. One calf had slight bulging of the forehead. Exophthalmia or asymmetrical appearance of eyes has been reported (Table 4.3). Birth defects included scoliosis and narrowing of the spinal cord. Most of these calves were euthanized and complete necropsies were performed. In 1 calf, there were areas of discoloration throughout the spinal cord (Figure 4.2b1–b3), severe extensive multifocal necrotizing nonsuppurative myelitis, severe perivascular cuffing, and degenerated parenchyma (Figures 4.12 and 4.13). It is noteworthy that lesions in this and other calves were confined to the CNS and muscle.

In an unusual case, a full-term calf was aymptomatic at birth but became weak at 2 weeks, and was eating mud. The calf was euthanized at 4 weeks. Lesions were confined to brain and muscles. In the brain, there were distinct areas of necrosis and cavitations of up to 6 mm in diameter.[550] There was marked gliosis throughout the gray and white matter. Only tachyzoites, and no tissue cysts, were seen.

4.1.2.3 Lesions in Weaned Calves and Adult Cattle

N. caninum has not been demonstrated histologically in tissues of cattle older than 8 weeks of age, although the viable parasite was isolated even from adult cows (Table 4.6). In 2 reports from Spain, viable *N. caninum* tachyzoites were isolated by bioassay of CNS of 9 asymptomatic 57-day-old calves; 6 of these 9 calves had scattered foci of mild gliosis and perivascular cuffing with mononuclear cells.[1674,1729] In an 8-month-old calf from Italy,[689] authors reported focal myocarditis but no encephalitis. They observed tachyzoites in sections of heart by IHC but the results need

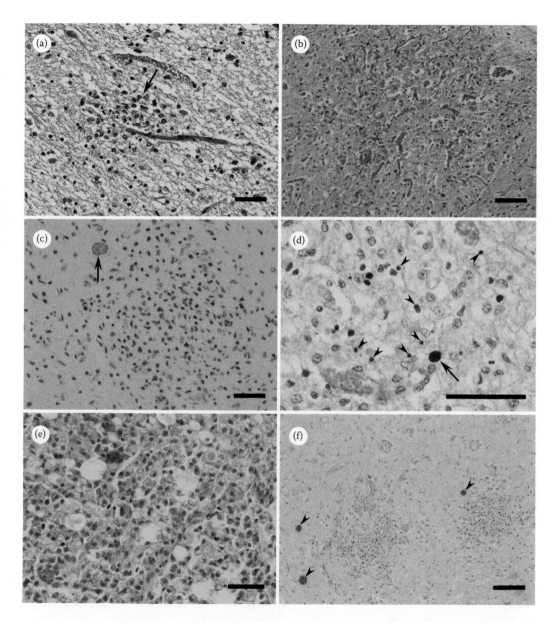

Figure 4.9 Encephalomyelitis in full-term fetuses and congenitally infected calves. a–e = full term, stillborn calf reported in Reference **546**, f = congenitally infected calf reported in Reference **132**. a–c, e = HE stain. (a) Focal inflammation adjacent to a blood vessel (arrow). (b) Focal necrosis and neo-vascularization. (c) Focal nodule, mainly mononuclear cells, and a group of tachyzoites (arrow). (d) Focal necrosis and individual bradyzoites (arrowheads), a tissue cyst (arrow) scattered in the lesion. IHC staining with BAG1 antibodies. (e) Macrophage accumulation in an area of malacia. (f) Perivascular inflammation and 2 glial nodules. Arrowheads point to 3 tissue cysts. IHC staining with polyclonal *N. caninum* rabbit antibodies. Bars are = 50 μm.

confirmation because the related *Sarcocystis* sarcocysts are common in bovine myocardium and could cross react with *N. caninum*. In another clinically normal heifer and her calf, there were no histological cerebral lesions, although *N. caninum* was isolated *in vitro*.[1507] Gliosis and perivascular cuffs, both in the CNS, focal myositis, and myocarditis, and infiltrates of mononuclear cells in the liver and kidney, were reported for a cow from which viable *N. caninum* had been isolated.[1786]

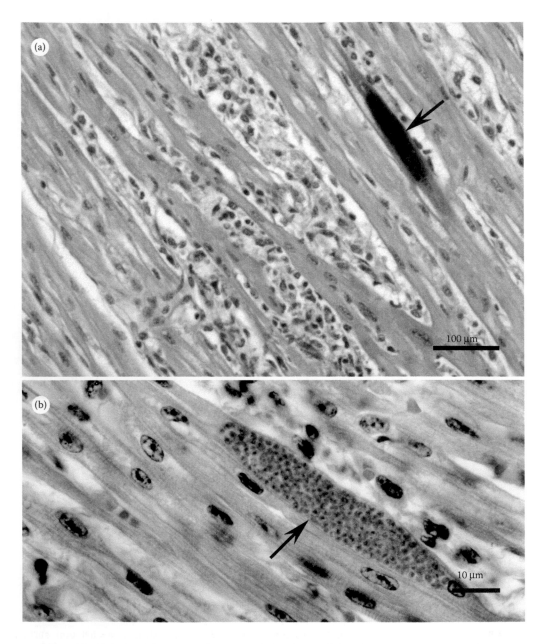

Figure 4.10   Necrosis and inflammation in the myocardium of the stillborn fetus. (a) IHC with *N. caninum* polyclonal antibodies. (b) HE stain. Note long intracellular group of tachyzoites (arrows).

4.1.3 Isolation of Viable *N. caninum* from Bovine Tissues

Information on isolation from bovine tissues is summarized in Table 4.6. Attempts at isolation
of viable *N. caninum* by bioassay in mice or cell culture have been largely unsuccessful. Many
attempts to isolate viable *N. caninum* have been unsuccessful because most parasitic stages die
within the fetus when it succumbs to the infection. Probably for this reason, *N. caninum* was only
recovered from 2 of 49 histologically confirmed *N. caninum*-infected fetuses (among more than 100
fetuses),[373] and in both instances, tissue cysts were present. This monumental effort illustrates the

Table 4.5   Distribution *N. caninum* in Tissues of Fetuses Aborted Due to Neosporosis

No. of Fetuses with Confirmed Neosporosis	No. of Fetuses Positive (%) for *Neospora*										
	Brain	Heart	Liver	Lung	Kidney	Skeletal Muscle	Spleen	Adre-nal	Intes-tine	Pla-centa	Refer-ence
80	71[a] (88.7)	11 (13.7)	21 (26.2)	ND	ND	ND	ND	ND	ND	ND	2121
29	21 (72.4)	24 (82.7)	18 (62.0)	16 (55.2)	17 (55.2)	17 (58.6)	12 (41.3)	10 (34.5)	8 (27.6)	5 (17.2)	1380
55	41 (74.5)	42 (76.4)	4 (7.3)	41 (74.5)	4 (7.3)	51 (92.7)	ND	ND	ND	ND	1595
41	21 (51.2)	24 (58.5)	25 (60.9)	ND	ND	ND	ND	ND	ND	ND	1397

Note: ND = not done/no data.
[a] Tachyzoites in 65, only tissue cysts in 3, and tachyzoites and tissue cysts in 17.

difficulty of isolating viable organisms from bovine fetuses. It is easier to isolate *N. caninum* from neural tissues of congenitally infected full-term calves, perhaps because tissue cysts are likely to be present and viable in tissues of a living animal.

The recovery of a few of the isolates of *N. caninum* listed in Table 4.6 was unusual. The Nc-LivB1 isolate was obtained from a stillborn calf that was seropositive but had no histologically demonstrable organisms in its tissues. The dam of the calf had no detectable antibodies with the MASTAZYME-ELISA at the time of calving and 2 weeks later, but she seroconverted during a subsequent pregnancy.[432] The NC-PV1 isolate was obtained from a congenitally infected calf with clinical signs[1253] that was euthanized when recumbant at 45 days of age. This is the longest time that an *N. caninum*-infected calf born with clinical signs has survived. The isolation of *N. caninum* from an 8-month-old calf is also interesting for several reasons.[690] First, the dam was diagnosed as having

Figure 4.11   Three-day-old recumbent calf with contracted forelimbs and slight bulge on the forehead. The calf was stunted, weak, and unable to stand. Many tissue cysts were found in lesions of encephalo-myelitis. (From Barr, B. C. et al. 1991. *J. Vet. Diagn. Invest.* 3, 39–46.)

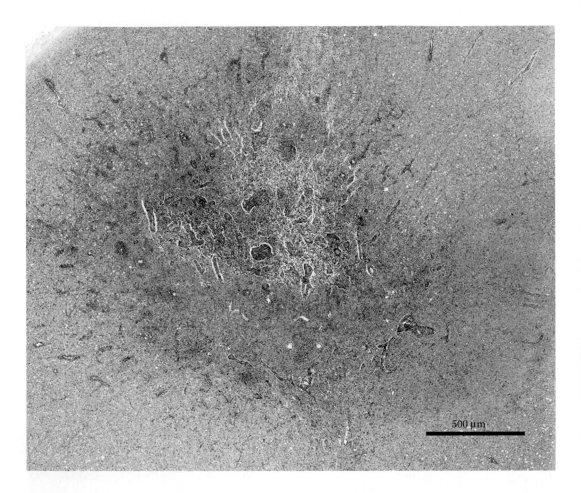

Figure 4.12   Severe myelitis with malacia in the spinal cord of a congenitally infected 5-day-old calf described
in References **530** and **1495**. This lesion was grossly visible as discoloration and softening of c4
segment. HE stain.

recently acquired *N. caninum* infection based on the detection of high IgG and IgM antibodies at
230 days of gestation. Second, although a healthy calf was born at 280 days of gestation, the placenta contained inflammatory foci and viable *N. caninum* was isolated by bioassay in mice. Third,
thick-walled *N. caninum* tissue cysts were found directly in the brain of the calf after it was killed
at 8 months of age and viable *N. caninum* was recovered in mice and in cell culture, even though
the calf was clinically normal.

Most of the isolates of *N. caninum* were obtained from cattle with clinical signs. As mentioned
earlier, in 2 studies from Spain, *N. caninum* was isolated from 9 healthy calves up to 57 days old.
These calves were selected for isolation of *N. caninum* based on seropositivity before the ingestion
of colostrum.

Viable *N. caninum* was isolated from 2 two-year-old asymptomatic cows from Japan and New
Zealand; both had transmitted neosporosis to their fetuses. The cow in Japan had twice aborted
N. caninum-infected fetuses, and was euthanized 24 days after the second abortion.[1786] Although
protozoa were not demonstrable in histologic sections of the infected brain, it had a mild nonsuppurative encephalitis. The cow in New Zealand had given birth to an asymptomatic congenitally infected
calf.[1507] The cow and her calf were euthanized 2 days after the birth of the calf. Viable *N. caninum*
was isolated from both the cow and her calf; there were no histologic lesions in both animals.

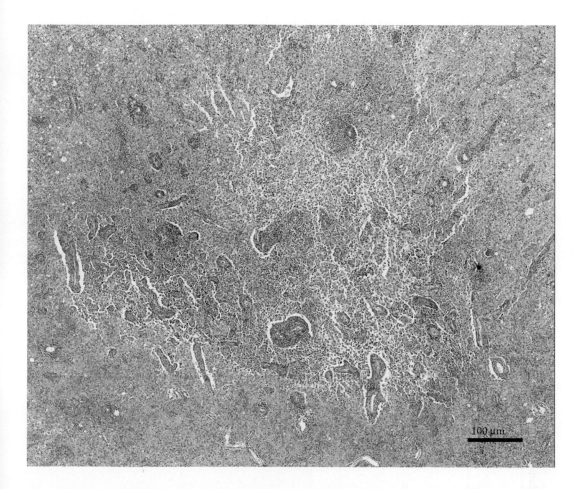

Figure 4.13 Higher magnification of lesion in Figure 4.12. Note extensive myelitis with vasculitis and necrosis of all elements of neuropile. HE stain.

In addition to data in Table 4.6, *N. caninum* was recovered in mice inoculated with neural tissue from a calf in Canada[251]; this result is of interest because it was reported that the bovine tissue had been frozen at −52°C for approximately 4 months before inoculation into mice. Similarly, viable *Neospora* was isolated from the brain of an aborted fetus in Australia.[1497] No other details are available for these isolates, apparently both isolates were not cryopreserved.

Viable *N. caninum* was isolated by bioassay in mice from placentas of 3 cows and their calves born in 3 subsequent pregnancies.[690] The cows and calves remained asymptomatic and the calves were euthanized in good health at 2 months of age.[690] These isolates were apparently not cryopreserved.

4.2 EXPERIMENTAL INFECTIONS

4.2.1 Calves and Adult Cattle

Calves and cattle inoculated orally[434,2009] or parenterally with tachyzoites[52,672,1076,1234,1278,1831] and orally with oocysts[445] became infected, developed cellular and humoral responses, but remained asymptomatic. In 1 study, heifers inoculated intraconjunctivaly with tachyzoites became seropositive temporarily but became seronegative by 4 months PI.[470]

Table 4.6 Isolation of Viable *N. caninum* from Neural Tissues of Cattle

Country	Source	Bioassay Animals	Bioassay Cell Culture	Isolate Designation	Museum Deposition	Molecular Data	Remarks	Reference
Argentina	Asymptomatic, 7 days old	KO mice	Vero	*NC-ArgentinaLP1*	—	Multilocus microsatellite (MS1B, 2, 3, 5, 6A, 6B, 7, 10, 12, 21); KJ700413 [MS10] and KJ700414 [MS2]	Brains of 5 seropositive calves were bioassayed; only 1 was positive in bioassay	279
Australia	Asymptomatic, 7 days old	KO mice	Vero[a]	NC-Nowra	—	Multilocus microsatellite (MS10, GU1289551; Cont-6, -14, 16, FJ883901, FJ883930, FJ883945; Tand-3, -4, -8, -9, -12, -13, -15, -16, -23, -24, -25, -26, -30, -31, -32, -34, FJ824930, FJ824996, FJ824942, FJ830458, FJ824959, FJ824984, FJ825000, FJ825006, FJ825012, FJ830463, FJ830467, FJ830471, FJ830479, FJ830485, FJ825016, FJ830489; ITS1 (AF029702, AF338411); pNc5 gene (JF937547.1, JF937546, JF937545)	Pathogenic to BALB/c mice	29, 1353
Brazil	Aborted fetus, 7 months of gestation	Mice, gerbils	Vero	BCN/PR1	—	Not provided	Initial isolation in immunosuppressed mice. Tachyzoites infective to immunocompetent mice and gerbils	1202
	Clinical, blind, neurological calf, 3 months old	SW mice	Vero	BCN/PR3	—	Not provided	Tissue cysts were found in the brains of SW mice that were immunocompetent	1201
	Asymptomatic, 4 months old	KO mice	MARC-145	Nc-Goiás 1	—	Multilocus microsatellite (MS1A, 1B, 2, 4, 5, 6A, 6B, 7, 8, 10, 12, 21)	Not pathogenic to BALB/c mice	729, 1679
China	Cow 3 years old	NI	Vero	NC-Bj	—	Multilocus microsatellite (MS10, identical to MS10 pattern of NC-1), KC832778, Cont-14 (KC832779)	Isolation from peripheral cow blood of the that had aborted an *N. caninum*-infected fetus[c]	849

(Continued)

Table 4.6 (Continued) Isolation of Viable N. caninum from Neural Tissues of Cattle

Country	Source	Bioassay		Isolate Designation	Museum Deposition	Molecular Data	Remarks	Reference
		Animals	Cell Culture					
Iran	Aborted fetus, 4 months gestational age	BALB/c	Vero	Nc-Iran	–	Not provided	–	1766
Israel	Aborted fetus, second trimester	NI	Vero	Ncls491	–	Multilocus microsatellite (Tand-3, -12, -13, FJ824933, FJ824964, FJ824989; Cont-6, -14, -16, FJ883910, FJ883927, FJ883950)	Nonpathogenic to gerbils	28, 29
	Aborted fetus, third trimester	NI	Vero	Ncls580	–	Not provided	Nonpathogenic to gerbils	692
Italy	Clinical, 45 days old	NI	Vero	Nc-PVI	–	SSU-rDNA multilocus microsatellite (MS4, 5, 6A, 6B, 7, 8, 10, 12, 21)	–	1253, 1255, 1679
	Calf, 8 months old, brain	SW mice	Vero	NCPG1	–	Not provided	Tissue cysts seen in brain of immunosuppressed mice. Tissue cysts seen directly in brain of calf	689
	Placenta	SW mice	Vero	NCPG1	–	Not provided	Tissue cysts seen in brain of immunosuppressed mice	689
Japan	Calf, 2 weeks old	No	CPAE	JPA1	–	Multilocus microsatellite (Cont-6, -14, -16, FJ883911, FJ883919, FJ883951; Tand-3, -4, -8, -9, -12, -13, -16, -23, -24, -25, -30, -31; FJ824928, FJ824993, FJ824941, FJ830454, FJ824957, FJ824982, FJ825004, FJ825010, FJ830461, FJ830465, FJ830476, FJ830483)	Precolostral IFAT 1:3200	28, 29, 2144
Japan	Calf	Nude	CPAE[a]	BT2	–	Not provided	Tissue cysts in brain and spinal cord of calf. Precolostral IFAT 1:3200	2145
Japan	Calf	Nude	CPAE[a]	JPA2	–	Not provided	Precolostral IFAT 1:1600	2145

(Continued)

Table 4.6 (Continued)   Isolation of Viable N. caninum from Neural Tissues of Cattle

| Country | Source | Bioassay | | Isolate Designation | Museum Deposition | Molecular Data | Remarks | Reference |
		Animals	Cell Culture					
	Calf	Nude	CPAE[a]	JPA5	–	Not provided	Precolostral IFAT 1:6400	2145
	Stillborn calf	Nude	CPAE[a]	JPA4	–	Not provided	Precolostral IFAT 1:400	2145
	Cow, 2 years old	Nude mice	Vero[a]	BT-3	–	Not provided	Cow had aborted 2 N. caninum-infected fetuses in 2 pregnancies	1786
South Korea	Clinical, tetraparetic, 1 day old	NI	Vero	KBA-1	–	Multilocus microsatellite (MS1A, 1B, 2, 3, 4, 5, 6A, 6B,78, 10, 12, 21; AY937146, AY937148, AY937149, AY937151, AY937152, AY937153, AY937154, AY937155, AY937156, AY937178, AY937157, AY937158)	IFAT titer 1:3200 in precolostral serum	1059, 1060, 1679
Korea	Fetus, 8 months old	NI	Vero	KBA-2	–	Multilocus microsatellite (MS1A, 1B, 2, 3, 4, 5, 6A, 6B, 7, 8, 10, 12, 21; AY937159, AY937160, AY937161, AY937164, AY937165, AY937166, AY937167, AY937168, AY937169, AY937170, AY937171, AY937172)	IFAT titer 1:1600 in thoracic fluid	1059, 1060, 1679
Malaysia	Clinical, 1 day old	BALB/c mice	Vero[a]	Nc-MalB1	–	Not provided	IFAT titer 1:3200 in precolostral serum	335
New Zealand	Cow 2 years old	NI	Vero	NcNZ 1	–	Multilocus microsatellite (Cont-6, -14, -16, FJ883913, FJ883931, FJ883947; Tand-3, -12, -13, FJ824934, FJ824965, FJ824990); ITS1 (AY601347)	IFAT titer 1:2000. 10[5] tachyzoites pathogenic to BALB/c mice	28, 29, 1507
	Asymptomatic 2 days old	NI	Vero	NcNZ 2	–	Multilocus microsatellite (Cont-6, -14, -16, FJ883914, FJ883932, FJ883948; Tand-3, -12, -13, FJ824935, FJ824966, FJ824991); ITS1 (AY601349)	IFAT titer 1:2000, precolostral. 10[5] tachyzoites pathogenic to BALB/c mice	28, 29, 1507
	Stillborn calf	NI	Vero	NcNZ 3	–	Multilocus microsatellite (Cont-6, -14, -16, FJ883915, FJ883933, FJ883949; Tand-3, -12, -13, FJ824936, FJ824967, FJ824992); ITS1 (AY601348)	IFAT titer 1:8000. 10[5] tachyzoites pathogenic to BALB/c mice	28, 29, 1507

(Continued)

Table 4.6 (Continued) Isolation of Viable *N. caninum* from Neural Tissues of Cattle

Country	Source	Bioassay		Isolate Designation	Museum Deposition	Molecular Data	Remarks	Reference
		Animals	Cell Culture					
Poland	Asymptomatic calf, 12 hours old	NI	Vero	NcPolB1	–	Not provided	No isolation from 8 other seropositive calves	801
Portugal	Fetus, 4 months gestation	SW mice-dexamethasone-treated	–	NC-porto1	–	Not provided	–	281
Slovakia	Cow 4 years old	Gerbils	Vero[a]	Nc-SKB1	–	Not provided	Tachyzoites were first seen 77 days after inoculation with gerbil brain. No direct isolation from cow brain	1702
Spain	Fetus, 6 months gestation	SW mice	Vero	NCSP-1	–	Not provided	Cortisonized mice	284
	Asymptomatic calf, 14 days old	Nude mice	MARC-145	Nc-Spain 6	–	Multilocus microsatellite (MS1A, 1B, 2, 3, 4, 5, 6A, 6B, 7, 8, 10, 12, 21; EU16099.1, EU816107, EU816115, EU816123, EU816131, EU816139, EU816147, EU816155, EU816163, EU816171, EU816180, EU816188)	Brain of calf PCR positive for *N. caninum*	1674
	Asymptomatic calf, 57 days old	Nude mice	MARC-145	Nc-Spain 7	–	Multilocus microsatellite (MS1A, 1B, 2, 3, 4, 5, 6A, 6B, 7, 8, 10, 12, 21; EU816100, EU816108, EU816116, EU816124, EU816132, EU816140, EU816148, EU816156, EU816164, EU816172, EU816181, EU816189); ROP40 gene (KP731806.1)	Brain of calf PCR-positive for *N. caninum*	1674
	Asymptomatic calf, 29 days old	Nude mice	MARC-145	Nc-Spain 8	–	Multilocus microsatellite (MS1A, 1B, 2, 3, 4, 5, 6A, 6B, 7, 8, 10, 12, 21; EU816101, EU816109, EU816117, EU816125, EU816133, EU816141, EU816149, EU816157, EU816165, EU816173, EU816182, EU816190)	Brain of calf PCR-positive for *N. caninum*	1674

(Continued)

Table 4.6 (*Continued*) Isolation of Viable *N. caninum* from Neural Tissues of Cattle

| Country | Source | Bioassay | | Isolate Designation | Museum Deposition | Molecular Data | Remarks | Reference |
		Animals	Cell Culture					
	Asymptomatic calf, 7 days old	Nude mice	MARC-145	Nc-Spain 9	–	Multilocus microsatellite (MS1A, 1B, 2, 3, 4, 5, 6A, 6B, 7, 8, 10, 12, 21; EU816102, EU816110, EU816118, EU816126, EU816134, EU816142, EU816150, EU816158, EU816166, EU816174, EU816183, EU816191)	Brain of calf PCR-positive for *N. caninum*	1674
	Weak calf, 2 days old	Nude mice	MARC-145	Nc-Spain 10	–	Multilocus microsatellite (MS1A, 1B, 2, 4, 5, 6A, 6B, 7, 8, 10, 12, 21; Genbank: MS10: EU816175)	Brain of calf PCR-positive for *N. caninum*	1674
	Asymptomatic calf, 2 days old	Nude mice	MARC-145	Nc-Spain 2H	–	Multilocus microsatellite (MS1A, 1B, 2, 3, 4, 5, 6A, 6B, 7, 8, 10, 12, 21; EU816095, EU816103, EU816111, EU816119, EU816127, EU816135, EU816143, EU816151, EU816159, EU816167, EU816176, EU816184)	Brain of calf PCR-positive for *N. caninum*	1674
	Asymptomatic calf, 57 days old	Nude mice	MARC-145	Nc-Spain 3H	–	Multilocus microsatellite (MS1A, 1B, 2, 3, 4, 5, 6A, 6B, 7, 8, 10, 12, 21; EU816096, EU816104, EU816112, EU816120, EU816128, EU816136, EU816144, EU816152, EU816160, EU816168, EU816177, EU816185)	Brain of calf PCR-positive for *N. caninum*	1674
	Asymptomatic calf, 22 days old	Nude mice	MARC-145	Nc-Spain 4H	–	Multilocus microsatellite (MS1A, 1B, 2, 3, 4, 5, 6A, 6B, 7, 8, 10, 12, 21; EU816097, EU816105, EU816113, EU816121, EU816129, EU816137, EU816145, EU816153, EU816161, EU816169, EU816178, EU816186)	Brain of calf PCR-positive for *N. caninum*	1674
	Asymptomatic calf, 14 days old	Nude mice	MARC-145	Nc-Spain 5H	–	Multilocus microsatellite (MS1A, 1B, 2, 3, 4, 5, 6A, 6B, 7, 8, 10, 12, 21; EU816098, EU816106, EU816114, EU816122, EU816130, EU816138, EU816146, EU816154, EU816162, EU816170, EU816179, EU816187)	Brain of calf PCR-positive for *N. caninum*	1674

(*Continued*)

Table 4.6 (Continued) Isolation of Viable *N. caninum* from Neural Tissues of Cattle

Country	Source	Bioassay		Isolate Designation	Museum Deposition	Molecular Data	Remarks	Reference
		Animals	Cell Culture					
	Asymptomatic calf	Nude mice	MARC-145	Nc-Spain 1H	–	Multilocus microsatellite (MS1A, 1B, 2, 3, 4, 5, 6A, 6B, 7, 8, 10, 12, 21); ITS1 (EU564165); POP40 gene (KP731807)	Low pathogenicity for BALB/c mice	1729
Sweden	Stillborn	No	Vero	NC-SweB1	–	Multilocus microsatellite (MS4, 5, 6A, 6B, 7, 8, 10, 12, 21, AY935177, AY935178, AY935179, AY935180, AY935181, AY935182, AY935183, AY935184, AY935185, AY935186.2, AY935187, AY935188; Cont-6, -14, -16, FJ883917, FJ883934, FJ883946; Tand-3, -4, -9, -12, -13, -15, -16, -23, -24, -25, -26, -30, -31, -32, FJ824931, FJ824998, FJ830457, FJ824960, FJ824985, FJ825002, FJ825007, FJ825008, FJ830464, FJ830468, FJ830472, FJ830480, FJ830486, FJ825017); ITS1 (AF029702, EU564167, AY259039)	–	28, 29, 1679, 1907
UK	Stillborn	No	Vero	Nc-LivB1	–	Multilocus microsatellite (Tand-3, -12, -13, FJ824968, FJ824943, FJ824912)	Calf seropositive but dam was seronegative	28, 432
	Fetus, aborted	NI	NI	Nc-LivB2	–	Multilocus microsatellite (Tand-3, -12, -13, FJ824969, FJ824944, FJ824913)		28, 1993
USA	Fetus, 4 months gestation	NI	CPAE	BPA1	ATCC,75710	Multilocus microsatellite (Cont-6, -14, -16, FJ883904, FJ883921, FJ883938; Tand-3, -8, -9, -12, -13, -23, -15, -24, -26, -30, -31, -32, FJ824924, FJ824939, FJ830453, FJ824947, FJ824972, FJ825009, FJ825003, FJ830459, FJ830469, FJ830474, FJ830482, FJ825013); ITS1 (AF038860); SSU-rDNA (U17345); SRS2 gene (AY940481)	8–10 µm tissue cysts in fetal brain	28, 29, 373
	Fetus, 6 months gestation	NI	CPAE	BPA2	–	Not provided	8–13 µm tissue cysts in fetal brain	373
USA	Clinical, calf	NI	NI	BPA3	–	Not provided	–	1281

(Continued)

Table 4.6 (Continued) Isolation of Viable *N. caninum* from Neural Tissues of Cattle

Country	Source	Bioassay		Isolate Designation	Museum Deposition	Molecular Data	Remarks	Reference
		Animals	Cell Culture					
USA	Clinical, calf	NI	NI	BPA4	–	Not provided	–	1281
	Fetus	NI	Vero	BPA 6	ATCC,75711	Multilocus microsatellite (Cont-6, -16, -14, FJ883905, FJ883939, FJ883922; Tand-3, -12, -13f, FJ824925, FJ824948, FJ824973); SRS gene (AY940482)	–	28, 29
	Clinical, 39 days old	KO	Vero[a]	NC-Illinois	ATCC,PRA-139	Multilocus microsatellite (Tand-3, -8, -9, -12, -13, -15, -26, -30, -31, FJ824927, FJ824940, FJ830455, FJ824961, FJ824988, FJ825001, FJ830470, FJ830475, FJ830487); ITS1 (AY259041)	No isolation in gerbils inoculated with calf brain. Cultures infectious to a rabbit	28, 29, 780, 1315
	Clinical, newborn	SW	M617[a]	NC-Beef[b]	ATCC,PRA-140	Multilocus microsatellite (Cont-6, -14, -16, FJ883903.1, FJ883920.1, FJ883937.1; Tand-3, -4, -8, -12, -13, -24, -31, -32, FJ824926, FJ824997, FJ824938, FJ824946, FJ824971, FJ830460, FJ830481 FJ825014); ITS1 (AY2590401, AF249968)	SW mice were immunosuppressed with corticosteroids	28, 29, 1309, 1314
		NI	Cell culture	S-197	NI	Not provided	–	1314
	Fetus, aborted beef cow	NI	Vero	VMDL1	–	PCR-RFLP	–	963

[a] Inoculated with brain homogenate of KO mouse.

[b] Details of the isolation of the NC-Beef1 isolate are given here for the record for clarity. The isolate came from a clinical calf from a beef cattle herd that had suffered neosporosis-associated abortions.[1314] Unfixed brain from the calf was sent by Dr. M. McAllister and received by J. P. Dubey on February 18, 1998. Homogenized calf brain was inoculated into SW mice that were orally medicated with dexamethasone from the day of inoculation. Twenty-two days later homogenate of the liver of an ill mouse was seeded onto M617 culture; 5 days later *Neospora* tachyzoites were seen. The cortisonized mice inoculated with cell culture-derived tachyzoites were used for feeding to dogs; dogs fed tissue cysts of this strain excreted oocysts as reported in References **1180 and 1309**. Dr. McAllister sent the brain from another calf to Dr. David Brake (company Pfizer) and the isolate obtained in cell culture was designated S-197 (Personal communication from Dr. M. McAllister to J. P. Dubey, March, 2016).

[c] Possible laboratory contamination with NC1 strain, based on microsatellite results.

Abbreviation: NI = No information; KO = Gamma interferon gene knockout mice; SW = Swiss Webster outbred mice; Clinical = Clinical neosporosis.

4.2.2 Cows during Pregnancy

Many aspects of neosporosis induced abortion in cattle are unknown. Numerous experiments have been performed by different research groups in dairy, beef, and cross breeds. Dairy cattle were mostly Holstein-Friesian and the beef breed was mostly Angus. Because breed was not a determinant, these data are omitted from Table 4.7. Most cattle were inoculated IV with culture-derived 10^6–10^8 tachyzoites; the dose was also omitted from Table 4.7 for simplicity. The route of inoculation could affect the outcome of infection and should be considered while evaluating results. After SC inoculation, tachyzoites multiply locally and in draining lymph nodes, allowing the host to react to invading parasites. However, there is great variability in the course of infection between animals inoculated SC. After IV inoculation, tachyzoites can reach the placenta and fetus quickly but the process is unnatural because the host is presented with a large bolus of parasites at 1 time. It is noteworthy that, except for mild pyrexia, all adult cattle inoculated with *Neospora* remained healthy, irrespective of the route of inoculation, stage of the parasite and dose inoculated, and age of cattle.

4.2.2.1 Early Studies

In 3 early studies,[136,374,552] only a few cows were used. The main purpose was to fulfill Koch's postulates using dog- or cattle-derived isolates of *N. caninum*. Of the 3 cows inoculated SC and IM with a mixture of tachyzoites and bradyzoites of the 3 canine strains of *N. caninum* available at that time, 1 cow aborted an autolyzed fetus, 1 had a mummified fetus, and the third cow had a severely diseased but live fetus when removed 32 DPI.[552] This live fetus had disseminated neosporosis with severe necrosis. In the second report, 2 cows were inoculated IV with a bovine *Neospora* isolate at 120 days of gestation.[374] Congenital infection was induced in both cows; 1 fetus removed by Caesarean section was infected, *Neospora* was demonstrated in sections of brain and isolated by bioassay, and the fetus had *N. caninum*-specific antibodies. The second cow delivered a congenitally infected calf.

In the third study, 2 fetuses directly inoculated intra uterus with a bovine strain of *N. caninum* in the fetal leg died of disseminated neosporosis between days 16 and 17 PI.[136] The sonogram detected a heart beat on day 16 but the fetuses were dead the next day. It is noteworthy that although there were disseminated infections with demonstrable tachyzoites, viable *N. caninum* could not be recovered by bioassay in cell culture. This study described lesions in 6 other fetuses whose dams were inoculated with *Neospora* at 85–161 days gestational age.

4.2.2.2 Pathogenesis of Neosporosis

The following information is derived from Table 4.7.

4.2.2.2.1 Parasitemia

Parasitemia has been demonstrated sporadically between 2 and 41 days in inoculated cows by detecting parasite DNA in peripheral blood (white blood cells), depending on the route and the strain of the parasite (Table 4.8). The very small amount of blood sample could account for the sporadic nature of results obtained.

4.2.2.2.2 Invasion of Placenta and Fetal Tissues

In studies from a laboratory in Scotland, 34 cows were euthanized at 14, 28, 42, and 56 DPI with NC1 strain; 20 cows were inoculated at day 70, and 14 cows were inoculated at day 140 of gestation.[1244,1264] Complete necropsies were performed on cows and their progeny. Of the 7 live fetuses

Table 4.7 Outcome of Pregnancy in *Neospora*-Seronegative Cows Inoculated with *N. caninum* during Pregnancy

No. of Cows	*N. caninum* Strain, Route[a]	Day of Gestation	Outcome of Pregnancy	Main Focus	Reference
3	NC1, 2, 3 SC, IM	129	Euthanized day 32 DPI, fetus alive, infected. Disseminated lesions	Histology	552
		126	Aborted macerated fetus 101 DPI, not examined		
		81	Mummified fetus 74 DPI, not examined		
2	BPA1, IV, IM	120	Caesarean section, fetus removed 32 DPI, infected, lesions. *Neospora* detected histologically and viable *N. caninum* recovered by bioassay	Antibody responses	374
			Infected calf born with mild neurologic signs. Mild lesions in brain		
8	BPA-1, IV, IM	118 (IU)	2 dead fetuses removed 17 DPI, disseminated lesions	Histology	136
		161	Fetus removed 29 DPI, infected		
		138	Fetus removed 30 DPI, infected		
		120	Fetus removed 31 DPI, infected		
		85	Fetus removed 26 DPI, not infected		
		95	Mummified fetus removed 67 DPI, infected		
		115	Live, infected calf, no lesions		
		120	Live calf born infected (precolostral *N. caninum* antibodies). No lesions		
6	Nc-Liv, IM	70	Fetuses in 5 of 6 cows died *in utero* at 3 weeks, resorbed. The sixth cow delivered a normal uninfected calf	Immunity	1995, 2097, 2099
6		210	All 6 cows had normal calves infected with *N. caninum*		
7	BPA-1, IV, IM	113–122	No viable fetus from these cows. 2 aborted 24 and 33 DPI, fetuses not found, placentas were infected. Fetuses from all 5 cows died *in utero*, 26, 26, 27, 29, and 32 DPI. All 5 were infected	Immunity	76
5	BPA-1, IV, IM	159–169	Fetuses removed 9 weeks PI, all were live but infected. Mild lesions in all fetuses but *N. caninum* seen in only 1 fetus	Immunity	77
3	Nc-Liv oocysts, PO	70	All 3 cows had live calves serologically negative	Pathogenesis	1994
6	NC1, SC	140	All calves born alive, killed 6 weeks after birth % of 6 calves PCR positive	Immunity	976
4	Nc-Illinois, IV	110	Cows killed 3–4 weeks, PI. All fetuses alive but infected	Immunity	38, 43
14	NC1, SC	140	Cows killed at 14, 28, 42, and 56 DPI. All fetuses were live. Fetuses PCR positive in 10 fetuses, including 1 at 14 DPI	Pathogenesis, immunity	147, 1264

(Continued)

Table 4.7 (Continued) Outcome of Pregnancy in Neosporosis in Neospora-Seronegative Cows Inoculated with N. caninum during Pregnancy

No. of Cows	N. caninum Strain, Route[a]	Day of Gestation	Outcome of Pregnancy	Main Focus	Reference
4	Nc-Liv, IV	70	All 4 fetuses died in utero, 3–5 weeks PI, Fetuses were recovered following prostaglandin injection. All fetuses were PCR positive with histologically demonstrable organisms	Immunity	2098
8	NC1, IV	70	Cows were killed 14, 28, 42, and 56 DPI. At 14 DPI. 2 fetuses were alive, at 28 DPI, no live fetuses, at 42 and 56 DPI, fetuses not found. Placental lesions in all cows, N. caninum not found in fetal tissues	Pathogenesis, immunity	151, 1244
8	NC1, SC	70	Cows were killed 14, 28, 42, and 56 DPI. At 14 DPI, 2 live fetuses, at 28, 42, and 56 DPI, only 3 fetuses were detected from 6 cows. Placental lesions, N. caninum not found in fetal tissues		
3	NC2 oocysts, PO	141–176	Fetal infection in 1 of 3 cows. N. caninum isolated from fetal tissues	Pathogenesis	222, 780
14	NC-Beef, via	70–130	Fetal infection in 4 cows, 1 aborted calf, 1 stillborn calf, 2 healthy full-term calves. N. caninum shown by histology and by PCR		
2	NC-Illinois, via	120	Fetal infection in 1 fetus. N. caninum isolated and shown by PCR		
9	NC1, IV	150	Healthy, full-term uninfected calves, post colostral antibodies	Immunity	470
6	Nc-Liv oocysts, 40,000, PO	70	Healthy, full-term uninfected calves	Pathogenesis	1317
6	Nc-Liv oocysts, 40,000, PO	120	1 stillborn, 5 healthy, full-term uninfected calves		
6	Nc-Liv oocysts, 40,000, PO	210	1 aborted, 4 had healthy, infected calves		
6	Nc-Liv, IV	70	Euthanized 3 weeks PI, fetuses died. Fetal death associated with placental lesions	Pathogenesis, immunity	766, 1743, 1744, 1745
6	Nc-Liv, IV	210	Euthanized 3 weeks PI, fetuses alive. Mild lesions in placenta		
5	NC1, IV	70	Monitored for 45 days. Euthanized when fetuses died in utero. Fetal death between 26–34 DPI	Pathogenesis	1728
5	Nc-Spain1H, via	70	Monitored for 45 days. Euthanized when fetuses died in utero. Fetuses remained viable. Lesions and parasites		

(Continued)

Table 4.7 (Continued)   Outcome of Pregnancy in Neospora-Seronegative Cows Inoculated with N. caninum during Pregnancy

No. of Cows	N. caninum Strain, Route[a]	Day of Gestation	Outcome of Pregnancy	Main Focus	Reference
3	NC-Illinois, IV	110	Euthanized 3 weeks PI, 3 fetuses alive, 1 fetus had disseminated infection	Pathogenesis, immunity	42, 47
3	NC-Illinois, IV	110	Euthanized 6 weeks PI, 2 fetuses alive, 1 fetus dead		
3	NC-Illinois, IV	110	Euthanized 9 weeks PI 3 fetuses alive, fetus had lesions		
3	Nc-K9WA, SC	150	Euthanized 119, 119, 99 DPI, 2 uninfected live calves, 1 uninfected dead fetus	Pathogenesis	2089
12	NC1, SC	210	Cows in groups of 3 euthanized at 14, 28, 42, 56 DPI. No abortion. All fetuses viable. Fetuses and placenta negative at 14 DPI	Pathogenesis, immunity	153, 189, 293, 295, 296
7	Nc Spain7, IV	65	4 cows aborted between 28 and 35 DPI	Pathogenesis	313
4	NC1, IV	65	1 cow aborted, between 28 and 31 DPI, 1 cow had dead fetus 42 DPI		
9	Nc-Nowra, IV	65	1 fetus infected, 8 calves not infected. None aborted	Immunity	2078
7	NC6 Argentina, NC1, IV	65	Cows euthanized 7 weeks later. 1 cow had severely autolyzed fetus, 1 cow aborted-both fetus not suitable for diagnosis. The other 5 fetuses had lesions and parasites	Pathogenesis, immunity	106, 866
6	Nc-Spain7, IV	70	Fetal death, median day 34. Lesions in fetal tissues and placenta	Pathogenesis	1680
6	Nc-Spain8, IV	70	Fetal death, median day 41. Lesions in fetal tissues and placenta		
5	Nc-Spain1H, IV	70	4 fetal dead, 26–83 DPI, 1 alive	Immunity	1732
5	NC-Bahia, IV	70	1 aborted 42 DPI	Pathogenesis	348
2	NC1, IV	70	Fetal death in all 3 cows, 35 DPI		
2	NC1, IV	161	No abortion. Infected fetuses	Pathogenesis	1392
6	Nc-Spain7, IV	110	Euthanized 42 DPI, 1 aborted 14 DPI, 1 aborted 21 DPI, and 1 mummified fetus. The placentae were infected	Pathogenesis	1435, 1436, 1833

[a] Administration route: SC = subcutaneous, IM = intramuscular, IV = intravenous, IU = intrauterine; PO = per os.

Table 4.8   Detection of _N. caninum_ DNA in Peripheral Blood of Experimentally Infected Cows

Parasite Strain	Route	No. of Cows	Tachyzoite Dose	Gestational Day	Days Tested	Result	Reference
NC1	IV	8	5×10^8	70	Daily for 2 weeks or more	Positive on 2 and 5 DPI	**1244**
	SC	8				Negative	
	SC	8	10^7	140		Negative	**1264**
	SC	6	5×10^8			Negative	
	SC	11	5×10^8	210	Daily for 2 weeks	Positive in all 11 cows, between 8 and 14 DPI[a]	**189**
Nc-Spain 1H	IV	5	10^7	70	2 times per week for 7 weeks	Only once in all 5 cows[b]	**1728**
	IV	5				Positive in 3 of 5 cows[c]	

[a] On day 1 only in 3 cows, on day 2 in 4 cows, on day 3 in 2 cows, and on day 4 in 2 cows.
[b] On day 5 in 2 cows, on day 12 in 1 cow, on day 19 in 1 cow, and on day 24 in 1 cow.
[c] On days 3 and 5 in 1 cow, on day 19 in 1 cow, and on day 41 in 1 cow.

examined 14 DPI, lesions and _N. caninum_ were detected in the placenta of 4, and in the brain of 1 fetus. This is the earliest demonstration of _N. caninum_ in fetal tissues. The intensity of infection increased in fetuses examined at 28 DPI.

In another study, fetal death occurred in 5 of 6 cows inoculated on day 70 of gestation.[2097] The outcome of pregnancy varied a lot, from abortion, birth of congenitally infected calves to birth of uninfected calves among different studies summarized in Table 4.7.

The following account of pathogenesis of neosporosis is derived largely from a previous review.[605] Lesions vary with infection of fetus at different stages of gestation. At around 100 days of gestation, the bovine fetus starts to become immunocompetent. Thus, in the first trimester, the fetus is exceptionally vulnerable to the effects of _N. caninum_ infection if it should occur, although transplacental transmission is less likely at this stage. Later in the middle third of pregnancy, fetuses may be able to mount an immune response to _N. caninum_ infection[38,77,147,979] but this may or may not be sufficient to mediate protection.

In experimental infections, the most severe lesions were found in the placenta and fetal brain. Earliest lesions were found in placenta (Table 4.9). _N. caninum_ invaded cells in the maternal placenta, it caused focal destruction by multiplying in both maternal and fetal tissue at the materno-fetal interface and elicited a largely nonsuppurative inflammatory response[147,1244] (Figures 4.14 and 4.15). As early as 14 DPI, _N. caninum_ multiplied in maternal and fetal villi with villous necrosis, sometimes with serum leakage between fetal villus and maternal septum, and nonsuppurative inflammation in the maternal septa of cows inoculated at 70 days of pregnancy.[295,1680,1728] At this early stage of gestation, fetal inflammation was largely absent. In cows inoculated IV at 70 days of gestation, all fetuses were lost after 14 days. In cows inoculated SC, lesions were only seen in half the cows infected. At 28 DPI, there was breakdown of the placentome with separation of fetal cotyledons from maternal caruncles.[1244] At later time points, autolysis of the maternal caruncular tissues and the fetal elements of the placenta were found to have been rapid, and maternal uterine tissues were returning to normal, with re-epithelization of the resolved caruncles.

Coincidental with the onset of placental infection, the parasite enters the fetal bloodstream and invades fetal tissues, with a predilection for the brain (Figure 4.16).[1244] Here, _N. caninum_ initially locates in and around blood vessels[130,132,552] and, in the younger fetus, its uncontrolled multiplication can cause lethal widespread destruction of the neuropil, with little or no inflammation (Figure 4.16a and b). In aborted fetuses, the hemmorhages and focal malacia are seen in experimental infections in live fetuses removed from experimentally infected cows after euthanasia, and also are often masked by autolysis.[261,1503] Older fetuses have a higher degree of

Table 4.9   Lesions Observed in Placentomes and Tissues of Fetuses after Experimental Infection of Dams with *N. caninum*

Time after Infection	Placentome	Brain, Spinal Cord	Liver	Other Tissues
Early (2 weeks PI)	Focal necrosis of maternal septae and fetal villi serum leakage between maternal and fetal tissues; occasional degenerative fetal villi, with necrosis of fetal trophoblast cells and surrounding maternal epithelial cells[189,1244,1264]	Focal malacia,[1244] few focal lesions with perivascular inflammatory cells (lymphocytic and microglial cells), no lesions[189,1264]	No[1264]	No[1264]
Medium (3–5 weeks PI)	Mixed inflammatory response in maternal septa (infiltrations by lymphocytes, histiocytes, neutrophils, and focally eosinophils); focal stromal necrosis affecting the maternal septa; autolysis and necrosis of fetal trophoblast cells; acute degeneration (nuclear pyknosis and karyorrhexis), coagulative necrosis and autolysis of fetal villi; breakdown of the placentome; separation of fetal chorion from the caruncle, with coalescing areas of serum leakage and hemorrhages[136,189,766,1244,1264,1680]	Focal areas of coagulative necrosis, focal encephalomyelitis, focal microgliosis, perivascular mononuclear cell infiltrations (lymphocytes, macrophages, plasma cells), mild focal mononuclear meningitis[136,189,766,1264,1317,1680]	Multifocal to coalescing necrosis of hepatocytes associated with non-purulent inflammatory infiltrates, periportal inflammatory infiltrates of mononuclear cells[136,189,766,1264]	Scattered individual necrotic parenchymal cells and mild inflammation, multifocal interstitial infiltration of macrophages, lymphocytes, and plasma cells (myocytes in skeletal muscle, renal tubular epithelial cells, pulmonary and pancreatic parenchymal cells and cells in thymus, spleen, and bone marrow[136,189,766,1264,1317,1680]
Late (≥6–7 weeks PI)	Accumulations of mononuclear inflammatory cells in the maternal, caruncular septa adjacent to lesions; fetal villi adjacent to focal lesions are replaced by fibrous tissue with disappearance of trophoblast cells[189,1244,1264,1680]	Multifocal nonsuppurative encephalomyelitis, focal encephalitis with gliosis, focal nonsuppurative myelitis with clusters of tachyzoites[77,784,1264,1680]	Focal of non-purulent hepatitis[189,1680]	Non-purulent interstitial pneumonia characterized by alveolar walls thickened by the infiltration of macrophages, lymphocytes, and plasma cells, infiltrates of mononuclear leukocytes in muscular tissues, non-purulent nephritis and myocarditis[189,784,1680]

immunocompetence and are better able to respond to the parasite. They are able to restrict multiplication, and necrosis is confined to small foci of damage surrounded by a relatively intense fetal inflammatory responses involving microglia, reactive astrocytes, and cells of the monocyte and lymphoid system (Figure 4.16c)[136,1536,1812,2121] that may become mineralized.[235,790] Associated mild meningitis may also be present. Lesions may be present in any part of the brain but are more common in the cerebrum.[1481]

Aborted fetuses infected with *N. caninum* have multifocal necrosis and widespread mononuclear infiltrations in many tissues. Destruction of fetal cells and associated lymphoid inflammation may occur in several tissues including the heart, skeletal muscle, lung, and liver.[62,130,131,2121]

In some fetuses, *N. caninum* may cause characteristic lesions of inflammation and necrosis, with demonstrable parasite in tissues such as the liver and heart, while in the brain focal leucomalacia, indicative of fetal hypoxia just prior to birth, may be seen. Thus, *N. caninum* is a primary pathogen

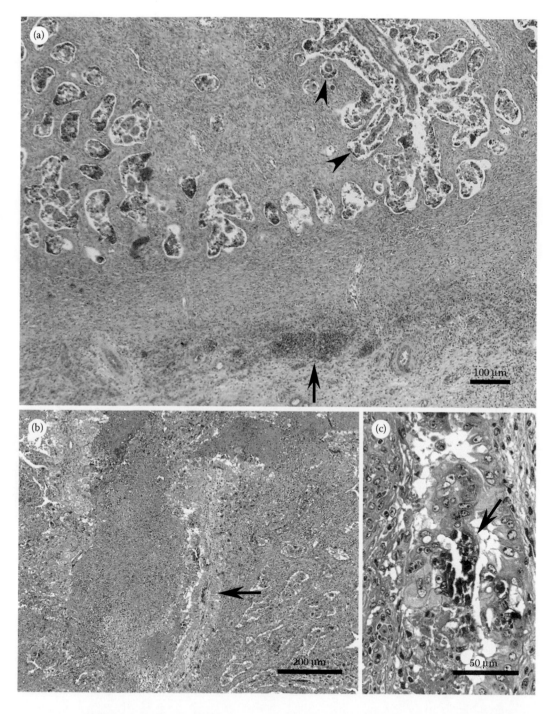

Figure 4.14 Placentitis caused by experimental neosporosis. Fetal placenta has not separated from the maternal part. HE stain. (a) Histological section of a placentome from a cow 28 days after subcutaneous inoculation with NC1 tachyzoites at 70 day of gestation. Note the maternal inflammation at the base of the caruncle (arrow), maternal septal inflammation and fetal villous necrosis (arrowheads). (Courtesy of Moredun Research Institute, see Reference **1265**.) (b,c) Focal necrosis, inflammatory foci, and mineralization (arrows) in placentas of cows inoculated intravenously with Nc-Spain7 strain at 110 days of pregnancy, and euthanized 6 weeks later. (From Almería, S. et al. 2016. *Exp. Parasitol.* 168, 62–69.)

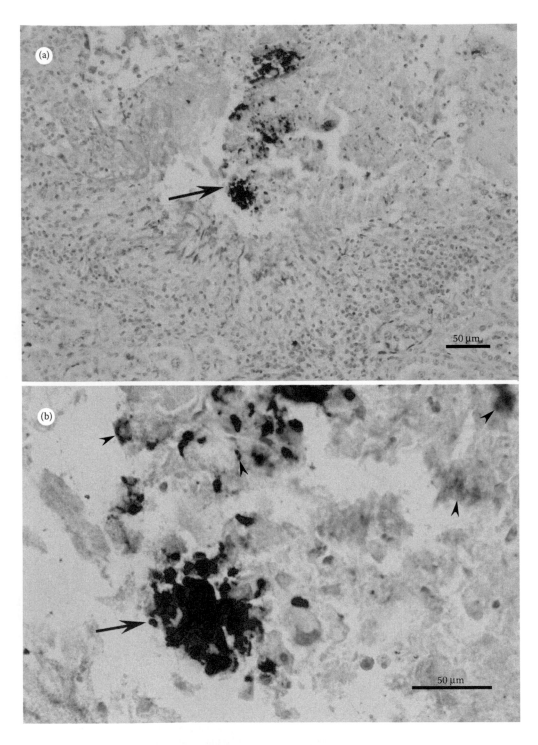

Figure 4.15   Placentitis caused by experimental neosporosis. Lesions in maternal and fetal placenta in a cow, 21 days after intravenous inoculation of Nc-Illinois tachyzoites. (From Dubey, J. P. et al. 1998. *J. Comp. Pathol.* 118, 169–173.) Fetal placenta has not separated from the maternal part. IHC staining with polyclonal rabbit *N. caninum* antibodies. (a) Large focal necrosis at maternal-fetal junction. Many tachyzoites are present (arrow). (b) Higher magnification showing intact tachyzoites (arrow), and debris (arrowheads) reacting to antibodies.

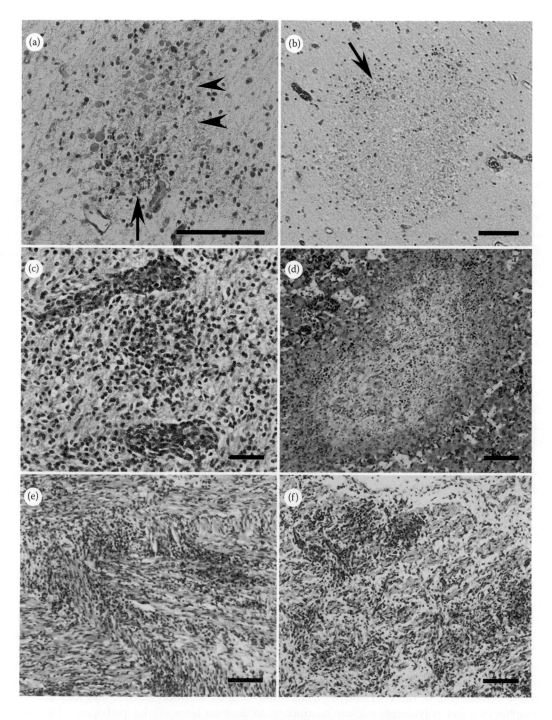

Figure 4.16   Lesions in live fetuses removed after euthanasia of experimentally infected cows. (a, b) 32 days postinoculation [Adapted From Dubey, J. P. et al. 1992. *J. Am. Vet. Assoc.* 201, 709–713]; (c–f) 21 days post inoculation [Adapted from Almeria et al. 2010. *Vet. Parasitol.* 169, 304–311.]. HE stain. (a) Hemorrhage (arrow) and necrosis (arrowheads) in cerebrum. (b) Focal necrosis (arrow) of neuropile with few mononuclear cells at the periphery. (c) Vasculitis and an inflammatory focus in the spinal cord. (d) Focal necrosis in liver. (e) Mononuclear cell infiltration in tongue. (f) Myositis in quadriceps. Bars = 200 μm.

that is capable of causing abortion either through maternal placental inflammation, maternal and fetal placental necrosis, fetal damage, or a combination of all three.

4.2.2.2.3 Oocyst-Induced Infection

In 3 similar studies, out of the 23 cows that became infected after feeding oocysts and followed until abortion or birth (Table 4.7), clinical infections were found only in cows inoculated during mid-gestation (120–130 days).[1316] All 7 cows inoculated in early pregnancy (70 days) delivered uninfected calves. Six of 7 cows infected late in pregnancy (162–210 days) delivered infected but otherwise healthy calves. In contrast, 3 of the 9 cows infected at mid-gestation aborted or had still-born calves. These effects are probably related to the permeability of the placenta at different gestation periods; the placenta is most permeable in the last trimester of pregnancy as documented for toxoplasmosis in sheep and humans.[613] In contrast to these laboratory findings, neosporosis-associated abortion storms occurred in cattle herds where as many as about 30% of the pregnant dams aborted within a short-time frame. It is highly unlikely that in nature cows ingest more oocysts than used in the experiments shown in Table 4.7. Other factors involved in natural infections have not been elucidated.

4.2.2.2.4 Pathogenesis of Abortion

4.2.2.2.4.1 *General Observations*—Abortion can be caused by cell death due to multiplication of *N. caninum* in the placenta, by cytokines that are detrimental to the maintenance of pregnancy, soluble mediators secreted locally that allow the producing cell to exert a powerful local effect on certain other cells of lymphoid and non-lymphoid origin, and hormonal regulation.[38,42,43,47,151,261,293,296,313,866,1743–1745,2097,2098] Similarly, it has been suggested that placental infection and inflammation may trigger prostaglandin-induced luteolysis causing premature uterine contraction and fetal expulsion.[296,605,1023,1652,1833]

There is speculation as to what extent antibody titers might be associated with abortion risk or protection. Immunity to neosporosis is cell mediated rather than humoral. Thus, serologic responses are just indicative of parasite exposure. In addition, antibody titers can fluctuate, especially during pregnancy, and might be affected by the virulence of the infecting *N. caninum* strains, and the subclasses of antibodies.[41,78,725,792,825,1039,1217,1305,1387,1484,1657,1774,1776,1792,1794,1908,2063,2125,2150]

4.2.2.2.4.2 *Cytokines*—Cytokines are a large group of proteins, peptides, or glycoproteins secreted by a broad range of cells of the immune system: macrophages, B lymphocytes, T lymphocytes, and mast cells, as well as endothelial cells, fibroblasts, and various stromal cells. Cytokines function as intercellular messenger molecules. There are many types of cytokines (chemokines, interferons, interleukines, lymphokines, and tumor necrosis factor). A given cytokine may be produced by more than 1 cell type. T-lymphocytes are divided into CD4+ and CD8+ T cells. The majority of CD4+ T cells are classified as either Th1, Th2, Th17, or T regulatory (Treg) cells, based on the cytokines they produce and the type of immune activity they promote. Th1 cells are the primary source for the proinflammatory cytokines IFN-γ, IL-2, and TNF-α. Th1 cytokines play a critical role in the host defense against intracellular microorganisms, and they activate lymphocytes and polymorphonuclear neutrophils to destruct intracellular pathogens. Th2 cells secrete anti-inflammatory cytokines, such as IL-4, IL-5, IL-6, IL-9, IL-13, and IL-25 (IL-17E). Th2 cells mediate the activation and maintenance of the humoral, or antibody-mediated, immune response. They induce strong antibody production, eosinophil activation, and they inhibit several macrophage activities. Treg cells are essential to maintain the homeostasis of cellular subsets involved in the adaptive immune response, either by contact-dependent suppression or by releasing anti-inflammatory cytokines such as IL-10 and TGF-β1. More recently, a third group, Th17 helper cells have been identified, which produce the signature cytokines IL-17, IL-21, and IL-22. This has fundamentally changed the previously established paradigm of Th1/Th2 dichotomy (http://www.sinobiological.com/Th17-cytokines.html).

Inflammatory cytokines, however, are partly responsible for placental damage, and their expression has been considered as a possible cause of abortion in early gestation[1265,1744] or in mid-gestation.[42,43,49] Activation of inflammation at the materno-fetal interface and in the lymph nodes is necessary to halt tachyzoite proliferation, but can be detrimental to the fetus, and thus the success of the pregnancy. High IFN-γ levels have been observed in both aborting and non-aborting cows,[2097] and transplacental transmission occurs even if high levels of IFN-γ are produced.[38,50] Fetal death may occur because of the adverse effects of cytokines,[42,50,147,296,766,979,1265,1652] particularly in the placenta.[1680] The severity of placental lesions and the strong IFN-γ response in some fetuses, possibly as part of the immune response trying to control the high parasitemia, could have been the cause of fetal death. These results showed that IFN-γ is required to limit parasite proliferation, but a critical threshold of the IFN-γ response is required to limit adverse effects on pregnancy.[47,48]

Proinflammatory cytokines may be counter regulated by IL-10, IL-4, and TGF-β, which are expressed at the materno-fetal interface to avoid fetal rejection.[974,981] However, they at the same time render the pregnant host unable to control Neospora infections by allowing parasite proliferation and transplacental transmission.[38,1745] Several studies have demonstrated the concomitant production and/or upregulation of both Th1 cytokines (mainly IFN-γ) and Th2- and Treg-cytokines during the course of experimental N. caninum infection in both aborting and non-aborting dams. This can occur upon infection early in pregnancy,[151,1680,1743,1744] at mid-gestation,[40,43,47,50] or at late gestation.[153] The upregulation of Th1, Th2, and Treg cytokines such as IL-2, IFN-γ, IL-12p40, TNF-α, IL-18, IL-10, and IL-4 was more marked when fetal death was observed in early gestation.[1744]

The upregulation of Th1, Th2, and Treg cytokines has also been frequently described in placenta.[43,47,296,868,1680,1743,1744] Mixed Th1–Th2 patterns were evident in the placenta of naturally infected pregnant dams that experienced recrudescence of a persistent infection between 20 and 33 weeks of gestation, with upregulated expression of IFN-γ and IL-4, but also increased expression of IL-12p40, IL-10, TNF-α, and increased MHC II antigen expression.[1745] The same was found in a study that analyzed cytokine mRNA expression levels in PBMC during the course of pregnancy in naturally infected cows compared to noninfected cattle. In addition to increased IFN-γ production, PBMC of Neospora infected dams had upregulated mRNA levels of TNF-α and IL-12p40, along with upregulation of the Treg cytokine IL10. In contrast, expression levels of the Th2 cytokine IL-4 were increased but did not differ significantly among infected and noninfected animals throughout the study period.[47] These results indicate that a partially protective immune response encompasses increased IFN-γ expression, which has to be counterbalanced by other cytokines such as IL-12 and IL-10, especially toward the end of pregnancy.

Besides IFN-γ, a second major inflammatory cytokine, IL-17A, could exert a cumulative effect in promoting either the pathological or protective role of IFN-γ.[694] IL17 CD4+ T cells and WC1+ T cells produce IL-17 in response to N. caninum infection in vitro,[1577] and IL-17 has been detected in both nonpregnant and pregnant N. caninum-infected mice.[478] However, a recent study in experimentally infected pregnant cattle[427] studied IL-17A production in aborting and non-aborting dams and their fetuses, and found that levels for this cytokine were low. IL-17 was only occasionally detected in the dams, and only negligible levels were found in their fetuses. Hence, in this experiment, IL-17A did not seem to be a major factor regulating IFN-γ production, but significantly lower IFN-γ/IL-4 ratios were observed in the dams with live fetuses, which indicate the importance of Th1/Th2 positioning in the course of N. caninum protection against abortion in cattle.

As discussed in Chapter 2 (Section 2.8.1.2.3), considerable progress has been made understanding the role of humoral and cellular immunity to neosporosis using mouse models. However, caution is needed to apply the findings to cattle. For instance, while antibodies can pass through placental tissue in mice, this is not the case in cattle. In mice, an increase of IFN-γ production is correlated to increases of IgG2a- and IgG3-synthesis, increased IL-4 production is associated with increased IgG1- and IgE-levels, and elevated TGF-β is associated with higher IgA responses. In cattle, the situation is less polarized and the classical roles of many cytokines in the laboratory mouse do not extrapolate entirely, or at all, to the situation in cattle.[656] Compared to mice, cattle have a higher number of circulating γ-δT-cells. These cells can respond to antigens and PAMPs from different disease agents, and may play a role in *Neospora* infections. In addition, in cattle, *Neospora*-specific cytotoxic lymphocytes are CD4+, while cytotoxic T cells are CD8+ in the murine model.[1900]

Defining the maternal and fetal immune responses against *N. caninum* infection in cattle at different stages of gestation will help to understand the immunological hallmarks of the complex host–parasite interplay that may result in disease of the fetus. Cell-mediated immune responses play an important role in the development of protective immunity to *N. caninum*,[607] and understanding of the pathogenesis of neosporosis in pregnant cattle requires knowledge of the role of cytokines, in particular at the materno-fetal interface.[1265]

4.2.2.2.4.3 *Immune Response of the Dams*—It has been shown that the initial timing and priming of the immune system by *N. caninum* determines the clinical consequences of subsequent infections, with different outcomes depending on whether the infection occurs *in utero* prior to birth, or whether postnatal infection takes place. First, postnatal infection of dams during gestation results in either abortion or infected calves (caused by exogenous transplacental transmission). Endogenous transmission does not require that the dam herself was infected congenitally. In a herd of beef cows that had suffered a neosporosis abortion outbreak, avidity serology indicated that most cows in the herd had acquired infection recently, consistent with a point source exposure event.[1314] However, in the following 2 years, the calculated rate of endogenous transmission in that same herd was 85%. This finding is evidence that the initial episode of horizontal infection of pregnant dams, and the occurrence of exogenous transmission, later resulted in a high rate of endogenous transmission in those same cows.[221] Transmission rate to progeny from chronically infected cows varies.[146,500,564] In contrast, cows infected with *N. caninum* prior to gestation give birth to seronegative calves without evidence of *N. caninum* infection.[976,1584,2097] Thus, these nonpregnant postnatally infected animals can clear the infection and develop the immunity that protects against abortion or transmission to successive generations. On the other hand, animals infected *in utero* through endogenous transplacental transmission remain persistently infected and can experience recrudescence and abort or transmit the infection to successive progeny.[64,134,1964] Thus, cattle infected during gestation and cattle infected *in utero* do not easily develop effective immunity to the parasite.[979]

In the field, most *N. caninum*-associated abortions occur at mid-gestation.[607] Studies performed in pregnant cattle at this time point (e.g., at 110 days of pregnancy) have reported a transient state of T-cell hyporesponsiveness to the parasite with downregulation of cell proliferation, reduced IFN-γ responses, and decreased lymphocyte subpopulations.[38] The first studies showing that IFN-γ inhibited the intracellular proliferation of *N. caninum* tachyzoites were performed *in vitro* employing infected ovine fibroblast cells treated with ovine recombinant IFN-γ.[973] Growth-inhibitory effects of IFN-γ on *N. caninum* were also found in murine macrophages and rat glial cells.[1013,1936] Similar effects were observed in IFN-γ and TNF-α treated bovine brain cells[2146] and growth-inhibitory effects of canine IFN-γ, β, and α were also found on canine cells infected with *N. caninum* tachyzoites.[1467] Enhanced gene expression of proinflammatory and immunomodulatory molecules was observed in bovine umbilical vein endothelial cells (BUVEC) infected with *N. caninum* tachyzoites.[1938–1940] Infection by *N. caninum* and presentation of *N. caninum* antigens was also demonstrated for bovine dendritic cells.[806]

The first line of immunological defense against invading pathogens in cattle is mediated by natural killer (NK) cells, which were shown to exhibit the capacity to lyse *N. caninum*-infected fibroblasts and to produce IFN-γ.[240] The NK cells show a decrease at 4–6 DPI in calves, but reach a peak at 11

DPI, simultaneous to that of CD8+ T-cells and γ-δ T-cells.[1076] This is associated with increased levels of antioxidant activity, malondialdehyde, and NO in cows naturally infected with *N. caninum*.[681] In addition, CD4+ cytotoxic T-cells are important for the killing of *N. caninum*-infected cells.[1900] Infected cattle elicit a Th1-type response associated with CD4+ T-cell activation and IFN-γ expression.[37,38,44,1265,1743,1999,2001,2097] CD4+ T cell lines from PBMC from infected cows also produced IFN-γ upon stimulation with *N. caninum* cyclophilin (Nc CyP).[1999,2000] In naturally infected cattle, IFN-γ production during pregnancy was shown to be effective in preventing abortion.[40,1220]

4.2.2.2.4.4 Immune Response in the Fetus—The immune response of the fetus also plays a crucial role for the outcome of the infection, and thus the time point of infection is an important parameter. The capacity of the fetus to mount an immune response to an invading pathogen increases progressively with time as the immune system develops. Fetuses from experimentally infected dams are able to induce significant cellular and humoral immune responses to *N. caninum* infection from 3 to 4 months of gestational age, depending on the isolate used for infection.[367,1680] In experimentally infected pregnant cows at day 70 of gestation, higher IFN-γ levels and lymphocyte proliferation were observed in the cows carrying live fetuses compared to those carrying dead fetuses.[151] Significant cytokine expression was observed in fetuses of experimentally infected heifers at 110 days of gestation.[38] Another study, however, failed to detect consistent inflammation in the uteri of infected animals.[1522] Moreover, when cows were infected at day 210 of pregnancy, abortion and lesions were not observed in the fetuses, suggesting a control of the infection by the maternal as well as fetal immune responses.[189] Although a high rate of vertical transmission was observed, parasites were observed in the placenta only from 28 DPI, while they were already present at 14 DPI when the cattle were infected earlier in pregnancy.[367,1244,1265]

A specific antibody response of the fetus against *Neospora* was found at least from day 100 of gestation onward.[151,153] Mitogenic cellular responses were detected after 12 weeks of gestation in fetal blood cells and after 14 weeks in the thymus and the spleen.[147] Infection in early pregnancy was associated with dissemination of the parasite and *Neospora*-specific necrotic lesions, but no inflammatory reactions were detected,[766,1244] while infection in mid and late pregnancy results in fewer lesions and reduction of parasite dissemination, also without inflammatory lesions.[766,1265] These findings suggest the fetal immune response contributes to the outcome of the infection.

4.2.2.2.4.5 Immune Response in Bulls—Cytokine responses to *N. caninum* infection have also been studied in bulls. Similar to calves and heifers, naturally infected seropositive bulls also had higher IFN-γ responses than controls, although high individual variations were observed over time.[671,1831] When bulls were reinfected with the parasite, mean IFN-γ levels of re-infected bulls were significantly higher than in other bull groups as early as 3 and 7 DPI.[672] Therefore, in bulls, measurement of the plasma IFN-γ level seems to be a good indicator of recent *N. caninum* infection.[672] Heifers intrauterine infected with semen contaminated with *N. caninum* seroconverted and showed specific IFN-γ responses[1830] but only in some animals (9–65 days after infection). Only *N. caninum* DNA has been detected in semen[1524] but no viable parasites were observed,[671,672] thus, there is no evidence for *N. caninum* venereal transmission in natural infection.[46]

4.3 DIAGNOSIS

4.3.1 General Considerations

Subclinical infection is the most common result. Neonatal mortality is the main clinically important outcome of neosporosis. One route by which *N. caninum* can be introduced into a naïve herd is via purchasing infected replacement cattle. Persistently infected but actually non-diseased dams heifers or calves may suffer from neosporosis in future pregnancies. Consequently, not only the diagnosis of diseased cattle but also the detection of asymptomatic but infected cattle is of importance. Diagnosis of bovine neosporosis should optimally be based on several components: (i) histological examination, (ii) detection of the parasitic genome by PCR, (iii) detection of specific antibodies in individual animals, and (iv) epidemiological evidence (Figure 4.19). Results of fetal histopathology and immunohistochemical examination are expected to have low positive predictive value and it was recommended to base final diagnosis on further information that includes lesion severity, fetal age, herd status, and exclusion of other possible abortifacients.[1968]

4.3.2 Submission of Samples to a Diagnostic Laboratory

Submission of aborted fetus, placenta, and serum/blood from the dam to a diagnostic laboratory can provide initial diagnosis.[603,607,1316,1525] Ideally, the entire chilled fetus should be submitted to a diagnostic laboratory for complete necropsy. If this is not practical, it is important that properly fixed specimens are submitted to a diagnostic laboratory. However, it has to be kept in mind that the sensitivity of the DNA extraction methods of the formalin-fixed tissues is substantially reduced. As explained earlier, examination of multiple tissues increases chances for diagnosis. If a choice has to be made by the farmer or practicing veterinarian, then brain, heart, tongue, diaphragm, liver, and placental cotyledons are most suitable for providing provisional diagnosis. These tissues should be fixed in 10% buffered formalin. Brain tissue in aborted fetuses is often soft and sometimes almost runny; this specimen can be poured in a jar with formalin (pathologists can find bits of solid tissue sufficient enough for diagnosis). A cross section of heart at midpoint and a slice of liver (away from gallbladder) can be fixed in formalin; these tissues should not be more than 0.5 cm thick. Placenta should be rinsed with saline before fixation to remove non-host materials (straw and dirt), and specimens should include both cotyledons and intercotyledonary tissue. Fixation is ideal if the amount of formalin solution is at least 10 times the volume of the tissues. After 1–2 days of fixation, specimens can be sealed in a biological specimen container, and mailed in a secure box with a biohazard sign. Volume of the sample can be reduced by pouring off excess formalin; only enough formalin is needed to keep the tissues moist. Advice for proper submission of tissues for diagnosis has been recently provided.[1316] Together with tissue samples, blood samples of aborting cattle for serological analyses should also be submitted. Bovine IgG remained stable for 48 or 118 days even at room temperature when taken sterile by venipuncture.[990]

4.3.3 Routine Histopathological Examination

Routine histopathological examination of sections of fetal tissues after staining with HE may provide initial evidence for etiology, and is most economical.[1968] Focal necrosis and inflammatory areas in the brain are the hall marks of neosporosis (Figure 4.3a). Nonsuppurative myocarditis and myositis is the second most common lesion of neosporosis (Figure 4.3e). The heart in aborted fetuses is often autolytic, but the epicardium is better preserved than the deeper myocardium. With respect to the placenta, the lesions are patchy and confined to cotyledons (Figure 4.3d), therefore inclusion of several cotyledons in specimens is recommended. As explained earlier, fetal lesions in the liver can vary in endemic (sporadic) and epidemic abortions. Hepatitis is more severe and common in epidemic versus endemic neosporosis but lesions in the brain, heart, and placenta are similar in both types of abortions.

Besides *Neospora* and *Toxoplasma*, *Sarcocystis* and *Tritrichomonas* are other protozoans that are abortifacients in cattle. *trichomonas foetus* is a flagellate and morphologically distinct from *Toxoplasma* and *Neospora*. *Neospora*, *Sarcocystis*, and *Toxoplasma* are related apicomplexans and their differential diagnosis was discussed in Chapter 2. If apicomplexan-like protozoa are found in the brain of bovine aborted fetuses, they can be assumed to be *N. caninum*. Although *T. gondii* DNA

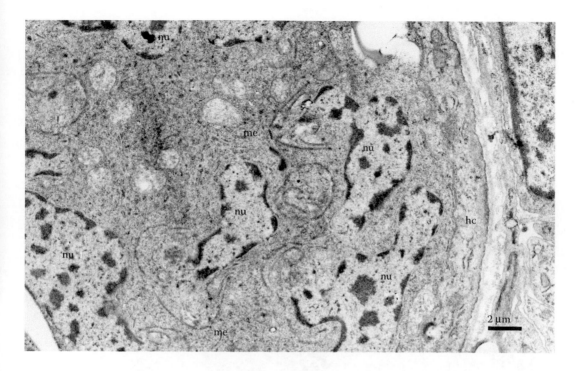

Figure 4.17 TEM of an immature *S. cruzi* schizont in a vascular endothelial cell of an experimentally infected calf. Note lobulation of the nucleus (nu) and formation of merozoite anlagen (me); hc = host cell cytoplasm. (Courtesy of Dr. C. A. Speer.)

was detected in cattle[193,640,797,1520,1757] and *T. gondii* could be isolated from aborted bovine fetuses,[280] there is no documented case of a proven *T. gondii*-associated abortion in cattle.

Although cattle are intermediate hosts to several *Sarcocystis* species,[622] only *S. cruzi* is known to be abortifacient. *S. cruzi* schizonts occur in vascular endothelial cells (Figure 4.17) and have immature stages (without merozoites). Unlike *Neospora*, *Sarcocystis* schizonts multiply by endo-polygeny where the nucleus becomes lobed and lobes remain interconnected. At the last division of the nuclear lobe, merozoites are formed (Figure 4.18). As stated earlier, *Sarcocystis* merozoites lack rhoptries (Figure 2.31).

4.3.4 Immunohistological Examinations

The number of *N. caninum* in aborted fetal tissues is very low and most of the tachyzoites are dead and autolytic. Therefore, a search for tachyzoites in the HE-stained section is unrewarding. Although tissue cysts are better preserved than tachyzoites, their number is low and often not associated with lesions. IHC staining with antibodies specific for *N. caninum*, greatly facilitates diagnosis. However, a negative reaction does not rule out neosporosis because of the low numbers of parasites. Greater than 10,000 organisms per gram of tissue are needed to find them in a 5 μm histologic section (J. P. Dubey, own observations). Usually, polyclonal antibodies (e.g., rabbit hyper-immune sera) are used for parasite detection in IHC. These polyclonal sera might not be entirely specific. The use of monoclonal antibodies might increase specificity of IHC as compared to the application of polyclonal antibodies. However, in 1 study using a monoclonal antibody specifically detecting a dense granule antigen, many PCR positives were not diagnosed positive during IHC examination.[1704] The diagnostic sensitivity of an IHC based on monoclonal relative to PCR may increase once several monoclonal antibodies are used in combination.[2015]

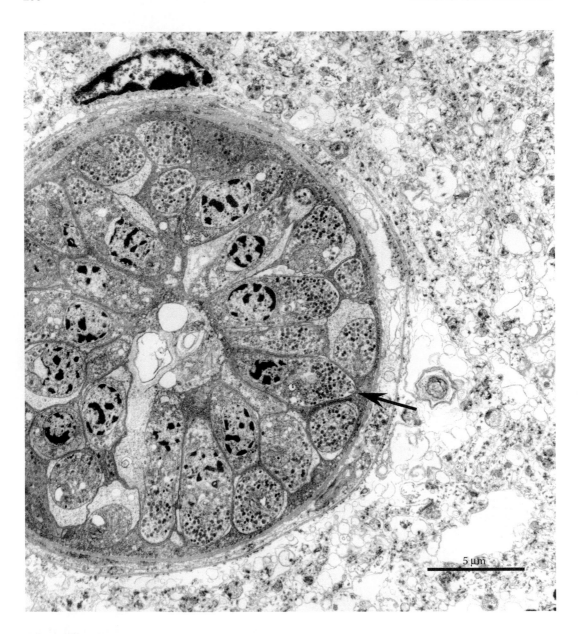

5 μm

Figure 4.18 TEM of a mature *S. cruzi* schizont in placenta of a naturally aborted fetus. Although the schizont is poorly preserved, the merozoites lack rhoptries (arrow) but have many micronemes. (Courtesy of Dr. Bradd Barr, see References **130** and **132**.)

Proficiency testing among a number of European laboratories specialized for the diagnosis of neosporosis revealed a low level of agreement between the results in IHC staining in different laboratories.[2022] It seems to be a general problem that IHC is less sensitive than PCR in detecting infection, in both situations, when monoclonal antibodies[1671] or polyclonal antibodies are applied.[2022] Monoclonal antibodies against *N. caninum* are commercially available, although their suitability for IHC examinations needs to be confirmed (e.g., *Neospora caninum* mAb gp65 IgG1 Isotype, VMRD, Pullman, Washington, USA). Polyclonal antibodies to *N. caninum* prepared in goats may react with other related parasites (*Toxoplasma* and *Sarcocystis*) if the goat used for *Neospora* immunization had been exposed to other protozoan parasites.

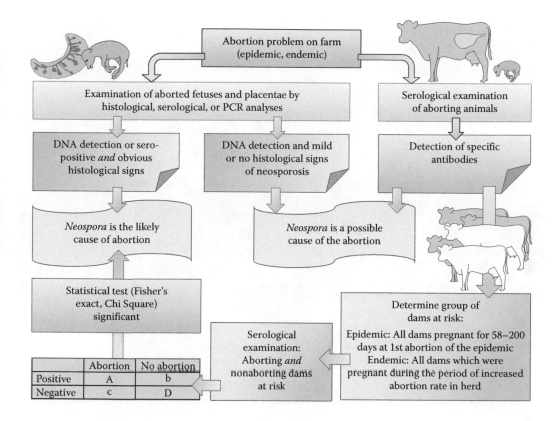

Figure 4.19   Diagnosis of *N. caninum*-associated abortion. The serologic examination of maternal sera and fetal body fluids as well as the histological and the PCR analysis may provide first but not yet definitive evidence for an *N. caninum*-associated abortion. If the lesions are severe enough to kill a fetus and if these lesions are immunohistochemically linked to *N. caninum* it may be justified to conclude that the abortion is caused by *N. caninum*. The involvement of *N. caninum* in bovine abortion may also be confirmed by the observation of a statistically significant association between seropositivity and abortion within the group of dams with an abortion risk ("dams at risk"). Definitions for "dams at risk" are provided in the text.

IHC procedure using rabbit antibodies is described in Chapter 3.

4.3.5 Demonstration of Viable *Neospora*

Isolation of viable *N. caninum* is possible as has been shown several times (Table 4.6) but has no practical diagnostic importance.

4.3.6 Detection of *N. caninum* DNA

The specific detection of *N. caninum* DNA by PCR represents one of the most important methods to determine *N. caninum* infection in aborted fetuses, stillborn calves, or calves suffering from neonatal neosporosis. PCR on DNA extracted from tissues are standard but also an *in situ* PCR has been established to perform PCR on histological sections or cell cultures[1226]; however, we are not aware that this technique which combines the sensitivity of PCR with the possibility to link positive reaction with histological observations (i.e., lesions) has been further used for *N. caninum* diagnosis.

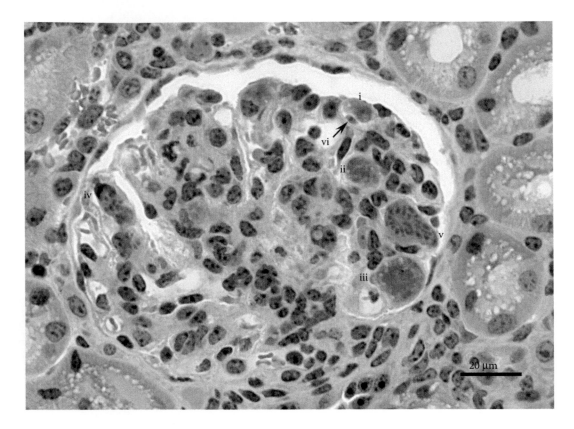

Figure 4.20 *Sarcocystis cruzi* schizonts in renal glomerulus of an experimentally infected calf. Note development of schizonts. (i) Uninucleate, macrophage-like schizont, (ii–iii) immature schizonts with division of nucleus, (iv) developing merozoites, (v) mature schizont with merozoites, and (vi) extracellular merozoite. HE stain.

Advantages of PCR are a high specificity and sensitivity of this methodology. However, the high sensitivity of the PCR method is also a disadvantage at the same time, as this technique is prone for cross-contamination and carry-over which may cause false positive results if not efficiently controlled.[378] A trial among a number of European laboratories specialized on the diagnosis of neosporosis revealed false positive findings, even in these specialized laboratories.[2022] Although highly sensitive, nested PCR protocols (if not used in a single tube format) are extremely vulnerable. On the other hand, PCR is also vulnerable to inhibitory effects if the DNA extracted is contaminated by inhibitory components and this effect needs to be efficiently controlled by internal control DNA which is amplified along with the specific targets.[378] All targets mentioned in Table 2.13 from Chapter 2 are suitable for diagnostic purposes in cattle. In the above-mentioned trial among European laboratories, the observed differences in the diagnostic performance between laboratories could not be explained by the target gene or the type of PCR applied, and it was concluded that other variables, for example, protocols followed to extract DNA, have an impact on the outcome of the PCR analysis.[2022]

4.3.6.1 Clinical Cases

PCR is an important tool in the diagnosis of clinical neosporosis. PCR was superior in terms of sensitivity to detect infection compared to other diagnostic tests (histological examination, IHC) in most studies on aborted or bovine fetuses that died *in utero* after experimental infection of dams.[266,313,784,1335,1389,1670,1757,1880] However, the concordance among different methods (PCR vs. serologic results and PCR vs. histologic results) was fair[1335,1880] or even poor.[1330,1389] In 1 study, histological examination of fetal tissues was superior to PCR examination.[1590] In some studies, sensitivity of PCR was improved by employing a nested PCR to diagnose *N. caninum* infection in bovine fetuses.[1572] One study compared 2 nested PCRs published[138,954] and observed a higher apparent sensitivity in one of these PCRs.[954] However, nested PCR protocols are labor-intensive and prone to cross-contamination problems.

Several studies in cattle indicate that placenta and brain are the most suitable for the detection of *N. caninum* DNA by PCR followed by the lung, heart, liver, and kidneys.[42,106,174,367,766,797,912,976,1264,1330,1584,1680,2151] One study suggests that the likelihood of positive findings is higher in placentomes than in the interplacentome area.[766]

DNA has been detected in fetal amniotic, allantoic,[766,912,1330] or other body fluids.[1680,2086] Amniotic and allantoic fluids tested in parallel, revealed a lower proportion of positive findings as compared to fetal tissues.[766] PCR analysis of amniotic fluid seemed to be superior relative to allantoic fluid.[766] The parasite DNA load and diagnostic efficacy can also be influenced by the stage of gestation at which a fetus was aborted.[367] In the last trimester, the parasite was only rarely detected in fetal tissues.[766]

Autolysis influences the diagnostic sensitivity of *N. caninum*-specific PCR. It is assumed that specific PCRs based on the amplification of small fragments are less affected by autolytic processes than PCRs based on larger amplicons.[643] *N. caninum* DNA has been detected even in mummified fetuses.[760,1924,2176]

In neonatally affected calves, the parasite and its DNA is localized mainly in the CNS (Table 4.5). *N. caninum* DNA was also found in brain and blood of seropositive cows that had congenitally infected fetuses.[1508]

4.3.6.2 Asymptomatic Cattle

Little is known of the tissue distribution of *N. caninum* in juvenile and adult cattle. Chances to detect infection are extremely low. In a slaughterhouse study in Brazil, only 5 of 100 beef cattle tested positive by PCR and sequencing. Both heart and brain were examined but only the brain tested positive.[177] Histopathology and quantitative real-time PCR were used to determine the tissue distribution of *N. caninum* in calves at 80 days after experimental infection and many tissues were tested (liver, spleen, kidney, heart, lung, adrenal gland, thyroid gland, pancreas, thymus, tongue, parotid gland, mandibular salivary gland, skeletal muscle, brachial and sciatic plexus, sympathetic trunk, cerebrum, cerebellum, spinal cord, eye, optic nerve, pituitary gland, gastrointestinal tract, and lymph nodes) by real-time PCR; parasite DNA was detected only in the CNS. The highest burden of *N. caninum* DNA was observed in the cerebellum, especially in the amygdala and hippocampus.[1481]

N. caninum DNA has also been detected in the blood, serum, milk or colostrum of cows,[671,1330,1411,1412,1506] in semen, and seminal cell fraction of bulls.[268,671,672,1524,1532,1831] However, clinical significance of these findings is unknown and there are studies which could not confirm the findings mentioned above.[1904,2030]

4.3.7 Detection of Antibodies

4.3.7.1 Antibody Types and Isotypes

In cattle, as in other hosts, IgM antibodies appear prior to IgG antibodies. Specific IgM levels peaked 2 weeks after *N. caninum* infection, but declined by 4 weeks[445] whereas IgG levels increased during the first weeks up to 3–6 months after experimental primary infection.[172,374,445,1792,1794,1994,2009,2097] An initial rise in specific IgG1 was followed by a slightly delayed surge of IgG2.[77,445,2097] No elevated IgA levels were observed in calves experimentally infected with oocysts.[445] Infected bovine fetuses can also mount a specific IgM and IgG response.[77,147,256,1867,1981] Immunoblot analyses revealed that in infected cattle not all antibody types recognize the same set of antigens.[1847]

4.3.7.2 Ratio of Specific IgG1 and IgG2

In persistently infected cows that gave birth to *N. caninum* positive calves, the reciprocal IgG2 titers were markedly higher than those of IgG1 7 months after they had been inseminated.[825] In contrast, in nonpregnant cows, in cows with uninfected calves, and in a cow that aborted after infection, there was no significant difference between IgG1 and IgG2 titers.[825] An examination of the specific IgG1/IgG2 ratio in 31 dams with a naturally acquired chronic *N. caninum* infection revealed a predominant IgG2 pattern in the vast majority of aborting dams. In many of the non-aborting dams, an IgG1 pattern was observed throughout the gestation (i.e., from day 40 to 210 of gestation).[40] The prominent IgG2 response suggested a T-helper cell type 1 bias in those dams in which transplacental transmission occurred although, in pregnant dams a T-helper cell type 2 bias was expected due to the inherent immunomodulation during pregnancy.[825,979,2079] For a protective immune response against *N. caninum*, a Th1 cell-biased response (associated with IgG2) is regarded as essential.[173,979,1051] However, an overwhelming Th1-biased response in pregnant cattle could affect the feto-maternal interface and cause rejection or abortion of the fetus.[42,296,979,1652,1680,1744] For further details, see Section 4.2.2.2.4 on pathogenesis of abortion.

In fetuses, specific IgG1 is dominating[77,147,256,1867,1981] and also in newborn calves, specific IgG1 dominates the IgG response in most of the calves.[1903]

4.3.7.3 Fluctuating Antibody Levels in Relation to Risk of Abortion and Vertical Transmission

Levels of specific antibodies may persist for life but are influenced by the age and stage of gestation.[301] Antibody levels fluctuate especially in pregnant dams and sometimes are below the detection limits of serological tests, depending on test sensitivity.[1484] In a study from Israel using stringent IFAT cut-offs (1:400 for positive and 1:200 for borderline), a small proportion of dams (7.7%) tested positive between 100 and 130 days of gestation, but tested negative at the end of pregnancy; 10% of the calves born to seronegative dams were found seropositive.[1301]

Reactivation of chronic infection as a putative factor responsible for endogenous transplacental transmission became a likely option to be considered when several independent studies observed an increase of specific antibody levels during bovine gestation.[78a,374,422,690,791,825,908,1484,1490,1657,1724,1908] The observation that this antibody rise usually occurs in the second half of gestation[825,1484] or after the second half of gestation[791,1484] suggested that both immunological changes in the cow and also a parasitic recrudescence could have contributed to the elevated titers. A transient immune suppression of T lymphocytes was observed in cattle experimentally infected with *N. caninum* after about 18 weeks of gestation.[976] This could increase the susceptibility of animals, and favor recrudescence probably causing parasitemia.[978,979] In 1 cow, a rise of specific antibodies was only observed around the time of abortion,[825] which suggest that parasitic multiplication affected the calf and the dam almost at the same time, which caused fetal death and triggered the antibody response in the dam. These studies indicate that the kinetic of a serological response to *N. caninum* during pregnancy may help to predict abortion or vertical transmission in chronically infected dams.[41,825,1558,1589,1657] The likelihood of abortion was 3.2 times lower in parous cows with low antibody titers to *N. caninum* than in cows with elevated antibody levels at the last screening prior to 90 days.[2150] In contrast to parous cows, levels of antibodies in heifers at the last screening prior to 90 days pregnancy diagnosis could not be linked to abortion risk.[2150] Antibody levels can fluctuate during pregnancy without effect on the clinical outcome. Cows that experienced a rise in *N. caninum* antibody levels during the third trimester delivered healthy calves.[1891a]

Not all persistently infected dams experience a rise of antibodies during gestation. In a study of 65 persistently infected non-aborting cows, 32 cows had a marked and consistent increase in specific antibodies during the second half of gestation; antibody levels in the remaining 33 non-aborting cows (also persistently infected) were almost constant throughout gestation and rose only after 210 days of gestation.[1484] Only 11 of the 21 cows (52.4%) that had aborted had an antibody peak prior to abortion. In 2 cows, no specific antibodies were detected at the time of abortion, although histological examination of aborted fetuses suggested an *N. caninum*-associated abortion and in 1 of these cows, specific antibodies had peaked in the second half of gestation.[1484] Although these 2 cows may be an exception, results indicate that diagnosis based on the assumption that abortion is always linked to a rise in specific antibodies can be false.

In cattle of different age groups in 14 British herds which had been affected by *N. caninum*-associated abortion, vertical transmission of *N. caninum* was dominating because seroprevalence did not increase by age.[429] However, there was a significantly lower prevalence in 13- to 24-month-old cattle which suggested that in cattle infected by endogenous transplacental transmission antibody levels may drop during the first months of life.[429]

4.3.7.4 Avidity Maturation

After primary *N. caninum* infection, the avidity of specific antibodies increases over time[219,222] and this can provide additional information regarding the time point and duration of a primary infection. Several avidity assays have been developed to differentiate low avidity IgG responses (indicative for a recent primary infection, approximately of 2 months duration) from high avidity IgG responses (indicative for a chronic infection) (discussed in Chapter 2). Usually, high avidity IgG responses are observed in cattle naturally infected for more than 6 months.[219,221] In field studies, low avidity IgG responses could be linked to *N. caninum*-associated abortion epidemics suggesting that a recent primary infection was the cause of abortion.[1004,1314,1759,1800] Contamination of fodder or drinking water with *N. caninum* oocysts was thought to be responsible for these infections.[1314,1315]

4.3.7.5 Antibodies in Adult Cattle due to Vaccination

After vaccination of cattle with inactivated *N. caninum* or live vaccines, the interpretation of serologic data may become difficult.[128,866,868,1302,1387,1739,2087] Animals inoculated with these vaccines developed antibody responses which were similar to those of cattle naturally infected with *N. caninum*.[77,346,866,1387] Therefore, it was proposed to develop marker vaccines together with companion serological tests to distinguish the antibody responses in vaccinated cattle from those naturally exposed.[607]

4.3.7.6 Antibodies in Fetuses

Ruminants have a syndesmochorial placentation, which does not allow maternal immuno-globulins to reach fetal tissues.[1529] Because bovine fetuses develop immunocompetence after around 120 days of gestation,[1926] many but not all infected fetuses are able to develop specific antibodies to the transplacentally invading *N. caninum* tachyzoites. However, also fetuses younger than 120 days may already be able to mount specific antibodies. At day 28 after experimental infection (i.e., at day 98 of gestation), a fetus mounted a specific IgG response (IFAT titer 1:28) but no IgM response.[151] There are numerous reports on fetal antibodies after experimental infection[77,136,147,151,784] and also in field studies, the detection of specific fetal antibodies were repor ted.[108,137,256,450,775,814,1590,1682,1867,1880,2120] The fetal antibody response is a predominant IgG1 response but also specific IgM or IgG2 are detected.[77,147,256,1867,1981]

The low sensitivity of fetal serology may be due to lack of fetal immunocompetence, especially in bovine fetuses younger than 6 months.[2120] In experimental infection of pregnant dams at day 140 of gestation, that is, after a bovine fetus is regarded as at least partially immunocompetent, specific IgM or IgG were not observed until day 42 after inoculation.[147] In another experimental infection with oocysts at day 120 of gestation, the crown-rump length of the fetus suggested that it died at day 130 of gestation, that is, 10 DPI and no specific antibodies were observed.[784] Thus, also a short interval between infection and fetal death may account for fetuses failing to mount a specific immune response.[2120] In addition, autolysis may cause degradation of fetal immunoglobulins[2120] and may lead to low levels of specific antibodies. Thus, a negative serological result in an aborted fetus does not rule out *N. caninum* infection. Western blot-based assays are shown to increase sensitivity and specificity of fetal serology.[53,1880] However, it has to be stressed, that the demonstration of specific antibodies against *N. caninum* in an aborted fetus does not allow the conclusion that the parasite was responsible for disease because the vast majority of *N. caninum*-infected fetuses infected later in gestation develop normally and are born as healthy calves.[1557] In 3 experimentally infected cows, 1 mummified fetus from a dam inoculated at 95 days of gestation had specific antibodies (IFAT 1:160) while 2 dead fetuses from the dams inoculated at day 118 of gestation remained seronegative.[136]

Serological tests can be used to study the infection status of an aborted fetus because fetal blood, serosanguinous fluids in the pleural or peritoneal body cavities, pericardial fluid, and abomasal content may contain specific antibodies against *N. caninum*.[1382] In a study from Denmark, it was concluded that the IFAT (with a cut-off titer of 1:20) was a specific method for diagnosis of neosporosis in fetuses older than 4.5 months of gestation.[1867] Others used IFAT titers of 1:32,[53] 1:16,[54] or 1:64[151] as cut-offs or validated various titers from undiluted up to 1:800.[1880,2120] There are several reports indicating a low sensitivity when fetal serology is performed with IFAT or ELISA[54,137,797,1682,1812,1867,1880,2120] even when low cut-offs were applied. Using a commercially available ELISA (i.e., CIVTEST BOVIS NEOSPORA; Table 2.11 in Chapter 2) to detect specific antibodies in fetal fluids, a less stringent optimal cut-off was established as compared to the one for diagnosis in adult cattle.[54]

In some fetuses, there is a low agreement among serological results and those obtained with PCR and histology,[1389,1590] probably due to autolysis. The main problem with IFAT for fetal serology seems to be test sensitivity, not specificity, irrespective of the serum dilution factor.[1880] Using an immunoblot did markedly increase the agreement between fetal serology and histological/IHC/PCR results.[1880]

4.3.7.7 Antibodies in Newborn Calves

Fetal infection may lead to the birth of full-term congenitally infected calves that are clinically normal, but some calves may develop neurological disease (Table 4.3). Intrauterine infection with *N. caninum* seems to induce the development of specific antibodies against the parasite in most infected calves.[65,912] The appearance of specific antibodies is not generally associated with high total serum IgG concentrations.[340] Although pathogen-specific tolerance in congenital *N. caninum* infection has been suspected,[66,815] the absence of antibodies to the parasite in stillborn or newborn calves makes an *N. caninum* infection unlikely. Newborn calves must be tested before suckling because ingested colostral IgG antibodies may cause false positive test results.[1003,1557] The magnitude of antibody titer after the ingestion of colostrum by the calf could be higher than the titer in dam serum because antibodies are concentrated in the colostrum. It was shown that colostral antibodies in calves persisted for several months.[300,906,2123] A precolostrally seropositive calf was followed for 28 days after birth. The initial high (1:320) IgM titer in the calf declined to undetectable levels by 24 days after parturition whereas the IgG titer increased from 1:6400 to 1:25,600 (weeks

5–11 post parturition) and dropped to 1:6400 again.[689] Study results from Belgium in 6 cases of calves with neurological signs suggested a limited sensitivity of IHC and PCR in neonatal bovine neosporosis but a reasonable performance of serological testing.[451]

4.3.7.8 In-House and Commercial Serological Assays

A large number of serological assays have been reported for detecting antibodies to *N. caninum* (Table 2.10). In addition to in-house tests, a number of commercial serological tests are available for the diagnosis of *N. caninum* infection (Table 2.11). Most of the commercial tests are based on previously published in-house tests and were developed to detect specific antibodies in cattle. Almost all commercial tests are ELISAs using ELISA plates; we are aware of 1 test also following the solid phase immunoassay principle, but applying a plastic card, shaped like a comb, on which purified *N. caninum* antigen is attached (ImmunoComb).[1034a]

Most serological assays used native or recombinant antigens, mostly from tachyzoites (Table 2.10). A single ELISA assay used bradyzoite-specific recombinant NcSAG4.[14] Other recombinant bradyzoites antigens (such as NcBSR4 or NcSRS9) were recognized only by a small proportion of *N. caninum*-infected cattle and seem to be unsuitable for the establishment of serological assays.[1715]

Several ELISAs have been described to test bovine sera for *N. caninum* antibodies (Table 2.10). These ELISAs utilize either whole or fixed *N. caninum* tachyzoites, aqueous or detergent-soluble total tachyzoite extracts, single native antigens or recombinant tachyzoite antigens. Among recombinant antigens, those based on parasite surface (such as NcSRS2 and NcSAG1) or dense granules (NcGRA7) dominate (Table 2.10). Recombinant antigen technique can be used to define antigenic domains on particular proteins as it has been shown for NcSRS2.[1877] Different methods have been applied to solubilize tachyzoite antigens. Some of these methods were comparatively evaluated.[2183] Based on monoclonal and polyclonal antibodies, cELISAs have been developed which detect antibodies directed against particular *N. caninum*-specific epitopes[172,176,462,1323,1863] (Table 2.10).

Most of the commercialized ELISAs for the detection of *N. caninum*-specific bovine antibodies are based on total tachyzoite lysate antigen (Table 2.11).[972,1096] However, a test using fixed *N. caninum* tachyzoites was also marketed but is no longer available.[2095] In addition, ISCOM-incorporated tachyzoite antigens, native antigen captured by monoclonal antibodies against a 65 kDa *N. caninum* tachyzoite antigen[176] or NcSRS2[755] or native affinity-purified NcSRS2[2049] are used in ELISAs currently marketed for serological diagnosis in cattle. One commercialized blocking or cELISA uses a recombinant antigen (rNcSRS2 expressed in *E. coli*) in combination with a specific monoclonal antibody labeled with peroxidase (Table 2.11).[972,1096] In most cases, commercial assays are superior in terms of reproducibility of test results, even when different batches of test kits are used. However, there are examples in the literature which show that massive variations in test performance occurred when different batches of commercial tests were used for *N. caninum* research.[145,2074]

4.3.7.8.1 Fit for Purpose Selection of Appropriate Test Cut-Offs in Serological Assays

Each of the different serological methods can be applied for different purposes. However, it has to be stressed that it is not advisable to use serological tests before evaluating them for the application in which they will be used (http://www.oie.int/international-standard-setting/terrestrial-manual/access-online/). One of the parameters to be evaluated is diagnostic specificity, that is, the proportion of test negatives among all animals being true negative. Cross-reactivities among related parasites, for example, between *N. caninum, B. besnoiti*, and *Sarcocystis* sp., or between *N. caninum* and *T. gondii* have been reported[728,869] affecting diagnostic specificity of serological tests and it is necessary to reevaluate tests for diagnostic specificities when used in new hosts or in other regions of the world.[378] Selecting an appropriate cut-off[54,56,603,1000,2049] is critical for any serological assay used for bovine neosporosis. The diagnostic sensitivity, that is, the proportion of test

Table 4.10   Characteristics and Performance of ELISA Tests Established to Test Bovine Milk for Specific Antibodies against *N. caninum*

Test (In-House or Commercial)	Type of Milk	Milk Dilution	Optimal Cut-Off (Milk)	Reference Test on Serum	Kappa (Relative to Reference)	Sensitivity (%, Relative to Reference)	Specificity (%, Relative to Reference)	Correlation Coefficient (R)	Bulk Milk (No. Herds or Animal Groups Tested; Expected Seroprevalence in Positive Herds)	Reference
IH ELISA for individual milk ISCOM ELISA	Skim milk, centrifugation 1000 × *g*, stored at −20°C	Undiluted	0.15	ISCOM ELISA, IFAT	ND (agreement 95%)	ND	ND	ND	NA	216
p38 (native NcSRS2) ELISA	Whole milk, sodium acid preserved, frozen −20°C	Undiluted	0.15	p38 (native NcSRS2) ELISA	0.85	93	93	NA	NA	1804
Commercial ELISA for individual milk IDEXX Herdcheck (now IDEXX Bov)	Whole milk, sodium acid preserved, frozen −20°C	1:2	0.216	IDEXX Herdcheck (serum)	0.80	90	90	0.86	NA	1802
SVANOVIR *Neospora*-Ab ELISA (SVANOVIR *Neospora*, now marketed by Boehringer Ingelheim)	Skimmed milk, stored at −20°C	1:2	0.145	SVANOVIR *Neospora*	0.94	96	96	ND	NA	2033
Indirect Institut Pourquier ELISA P00511/01 (IP ELISA, no longer marketed)	Skimmed milk, stored at 4°C and −20°C, centrifugation at 1000 × *g*, stored at −20°C	Undiluted	28.8	IDEXX Herdcheck (now IDEXX Bov)	ND	96.9	96.9	0.83	NA	835
NEOSPORA CANINUM ELISA antibody KIT (BIO K 192, Bio-X Diagnostics)	Skimmed milk, centrifugation at 2000 × *g*, stored at −20°C	Not stated, 1:4 (according to manufacturer)	0.1	BIO K 192	0.91	93	99	0.93	NA	792

(Continued)

Table 4.10 (Continued) Characteristics and Performance of ELISA Tests Established to Test Bovine Milk for Specific Antibodies against *N. caninum*

Test (In-House or Commercial)	Type of Milk	Milk Dilution	Optimal Cut-Off (Milk)	Reference Test on Serum	Kappa (Relative to Reference)	Sensitivity (%, Relative to Reference)	Specificity (%, Relative to Reference)	Correlation Coefficient (*R*)	Bulk Milk (No. Herds or Animal Groups Tested; Expected Seroprevalence in Positive Herds)	Reference
IDEXX Herdcheck (now IDEXX Bov)	Whole milk	1:2	0.14	IDEXX Herdcheck	0.70	75	95	0.87	NA	262
IDEXX Herdcheck (now IDEXX Bov)	Skim milk (aqueous portion after freezing at −20°C)	1:2	0.3	IDEXX Herdcheck	0.72	77	95	0.85	NA	262
IDEXX Herdcheck (now IDEXX Bov)	Frozen milk	1:2	0.1	IDEXX Herdcheck	0.52	ND	ND	ND	NA	2067
IDEXX Herdcheck (now IDEXX Bov)	Centrifuged (1000 × *g*) milk, stored −20°C	1:2	0.704	IDEXX Herdcheck	0.723	ND	ND	ND	NA	650
IDVET ID Screen *N. caninum* indirect Multi-species (IDVET *Neospora*)	Centrifuged milk, stored −20°C	1:2	7.966	IDvet *Neospora*	0.77	ND	ND	ND	NA	650
Mastazyme ELISA (no longer marketed)	Centrifuged (1000 × *g*) milk, stored −20°C	1:5	15.5	Mastazyme ELISA	ND	>73	>73	ND	2 herds; NA	1355

(Continued)

Table 4.10 (Continued) Characteristics and Performance of ELISA Tests Established to Test Bovine Milk for Specific Antibodies against *N. caninum*

Test (In-House or Commercial)	Type of Milk	Milk Dilution	Optimal Cut-Off (Milk)	Reference Test on Serum	Kappa (Relative to Reference)	Sensitivity (%, Relative to Reference)	Specificity (%, Relative to Reference)	Correlation Coefficient (R)	Bulk Milk (No. Herds or Animal Groups Tested; Expected Seroprevalence in Positive Herds)	Reference
In-house bulk or tank milk ELISA	ISCOM ELISA	Skimmed milk, inactivated at 56°C, 30 min, stored frozen −20°C	1:2	0.2	ISCOM ELISA	ND	ND	ND	ND	11 herds; NA	322
	p38 (native NcSRS2) ELISA	Whole milk, frozen −20°C	Undiluted	0.9	p38 (native NcSRS2) ELISA	ND	80	100	ND	60 artificial bulk milks; aimed at the detection of ≥20% seroprevalence	1801
	AHS in-house	Defatted by storage at 4–8°C and centrifugation 2000 × g, storage at −20°C	1:2	0.1	AHS IH	0.47 (mean, 3 sampling dates)	63 (2nd sampling)	70 (2nd sampling)	0.26–0.73 (3 sampling dates per herd)	162 herds; aimed at the detection of ≥15% seroprevalence	142
	ISCOM ELISA	Skimmed milk, centrifuged 1000 × g, stored frozen −20°C	1:2	0.1	ISCOM ELISA	ND	50	81	ND	98 herds; aimed at the detection of 2 individuals cattle with a serum ISCOM ELISA result of OD 0.35	706
Commercial bulk or tank milk ELISA	IDEXX Herdcheck (now IDEXX Bov)	Defatted by storage at 4–8°C and centrifugation 2000 × g, storage at −20°C	1:2	0.6	AHS IH	0.78 (mean, 3 sampling dates)	61 (2nd sampling)	92 (2nd sampling)	0.71–0.80 (3 sampling dates per herd)	162 herds; aimed at the detection of ≥15% seroprevalence	142
	LSI *N. caninum* milk, Laboratoire Service International, Lissieu, France (no longer marketed)	Defatted by storage at 4–8°C and centrifugation 2000 × g, storage at −20°C	1:10	0.18	AHS IH	0.72 (mean, 3 sampling dates)	53 (2nd sampling)	91 (2nd sampling)	0.54–0.64 (3 sampling dates per herd)	162 herds; aimed at the detection of ≥15% seroprevalence	142

(Continued)

Table 4.10 (Continued) Characteristics and Performance of ELISA Tests Established to Test Bovine Milk for Specific Antibodies against *N. caninum*

Test (In-House or Commercial)	Type of Milk	Milk Dilution	Optimal Cut-Off (Milk)	Reference Test on Serum	Kappa (Relative to Reference)	Sensitivity (%, Relative to Reference)	Specificity (%, Relative to Reference)	Correlation Coefficient (R)	Bulk Milk (No. Herds or Animal Groups Tested; Expected Seroprevalence in Positive Herds)	Reference
SVANOVIR *Neospora-Ab* ELISA (SVANOVIR *Neospora*, now marketed by Boehringer Ingelheim)	Skimmed milk, centrifugation at 1000 × g, stored at −20°C	1:2	0.2	SVANOVIR Neospora	ND	100	ND	ND	4 herds with known seroprevalence; aimed at the detection of ≥10%	957
IDEXX Herdcheck (now IDEXX Bov)	Skimmed milk, centrifugation at 1000 × g, stored at −20°C	1:2	0.6	IDEXX Herdcheck (now IDEXX Bov)	ND	83.3 (5/6)	100 (5/5)	0.87	11 herds; aimed at the detection of ≥15% seroprevalence	2073
NEOSPORA CANINUM ELISA antibody KIT (BIO K 192, Bio-X Diagnostics)	Skimmed milk, centrifugation at 2000 × g, stored at −20°C	Not stated, 1:4 (according to manufacturer)	0.1	BIO K 192	ND	ND	ND	0.87	38 herds; aimed at the detection of ≥10% seroprevalence	792
IDEXX Herdcheck (now IDEXX Bov)	Milk, centrifuged at 1000 × g, stored −20°C	1:2	0.398	IDEXX Herdcheck	ND	74.1	90.9	ND	3; aimed at the detection of ≥15% seroprevalence	650

Note: ND = not done; NA = not available; AHS = Animal Health Service, The Netherlands; ISCOM = Detergent extracted tachyzoite antigen incorporated in immune stimulating complex particles.

Table 4.11 Asymptomatic Congenital Transmission of *N. caninum* in Cattle from Farms with No Indication of a Point Source Exposure of Dams to *N. caninum* during or after Pregnancy

Study Type	Country	Region	Seropositive Dams or Pregnancies (% Seropositivity in Progeny)	Seronegative Dams or Pregnancies (% Seropositivity in Progeny)	Test, Titer, (Cut-Off or Provider)	Remarks	Reference
Dam-calf (precolostral)	Argentina	–	140 (73)	33 (3)	IFAT, 1:25	–	1403
	Brazil	Goias	25 (24)	65 (0)	IFAT, 1:200	Embryo transfer	74
		Bahia	11 (100)	1 (0)	IFAT (1:200 adult cattle; 1:25 precolostral calves)	HT studied (6 per 100 animals/year), familial analysis	444
		Pernambuco	Cows 18 (72.2)	95 (3.2)	IFAT, 1:200	–	1666a
			Cows 39 (43.5)	74 (8.1)	IH-ELISA		
			Heifers 13 (69.2)	32 (9.4)	IFAT, 1:200		
			Heifers 18 (50.0)	27 (11.1)	IH-ELISA		
	The Netherlands	–	36[a] (89)	14[a] (14)	iELISA-IH	–	2123
			14[b] (100)	3[b] (0)			
	Spain	Northwest	98 (50)	192 (7)	IFAT (1:1024)	–	1589
		Northwest	25 (48)	73 (0)	IFAT (1:1024)		
	UK	–	124 (95)	248 (2)	WT-iELISA (MAST Diagnostics)	HT studied	430
	USA	California	51 (88.2)	No data	WT-IHCA, ≥0.45	–	1557
Dam-progeny	Argentina	–	10 (90)	2 (50)	IFAT, 1:25	Effect of ingestion of colostrum examined	1403
		–	16 (100)	15 (0)	IFAT (1:200 adult cattle; 1:25 fetal fluids)	Embryo transfer	276
	Brazil	Mato Grosso do Sul	194 (68)	206 (30)	IFAT, 1:50	–	74
	China	–	33 (93.9)	35 (20)	iELISA	–	845
	Germany	–	15[c] (94)	43[c] (2)	IFAT, 1:50; IB; iELISA (IDEXX HerdCheck); at least 2 of 3 tests positive	Familial analysis	1792

(Continued)

Table 4.11 (*Continued*) Asymptomatic Congenital Transmission of *N. caninum* in Cattle from Farms with No Indication of a Point Source Exposure of Dams to *N. caninum* during or after Pregnancy

Study Type	Country	Region	Seropositive Dams or Pregnancies (% Seropositivity in Progeny)	Seronegative Dams or Pregnancies (% Seropositivity in Progeny)	Test, Titer, (Cut-Off or Provider)	Remarks	Reference
	Israel	–	59 (70)	125 (14)	IFAT (1:200)	–	1301
		–	59 (61)	125 (11)	IFAT (1:400)	–	1301
	The Netherlands	–	500 (73)	No data	WT-AHS-iELISA	HT studied (5% positive seroconversion)	498
		–	180 (80)	No data	WT-AHS-iELISA	Dam (1st parity)	498
		–	133 (71)	No data	WT-AHS-iELISA	Dam (2nd parity)	498
		–	94 (67)	No data	WT-AHS-iELISA	Dam (3rd parity)	498
		–	93 (66)	No data	WT-AHS-iELISA	Dam (>3rd parity)	498
		–	24[d] (58)	No data	WT-iELISA (GD-AHS ELISA)	HT studied	500
	Spain	Northeast	32 (91)	No data	WT-iELISA (IDEXX Herdchek)	Dam-progeny	1215
		Galicia	35 (97)	32 (9)	WT-iELISA (CIVTEST)	–	791
	Thailand	–	52 (58)	360 (5)	ISCOM-iELISA (in-house)	HT studied (apparent HT 5%)	326
	UK	–	Only statistics	Only statistics	iELISA (IDEXX HerdChek)	HT studied (apparent HT 2%)	2111
	USA	California	25 (100)	25 (0)	IFAT (1:80)	–	65
		Nebraska	150 (89)	41 (22)	ISCOM-iELISA (in-house)	–	221
Dam-daughter	Australia	–	74 (60)	130 (21)	WT-iELISA (IDEXX Herd-check)	–	1116
	Brazil	São Paulo	18 (50)	42 (36)	WT-iELISA	HT studied (apparent HT 20%)	1604
		São Paulo	36 (78)	215 (2)	IFAT (1:100)	HT studied (0.4–7 positive seroconversions per 100 animals/year)	302
	Canada	Ontario	619 (41)	2490 (7)	WT-IHCA (≥0.45)	HT studied (apparent HT 4%–5%)	1549
		Ontario	307 (45)	2802 (4)	WT-IHCA (≥0.70)	–	1549
		Quebec	144 (44)	No data	WT-iELISA (BIOVET)	Familial analysis	194
	Costa Rica	–	249 (68)	498 (24)	IH-WT-iELISA	–	1738
	Germany	–	31 (55)	103 (5)	P38(NcSRS2)-IH-iELISA	–	1800

(*Continued*)

Table 4.11 (Continued)   **Asymptomatic Congenital Transmission of N. caninum in Cattle from Farms with No Indication of a Point Source Exposure of Dams to N. caninum during or after Pregnancy**

Study Type	Country	Region	Seropositive Dams or Pregnancies (% Seropositivity in Progeny)	Seronegative Dams or Pregnancies (% Seropositivity in Progeny)	Test, Titer, (Cut-Off or Provider)	Remarks	Reference
	The Netherlands	–	204[c] (80)	248[c] (17)	WT-AHS-IH-iELISA	–	492
		–	526 (62)	3565 (3)	WT-iELISA (GD-AHS ELISA)	Raw data, HT studied (apparent HT 3%)	146
		–	424 (47)	3658 (4)	WT-iELISA (GD-AHS ELISA)	Adjusted for imperfect test HT studied (adjusted HT 4%)	146
		–	405 (47)	3687 (5)	WT-iELISA (GD-AHS ELISA)	Adjusted for postnatal infection of dam after birth of daugter, HT studied (adjusted HT 5%)	146
		–	405 (45)	3687 (5)	WT-iELISA (GD-AHS ELISA)	Adjusted for postnatal infection of daughter after birth, HT (adjusted HT 5%)	146
	USA	Maryland	16[d] (56)	No data	WT-iELISA	HT studied	500
		–	74 (43)	No data	IFAT	–	630
Dam-daugher/ grand mother-daughter	Australia	–	27 (74)	27 (15)	cELISA (POURQUIER)	Familial analysis	833

Note: HT = horizontal transmission paths, iELISA = indirect ELISA, cELISA = competitive ELISA, WT = Whole tachyzoite, IHCA = Kinetic ELISA-California.[1555]

[a] F1 progeny of cows which had aborted previously during an outbreak.
[b] F2 progeny of cows which had aborted previously during an outbreak.
[c] From herds with no evidence of point source exposure to N. caninum.
[d] Dams became pregnant after seroconversion during an episode of postnatal infection in the herd.

positives detected among all true positives, may change when a cut-off is changed. A less stringent cut-off enables to make a test more sensitive. Applying a lower cut-off needs careful consideration, because usually a test with a less stringent cut-off becomes less specific and the number of false positive test results increase.[378] For some application, it might be advisable also to change the test protocol (e.g., the type of antigen, antigen concentration, serum and conjugate dilution, isotype or subtype specificity of conjugate). Most of the serological tests described were developed to diagnose bovine abortion (Tables 2.10 and 2.11).

Another important application of serology is the detection of infected cattle (e.g., calves, replacement heifers, and bulls) in order to cull *N. caninum*-infected animals from herds,[833] to prevent the new introduction of infected animals into herds or other regions, to exclude infected dams from embryo-transfer,[111,1115] or to evaluate the potential adverse effect of *N. caninum*-infected donors or recipients on the success of embryo transfer.[276,461] Serological tests for the identification of infected cattle may require a higher sensitivity (i.e., lower cut-offs) than those meant to diagnose bovine abortion[54,56,1790] and not all commercially available diagnostic tests were equally suitable to detect infected animals.[56,2049] Different cut-offs might be necessary to ensure maximum sensitivity or specificity.[54] One of the constraints to establish tests to detect infection in dams (and not only disease, i.e., neosporosis) is that a reference is missing to define in population of cattle those which are true positive and those which are true negative. Different approaches are followed to overcome this problem and to validate serological tests for the purpose of detecting *N. caninum* infection or to diagnose bovine neosporosis, including gold-standard-free approaches.[54,176,283,345,703,1726,2040,2049,2095] As a potential way to define true positive dams, precolostral seropositivity of calves born to dams was proposed as an indication.[2096] Using *N. caninum* infections in cattle as an example, an alternative diagnostic procedure, called "probability diagnostic assignment (PDA)" was evaluated that used continuous serologic measures and infection prevalence to estimate the probability of an animal being infected.[1969]

IB, IFAT, and ELISA were validated to detect specific antibodies in fetuses and these examinations revealed less stringent cut-offs as compared to those established for adult cattle.[54]

In the past, there was some debate on appropriate cut-off titers for IFATs. However, a specific situation exists regarding the selection of appropriate cut-offs for this assay. Because the IFAT titers are largely dependent on the quality of the equipment used for fluorescence microscopy, it is often hard to standardize the IFAT test results among different laboratories. Consequently, a cut-off titer appropriate in one particular laboratory might not be suitable in another. Reference sera from an experienced laboratory with defined titers help establish IFAT in an as yet not experienced laboratory.

4.3.7.8.2 Suitability of Milk to Examine the Serological Status of Individual Cattle

Milk of infected cows contains specific antibodies against *N. caninum*. The major IgG isotype in milk is IgG1.[254] Some of the in-house and commercialized ELISAs (initially validated for use with bovine serum or plasma) were modified and shown to be suitable for the examination of individual bovine milk samples as well as bulk milk samples.

In-house tests[142,216,322,706,1801,1804] as well as commercialized ELISAs[262,650,792,835,1355,1802,2033,2067] (Tables 2.10 and 4.10) were modified and used to test milk. In general, comparative analyses showed that not exactly the same results can be achieved when replacing serum by milk. One exception is a report on 35 serum/milk pairs of a single herd examined by immunoblot, which revealed complete concordance.[1523]

A number of in-house tests to assay individual milk samples have been reported.[216,324,1804] A high level of agreement between serum and milk was observed when using an ISCOM ELISA[216] and also in a native surface antigen (NcSRS2 or p38) ELISA, as characterized by high kappa values.[1804] The antibody response in the milk of an individual dam largely reflects the antibody response in serum. However, relative to the serum antibody levels the specific milk antibody levels seem to be affected by other variables such as the time between last calving and sampling,[324,1802] number of lactations,[324] or age of a dam.[1804] However, several studies revealed that a significant proportion of serologically negative animals tested positive in milk, while animals which tested negative in milk were serologically positive (Table 4.10). In individual milk samples, the sensitivity of milk relative to serum testing was usually in the range of 75%–97%, while specificity was in the range of 90%–99%. Kappa values, characterizing agreement, were in the range of 0.52–0.94. Some commercial tests performed better than others (Table 4.10). This shows that serological tests should not be used for milk testing without a thorough validation. The establishment of test modifications is essential when used for milk including, for example, sample dilutions and cut-off selection (Table 4.10). However, for the same tests, different studies established different optimal cut-offs (Table 4.10). The reason is most likely that different studies applied different sets of samples. Some studies used only few herds for creating their sample set (Table 4.10). The more herds were enrolled for a respective study, the better the established cut-off was validated.

4.3.7.9 *Testing of Individual Aborting Dams or Dams at Abortion Risk*

After the occurrence of bovine abortion, stillbirth or neonatal mortality on a farm, a serological examination may provide information if these dams are infected with *N. caninum* (Figure 4.20). All of the serological tests listed in Table 2.10 can be applied for this purpose. However, in contrast to most other methods, ELISA is the most convenient to test a larger number of sera in a short period of time.

Most cows who delivered an *N. caninum*-infected fetus or calf, had specific antibodies against *N. caninum* at the time of abortion[450,1538,1880,2123] or after calving.[65,430,912] Consequently, a negative serological test result for the dam makes it unlikely that *N. caninum* was involved in abortion, stillbirth, or neonatal mortality.

In cases of epidemic and endemic abortion, antibody levels as determined by 2 different ELISAs were higher in aborting than in non-aborting dams at risk,[1792,1794] suggesting that aborting cows were more exposed to parasitic antigens than non-aborting infected herd mates. Also, many other working groups observed that dams that abort due to neosporosis often have higher *N. caninum*-specific antibody levels than infected but non-aborting dams.[41,569,792,898,1039,1217,1301,1305,1657,1774,1792,1794,1908,2063,2125,2150]

The same is true for cows that transplacentally transmit the infection to their calves.[825,35799,1301,1403,1558] Those serological tests that have a cut-off not adjusted to detect dams that are infected with *N. caninum* but adjusted to detect those animals with high *N. caninum*-specific antibody levels are useful to demonstrate the association between seropositivity and abortion[1790] or the relationship between seropositivity and vertical transmission.[430]

By IFAT, aborting dams from herds with endemic bovine abortion appeared to have lower antibody levels than dams from herds affected by recent epidemic abortion.[1792,1794] Surprisingly, the opposite was the case when ELISAs were used to examine the same animals; aborting dams from herds with endemic bovine abortion appeared to have higher antibody levels than dams from herds afflicted by recent epidemic abortion.[1792,1794] This shows that not all serological assays show the same diagnostic capability to diagnose *N. caninum* infection depending on time of infection, which appears to be recent in epidemic cases and is regarded as latent or chronic in endemic abortion.

As stated earlier, fluctuating antibody levels are observed during pregnancy[78a,374,422,690,825,1558,1657,1908] around abortion[825,1792,2124] or after calving.[1908] Serial sampling of cows that aborted *N. caninum*-infected fetuses revealed a decline of antibody titers 20–140 days post abortion.[374]

The antibody levels may drop below the detection limit of less sensitive tests and cause false negative results.[422,498,1757,2124] This is eventually the reason for cases in which *N. caninum*-infected fetuses were aborted by or positive calves were born to seronegative dams.[430,432,791,1301,1757] In 1 case viable *N. caninum* was isolated from a dam serologically negative by NAT, a commercial ELISA and IFAT.[791] In 2 chronically infected cows, no specific antibodies were detected at the time of abortion, although histological examination of aborted fetuses suggested a *N. caninum*-associated abortion. However, in 1 of these cows, specific antibodies had peaked in the second half of gestation[1484] which may indicate that antibody levels may have dropped or even disappeared after fetal death but prior to expulsion. Although infected, some animals do not develop antibodies at all, or the antibodies are not detectable by particular tests as it was reported in a calf experimentally infected by oocysts.[445] In an experimental study, *N. caninum* was detected by PCR in several tissues (midbrain, uterine lymph node, heart, skeletal muscle, and kidneys) of a dam that tested several times serologically negative using a commercial ELISA.[153,189]

The presence of antibodies to *N. caninum* in the serum of a dam allows no definitive diagnosis regarding the cause of abortion because only a low proportion of infected dams abort, and most of their calves are born infected but healthy.[825,1558,1964,2123] Therefore, a dam may have antibodies against *N. caninum*, although abortion, stillbirth, or birth of a weak calf may have had another cause. Positive serological testing of individual dams only allows one to suspect *N. caninum* infection but is no proof that *N. caninum* was involved in the reproductive failure (Figure 4.19).

4.3.7.10 Serological Testing on a Herd Level

A definitive diagnosis of bovine abortion due to neosporosis can be achieved when all dams at risk are tested and this testing is called herd-based approach.[1961] The rationale is to determine by statistical methods if the proportion of seropositive aborting dams is higher than the proportion of serological positives among non-aborting dams at risk (i.e., to determine whether abortion is statistically associated with seropositivity to *N. caninum*). Positive and negative serological results of aborting and non-aborting dams are usually entered into a 2×2 table (Figure 4.19) and examined by standard statistical tests, like Fisher's exact test or chi-square test (i.e., statistical tests for which also on-line test calculators are available).

It must be stressed that only serological results that have been obtained from dams at risk (i.e., those dams that were pregnant during the period of time when the abortion problem occurred) should be included in the analysis. The selection of dams at risk depends on the pattern of the within-herd abortion problem (Figure 4.19). In endemic (sporadic) cases, the period during which pregnant dams have an abortion risk may last several months[429,1789,1800,1966] and all dams that had been pregnant during that time are regarded at risk. In contrast, in epidemic cases, the period for pregnant cattle being at risk may last only a few weeks.[1305,1960,2125,2136] In abortion storms, dams were regarded at risk if they had been pregnant for at least 58–260 days at the day the epidemic (i.e., abortion storm) started.[1800]

By comparing the *N. caninum* seroprevalences of different age groups or groups of animals which were housed together, additional information on the route by which an infection came into a herd may be provided.[492] Once only particular age-groups housed or pastured together are affected, a postnatal, oocyst-mediated infection is most likely.

The analysis of dam-daughter or dam-calf pairs may provide information whether the infection is predominantly transmitted vertically in a herd, that is, by endogenous transplacental transmission.[1966] It is important to note that if calves are not sampled presuckling, maternal antibodies are going to interfere with the analyses. However, maternal antibodies are no longer detectable in 6-month-old calves.[906,2123] In herds in which there is a positive association regarding the seropositivity of dams and their daughters or progeny, the predominant route of infection seems to be by endogenous transplacental, but not via postnatal infection. For example, in a herd with a high number of seroconversions from negative to positive over a period of 3 years, there was no statistical association between the serological response in dams and their respective daughters confirming that a high rate of postnatal infections occurred during that time.[1604]

In general, all of the serological tests listed in Table 2.10 can be applied for this purpose. However, again ELISAs represent the most suitable tests for this purpose, because ELISAs allow the examination of a larger number of sera at 1 time.

4.3.7.11 Avidity Tests to Define the Time Point of Infection

It may become necessary to make supplementary analyses to find indication whether the infection in cattle occurred recently or is chronic. These analyses can be done in individual animals but also in groups of animals (e.g., in entire herds or subgroups of herds, e.g., in aborting dams) to identify the predominant route by which the cattle became infected with *N. caninum*. In case there are indications of recent infection, a postnatal infection via oocysts is the most likely route.

To define the stage of infection, a number of avidity ELISAs have been developed (Table 2.10; Chapters 2 and 3).[13,219,222,223,1758,1800] There seem to be differences in avidity maturation between different antigens. While an ISCOM and NcSRS2-based ELISAs could follow avidity maturation up to 20 weeks p.i., this was hardly or not possible in assays based on a crude-antigen extract as shown by an inter-laboratory comparison.[223]

Optimally, these avidity tests are applied on a herd level. Avidity test results for individual animals should be interpreted with caution, because individual animals can maintain a low avidity antibody response although infected for several years.[221] For avidity testing of an antibody response, not only ELISAs but also other serological methods are suitable. In 1 study, immunoblot was used for avidity testing[13] and by this approach differences in avidity for particular antigens were examined.[13]

Other approaches to differentiate recent from chronic infection are based on the comparative examination of antibody reactions against a limited set of antigens, that is, antigens recognized during early phase of infection and antigens more often recognized late phase of infection or during chronic infection. Employing an antigen specific for bradyzoites, for example, a recombinant NcSAG4 antigen[14] in ELISA generally revealed only low levels of specific antibodies. Reactions against NcSAG4 seem to be an indicator for chronic infection only when used in combination with an ELISA detecting antibodies elicited in the early phase of infection, as, for example, anti-NcGRA7 antibodies.[14] Other recombinant bradyzoite antigens, such as NcBSR4 or NcSRS9 were recognized only by a small proportion of *N. caninum*-infected cattle and do not seem to be suitable for the establishment of serological assays.[1715] Others observed that recombinant NcSAG1 antigen is suitable to detect chronically infected cattle but antibodies against recombinant NcGRA7 were only detected during pregnancy (most often between 6th and 7th months of gestation), eventually indicating recrudescence.[1930] In another study, antibodies levels against NcGRA7 were higher in aborting than in non-aborting seropositive dams and a possible explanation is that antibodies against NcGRA7 are indicative for a reactivated infection.[950] In experimentally tachyzoite-infected cattle, antibodies (IgG1) to recombinant NcGRA7 peaked shortly after infection, while antibodies to recombinant NcSAG1 (Ig1 and IgG2) increased steadily until 56 DPI,[904] which again suggested that NcGRA7 is a useful antigen to detect recrudescence.

To our knowledge, there is so far only 1 commercial test that offers the possibility to do avidity analyses (i.e., CIVTEST BOVIS NEOSPORA; Table 2.11). The adaptation of other commercial ELISAs to examine the avidity maturation of *N. caninum*-specific IgG in bovine sera seems to be possible, because a commercial assay for the detection of specific antibodies in ruminants was successfully validated for avidity analyses in experimentally and naturally infected sheep.

4.3.7.12 Serological Tests to Estimate the Herd Seroprevalence

Several working groups adapted ELISAs for analysis of bulk milk samples (Table 4.10). An analysis with these ELISAs provides an estimate of the seroprevalence within the group of animals that contributed to the bulk milk or tank milk sample[142,145,322,324,706,792,1801,2033,2073] (Tables 4.10 and 4.11).

Several in-house tests to estimate the seroprevalence in a group of animals or a herd (i.e., by testing of bulk or tank milk) were published (Table 4.10). Tests were able to detect herds or groups of animals with a prevalence of at least 15%–20% with sensitivities of 63%–80%.[142,1801] Test specificities reported ranged from 70%[142] to 100%[1801] which shows huge differences in test performance. Especially in small herds, the outcome of bulk milk testing is extremely variable among different sampling dates as shown in a study on 418 herds in Thailand,[325] most likely due the small and variable number of seropositive animals contributing to a bulk or tank milk sample and variations in antibody levels in milk of individual animals during gestation and between different lactations.[324] Using the same test, that is, the in-house ISCOM ELISA, the effect of milk yield and different levels of specific antibodies in the serum of individual animals on the composition of tank milk was studied.[706]

Commercialized tests were also validated to estimate seroprevalence in herds (Table 4.10). Studies aimed at the correct identification of herds with seroprevalences ≥10% or ≥15%. Correlation coefficients ranged from 0.54 to 0.87, which shows that not all tests have the same capacity to estimate herd prevalence.[142,792,957,2073] A large study with 2 commercial and 1 in-house test showed that there is variation between different sampling dates which in the case of the in-house tests were related to differences in test performance.[142] Another study showed that similar problems related to test performance or test stability may also occur in commercialized tests.[145] Further variability influencing the reliability of estimating herd prevalence is the extent to which strong or weak seropositive cattle contribute to the bulk milk sample.[145] Differences between whole milk or skimmed milk seem to be negligible.[262] In addition to their application in epidemiological studies,[792,1490,1491,1801,1803] bulk milk ELISAs are potentially important to support farmers in maintaining bovine herds free of *N. caninum* infections and to identify herds already infected with this parasite.[142,145,325]

4.4 EPIDEMIOLOGY

The following is based on a previous review[607] but has been updated.

4.4.1 Transmission and Risk Factors in Cattle

Long-term serological studies in cattle suggest that *N. caninum* infection persists lifelong. In many regions of the world, transplacental transmission represents the main route of infection in cattle[607] (Figures 4.21 and 4.22). In persistently infected dams, a reactivation during gestation is responsible for endogenous transplacental transmission.[605] In as yet naïve dams a postnatal infection during gestation, most likely caused by ingested oocysts, is responsible for exogenous transplacental transmission (Figure 4.21). Once the female calf is born healthy, it remains persistently infected and is able to transmit *Neospora* infection again (endogenously) to its offspring with the result that 1 postnatal infection event may cause several infected cattle generations[607] (Figure 4.22). Mathematical modeling was used to study how a certain prevalence is maintained in a population considering different rates of endogenous transplacental transmission.[698,699]

4.4.1.1 Prenatal (Transplacental) Transmission

N. caninum is one of the most efficiently transplacentally transmitted parasites in cattle. In some herds virtually all calves were born infected but were asymptomatic (Table 4.11). Evidence for this efficient transplacental transmission is derived from natural and experimental studies. In 1 type of field studies, the familial relationship between serologically positive and negative animals in herds was analyzed (Table 4.11). Other studies compared antibody status of dams and the precolostral antibody response of newborn calves and there are further reports on the association between the antibody response of dams and their older progeny or daughters (Table 4.11).

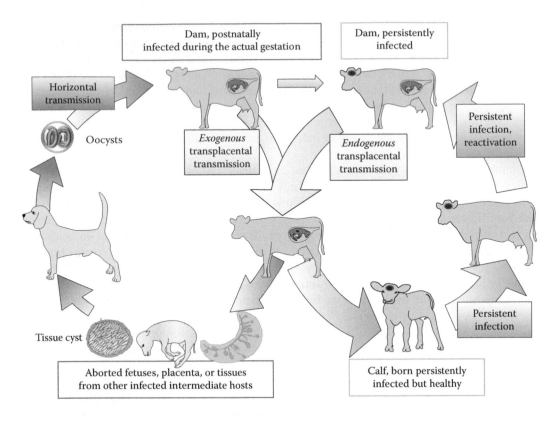

Figure 4.21 Endogenous and exogenous transplacental transmission of *Neospora* in cattle.

A study of dairy cows in a herd in Sweden[215] traced the familial history of *N. caninum* seropositivity and found that all infected animals were the progeny of 2 cows that were bought when the herd was established 16 years earlier. Insemination records suggested that venereal transmission was not a factor. Similar results, that is, the demonstration of family trees with a high proportion of positive cattle were obtained in other studies performed in Germany,[1789] Canada,[194] Australia,[833] Sweden,[705] and Brazil.[444]

A study in California, USA[65] provided convincing evidence that chronic persistent infection can be passed to progeny via endogenous transplacental transmission. Twenty-five seronegative heifers were housed with 25 seropositive heifers beginning at birth, and their progeny were evaluated for *N. caninum* infection. The seronegative heifers remained seronegative and gave birth to calves not infected with *N. caninum*. The seropositive heifers remained clinically normal but gave birth to congenitally infected calves. Seven of these congenitally infected calves were necropsied (Table 4.3); all had histologic evidence of *N. caninum* infection and 4 were recumbent.[65] Transplacentally infected and precolostrally seropositive calves had lower mortality than seronegative calves[1557] which is a clear indication that persistently infected calves are born infected but asymptomatic.

Strong evidence for transplacental transmission of *N. caninum* was obtained by finding statistically significant associations between the antibody responses in dams and their newborn calves. In several studies, it was observed that most of the newborn progeny of positive dams are serologically positive, too, while in seronegative dams most of the progeny are negative. In cattle and other ruminants, there is no transfer of antibodies from the dam to the fetus, not even through a placenta that has been damaged by infection.[526] Therefore, detection of specific antibodies in precolostral serum indicates *in utero* synthesis of antibodies by the fetus and is a strong indication of transplacental

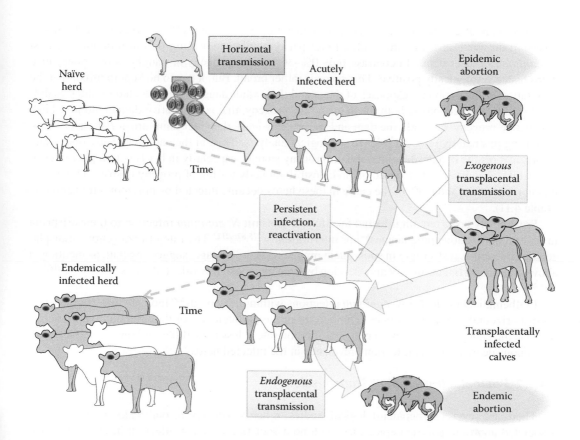

Figure 4.22 Dynamics of vertical transmission of *Neospora* in cattle.

transmission. Usually, a high proportion of precolostral positive calves are found in bovine herds ranging from 24% up to 100% (Table 4.11). In a few studies, one of them conducted in an embryo transfer center,[74] low rates of vertical transmission were observed, ranging from 24% to 50%[74] (Table 4.11). However, a finding of no specific antibodies in a calf is not conclusive for the absence of infection because the fetus might have been infected late in gestation and antibody synthesis may not have started. It was hypothesized that transplacentally infected calves may become immuno-tolerant[1330]; and this hypothesis was based on the non-confirmed detection of parasite DNA in the serum of serologically negative cattle and thus needs further studies for confirmation. In most studies comparing the serological results in seronegative dams and the precolostral antibody response in calves, the proportion of positive calves was low and ranged from 0% to 7% (Table 4.11). There was a single study on the F1 progeny of cows which had aborted previously during an outbreak in which 14% of the descendants of seronegative dams were precolostrally positive.[2123] Rarely, it is possible that a seronegative dam might give birth to a seropositive calf; this may be because the cow had been infected for some time and that the level of antibodies has declined to an undetectable level in the dam.[374,705,1215,1757] In addition to low sensitivity, also low specificity of serological tests has to be taken into account,[146,498] that is, it is possible that in some of these studies, a proportion of true negative calves had been tested false positive.

Further results obtained in studies comparing the serological responses in dams and their several month old or adult progeny or daughters are summarized in Table 4.11. In these studies, the proportion of serologically positive progeny of positive dams ranged between 41% and 100% (Table 4.11).

Parity may affect transmission rate as evidenced by a Dutch study[498]; the transplacental trans-mission rate was higher in dams with a lower parity number indicating partial immunity against *N. caninum* in older dams. In contrast, only 0%–36% of the progeny or daughters of seronegative dams tested serologically positive. This range is rather broad. But, it has to be kept in mind, that the correlation of the antibody responses of dams and their adult daughters is affected by the possibil-ity of postnatal infections in both the dam and the progeny, that is, a negative dam could give birth to negative progeny, but both, the dam and the progeny (calf, daughter dam) are at risk, later on, of becoming postnatally infected prior to serological analysis. This, in the end, may cause discordant serological results in dam-daughter or dam-progeny studies, especially in herds with higher rates of postnatal transmission. Data based on precolostral antibody testing of progeny (Table 4.11) suggest that not more than about 5% of the cattle in these herds became infected by horizontal transmission (Table 4.11).

Presumably cows remain infected for life and transmit *N. caninum* infection to their offspring in several consecutive pregnancies[690] or intermittently.[235,825,2123] The rate of endogenous transpla-cental infection may decrease in subsequent pregnancies or with age, suggesting that, as mentioned above, at least partial immunity may develop by time in persistently infected dams.[498,898,1215,1218,1737, 1965,2063]

Mathematical models of *N. caninum* infections within dairy herds[699] indicate that even low lev-els of horizontal transmission may be important in the maintenance of the infection within herds, because the endogenous transplacental transmission rate is less than 100% and thus would lead to a continuous decrease in the infection prevalence in the infected herds.

4.4.1.2 Postnatal (Horizontal) Transmission

There are several seroepidemiological studies in herds with severe outbreaks of *N. caninum*-associated abortion (abortion epidemics) with no association between infection in dams and their older progeny, indicating high rates of postnatal transmission (and as a consequence exogenous transplacental transmission later on, Table 4.11).[492,1568,1603,1792,1800,2058] In some studies, it was shown that only particular age groups within a herd were affected by infection[495,1603] which was explained by common housing and feeding of these animals.[495] In a dairy herd from Taiwan, a large propor-tion of dams had seroconverted and more than half of these dams had aborted almost at the same time.[948] These observations were in accord with the hypothesis that food or water provided to these animals might have been contaminated by oocysts of *N. caninum* and that this contamination was the source of infection (Figure 4.22).

The ingestion of sporulated *N. caninum* oocysts from the environment is the only demonstrated natural mode of infection in cattle after birth.[445,784,1317,1994] Cow-to-cow transmission of *N. caninum* has not been observed.[65] There is no evidence that placentophagia plays a role in transmission.[1795] At present, there is no definitive evidence that live *N. caninum* is present in the excretions or secretions of adult asymptomatic cows. Neonatal calves became infected after the ingestion of milk experi-mentally contaminated with tachyzoites.[2009] *N. caninum* DNA in milk, including colostrum, has been demonstrated.[1411,1412] However, there is no evidence that lactogenic transmission of *N. caninum* occurs in nature.[434,493,1991] The inoculation of high doses of *N. caninum* via the conjunctival route of cattle resulted in seroconversion[470] but no vertical transmission was observed[1392] and a natural infection via this route appears to be unlikely.

Venereal transmission may be possible, but unlikely, as evidenced in heifers experimentally infected by intrauterine inoculation of semen contaminated with tachyzoites[1827] and a dose response has been observed in a titration experiment with seroconversion and maintained antibody levels in heifers inoculated with semen contaminated with 5×10^4 tachyzoites.[1830] Although *N. caninum* DNA has been found in the semen of naturally exposed bulls[268,671,1524] or experimentally infected bulls,[1830] results suggest that viable organisms, if present, are few and infrequent. In addition, cows inseminated with frozen-thawed semen contaminated with *N. caninum* tachyzoites failed to acquire infection.[285]

4.4.1.3 Epidemic and Endemic N. caninum-Associated Abortion

N. caninum-associated abortion in bovine herds may have an epidemic or an endemic (sporadic) pattern. There are reports that in the years after an epidemic abortion outbreak, the affected herd may experience endemic (sporadic) abortions.[221,1370,1603] Abortion outbreaks have been defined as epidemic if the abortion outbreak is temporary and if 15% of cows at risk abort within 4 weeks, 12.5% of cows abort within 8 weeks, or 10% of cows abort within 6 weeks.[1370,1800,2125] In contrast, an abortion problem is regarded as endemic if it persists in the herd for several months or years. It is likely that these 2 patterns of *N. caninum*-associated abortion are related to the 2 routes by which *N. caninum* infections can cause abortion[1996,2101], that is, by endogenous or exogenous transplacental transmission (Figure 4.22). However, these 2 patterns represent extremes and intermediate abortion patterns (related to neosporosis, too), representing mixed patterns, possibly caused by endogenous and exogenous transplacental transmission are observed in the field.

4.4.1.3.1 Epidemic Abortions Caused by Exogenous Transplacental Transmission

Epidemic abortions or abortion storms are thought to be due to a primary infection of naïve dams with *N. caninum*, probably due to ingestion of feed or water contaminated with oocysts[1314,1315] (Figure 4.22). Because pregnant dams may be exposed to contamination with oocysts almost at 1 time (point source exposure), exogenous transplacental fetal infection and the resulting abortions occur within a short period of time (termed exogenous transplacental transmission, because the infection of the dam came from outside the animal during gestation[1996,2101]). Finding low avidity IgG responses in experimentally infected cattle, indicate a recent infection.[13,221,223,507,1800] Low avidity *N. caninum*-specific antibodies in the majority of aborting dams from herds with epidemic abortion, confirmed that the aborting animals became infected recently.[14,169,221,1004,1314,1759,1800]

4.4.1.3.2 Endemic Abortion Caused by Endogenous Transplacental Transmission

Recrudescence or reactivation of a latent or persistent infection in the dam during gestation may cause abortion.[825,1558,1908,1996,2085,2101] In many cattle herds with endemic abortion due to neosporosis, there is a strong positive association between the serostatus of mothers and their adult daughters, that is, there is evidence that the major route of vertical transmission in these herds is endogenous[194,215,221,302,492,833,1301,1786,1800,1966,2123] and that the rate of postnatal infection of pregnant dams, and the rate of exogenous transplacental transmission is very low.[146]

Several studies demonstrate that chronically infected seropositive cows have a 2-fold or higher increased risk of abortion compared to seronegative dams.[1214,1215,1542,1558] In a dairy herd from Spain, seroprevalence remained between 18% and 21% over a period of 3 years. Of 13 abortions recorded in seropositive animals 8 (62%) were repeat abortions and no abortions had been recorded in seronegative animals.[1542] Also in a study from Israel, the proportion of repetitive abortion in seropositive dams was about 5 times higher than in seronegative dams.[1301]

Parity of the dams or gestation number may influence the risk of abortion caused by endogenous transplacental transmission.[1010,1218,1964] In 1 study, a markedly increased abortion risk in congenitally infected heifers was observed during their first gestation but not in later gestations compared to the abortion risk in seronegative controls.[1964] In contrast, in a study from Israel, the proportion of abortion in pregnant seropositive dams appeared to be higher in elder cattle.[1301] However, it remained uncertain whether only endogenous or in addition also exogenous infection of dams had caused the abortions.[1301] In a study from The Netherlands which analyzed herds with abortion storms, it was assumed for some of these herds that many cattle became infected postnatally, that is, in many dams an exogenous transplacental transmission could have contributed to abortion.[2125]

Latent or persistent infection in dams may have been acquired vertically[65] or postnatally.[500,1370] The mechanism of reactivation of latent *N. caninum* infection is unknown. Whether immune suppression induced by ingestion of toxic feeds or other concurrent infections can cause reactivation has been debated but not supported by data.[141,1603,2125] Several others suspect "stress" as a trigger of endemic *N. caninum*-associated abortion. A period with increased rainfall in a usually dry region was found associated with *N. caninum*-abortions in herds with endemic infection and it was hypothesized that these abortions had been induced by stress especially in dams in the second trimester of gestation.[1215,1218,2150] In another study, it was shown that progesterone supplementation during mid-gestation increased the risk of abortion in those *N. caninum*-infected dairy cows, which, at the same time, had high titers of *N. caninum*-specific antibodies[179,180]; the effect of progesterone could not be explained.

4.4.1.4 Persistent N. caninum Infection and Fertility

A persistent *N. caninum* infection seems to have no effect on bovine embryos[1215] and after abortion, persistently infected dams had better fertility as compared to seronegative dams.[1775]

4.4.1.5 Effect of Infection on Fetal Survival

Irrespective of the origin of infection (exogenous or endogenous), not all congenitally infected fetuses become ill.[500] In abortion epidemics, up to 57% of aborting dams (among all dams at risk) have been reported.[1800,2125] However, in the Netherlands, a high rate of seroconversion together with low avidity responses were observed in a dairy herd, suggesting a recent exposure of this herd to *N. caninum,* though no increased abortion incidence was observed in the herd.[496] If epidemic abortion is caused by an exposure to oocyst-contaminated feed or water the observed variability regarding abortion risk may be explained by factors such as the infection dose,[784] the pathogenicity of the parasite strain by which the animals became infected,[730,1729] and by the susceptibility of the dams (e.g., immune status, which is also related to the stage of gestation).[784]

4.4.2 Risk Factors Associated with Infection

Knowledge of risk factors for herds to acquire *N. caninum* infection and have *N. caninum*-associated abortions is important for the development and implementation of measures to control bovine neosporosis. Knowledge of risk or protective factors with respect to bovine neosporosis is based largely on retrospective cross-sectional or case-control studies. Retrospective assessment generally allows the identification of putative risk or protective factors, but conclusive data can only be obtained by prospective cohort or experimental studies. However, the repeated identification of the same risk or protective factor in several independent retrospective cross-sectional or case-control studies increases the evidence that this factor is a "true" risk or protective factor for an infection or a disease.

Serologic prevalences of *N. caninum* (Table 4.1) indicate that there are considerable differences among countries, within countries, between regions, and between beef and dairy cattle.[144,448,634,1086,1380,1390,1656,2028] However, these results should be interpreted with caution because of differences in serologic techniques, study design, and sample size between studies. Data reported on the European situation[144] are noteworthy because sera were tested by standardized serological techniques[2049] and a similar study design was followed in different countries. From the data, it is evident that seroprevalence of *N. caninum* was lowest in Sweden compared with the prevalence in other European countries. Similar observations were made in Canada where also marked differences in seroprevalence between regions were found in randomly sampled dairy cows and herds.[2028] Results suggest that there are differences in the infection risk among different regions, within a particular region, and among different management systems. Therefore, caution should be used when transferring the results of a risk factor analysis obtained in a particular region or management system to another.

There are numerous risk factor studies assessing either the risk of individual cattle or herds, which either became infected with *N. caninum* or experienced *N. caninum*-associated abortions. These risks (infection risk and abortion risk) are positively associated with each other but influenced differently (Figure 4.22). After exogenous transplacental transmission, abortion risk might be influenced by, for example, the number of oocysts ingested by the dam and the gestational stage,[784] whereas the occurrence of abortions in endogenous transplacental transmission might be influenced by as yet unknown factors, for example, the immune status of the dam or simply stress.[1215,1218,2150]

Several studies have examined *N. caninum* infection risk at the herd level or animal level with the serostatus of herds or individual cattle (dams, calves) as a dependent variable, that is, as the target or outcome variable. The results of these studies have been influenced by the sensitivity and specificity of the serological tests used and fluctuations of antibodies during pregnancy as discussed earlier. Demonstration of seropositivity does not provide information on the viability of infection. Furthermore, rarely, an animal may be infected but seronegative.[188,791]

As yet, there is no serological test available providing information on the route of infection (horizontal by oocysts or vertical by transplacental transmission). Only a few tests, but not those routinely used in seroepidemiological studies provide reliable information on how recently the infection occurred. To partially overcome the later problem, some risk factor studies focused on herds with epidemic abortion[141,495] because in these it is likely that the infection of most of the aborting dams was postnatal and these herds could provide better information on the factors that favor on-farm postnatal infection of cattle with *N. caninum* than herds in which infection occurred a long time ago. However, it was regarded as doubtful that all *N. caninum* abortion storms are the immediate result of a recent exposure to oocyst shedding farm dogs.[2128]

No association between the allele frequency distribution for BoLA DRB3 and DQA1 (i.e., genes which are part of the MHC) and infection with *N. caninum* were identified, that is, there is no evidence for a genetic predisposition of particular cattle as yet.[1816]

Results of studies that assessed risk factors for infection on either the animal or herd level follow. A summary of the actual data available is provided in Table 4.12.

4.4.2.1 Age

The risk of being seropositive may increase with the age or the gestation or parity number in beef and dairy cattle (Table 4.12) and suggests that postnatal infection by horizontally transmitted *N. caninum* is of particular importance in some herds. In contrast, a negative age effect on the prevalence of seropositive animals was reported in beef cattle in Canada.[2057] In the same study, it was observed that the risk of being culled was significantly greater in seropositive than in seronegative cows, suggesting that selective culling could be a possible reason for the age effect. Based on these findings it is not surprising, that both the mean age of cows in a herd and the proportion of heifers in a herd could explain differences in the level of herd seropositivity as demonstrated in beef cattle in the northwestern USA (Idaho, Montana, Oregon, and Washington).[1772] Age effect on seropositivity in dairy cattle may vary in different study areas. In Spain, for instance, the risk of being seropositive increased with age, while in Sweden the situation was opposite.[144] In another study, in beef and dairy cattle from the same area in Spain (Galicia), the seroprevalence also increased with age.[634] It was hypothesized that the age effect might be influenced by variations in the probability of horizontal transmission (e.g., by the risk of ingesting oocysts), by regional differences regarding replacement rate (influencing the time

Table 4.12 Risk and Protective Factors for *N. caninum* Infection in Cattle

Factor Description			References of Studies Which Identified Putative Risk or Protective Factors[a]	
			Risk	Protect
Age, parity, gestation, and lactation number	Age of cattle, heifers vs. cows		84[*], 93, 144[**], 429, 458a, 630, 634[**], 758[**], 791, 818, 930[**], 1393[*], 1737, 1824, 1835, 2111	144[**], 1072[b], 2057[*]
	Mean age of cows in a herd		1772[*]	–
	Proportion of heifers in a herd		–	1772[*]
	Heifers, adult cattle vs. calves		1711[**]	–
Parity number of the mother of an animal	Gestation/parity number/number of pregnancies		1010	1072[b]
			–	498, 1737
Definitive hosts (dogs, coyotes) and other canids	Dogs	Presence of farm dogs	754, 758, 764, 791, 1456, 1541, 1559, 1770, 2028, 2050[**]	127[*]
		Presence of farm dogs in the past 10 years	2050	–
		No. of farm dogs	93, 384, 972, 1260, 1559, 1803, 2050	–
		N. caninum seroprevalence in farm dogs	818, 1380	–
		Exposure to *N. caninum* seropositive dogs	1498	–
	Behavior of farm dogs	Dog has contact to herd	764	–
		Dogs kept loose	249	–
		Defecating on feeding alley	497	–
		Defecating on grass/corn silage storage	497	–
		Dog feeding on raw meat, placenta, fetuses, uterine discharge, colostrum, or milk	249, 497, 2028	–
		Presence of dogs but not known to eat placentas and/or fetuses	2028	–
		Density of dogs in the district or municipality of farm location	1801, 1803, 2050	–

(Continued)

Table 4.12 (*Continued*) Risk and Protective Factors for *N. caninum* Infection in Cattle

Factor Description		References of Studies Which Identified Putative Risk or Protective Factors[a]	
		Risk	Protect
Wild carnivores, incl. coyotes, hyenas, foxes	Abundance in the ecological region of farm location	125(*)	—
	Wild carnivores have access to farm, presence of coyotes	93, 1854	—
	Presence of canids/stray canids	764, 1854, 1919(**)	—
Canids			
Cats	Presence of cats	—	1541, 1075
Other (potential) intermediate hosts	Presence of rabbits and/or ducks	1075, 1541	—
	Presence of poultry by increasing number of dogs	1533(**)	—
	Presence of domestic poultry	1854	—
Grazing and fodder	No grazing	1533(**), 2068	—
	Early fall housing of cattle	1491	—
	Feed is produced on farm	1287	818
	Silage feeding		818
	Using monensin in dry cows		2028
	Grazing on rangeland during summer		1772(*)
	Use a hay ring with round bales of hay	127(*)	—
	Early housing/ending of grazing period	1491	—
	Use of self-contained feeder for cow supplement		127(*)
	Wildlife contact with the weaning ration, cattle feed	127(*), 1287	—
	Wildlife has access to farm facilities	1287	—
Source of drinking water	Pond/well vs. well or public water supply	1075, 1541, 1919(**)	—
Colostrum or milk	Feeding pooled colostrum to calves	384	—
Calving management	Spring calving vs. fall calving	127(*)	—
	Year round calving vs. split calving, Year round calving vs. spring calving	1491	—
	Calving period of >3 months		1541

(*Continued*)

Table 4.12 (Continued) Risk and Protective Factors for *N. caninum* Infection in Cattle

Factor Description		References of Studies Which Identified Putative Risk or Protective Factors[a]	
		Risk	Protect
History of reproductive problems	Abortion/stillbirth	94, 1854	—
	Prolonged calving interval	94, 95, 2029	—
	Retained fetal membranes	94, 95	—
	Uterine infection, endometritis	94, 95	—
	Repeated breeding	95	—
Cattle density, cattle stocking density	Stocking density	127(*), 1772(*)	—
	Cattle stocking density during winter	1772(*)	—
	Size of farmland, area (acres) used for forage production	—	385, 2028
Herd size	Large herds	93, 764, 1533(**)	1107
	Large herds by number of dogs	1533(**)	—
	Herd size	1803	791
	Number of lactating cows	—	2028
Source of replacement heifers	Ranch-raised replacement females, closed management	94, 127(*), 732, 1640(*)	93, 2111
Breed	Holstein-Friesian-local breed crosses vs. Holstein-Friesian	—	93
	Jersey vs. Holstein-Friesian	1737	—
	Zebu or crossbred vs. Holstein-Friesian	—	818
	Native breed vs. Holstein-Friesian, Spanish beef cattle Rubia Gallega, mixed	—	144(**)
	Swedish red and white vs. Swedish Friesian, other	144(**)	—
	Local breeds vs. exotic European breeds (including Holstein-Friesian, Fleckvieh, Montebiliarde, Jersey, Girolando)	757(**), 758(**), 1004(**)	—
	Limousin-mixed/Limousin vs. other European beef breeds	—	84
Dairy vs. beef	Mixed vs. pure breeds	458[a]	—
	Dairy vs. beef	1380(***), 1390(***), 1656(***), 2104(***)	144(***), 634(***)

(Continued)

Table 4.12 (Continued)   Risk and Protective Factors for *N. caninum* Infection in Cattle

Factor Description		References of Studies Which Identified Putative Risk or Protective Factors[a]	
		Risk	Protect
Type of housing	Tethered vs. loose	1541	–
	Presence of calving pen	–	1931
Hygienic status, biosecurity	Low vs. high	93, 758	–
	Aborted dams are not routinely serologically tested for abortion causes	1737	–
	Asking for a BVDV-negative test before introducing an animal		2028
	Workers visiting other farms	1931	–
Indicators for other diseases or infections	Antibodies against *T. gondii*	791	–
	Antibodies against BVDV	220	–
	Antibodies against BHV-1	1712[(**)]	–
	BT vaccination	316	1541
	Average herd somatic cell count $200-400 \times 10^3$ vs. a somatic cell count $<200 \times 10^3$ in the past 6 months	–	–
Climate	Mean temperature in July (Summer, Germany)	1803	–
	Mean temperature in spring (Italy)	1711[(**)]	–
Climatic season	Spring	758[(**)]	–
	November vs. August sampling	1490	–
Vegetation	NDVI	1711[(**)]	–
Demographic factors	Human population	1794, 1801	–
Geographic cluster		707, 1077, 1210, 1485[(*)]	–

[a] Studies on beef cattle are marked with [(*)] and those not differentiating between beef and dairy are marked with [(**)]. Studies on dairy cattle are not marked. Studies differentiating between beef and dairy [(***)].
[b] In cows with a history of abortion.

cattle may be exposed to horizontal transmission), and by management practices such as selective culling of seropositive animals.[144] Nonselective culling of animals in a herd with a high seroprevalence could result in a positive relationship between age and prevalence, if the population from which successive external replacement heifers are purchased has a lower seroprevalence than the herd itself. This effect (lower prevalence in younger age groups) is further strengthened by the fact that the proportion of transplacental transmission is often much lower than 100%.[430] As discussed earlier, recrudescence during gestation may have further contributed to a higher seroprevalence in older age groups.[429]

4.4.2.2 Parity Number of Dam

The higher the parity number of seropositive dams the lower was the chance of progeny to be prenatally infected by *N. caninum* as shown in the Netherlands.[498] Observations in Costa Rica were in accord with this finding, showing that daughters born by dams which had a parity number ≥6 at delivery, had a lower risk of being seropositive.[1737]

4.4.2.3 Definitive Hosts (e.g., Dogs, Coyotes, and Wolves)

In most epidemiological studies in dairy herds, the presence of farm dogs (as illustrated in Figure 4.22), either current or within the past 10 years,[2050] or the actual number of farm dogs (Table 4.12) was a risk factor for seropositivity in cattle. This is not surprising since dogs are definitive hosts of *N. caninum*. Three studies observed positive correlations between the *N. caninum* seroprevalence in dogs and cattle[754,818,1380] which further confirms that the occurrence of *N. caninum* in cattle is linked to infection in dogs.

Not only the presence of dogs but also the possibility that the dog has contact to the herd[764] or is kept loose[249] were identified as risk factors. Furthermore, the putative ways by which dogs may pose an infection risk to dairy cattle has been studied.[497] Defecation by farm dogs on feeding alleys and on stored grass or corn silage was reported more often by farmers of herds with evidence for postnatal bovine infection than by those herds with no such evidence.[497] Interestingly, in a study of herds with evidence of recent postnatal infection, seropositivity to *N. caninum* was more often associated with common housing than with common feeding of the seropositive age group.[495] Based on these results, it might be justified to assume that contamination in the feeding area is more closely related to infection than contamination of fodder during storage.

Farmers of herds with evidence of postnatal infections more often observed dogs feeding on raw meat, placenta, fetuses, uterine discharge, colostrum, or milk.[249,497,2028] This suggested that these materials may pose an infection risk to dogs; that is, these materials may facilitate dogs becoming infected with *N. caninum*. In an experimental study, placenta, but not colostrum, has been confirmed as an infection source for dogs.[493] However, oocysts were not excreted when aborted fetuses or brains of fetuses were fed to dogs.[195] These results were most likely influenced by the stage of autolysis in the fetus, killing the parasite along with the host cells. Most *N. caninum* in aborted fetuses die with the host cells and it is rare to find intact tachyzoites in such tissues.[605] Several studies reported on dogs that have shed oocysts after ingesting a variety of tissues, including neural, muscular, visceral, and fetal membranes (Chapter 5). Therefore, it can be assumed that fresh tissues (e.g., tissues of *N. caninum*-infected calves or cows that died, fresh placentas or remnants after slaughter) may cause excretion of *N. caninum* oocysts while autolyzed materials (e.g., aborted fetuses, retained placentas) seem to be less appropriate.

There is some evidence that recently introduced dogs pose a higher risk of transmission of *N. caninum* than resident dogs.[495] This could be explained in an analogy to the hypotheses for *T. gondii*, for which it is suspected that naïve definitive hosts are crucial for the life cycle.[428] In *N. caninum*, this situation seems to be similar as dogs shed no (or only a few) oocysts after being fed repeatedly with infectious material.[493,785,1798] In addition, higher numbers of oocysts are excreted by young dogs (10–14 weeks old) than by older dogs (2–3 years old).[785]

In addition to farm dogs, dogs kept in the neighborhood of farms may pose an infection risk. In a German cross-sectional study, dog densities in districts, cities, or municipalities were predictors of the prevalence of bulk milk positive herds[1801] or were identified as risk factors for herd seropositivity.[1803,2050] Coyotes were found to be additional definitive hosts of *N. caninum*.[781] This was suspected after epidemiological studies of beef calves had shown that the abundance of coyotes or gray foxes in different ecological zones of Texas was associated with the seroprevalence of *N. caninum* in beef calves.[125] Whether gray foxes are also definitive hosts of *N. caninum* remains to be determined. In addition to coyotes, wolves are also definitive hosts of *N. caninum*.[615] Especially in regions with a high abundance of wild-living definitive hosts (e.g., coyotes and wolves), the sylvatic cycle (e.g., from deer to wild-living canid species) might be important in maintaining the domestic cycle (from cattle to dogs) of the parasite.[783] Although 1 experimental study indicates that the red fox is not a definitive host for *N. caninum*,[1799] there is an ongoing discussion as to whether red foxes could be sources of postnatal infections with *N. caninum*. *N. caninum*-like oocysts were reported in the feces of naturally infected foxes from Canada[2072] but there is no unambiguous proof that foxes represent definitive hosts of *N. caninum*. It is suspected that other canids may represent definitive hosts for *N. caninum* and the access of wild canids to farms was identified as a risk factor of infection with *N. caninum* for Ethiopian cattle[93] and the presence of not further specified canids or stray canids were also reported as a risk factor.[764,1919] Available information related to wild canids is summarized in Chapter 16.

For beef cattle, there is as yet no clear evidence that farm dogs or dogs kept in the surroundings of farms pose an infection risk.[2050] A possible explanation for this is that on the less intensively managed beef farms, there is in general no close contact between the excretions of farm dogs and beef cattle.[127,1533,1772] An unusual observation was that the presence of farm dogs on beef farms in Texas, USA was a putative protective factor[127]; in the same region, it was demonstrated that wildlife contact with the weaning ration could explain seroprevalences in beef calves.[127] Possibly, the presence of dogs was inversely related to the presence of wild canids on farm land, as hypothesized in a Canadian study on *N. caninum* abortion.[918]

4.4.2.4 Carnivores Other than Canids

In experimental studies, cats failed to serve as definitive hosts for *N. caninum*.[1310] However, there is 1 epidemiological study of dairy cattle that observed a protective effect for cats being present on the farm.[1541] It is possible that this factor is a confounder related to the absence of dogs. However, another possible explanation for the protective effect of the factor "presence of cats" might be that cats are predators of putative intermediate hosts of *N. caninum* (e.g., rodents) which could reduce the frequency by which definitive hosts of *N. caninum* have access to tissues of infected intermediate hosts.

4.4.2.5 Intermediate Hosts Other than Cattle

Not only cattle but also other intermediate hosts of *N. caninum* may be a source of infection for dogs and other canids. The presence of *N. caninum* DNA in naturally infected mice and rats suggests that these animals may be important sources of infection for carnivore hosts of *N. caninum* (Table 2.2). A study from France reported the presence of rabbits and/or ducks as a putative risk factor for seropositivity in dairy cattle.[1541] In Mexico, the presence of domestic poultry was found as a risk factor for seropositivity.[1854] In another study from northern Italy, the risk for seropositivity in individual cattle increased with the number of farm dogs when poultry were present on the farm.[1533] In The Netherlands, the presence of poultry on the farm was also found to be a risk factor for the occurrence of *N. caninum*-associated abortion; their possible role as vector of oocysts was discussed.[141,2127] The susceptibility of poultry for *N. caninum* is a matter of debate. Although antibodies against *N. caninum* have been observed in chicken[390,968,1288] and DNA was observed in seropositive chicken[390,773] all attempts to date have failed to isolate viable *N. caninum* from chicken. Chicken seems to be a better host for *T. gondii* than for *N. caninum*.[773] Although experimental infection in embryonated eggs is possible[110,715,1054] and material from infected embryonated chicken eggs induced oocyst excretion in dogs,[715] evidence is lacking that chickens are important natural intermediate hosts of *N. caninum*. Experimentally inoculated pigeons developed antibodies against *N. caninum*[1360] and researchers from China, Israel, and Iran reported *N. caninum* DNA in naturally infected pigeons, crows and sparrows, respectively.[2,520,1762] In addition, in Spain, *N. caninum* DNA was observed in 2 magpies (*Pica pica*) and 1 common buzzard (*Buteo buteo*)[426] but the importance of wild birds in the life cycle of *N. caninum* remains to be clarified. These results warrant further examination of the susceptibility of other livestock to *N. caninum* infection, such as rabbits, ducks, chicken, and other poultry as well as wild animals, and whether these potential intermediate hosts could pose an infection risk to definitive hosts.

4.4.2.6 Grazing, Fodder, and Drinking Water

Oocyst-contaminated pastures, fodder, and drinking water are regarded as the potential sources for postnatal infection of cattle. Therefore, it is important to know which feeding practices pose an increased infection risk.

In Italy and northwestern USA, grazing of cattle on rangeland during summer seems to be a protective factor.[1533,1772] In Ireland, those farms that housed cattle early in fall had an increased risk of *N. caninum* antibodies, suggesting that housed cattle have a higher risk to acquire an *N. caninum* infection.[1491] Although wild canids and dogs have free access to rangeland, oocyst contaminations caused by definitive hosts may be too low to pose a significant infection risk or oocysts may not survive during summer months if these are very hot and dry. Unfortunately, information on the climatic conditions under which *N. caninum* oocysts are able to survive in the environment is lacking.

In beef herds, the use of a hay ring appeared as a putative risk factor for seropositivity.[127] This factor was explained by the observation that cows often calve, abort, or expel placentas near hay feeders. Because these feeders are seldomly moved it was hypothesized that fecal contaminations by definitive hosts which have fed on infectious material may concentrate close to the feeders.[127] In the same study, a procedure implemented to avoid the contamination of fodder, that is, the use of a self-contained feeder for cow supplements, was identified as a probable protective factor.[127] Related to this is the observation that ranches with wildlife access to the weaning supplement had an increased risk of their calves being *N. caninum* positive[127] and another study reported an increased risk of seropositivity in Brazilian dairy cattle when wildlife had access to farm facilities.[1287] Since farm dogs play a major role, contamination could occur especially when feed is produced on the farm. However, this was observed as a risk factor in 1 study[1287] but identified as a protective factor in another study,[818] both from Brazil. Silage is stored on the farm and is regarded as 1 of the feeds potentially contaminated by *N. caninum* oocysts.[1316] In addition, rodents as intermediate hosts of *N. caninum* may concentrate close to silage. However, silage feeding was identified as a protective variable in a Brazilian study.[818]

In a Canadian study, it was observed that feeding monensin to dry cows was a herd-related factor reducing the risk of *N. caninum* seropositivity.[2028] Monensin is used in dairy herds to assist in the management of negative-energy balances during milk production. Monensin was also shown to reduce the population of *N. caninum* tachyzoites in cell culture[1171] and the infectivity and viability of bradyzoites of *T. gondii* in cell culture was inhibited by monensin at low concentrations (0.1 ng/mL).[392] In 2 groups of cattle infected with *N. caninum* tachyzoites, the 1 receiving monensin had significantly lower antibody responses than the one not given monensin. However, including also other sampling dates PI into the analysis, led to nonsignificant serological differences between both groups of cattle.[2030] Nevertheless, further studies are necessary to elucidate whether feeding monensin has a direct effect on *N. caninum* or whether "feeding monensin to dry cows" represents a confounding variable (i.e., can be explained by unknown variables associated to both infection risk and "feeding monensin to dry cows").

In a study conducted in France, the use of ponds rather than the use of a well or public water supply for drinking water was a risk factor for *N. caninum* infection in dairy cattle.[1541] Seroprevalence data in feral marine mammals suggest that *N. caninum* oocysts may contaminate surface water and subsequently contaminate sea water.[593] In 1 study from Mexico the presence of *N. caninum* DNA was reported in drinking water collected on dairy farms, and the presence of oocysts suspected but not confirmed.[1854]

4.4.2.7 Feeding Colostrum or Milk

Experimental studies have demonstrated that neonatal calves may become infected by the ingestion of milk containing tachyzoites.[434,2009] However, cross-suckling of calves born to seronegative mothers on seropositive cows has not led to an infection.[434] Because *N. caninum* DNA was demonstrated in bovine milk,[1411,1412] there is an ongoing debate whether or not the lactogenic transmission of *N. caninum* is possible. With respect to this, it is interesting that 1 study in dairy cattle suggested that feeding of pooled colostrum is a putative risk factor for seropositivity.[384]

4.4.2.8 Calving Management

In a risk factor analysis of beef calves in Texas, the effect of seasonal calving during spring was profound, that is, the risk of calves of being seropositive was higher than it was on ranches with a fall-calving season.[127] There was no explanation offered for this observation. Possibly, there are seasonal effects in these beef herds on the risk for calves to become infected, either by transplacental or by horizontal (postnatal) transmission. This seasonality may be biologically linked to the whelping season of the putative definitive hosts in Texas, coyotes and gray foxes. Since, naïve or young dogs are more likely to excrete *N. caninum* oocysts than older or immune dogs,[493,785,1798] the same may also be true for young coyotes and gray foxes. Further studies are needed to explain the observations in Texas beef calves. Interestingly, in a French study, prolonged herd calving periods of 3–6 or 6–12 months reduced the risk of herd seropositivity compared to herd calving periods of up to only 3 months.[1541] There was no explanation for this observation. A study in Ireland revealed that year-round calving (in contrast to split or spring calving) was statistically associated with the presence of *N. caninum*-specific antibodies in bulk milk collected on farms.[1491] Two explanations were offered for this observation: (i) on farms with year-round calving, those dams that have aborted are made pregnant again and are most likely not culled, which may prevent *N. caninum*-infected dams from being removed from the herd and (ii) on farms with year-round calving definitive hosts are exposed continuously to potentially infectious material, which may perpetuate the on-farm lifecycle of *N. caninum*.

4.4.2.9 History of Reproduction Problems

History of reproductive problems (abortion/stillbirth, prolonged calving interval, retained fetal membranes, uterine infection, endometritis, and repeated breeding) were reported as risk factors for seropositivity in 2 related studies in Ethiopia.[94,95] The history of abortion on the farm was a risk factor for seropositivity in dairy cattle in Mexico.[1854]

4.4.2.10 Cattle Stocking Density and Size of Farm Land

In 2 studies on beef calves in Texas, a high stocking density was identified as a potential risk factor for seropositivity.[125,127] A similar effect was observed for stocking density of beef cows during winter in northwestern USA.[1772] This effect was explained by the observation that ranches with a high density of cattle (both beef and dairy) are more likely to use supplemental feeding practices.[85,125,127] Places on farms where supplemental feed are stored or fed to cattle may attract rodents that are potential prey for definitive hosts of *N. caninum*. This could cause these places to have an increased risk for being contaminated with the feces of definitive hosts and the presence of infected rodents, thus increasing the risk of postnatal infections.[125]

In studies on dairy cattle in southern Brazil and Canada, it was observed that with increasing size of farm land or area (acres) used for forage production the seroprevalence in herds decreased.[384,2028] However, in case of the Brazilian study, this protective effect was not linked to the stocking density.[384] It was hypothesized that on small farms, it is easier for farm dogs to have access to bovine carcasses, aborted fetuses, placenta, and uterine discharge than on larger farms.

4.4.2.11 Herd Size

In several studies, the risk of individual cattle becoming seropositive increased with the size of the herd.[93,764,1533] When in a study from Italy the analysis was restricted to data from northern Italy, the number of dogs per farm interacted significantly with herd size, that is, the risk of being seropositive increased in larger herds with an increasing number of dogs per farm.[1533] In a study conducted in Germany, larger herds had an increased risk of being serologically bulk milk positive.[1803] In a Canadian study, the number of lactating cows was one of the herd-level variables associated with the risk of seropositivity in dairy cattle.[2028] Possible explanations are that with increasing size of the herd there is an increasing chance to acquire an *N. caninum* infection by, for instance, purchasing external replacement heifers. Another explanation for herd size as a risk factor could be that hygienic measures to prevent dogs from feeding on infectious material are more difficult to follow in large herds than in small herds.[1803] In addition, the open storage of feed outside barns and silos (typical for large farms) may favor contamination.

4.4.2.12 Source of Replacement Heifers

Rearing replacement heifers on the farm rather than purchasing them from outside sources may foster the trend that an existing *N. caninum* infection in a herd may persist for many years.[705,1909] This could explain why, in one of the risk factor studies on beef cattle, "rearing of own replacement heifers" or "closed management" was identified as a potential risk factor for high seroprevalence in calves[127] or in dams from cow–calf operations in Alberta, Canada being serologically positive.[1640] If the seroprevalence is lower in the recipient herd than in the population from which the replacement heifers are obtained, the purchase of replacement heifers should increase infection in the recipient herd or even introduce infected animals into a naïve herd. This could explain the results of other studies which suggest that purchasing replacement represents a risk.[93,2111] In a Canadian study, no association was observed between purchasing dairy cattle and seroprevalence for *N. caninum* in dairy herds.[337]

4.4.2.13 Breed

There are indications from several countries that the *N. caninum* seroprevalence differs according to the cattle breed.[84,93,144,757,758,1028,1737] However, these results have to be interpreted with caution, because the differences observed might have been caused by differences in the production system or farm characteristics and not by differences in the breed-related susceptibility to infection. For example, native Spanish breeds were less likely to be seropositive than Holstein-Friesian or mixed breeds. This was explained by differences in the intensity of management.[144]

In contrast to Holstein-Friesian native breeds are predominately located on highland pastures with very low stocking densities. In the same study, breed-associated differences were reported from Sweden.

4.4.2.14 Type of Cattle

There are several studies reporting significant differences in the prevalence between dairy and beef cattle. In some studies, dairy cattle showed a higher risk to be seropositive[1380,1390,1656] and in other studies, the opposite was the case.[144,634] Obviously, there is no general rule which type of cattle has a higher risk of infection (dairy or beef). However, differences in abortion risk between dairy and beef cattle were observed (Section 4.4.3).

4.4.2.15 Type of Housing

In a French study, tethered dairy cattle had an increased risk of being seropositive compared to dairy cattle kept untethered indoors.[1541] No explanation for this effect was offered, but it shows that risk of infection is most likely not uniform among animals belonging to a particular cattle type. In a study conducted in Jordan, the presence of a calving pen was identified as a protective variable.[1931] A possible explanation for this observation is that hygienic conditions on farms with a calving pen are higher, for example, possibly in these farms placentas which could be infectious to dogs are regularly and adequately removed.

4.4.2.16 Hygienic Status and Biosecurity

Variables indicating a low level of biosecurity[1737,1931] or a low hygienic status of a farm[93,758] are related to increased *N. caninum* seropositivity in cattle. It is unlikely that these factors have a direct effect. The same can be assumed for a variable "asking for a bovine viral diarrhea virus (BVDV)-negative test before introducing an animal" that most likely also has no direct effect on *N. caninum* seropositivity, but characterizes the general hygienic status or biosecurity level of a cattle farm.

4.4.2.17 Factors Related to Concurrent Infections

T. gondii and *N. caninum* are closely related and both parasites are able to infect cattle post-natally via the oral ingestion of oocysts. In a study from Spain, the probability of infection by *N. caninum* was 6.1 times higher in cattle that were positive for *T. gondii*.[791] Common risk factors which may favor postnatal infections with coccidian infections may have contributed to this effect. This finding was surprising, because of the differences in life-cycles and in routes of transmission to cattle between *N. caninum* (mainly vertical) and *T. gondii* (mainly horizontal). Thus, this relation seems to be associated to biosecurity measures.

In Swedish cows, a statistically significant association between antibodies against *N. caninum* and BVDV were observed.[220] From this result, it was assumed that risk factors supporting the introduction and spread of BVDV in cattle, such as high cattle density and frequent purchase of animals, also increase the risk for *N. caninum* infection. However, a study in Spain found no such association.[1261] In an Italian study, a positive association between antibodies against bovine herpes virus 1 (BHV-1) and antibodies against *N. caninum* was demonstrated.[1712] The possibility whether BHV-1 induced immunosuppression after natural infection or vaccination could increase the susceptibility of cattle to secondary infection with *N. caninum* was discussed. Also, a Brazilian study suspected effects of coinfection between *N. caninum* and BVDV of BHV-1.[455] However, to prove this hypothesis, experimental or follow-up studies after infection or vaccination are necessary.[1712] Another

Italian study reported an increase in *N. caninum*-specific antibodies in dairy cows after blue tongue (BT) vaccination (serotype BT serotype 2 + 9 vaccine) in southern Italy.[316] Surprisingly, this was not observed in dairy cows from central Italy (vaccinated with BT serotype 2) and northern Italy (no vaccination)[316] and unexplained immunosuppressive effects and reactivation of *N. caninum* in case of the combined BT serotype 2 + 9 vaccine was hypothesized. In a Canadian study of 78 dairy herds in Ontario, no significant association between antibodies against *N. caninum* and the serostatus to *Leptospira interrogans* serovar *hardjo*, *icterohaemorrhagiae*, or *pomona* was observed.[1583] In a study from France, a slightly elevated somatic cell count $200–400 \times 10^3$ versus a somatic cell count $<200 \times 10^3$ (as averaged in the herd over the last 6 months) was reported as a protective variable but no explanation was offered for this observation.[1541]

4.4.2.18 Climatic Season

In 2 European studies that analyzed climate effects on the risk of seropositivity in herds or individual cattle, the factors "mean temperature in spring in a buffer zone around farm location" and "mean temperature in July in the municipality where the herd is localized" were identified as putative risk factors.[1711,1803] These observations can be explained by the effects of climate on sporulation or survival of oocysts. For example, a higher temperature (up to not yet defined limits) may favor a faster sporulation of oocysts in the fodder or in the environment surrounding cattle.

Studies reporting seasonal effects on seropositivity (Table 4.12) have to be interpreted with caution because there is reactivation during gestation causing higher levels of antibodies in the third trimester of gestation and seasonal calving could bias studies on seroprevalences during different climatic seasons. In Algeria, higher seroprevalences were observed in cattle sampled during spring.[758] Another study from Ireland also revealed significant differences in the prevalence of bulk milk positive herds between 4 different sampling periods (March, June, August, and November) with the highest prevalence in November and the highest incidence from August to November.[1490] Therefore, the existence of seasonal effects on the risk of *N. caninum* infection for cattle should not be excluded but needs further study.

4.4.2.19 Vegetation

An Italian study observed that the risk of seropositivity in individual cattle decreased with increasing summer normalized difference vegetation index (NDVI) values determined for the 3 km buffer zones around farm locations.[1711] A high summer NDVI is indicative for forests or broad-leaved trees. It was assumed that cattle from the respective farms were not pastured and thus had a lower chance to ingest *N. caninum* oocysts. However, this interpretation is not supported by the finding of another Italian study in which "no grazing" was identified as a risk factor for seropositivity in individual cattle.[1533]

4.4.2.20 Human Population Density

In Germany, the human population, similar to dog density, could also be used to predict the prevalence of bulk milk positive herds in district and cities.[1801,1808] Because dog density was identified as a putative risk factor for infection, it is not surprising that human population density also seems to have the same effect, due to correlation between population and dog density.

4.4.2.21 Geographic Clustering

Several well-controlled studies identified geographic clusters in which significantly higher levels of bovine infection with *N. caninum* were observed than in other parts of the study area.[707,1077,1210,1485] In none of the studies, these geographic clusters were explained by accompanying data.

4.4.3 Risk Factors Associated with Reproduction

Soon after the first description and isolation of *N. caninum* from dogs,[528,529] it became obvious that the same parasite is also associated with bovine abortion.[1954] In addition, *N. caninum* was also made responsible for stillbirth and perinatal mortality.[804] A few reports suggest an effect on fertility as well. A study from Senegal revealed that *N. caninum* seropositive cattle needed a higher number of inseminations to conception than seronegative cattle independent of age and breed.[1028] Significant effects on fertility were also reported in other studies. Seropositive cows required 3.7 inseminations to become pregnant in the most recent pregnancies, compared to 2.4 inseminations for seronegative cows.[833] Another study found that *N. caninum*-infected heifers were 1.8 times more likely not to conceive after 1 insemination than their noninfected herd mates.[1434] In contrast, others did not observe any adverse effect of *N. caninum* infection in early pregnancy.[215,1010,1216,1218,1740] In a Canadian study, positive *N. caninum* serological results were associated with increased calving intervals of >484, 534, or 584 days in dairy cattle.[2029]

Transplacental transmission of *N. caninum* from a chronically infected dam to the fetus is an important mode of infection and can occur in consecutive pregnancies via endogenous transplacental transmission[65,215] (Figure 4.22). This mode of transmission contrasts to the exogenous transplacental transmission to the fetus, when a previously naïve dam becomes infected postnatally most likely via ingestion of *N. caninum* oocysts from fodder or water and these contaminations with oocysts are related to definitive hosts (Figure 4.22). High rates of exogenous transplacental transmission in a bovine herd may cause an epidemic abortion outbreak, affecting a large proportion of the pregnant herd. However, this is not always the case. Exogenous transplacental transmission may cause only prenatal infections in fetuses and calves but neither abortion nor stillbirth may occur.[496]

Factors having an effect on the occurrence of epidemic abortion outbreaks may completely differ from those influencing the risk of endemic abortions. Risk factor analyses often have the disadvantage that there is no information regarding the epidemiological context (epidemic or endemic abortion problem). Consequently, it is not possible to clearly assign the risk or protective factors identified in epidemiological studies to the occurrence of epidemic or endemic abortions. However, some risk factor analyses restricted their analyses to herds with epidemic outbreaks[141,2125] or stratified analysis for herds with postnatal *N. caninum* infection and endemically infected control herds[497]; therefore, the risk factors identified in such studies can be related to the occurrence of abortion due to high rates of exogenous transplacental transmission. In many other studies, risk factors cannot be attributed clearly to endogenous and exogenous *N. caninum* transmission.

4.4.3.1 Abortion Risk Associated with Seropositivity of Individual Cattle

Seropositive cows are more likely to abort than seronegative cows as demonstrated in many studies, including retrospective and prospective cohort studies (Table 4.12). The strength of the association between seropositivity and abortion in a single group of animals may vary considerably if different serological assays are used or if for the same assay different cut-offs are applied.[1790,2057] Consequently, the estimates for odds ratio or relative risk may vary in relation to the serological test applied.

The abortion risk increases with increasing levels of *N. caninum*-specific antibodies in individual animals. A strong association between the level of antibodies in the dam and the occurrence of lesions in aborted fetuses consistent with *N. caninum* infection was found in 1 study from Belgium.[450] Herds with a high *N. caninum* seroprevalence had an increased risk for repeated abortion.[249] The same was observed in individual dams with a positive *N. caninum* serological status.[93,1301,1542] A few studies observed that seropositivity in dams was associated with stillbirth and perinatal mortality.[228,246,2057] However, in 1 study, calves born seropositive performed better than seronegative calves.[1557]

As stated above, there are 2 potential reasons for abortion, postnatal infection during pregnancy (causing exogenous transplacental transmission), or reactivated persistent infection (causing endogenous transplacental transmission) (Figure 4.22). With respect to postnatal infection, a high antibody level in the individual animal could be indicative of a high infection dose and/or for an efficient multiplication of the parasite in the infected host. In the case of a latent infection, a high antibody level or titer could also reflect the presence or intensity of recrudescence of an existing chronic infection. There is evidence from prospective studies in latently infected dams that the intensity and duration of the increase of specific antibodies during gestation could be related to the risk of fetal infection.[825,950,1908] Thus, it was proposed to use information on individual *N. caninum*-specific antibody levels or antibody titers (and not only seropositivity) as a predictive tool to identify animals with a high risk of abortion in herds with a high seroprevalence for *N. caninum*.[1657] It was also attempted to use the ratio of specific IgG1 and IgG2 (IgG1/IgG2 ratio) to predict or understand the outcome of bovine pregnancy in dams with a chronic infection because increased specific IgG2 is regarded as an indicator for a proinflammatory immune response (Section 4.2.2.2.4).[40] However, as yet no reliable protocol is established to predict abortion by serological tools.

4.4.3.2 Abortion Risk Associated with N. caninum Seroprevalence in the Herd

There are several case-control and cross-sectional studies which observed that a high *N. caninum* seroprevalence in herds is associated with an increased risk of abortion at the herd level.[141,691,757,918,1559,1757,1803,2125] This is explained by the increased abortion risk in latently infected as well as in recently infected individual dams (discussed above). However, not all herds with a high seroprevalence suffer from *N. caninum*-associated abortion.[496,1010,1559,1803] Long-term studies in herds which had experienced abortion outbreaks revealed no (or only slightly) elevated abortion rates in the years after the outbreak.[221,1603] Recent exposure to *N. caninum* infection, as evidenced by seroconversion and low avidity antibodies does not necessarily result in an increased abortion rate.[496] This supports that, in addition to infection, other factors may influence the abortion risk.

4.4.3.3 Factors Related to Infection Risk

Several factors putatively related to *N. caninum*-associated abortion were already discussed above with respect to infection risk. Moreover, several factors identified as putative risk or protective factors for *N. caninum* infection in cattle also seem to influence the risk of *N. caninum*-associated abortion (Table 4.13).

4.4.3.3.1 Effect of Age on Abortion Risk

A number of studies on herds with *N. caninum*-associated abortion reported an increased abortion risk with increasing age, parity number, number of pregnancies, or gestation number.[1010,1039,1301,2122,2125] However, in herds with clear evidence for endemic *N. caninum*-associated abortion, the association with age seems to be reversed. For example, in a study on the abortion risk in *N. caninum* seropositive dairy cows, lactation number was identified as a putative protective factor.[1218] Also, others found a lower risk of *N. caninum*-associated abortion in older cattle.[1669] These findings confirm previous reports on a 7.4-fold increased abortion risk in congenitally infected heifers during their first gestation but only a 1.7-fold higher risk of abortion in the first pregnancy of first lactation in comparison to the abortion risk in seronegative controls. In the first pregnancy of the second lactation, congenitally infected cows had the same abortion risk as seronegative cows.[1964] In another study conducted in a herd with endemic *N. caninum*-associated abortion where endogenous transplacental infection was the main mode of transmission, a 2.8-fold increased abortion risk during the first pregnancy of the second lactation in seropositive

Table 4.13 Risk and Protective Factors for *N. caninum* Abortion in Cattle

Factor Description		References of Studies Which Identified Putative Risk or Protective Factors[a]	
		Abortion	
		Risk	**Protect**
N. caninum-specific antibodies	Antibody positivity (serum, milk, bulk milk)	41, 64, 93, 100, 143, 228, 246, 249, 271, 276, 286, 339, 383, 393, 429, 431, 448[**], 670, 726, 731, 732[**], 764, 791, 792, 872, 898, 969, 1010, 1085, 1086, 1116, 1125, 1214, 1215, 1217, 1260, 1301, 1370, 1381[**], 1390, 1396, 1530, 1538, 1542, 1558, 1591, 1596, 1641a, 1655, 1669, 1700, 1774[*], 1794, 1800, 1824, 1892, 1899, 1966, 1994, 2016, 2057[*], 2058[*], 2063[*], 2085, 2088, 2123	1287, 1390
	Level of specific antibodies (titer, ELISA value)	41, 143, 792, 898, 1039, 1217, 1301, 1305, 1657[*], 1792, 1794, 1908, 2063[*], 2125[*], 2150	—
	Level of specific antibodies (titer, ELISA index) in combination with other markers		—
	Seropositive and no IFN-γ detected in plasma during gestation	1213	—
	Seropositive and low bovine pregnancy-associated-glycoprotein 2 level (PAG2 ≤ 4.5 ng/mL)	726[**]	—
	High level of specific antibody titer and progesterone supplementation in early gestation	179	—
	High level of specific antibody titer and progesterone supplementation in mid-gestation	180	—
Age, parity, gestation, and lactation number	Seroprevalence in the herd	141, 691, 757, 918, 1559, 1757, 1803, 2057	—
	Age	1039, 2063[*]	1669, 2063[*]
	Parity number, number of pregnancies, gestation number ≥2	1010, 1301, 2125	—
	Lactation number	—	1218, 1964, 2150
	2nd lactation vs. 1st, 3rd or ≥4th lactation	898	—
	Bred heifer vs. mature cow	2063[*]	—

(Continued)

Table 4.13 (Continued)   Risk and Protective Factors for *N. caninum* Abortion in Cattle

Factor Description			References of Studies Which Identified Putative Risk or Protective Factors[a]	
			Abortion	
			Risk	Protect
Definitive hosts (dogs, coyotes) and other canids. Contact with sick cattle	Dogs	Presence of farm dogs	141	—
		Number of farm dogs	918	—
		Farm dog serologically *N. caninum* positive	757	—
		Behavior of farm dogs: frequency of defecating in a feed manger	918	—
	Wild canids	Frequency of observation on farm premises	918	—
	Other animal species	No. of poultry >10	141	—
		Number of horses	918	—
		Frequency of stray cats are observed	—	918
	Contact to sick cattle	Calving pen used to hospitalize sick animals	141	—
Grazing and fodder		Feeding of moldy maize-silage to dairy cows during summer	141	—
		Feeding of remnant fodder to heifers during summer	141	—
Source of replacement heifers		Maternal relationship	860	—
Breed		Crossbreeding: using beef bull semen to inseminate Holstein-Friesian dairy cattle	—	1217, 2150
		Continental vs. British beef breeds	2063	—

(Continued)

Table 4.13 (Continued) Risk and Protective Factors for *N. caninum* Abortion in Cattle

Factor Description		References of Studies Which Identified Putative Risk or Protective Factors[a]	
		Abortion	
		Risk	Protect
Utilization	Beef vs. dairy	–	1380[***], 2104[***]
Body condition	Body condition score ≤4	2063[*]	–
Failures during and around reproduction	Previous abortion in congenitally infected cattle	1965	–
	Annual rate of cows returning to estrus post pregnancy check	918	–
	Annual rate of retained fetal membranes in herd	918	–
	Prevalence of retained after births in previous year >10%	141	–
Indicators for other diseases or infections	Antibodies against BVDV	860	–
	Antibodies against *Coxiella burnetii*	860	–
	Antibodies against *Chlamydia psittaci*	–	860
	Antibodies against *Leptospira* sp.	–	860
Type of housing	Loose housing	860	–
	Heifers housed on a loafing pack	–	918
Climate	Rainfall	1218, 2150	–
Climatic season	Summer (The Netherlands)	2125	–
	Winter (California, USA)	1961	–
	September–November (British Columbia, Canada)	2104[**]	–
	Summer and winter vs. fall and spring (Israel)	1301	–
Demographic factors	Proximity to a town or village	860	–

[a] Studies on beef cattle are marked with [*] and those not differentiating between beef and dairy are marked with [**]. Studies on dairy cattle are not marked. Studies differentiating between beef and dairy [***].

dams but not in the first pregnancies of the first, third, and later lactations were observed in another study.[898] In a Canadian study on beef cattle to find potential reasons for non-detectable pregnancies, both increased levels of *N. caninum* antibodies and age (bred heifer, >10 years old) were associated with risk.[2063]

4.4.3.3.2 Abortion Risk in Cattle and Farm Dogs

There are large number of studies confirming that the presence of dogs and other dog-related factors as a major risk factors for *N. caninum* infection in cattle (Table 4.12). As shown in Table 4.13, the presence of dogs, their number, their seropositivity, and the frequency of observing farm dogs defecating in a feed manger were associated with an increased abortion risk of the herd.[141,757,918] One of the studies identifying a positive association between the presence of farm dogs and *N. caninum*-associated abortion had selectively analyzed risk factors for epidemic abortion. Because epidemic abortion is possibly caused by oocyst-mediated horizontal transmission, the identification of the presence of potential definitive hosts, that is, farm dogs as a putative risk factor is not unexpected.[141,918] In addition, a positive correlation between seropositivity of farm dogs and increased seroprevalence in cattle, indicating a relationship between infection in dogs and in cattle was observed in another study.[2125] Investigated dogs were present both at farms with epidemic and endemic neosporosis.[2125]

4.4.3.3.3 Abortion Risk in Cattle and Wild Canids

In 1 study, the frequency with which wild canids (including coyotes, wolves, foxes, and wild dogs) were observed on the premises seemed to have a protective effect on the likelihood that farms experienced *N. caninum*-related abortion.[918] The protective effect was explained by hypothesizing a negative interaction between the presence of farm dogs (which seem to pose an infection risk) and wild canids. It was assumed that the more farm dogs are present on a farm, the lower the likelihood that wild canids are observed on the premises.

4.4.3.3.4 Abortion Risk in Cattle and Cats

The frequency with which stray cats were observed on the premises was identified as a putative protective factor for *N. caninum*-associated abortion.[918] As mentioned above, a study from France reported a reduced risk of infection in cattle when cats were present on farm.[1541] It was assumed that the presence of cats might be an indicator for the absence of dogs, resulting in a reduced risk of horizontal transmission.[918] In the same study, the frequency with which wild canids (including coyotes/wolves, foxes, and wild dogs) were observed on premises seemed to have a protective effect on the likelihood of *N. caninum*-related abortion on the farm.[918]

4.4.3.3.5 Other Potential Intermediate Hosts Such as Poultry and Horses

Case herds having experienced *N. caninum*-associated abortion outbreaks in The Netherlands more often kept, in addition to cattle, an increased number of poultry (more than 10).[141] Interestingly, in 1 study, the infection risk increased with the number of farm dogs if poultry was present on a farm suggesting an interaction between these intermediate and definitive host-related variables.[1533] Antibodies against *N. caninum* have been observed in chickens[390,968,1288] and DNA was observed in seropositive chickens.[390,773] However, all attempts so far failed to isolate viable *N. caninum* from chickens. Therefore, further studies are necessary to establish that poultry or which poultry species represent important intermediate hosts for *N. caninum*.

Unexpectedly, a Canadian study observed an association between the number of horses on a farm and the occurrence of *N. caninum*-related abortion.[918] The reason for this association is not clear. Horses are known as intermediate hosts of *N. hughesi* which seem to represents a species different from *N. caninum*.[1284] As yet *N. hughesi* (see Chapter 18) has not been isolated from cattle. Thus, it is unknown whether *N. hughesi* could be involved in bovine abortion. In addition, there is no definitive evidence that horses act as intermediate hosts for *N. caninum*.

4.4.3.3.6 Fodder

Feeding fodder of inferior quality, for example, "Feeding of moldy maize-silage to dairy cows during summer" or "Feeding of remnant fodder to heifers during summer" seemed to be risk factors for epidemic *N. caninum*-associated abortion in The Netherlands.[141] The effect of feeding fodder of inferior quality may involve a suspected negative impact of fungal toxins on the immune system of cattle.[141,1962,2125] In addition, remnant fodder may contain a higher proportion of contaminants, thus possibly also fecal contaminations of definitive hosts. A further explanation could be that inadequate rations may stress cattle.

4.4.3.3.7 Climate, Climatic Season, Stress, and Body Condition

A highly significant seasonal pattern regarding the submission of *N. caninum* positive aborted fetuses was observed in California. The highest number of positive cases was submitted during winter, which in California is mild and humid in contrast to the summer, which is hot and dry.[1962] In Canada, the proportion of *N. caninum* diagnoses peaked in the autumn months (i.e., in September–November)[2104] but no explanation for this observation was provided. In The Netherlands, abortion epidemics most often occurred in summer,[2125] which in The Netherlands is warm and humid. There are several possible explanations for these phenomena. Mild temperatures and humidity favor sporulation and survival of coccidian oocysts which may increase the risk of postnatal infection. A further explanation is that mild temperatures and humidity support the growth of fungi. Fungal toxins are suspected to cause immune suppression in cattle which may favor the recrudescence of *N. caninum* infections in persistently infected dams.[141,1962,2125]

Heat stress may have an effect on fertility and pregnancy maintenance.[724,1213] But a hot environment alone seems not to affect the risk of *N. caninum*-associated abortion.[1214,1218,2150] However, in a dry environment, rainfall during the second trimester of gestation was found to increase the risk of abortion in seropositive cows.[1218,2150] The effect of rainfall is regarded as nonspecific. This observation may reflect the general susceptibility of pregnant cows to any type of stress during the second trimester of gestation, that is, in a period of gestation when cows are immunocompromised.[976,2150] A risk factor analysis on abortion risk in *N. caninum* seropositive dams in 2 Spanish dairy herds suggested that there was a significant relationship between rainfall and abortion. It was suspected that increased rainfall may pose direct and indirect stresses to cattle by elevated heat production in response to cold temperatures, behavioral stress, impaired food quality, and diminished hygiene.[1218]

A Canadian study revealed that a thin body condition (body condition score 4 or less) increased, together with the level of *N. caninum*-specific antibodies, the chance of beef cows to experience a reproductive failure.[2063]

4.4.3.3.8 Farm-Raised Replacement Heifers

Rearing dams affected by abortion and the replacement heifers on the same farm was identified as a putative risk factor for *N. caninum*-associated abortion in a case-control study conducted in Switzerland.[860] This finding is in accord with previous findings on the infection risk in beef calves.[127]

4.4.3.3.9 Demographic Factors, Proximity to a Town and Village

In a Swiss case-control study "proximity to a town or village" was observed as a putative risk factor for *N. caninum*-associated abortion.[860] This observation is in accord with the findings of a German study which showed that herds had an increased risk of being positive in *N. caninum* bulk milk ELISA if they were located in districts or cities with a high human population density.[1801] An increased population density is correlated with a high dog density[1801] which may lead to an increased infection risk of herds located closer to towns or cities. This may cause an increased risk for *N. caninum*-associated abortion in these areas.

4.4.3.3.10 Factors Related to Antibodies against Other Infectious Agents

Infections with agents other than *N. caninum* could pose stress or immune suppression to animals thus supporting the recrudescence of chronic infections or postnatal transmission.[220,1962] In contrast, vaccination against other infectious agents could reduce the level of stress in a herd and thus reduce also the likelihood of *N. caninum*-associated abortions if stress triggers such abortions.[918] The effect of other infections or vaccination against other infectious agents on the risk of *N. caninum*-associated abortion is not clear. Both vaccination and infection induce antibodies against infectious agents and these serological responses can be used to address this question in epidemiological studies. However, the results of risk factor studies based on the serological responses to other infectious agents are often difficult to interpret because typically there is no (or only limited) information whether the antibodies are present because of infection or because of vaccination.

In an univariate analysis, a Swiss case-control study observed that herds with *N. caninum*-associated abortions were more often positive for antibodies against *Coxiella burnetii* and less often positive for antibodies against BVDV, *Chlamydia psittaci*, and *Leptospira* species than the control herds.[860] However, in a final multivariate model, positive BVD serology appeared to be the only putative serological risk factor for *N. caninum*-associated abortion at the herd level. The serostatus to *Coxiella, Chlamydia,* and *Leptospira* was eliminated from this final model because of the lack of statistical significance.

In a Dutch case-control study, no significant relationship was observed between the herd-level seropositivity for BVDV, BHV1, *Leptospira interrogans serovar* Hardjo and *Salmonella enterica serovar* Dublin, and the risk of epidemic *N. caninum*-associated abortion. However, among the aborting dams, there was a negative relationship between seropositivity to BVDV and seropositivity to *N. caninum*.[141]

4.4.3.3.11 Housing

In 2 studies, the type of housing had an effect on the risk of *N. caninum*-associated abortion. In a Swiss study, loose housing of cattle was identified as a putative factor increasing the abortion risk.[860] Apparently, loose housing is related to unknown management practices that increase the risk of *N. caninum*-associated abortion. For example, an association between loose housing and herd size was identified in a German study, because in large herds, cattle were more likely to be kept in pen barns.[1803] However, in another study conducted in France, loose housing was identified as a factor that reduced the infection risk.[1541]

In a Canadian study, the housing of heifers on a loafing pack (a housing pen divided into feed manger, scrape alley, and bedded pack areas) reduced the abortion risk.[918] It was assumed that some designs of loafing packs may hinder the access of farm dogs and that the effect is most likely associated with oocyst-mediated horizontal transmission of *N. caninum* to cattle.

4.4.3.4 Factors Associated with Reproduction

The results of several studies suggest that variables related to reproduction could affect the risk for bovine *N. caninum*-associated abortion.

4.4.3.4.1 History of Previous Abortions

Congenitally infected cows which had previously aborted, had a 5.6-fold increased abortion risk compared to congenitally infected cows which had not experienced an abortion before.[1964] Others also observed that dams with a positive *N. caninum* status had an increased risk for repeated abortion.[93,1301,1542]

4.4.3.4.2 Annual Rate of Cows Returning to Estrus Post Pregnancy Check

A Canadian case-control study revealed that there was a positive association between the occurrence of *N. caninum*-related abortions in a herd and the annual rate of cattle returning to estrus after pregnancy confirmation.[918] A high rate of early pregnancy losses could increase the chance for definitive hosts to have access to infectious material, increasing the rate of oocyst-mediated horizontal transmission. On the other hand, this result could indicate that *N. caninum* is not only associated with abortion but also with early pregnancy losses. Indeed, there are 4 other studies, 3 from Canada with results supporting this view.[1434,2057,2059,2063] In this context, cattle experimentally infected at day 70 of gestation by intravenous inoculation with high doses of *N. caninum* tachyzoites were more susceptible to abortion than those infected with the same dose at day 140 or 210 after insemination.[2097] Whether *N. caninum* contributes to early pregnancy losses remains open because a number of other epidemiological studies observed no indication that *N. caninum* is able to cause early pregnancy losses.[215,1010,1214,1218,1740]

4.4.3.4.3 Retained Afterbirths

There are 2 studies indicating that the risk of *N. caninum*-associated abortion may increase with an increasing annual rate of retained afterbirths.[141,918] Firstly, more retained afterbirths could provide more sources of infection for definitive hosts and thus increase the chance of oocyst-mediated horizontal transmission. Secondly, and more likely, *N. caninum* could not only be associated with abortion but also be involved in the pathogenesis of retained afterbirth.[85,94] Further studies are necessary to clarify this point.

4.4.3.4.4 Using Beef Bull Semen to Inseminate Dairy Cattle, Lower Susceptibility of Beef Breeds

In a prospective cohort study using dairy or beef bull semen to inseminate *N. caninum* seropositive dairy cows, it was observed that the use of beef bull semen reduced the risk of abortion[1217] which was confirmed in another study.[1218] Further studies from Spain demonstrated that crossbreed pregnancies efficiently reduced the abortion risk in *N. caninum*-infected dairy cows, especially if Limousin semen was used.[41,2150] But also crossbreed pregnancies of seropositive Holstein-Friesian cattle inseminated with semen from other beef breeds (Charolais, Belgium Blue, and Piedmontese) led to abortion less frequently than in seropositive Holstein-Friesian cows inseminated with semen from Holstein-Friesian bulls.[41] The positive effects of crossbreeding were stronger in cows with high levels of *N. caninum*-specific antibodies.[41] A study from Belgium observed that the distribution of intracerebral lesions was more extensive in aborted dairy than in beef fetuses.[450] A Canadian study revealed that the breed of dam was significantly associated with *N. caninum* diagnosis in aborted fetuses, with a higher prevalence in dairy versus beef breeds.[2104] These results could suggest that beef breeds in general or certain beef breeds (e.g., Limousin) are less susceptible to *N. caninum* infection than others. Results of a further study in Argentina support this view.[1380] A more recent review, which compiled findings on *N. caninum*-associated abortions in beef and dairy cattle, concluded no significant differences in *N. caninum*-specific-abortion risk between beef and dairy cattle. This observation has to be interpreted with care because most dairy and beef cattle compared in this study did not come from the same herds or regions.[1694]

It was hypothesized that placental function might be favored in crossbreed pregnancies possibly via an increased concentration of pregnancy-associated glycoproteins (PAG, i.e., PAG-1 and PAG-2).[48] PAG-1 and PAG-2 are expressed in the ruminant trophectoderm from the time the placenta attaches to parturition.[2064,2112] The precise role of these glycoproteins is unknown. Eventually, these PAGs interact with uterine serpins which are involved in the progesterone-mediated uterine immunosuppression during gestation.[1833] Both PAG-1 and PAG-2 levels undergo a sudden drop in aborting cows, thus acting as useful markers of the feto-placental status.[727,1219,1435] In 1 study, it was shown that in *N. caninum* seropositive, chronically infected but non-aborting cows, PAG-1 concentrations during the course of gestation were not affected and were similar to concentrations observed in *N. caninum* negative cows.[1219] However, PAG-1 concentrations were affected by breed and fetal genotype irrespective of *N. caninum* infection status.[724,727] PAG-1 plasma level seems to be a useful marker to monitor the feto-placental status in aborting animals; high levels seem to be beneficial.[1219] However, PAG-1 seems to have no predictive capacity related to *N. caninum*-associated abortion.[48] In contrast to PAG-1, PAG-2 levels might be a predictive marker regarding the fate of pregnancies in persistently *N. caninum*-infected pregnant dams.[48] In 1 study, both *N. caninum* seropositivity and low plasma PAG-2 concentrations (<4.5 ng/mL) on day 120 of pregnancy were associated with the likelihood of abortion.[726]

4.4.3.4.5 Use of Calving Pen to Hospitalize Sick Animals

In a Dutch case-control study, it was observed that herds on farms where the calving pen was also used to hospitalize sick animals had a higher risk of having a recent *N. caninum*-associated abortion epidemic than other herds.[141] The biological significance of this finding is not clear. It is very unlikely that *N. caninum* is transmitted horizontally among adult cattle, for instance via the exposure to placenta or uterine effusions.[1795] To date, all experiments aimed at infecting adult cattle or calves via oral ingestion of placental material from seropositive animals failed.[434] Therefore, it has to be assumed that the factor "calving pen used to hospitalize sick animals" was possibly linked to another as yet not identified risk factor.

4.4.3.4.6 Attendance of Cattle Shows

In a Dutch case-control study, it was observed that herds which had attended cattle shows during the previous 2 years had a reduced risk of *N. caninum*-associated abortion epidemics.[141] Possibly this factor is negatively associated with the factors "rearing of own replacement heifers"[127] or "rearing the dams affected by abortion and replacement heifers on the same farm"[860] because the attendance of cattle shows could indicate that a higher proportion of replacement heifers come from external sources. "Rearing own replacement heifers" was identified as a potential risk factor for a high *N. caninum* seroprevalence in beef cattle[127] and "rearing the dams affected by abortion and replacement heifers on the same farm" was identified as a putative risk factor for *N. caninum*-associated abortion in a Swiss case-control study.[860]

4.5 ECONOMICS AND CONTROL OF BOVINE NEOSPOROSIS

4.5.1 Economic Considerations

The major economic impact of neosporosis is caused by reproductive failure in cattle. Fetal death represents direct costs. However, other indirect losses increase the overall expenses for farmers, such as expenses for professional help and diagnosis, or lengthened intervals for rebreeding and replacement of culled cows, reduction of milk yield and reduced weight gain in infected animals.[603,607,1525,1992] However, the latter 2 are controversial: with respect to milk yield, different studies indeed showed a reduction,[897,1973] whereas others indicated increased[1603] or not significantly different[833,917] milk production. The same accounts for weight gain: reduced post-weaning weight gain in seropositive calves due to poor feed efficiency was observed,[126,127a] but no difference in weight gain between seropositive and seronegative calves was observed in more recent studies.[916,1404] Other factors that may impact on the economic effects of *N. caninum* may include differing genetics and susceptibilities of cattle, different herd management practices, nutrition, and the possibility of concurrent infection with other infectious agents that has not been fully evaluated nor consistently presented in published studies.[127,1691,1694,1695]

It is difficult to have a reliable global overview of the economic losses caused by abortions due to *N. caninum*. A comprehensive calculation based on the review of 99 publications from 10 countries estimated the median losses of *N. caninum*-induced abortion to be in excess of US$ 1.298 billion per year, with two-thirds incurred by the dairy cattle, and one-third by the beef cattle industry.[1694] The global costs were estimated to US$ 852.4 million in North America (65.7%) (USA, Canada, and Mexico), US$ 239.7 million in South America (18.5%) (Brazil and Argentina), US$ 137.5 million in Australasia (10.6%) (Australia and New Zealand), and US$ 68.7 million in Europe (5.3%) (The Netherlands, Spain, and UK).[1694]

Different patterns of *Neospora*-associated abortion occur: epidemic and endemic (including sporadic). The epidemic pattern is defined by temporary abortion outbreaks.[431] Epidemic abortions are due to primary infection of previously uninfected dams exposed at almost the same time to a single source of contamination.[1314] This pattern can result in a large proportion (>10%) of pregnant cows aborting over a short period of time.[607] These abortion storms are generally viewed as very costly and sometimes devastating to the farmer. In the endemic pattern, the abortions happen intermittently for months or years and are due in part to persistently infected dams that transplacentally transmit the parasite to their progeny.[833] A low background level of postnatal (horizontal) transmission continuously occurring in endemically infected herds also may contribute to *N. caninum*-associated abortion in these herds and this background level of horizontal transmission seems to be favored by the presence of dogs.[146] The prevalence in dogs shows a correlation with the prevalence in cattle.[575] Although a sylvatic cycle for *N. caninum* has been demonstrated,[786,1747] its importance as a reservoir for the transmission to domestic animals has not been definitely elucidated, but seems to be of minimal significance at a large scale, although it can be significant locally.[1068,1070]

Other factors that could have an influence on the infection risk are, for example, the presence of other intermediate hosts, the feeding mode of cattle, coinfections, vegetation index, climate, size of farmland, calving management, and feeding colostrum or milk.[607,614] Seroprevalences in beef herds are usually lower than in dairy herds, but this may be due to different farm management practices[144,1390] rather than breed-related susceptibility.[634] However, some studies found different rates of abortion and immune responses between different breeds of cattle.[84,1757,1776]

In order to diminish the costs of *N. caninum* infections within herds, different strategies have been proposed, depending on the country or region, infection rate, and associated risk factors.[607] Culling productive cattle is expensive. In farms with endemic abortions, the identification of infected animals and to cull or selectively breed them may represent an option to avoid further costs due to neosporosis. To avoid epidemic abortions, the contact between definitive hosts and cattle has to be avoided and contamination of food and water by feces containing oocysts must be carefully controlled.[607] There are also standard measures that can be applied to prevent the introduction of new infection sources in a herd. Replacement heifers or cows should be tested for *N. caninum* or purchased from disease-free herds. Access of dogs and wild animals to the housing zone, barn, water where cattle drink, and feed storage as well as to potentially infected tissue from intermediate hosts has to be avoided. A control of rodents in the farm area could also reduce the dogs' infection risk although the importance of rodents as sources for canine infection needs to be further studied. Reproductive management could also reduce the costs of infection in a herd (reviewed in Reference **607**). The 2 methods of reproductive management that have proven to reduce vertical transmission are the transfer of embryo from infected dams to uninfected recipients,[111,1115] although care should be taken to avoid contamination from the ovarian follicle,[1856] and the insemination of seropositive dairy dams with beef bull semen.[41,1217,1856] Indeed, crossbreeding was shown to have a favorable effect on the placental protective function, presumably through higher concentration of PAG and other associated factors.[1217,1435,1436,1833] However, the most effective option is not always the most economic one and a detailed and adapted economic study as to be made specifically for each case before deciding on a strategy.[858,860,1121,1689] Moreover, it is important to constantly recalculate the economics of the strategy to be in agreement with fluctuating prices.[1691] With all control strategy studies, it was never possible to reach a seroprevalence of zero, because of the existence of the horizontal transmission. In order to eradicate *N. caninum*, it would be necessary to control both transplacental transmission within a herd as well as horizontal transmission.[859,1316] Chemotherapeutic treatment of *Neospora*-seropositive animals has not been regarded as an economically viable option. Depending on the compounds used, milk or meat from drug-treated animals would remain unacceptable for consumption for some time.[607] Nevertheless, experimental studies have revealed potentially interesting effects of several compounds *in vitro* and *in vivo* in laboratory animal models. Many of the promising compounds originate from repurposing approaches and have been studied by cell culture-based screening methods.[1421,1422] A target-based screening approach has been used to identify inhibitors of calcium-dependent protein kinase 1 (CDPK1), which is conserved almost exclusively within the group of apicomplexan parasites including also *Plasmodium, Toxoplasma, Cryptosporidium, Sarcocystis*, and others, but most studies have been performed in small laboratory animals and only few in the actual target hosts such as cattle. On the other hand, vaccines have the advantage of presenting no risk of long-lasting residues in the meat or milk. It was also demonstrated that vaccination makes economic sense in case of high prevalence of the disease.[856–859,1687,1693,1695,1697] Therefore, an efficient vaccine that prevents *N. caninum* infection is needed and would fill an empty market field.[928]

4.5.2 Vaccination against Neosporosis in Cattle

The only licensed *Neospora* vaccine was Bovilis Neoguard®, which was composed of a tachyzoite lysate, and was available in selected countries for several years. However, this vaccine had only moderate efficacy in field trials,[1739] and 1 study suggested that vaccination itself could increase the risk of early embryonic death.[2087] Thus, this vaccine was withdrawn from the market. Vaccine trials against neosporosis in cattle, employing either live-attenuated strains or subunit vaccines, are compiled in Table 4.14.

4.5.2.1 Live-Attenuated Vaccines

Nc-Nowra was the first live-attenuated *N. caninum* strain that was experimentally assessed. It was demonstrated that live vaccines can protect against fetal death.[2100] For this, animals were

Table 4.14   Vaccine Studies against Neosporosis in Cattle

Vaccine	Set Up	Results	References
N. caninum tachyzoites (killed)	Pregnant cattle, vaccinated twice by combined IV/IM inoculation of tachyzoite extract formulated in POLYGEN adjuvant. Challenge during pregnancy	High humoral immune responses in vaccinated cattle, no protection against abortion and fetal infection	**75, 76**
N. caninum tachyzoites (Bovilis Neoguard™)	Field trial with dairy cattle. No challenge	Reduction of abortion from 20% in the placebo group to 11% in the vaccinated group	**1739**
NC-Nowra (naturally attenuated live vaccine strain)	Cattle vaccinated IV 90 days prior to pregnancy, challenged with Nc-Liv at day 70 of gestation	Cattle immunized with live tachyzoites had strong cellular and IFN-γ responses prior to challenge, and this correlated with protection against fetopathy. Dams vaccinated with Nc-extract were not protected	**2100**
N. caninum tachyzoites (Bovilis Neoguard)	Clinical trial on 5 dairy farms (SC, 2x, 4-week intervals)	Vaccination increases the risk of vertical transmission. In 1 of 5 herds, vaccination reduced abortion	**2087**
Native *N. caninum* antigens formulated in ISCOMs and live Nc6 vaccine strain	Pregnant heifers, immunized with live Nc-6 Argentina or antigen extract formulated with ISCOM (SC, 2x). Challenge with NC1 at day 70 of gestation. Termination of experiment at day 104 of gestation	Reduced vertical transmission through live vaccination, but not through vaccination with antigen extract	**866**
N. caninum tachyzoite extract	Cattle, aqueous tachyzoite extract at various concentrations with soybean-based adjuvant (2x). No challenge in first experiment, or infection at days 78 or 225 of gestation in second study	Increased IgG1 and IFN-γ levels in vaccinated animals as compared to controls. Stimulation of CD4(+)-T-cells. High systemic IFN-γ levels did not interfere with pregnancy	**1270, 1271**
N. caninum tachyzoites (naturally attenuated live vaccine strain NcSpain H1)	Seronegative heifers, vaccinated with Nc-Spain H1 (2x), challenge with NC1 post mating (2x)	Strong IgG and IFN-γ responses post immunization. No fetal loss in immunized but not challenged heifers. In challenged heifers, 50% protection against fetal loss	**1732**
N. caninum tachyzoites (live or frozen Nc-Nowra)	Cattle, 96 seronegative animals, immunized with Nc-Nowra (SC or IV, 1x), either live or from frozen stocks, prior to mating. Pregnant heifers were challenged with Nc-Liv.	Protection against abortion by vaccination, with live tachyzoites IV 85%, with cryopreserved tachyzoites 30%	**2078**
GRA7 (recombinant antigen)	Nonpregnant cattle, NcGRA7 (50 and 200 µg) entrapped in oligomannose microsomes (sc, 2x). Challenge with NC1 isolate 27 days after last boost. Euthanasia at 85–87 DPI	IgG and IFN-γ levels increased as compared to controls. Lower parasite load in brains in cattle immunized with 50 µg	**1482**
SAG1+HSP20+GRA7 (recombinant antigens)	Pregnant heifers, immunized with recombinant proteins formulated with ISCOM (SC, 2x). Challenge with NC1 at day 70 of gestation. Experiment terminated at day 104 of gestation	Immune responses against antigens. No IFN-γ response. No protection against vertical transmission	**867**

(Continued)

Table 4.14 (*Continued*)   Vaccine Studies against Neosporosis in Cattle

Vaccine	Set Up	Results	References
SAG1+HSP20+GRA7 or live NC-6 Argentina vaccine strain	Pregnant heifers, immunized with recombinant proteins formulated with ISCOM (SC 2x) or with live NC-6 Argentina strain. Challenge with NC1 at day 70 of gestation	Analysis of CD3(+), CD4(+), $\gamma\delta$-T cells, CD8(+) cells and macrophages, and expression levels of IFN-γ, IL-4, IL-10, IL-12, and TNF-α. Strongest cellular immune responses were observed in the placentomes of non-vaccinated animals and those that were immunized with inactivated vaccines	868
N. caninum tachyzoites (live NcIs491 vaccine strain)	Field study with 520 pregnant, seropositive heifers; 146 were vaccinated, 374 controls	Lower incidence of abortion in vaccinated as compared to control cows, with overall vaccine efficacy of 39%	1302
Soluble fraction of *N. caninum* tachyzoite lysate and a soy-based aqueous adjuvant (sNcAg/ AVEC)	10 pregnant heifers were vaccinated twice during the first trimester of gestation and 8 remained unvaccinated. No challenge infection	High antibody and IFN-γ responses in vaccinated animals without any influence on pregnancy outcome	1271

IV inoculated with 10^7 Nc-Nowra tachyzoites 9 weeks prior to artificial insemination in order to establish a chronic infection. Strong antibody responses and strong cellular and IFN-γ responses were noted prior to challenge. Experimental challenge was done IV with 10^7 Nc-Liv tachyzoites at day 70 of gestation. The protective efficacy of a naturally attenuated *N. caninum* isolate, the Nc-Spain1H isolate, was demonstrated.[1732] In these experiments, heifers were immunized SC twice with 10^7 live Nc-Spain1H tachyzoites prior to artificial insemination. Upon challenge with NC1 at day 70 of gestation, protection of 50% against fetal death was noted. When challenge was performed at mid-gestation (day 135), calves from immunized heifers had significantly lower precolostral *Neospora*-specific antibody titers compared to calves from the non-immunized/challenge group. An Argentinian isolate, Nc-6 Argentina, was compared with native Nc-6 Argentina antigen extract formulated in ISCOMs (both applied prior to mating), followed by IV challenge with the NC1 isolate also on day 70 of gestation.[866] A significant increase in *N. caninum* antibody responses was detected in the heifers of both groups prior to challenge, and IFN-γ responses were similar as well. The experiment was terminated at day 104 of pregnancy, and all fetuses were viable at that time point. PCR detected transplacental transmission in 1 of 4 fetuses from the live-vaccinated group, and in 3 of 4 fetuses of the group vaccinated with the ISCOM formulation. More recently, a live vaccine isolate (NcIs491) was assessed in a field trial comprised of 520 pregnant and *N. caninum* seropositive cows, of which 146 were vaccinated at mid-gestation and 374 served as controls.[1302] A significantly lower incidence of abortion was observed in vaccinated (16%) compared to non-vaccinated cows (26%), resulting in a vaccine efficacy of 39%. However, the number of seropositive offspring remained similar in both groups and it remains to be clarified to which extent the vaccine strain caused this infection in the offspring of vaccinated dams.

4.5.2.2 *N. caninum Tachyzoite Lysate-Based Vaccines*

The first antigen formulation to be assessed as a vaccine against experimental *N. caninum* infection in cattle was a POLYGEN-adjuvanted killed *Neospora* tachyzoite preparation.[76] The rationale behind this experiment was that cattle immunized with this formulation were previously shown to produce interferon IFN-γ at levels similar to those of tachyzoite-infected cattle. Heifers were immunized at days 35 and 63 of gestation and were challenged at day 91 of gestation. Immunization appeared to induce high IgG1 responses, elevated lymphoproliferative responses and IFN-γ production. However, following tachyzoite challenge, the cellular immune response dropped to undetectable levels, and fetal infection was not inhibited.

A study in Costa Rica investigated the effect of Bovilis Neoguard, on the crude abortion rate of dairy cows under field conditions, and results indicated a reduction of abortion from 20% in the placebo group to 11% in the vaccinated group.[1739] Subsequently, the efficacy of Bovilis Neoguard™ was reinvestigated by carrying out clinical trials in 5 dairy cattle farms in New Zealand, all of which had a history of *N. caninum* abortions. Cattle enrolled in these trials at days 30–60 of gestation were vaccinated at 4 week intervals. The conclusion of this study was that vaccination after conception prevented 61% of abortions in 1 of 5 herds, but increased transplacental transmission occurred, and results indicated that vaccination may have increased the risk of early embryonic death.

Another approach has been to formulate total antigen extract into ISCOMs. This formulation was injected into seronegative calves and the immune response and blood parasitemia was assessed following challenge with *N. caninum* tachyzoites. The protective effect of the vaccine was not assessed, but higher antibody titers and similar IFN-γ production were observed in those calves that had received the antigens with ISCOMs compared to those that had received live parasites as an immunization dose.[866]

Recently, the safety and immunogenicity of a soluble native *N. caninum* tachyzoite-extract vaccine formulated with a soy lecithin/β-glucan adjuvant in pregnant cattle was reported.[1271] This formulation had previously been shown to be protective in the mouse model and induced strong IFN-γ responses and high avidity antibodies in nonpregnant cattle.[1270] The vaccine was used during the first trimester of pregnancy, and anti-*N. caninum* immune responses were efficiently primed by vaccination. Infection was carried out at days 78 or 225 of pregnancy, and the high systemic IFN-γ levels that were induced did not interfere with pregnancy. However, protection against vertical transmission was not reported in this study.

4.5.2.3 Subunit Vaccines

In 1 study, nonpregnant cattle were immunized SC with recombinant NcGRA7 entrapped in oligomannose microsomes (M3-NcGRA7) twice with a 4 week interval.[1482] Challenge infection with 10^7 NC1 tachyzoites was done 27 days after the second immunization, and animals were euthanized at days 85–87 PI. NcGRA7-specific antibody production and IFN-γ production in PBMC was induced in vaccinated animals prior to challenge, and the parasite load in the brain was significantly decreased in cattle immunized with 50 μg M3-NcGRA7 compared with controls.

Another study was performed in pregnant animals and addressed transplacental transmission.[867] The bacterially expressed and purified recombinant antigens rNcSAG1, rNcHSP20, and rNcGRA7 were formulated with ISCOMs, and inoculated into heifers prior to mating. Immunogens were SC administered twice, with 1 group receiving all 3 antigens, 1 only ISCOMs, and 1 receiving sterile PBS only. Challenge was done at day 70 of gestation employing the NC1 isolate. The vaccinated groups had high antibody responses prior, and even more increased responses after, challenge, but there were no differences in IFN-γ production among the experimental groups at any point in time. Analysis of the fetal tissues showed that despite a clear humoral immune response, vaccination had failed to prevent fetal infection.

In a follow-up study,[868] the cell-mediated immune responses in the placenta of heifers that were vaccinated with this combined recombinant antigen ISCOM formulation were analyzed and compared with the corresponding immune responses of heifers vaccinated with either live Nc-6 Argentina tachyzoites or Nc-6 Argentina antigen extract ISCOM formulation. CD3(+), CD4(+), γδ-T cells, CD8(+) cells, and macrophages were analyzed,[868] as well as expression levels of IFN-γ, IL-4,

IL-10, IL-12, and TNF-α. Strongest cellular immune responses were observed in the placentomes of non-vaccinated animals and those that were immunized with inactivated vaccines. Animals vaccinated with live tachyzoites showed a milder immune cell infiltration to the placenta, possibly due to the existence of a protective systemic maternal immune response that helped to minimize *N. caninum* infection at the maternal-fetal interface.

Based on studies in mice which had demonstrated that vaccination with recombinant *N. caninum* profilin (rNcPRO) resulted in limited protection and a profound regulatory T cells response,[1273] the immune response against rNcPRO and rNcPRO fused to functional T cell epitopes of vesicular stomatitis virus glycoprotein G (rNcPRO/G) was analyzed in cattle.[1272] In addition, the 2 recombinant antigens were also mixed with TLR2 and TLR9 agonists. Vaccination of cattle with rNcPRO elicited only IgM antibodies, while antibodies in rNcPRO/G-vaccinated animals switched to IgG1 after the booster. The vaccine formulated with rNcPRO/G and TLR agonists improved the production of systemic IFN-γ and induced long-term recall B cell-responses, rendering this formulation interesting for further studies in pregnant cattle.

4.5.3 Reproductive Management

The following measures have been suggested to reduce the economic losses related to endogenous transplacental transmission in infected herds:

1. Embryo transfer. Transferring embryos that are taken from infected dams into uninfected animals will prevent endogenous transplacental transmission of *N. caninum*.[111] Only seronegative cows can act as recipients, thus pretransfer testing of recipients for infection with *N. caninum* is highly recommended. None of the 70 fetuses or calves born to seronegative cows that received embryos from seropositive donors were infected with *N. caninum*. In contrast, 5 of 6 calves resulting from embryo transfer from seronegative animals to seropositive cows were infected with *N. caninum*.[111] These findings were confirmed in another study[1115] that used commercial embryo transfer procedures. Another report showed that the zona pellicula protects preimplantation stage bovine embryos against *N. caninum* invasion,[199] which opens up the possibility to use this technique to recover uninfected calves from genetically valuable but *N. caninum*-infected dams. This is a great strategy for those uncommon instances in which there is a genetically highly valued cow with a neosporosis problem. This can allow them to recover the uninfected blood line. But this is an impractical very expensive strategy for the average dairy cow.
2. Artificial insemination of seropositive dairy cows with semen from beef bulls could have a beneficial effect on placental function due to crossbreeding. As shown in a study conducted in Spain in 2 high-producing dairy farms with a mean seroprevalence of 28%, insemination of these cows with beef bull semen resulted in a marked reduction of abortion.[1217]
3. "Test-and-cull" as an approach to eliminate seropositive cattle. *N. caninum*-infected cows within a herd represent a reservoir that may allow parasite transmission to other cattle, either vertically to their offspring, or by horizontal spread through the presence of a definitive host, such as via ingestion of feedstuff or water contaminated with oocysts. As a consequence, farmers may decide to remove infected cows or their progeny from the herd. The culling of infected cows is a control option that is effective, but not always economically viable. The following options are included in this strategy: (i) test and then remove seropositive dams or seropositive aborting dams from the herd, (ii) testing for seropositivity, followed by insemination of the progeny of seropositive dams only with beef bull semen, and (iii) test and exclude the progeny of seropositive dams from further breeding. Success was reported using these options.[833] Moreover, simulation models in endemically infected herds of beef cattle provided estimates on the economic return after using different "test-and-cull" strategies such as culling females that fail to calve, selling seropositive females and purchasing seronegative replacements, and excluding the female offspring of seropositive dams as potential replacements. In this model, testing of the entire herd and excluding the female offspring of seropositive dams as potential replacements appeared as the economically most factorable option return.[1121] Importantly,

these approaches can only be considered for herds with predominantly endogenous transplacental transmission of *N. caninum*, and only seronegative cattle should be used as replacements. Prior to adopting a "test-and-cull" strategy, it is important to analyze the risk factors for infection, such as whether endogenous transplacental transmission or the presence of dogs or other domestic or wild-life reservoirs,[826] is the main route of transmission, and a cost-benefit analysis for each farm should be performed prior to choosing any of these options. Cost-benefit analyses are to be performed using specialized computer programs.

4. Protection of feedstuffs and water from contamination by canid feces is the most economical control factor for housed cattle.

BIBLIOGRAPHY

Serologic Prevalence: 20, 21, 22, 23, 24, 71, 74, 83, 84, 89, 93, 94, 95, 100, 108, 111, 115, 126, 144, 160, 181, 192, 194, 204, 220, 228, 246, 249, 263, 269, 270, 282, 302, 320, 322, 326, 328, 375, 377, 383, 385, 389, 393, 394, 419, 422, 429, 448, 450, 452, 455, 498, 501, 504, 623, 625, 629, 630, 634, 650, 659, 669, 670, 671, 685, 705, 706, 707, 722, 731, 733, 739, 741, 751, 755, 758, 759, 764, 776, 791, 797, 798, 804, 814, 818, 845, 855, 871, 916, 917, 929, 930, 955, 956, 961, 966, 967, 969, 970, 972, 984, 1010, 1022, 1028, 1038, 1039, 1044, 1050, 1061, 1072, 1074, 1075, 1077, 1086, 1087, 1089, 1096, 1104, 1106, 1107, 1109, 1115, 1124, 1134, 1191, 1193, 1194, 1196, 1199, 1200, 1204, 1210, 1214, 1215, 1216, 1218, 1240, 1241, 1253, 1254, 1257, 1260, 1280, 1287, 1299, 1301, 1314, 1334, 1337, 1340, 1342, 1358, 1364, 1366, 1380, 1381, 1383, 1385, 1390, 1393, 1396, 1401, 1403, 1430, 1431, 1452, 1455, 1457, 1459, 1501, 1516, 1518, 1523, 1528, 1530, 1533, 1541, 1549, 1550, 1552, 1553, 1557, 1559, 1568, 1569, 1576, 1582, 1602, 1621, 1622, 1640, 1643, 1656, 1658, 1666a, 1669, 1683, 1686, 1688, 1700, 1712, 1725, 1738, 1740, 1741, 1753, 1765, 1770, 1772, 1774, 1779, 1780, 1783, 1789, 1790, 1801, 1808, 1817, 1818, 1822, 1834, 1835, 1837, 1843, 1854, 1859, 1862, 1882, 1899, 1908, 1909, 1911, 1919, 1922, 1930, 1931, 1951, 1957, 1960, 1972, 2016, 2023, 2027, 2039, 2040, 2048, 2050, 2055, 2056, 2060, 2061, 2063, 2068, 2077, 2085, 2091, 2111, 2134, 2135, 2151, 2155, 2161, 2163, 2170, 2180.

Clinical Infections: 10, 11, 40, 41, 57, 59, 62, 64, 65, 83, 85, 93, 96, 100, 107, 130, 131, 132, 134, 169, 202, 228, 230, 235, 246, 247, 249, 251, 266, 271, 275, 282, 339, 353, 367, 368, 371, 373, 381, 383, 384, 393, 395, 421, 431, 432, 448, 450, 513, 514, 530, 535, 542, 546, 550, 554, 557, 571, 624, 626, 633, 670, 689, 690, 691, 695, 717, 726, 731, 732, 757, 758, 760, 764, 775, 790, 791, 792, 820, 833, 845, 853, 854, 860, 862, 872, 874, 898, 923, 948, 969, 995, 997, 1004, 1053, 1057, 1058, 1062, 1085, 1086, 1104, 1110, 1116, 1214, 1215, 1217, 1220, 1253, 1260, 1263, 1277, 1301, 1305, 1331, 1333, 1335, 1336a, 1349, 1370, 1380, 1381, 1389, 1390, 1395, 1397, 1462, 1495, 1497, 1503, 1507, 1530, 1535, 1538, 1542, 1560, 1590, 1591, 1595, 1596, 1599, 1655, 1669, 1670, 1671, 1674, 1700, 1704, 1727, 1729, 1755, 1757, 1759, 1764, 1771, 1774, 1786, 1786, 1788, 1794, 1800, 1812, 1849, 1880, 1881, 1892, 1924, 1954, 1955, 1957, 1959, 1960, 1962, 1966, 1991, 2016, 2040, 2058, 2085, 2088, 2114, 2115, 2118, 2120, 2121, 2122, 2123, 2125, 2136, 2150, 2151, 2176.

Parasite Isolation: 28, 29, 279, 281, 284, 335, 373, 432, 689, 692, 729, 780, 801, 849, 963, 1059, 1060, 1180, 1201, 1202, 1253, 1255, 1281, 1309, 1314, 1315, 1353, 1507, 1674, 1679, 1702, 1729, 1766, 1786, 1907, 1993, 2143, 2144, 2145.

Experimental Infections: 37, 38, 40, 44, 46, 50, 52, 62, 64, 76, 77, 78, 106, 130, 132, 134, 136, 146, 147, 151, 153, 189, 222, 235, 240, 261, 293, 295, 296, 313, 348, 367, 374, 427, 434, 445, 470, 478, 480, 500, 552, 564, 605, 607, 613, 656, 671, 672, 681, 694, 725, 766, 780, 784, 790, 792, 806, 825, 866, 868, 973, 974, 976, 979, 981, 1023, 1039, 1076, 1217, 1220, 1234, 1244, 1264, 1265, 1278, 1305, 1316, 1317, 1387, 1391, 1392, 1435, 1436, 1467, 1484, 1503, 1522, 1524, 1536, 1577, 1584, 1652, 1657, 1680, 1728, 1732, 1743, 1745, 1774, 1776, 1792, 1794, 1812, 1830, 1831, 1833, 1900, 1908, 1938, 1939, 1940, 1964, 1994, 1995, 1999, 2000, 2001, 2009, 2063, 2078, 2089, 2097, 2099, 2121, 2125, 2146, 2150.

Diagnosis: 13, 14, 40, 42, 53, 54, 56, 65, 66, 77, 99, 106, 108, 111, 128, 136, 137, 142, 145, 147, 151, 153, 172, 174, 176, 177, 189, 193, 216, 219, 221, 223, 254, 256, 262, 266, 268, 276, 278, 280, 283, 296, 300, 301, 313, 322, 324, 325, 340, 345, 346, 367, 374, 378, 422, 429, 430, 432, 445, 450, 451, 461, 462, 492, 498, 569, 603, 607, 622, 640, 643, 650, 671, 672, 689, 690, 703, 706, 728, 755, 760, 766, 775, 784, 791, 792, 797, 814, 815, 825, 833, 835, 840, 866, 868, 869, 898, 904, 906, 912, 950, 954, 957, 976, 978, 979,

990, 1000, 1003, 1004, 1014, 1039, 1051, 1115, 1131, 1217, 1226, 1256, 1264, 1301, 1302, 1305, 1314, 1316, 1323, 1330, 1335, 1355, 1382, 1387, 1389, 1403, 1411, 1412, 1463, 1481, 1484, 1490, 1491, 1506, 1508, 1520, 1523, 1525, 1529, 1532, 1538, 1557, 1558, 1572, 1581, 1584, 1589, 1590, 1604, 1606, 1652, 1657, 1670, 1671, 1680, 1682, 1704, 1715, 1724, 1726, 1739, 1744, 1757, 1759, 1774, 1777, 1789, 1790, 1792, 1794, 1800, 1804, 1812, 1823, 1826, 1831, 1863, 1867, 1874, 1877, 1880, 1903, 1904, 1908, 1924, 1926, 1930, 1958, 1960, 1961, 1964, 1966, 1968, 1969, 1981, 1994, 2009, 2015, 2022, 2030, 2033, 2040, 2049, 2063, 2067, 2069, 2073, 2074, 2079, 2086, 2087, 2095, 2096, 2097, 2120, 2123, 2125, 2136, 2149, 2150, 2151, 2176.

Epidemiology: 13, 14, 40, 41, 48, 64, 65, 74, 84, 93, 95, 100, 125, 127, 141, 143, 144, 146, 169, 179, 180, 188, 194, 195, 215, 220, 221, 223, 228, 235, 246, 249, 268, 271, 276, 285, 286, 303, 302, 316, 326, 337, 339, 351, 374, 383, 384, 385, 390, 392, 393, 426, 428, 429, 430, 431, 444, 445, 448, 450, 455, 458a, 470, 492, 493, 495, 498, 500, 507, 520, 526, 528, 529, 593, 605, 607, 615, 630, 634, 666, 670, 671, 690, 691, 698, 699, 705, 707, 715, 723, 724, 726, 727, 730, 732, 754, 757, 758, 764, 773, 781, 783, 785, 791, 792, 804, 818, 825, 833, 845, 860, 872, 898, 918, 930, 948, 950, 968, 969, 972, 975, 976, 1004, 1010, 1028, 1039, 1054, 1072, 1075, 1077, 1085, 1086, 1107, 1116, 1125, 1171, 1210, 1213, 1215, 1217, 1219, 1260, 1261, 1284, 1287, 1288, 1301, 1305, 1310, 1314, 1315, 1317, 1330, 1360, 1370, 1380, 1381, 1384, 1390, 1392, 1393, 1396, 1403, 1411, 1412, 1434, 1435, 1456, 1485, 1490, 1491, 1498, 1524, 1530, 1533, 1538, 1541, 1542, 1549, 1557, 1559, 1568, 1583, 1589, 1591, 1596, 1603, 1604, 1640, 1656, 1657, 1666a, 1669, 1700, 1711, 1712, 1729, 1737, 1738, 1740, 1757, 1759, 1770, 1772, 1774, 1775, 1786, 1789, 1790, 1792, 1794, 1795, 1798, 1803, 1808, 1816, 1824, 1827, 1830, 1832, 1833, 1835, 1854, 1892, 1899, 1908, 1909, 1919, 1931, 1954, 1961, 1962, 1964, 1966, 1991, 1994, 1996, 2009, 2016, 2028, 2030, 2049, 2050, 2057, 2059, 2063, 2064, 2068, 2072, 2085, 2088, 2097, 2101, 2104, 2111, 2112, 2122, 2123, 2125, 2127, 2128, 2150.

Economics and Control: 41, 75, 76, 84, 111, 126, 144, 199, 336, 338, 379, 407, 431, 494, 499, 503, 603, 607, 614, 634, 786, 826, 833, 856, 860, 866, 868, 897, 916, 917, 928, 983, 1037, 1068, 1070, 1115, 1121, 1216, 1217, 1270, 1271, 1273, 1302, 1314, 1316, 1373, 1379, 1390, 1404, 1421, 1422, 1435, 1436, 1482, 1492, 1525, 1687, 1689, 1691, 1693, 1694, 1695, 1697, 1732, 1739, 1747, 1757, 1776, 1833, 1856, 1963, 1973, 1992, 2078, 2087, 2100.

Additional References: 46, 48, 59, 61, 63, 80, 190, 231, 237, 271, 272, 321, 323, 372, 376, 388, 437, 454, 499, 512, 560, 576, 577, 582, 584, 587, 591, 595, 596, 600, 604, 623, 696, 698, 704, 742, 746, 752, 769, 798, 812, 815, 851, 901, 928, 1005, 1024, 1066, 1120, 1169, 1170, 1179, 1227, 1249, 1251, 1252, 1314a, 1348, 1357, 1369, 1386, 1410, 1416, 1496, 1510, 1525, 1540, 1543, 1554, 1597, 1627, 1685, 1691, 1695, 1701, 1719, 1736, 1778, 1811, 1825, 1912, 1967, 2010, 2020, 2036, 2044, 2117, 2119, 2169.

CHAPTER 5

Neosporosis in Dogs

5.1 NATURAL INFECTIONS

5.1.1 Sources of Infection and Transmission of *N. caninum* for Dogs

As reviewed in Chapter 2, the main sources of *N. caninum* infections in dogs are related to ingestion of tissue cysts in prey or food. How dogs acquire *N. caninum* infections in nature is not fully understood.

Vertical transmission of neosporosis was first recognized in 3 successive litters from a bitch in Norway in 1984.[205,529] In a retrospective study, a congenitally infected German Shorthaired Pointer bitch from Ohio, USA transmitted infection to its progeny in 1957[541] resulting in the description of the most severe outbreak of neosporosis identified to date.[541] Transplacental transmission has also been confirmed in experimentally infected dogs.[362,533] In most cases of neonatal neosporosis, clinical signs are not apparent until 5–7 weeks after birth.[564] These data suggest that *N. caninum* is transmitted from the dam to the neonates during the terminal stage of gestation. A postnatal transmission via milk is not excluded. Overall, only a small proportion of dogs acquire *N. caninum* infection prenatally.[124,564,602] Recently, *N. caninum* DNA was detected in tissues of 22 of 41 (53.6%) stillborn puppies from 5 seropositive bitches.[1936a] This report needs confirmation because stillbirths due to neosporosis has not been confirmed in natural infections in dogs.

Feeding on infected intermediate host tissues is most likely responsible for postnatal infections in carnivores. In 1 report, 51% of 300 foxhounds fed bovine carcasses were found to have *N. caninum* antibodies.[1993] While consumption of aborted bovine fetuses does not appear to be an important source of *N. caninum* infection in dogs,[195,318,497] the consumption of bovine fetal membranes was shown to be a source of *N. caninum* infection. The parasite has been found in naturally infected placentas[196,689,1849] and dogs fed placenta from freshly calved seropositive cows excreted *N. caninum* oocysts.[493] It has been amply demonstrated that dogs can become infected by ingesting *N. caninum*-infected tissues. A large variety of tissues from naturally and experimentally infected ruminants including tissues from cattle, water buffaloes, goats, sheep, rodents, and also embryonic chicken can be infectious.[116,314,493,715,780,785,1067,1187,1309,1579,1797] Whether they can become infected by the ingestion of oocysts is not completely known but likely; dogs inoculated with oocysts seroconverted but did not excrete oocysts and *N. caninum* was not demonstrated in their tissues.[116]

5.1.2 Serologic Prevalence

Serological prevalences of *N. caninum* in the general canine population indicate that subclinical neosporosis is common worldwide (Table 5.1). Many of these serological studies were conducted in relation to dogs as a source of infection for livestock. Thus, data are biased. Risk factors for *N. caninum* infection are discussed as follows.

5.1.2.1 Risk Factors for Infection in Dogs

Several studies have identified factors that may influence the seroprevalence found in dogs (Tables 5.1 and 5.2); main factors are summarized below:

1. *Age*: The fact that *N. caninum* seroprevalence increases with age in dogs (Tables 5.1 and 5.2) suggests that most of these dogs became infected postnatally, most likely by the ingestion of tissue cyst containing material from intermediate hosts. However, only a small portion of dogs seroconvert and develop antibodies against *N. caninum* tachyzoites (Table 5.8).

2. *Gender*: Gender effects have been only observed in a few studies (Tables 5.1 and 5.2). Since *N. caninum* is well adapted to vertical transmission, it is not unlikely that also in female gravid dogs, there is a reactivation of *N. caninum* during gravidity. This was addressed in a study from Denmark. *Neospora* antibodies increased markedly during gestation in a pregnant bitch, suggesting that there is parasitic recrudescence during pregnancy.[1667] Other studies have shown that non-spayed female dogs were significantly more often seropositive than males,[610,803,1103,2126] but there is also 1 reference reporting the opposite.[2156]

3. *Breed*: A few studies showed that crossbred or mongrel dogs have a higher risk of being seropositive.[248,410,756] Others observed higher seroprevalences in particular breeds (Siberian Huskies in Japan[1103] and the Boxer breed in Italy[397]) or an increased risk in purebred versus crossbred.[298] These findings remained unexplained and it is possible that these singular observations are not related to the higher susceptibility of particular breeds.

4. *Presence of intermediate hosts of N. caninum*: Cattle farm dogs have a significantly higher risk of being positive compared to household, urban, rural, and guard dogs or even rescue kennel dogs (Tables 5.1. and 5.2), and dairy farm dogs have a higher infection risk than beef farm dogs,[410] possibly due to the closer contact with animals. Higher seroprevalence ratios in farm dogs are related to farms with an abortion history,[757,1486] or a higher *N. caninum* seroprevalence in cattle.[1676,2126]

   Also dogs from sheep farms have an increased risk of being seropositive relative to household or urban dogs[827]; but still continue to be less so than in those from cattle farms.[679] There is an exception in which farm dogs were significantly less often positive than dogs from sheep–beef farms in New Zealand.[81]

   Cats are not important intermediate hosts of *N. caninum*[1310]: Thus, there is no explanation for observations that the presence of cats[1667] significantly increased the risk of dogs becoming seropositive for *N. caninum*. Chickens are regarded as intermediate hosts of *N. caninum* based on serological and PCR examinations.[390] The importance of chicken and other domestic poultry in the life cycle of *N. caninum* is unknown. However, the presence of domestic poultry significantly increased the risk of dogs becoming seropositive for *N. caninum*[248] (Table 5.2).

5. *Habitat*: Not unexpectedly dogs from an urban area have a generally lower risk of being infected as compared to dogs from a rural or periurban area (Tables 5.1 and 5.2). Interestingly, in the case of urban dogs from Araçatuba, São Paulo state, Brazil dogs that were kept on soil or lawn had a higher risk of being seropositive as compared to dogs which were kept on a concrete floor.[245] Possibly, dogs kept on soil or lawn have a better chance to have access to prey, and contamination with *N. caninum* oocysts may persist for longer on soil.

6. *Type of dogs*: Street dogs as well as dogs which have access to the street are at higher risk (Table 5.2),[191,745] directly because they have access to food outside (point "7" of this list), including waste or prey animals. It is likely that dogs that hunt, or feed on tissues of domestic or wild animals become exposed to *N. caninum*. Many wild animal species are suspected as intermediate hosts of *N. caninum* (summarized in Chapters 9 through 18).

7. *Feeding/access to food*: Dogs with access to or fed on raw material from intermediate hosts of *N. caninum* had an increased risk for seropositivity (Tables 5.1 and 5.2). This included aborted fetuses or placenta,[1486,1676] raw meat including also beef[1099,1570,1676] and milk,[1853] which represent sources for which

Table 5.1 Prevalence of *N. caninum* Antibodies in Dogs (For more Details on Risk Factors, see Table 5.2)

Country	Location	Type	No. Tested	No. Positive	% Positive	Test	Titer/Company/Type of Test/Antigen	Risk Factors Studied[a]	Reference
Albania	Tirana	Urban	602	110	18.3	IFAT	1:50	–	837
Algeria	Algiers	Pound, urban	100	23	23.0	ELISA	BioX	–	755
	Algiers	Pound, urban	100	21	21.0	IFAT	1:100		755
	Algiers	Pound, urban	337	76	22.5	IFAT	1:50	Age, sex, **breed, habitat**, others	756
	Algiers	Pound, urban	91	6	6.5	IFAT	1:50		756
	Algiers	Breeder	209	25	11.9	IFAT	1:50		756
	Algiers	Cattle farm	144	64	44.4	IFAT	1:50		756
Argentina	Buenos Aires, South	Urban	160	42	26.2	IFAT	1:50	**Age, area**, clinical signs present	165
	Buenos Aires	23 dairy farms	125	60	48.0	IFAT	1:50		165
	Buenos Aires Province	10 beef farms	35	19	54.2	IFAT	1:50		165
	La Plata	Pet	97	46	47.4	IFAT	1:50	*T. gondii* infection	489
Australia	Melbourne	Pet	207	11	5.3	IFAT	1:50	–	123
	Sydney	Pet	150	18	12.0	IFAT	1:50	–	123
	Perth	Pet	94	13	13.8	IFAT	1:50	–	123
	Aboriginal communities	Community dogs	263	110	41.8	cELISA	VMRD	**Age**, sex	1070
	Aboriginal communities	Community dogs	194	40	20.6	IFAT	1:50	**Age**, sex	1070
Austria	Vienna, Lower Austria, and Styria	Rural	433	23	5.3	IFAT	1:50	Sex, breed, age	2071
	Vienna, Lower Austria, and Styria	Urban	381	8	2.1	IFAT	1:50	Sex, breed, age	2071
	Vienna, Lower Austria, and Styria	Unknown	956	32	3.3	IFAT	1:50	Sex, breed, age	2071
Belgium	Antwerp	Urban	100	11	11.0	IFAT	1:50	**Age**, sex, breeds	122
	Flanders	Clinics-neurological	100	11	11.0	IFAT	1:50	**Age**, sex, breeds	122
	Ghent	Urban	100	12	12.0	IFAT	1:50	**Age**, sex, breeds	122
	–	Farms	56	26	46.4	cELISA	VMRD	–	1123
	–	Farms	56	15	26.8	IFAT	1:100	–	1123

(Continued)

Table 5.1 (*Continued*) Prevalence of *N. caninum* Antibodies in Dogs (For more Details on Risk Factors, see Table 5.2)

Country	Location	Type	No. Tested	No. Positive	% Positive	Test	Titer/ Company/ Type of Test/ Antigen	Risk Factors Studied[a]	Reference
	—	Healthy	84	15	18.4	cELISA	VMRD	—	1123
	—	Healthy	84	8	9.7	IFAT	1:100	—	1123
	—	Neurological signs, urban	71	15	22.2	cELISA	VMRD	—	1123
	—	Neurological signs, urban	71	8	11.3	IFAT	1:100	—	1123
Brazil	Alagoas	128 urban, 99 rural	237	10	4.2	IFAT	1:50	Age, sex, breed, access to abortions (4.8% rural; 3.8% urban)	466
	Bahia	Urban	156	4	2.6	IFAT	1:50	**Habitat, contact with other dogs, feeding habits**, others	1853
	Bahia	Rural	41	2	14.6	IFAT	1:50	**Habitat, contact with other dogs, feeding habits**, others	1853
	Bahia	Periurban	214	28	13.1	IFAT	1:50	**Habitat, contact with other dogs, feeding habits**, others	1853
	Bahia	Stray	250	28	11.2	IFAT	1:50	Age, sex, others	440, 1012
	Bahia	Pet	165	22	13.3	IFAT	1:50	Age, sex, others	440
	11 villages in Amazon Region (Mato Grosso and Pará)	Stray	325	31	9.8	IFAT	1:50	**Age**, sex, *T. gondii*	1365
	Espírito Santo	Rural	187	22	11.7	IFAT	1:50	Sex, *T. gondii*	6a
	Goiânia	Urban, clinics	197	65	32.9	IFAT	1:50	Sex, 31.2% of 125 dogs positive from clinics	229
	Maranhão	Stray	100	45	45.0	IFAT	1:50	Sex	1949
	Mato Grosso	Clinics, pet	60	27	45.0	IFAT	1:50	Age, sex, diet, **access to streets**	191
	Mato Grosso	Dairy farms	37	25	67.6	IFAT	1:200	Cattle herds	192
	Mato Grosso do Sul	Urban pets	345	93	27.2	IFAT	1:50	**Age**, sex, *Leishmania* sp.	72

(*Continued*)

Table 5.1 (Continued) Prevalence of *N. caninum* Antibodies in Dogs (For more Details on Risk Factors, see Table 5.2)

Country	Location	Type	No. Tested	No. Positive	% Positive	Test	Titer/ Company/ Type of Test/ Antigen	Risk Factors Studied[a]	Reference
	Mato Grosso do Sul	Pet	245	65	26.5	IFAT	1:50	**Age**, sex	460
	Mato Grosso do Sul	Rural, urban	40	12	30.0	IFAT	1:100	Cattle herds	71
	Minas Gerais	Urban	300	32	10.7	IFAT	1:50	**Age**, breed, **sex**, **habitat**	667
	Minas Gerais	Periurban	58	11	18.9	IFAT	1:50	**Age**, breed, **sex**, **habitat**	667
	Minas Gerais	Rural	92	20	21.7	IFAT	1:50	**Age**, breed, **sex**, **habitat**	667
	Minas Gerais	Clinical	163	11	6.7	IFAT	1:25	3.1% of coinfection with *T. gondii*	1356
	Minas Gerais	Clinic	275	22	7.9	ELISA	ISCOM	*T. gondii*, age, breed, sex, others	1357
	Minas Gerais	Stray	94	12	12.8	ELISA	ISCOM	*T. gondii*, age, breed, sex, others	1357
	Minas Gerais	Clinic, stray	300	32	10.7	IFAT	1:50	Sex, age, breed	1858
	Minas Gerais	Clinic, stray	300	105	35.0	ELISA	1:25	Sex, age, breed	1858
	Minas Gerais	Clinics	228	7	3.1	IFAT	1:50	Age, sex, breed, *T. gondii*, *Leishmania chagasi*, *Babesia canis*	817
	Minas Gerais	Urban	182	15	8.2	IFAT	1:50	**Age**, diet, **hunting**,	1486
	Minas Gerais	Rural	421	58	13.7	IFAT	1:50	ingestion of placenta, others	1486
	Minas Gerais	Rural	240	36	15.0	IFAT	1:50	Age, sex, **breed**, others	248
	Pará	Rural	72	8	11.1	IFAT	1:50	*T. gondii*, **habitat**,	2019
	Pará	Urban-stray	57	8	14.0	IFAT	1:50	*Leishmania* sp., sex	2019
	Paraíba	Domestic, urban	286	24	8.4	IFAT	1:50	**Age**, sex, breed, others, **area**, *T. gondii* in 45.1%	105
	Paraná	Urban	181	23	12.7	IFAT	1:50	**Habitat**, lifestyle	700
	Paraná	Periurban	178	28	15.7	IFAT	1:50	–	700
	Paraná	Rural	197	50	25.3	IFAT	1:50	–	700
	Paraná	Dairy farms	134	29	21.6	IFAT	1:50	**Age**, sex, breed	468
	Paraná	Neurological	98	0	0.0	IFAT	1:50	*T. gondii* in 63%	768
	Paraná	Sheep farms	24	7	29.1	IFAT	1:50	Age, sex, breed	1734

(Continued)

Table 5.1 (Continued) Prevalence of N. caninum Antibodies in Dogs (For more Details on Risk Factors, see Table 5.2)

Country	Location	Type	No. Tested	No. Positive	% Positive	Test	Titer/ Company/ Type of Test/ Antigen	Risk Factors Studied[a]	Reference
	Paraná	Owned	127	14	11.0	IFAT	1:50	*T. gondii* in 21.1%, neurologic signs	1630
	Paraná	Stray	20	3	15.0	IFAT	1:50	neurologic signs	1630
	Paraná	Stray	26	3	11.5	IFAT	1:50	*T. gondii, Leishmania* spp., *Trypanosoma cruzi*	379a
	Paulista, Pernambuco	Domiciled	289	75	26.0	IFAT	1:50	*T. gondii* in 56.8%	684
	Amaraji, Pernambuco	Domiciled	168	44	26.2	IFAT	1:50	*T. gondii* in 64.1%	684
	Garanhuns, Pernambuco	Domiciled	168	58	34.5	IFAT	1:50	*T. gondii* in 54.4%	684
	Pernambuco	56	Rural villages	0	0	IFAT	1:50	*T. gondii* antibodies surveyed	85a
	Piauí	Urban	530	17	3.2	IFAT	1:50	**Age**, sex, breed, *L. infantum, T. gondii*	1211
	Piauí	71	Rural villages	5	(7.0)	IFAT	1:50	*T. gondii* antibodies surveyed	85a
	Rio de Janeiro	Urban, clinics	402	34	8.5	IFAT	1:50	**Age**	114
	Rio Grande do Norte	Urban	102	3	2.9	IFAT	1:50	***T. gondii***	1211
	Rio Grande do Sul	Rural	230	47	20.4	IFAT	1:50	**Age**, habitat, sex, others	410
	Rio Grande do Sul	Urban	109	6	5.5	IFAT	1:50		410
	Rio Grande do Sul	—	65	21	32.3	ELISA	NcSRS2	—	1608
	Rondônia	Domiciled, street access	157	13	8.3	IFAT, NAT	1:50, 1:25	**Age**, sex, diet, **street access**	289, 290
	Rondônia	Beef and dairy cattle farms	174	22	12.6	IFAT	1:50	Feed, abortion, stillbirth, others	20
	São Paulo	Urban	108	17	15.7	IFAT	1:50	Age, sex, diet, others, *T. gondii*	245
	São Paulo	Urban, rural, periurban	963	245	25.4	IFAT	1:50	223 of urban, 11 of rural, and 11 of periurban were positive. Sex, age	457
	São Paulo	Beef farms	39	23	58.9	IFAT	1:50	—	855

(Continued)

Table 5.1 (*Continued*) Prevalence of *N. caninum* Antibodies in Dogs (For more Details on Risk Factors, see Table 5.2)

Country	Location	Type	No. Tested	No. Positive	% Positive	Test	Titer/Company/Type of Test/Antigen	Risk Factors Studied[a]	Reference
	São Paulo	Urban	100	14	14.0	IFAT	1:25	No association with *Leishmania* sp. antibodies	808
	São Paulo	Urban	342	17	4.9	IFAT	1:25	Age, sex, breed, *T. gondii*	1119
	São Paulo	Pet	500	49	10.0	NAT	1:25	Sex, age, breed, **habitat**	745
	São Paulo	Street	611	151	25.0	NAT	1:25	Age, sex, breed, **habitat**	745
	São Paulo	Rural and urban	295	25	8.5	IFAT	1:50	Age, sex, *T. gondii*	2031
	São Paulo	Urban	204	36	17.6	IFAT	1:50	Age, sex, **Leishmania sp.**, *T. gondii*	748
	São Paulo	16 sheep farms	42	2	4.8	IFAT	1:25	—	1248
	São Paulo	Neurologic	50	7	14.0	IFAT	1:25	*T. gondii* in 22%	1118
	São Paulo	Rural	93	6	6.5	IFAT	1:50	No association with *Leishmania* sp. Other pathogens investigated: *Erhlichia* sp., *Babesia canis*, and *T. gondii*	1573
	São Paulo	Hospital, urban	203	44	21.6	IFAT	1:40	*T. gondii* in 81.3% (ELISA)	907
	São Paulo	Kennel, urban	167	37	22.1	IFAT	1:25	*T. gondii* (45.4%), *Leishmania* sp. (6.2%)	464
	São Paulo	Clinic	133	36	27.0	IFAT	1:25	*T. gondii* (44.3%), *Leishmania* sp. (1.5%)	464
	Tocantins	Rural	99	43	43.4	ELISA	IH	Age, **breed**, *T. gondii* in 62.6%	1659
				31	31.3	IFAT	1:25		
	Tocantins	Urban	105	45	42.9	ELISA	IH	Age, **breed**, *T. gondii* in 62.4%	1659
				31	29.5	IFAT	1:25		
Canada	Northwest Territories	Clinics	108	4	3.7	IFAT	1:25	Age, sex, feed, others	1763
	Alberta, Prince Edward Island, Ontario	—	88	11	12.5	IFAT	1:50	Age, sex, breed	332
Chile	IX Región	Rural	81	21	25.9	IFAT	1:50	Age, sex, feed, **habitat**, breed	1570
	IX Región	Urban	120	34	28.3	IFAT	1:50		1570

(Continued)

Table 5.1 *(Continued)* Prevalence of *N. caninum* Antibodies in Dogs (For more Details on Risk Factors, see Table 5.2)

Country	Location	Type	No. Tested	No. Positive	% Positive	Test	Titer/Company/Type of Test/Antigen	Risk Factors Studied[a]	Reference
China	–	Unknown	95	9	9.5	ELISA	NcGRA7t	–	850
	Jilin, Henan, Anhui	Countryside, slaughterhouse	96	6	6.2	NAT	1:25	Age, region, *T. gondii* (8.2%)	2148
	Henan	Urban, rural	1176	172	14.5	IFAT	1:50	Significant: habitat (urban–rural), age. Not-breed, gender, region	2070
Colombia	Codoba	Farms	22	11	50.5	ELISA	IDDEX	–	299a
Czech Republic	–	Pet	80	1	1.3	ELISA	ISCOM	–	1095
	Bohemia	Police	115	0	0.0	IFAT	1:50	**Age**, sex, breed	2017
		Pet	195	5	2.6	IFAT	1:50	**Age**, sex, breed	2017
	All across the country	Shelter	78	15	19.2	IFAT	1:50	**Age**, sex, breed	2017
	All across the country	Army	470	22	4.7	IFAT	1:50	**Age**, sex, breed	2017
Denmark	Jutland, Funen, Sealand, Lolland-Falster, Copenhagen	Pet	98	15	15.3	IFAT	1:160	Age, sex, diet, others	1667
Falkland Islands (UK)	–	Farm, pets	500	1	0.2	IFAT	1:50	–	123
France	West	Dairy farm	22	5	22.7	NAT	1:100	Abortions in cattle	1622
Germany	–	Clinic	200	26	13.0	IFAT	1:50	*T. gondii*, **sex**	1073
	–	Normal	50	2	4.0	IFAT	1:50	*T. gondii*, sex	1073
Grenada	All across the country	Stray, pets	107	2	1.8	IFAT	1:100	**Age**, sex, health status, *T. gondii* in 48.5%, others	610
	All across the country	Stray	368	6	1.6	ELISA	IDvet	Sex, habitat	1838
		Pets	257	3	1.2	ELISA	IDvet	Sex, habitat	1838
Hungary	11 locations	Rural	249	15	6.0	IFAT	1:80	Age, sex, breed, habitat	931
		Urban	402	4	1.0	IFAT	1:80	Age, sex, breed, habitat	931
India	Punjab	Rural	126	27	21.4	cELISA	VMRD	Age, sex, breed, **habitat**	1841
		Urban	58	4	6.9	cELISA	VMRD	Age, sex, breed, **habitat**	1841

(Continued)

Table 5.1 (*Continued*) Prevalence of *N. caninum* Antibodies in Dogs (For more Details on Risk Factors, see Table 5.2)

Country	Location	Type	No. Tested	No. Positive	% Positive	Test	Titer/ Company/ Type of Test/ Antigen	Risk Factors Studied[a]	Reference
Iran	Chaharmahal va Bakhtiari, Isfahan, Khoozestan	Random	548	159	29.0	ELISA	IH	**Age, sex, habitat,** *T. gondii* in 26.8%, others	937
	East Azerbaijan	Shepherd dogs	384	41	10.6	IFAT	1:50	Age, location, breed	1052
	Tehran	Farm	50	14	28.0	IFAT	1:50	**Age, sex**	827
	Tehran	Urban	53	6	11.3	IFAT	1:50	**Age, sex**	827
	Tehran	Clinics, healthy	233	24	10.3	ELISA	P38	–	936
	Tehran	Urban, pet	50	10	20.0	IFAT	1:50	**Age, sex, habitat**	1266
		Beef cattle farm	50	23	46.0				
	Urmia	Stray	135	36	27.0	IFAT	1:50	**Age, sex**, others	2140
Italy	Abruzzo	Cattle farms	50	23	46.0	IFAT	1:50	Habitat, age, sex, breed, others	1718
		Breeding facilities	50	9	18.0	IFAT	1:50	Habitat, age, sex, breed, others	1718
	Apulia	Kennel	144	21	14.6	ELISA	MASTAZYME	*Leishmania* spp., **age,** sex, feeding, others	1552
		Farm	162	43	26.5	ELISA	MASTAZYME		1552
	Campania	Pet	1058	68	6.4	IFAT	1:50	Age, sex, **breed,** *Leishmania* spp., others	397
	Campania	Pet	194	56	28.9	IFAT	1:50	**Sex, *Leishmania* spp.**	396
	Parma	Pet	282	51	18.1	IFAT	1:50	**Age, habitat, housing, feeding habits**	1099
	Piedmont	Urban	188	38	20.2	NAT	1:40	Sex, breed, **age,** lifestyle	679
		Rural	302	110	36.4				
	Trentino-Alto Adige and Veneto	Kennel and pet	707	77	10.9	cELISA	VMRD	**Age,** sex, others, negative for *Leishmania* spp.	298, 299
Japan	9 prefectures	Urban	198	14	7.1	IFAT	1:50	Age, breed, habitat	1785
		Dairy farm	48	15	31.3	IFAT	1:50	Age, breed, habitat	1785
		Sheepdog breeder	39	36	92.3	IFAT	1:50	Age, breed, habitat	1785
	30 prefectures	Clinics	1206	126	10.4	ELISA	IH tNcSAG1	**Sex, age, breed,** others	1103
	–	Unknown	135	23	17.0	ELISA	NcGRA7t	–	850

(*Continued*)

Table 5.1 (Continued)   Prevalence of *N. caninum* Antibodies in Dogs (For more Details on Risk Factors, see Table 5.2)

Country	Location	Type	No. Tested	No. Positive	% Positive	Test	Titer/ Company/ Type of Test/ Antigen	Risk Factors Studied[a]	Reference
Kenya	–	Rural (feral)	140	0	0.0	IFAT	1:50	–	123
	Nakuru District	Farm	84	15	17.9	ELISA	IDEXX	Age, sex, habitat, breed, others	1508a
Korea	–	Urban, clinics	289	24	8.3	IFAT	1:50	Sex, age	1063
	–	Dairy farm	51	11	21.6	IFAT	1:50	Sex, age	1063
	9 provinces	Domestic	553	20	3.6	IFAT	1:50	Age, sex, breed, *T. gondii* in 51.5%, **habitat**	1461
Mexico	Hidalgo	Farm	27	14	51.0	ELISA	IDEXX	Sex, **habitat**	1770
		City	30	6	20.0	ELISA	IDEXX	Sex, **habitat**	1770
	Durango	Pound	101	2	2.0	IFAT, NAT	1:25, 1:25	*T. gondii* (12.8%), age, breed, sex, others	609
	Aguascalientes	Urban	116	23	19.8	ELISA	IDEXX	**Age**, sex, **habitat**, size	401
		Dairy farms	152	62	41.0	ELISA	IDEXX		
New Zealand	–	–	200	44	22.0	IFAT	1:40	–	1683
	Taranaki, Manawatu, Rangitikei, Wairarapa	Dairy farms	161	157	97.5	IFAT	1:50	47 (29.2) positive at 1:400. **Habitat**	81
		Sheep/beef farms	154	154	100.0	IFAT	1:50	32 (20.7) positive at 1:400. **Habitat**	
		Urban	150	114	76.0	IFAT	1:50	12 (8.0) positive at 1:400. **Habitat**	
Nigeria	Oyo, Ogun	Caged	44	0	0.0	WB	NA	Age, sex, breed, *T. gondii*, others	104
		Roaming	189	5	2.6				
Pakistan	Punjab	Stray	92	37	40.2	cELISA	VMRD	**Age**, breed, **habitat**	1456
		24 dairy farms	138	51	36.9				
		Pets, healthy	294	31	10.5				
		Neurological	76	22	28.9				
Peru	Lima	22 dairy farms	104	34	32.7	IFAT	1:50	Age, sex	479
	Puno	Farm	122	18	14.8	IFAT	1:50	Age, sex, habitat	2037
	Lima	Pets, healthy	120	2	1.7	IFAT	1:50	*T. gondii* (3.3%), neurologic signs	1750

(Continued)

Table 5.1 (Continued) Prevalence of *N. caninum* Antibodies in Dogs (For more Details on Risk Factors, see Table 5.2)

Country	Location	Type	No. Tested	No. Positive	% Positive	Test	Titer/Company/Type of Test/Antigen	Risk Factors Studied[a]	Reference
	Lima	Pets, neurological	96	5	5.2	IFAT	1:50	*T. gondii* (24.0%), neurologic signs	1750
Poland	South-western	Clinics	110	18	16.4	IFAT	1:50	*T. gondii*, habitat	1629
Poland	Warsaw	Urban, clinics	257	56	21.7	ELISA, IB	IH	**Sex**, age	803
Portugal	North, Center, Alentejo, Lisbon, Algarve	Urban	132	9	6.8	cELISA	VMRD	Sex, breed, age, others	1259
		Rural	227	19	8.4				
Romania	Cluj Napoca	Stray	56	7	12.5	IFAT	1:100	–	1923
	Southern	Stray	38	7	18.4	IFAT	1:50	**Age, lifestyle**	1367
		Guard, rural	25	2	8.0				
		Cattle farms	21	8	38.1				
	8 regions	Different categories	1114	364	32.7	IFAT	1:50	Breed, **age**, sex, area	738
Senegal	Saint Louis	Pets	100	48	48.0	ELISA	LSIVet	*T. gondii*, sex, age, others	1028, 1029
Serbia	Vojvodina	Pet, kennel	31	4	12.9	IFAT	1:50	Age	1574
Spain	Catalonia	Pet	139	17	12.2	IFAT	1:50	**Age**, sex, breed, others	1526
	Andalusia	Feral	28	4	17.0	cELISA	VMRD	–	1351
	Galicia	Farm	141	67	47.5	IFAT	1:50	**Age, housing**	1676
		Stray	134	53	39.5				
	Majorca island	Kennel	44	0	0.0	cELISA	VMRD	*Leishmania* spp. (56.3%), *T. gondii* (58.7%)	265
	Several areas	Household	102	3	2.9	IFAT	1:50	**Age, sex, breed**	369
		Kennel-stray	94	23	24.5				
		Hunting	100	23	23.0				
		Dairy farms	100	51	51.0				
Sweden	Several areas	Pet-clinics	398	2	0.5	ELISA	IH-ISCOM	Age, sex, breed, *T. gondii*	212
	Several areas	Pet clinics	398	1	0.2	IFAT	1:80	Age, sex, breed, *T. gondii*	212
	Several areas	Pet clinics	12	0	0.0	DAT	1:40	Age, sex, breed, *T. gondii*	212
	Also from Britain	–	194	43	22.1	IFAT	1:40	Clinic	213

(Continued)

Table 5.1 (Continued) Prevalence of *N. caninum* Antibodies in Dogs (For more Details on Risk Factors, see Table 5.2)

Country	Location	Type	No. Tested	No. Positive	% Positive	Test	Titer/Company/Type of Test/Antigen	Risk Factors Studied[a]	Reference
Switzerland	Bern	Pet	1080	78	7.3	ELISA	IH	**Habitat**	**1760**
Taiwan	–	Dairy farm	30	6	20.0				
Taiwan	–	Dairy farm	13	3	23.0	IFAT	1:50	–	1518
Tanzania	–	Rural (domestic)	49	11	22.5	IFAT	1:50	–	123
Thailand	Nakhon Pathom	Dairy farm	82	1	1.2	cELISA	VMRD	–	1109
Thailand	Nakhon Pathom, Rachaburi, Kanchanaburi	Dairy farm	114	8	7.0	cELISA	VMRD, 5 positive confirmed by IFAT 1:100	Age, sex, others	90
Turkey	Bursa, Adana	Kennels, pet	150	15	10.0	IFAT	1:50	Age, sex	387
Turkey	Kirikkale	Stray	121	35	28.9	IFAT	1:16	Age, **sex**, others, *T. gondii* (54.3% of *N. caninum* seropositive dogs)	[a] 2156
	Konya	Pet, stray	187	31	16.6	ELISA	rNCSAG1	–	2180
UK	North Mymms	Pet, urban	104	6	5.8	IFAT	1:50	1 dog with clinical neosporosis	1126
USA	Liverpool	Pet	163	27	16.6	IFAT	1:50	Sex, age, diet, others	1990
USA	Kansas	Pet, clinics	229	5	2.0	IFAT	1:50	*T. gondii* (25.0%)	1165
	35 states	Pet	1077	75	7.0	IFAT	1:50	Age, sex, breed	332
	Maryland	Pet	20	5	25.0	IFAT	1:20	Described clinical cases	606
	Wisconsin, Colorado	Pet	88 (44 with ACP[c])	4	4.5	IFAT	1:100	6.8% dogs with ACP, 2.3% without ACP. No significant	925
Uruguay	–	Farm, domestic	414	82	19.8	IFAT	1:50	–	123

a Modified from Dubey, J. P. 2013. *Commonwealth Agricultural Bureau Reviews* 8, e55.

b Risk factors, statistically significant are in bold. For more details of risk factors see Table 5.2.

c Acute canine polyradiculoneuritis.

Abbreviations: ELISA = enzyme-linked immunosorbent assay, IDEXX = IDEXX HerdChek *Neospora caninum* antibody, indirect ELISA, sonicate lysate of tachyzoites, IDEXX Laboratories, USA, IFAT = indirect fluorescent antibody test, IH = in-house, ISCOM = Detergent extracted tachyzoite antigen incorporated in immune stimulating complex particles, MASTAZYME = MASTAZYME *NEOSPORA*, indirect ELISA, formaldehyde-fixed whole tachyzoites, MAST GROUP UK, NAT = *Neospora* agglutination test, NS = not stated, WH = whole tachyzoite extract, VMRD = *Neospora caninum* cELISA Competitive ELISA GP65 surface antigen of tachyzoites VMRD, USA, BioX = NEOSPORA CANINUM Antibody Detection Kit (BIO K 192), Bio-X Diagnostics, Belgium, IDvet = Neospora caninum Indirect Multispecies, ID.vet, France, LSIVet = Neospora caninum ANTIBODY TEST KIT C-ELISA.

Table 5.2 Risk and Protective Factors for Postnatal *N. caninum* Infection

Factor Description				References of Studies Which Identified Putative Risk or Protective Factors	
				Risk	**Protective**
Age	Pup, young vs. old			20, 72, 122, 165, 298, 369, 396, 401, 410, 415, 457, 460, 468, 679, 827, 937, 1070, 1099, 1103, 1266, 1365, 1367, 1526, 1676, 2017, 2126, 2140	—
Gender	Female			610, 748, 803, 1103, 2126	2156
Breed	Crossbred/mongrel vs. purebred dogs			248, 410, 756, 1659	—
	Particular breeds vs. other breeds (e.g., Boxer and Siberian Husky)			397, 1103	—
	Dog size			401	—
Presence of intermediate hosts of *N. caninum*	Ruminants	Cattle	Cattle farm dogs vs. household/urban dogs	81, 165, 369, 401, 931, 1063, 1123, 1266, 1486, 1770, 1785, 2126	—
			Cattle farm dogs vs. rural guard dogs	1367	—
			Cattle farm dogs vs. rescue kennel dogs	1552	—
			Dog with contact with bovines	1099, 1853	—
			Dairy farm dogs vs. beef farm dogs	410	—
			Dogs from farms with higher prevalence of *N. caninum* positive cattle	1676, 2126	—
			Dogs from farms with abortion history	757, 1486	—
			Carcasses of dead cattle are collected		410
			Carcasses of dead cattle are removed from pasture and brought to rendering		410
			Farm dogs with free access to cattle farm vs. dogs controlled farm dogs	1676	—
			Pure breed dogs living in a cattle farm vs. other dogs	1718	—
		Sheep	Sheep farm dogs vs. household/urban dogs	827	—
		Cattle and sheep	Cattle or sheep farm dogs vs. household/urban dogs	81, 679	—
			Cattle farm dogs vs. sheep farm dogs	679	—
			Dairy farm dogs vs. sheep–beef farm dogs		81
	Carnivores	Cats		1667	—
	Birds	Presence of domestic poultry		248	—

(Continued)

Table 5.2 (*Continued*) Risk and Protective Factors for Postnatal *N. caninum* Infection

Factor Description		References of Studies Which Identified Putative Risk or Protective Factors	
		Risk	Protective
Habitat of dog	Rural vs. urban	410, 667, 700, 931, 1486, 1841, 1853, 2071	—
	Periurban *vs.* urban	667, 1853	—
	Dog kept on soil or lawn vs. on concrete	245	—
Type of dog	Stray/shepherd dogs vs. household dogs	937	—
	Street dog vs. owned dogs	745	—
	Dog with access to street	191, 1508a	—
	Guard dogs	1486	—
Feeding habits	Contact with or ingestion of aborted fetuses or placenta	1486	—
Access to/ feeding with raw material from potential intermediate hosts	Raw meat, raw bovine meat	1099, 1570	—
	Ingestion of milk	1853	—
	Dog fed with noncommercial food	1486	—
	Access to food outside home	1853	—
	Dog hunts	1486, 1853	—
Other infections	*Leishmania infantum* antibodies	369, 396, 397	—
Level of attention for the dog	Vaccinated vs. non-vaccinated	756	—
Seasonal effects	Sampling in summer	756	—

a possible transmission has been shown by experimental studies. In farms in which carcasses of dead cattle are properly removed, dogs presented a significantly lower risk of seropositivity.[410]

In addition, also dogs fed with noncommercial food, dogs which have access to food outside their home, and dogs which are hunting had an increased risk to be *N. caninum* seropositive.[1486,1853] Dogs that hunt or feed on animal species which are suspected to be intermediate hosts of *N. caninum* may become infected and develop antibodies. Rodents (rats and mice),[679,949,960,1008,1336,1338] lagomorphs (rabbits),[954,2166] insectivores (shrews),[1338] or domestic and wild birds[2,390,426,520,789,1762] are suspected to be intermediate hosts of *N. caninum* based on PCR results or on PCR and IHC results[1336] (see Table 2.2 in Chapter 2).

8. *Coinfections*: Several studies on *N. caninum* seroprevalence in pet dogs observed that the likelihood of *Leishmania infantum* positive dogs being *N. caninum* antibody positive was higher than in *L. infantum* negative dogs.[396,397,748] Such an association has been also reported for hunting and stray dogs.[369] There is no evidence that an infection with *L. infantum*, or other infectious agents (Table 5.1) increases the susceptibility for *N. caninum* or vice versa. In such cases, most likely there are common risk factors associated to habitat which favor coinfections.

9. *Other factors*: In an observation from Algeria,[756] vaccinated dogs had a significantly lower seroprevalence as compared to non-vaccinated dogs, this is related to the level of attention provided by owners, so such dogs are fed regularly with a known source. Also, in the same study,[756] a seasonal effect is reported, with higher seroprevalence in dogs sampled in summer, climatic conditions might influence the survival time of excreted oocysts, but also the availability and the abundance of infected intermediate hosts. Similar effects were observed for *T. gondii* in cats.[9,1809]

5.1.3 Prevalence of *N. caninum* Oocysts in Dog Feces

As dogs play both the role of definitive and intermediate host in the life cycle of *N. caninum*, they are the key element in the epidemiology of the disease. Although serological prevalence is high in both dogs and intermediate hosts, prevalence of dogs excreting oocysts is low at a moment. Surveys on the prevalence of *N. caninum*-like oocysts in feces of dogs are summarized in Table 2.3 (see Chapter 2). *N. caninum*-specific methods should be used to differentiate oocysts from those of *Hammondia heydorni* and other related parasites.

5.1.4 Isolation of Viable *N. caninum* from Dogs

Current information regarding isolation of viable *N. caninum* from either tissues (clinical cases and asymptomatic dogs) or feces has been summarized in Tables 2.3 (see Chapter 2) and 5.3. Molecular data are provided for future epidemiological studies.

5.1.5 Clinical Infections

Worldwide reports of clinical canine neosporosis are summarized in Tables 5.4 through 5.6. Serological prevalence data indicate that *N. caninum* infections are common but clinical disease is relatively rare. Why some dogs develop clinical neosporosis whereas most remain asymptomatic needs clarification. Whether the severity of clinical disease is related to parasite virulence or host susceptibility, or a combination of both, is unknown. Currently, 26 viable isolates of *N. caninum* have been cultured: 11 from feces (Table 2.3, Chapter 2) and 15 from tissues (Tables 5.3 through 5.5). Most of them are derived from dogs with clinical illness. At present, there are no pathogenicity-related genetic markers to distinguish different strains of *N. caninum*.[167,614] Clinical neosporosis is most common in young, very old, and immunosuppressed dogs (Tables 5.4 through 5.6). Administration of exogenous glucocorticoids can

Table 5.3 Isolation of Viable *N. caninum* from Dogs

Country	Breed, Age	Bioassay Source	Animals	Cell Culture	Isolate Designation	Museum Deposition	Molecular Data	Remarks	Reference
Argentina	Rottweiler, 45 d	Feces	Gerbils, KO mice	M617	NC-6 Argentina	–	Multilocus mini-, microsatellites (MS1B, 2, 3, 4, 5, 6A, 6B, 10, 12, 21, JN642712, JN642713, JN642714, JN642715)	Primary isolation in gerbils	106, 164
Australia	White Highland White Terrier, 13 m	Skin	No	Vero	WA-K9	–	Multilocus microsatellites (Cont-14, -16, FJ883935.1, FJ883953.1; Tand-3, -12, -13, FJ824923, FJ824962, FJ824986)	Parasite DNA in feces of dog 2.5 years later	28, 29, 1329
Brazil	Collie, 7 yr	Brain	Gerbil	Vero (COS-1)	NC-Bahia	ATCC, PRA-138	Multilocus mini-, microsatellites (MS1A, 1B, 2, 3, 4, 5, 6A, 7, 8, 10, 12, 21; AY935189, AY935190, AY935191, AY935192, AY935193, AY935194, AY935195, AY935196, AY935197, AY935198, AY935199, AY935200; Cont-6, -14, -16, FJ883902, FJ883918, FJ883936; Tand-3, -12, -13, FJ824914, FJ824945, FJ824970; ITS-1, AY259042, AY259043)	Isolate pathogenic to gerbils	28, 29, 779, 782, 1679

(Continued)

Table 5.3 (Continued) Isolation of Viable *N. caninum* from Dogs

Country	Breed, Age	Source	Bioassay Animals	Cell Culture	Isolate Designation	Museum Deposition	Molecular Data	Remarks	Reference
Germany	Kleiner Müsterländer, 11 w	Brain, spinal cord	Nude	Vero	NC-GER1	–	Multilocus mini-, microsatellites (MS1A, 1B, 2, 3, 4, 5, 6A, 6B, 7, 8, 10, 12, 21, AY937131, AY937132, AY937133, GU597071, AY937134, AY937135, AY937136, AY937137, AY937138, AY937139, AY937141, GU597088, AY937142, AY937144; Cont-6, -14, -16, FJ883906, FJ883923, FJ883940; Tand-3, -12, -13, FJ824915, FJ824949, FJ824974)	No pathogenic to gerbils	**28, 29, 169, 1598**
	Unknown	Feces	Goat, sheep, gerbil, Guinea pig, multimammate rat, BALB/c mouse	Vero	HY-Berlin-1996[a]		Multilocus mini-, microsatellites (MS1A, 1B, 2, 3, 4, 5, 6A, 7, 8, 10, 12, 21; AY937119, AY937120, AY937121, AY937122, AY937123, AY937124, AY937125, AY937126, AY937127, AY937128, AY937129, AY937130)	Primary isolation	**1673, 1679, 1798**
	Dutch Shepherd, 2 m	Feces	Gerbils, KO mice	Vero	NC-GER2	–	Multilocus mini-, microsatellites (Cont-6, -14, -16, FJ883907, FJ883924, FJ883941; Tand-3, -12, -13, FJ824916, FJ824950, FJ824975)	No pathogenic to gerbils	**28, 29, 1805**
	Wirehaired Vizsla, 4 m	Feces	Gerbils, KO mice	Vero	NC-GER3	–	Multilocus mini-, microsatellites (Cont-6, -16, -14, FJ883908.1, FJ883942, FJ883925; Tand-3, -12, -13, FJ824917, FJ824951, FJ824976)	No pathogenic to gerbils	**28, 29, 1805**

(Continued)

Table 5.3 (*Continued*) Isolation of Viable *N. caninum* from Dogs

Country	Breed, Age	Source	Bioassay Animals	Cell Culture	Isolate Designation	Museum Deposition	Molecular Data	Remarks	Reference
	Hovawart, 7 yr	Feces	Gerbils, KO mice	Vero	NC-GER4	–	Multilocus mini-, microsatellites (Tand-3, -12, -13, FJ824918, FJ824952, FJ824977)	No pathogenic to gerbils	**28, 1805**
	German Spitz, 2 yr	Feces	Gerbils, KO mice	Vero	NC-GER5	–	Multilocus microsatellites (MS1B, 2, 3, 4, 5, 10, EU872323, EU872322, EU872338, EU872337, EU872353, EU872352, EU872368, EU872367, EU872383, EU872382, EU872398, EU872397; Tand-3, -12, -13, FJ824919, FJ824953, FJ824978)	No pathogenic to gerbils	**28, 168, 1805**
	Wirehaired Daschshund, 13 yr	Feces	Gerbils, KO mice	Vero	NC-GER6	–	Multilocus mini-, microsatellites (MS1B, 2, 3, 4, 5, 10, EU872326.1, EU872325, EU872324, EU872341, EU872340, EU872339, EU872356, EU872355, EU872354, EU872371, EU872370, EU872369, EU872386, EU872385, EU872384, EU872401, EU872400, EU872399; Tand-3, -12, -13, FJ824920.1, FJ824954.1, FJ824979.1)	No pathogenic to gerbils	**28, 168, 1805**

(Continued)

Table 5.3 (Continued) Isolation of Viable *N. caninum* from Dogs

Country	Breed, Age	Source	Bioassay		Isolate Designation	Museum Deposition	Molecular Data	Remarks	Reference
			Animals	Cell Culture					
	Unknown, 4 yr	Feces	KO mice	KH-R	NC-GER7	–	Multilocus mini-, microsatellites (MS1B, 2, 3, 4, 5, 10, EU872328, EU872327, EU872343, EU872342, EU872358, EU872357, EU872373, EU872372, EU872388, EU872387, EU872403, EU872402)	–	168
	Flat-Coated Retriever, 2.5 m	Feces	KO mice	KH-R	NC-GER8	–	Multilocus mini-, microsatellites (MS1B, 2, 3, 4, 5, 10, EU872330.1, EU872329, EU872345, EU872344, EU872360, EU872359, EU872375, EU872374, EU872390, EU872389, EU872405, EU872404; Tand-3, -12, -13, FJ824921, FJ824955, FJ824980)	–	28, 168
	Shih Tzu–Jack Russell Terrier-mix, 2.5 yr	Feces	KO mice	KH-R	NC-GER9	–	Multilocus mini-, microsatellites (MS1B, 2, 3, 4, 5, 10, EU872333, EU872332, EU872331, EU872348, EU872347, EU872346, EU872363, EU872362, EU872361, EU872378, EU872377, EU872376, EU872393, EU872392, EU872391, EU872408, EU872407, EU872406; Cont-6, -14, -16, FJ883909, FJ883926, FJ883943; Tand-3, -12, -13, J824922.1, FJ824956.1, FJ824981.1)	–	28, 29, 168

(Continued)

Table 5.3 (*Continued*) Isolation of Viable *N. caninum* from Dogs

Country	Breed, Age	Source	Bioassay		Isolate Designation	Museum Deposition	Molecular Data	Remarks	Reference
			Animals	Cell Culture					
Portugal	Boxer, 8 yr	Feces	KO mice	Vero	NC-P1	—	Multilocus mini-, microsatellites (MS1B, 2, 3, 4, 5, 10, EU872335, EU872334, EU872350, EU872349, EU872365, EU872364, EU872380, EU872379, EU872395, EU872394, EU872410, EU872409)	—	167, 168
UK	Boxer, 5 w	Brain	No	Vero	Nc-Liv	ATCC, 50845	Multilocus mini-, microsatellites (MS1A, 1B, 2, 3, 4, 5, 6A, 7, 8, 10, 12, 21, AY935165, AY935166, AY935167, AY935168, AY935169, AY935170, AY935171, AY935172, AY935173, AY935174, AY935175, AY935176; Cont-6, -14, -16, FJ883900, FJ883928.1, FJ883944.1; Tand-3, -4, -8, -9, -12, -13, -15, -16, -23, -24, -25, -26, -30, -31, -32, -34, FJ824929, FJ824995, FJ824937, FJ830456, FJ824958, FJ824983, FJ824999, FJ825005, FJ825011, FJ830462, FJ830466, FJ830473, FJ830477, FJ830484, FJ825015, FJ830488), rDNA (SSU, AF304319, AF304315; LSU, AF304316, AF001946, AF304322; ITS-1, EU564166, AY259038), Cytochome b, Cytochome C oxidase subunit 1 (JX473252, JX473256, JX473271, JX473267)	First out of USA	28, 29, 118, 119, 642, 770, 771, 782, 1679, 1729

(*Continued*)

Table 5.3 (Continued) Isolation of Viable *N. caninum* from Dogs

Country	Breed, Age	Source	Bioassay Animals	Cell Culture	Isolate Designation	Museum Deposition	Molecular Data	Remarks	Reference
USA	Labrador Retriever, 5 w	Brain, spinal cord	SW mice	BM, CPA	NC1	ATCC, 50977	Multilocus mini-, microsatellites (MS1B, 2, 3, 4, 5, 6A, 6B, 7, 8, 10, 12, 21, EU872336, KC542883, JN651092, KM241868, KP845096, KP890250; Cont-6, -14, -16, FJ883912, FJ883929, FJ883952.1; Tand-3, -4, -12, -13, -30, FJ824932, FJ824994, FJ824963, FJ824987, FJ830478), rDNA (SSU, U93870, AY155364, AB586643, AF094762; ITS-1, AB526831), NcHSP70 (U82229, AF094762), Cytochrome C oxidase subunit 1 (HM771688)	—	**28, 29, 528, 662, 782, 1502, 1679**
	Labrador Retriever, 12 w	Muscle biopsy	SW mice	M617	NC2	ATCC, 50844	ITS-1 (AY259037, AF249969)	No isolation in SW mice	**585, 863**
	English Springer Spaniel, 12 w	Brain, spinal cord, muscle	Gerbils	BM	NC3		rDNA (SSU, U25044)	Pathogenic to gerbils. No isolation in SW mice, rabbits, hamsters, and guinea pigs	**402, 1166, 1281**
	Labrador Retriever, 53 d	Brain	KO mice	HS68	NC4	—	NA	—	**573**
	Labrador Retriever, 113 d	Brain	KO mice	HS68	NC5	—	rDNA (LSU, AF249972, ITS-1, F249970)	—	**573, 585**

(Continued)

Table 5.3 (Continued) Isolation of Viable N. caninum from Dogs

Country	Breed, Age	Source	Bioassay Animals	Cell Culture	Isolate Designation	Museum Deposition	Molecular Data	Remarks	Reference
	Labrador Retriever, 116 d	Brain	Gerbils, KO mice	CV1	NC6	–	rDNA (ITS-1, AY665716), pNc5 (AY665720)	Pathogenic to gerbils	573, 598, 602
	Labrador Retriever, 116 d	Brain	Gerbils, KO mice	CV1	NC7	–	rDNA (ITS-1, AY665717), pNc5 (AY665721)	Not pathogenic to gerbils	
	Labrador Retriever, 135 d	Brain	Gerbils, KO mice	CV1	NC8	–	rDNA (ITS-1, AY665718), pNc5 (AY665722)	Infective to gerbils, but not pathogenic	
	Labrador Retriever, 13 w	Brain, spinal cord, muscle	KO mice	M617, CV1	NC9	–	rDNA (ITS-1, EF219139)	Parasites isolated from each tissue in KO mice. Gerbils seroconverted but Neospora not found in their tissues	608
	Rhodesian Ridgeback, 6 w	Brain, spinal cord	No	Vero	CN1	–	rDNA (ITS-1, AF038861)	–	1284

Note: d = day, m = month, w = week, yr = year.

a *N. caninum* isolate originally named *Hammondia heydorni* Berlin-1996 (HY-Berlin-1996), because at the time of isolation the dog had not yet been established as a definitive host of *N. caninum*.

exacerbate neosporosis[708,719,735,927,1111,1133,1521] or neosporosis is associated to other disease such as lymphosarcoma,[562] or traumas.[920]

Although overlaps exist, 4 forms of canine neosporosis can be differentiated (Table 5.6)[617,1098,1749,1956]:

1. The most severe signs are seen in transplacentally infected puppies that develop ascending rearlimb paresis and paralysis, with rigid contracture of the muscles of the pelvic limbs, often including muscle atrophy (Figures 5.1a and 5.2). Initial symptoms often start in the first weeks of life. The majority of these dogs are 2 months old at diagnosis, most likely prenatally infected dogs, *N. caninum* has a tropism for the lumbosacral spinal nerve roots. Polyradiculoneuritis and polymyositis, eventually including meningoencephalomyelitis, are typical.

2. Adult dogs may suffer from a non-generalized neosporosis and may develop a wider range of manifestations, predominantly including encephalitis, meningoencephalomyelitis, or myositis. Paraparesis is the most common neurological sign, but seizures, abnormal behavior, and vestibular dysfunction can also occur. Signs of myositis include stiff gait, weakness, and pain or atrophy of skeletal muscles. Almost any other organ can be involved, including the lungs, liver, eyes, and heart (Figure 5.1c).

3. In some dogs, a multifocal, systemic dissemination affecting a variable number of organs, other than the CNS, is observed.

4. In addition, there are several case reports on polygranulomatous *N. caninum* dermatitis with or without a systemic dissemination of the parasite into other organs. Dermal neosporosis affects mainly adult dogs (Figure 5.1b).

Some of the factors that may affect clinical infections are

1. *Age*: The most severe cases of neosporosis occur in young congenitally infected pups (Tables 5.5 and 5.6). Although stillbirths and neonatal deaths have been reported in littermates of pups with confirmed neosporosis (Table 5.5), the youngest pups with verified neosporosis were 8 and 12 days old.[1636] Necropsy examinations were performed on 2 of the 4 pups born by Caesarean section. Both pups had generalized neosporosis with *N. caninum* demonstrated in histological sections of their lungs, livers, hearts, adrenal glands, and kidneys. Among the 27 *Neospora*-infected dogs reported in Reference **120**, there was a 2-day-old pup diagnosed by IHC, but no additional details were provided. Young dogs develop pelvic limb paresis (Figure 5.2) that becomes progressive paralysis. Rarely, all littermates are affected. The earliest clinical sign may be limping or dragging of 1 limb or sudden lameness.[573,608] The pelvic limbs are more severely affected than the thoracic limbs and often have rigid hyperextension.[120,402] Some dogs learn to sit and to use their thoracic limbs, and are able to hop for months. A Labrador[617] lived for 5 years after initial clinical signs noticed at 3 months of age. During the 5 years, the dog had persistent *N. caninum* IFAT titers of 800 or higher.[617]

   Subclinically infected bitches can transmit the parasite to their fetuses, and successive litters from the same bitch may be born infected or skip a generation among infected litters.[124,209,210,541,870] At what stage of gestation *N. caninum* is transmitted from bitch to the fetus is unknown. In most cases of neonatal neosporosis, clinical signs are not apparent until 5–7 weeks after birth (Tables 5.4 through 5.6). These data suggest that *N. caninum* is transmitted from the dam to the neonates toward the terminal stages of gestation or postnatally via milk. Vertical transmission of *N. caninum* in dogs is considered highly variable and not likely to persist in nature in the absence of horizontal transmission of infection.[124] Pups may have a variety of neurological signs from hyperexcitability to complete stupor and the dysfunction reflects the area of CNS involvement. Other dysfunctions which occur include difficulty in swallowing, paralysis or stiffness of the jaw,[120,166,863] muscle flaccidity, and muscle atrophy. In some cases, the severe stiffness of the jaw causes a small (1–2 cm) opening of the mouth. In these pups, the food may stay in the pharynx, indicating weakness of reflexes or muscles or both. Dogs with pelvic limb paralysis may be alert and survive for months. Tetraplegia may follow.

   The cause of the pelvic limb hyperextension, which is predominantly observed in young congenitally affected dogs, is likely due to a combination of lower motor neuron paralysis caused by radiculoneuritis and myositis, which results in rapidly progressive fibrous tissue scarring of the muscles

Table 5.4 Summary of Reports of Clinical Neosporosis in Single Dogs

| Country | No. of Dogs | Breed | Age | Necropsy | Results | | | |
					Tissues Positive	IFAT	Other Findings, *N. caninum* Strain Isolated	Reference
Argentina	1	Boxer	2 m	Yes, complete	B, M	1:12,800	Megaesophagus. Bioassay positive in gerbils. PCR positive in pup and gerbils	166
Argentina	1	Rottweiler	2 yr	No	M	1:400	Megaesophagus, tetraparesis, and muscle atrophy	772a
Australia	1	Unknown	>6 m	Yes	B, H	–	Retrospective	1429
	3	Unknown (same litter)	<3 m	Yes	B, S	–	Retrospective	736
	1	Boxer	6 m	Yes, only CNS	B, S	–	Retrospective	
	1	West Highland White Terrier	13 m	Skin biopsy	Skin	1:400,000	PCR-skin positive, bioassay, WA-K9	1329
	1	Bulldog	17 d	Yes	B, H, Li, Lu, A, Kidney		Severe myocarditis giving appearance of infarcation	1316a
Austria	1	Rhodesian Ridgeback	5 yr	Yes	B, S	1:5120 (1:1280 in CSF)	PCR positive in blood and CSF, chemotherapy	1225
Brazil	1	Collie	7 yr	Yes, complete	B	1:1600	NC-Bahia	779
	1	Cocker Spaniel	10 yr	Biopsy	Skin	1:6400	Chemotherapy	1268
Belgium	1	Labrador	4 w	Yes	S	–	TEM	1633, 1634
	4	1 Labrador	6 m	Yes	B, S	1:800	–	2021
		1 Labrador	7.5 m	No	–	1:800	Treated, recovered	
		1 Boxer	4 m	Yes	B, S	1:3200	–	
		1 Rottweiler	2 yr	Yes	B	1:12,800	–	
	1	Welsh Springer Spaniel	6 yr	Yes	S	–	Polyradiculitis	1756
	1	Greyhound	12 yr	No, skin biopsy	Skin	1:80	Chemotherapy	463
Canada	1	Bullmastiff	10 m	Yes, complete	H	–	Severe myocarditis with numerous tachyzoites	1500
	1	Golden Retriever	12 w	Yes, complete	M, B, Li, Lu, H	<1:64	Cysts in brain	357
	1	Unknown	7 m	Yes	H	–	–	606
	1	Basset Hound	5 yr	Yes	B, S	1:800	–	

(Continued)

Table 5.4 (Continued) Summary of Reports of Clinical Neosporosis in Single Dogs

Country	No. of Dogs	Breed	Age	Necropsy	Tissues Positive	Results IFAT	Other Findings, N. caninum Strain Isolated	Reference
	1	Rottweiler	16 m	Yes	B, S, M	1:800	Myositis, meningoencephalomyelitis	1192
Costa Rica	1	Cocker Spaniel	2 yr	Yes	B	–	TEM	1398
Czech Republic	1	German Wirehaired Pointer	6 yr	Skin	Skin biopsy	1:160	–	1095
Finland	1	Great Dane	8 m	Yes	B, S, M	–	TEM, meningoencephalomyelitis, polyradiculoneuritis, myositis	1748
France	1	Siberian Husky	6 yr	No, skin biopsy	Skin, bone marrow	1:1280	TEM	702
Germany	1	Blue Picardy Spaniel	4 w	Yes	S, B	–	TEM	1633, 1634
Germany	2	1 Boxer	2 m	Yes	B, S	–	–	253
		1 Labrador Retriever	5 m	Yes	B, S, M	–	–	
	1	Airedale	4 yr	–	–	1:400	Chemotherapy	1093
	1	Irish Wolfhound	21 m	No	–	1:800	Chemotherapy	1551
Hungary	1	Basset hound	2 m	Yes	B	–	TEM	1895
Ireland	3	1 Boxer	6 m	Yes	B, S	–	TEM	1842
		1 Greyhound	7 w	Yes	B, S	–	–	
		1 Pointer	10 m	Yes	B, S	–	–	
Israel	1	Boxer	11 yr	Yes	Skin	–	–	1593
Italy	1	Bernese Mountain Dog	5 yr	Skin biopsy	Skin	1:640	TEM, chemotherapy	1631
	1	Pit Bull Terrier	2 m	Muscle biopsy	M	1:3200	–	1563
	1	Labrador Retriever	14 yr	Yes, complete	B	–	No cysts observed	292
	1	Argentine Dogo	9 m	Yes, complete	Skin	1:800	TEM, coinfection with Leishmania sp.	1937
	1	Argentine Dogo	2 m	Muscle biopsy	M	1:25,600	–	1544
	1	Golden Retriever	10 yr	Skin biopsy	Skin	1:800	PCR positive from blood	1133
Japan	1	Shetland Sheepdog	5 m	Yes	B, S, H	–	TEM	2011
	1	Greyhound	1 m	Muscle biopsy	M	1:3200	PCR positive in CSF	986

(Continued)

Table 5.4 (Continued)   Summary of Reports of Clinical Neosporosis in Single Dogs

Country	No. of Dogs	Breed	Age	Necropsy	Tissues Positive	Results IFAT	Results Other Findings, *N. caninum* Strain Isolated	Reference
Mexico	1	Toy Poodle	4 yr	Yes	B, Li, P, intestine	–	–	1250
New Zealand	1	Standard Poodle	4 yr	Liver biopsy	Li	–	PCR positive in liver	708
	1	West Highland White Terrier	1 yr	Yes, complete	B, S	–	TEM, retrospective	1567
	1	West Highland White Terrier	6 yr	Yes, complete	B, S	–	Retrospective	
	1	Labrador	19 m	Yes, complete	B, S	–	Retrospective	
The Netherlands	1	Rottweiler	3 yr	Muscle biopsy	M	–	Chemotherapy	1953
Norway	1	Saluki	4 m	Yes	B, S, M	–	Retrospective	206, 207, 210
	6	Boxer	5 m to 2 yr	Yes	B, S, M	–	Bioassay in mice and blue fox	Personal communication to JPD
South Africa	3	1 Great Dane	12 w	Yes	B, S	–	TEM	994, 998
		1 Labrador Retriever	6 m	Yes	B, S	–	TEM	
		1 Scottish terrier	2 w	Yes	H, Li, Lu	–	Severe myocarditis	
Spain	1	Napolitan Mastiff	4 m	Yes	B, S	–	–	1642
	1	Labrador Retriever	4 yr	Yes, complete	B, S	–	Tachyzoites in CSF	718
	1	Alaskan Malamute	4 yr	Yes	B, S, M	1:1500	Chemotherapy	1224
	1	Rottweiler	4 yr	No, skin biopsy	Skin	1:1600	TEM, immunodepressed, chemotherapy	1521
Sweden	1	Greyhound	4 m	Yes	S	–	TEM, retrospective	908
	1	Boxer	10 w	Yes	B, M	–	TEM, retrospective	2007
	1	Riesenschnauzer	4 yr	Yes	S, M	–	Retrospective	2008
Switzerland	9	See footnote[a]	"Pups"	Yes	B, S, M in 6 of 9 pups	–	–	2109
Turkey	1	Doberman Pinscher	9 w	No	–	1:800	Chemotherapy	178
UK	1	Shetland Sheepdog	9 m	Yes	B	–	Retrospective	547

(Continued)

Table 5.4 (Continued) Summary of Reports of Clinical Neosporosis in Single Dogs

Country	No. of Dogs	Breed	Age	Necropsy	Tissues Positive	IFAT	Results — Other Findings, *N. caninum* Strain Isolated	Reference
	3	1 Labrador retriever	15 w	No	—	1:3200	Chemotherapy	1080
		1 Boxer	18 m	Yes	B,S,M	1:12,800	—	
		1 Boxer	5 w	Muscle biopsy	M	1:3200	Chemotherapy	
	27	8 Labrador retriever, 7 Boxers, 2 Greyhounds, 10 others[b]	2 days to 7 yr	Yes, in 17	CNS of 15, M of 11 of 12	20 dogs positive (all tested)	Isolation, Nc-Liv	118, 119, 120, 121
	1	Boxer	16 w	Yes, also biopsy skin	Skin, B, H, M, P,	1:800	Tachyzoites in needle aspirate of skin lesion	238
	1	Collie (crossbred)	7 yr	No	—	1:800	Normal by computed tomography, chemotherapy	239
	1	See footnote[c]	18 m to 10 yr	Yes, 1 dog (4 yr)	B	—	MRI, PCR positive on CSF of 4 dogs	735
	4	1 Greyhound	9.5 yr	No	—	1:1600	PCR positive in CSF, MRI, chemotherapy	1562
		1 Labrador Retriever	2.5 yr	No	—	1:1600	PCR positive in CSF, MRI, chemotherapy	
		1 Labrador Retriever	5 m	No	—	1:1600	PCR positive in CSF, MRI, chemotherapy	
		1 Cavalier King Charles Spaniel	3 yr	No	—	1:800	PCR positive in CSF, MRI, chemotherapy	
USA	1	Doberman	12 w	Yes	B, M	—	Retrospective[d,e]	See Reference 617
	1	Weimaraner	6 yr	No, biopsy muscle	M	—	Retrospective[f], chemotherapy	809
	2	1 Bloodhound	3.5 yr	Yes	B, M	—	Retrospective[g], TEM	242
		1 Borzoi	6 yr	Yes	B, M	—		
	10	See footnote[g]	6 w to 15 yr	Yes, complete	Many[h]	—	TEM	529
	1	Labrador Retriever	12 w	No, muscle biopsy	M	1:3200	Chemotherapy, viable *N. caninum* isolated by bioassay (NC2)	863
	1	Basset hound	10 yr	Yes	H, A	—	—	933

(Continued)

Table 5.4 (Continued) Summary of Reports of Clinical Neosporosis in Single Dogs

Country	No. of Dogs	Breed	Age	Necropsy	Results — Tissues Positive	Results — IFAT	Results — Other Findings, N. caninum Strain Isolated	Results — Reference
	1	Boxer	6 m	Yes	B, S	–	–	736
	1	Vizsla	3 yr	Yes	B, S	–	–	1749
	1	Golden Retriever	3.5 yr	Yes	B, M, S	1:400	–	987
	1	Golden Retriever	12 yr	Yes, complete	Skin	1:3200	Chemotherapy, TEM, concurrent lymphosarcoma	562
	1	West Highland White Terrier	11 yr	Yes, complete	Lu, M	–	Tachyzoites in lung aspirate	811
	1	Great Pyrenees	10 w	Muscle biopsy	M	1:6400	Chemotherapy	1209
	1	Rhodesian Ridgeback	7 yr	No, peritoneal tap	–	1:20, 480- 1:163, 480	Peritoneal fluid positive for intact tachyzoites and DNA by PCR, chemotherapy	920
	2	1 Italian Greyhound	9 yr	Skin biopsy	Skin	–	Chemotherapy	1111
		1 Labrador Retriever	7 yr	Skin biopsy	Skin	–	Concurrent lymphosarcoma, chemotherapy, euthanized	
	1	Maltese	6 yr	Yes, skin biopsy	Skin, Lu, M	1:1600	Prednisone therapy	822
	1	Shetland Sheepdog	9 yr	Yes, complete	B	1:50	Tachyzoites in CSF	719
	1	Border Collie	4 yr	Yes	Li	–	Severe hepatitis	927
	1	Mastiff	3.5 yr	Yes	H	–	Severe myocarditis, likely immune mediated	1346

Source: From Dubey, J. P. 2013. Neosporosis in dogs. *Commonwealth Agricultural Bureau Reviews* 8, e55.

a 5 Boxers, 1 Labrador Retriever, 1 Golden Retriever, 1 St. Bernard, and 1 Irish Wolfhound.

b 1 each of Bloodhound, Border Collie, Bull Mastiff, Cavalier King Charles Spaniel, Flat-Coated Retriever, Golden Retriever, Hamiltonstövare, Lucher, West Highland White Terrier, and Irish Wolfhound.

c 3 Labrador Retriever (18 m, 6 yr, and 4 yr), 2 West Highland White Terrier (5 yr and 9 yr), 1 Greyhound (7 yr), and 1 Dachshund (10 yr).

d 6 dogs with protozoal-associated encephalitis from USA were diagnosed by IHC and confirmed by specific PCR assay.[1810]

e *Toxoplasma*-associated neuropathology was described in 63 cases in dogs,[1084] those cases were revised[529] and resulted in the first cases of neosporosis in dogs described in USA and in the literature.

f Originally reported as *Toxoplasma*; however, subsequently confirmed as *Neospora* by J. P. Dubey.

g 1 Shetland Sheepdog (8 m), 3 Basset Hounds (8 m, 5 yr, and 5 m), 1 Bullmastiff (3 m), 1 Collie (4 yr), 1 Golden Retriever (6 m), 2 Poodles (2 yr, 6 w), and 1 mixed breed (15 yr).

h Brains of 4, spinal cords of 2, heart of 1, muscles of 2, livers of 2, lungs of 2, pancreas of 1, skin of 1, and ureter of 1.

Abbreviations: m = months, w = weeks, yr = year, IFA = indirect fluorescent antibody, TEM = transmission electron microscopy, A = Adrenal glands, B = Brain, H = Heart, Li = Liver, Lu = Lung, M = Skeletal muscle, P = pancreas, MRI = magnetic resonance imaging.

Table 5.5 Neosporosis in Littermate Pups

Country	Breed (No. of Pups)	Clinical Outcome, Remarks	Histopathology	Tissues Positive	IFAT	Viable N. caninum Isolated[a]	Reference
Australia	Greyhound (?)[b]	2 pups died, 2 months apart	2	B, B	—	—	1429
	Bernese (4)	2-w-old, 3 out of 4 died	2	L, H, K, A	—	—	1636
Austria	American Staffordshire Terrier (5)	2 pups ill at 5 and 8 w, euthanized	2	B, B	1500, 1500 in survival asymptomatic pups	—	2081
Belgium	Labrador Retriever (9)	4 pups ill—3 at 6 w, 1 at 4 m. 1 dead at birth. Chemotherapy	1	M	50, 50, 200, 200	—	2021
Canada	Irish Wolfhound (8)	A 7-w-old pup ill; biopsy. Chemotherapy	Muscle biopsy	M	800	—	400
Denmark	Labrador Retriever (8)	6 pups ill, 12 w to 6 m, 1 euthanized 19 w	1	B, S, M	All 6 at 1:600	—	693
Germany	Kleiner Münsterländer (8)	2 pups ill, 1 at 7 days died, 2nd at 7 w, euthanized	1	Many[c]	1600	NC-GER1	1598, 1599
	Doberman (3 litters: 9 + 5 + 8)	Of 22 pups in 3 successive litters (A, B, C) 6 pups (3A + 1B + 2C) were seropositive but only 1 from litter C was ill, euthanized at 16 w	1	B	5120 (other 5 were seropositive)	—	870
Ireland	Greyhound (5)[b]	3 pups ill, 1 examined, 2nd successive litter. Previous litter similar signs but not tested	4	B, B, B, B	—	—	1842
	Greyhound (7)	3 pups ill					
	Greyhound (NR)	2 pups ill					
New Zealand	Boxer (5)	3 ill, 2 littermates died but not necropsied. A 13-w-old pup examined	1	B, M	51,200 (pup examined), 1600 (dam), 1600 (other pup)	—	1684
Norway	Boxer (6 in 3 litters)[b]	6 pups ill from 3 litters at 2–5.5 m	6	Neo in sections of CNS of all 6 pups. Tachyzoites in M of 5	—		205, 207, 208, 209

(Continued)

Table 5.5 (Continued)　Neosporosis in Littermate Pups

Country	Breed (No. of Pups)	Clinical Outcome, Remarks	Histopathology	Tissues Positive	IFAT	Viable N. caninum Isolated[a]	Reference
The Netherlands	Labrador Retriever (10)	1 stillborn. 4 pups ill 3 w to 5 m	3	CNS and M of all 4 pups	–	–	2117
South Africa	Labrador Retriever (11)	3 pups ill 4 w, 2 pups necropsied	2	CNS and M of both	–	–	988
UK	Bloodhound (3 + 9)	2 successive litters, all 3 ill 1st litters, 3 of 9 ill 2nd litter. Chemotherapy	1 from 1st litter	B, S	100, 400 (2 of 3 investigated)	–	1298
	Labrador Retriever (?)[b]	2 littermates ill with similar signs,	1	CNS	–	–	1325
	Boxer (3)	1 sick, 2 seropositive	1	CNS, M	200	Nc-Liv	119
USA	German Shorthaired Pointer (39)[b]	39 pups in 4 litters; 29 had limb paralysis in 1957	6	CNS of all 6, eyes of 5	–	–	541
	Labrador Retriever (8 + 7)[b]	4 pups ill in each litter. Tetraplegia	2	CNS of both	–	–	404
	Labrador Retriever (7 + 7)	2 litters from different dams, 11 of 14 pups ill. 5–8 w after birth	5	CNS of 5, H in 1	50, 100, 200, 200, 800, 800	NC1	528
	English Springer Spaniel (5)	4 pups ill; 2 euthanized but not examined. 2 pups necropsied at 12 w	2	CNS and M of 1	1600, 1600	NC3	402, 1166
	Labrador Retriever (6)	3 pups ill starting day 34	2	B, B, M, M	100, 3200, 50,200	NC4, NC5	573
	Rhodesian Ridgeback (10)	3 born dead, 1 died at birth, 5 asymptomatic, 1 euthanized 6 w	1	A, B, S, M[d]	10,240	CN1	1284
	Labrador Retriever (11)	3 pups ill, euthanized 116, 116, 135 days. Chemotherapy	3	B, M of all 3	800, 800, 3200	NC6, NC7, NC8	598, 602
	Beagle (5)	4 pups ill, euth. 137 days. 1 pup necropsied	1	B, M	800, 400, 800, 400, <25	NC9	608

Source: Modified from Dubey, J. P. 2013. Commonwealth Agricultural Bureau Reviews 8, e55.

[a] N. caninum strain designation.

[b] Retrospective.

[c] Brain, spinal cord, retina, muscles, thymus, heart, liver, kidney, stomach, adrenal gland, and skin.

[d] Tissues and sera from the dam and pups were also received in J. P. Dubey's laboratory in 1995; the NAT titers were for the dam 6400 and 1600, 800, 400, 200 in 3, 100 in 2 pups. Viable N. caninum was also isolated in J. P. Dubey's laboratory by bioassay in mice and cell culture. The attending veterinarian suspected outbreak of neosporosis in his kennel and other client dogs. Of 36 dogs tested, 13 had NAT antibodies (9 with titers of 50 and 4 with titers of 100) (J. P. Dubey, unpublished).

Abbreviations: A = adrenal, B = brain, H = heart, M = muscle, CNS = central nervous system, K = kidney, Li = liver, Lu = lung, S = spinal cord. d = day, w = week, y = year.

Table 5.6   Reports on Different Presentations of Canine Neosporosis in Dogs ≤12 Months or Dogs >12 Months of Age

Age	Progressive Polyradiculitis and Polymyositis, Eventually Involving Meningoence-phalomyelitis	Encephalitis, Meningoence-phalomyelitis, Eventually with the Dissemination to Other Organs	Disseminated Neosporosis, Including, for Example, Myocarditis, Polymyositis, Hepatitis, Interstitial Pneumonia	Polygranulomatous Dermatitis, Eventually with Dissemination to Other Organs
≤12 months	118, 121, 205, 253, 357, 400, 402, 404, 528, 529, 541, 547, 573, 602, 863, 908, 986, 994, 1080, 1298, 1563, 1567, 1598, 1599, 1633, 1634, 1642, 1684, 1842, 1895, 2008, 2011, 2021	547, 2081	121[a,b], 529[c,d], 597[c,e], 994[c,d,e], 1500[c], 2081[c]	238, 1937
>12 months	121, 242, 529, 597, 1080, 1093, 1192, 1225, 1551, 1567, 1953, 2007	292, 529, 718, 735, 772a[a], 779, 987, 1224, 1250, 1398, 1749, 2021	239[b], 529[d,e,f], 708[e], 772a[a], 811[d]	529, 562, 702, 1111, 1133, 1329, 1521, 1593, 1631

Note: Organs affected in disseminated canine neosporosis.
[a] Myositis.
[b] Eye innervations.
[c] Myocarditis.
[d] Pneumonia.
[e] Hepatitis.
[f] Skin.

that leads to arthrogryposis (fixation of joints). The rigidity and hyperextended limb position may be explained by the opposing forces from scar contracture of muscles opposing the concurrent long bone growth. The disease may be localized or generalized and virtually all organs may be involved, including the skin. The dams show no clinical signs of infection.

2. *Breed*: Of 148 isolated cases summarized in Table 5.4, there were 29 Boxers, 24 Labrador Retrievers, 9 Greyhounds, 7 West Highland White Terriers, 7 Golden Retrievers, 6 Basset Hounds, 5 Poodles, 4 Bullmastiffs, 4 Shetland Sheepdogs, 3 Collies, 3 Irish Wolfhounds, 2 Great Danes, 2 Bloodhounds, 2 Border Collies, 2 Cavalier King Charles Spaniels, and 39 isolated breeds. For cases in littermate dogs, most reports were in Labradors (Table 5.5). Some of these reports were retrospective and thus details are incomplete. All *N. caninum*-infected pups were from 36 litters, some from the same dams. The breeds affected were: 11 Labradors, 5 Boxers, 4 German Shorthaired Pointers, 4 Greyhounds, 2 Bloodhounds, and 1 each of Rhodesian Ridgeback, Doberman, Kleiner Müsterländer, English Springer Spaniel, American Staffordshire Terrier, Irish Wolf Hound, Bloodhound, Bernese Mountain Dog, and Beagle (Table 5.5).

3. *Tissues parasitized*: Clinical signs varied with the organs affected. Many of the reports on neosporosis in dogs were retrospective (Tables 5.4 through 5.6). Clinical signs in most cases of canine neosporosis were neuromuscular, especially in pups.

Dermatitis is another common presentation of neosporosis in dogs.[238,529,562,702,822,1095,1111,1133,1268,1329,1593,1631,1937] Lesions may ulcerate with a purulent discharge. Most of these cases were in aged dogs or those on immunosuppressive drugs, but pups can also be affected. The youngest affected dog was a 16-week-old Boxer.[238] This dog had multiple ulcerative dermal nodules throughout the body and the dog died within 6 days of the first appearance of the dermal nodules. *N. caninum* was found in the skin, pancreas, muscles, and brain. A prior episode of *Salmonella* gastroenteritis and glucocorticoid therapy probably contributed to immunosuppression and demise of the pup. A 13-month-old West Highland White Terrier in Australia developed dermal neosporosis. Initially, non-painful, non-ulcerated nodules were noted under the right eye, right elbow, and in the perineum.[1329] This dog had a history of progressive weight loss associated with ulcerative

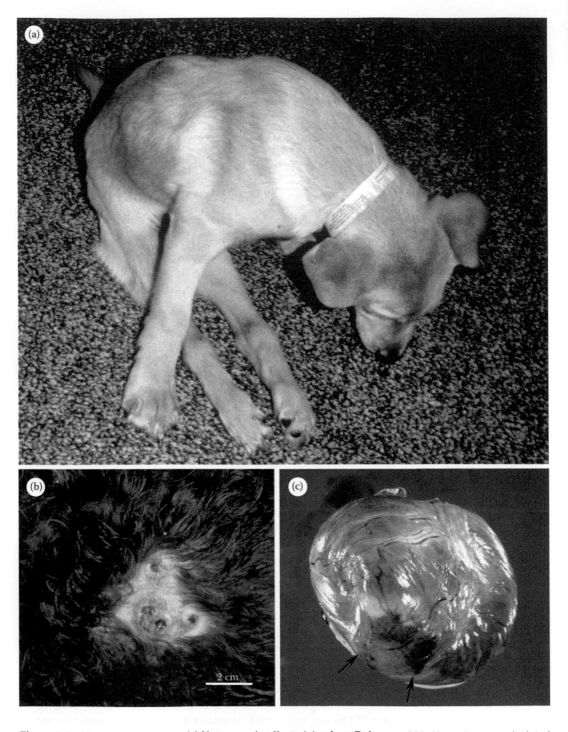

Figure 5.1 Neosporosis in dogs. (a) Neosporosis affected dog from Reference **863**. *N. caninum* was isolated from muscle biopsy of this animal. The dog is depressed and the hind limbs are in hyperextension. Its jaw was stiff allowing only 1 cm opening. The dog lived with hand feeding. (From Dubey, J. P. 2013. *Commonwealth Agricultural Bureau Reviews* 8, e55). (b) Ulcerative, hyperemic, nodules in the skin of a dog (From Dubey, J.P. et al. 1988. *J. Am. Vet. Med. Assoc.* 192, 1269–1285.) (c) Heart of a naturally infected dog from Canada (From Odin, M. and Dubey, J.P. 1993. *J. am. Vet. assoc.* 203, 831–833). The dog died suddenly. Note a large area of infarction (arrows).

Figure 5.2 Three-month-old Labrador Retriever with rear limbs paralysis and hyperextension. Viable *N. caninum* was isolated from this dog.

colitis, probably due to undiagnosed immune deficiency. The dog subsequently developed disseminated neosporosis despite therapy. In USA, 2 cases of dermal neosporosis in Golden and Labrador Retrievers were associated to lymphosarcoma,[562,1111] in another case it was associated to chronic immunosuppressive therapy.[562,1111] Concurrent dermal leishmaniasis was found in 1 dog from Italy.[1937] Neosporosis in a Maltese dog was associated with long-term prednisone therapy.[822] In a 9-year-old dog from Israel, there was generalized lymphadenomegaly and dermal nodules all over the body of the patient.[1593] The 10-year-old Golden Retriever from Italy[1133] had been given prednisone for immune-mediated myelofibrosis. In addition, the 10-year-old dog from Brazil had hyperadrenalocorticism; the patient was treated successfully with clindamycin but relapsed.[1268]

Severe neosporosis hepatitis was reported in 5 dogs (Table 5.4). In a 4-year-old Poodle, it was the result of a reactivated infection after drug-induced immunosuppression[708]; the extent of liver involvement was unknown because the diagnosis was based on liver biopsy, and other organs were not examined. A similar case was observed in a 4-year-old Border Collie treated with corticosteroids for pemphigus foliaceus. Severe hepatic necrosis was associated with intralesional tachyzoites; the brain was not parasitized.[927] Hepatitis was the most prominent in an 8-month-old female Basset Hound that had generalized neosporosis. Most (>75%) of its liver was necrotic and masses of tachyzoites were in the lesions (Figure 2 of Reference **529**). The dog had been vaccinated against canine distemper virus, canine hepatitis, leptospirosis, and rabies and had not been ill before this episode. No obvious immunosuppression was identified. This dog also had encephalomyelitis, myositis, granulomatous masses in its ureter (Figure 5.3), and vena cava. In addition to numerous tachyzoites in the liver and granulomatous masses, organisms were seen in the pancreas, esophageal, temporal, and pelvic limb muscles, stomach mucosa, lungs, and the retropharyngeal, bronchial, pancreatic, and hepatic lymph nodes. Tachyzoites were also seen in the myocardium

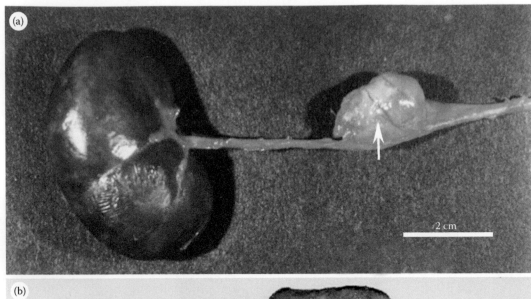

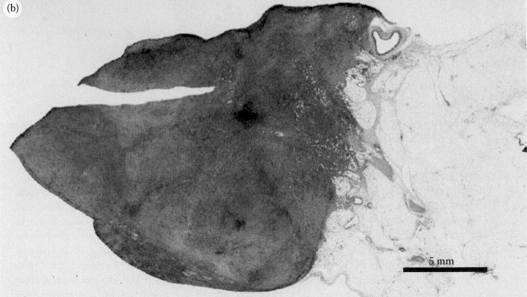

Figure 5.3   (a) Fibrotic mass (arrow) attached to ureter of dog (From Dubey, J.P. et al. 1988. *J. Am. Vet. Med. assoc.* 192, 1269–1285.) A similar mass was attached to the right adrenal gland and caudal vena cava. (b) Granulomatous mass in adrenal gland of another dog (JPD, unpublished).

but were not associated with inflammation. Perhaps, this is the most severe reported case of neosporosis in a dog. Two neonatal cases, 1 in a 12-week-old Golden Retriever and 1 in a 2-week-old Scottish Terrier with patent protozoal multifocal foci of necrosis were described in Canada and South Africa, respectively.[357,994]

Differential diagnosis needs to be carried out in cases of protozoal hepatitis in dogs.[622]

Dogs can die suddenly of severe myocarditis and heart failure without the obvious involvement of other tissues.[529,1346,1500]

An unusual case of diffuse peritonitis with peritoneal effusion has been reported in an 11-year-old Rhodesian Ridgeback from USA.[920]

Pneumonitis was suspected in an 11-year-old dog with persistent cough and tachypnea.[811] Tachyzoites were found in its lung aspirate antemortem (Figure 3.1a, Chapter 3). The dog died of respiratory failure. On necropsy examination, the diagnosis of neosporosis pneumonitis was confirmed. Few tachyzoites were also found in skeletal muscles but not in other organs.

5.2 DIAGNOSIS OF CANINE NEOSPOROSIS AND CANINE *N. CANINUM* INFECTION

5.2.1 General Considerations on Clinical Examination

Antemortem diagnosis of neosporosis is difficult. Clinical history, age of dogs, and serological examinations are helpful for the diagnosis of neosporosis in dogs. Most cases of canine neosporosis have been identified in congenitally infected dogs involving littermates (Table 5.5). In most instances, dogs begin to develop clinical signs 3 or more weeks after birth. Not all pups are affected to the same degree (Tables 5.5 and 5.6). Paralysis of pelvic limbs, often with contracture, is the most consistent sign of neosporosis in dogs. Features that distinguish neosporosis from other forms of paralysis are gradual muscle atrophy and stiffness, usually as an ascending paralysis, and the pelvic limbs are more severely affected than the thoracic limbs.[616] Paralysis progresses to rigid contracture of muscles of affected limbs. This arthrogryposis is a result of scar formation in the muscles from lower motor neuron damage and myositis. In some pups, joint deformation and genu recurvatum may develop. Differential diagnoses have to be carefully excluded. In a study on potential infectious reasons for acute canine polyradiculoneuritis (which is a putative model for Guillain–Barré syndrome in humans) antibodies against *T. gondii* but not *N. caninum* were associated with the disease.[925] Cervical weakness, dysphagia, megaesophagus, and ultimately death occur. In some dogs, the progression may become static. Dogs do not develop severe intracranial manifestations and maintain alert attitudes. They can survive for months with hand feeding and care but remain paralyzed with associated complications.[616] In dogs older than 6 months, the most common signs are related to multifocal CNS involvement with or without polymyositis resulting in ataxia, paraplegia, shivering, unsteadiness, seizures, stupor, abnormal gait, dragging of pelvic limbs, and bumping into objects.[529] Cranial nerve functions may be affected.[120] Hematologic and biochemical findings have been variable, depending on the organ system of involvement.[178,238,357,400,402,529,602,702,708,719,822,920,933,986,987,1080,1224,1268,1325,1684,1749,2021] With muscle disease, creatinine kinase and aspartate aminotransferase activities have been increased. Serum alanine aminotransaminase and alkaline phosphatase activities are increased in dogs that develop hepatic inflammation. CSF abnormalities have included mild increases in protein (over 20 but under 150 mg/dL) and nucleated cell counts (over 10 but <100 cells/dL). Differential leukocyte counts, in decreasing numbers, included lymphocytes, monocytes and macrophages, neutrophils, and eosinophils.[616] Thus, mononuclear cells seem to predominate. CSF analysis results can be within reference limits in some dogs. Eosinophilic pleocytosis has been observed in rare instances in dogs when *N. caninum* is present. Tachyzoites are occasionally observed in CSF. Electromyographic, radiographic, and magnetic resonance imaging (MRI) can aid diagnosis by locating lesions and conduction abnormalities.[120,357,400,402,718,735,863,2021] Nerve conduction velocities may be reduced in the most severely affected limbs, especially proximally as a result of radiculoneuritis, but they are often within reference range. Low evoked action potentials may be found with myositis.

5.2.2 Histopathologic Diagnosis

Relative to the number of infected dogs (Table 5.1), cases of canine neosporosis are rarely reported (Tables 5.4 through 5.6). *N. caninum* was isolated also from brains of non-diseased dogs which suggest that *N. caninum* causes a persistent infection in dogs.[1118] In a prospective study, it was observed that in the vast majority of infected dogs infection was asymptomatic.[124]

5.2.2.1 Antemortem Findings

Symptoms of neuromuscular disease are suspicious of canine neosporosis. In addition to neurological signs, seizures, abnormal behavior, and vestibular dysfunction are suspicious, too. Signs of myositis (stiff gait, weakness, and pain or atrophy of skeletal muscles) are indications for canine neosporosis. Heart failure, pneumonia, and hepatitis, as well as skin lesions could be other signs of less common forms of canine neosporosis. However, positive serological findings in symptomatic dogs are not sufficient to diagnose canine neosporosis because infection is widespread also in non-diseased dogs (Table 5.1). In cases of encephalomyelitis, but also in other cases (Tables 5.4 through 5.6), the examination of CSF can often but not always help to confirm diagnosis. In most cases, a pleocytosis[178,238,357,402,547,718,719,735,822,1598] is observed; however, only in rare cases, the presence of *N. caninum* could be confirmed by microscopic examination[529,718,719,1325] or indirect via PCR.[719,735,986,1225,1598,1810] In some confirmed cases, CSF presented normal[2021] and even in a case when *N. caninum* DNA was detected by PCR in CSF, the CSF protein-content and cellular counts presented normal.[735] In 1 study, a repeated CSF analysis was recommended to avoid false negative results.[719]

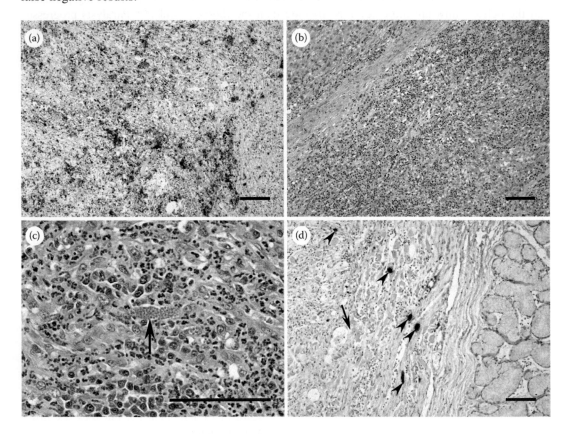

Figure 5.4   Higher magnification of granulomatous mass in Figure 5.3b. (a) Massive numbers of *N. caninum* tachyzoites (all dark/black areas). IHC with *N. caninum* polyclonal antibodies. (b) The adrenal gland architecture has been replaced with inflammatory mass. HE stain. (c) Neutrophils, mononuclear cells, and epitheloid cells obliterating adrenal architecture. Arrow points to a large group of tachyzoites. (d) Esophagitis in a 3-month-old Golden Retriever suffering from congenital neosporosis. (From case reported in Cochrane, S. M., Dubey, J. P. 1993. *Can. Vet. J.* 34, 232–233.) Note necrosis of smooth muscle (arrow) and numerous tachyzoites (all dark areas, arrows). IHC with *N. caninum* antibodies. Megaesophagus is one of the lesions of congenital neosporosis in dogs. Bars = 50 μm.

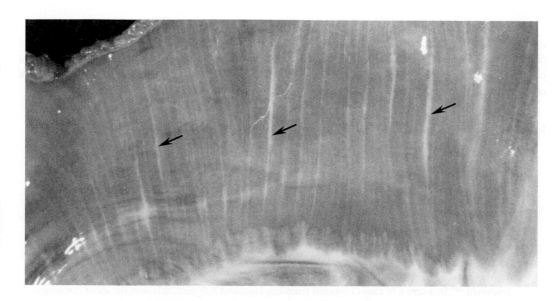

Figure 5.5 Foci of extreme pallor (arrows) in diaphragm of one of the congenitally infected 3-month-old dogs. (Reported in Dubey, J. P. et al. 1988. *J. Am. Vet. Med. Assoc.* 193, 1259–1263.) Unstained.

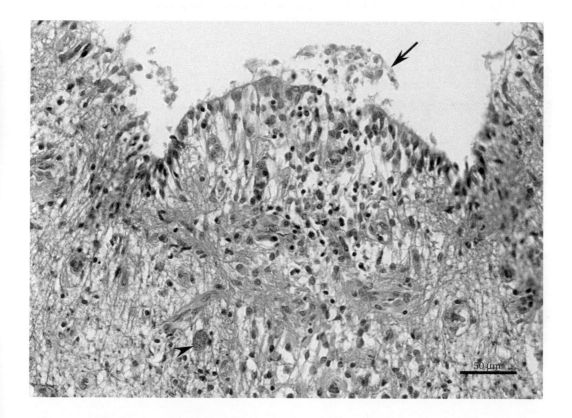

Figure 5.6 Section of medulla oblongata of case no. 7 from Reference **529**. Note inflammatory focus in the subependymal tissue and shedding of cellular elements (arrow) in the ventricle. Arrowhead points to a group of tachyzoites. HE stain.

Biopsies taken from affected organs (muscle, liver, skin, or aspirates from lung) were successfully used to detect the parasite or its DNA (Tables 5.4 through 5.6). Also, in cases of *N. caninum*-associated neuromuscular disease, however, muscle biopsies confirmed myositis but parasitic stages were not detectable in muscle aspirates.[122,400] In other cases, parasitic stages were observed microscopically[357,811,1209] or suspected after positive PCR findings in biopsies.[708]

5.2.2.2 *Lesions in Dogs with Canine Neosporosis, Postmortem Findings*

1. *Gross pathology*: Gross postmortem findings are generally nonspecific; gross lesions of neosporosis reported were cerebral atrophy,[1224] necrotic areas in the CNS[529,2021] and in the liver,[529] granulomas (≤1 cm in diameter) in visceral tissues (Figures 5.3 and 5.4),[529] yellowish-white streaks in muscles (Figure 5.5),[528,539,1598,2011] cerebellar atrophy,[208,735,987] brownish, reddish, or yellowish areas in the heart,[1500] myocardial infarcts (Figure 5.1c),[1346] megaesophagus,[121,166,357,1080,1093] hemorrhages in skeletal musculature,[2021] white greyish striations in limb muscles,[166] and ulcerative dermatitis (Figure 5.1b).[238,463,529,702,1111,1133,1329,1593,1631,1937]

   Muscle atrophy,[120,121,693,735,863,986,1298,1325,1563,1598,1633,1748,2007] including fibrosis[988,1684] of the pelvic limb musculature is frequently reported. In 1 case from USA, masseter muscle atrophy was described.[529] In a series of cases reported from England and Belgium, this was the most consistent finding observed in 21 out of 27 cases.[120] In case of gross lesions, *N. caninum* infection needs to be confirmed either by IHC or by PCR-derived methods.

2. *Histology*: In general, progressive meningoencephalomyelitis, polyradiculitis, and polymyositis is characteristic in transplacentally infected puppies with ascending pelvic limb paresis and paralysis (Figures 5.6 through 5.11 and Tables 5.5 and 5.6). In addition, the parasite may affect a number of other organs, causing lesions, characterized by inflammation and focal necrosis. In adult dogs,

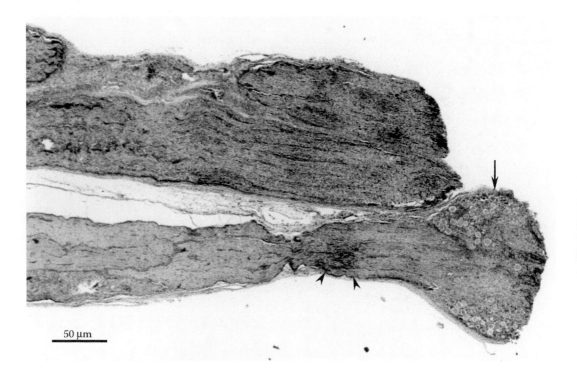

50 µm

Figure 5.7   Neosporosis of peripheral nerves in the dog in Figure 5.2. HE stain. Note severe inflammation of nerve (arrowheads) and of the ganglia (arrow). (From Dubey, J. P. et al. 1988. *J. Am. Vet. Med. Assoc.* 193, 1259–1263.)

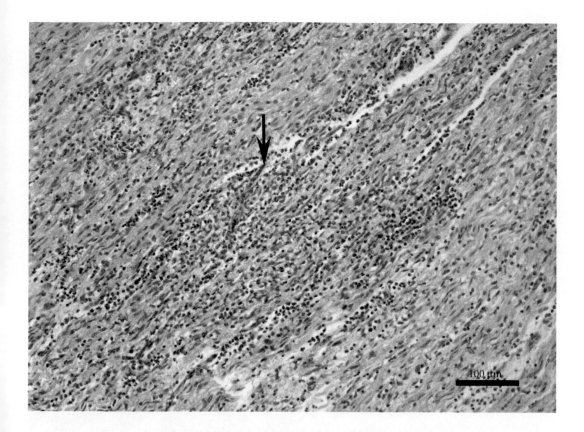

Figure 5.8 Sciatic nerve from dog in Figure 5.2. Severe mononuclear cell infiltration (arrow). HE stain.

multifocal encephalomyelitis, myocarditis, polymyositis, dermatitis, or multifocal dissemination to inner organs including lung and liver are observed (Tables 5.5 and 5.6).

In prenatally infected young dogs, lesions are seen in the spinal cord and brain stem with demyelination, neuronal degeneration with neurophagia and areas of malacia.[1598,2007] The brains of dogs with neuronal canine neosporosis had multifocal nonsuppurative encephalitis with neuronal degeneration and necrosis, perivascular cellular infiltration, and gliosis.[120,121,402,528,529,541,693,988,1567,1598,2021,2109] In 6 cases of canine neosporosis, the distribution of *N. caninum* brain infection was reported in more detail.[121] Lesions were most consistently found in the cerebrum although widespread in the entire CNS.[121] In 3 clinical cases in a litter of Labrador Retrievers, lesions were more severe in the cerebrum than in the midbrain.[602]

In HE-stained tissue sections, severe, diffuse, mixed inflammatory infiltrates affecting the leptomeninges and extending into the adjacent neural parenchyma (thalamus, optic chiasm, pituitary gland, rhinencephalon, and hippocampus) might be observed.[718] In the leptomeninges, the infiltrate consisted of plasma cells, lymphocytes, macrophages, and eosinophils. Gliosis, neuronal necrosis, and tissue cysts were observed in the neural parenchyma.[811] Large areas of neuronal necrosis, gitter cells, proliferating vessels, white matter spongiosis, and small hemorrhages might be observed.[718] In the spinal cord, lesions were widely distributed and often radiculoneuritis was observed (Figure 5.6).[121] In addition, inflammation with infiltrations of neutrophils, plasma cells, and macrophages were seen, even when *Neospora* was not detectable.[121] In dogs infected with *N. caninum*, meningoencephalomyelitis[718,735,1748] is most commonly reported, but also myocarditis,[120,529,597,994,1346,1429,1500,2081] polymyositis,[529,693,863,988,1080,1748] pneumonia,[529,539,811,994,1080] and hepatitis.[708,1080] Further disseminations into adrenal glands have been observed.[529]

N. caninum-associated dermatitis has been frequently reported.[238,463,529,562,702,822,1111,1133,1268,1329,1521,1593,1631,1937] Cutaneous lesions consisted of multifocal, ulcerated, and sometimes exudative

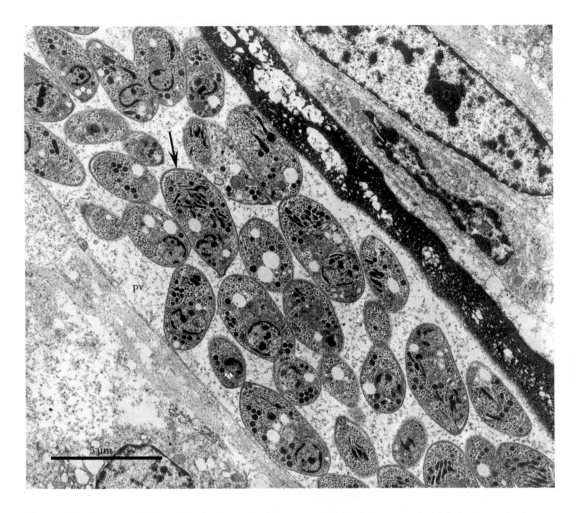

Figure 5.9 Numerous tachyzoites in a parasitophorous vacuole (pv) in a neural cell in lumbar spinal nerve of a dog. Arrow indicates a tachyzoite with several rhoptries. (From Dubey, J. P. et al. 1988. *J. Am. Vet. Med. Assoc.* 193, 1259–1263.)

nodules, measuring 0.5–5 cm in diameter with variable distribution (Figure 5.1b). Lesions were characterized by pyogranulomatous necrotizing inflammation.[529,562,1111]

In most studies, histologic examination detected only tachyzoites. Tissue cysts are rarely detected in cases of canine neosporosis and almost exclusively in the brain.[121,166,357,529,547,573,597,602,693,718,719,779,988,994,1080,1185,1224,1429,1567,1598,1642,1749,1842,1895,2008,2011,2021,2081,2116] There are only few reports of definitive identification of *N. caninum* tissue cysts in the muscle cells of infected dogs.[602,1599]

5.2.3 Demonstration of Viable *N. caninum* in Dogs

Specific protocols for isolation of viable parasites from fecal matter and tissues were reviewed in Chapter 3. Reported isolates in dogs are summarized in Table 5.3.

5.2.3.1 Demonstration of Viable N. caninum in Cases of Canine Neosporosis

1. *Postmortem*: There are several reports on the *postmortem* isolation of *N. caninum* from diseased (and/or euthanized) dogs.[119,528,573,598,602,608,779,1598] As a diagnostic tool, isolation has no practical importance because it is too laborious, and sensitivity of this method has not been validated.

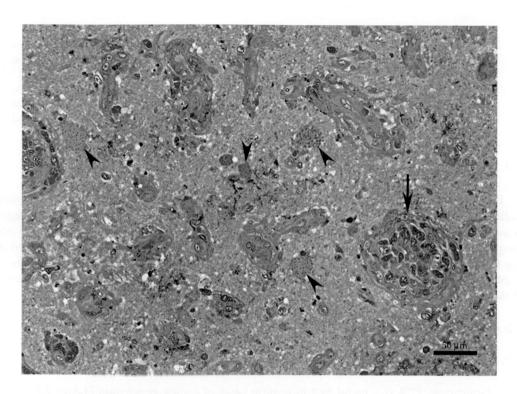

Figure 5.10 Severe vasculitis (arrow) and necrosis of neuropile and *Neospora* (arrowheads) in midbrain of a dog (case no. 7 of Reference **529**). Giemsa stain.

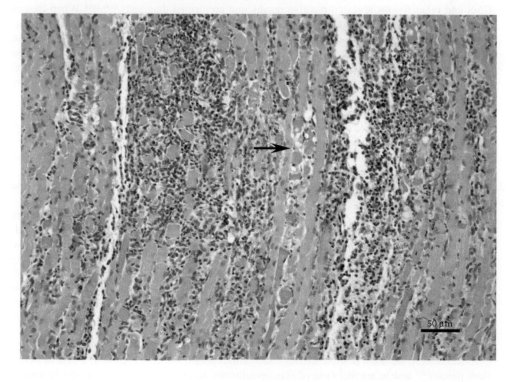

Figure 5.11 Severe myositis and necrosis (arrow) in gluteal muscle of dog represented in Figure 5.2. HE stain.

Isolations can be achieved either from the infected dog tissues into cell culture,[119,402,528,573] via mouse inoculations and passages,[528,573,598,602,1598] or gerbil inoculation and passages.[402,598,779] Homogenates of brain, muscle tissues, and peritoneal fluids were used for cell culture inoculation or sub-inoculation of rodents.

2. *Antemortem*: Cell culture isolation was achieved directly from muscle biopsy[863] or fine-needle aspiration from skin.[1329] An exploratory laparotomy of a dog with a 4-month history of peritonitis revealed a septic exudate including also protozoal zoites,[920] which could be *in vitro* isolated and diagnosed by using *N. caninum*-specific PCR.

5.2.3.2 Demonstration of Viable N. caninum Oocysts

Dogs rarely excreted large numbers of *N. caninum* oocysts (Table 2.3 in Chapter 2). To *in vitro* isolate *N. caninum* from fecal samples into cell culture a passage via inoculations and passages of gerbils or mice is the most promising method.[164,167,168,1798,1805] Because *N. caninum* oocysts are morphologically similar to those of *H. heydorni* and related parasites (e.g., *T. gondii* observed in dog feces most likely because of coprophagia) bioassay is the ultimate way to confirm the identity of viable oocysts present in canine feces.[1805] In addition, molecular tools can also discern the identity of the oocysts.[314,910,1798]

5.2.4 Detection of N. caninum DNA in Dogs

5.2.4.1 N. caninum DNA in Canine Neosporosis

1. *Postmortem*: PCR is a useful tool to confirm cases of canine neosporosis (Tables 5.4 and 5.5). *N. caninum* DNA was detected by PCR in brain,[166,573,1225,1598] spinal cord,[1598] lung,[166,1598] liver,[708,1598] spleen,[1598] kidney,[1598] heart,[1598] and striated muscle.[166,573,1598] Paraffin-embedded CNS and other tissues revealed positive results and confirmed diagnosis.[1636,1684] When tissues (lung, liver, kidney, and heart) of a dog with confirmed neosporosis were examined after treatment, a negative result was achieved although the brain tested positive in PCR.[1225] In a case in which immunosuppressive therapy may have been responsible for *N. caninum* reactivation in a dog, a "Tru-Cut" biopsy of liver contained numerous (20–100) cytoplasmic basophilic, oval, 2–3 μm diameter, globular to ovoid bodies (most likely tachyzoites) which were further diagnosed as *N. caninum* by PCR.[708]

   In the city of São Paulo, Brazil, sera and tissues of 50 dogs with neurological signs were tested; 7 dogs resulted seropositive, 2 dogs were positive by bioassay in gerbils, and 3 dogs were positive by PCR.[1118] In Algeria, *N. caninum* DNA was detected in liver and spleens of 32% (28 of 87) asymptomatic pound dogs by using Taqman real time PCR[753]; however, this technique is highly sensitive and there is always a chance of over estimation.

2. *Antemortem*: In cases of encephalomyelitis, but also in other cases, testing of CSF by PCR could confirm neosporosis.[719,735,986,1225,1598,1810]

   N. caninum or its DNA has been detected in biopsies or fine-needle aspirates taken from dermal lesions (Tables 5.4 and 5.5).[238,463,529,562,702,822,1111,1133,1268,1329,1521,1593,1631,1937] The finding of *N. caninum* DNA in skin lesions confirms diagnosis.[463,822,1133,1329] In cases of dermatitis, *Leishmania* sp. can coexist with *N. caninum*[1937] and amastigotes can be confused with dividing *N. caninum*. Historically, the first case of neosporosis was a dog with dermatitis case no. 10 of Reference **529** the dog was euthanized in 1987 (before the naming of *Neospora caninum*) because it was suspected to have leishmaniasis because of globular tachyzoites (J. P. Dubey, unpublished) (Figure 3.3b). On closer examination, *N. caninum* tachyzoites are crecenteric, whereas *Leishmania* amastigotes are globular, and these organisms are easily distinguished by TEM.

   In addition, *N. caninum* positive findings were also reported for whole blood in a case of pelvic limb paresis[1225] and in another case of skin neosporosis, together with positive results in serum, rectal scrapings, and feces.[1329]

5.2.4.2 N. caninum DNA in Canine Feces

N. caninum oocysts are morphologically very similar to those of *H. heydorni* and related parasites (e.g., *T. gondii* or *H. hammondi*)[899,1805,1866] and PCR is, in addition to bioassay, one of the ways to confirm the identity of oocysts in canine feces.[1805]

Studies on oocyst excretion of *N. caninum* in dogs should be interpreted with caution because in many studies oocyst excretion could not be confirmed unambiguously (Tables 2.3 in Chapter 2, 5.7, and 5.8). Only findings in which living parasites were isolated, that is, the viability of oocysts was shown and oocysts or *in vitro* isolated parasites typed, for example, by microsatellite typing should be regarded as definitive (Table 5.3).[164,167,168,615,781,1798,1805] *N. caninum*-like oocyst detection and properly controlled detection of *N. caninum* DNA constitutes an acceptable tool for epidemiological studies on already proven definitive hosts (Table 2.3 in Chapter 2). Studies in which just DNA of *N. caninum* is shown in feces are inconclusive since the demonstration of parasitic DNA could have many reasons, including laboratory cross-contamination or the presence of *N. caninum* DNA in the feed-diet.

5.2.4.3 Genotyping of N. caninum Strains Isolated from Dogs

Methodologies have been described in Chapter 3. In general, microsatellite typing[28,29,168,1578,1636,1673] has become a valuable tool to fingerprint isolates and to confirm the absence of laboratory cell culture contamination. In Table 5.3, available molecular data for *N. caninum* isolates are summarized. Microsatellite assays have been used to type *N. caninum* isolated from tissues of clinical cases[1636] or from oocysts excreted by naturally infected dogs.[1649] Other methodologies such as PCR sequencing of ITS1 or pNc5 genes have been also employed with the aim of detecting variability in *N. caninum* positive samples.[314,416,1649]

5.2.5 N. caninum-Specific Antibodies in Dogs with Canine Neosporosis

Based on the serological findings, the clinical signs, the application of diagnostic imaging procedures including MRI,[718,735] radiographic imaging (e.g., megaesophagus[120]), and electromyographic examination[402,657] may aid further diagnostic steps. Demonstrating serum antibodies to *N. caninum* by any of the different tests can help confirm the diagnosis of neosporosis. In general, canine antibody concentrations may decrease over time and some dogs eventually become seronegative.[124] Antibody titers can also vary among different laboratories. The test used in most clinical cases of neosporosis was IFAT; individual titers are listed in Tables 5.4 and 5.5. IFAT titers were generally 800 or higher. Although IFAT titers were generally high in cases with severe neosporosis, this should not be stressed because there is no correlation found between the magnitude of titer and clinical signs.

Low antibody titers have been reported in some dogs with histologically verified neosporosis.[573,608] The IFAT can be used to detect IgM antibodies; however, dogs with confirmed acute neosporosis may have negative IgM test results.[573] Some very young dogs might still have maternal antibodies. Maternal antibodies may be transferred from dam to offspring, resulting in false-positive titer results; however, these levels were gone in uninfected pups by day 32 of life.[124]

Toxoplasma gondii infections are also common in dogs and dual infections with *T. gondii* and *N. caninum* can coexist in the same dog.[562] Cross-reactivity between *T. gondii* and *N. caninum* was not demonstrable in verified infections.[573,602] Slight cross-reactivity with sera from dogs infected with *Babesia gibsoni* but not *Babesia canis* has been observed.[2142] There was no cross-reactivity between *N. caninum* and *Leishmania* sp.[72,298,397,748,1552]

Most dogs do not seroconvert for tachyzoite antigens after excreting *N. caninum* oocysts (Table 5.7). Antibody tests able to unambiguously identify dogs that have excreted oocyst are still

Table 5.7　Antibodies against *N. caninum* Tachyzoites in Dogs Having Excreted *N. caninum* Oocysts after Primary Experimental Infection

Country	Type of Inoculum	Strain	Number of Dogs Shedding *N. caninum* Oocysts	Blood Sampling (Days Post Feeding)	Number of IFAT Positive Dogs (Applied Cut-Off)	Number of NAT Positive Dogs (Applied Cut-Off)	Proportion of Serologically Positive Dogs	Reference
Brazil	Bubaline-infected brains	Naturally infected tissues	4	61	2 (≥1:50)	No data	50% (2/4)	1722
	Ovine-infected brains	Naturally infected tissues	1	49	0 (≥1:50)	No data	0% (0/1)	1579
	Infected embryonated eggs	NC1	2	30–34	0 (≥1:25)	No data	0% (0/2)	715
	Cattle tissues (brain, masseter, liver, heart)	Naturally infected tissues	8	30	0 (≥1:50)	No data	0% (0/8)	314
Germany	Bovine-infected brain	Naturally infected	1	1–135	0 (≥1:50)	No data	0% (0/1)	1510a
	Sheep-infected heart and skeletal muscle	HY-Berlin-1996	7	30–60	0 (≥1:20)	No data	0% (0/6)	1798
	Multimammate rat-infected carcass	HY-Berlin-1996	1	60–120	0 (≥1:20)	No data	0% (0/1)	1798
	Goat-infected brain, heart, and skeletal muscle	HY-Berlin-1996	3	30–120	0 (≥1:20)	No data	0% (0/3)	1798
	Guinea pig-infected carcass	HY-Berlin-1996	5	>120	0 (≥1:20)	No data	0% (0/2)	1798
The Netherlands	Placental cotyledonary tissues from seropositive cows	Naturally infected tissues	3	77	0 (≥1:20)	No data	0% (0/3)	493
USA	Mouse-infected tissues	NC2, NC-beef, Nc-Liv	3	37	2 (≥1:50)	No data	67% (2/3)	1309
	Mouse-infected brains	NC-beef	2	42	1 (≥1:25)	No data	50% (1/2)	1180
	Mouse-infected brains	CKO, PB-1-2C	3	36	ND	2 (≥1:25)	67% (2/3)	1187
	Bovine-infected tissues	NC-beef, Nc-Liv	14	25	6 (≥1:50)	No data	43% (6/14)	780
	Bovine-infected tissues	NC-beef, Nc-Liv	6	28	4 (≥1:50)	No data	67% (4/6)	785
Total			59	—	15	2	27.1% (16/59)	—

Note:　ND = not done.

needed. Nevertheless, tachyzoite antigen-based serological assays are important tools in epidemiological studies to examine the exposure of dogs to *N. caninum* (Table 5.1).

5.2.5.1 Antibody Development and Antibody Types

Dogs develop antibodies against *N. caninum* antigens after postnatal infection, by ingesting infectious tissue cysts containing feed-food, but also when puppies are infected prenatally via transplacental transmission. The uptake of infectious material may cause oocyst excretion in these dogs; however, there are a number of reports on experiments in which dogs and other canids were fed infectious material, did not excrete oocysts during the following days, but subsequently seroconverted, that is, developed antibodies against *N. caninum*[715,780,785,1067,1187,1309] (Table 5.8). The materials used to feed the experimental dogs were often tissues that contained tissue cysts or were suspected to contain tissue cysts and/or tachyzoites of *N. caninum*. In 1 study, a dog excreted oocysts 10 days after being fed a single meal of infected chicken embryos; the chicken embryos had been inoculated with 10^6 *N. caninum* tachyzoites by the chorioallantoic membrane 8 days previously.[715] In the same study, another dog excreted oocysts 34 days after ingesting infected chicken embryos; the embryos had been inoculated 8–12 days previously. It is not known whether tissue cysts were present in the inocula fed to dogs.[715] The results of experiments by feeding dogs with infectious tissues suggest that extra-intestinal stages develop after the ingestion of intermediate host tissues containing tissue cysts and/or tachyzoites. Whether the oral ingestion of tachyzoite-containing material is infectious to dogs has not been shown and seems to be unlikely. An experiment with milk spiked with cell culture-derived tachyzoites failed to induce oocyst excretion in dogs.[493]

In addition to carnivorism, fecal transmission via sporulated oocysts is a possibility.[116]

The development of specific IgG and IgE *N. caninum* antibodies was followed in experimentally and naturally infected dogs. While the IgG response was detected by IFAT, ELISA, and immunoblot, only weak specific IgE reactions were observed.[1012]

In addition to postnatal infection (with tissue cysts and oocysts), seroconversion can occur prenatally (Table 5.5). In 53–182-day-old prenatally infected puppies, no specific IgM but specific IgG was detected by IFAT.[573] Whether this is generally the case in prenatally infected dogs needs further investigation.

5.2.5.2 Persistence of Antibodies

A retrospective testing of naturally infected dogs of Hamiltonstövare breed, the use of IFAT revealed that *N. caninum*-specific antibodies persisted for up to 4 years in 9 dogs, but in 4 dogs, all with low titers on initial testing (1:50), antibody titers fell below the detection limit over a 1–2-year period.[124] Antibody titers declined after parturition in naturally infected seropositive bitches[573] or remained at almost the same level.[602,1684]

5.2.5.3 Antibodies and Vertical Transmission in Dogs

Persistently infected bitches may repeatedly transmit their infection vertically during several pregnancies to their offspring.[124,209,541,602,870,1298] In addition, *N. caninum* can be transmitted through several dog generations.[124]

1. *Puppies*: Not all puppies of a litter are seropositive[124,400,870,1298,1684] which is an indication that not all puppies of a litter become infected. After experimentally induced vertical transmission of *N. caninum*, 1 newborn puppy died of neosporosis but no specific antibodies were detectable by IFAT.[533] Congenitally infected puppies may have high antibody IFAT titers after birth and these titers decline later on.[1684]
2. *Bitches*: Transplacental transmission of *N. caninum* was prospectively studied in 17 naturally infected, privately owned bitches of the Hamiltonstövare breed using IFAT; puppies of these seropositive bitches (IFAT titers 1:50 or higher) were tested by IFAT to determine their infection status.

Table 5.8 Positive Serologic Reactions against *N. caninum* Tachyzoites in Dogs Failing to Excrete Oocysts but Fed with Infectious Material/Material Suspected to be Infectious

Country	Number of Dogs Failing to Excreted Oocysts after Feeding Infectious Material/ Material Suspected to be Infectious	Material Used to Feed Dogs	Serological Test (Cut-Off)	Reference
Australia	2	Tissue from calves (inoculated 10^7 or 10^8 tachyzoites, culture adapted Nc-Nowra, 12 weeks after infection; tissue: brain, spinal cord, heart, lung, kidney, and tongue)	VMRD-cELISA (>30% inhibition)	**1067**
Brazil	3	Repeated feeding with embryonated eggs 10 days after inoculated with 10^6 *N. caninum* tachyzoites/each; dogs were fed each with 2 eggs/day (excluding egg shells), during 3 consecutive days, with feeding procedures being repeated at 15 and 30 days after first feeding	IFAT (1:25)	**715**
USA	1	100 tissue cysts in mouse brains (NC2 isolate)	IFAT (1:50)	**1309**
	3	Brains from mice infected for 58 or 92 days with clones of CKO or PB-1-2C strains	NAT (1:25)	**1187**
	1 (reported)	Tissue of a seropositive calf inoculated with 500 oocysts of NC-beef (6 weeks—2 months after infection; tissue: brain, spinal cord, heart, liver, kidney, tongue, diaphragm, and other skeletal muscles)	IFAT (1:50)	**780**
	2	Infected calf tissues (NC-beef tachyzoites, 2×10^6–3.2×10^7; 6 weeks—2 months after infection; tissue: brain, spinal cord, heart, kidney, tongue, diaphragm, and other skeletal muscles)	IFAT (1:50)	**785**

Only 4/118 (3%) puppies from these 17 bitches were seropositive and only 2 of these developed disease compatible with neosporosis.[124] Retrospective studies in seropositive bitches (6 Hamiltonstövare, 3 Boxer, 2 Flat-Coated Retriever, and 2 Labrador Retriever) including 179 pups showed that the proportion of seropositive puppies was positively associated with the IFAT titers in the dam and ranged from 5% in dams with an IFAT titer of 1:50 to 89% in dams with an IFAT titer of 1:12,800.[124] This shows that the proportion of vertical transmission varies in female dogs and the efficiency of transmission can be considerably high in dams with very high IFAT titers.

5.2.5.4 Antibodies in Dogs with Myositis, Polyradiculoneuritis, and Encephalomyelitis

Most dogs in which myositis, polyradiculoneuritis, and encephalomyelitis predominate (i.e., predominantly young dogs most likely transplacentally infected) have IFAT titers of 1:400–1:1600 (Tables 5.4, 5.5, and 5.9). However, some dogs with *N. caninum*-associated myositis,

Table 5.9 IFAT Titers Reported in Clinical Cases of Canine Neosporosis

Canine Neosporosis	IFAT Titer Category	Age											References
		1 m	2 m	3 m	4 m	5 m	6 m	7–12 m	1–2 yr	3–6 yr	>6 yr	Total	
Myositis, polyradiculoneuritis, encephalomyelitis	>1:1600	3	3	3	1				2	1		13	121, 178, 400, 402, 528, 573, 597, 602, 693, 772a, 863, 986, 1067, 1080, 1093, 1192, 1209, 1225, 1298, 1563, 1598, 1684, 2021
	1:400–1:1600	1	4	8	2		1		2	3		20	
	1:50–1:200		3	1	3							7	
	Total	4	10	12	6	0	1	0	4	4	0	41	
Encephalitis, meningoencephalomyelitis with/without systemic distribution	>1:1600		1				1					2	735, 779, 987, 1224, 2021
	1:400–1:1600									4	5	9	
	1:50–1:200											0	
	Total	0	1	0	0	0	1	0	1	4	5	11	
Multifocal disseminated neosporosis affecting internal organs	>1:1600	1	1									2	121, 239, 1080
	1:400–1:1600							1				1	
	1:50–1:200											0	
	Total	1	1	0	0	0	0	1	0	0	0	3	
Cutaneous neosporosis with/without systemic distribution	>1:1600								1	1	1	3	9, 238, 463, 529, 702, 822, 1111, 1133, 1268, 1329, 1593, 1631, 1937
	1:400–1:1600				1			1		2	1	5	
	1:50–1:200								1			1	
	Total	0	0	0	1	0	0	1	2	3	2	9	

Note: m = months; yr = years.

polyradiculoneuritis, and encephalomyelitis had only IFAT titers of 1:50 up to 1:200. Cut-offs of 1:800 as suggested in the literature for the diagnosis of canine neosporosis should not be applied. However, titers in congenitally infected dogs may increase overtime.[573,863] In a case of natural congenital infection, 4 pups had initial titers of 1:50 to 1:400 at day 49 or 50 after parturition; in contrast, titers ranging from 1:100 to 1:3200 were reported at days 53–182 after parturition in the same dogs.[573]

5.2.5.5 Antibodies in Dogs with Other Forms of Canine Neosporosis

In elder dogs suffering from *N. caninum*-associated encephalitis, meningoencephalomyelitis with/without systemic involvement, in dogs with multifocal disseminated neosporosis affecting internal organs, and in dogs with cutaneous neosporosis with/without systemic involvement, the reported IFAT titers ranged between 1:400 and 1:1600 in most cases (Tables 5.4 and 5.9).

5.2.5.6 Antibody Detection in CSF

Antibodies against *N. caninum* can be detected in CSF.[779,863,987,1225,1598] While the serum IFAT titers ranged from 1:400 to 1:5120 in the examined dogs, the titers in CSF only ranged from 1:40 to 1:1280 and the quotient IFAT titer (in serum)/IFAT titer (in CSF) ranged from 4 to 16. A low quotient may indicate a destroyed blood-brain barrier. The examination of CSF for antibodies is able to confirm findings in serum, but (due to the low analytical sensitivity) a negative outcome is not suitable to exclude an *N. caninum* infection.

5.2.5.7 Effect of chemotherapeutical Treatment on Antibody Levels

After treatment, antibody levels may drop. Several months after treatments with trimethoprim/ sulfonamide and clindamycin were discontinued and a puppy with neosporosis had partially improved, titers of initial 1:3200 had returned to 1:200.[1080] No effect on IFAT titer after treatment for 12 days with trimethoprim and sulfadiazine was observed in a pup with congenital neosporosis when examined 3–4 weeks later[1598] and also in another dog treated with clindamycin, antibody levels as determined by IFAT did not change.[1551] In contrast, in a litter of Labrador Retrievers with canine neosporosis, IFAT titers declined within 4 weeks after treatment with clindamycin.[602]

5.2.5.8 Avidity Maturation of a Specific IgG Response in Dogs

There is 1 study reporting a low avidity IgG response in dogs.[1012] In dogs that excreted oocysts after ingesting tissue cyst-infected meat, serum samples were tested 30–35 DPI. In these sera, low avidity IgG antibodies were observed in an *N. caninum* in-house ELISA.[1012] In contrast, naturally infected dogs, suspected to have a chronic infection, had higher avidity values in this test. Further studies are needed to confirm these findings using well-defined sera from experimentally infected dogs, which are followed over several months, and avidity-ELISA protocols need to be established for dog samples similar to those validated for cattle.[223]

5.2.6 Serological Reactions in Dogs after Excreting Oocysts

5.2.6.1 Reactions against Conventional Tachyzoite Antigen

Seropositivity is indicative of exposure to the parasite but not of *N. caninum* oocyst excretion. Most experimentally infected dogs (40/58) that excreted oocysts did not seroconvert (Table 5.7). These results were irrespective of the serological test, including the IB tests.[1798] This observation was also confirmed by testing sera from naturally infected dogs found excreting *N. caninum* oocysts.[1805] The lack of seropositivity suggests that only in some of these dogs extra-intestinal stages occurred during the intestinal development of *N. caninum* preceding oocyst development.

Another potential reason for seroconversion in some of these dogs might be ingestion of oocysts from own feces or feces of other dogs. However, infection of dogs via ingestion of *N. caninum* oocysts may not be efficient or successful. Four dogs were fed with *N. caninum* oocysts (1 dog 1×10^3, 1 dog 5×10^3, and 2 dogs 1×10^4). None of the dogs excreted *N. caninum* oocysts but 2 of the dogs (infected by 1×10^4 oocysts) seroconverted as shown by IFAT and immunoblot.[116] Neither parasite DNA nor parasite stages were observed in dogs necropsied 6 months after inoculation with oocysts.

In summary, the presence of tachyzoite or bradyzoite-specific antibodies in a dog only indicates that the animal became exposed to *N. caninum* but provides no evidence for oocyst shedding.

5.2.6.2 Serological Reactions against a 152 kDa Tachyzoite Antigen in Oocyst Shedding Dogs

Reactions against a 152 kDa tachyzoite antigen were recognized in the sera of dogs after they excreted oocysts following feeding with infected intermediate host tissues.[1797] The reactions were observed in dogs between days 35 and 447 after feeding with infected tissue but not prior to oocyst excretion. Recognition of the 152 kDa tachyzoite antigen may thus be a marker for dogs having excreted *N. caninum* oocysts in the past. The 152 kDa tachyzoite antigen was also recognized by those dogs which neither showed positive IFAT reactions nor recognized *N. caninum* immunodominant antigens in the immunoblot. Possibly, this 152 kDa is an *N. caninum* antigen not only present in the tachyzoite stage but also in stages involved in the asexual or sexual development of *N. caninum* in the intestine of the definitive host. When this technique was applied to analyze sera from 4 naturally *N. caninum* shedding dogs, all these sera reacted with the 152 kDa antigen.[1805]

5.2.7 Serological Assays

5.2.7.1 Available Types of Assays

The IFAT is still the most often used test for serological analyses in dogs (Table 5.1). Although other tests, such as NAT[573,1545] or ELISA have also been used or validated for dogs,[213,573,602,759,936, 1012,1323,1472,1608] these tests are currently not broadly applied for diagnosis. Banding patterns in immunoblot are similar to those reported for other animal species[211,213,936,1012,1797] and immunoblot tests were used to confirm results of other serological tests (e.g., Reference **1598**). The advantage of IFAT is that this test is easy to perform and the IFAT titers provides more relevant information on antibody levels than ELISA tests, in which usually only a single serum dilution is analysed.

Generally, all tests available and developed for cattle should also be applicable for the examination of dog sera (Table 2.10 in Chapter 2). Tests based on parasite lysate,[568,1012,1858,2126] and ISCOM-incorporated antigens[213] have been published. In addition, a dot ELISA employing parasite lysate antigens after sonication has been described.[1609] An ELISA using tachyzoites fixed by formaldehyde to ELISA plates was modified as a competitive ELISA (using a polyclonal rabbit anti-*N. caninum* antibody) and the test showed high sensitivity and specificity in the examination of dog reference sera.[1323]

An ELISA based on native, affinity-purified NcSRS2 showed optimal sensitivity and specificity (100% and 97.9%) when validated relative to IFAT and immunoblot and showed no cross-reactivity with sera of *L. infantum*-infected dogs.[936] NcSRS2 also seems to be a suitable antigen for the development of recombinant antigens for serological testing. Recombinant NcSRS2 expressed in baculovirus[1472] or in *Pichia pastoris*[1607] was successfully used to establish serological tests to detect *N. caninum* antibodies in canine sera.[1472,1607] Another surface antigen, that is, seems to be a suitable candidate for serological assays in dogs, too. Recombinant NcSAG1 was used to establish an RIT[1148] and ELISA.[905,1020]

A number of other recombinant antigens were used to develop assays for testing canine sera. Anti-NcPF (i.e., anti-*N. caninum* profilin) and anti-NcGRA7 antibodies were mainly detected at the acute stage in experimentally infected dogs, while anti-NcSAG1 antibodies were produced during both acute and chronic stages.[905] Recombinant truncated NcGRA2t[1020] or NcGRA7t[850] were also used.

A recombinant GRA6-based LAT was validated for dog serology.[759] For dog sera, the recombinant GRA6-based LAT showed excellent specificity but low sensitivity relative to IFAT.[759]

5.2.7.2 Commercial Assays

We are aware of only 1 commercialized *N. caninum* assay which is explicitly marketed for use in dogs (ID-VET *Neospora*). Other commercialized indirect ELISA tests (NEOSPORA CANINUM ELISA antibody KIT BIO K 192; IDEXX *Neospora* X2 [IDEXX Bov]) were also validated or used to test canine sera[401,755,759,1770] (Table 5.1). Several competitive ELISAs have been validated for use in dogs including a monoclonal antibody-based test,[176] which is commercially available[299,1070,1123] (Table 2.11 in Chapter 2).

5.3 THERAPY

Clinical responses in naturally infected dogs are summarized in Tables 5.10 and 5.11. Clindamycin, sulfadiazine, and pyrimethamine alone or in combination with other drugs have been administered to treat canine neosporosis. In adult dogs with acute paralysis from myositis, dysfunction is often more amenable to early treatment because scar contracture is less common. Dermatitis and myositis responded to therapy with clindamycin (Table 5.10). In neonates, clinical improvement did not occur in the presence of muscle contracture or rapidly advancing paralysis (Table 5.11). To reduce the chance of illness, all dogs in an affected litter should be treated as soon as the diagnosis is made in 1 littermate. Older (over 16 weeks) puppies and adult dogs responded better to treatment (Tables 5.10 and 5.11).

With neurologic involvement, trimethoprim–sulfonamide or pyrimethamine and sulfonamide are better treatment options because of more efficient penetration of the CNS. Clindamycin is effective in suppressing the replication and dissemination of tachyzoites but does not appear to be effective against encysted bradyzoites. Despite clinical improvement, treatment with drugs such as clindamycin does not eliminate the *N. caninum* infection from the body (Tables 5.10 and 5.11).

In a study from UK, data on the treatment of 16 dogs with neosporosis with clindamycin, pyrimethamine, and sulfadiazine, either alone or in combination were reported.[120] Five dogs made a full recovery, and good response to treatment was seen in another 5 dogs. Treatment was less effective in younger pups with neuromuscular signs.

5.4 PREVENTION

In dogs, *N. caninum* can be transmitted repeatedly through successive litters and litters of their progeny. This tendency should be considered when planning the breeding of bitches infected with *Neospora*. Dogs should not be fed uncooked meat, especially beef. On farms, dogs should not be allowed to feed on offal or aborted materials. They should be prevented, when possible, from defecating in feed troughs, watering sources, pastures, or livestock holding pens where cattle are housed. No vaccine has been developed to combat neosporosis. No drugs are known to prevent transplacental transmission.

Table 5.10 Attempted Treatment of Neosporosis in Isolated Cases of Dogs

Medicine	Dosage/Day	Duration	Age	Clinical Improvement	Evaluation	Reference
		Clindamycin Alone				
Clindamycin	5–7.5 mg/kg IM/PO	4 w	6 yr	Yes	Clinical, biopsy	809
	7.5 mg/kg, PO, twice	45 d	12 yr	Yes[a]	Clinical, necropsy	562
	12.5 mg/kg, twice	Several months	6 yr	Yes	Clinical	1093
	7.5 mg/kg PO	3 w	5 yr	Yes	Clinical	1631
	22 mg/kg, PO, twice	2 w	7 yr	No, died	Clinical, histology, isolation	779
	25 mg/kg, PO	6 w	7 yr	Yes	Clinical	239
	10 mg/kg, PO, twice	6 w	21 m	Yes	Clinical	1551
	12.5 mg/kg, twice	More than 3 months	4 yr	Yes	Clinical	1521
	13 mg/kg, PO, twice	9 d	7 yr	Yes	Clinical	920
	6–15 mg/kg, PO, twice	3 w/2.5 yr	13 m	Yes/no[b], dog lived 2.5 yr	Clinical	1329
	20 mg/kg, twice	4 w	12 yr	Yes	Clinical, biopsy	463
	11 mg/kg, PO, twice	4 w	10 yr	No	Clinical	1133
	6 mg/kg, PO, twice	28 d	10 yr	Yes/no[a], lesions appeared 40 d after treatment had been discontinued	Clinical, biopsy	1268
		Combined				
Sulfadiazine + Trimethoprim	60 mg/kg + 1 mg/kg, twice	4 w	12 w	Mild, pelvic limb extension unaffected	Clinical	863
Sulfadiazine + pyrimethamine	31 mg/kg + 6 mg/kg	2 w				
Clindamycin	13.5 mg/kg, PO, 3 times	4 w				
Trimethoprim + sulfonamide	15 mg/kg, twice	2 w	15 w	Yes	Clinical	1080
Clindamycin	10 mg/kg, 3 times	6 w				
Sulfadiazine + Trimethoprim	15 mg/kg + 3 mg/kg, twice	11 w	5 yr	Yes, but unable to walk euthanized	Clinical, PCR	1225
Pyrimethamine	1 mg/kg, once					
Clindamycin	20 mg/kg, PO, 3 times					

(Continued)

Table 5.10 (Continued) Attempted Treatment of Neosporosis in Isolated Cases of Dogs

Medicine	Dosage/Day	Duration	Age	Clinical Improvement	Evaluation	Reference
Clindamycin (alone or with other drugs)	16 dogs (see original paper)			Yes in 10, no in 6	Clinical	**120**
Clindamycin (alone or with other drugs)	27 dogs (see original paper)			Several recovered	Clinical	**657**
Trimethoprim + sulphadiazine	15 mg/kg, PO, twice	18 d	9 w	No, died	Clinical	**178**
Pyrimethamine	1 mg/kg, PO, twice					
Clindamycin	10 mg/kg, IM, twice					
Trimethoprim + sulphadiazine	6 dogs (see original paper)			Yes in 3, 1 totally healed, 2 died	Clinical, PCR	**735**
Clindamycin						
Clindamycin	11 mg/kg, PO, twice	72 d	1 m	Yes, dog lived for 2.5 yr	Clinical	**986**
Sulfa-trimethoprim	14 mg/kg, twice					
Clindamycin	20 mg/kg, twice	6 w	9.5 yr	Yes	Clinical	**1562**
Trimethoprim + sulfonamide	20 mg/kg + 15 mg/kg twice					
Clindamycin	20 mg/kg, twice	8 m		Yes	Clinical	
Trimethoprim + sulfonamide	15 mg/kg, twice	8 w	2.5 yr	Again, released symptoms because owner stopped treatment	Clinical	
Pyrimethamine	0.9 mg/kg, once	4 w				
Clindamycin	20 mg/kg, twice	3 w				
Trimethoprim + sulfonamide	15 mg/kg, twice	More than 5 m	5 m	Yes	Clinical	
Clindamycin	20 mg/kg, twice	4 m				
Trimethoprim + sulfonamide	15 mg/kg, twice	3 m	3 yr	Yes	Clinical	
Clindamycin	20 mg/kg, twice	5 m				
Other Drugs						
Sulfadiazine + Trimethoprim	15 mg/kg, twice	3 w	10 w	No	Clinical	**1209**
Pyrimethamine + sulfadoxine	1 mg/kg + 20 mg/kg	4 w	3 yr	No	Clinical	**1953**

Note: d = day, m = month, w = week, yr = year.

a After 2 weeks dermal nodules disappeared. Therapy was continued for 30 days. Nodules reappeared 30 days after cessation of therapy. Dermal nodules resolved after 21-day therapy, and the results were verified by necropsy examination.

b Dermal nodules disappeared after 3 weeks treatment. Two weeks after cessation of therapy, neurological signs developed. The dose of clindamycin was increased from 6 to 15 mg/kg and continued off/on for 2.5 years. Neurological deficit was greatly improved.

Table 5.11 Attempted Treatment of Neosporosis in Congenitally Infected Pups

Medicine	Dosage	Duration	No. of Pups	Age of Pup	Clinical Improvement	Evaluation	Reference
Trimethoprim + Sulphadiazine	30 mg/kg, once daily	3 d	1	4 w	No	Necropsy, viable *N. caninum* isolated	119
Trimethoprim + Sulphadiazine	Not given	12 d	1	6 w	No	Necropsy at 11 w; viable *N. caninum* isolated	1598
Sulfadiazine + Trimethoprim	30 mg/kg, once daily	14 d	1	3 w	Yes	Clinical	1636[a]
Clindamycin	11 mg/kg, twice daily						
Trimethoprim	120 mg, 2 times daily	14 d	1	13 w	No	Necropsy	1684
Sulphamethoxazole	600 mg, 2 times daily						
Trimethoprim + Sulphadiazine	15 mg/kg, 2 times daily	4 w in first pup, 1 w in the second pup	2	17 w	Yes, ataxia and paresis resolved in the first pup, and partial recovery in the second	Clinical	1298
Pyrimethamine	1 mg/kg, daily						
Trimethoprim + Sulphadiazine	20 mg/kg, sulfadiazine 100 mg, 2 times daily	14 d	1	7 w	Yes	Clinical	1325
Pyrimethamine	1 mg/kg, day						
Trimethoprim + Sulphadiazine	30 mg/kg daily	3	1	4 w	No	Necropsy, *N. caninum* isolated	119
Clindamycin	12.5–18.5 mg/kg, PO, 2 times daily	10 d	1 (studied to the end)	40 d	Stable (3 remaining pups)	Necropsy at 131 days, viable *N. caninum* isolated	573
	150 mg total, PO for 3 w, then 300 mg at 13 w	8 m continuous	4	10 w	Yes	Viable *N. caninum* isolated after 54 day treatment	608
	10 mg/kg, PO, 3 times daily	2 w	2	9 w	No, 2 were euthanized	Histology	988
	50 mg/kg, PO, 2 times daily	62 d	3	54 d	No	Viable *N. caninum* isolated from all 3 pups after 51–62 days treatment	598, 602
	50 mg/kg, PO, 2 times daily	62 d		54 d	No		
	50 mg/kg, injectable, 2 times daily	51 d		54 d	Yes		
	25 mg /kg, 2 times daily	1 w	1	6 w	No	Clinical	2021
		4 w	1	3 m	Yes		
	12 mg/kg, PO, 3 times daily	18 w	1	7 w	Yes	Clinical, dog lived for 9 yr	400

Note: d = day, m = month, w = week, yr = year.
[a] Personal communication, J. P. Dubey on May 3, 2016.

5.5 EXPERIMENTAL INFECTIONS

5.5.1 Using Dogs as Intermediate Hosts

Clinical neosporosis has been induced postnatally by using high doses of tachyzoites. Three of the 8 5-day-old pups inoculated SC with 10^6 tachyzoites of the NC1, NC2, and NC3 isolates, developed clinical neosporosis.[363] One pup had myalgia, muscle atrophy of the pelvic limbs, and persistent cough and N. caninum was isolated from its brain and muscle tissue when the pup was killed 27 DPI. Histologically, the pups had evidence of encephalitis, myositis, hepatitis, and interstitial pneumonia, and N. caninum tachyzoites were seen in lesions. Tissue cyst-like structures were seen in microscopic sections of the brain. One pup was laterally recumbent and was euthanized 37 DPI. Microscopic lesions were seen in the brain, heart, tongue, skeletal muscles, lung, and liver. A third pup died of neosporosis day 12 PI with inflammatory and necrotic lesions and N. caninum tachyzoites in the brain, heart, tongue, limb muscles, diaphragm, lung, and liver. The results demonstrate that postnatally infected pups can develop clinical neosporosis.

The pioneer experiment[528] regarding experimental transmission of N. caninum in dogs was performed as follows. N. caninum infection was diagnosed in 3–8-week-old naturally infected pups and the first viable isolate of the parasite (NC1) was isolated from 3 of them. A dog (no. 12) was SC inoculated with 10^5 culture-derived tachyzoites obtained from the tissues of 2 infected pups (nos. 3 and 4). This dog (no. 12) was given methylprednisolone acetate on 22 and 29 DPI. The dog remained clinically normal for 4 weeks, and then 2 days before death appeared depressed. Lesions were found in the lungs, liver, myocardium, muscles (including ocular muscles), brain, and spinal cord; parasites were specially evidenced causing the important necrosis in the liver. Other dogs inoculated with dog-derived infected tissues remained asymptomatic and no organisms or lesions were found in their tissues.

Experimentally, 3 studies described the transplacental transmission of N. caninum in pregnant bitches inoculated with the NC1 isolate.[315,363,533] In the first experiment,[533] a laboratory-bred Beagle was inoculated SC and IM with 1.5×10^6 tachyzoites.[533] The bitch remained clinically normal and delivered 8 full term pups, 28 days later. Pup no. 1 was born dead, pup no. 2 died 2 days later, and pups no. 3 and no. 4 were euthanized at 2 and 3 days of age because they were not nursing. Pup no. 5 was euthanized in a moribund condition on day 20. N. caninum was isolated in cell cultures inoculated with tissues of each of the 5 pups. Histologically, although all 5 pups had lesions, encephalitis and myocarditis in association with tachyzoites were verified only in pup no. 5. Pups no. 6–8 and the bitch remained healthy. However, they had pneumonia and myositis following the administration of large doses of corticosteroids, indicating they were subclinically infected.[533] N. caninum tachyzoites were seen in tissues of pups no. 6–8 and the bitch, indicating that under certain circumstances subclinical neosporosis can be reactivated.

In the second study,[363] 6 mixed breed bitches were inoculated SC with 5×10^6 tachyzoites of NC1 strain at day 21 of pregnancy.[363] Four bitches had only macerated, mummified, or resorbed fetuses and it was not possible to demonstrate N. caninum in fetal tissues; 1 of these bitches died of disseminated neosporosis 17 DPI. One bitch gave birth to 3 live full term pups, with mild neurologic deficits but N. caninum was not demonstrated in the tissues of pups killed 9–13 weeks after birth. The last bitch was killed 18 days after N. caninum inoculation because it had ultrasonic evidence of fetal death. There were 12 fetuses or sites of fetal attachment. Three fetal sites were not investigated because the fetuses were resorbed. N. caninum was found in the tissues of the 9 remaining fetuses, 2 of which were alive at the time of necropsy. There were no inflammatory lesions although numerous tachyzoites were seen in fetal tissues. Many tachyzoites were degenerating along with fetal tissues. Necrosis and N. caninum tachyzoites were found in maternal and fetal placenta.[564]

In the third study,[315] 6 bitches in 2 groups were inoculated with a very high dose (10^8 tachyzoites). In group I, 3 bitches were inoculated during the third week of gestation, and in group II,

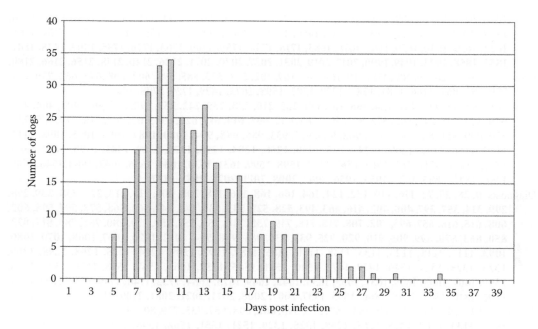

Figure 5.12   Summary of data from the literature showing the excretion of *N. caninum* oocysts by 52 experimentally infected dogs reported in References **116, 314, 493, 715, 780, 1180, 1187, 1309, 1579,
and 1798**. Excretion had started in most dogs by day 7 and ended 14 DPI.

3 bitches were inoculated at the sixth week of gestation. The bitches were allowed to whelp
naturally. Dams and their pups were examined by IHC, serology, and by PCR. In group I, 6 of
the 10 pups died within 48 h of birth. In group II, 7 of the 13 pups died between 5 and 10 days
of birth. *N. caninum* DNA was detected by nested PCR in 2 pups (hearts of both and liver of
1) from group I, and 1 pup (CNS and lymph node) from group II. The dams and the pups that
survived were clinically normal. *N. caninum* was not demonstrable in the tissues of any of the
pups and their dams.

5.5.2 Using Dogs as Definitive Hosts

This topic was discussed in Chapter 2, experiments were summarized in Tables 5.7 and 5.8;
some of the dogs fed infected tissues excreted oocysts but none developed clinical signs. In most
cases, the production of oocysts was low (Table 5.7) or absent.[195] The prepatent period was generally 7 days and patency was short. Graphical summary of the relation oocysts excreted/DPI of 52
experimentally infected dogs available in the literature is shown in Figure 5.12. In a recently published study (not included in Figure 5.12), oocyst shedding started at 7 DPI and ended at 21 DPI.[1510a]

BIBLIOGRAPHY

Sources of Infection and Transmission: **116, 124, 195, 196, 205, 314, 362, 493, 497, 529, 533, 541, 564, 602,
689, 715, 780, 785, 1067, 1187, 1309, 1579, 1797, 1849, 1993.**
Serologic Prevalence: **9, 20, 71, 72, 81, 90, 104, 105, 114, 122, 123, 165, 191, 192, 212, 213, 229, 245, 248,
265, 289, 290, 298, 299, 332, 369, 387, 390, 396, 398, 401, 410, 415, 426, 440, 457, 460, 464, 466, 468,
479, 489, 520, 606, 609, 610, 617, 667, 679, 684, 700, 738, 745, 748, 755, 756, 757, 768, 773, 803, 808,
817, 827, 837, 850, 855, 907, 925, 931, 936, 937, 949, 954, 960, 1008, 1012, 1028, 1029, 1052, 1063,**

1070, 1073, 1095, 1099, 1103, 1109, 1118, 1119, 1123, 1126, 1165, 1211, 1212, 1248, 1259, 1266, 1336, 1338, 1351, 1356, 1357, 1365, 1367, 1456, 1461, 1486, 1518, 1526, 1552, 1570, 1573, 1574, 1608, 1622, 1629, 1630, 1659, 1667, 1676, 1683, 1718, 1734, 1750, 1760, 1763, 1770, 1785, 1809, 1838, 1841, 1853, 1858, 1923, 1949, 1990, 2017, 2019, 2031, 2037, 2070, 2071, 2126, 2140, 2148, 2156, 2166, 2180.

Parasite Isolation: 28, 29, 106, 118, 119, 164, 167, 169, 402, 528, 573, 585, 598, 602, 608, 642, 662, 770, 771, 779, 782, 863, 1166, 1281, 1284, 1329, 1502, 1598, 1673, 1679, 1729, 1798, 1805.

Clinical Infections: 118, 121, 124, 166, 167, 178, 205, 210, 238, 239, 242, 253, 292, 357, 400, 402, 404, 463, 528, 529, 541, 547, 562, 573, 597, 598, 602, 606, 608, 614, 617, 622, 693, 702, 708, 718, 719, 735, 736, 779, 809, 811, 822, 863, 870, 908, 920, 927, 933, 986, 988, 994, 1080, 1084, 1093, 1095, 1098, 1111, 1133, 1166, 1192, 1209, 1224, 1225, 1250, 1268, 1284, 1298, 1325, 1329, 1346, 1398, 1429, 1500, 1521, 1544, 1551, 1562, 1563, 1567, 1593, 1598, 1599, 1631, 1633, 1634, 1636, 1642, 1684, 1748, 1749, 1756, 1842, 1895, 1937, 1953, 1956, 2007, 2008, 2011, 2021, 2081, 2117.

Diagnosis: 9, 28, 29, 72, 116, 119, 122, 124, 164, 166, 168, 176, 178, 208, 209, 211, 213, 223, 238, 239, 298, 299, 314, 357, 397, 400, 402, 416, 463, 493, 528, 529, 533, 539, 541, 547, 562, 568, 573, 597, 598, 602, 608, 615, 616, 657, 693, 702, 708, 715, 718, 719, 735, 748, 753, 755, 759, 779, 780, 781, 785, 811, 822, 850, 863, 870, 899, 905, 910, 920, 925, 933, 936, 986, 988, 994, 1012, 1020, 1067, 1068, 1070, 1080, 1093, 1111, 1118, 1123, 1133, 1148, 1180, 1185, 1187, 1192, 1209, 1224, 1225, 1268, 1298, 1309, 1323, 1325, 1329, 1346, 1429, 1472, 1500, 1521, 1545, 1552, 1563, 1567, 1578, 1579, 1593, 1598, 1599, 1607, 1608, 1609, 1631, 1633, 1636, 1642, 1649, 1673, 1684, 1722, 1748, 1749, 1770, 1797, 1798, 1805, 1810, 1842, 1858, 1866, 1895, 1937, 2007, 2008, 2011, 2021, 2081, 2109, 2116, 2126, 2142.

Therapy: 119, 120, 178, 239, 400, 463, 562, 573, 598, 602, 608, 657, 735, 779, 809, 863, 920, 986, 988, 1080, 1093, 1133, 1209, 1225, 1268, 1298, 1325, 1329, 1521, 1551, 1562, 1598, 1631, 1636, 1684, 1953, 2021.

Experimental Infections: 195, 315, 363, 528, 533, 564.

Additional References: 210, 236, 417, 447, 531, 537, 543, 548, 549, 553, 555, 561, 616, 617, 999, 1098, 1184, 1239, 1320, 1321, 1690, 1845, 1894, 1917, 1956, 1988, 2103, 2130, 2168.

CHAPTER 6

Neosporosis in Sheep

6.1 NATURAL INFECTIONS

6.1.1 Serologic and DNA Prevalence

Data are summarized in Table 6.1. Seroprevalence varied from 0% to 64% depending on the serological test and cut-off employed, age and type of sheep, and gender and management of animals. Conclusive data are lacking with relation to age and seropositivity to access prenatal and postnatal transmission. None of the studies reported sequential sampling of sheep and different age groups from a single farm. When bulk milk from 613 dairy sheep herds in Sardinia, Italy, was investigated, 44.2% of the herds were positive.[1934a]

In 1 study from Tehran, Iran, *N. caninum* DNA was detected in 1 of 150 brains and 120 of 180 hearts of sheep from an abattoir.[81a]

6.1.2 Clinical Infections

The economic, clinical, and epidemiologic importance of *N. caninum* infection in sheep remains uncertain. In one of the largest investigations of protozoal abortion in a diagnostic center in England, there was no evidence of congenital *N. caninum* infection in sheep.[1539] In this investigation, antibodies to *N. caninum* were not found in the pleural fluid of 179 fetuses and 141 lambs. *N. caninum* was also not identified in tissues from 281 fetuses/lambs by IHC.[1539] Contrary, to this report *N. caninum* was identified in some sheep flocks, including flocks from Italy, New Zealand, and Spain.[793,946,1293,1405]

Occasionally, *N. caninum* can cause abortion, birth of weak lambs, neonatal mortality, and even clinical signs in adult sheep. Available evidence is summarized in Table 6.2.

Although seroprevalence is generally higher in sheep that aborted versus sheep with normal pregnancy, the causal relationship needs further investigation.[945,946,1891,2083] In 1 study the lambing rate in ewes seropositive to *N. caninum* was significantly lower than in seronegative ewes.[793]

6.1.2.1 Histologically Confirmed Abortion and Stillbirth

There are 4 reports of histologically confirmed congenital neosporosis in neonatal sheep (Table 6.2); in the brain of an aborted fetus and a stillborn lamb,[793] in a ewe and 2 surgically delivered twin fetuses,[1083] and a lamb from a flock in England.[545,854] The lamb from England was born weak and had nonsuppurative encephalomyelitis with intact and degenerating tissue cysts (Figure 6.1).

Table 6.1 Prevalence of Antibodies to *N. caninum* in Sheep

Country	Region[a]	No. Examined	Type of Animals	No. Positive	% Positive	Assay	Cut-Off Titer IFAT or Test[b]	Remarks	Reference
Argentina	Pampa	704	6 dairy farms, dairy	21	3.0	IFAT	1:50	Also tested for *T. gondii*	865
Australia	New South Wales	232	5 farms, meat	5	2.2	cELISA	VMRD	1 farm had clinical neosporosis	203
Brazil	Alagoas	343	26 farms	33	9.6	IFAT	1:50	Adult sheep sampled. Water supply is a risk factor	661
	Federal District	1028	321 flocks	90	8.7	IFAT	1:50	Titers up to 51,200. Also tested for *T. gondii*	2006
	Maranhão	64	5 farms	3	4.7	IFAT	1:25	Food supplementation, reproductive problems. Also tested for *T. gondii*	1394
	Mato Grosso do Sul	441	1 farm, meat	136 141	30.8 32.0	IFAT ELISA	1:50 IH, rNcSRS2	Sensitivity (98.6%) and specificity (98.3%) of ELISA were excellent compared with IFAT	73
	Minas Gerais	488	63 farms	64	13.1	IFAT	1:50	No age difference, 6–36-month-olds surveyed	68
	Minas Gerais	155	2 farms	73 41	47.1 26.4	IFAT ELISA	1:64 IH	No relation with age. Also tested for *T. gondii*	1746
	Minas Gerais	334	12 flocks	27	8.1	IFAT	1:50	Herds with abortion problems had significantly more positive animals. No relation with age	1761
	Paraná	305	9 farms	29	9.5	IFAT	1:50	Gender, age, breed with no effect. Also tested for *T. gondii*	1734
	Pernambuco	179	Rural villages	39	21.8	IFAT	1:50	*T. gondii* antibodies surveyed	85a
	Pernambuco	81	23 farms	52	64.2	IFAT	1:50	Prevalence increase with age	1950
	Piauí	153	Rural villages	8	5.2	IFAT	1:50	*T. gondii* antibodies surveyed	85a
	Rio Grande do Norte	409	35 farms	7	1.8	IFAT	1:50	No relation with gender. Also tested for *T. gondii*	1869
	Rio Grande do Sul	110	Adult	37 39	33.6 NS	ELISA IFAT	IH, rNcSRS2 1:50	—	1607, 1608

(Continued)

Table 6.1 (Continued) Prevalence of Antibodies to N. caninum in Sheep

Country	Region[a]	No. Examined	Type of Animals	No. Positive	% Positive	Assay	Cut-Off Titer IFAT or Test[b]	Remarks	Reference
	Rio Grande do Sul	62	4 municipalities	2	3.2	ELISA	IDEXX	–	2048
	Rondônia	141	15 farms	41	29.0	IFAT	1:50	Titers up to 1:25,600	19
	São Paulo	597	30 farms, meat breeds	55	9.2	IFAT	1:50	No age association. 3.5% coinfected with T. gondii	683
	São Paulo	382	8 farms	49	12.8	IFAT	1:25	0%–24.2% prevalence in different farms. Also tested for T. gondii	1117
	São Paulo	1497	16 farms	120	8.0	IFAT	1:25	Water supply, domestic canids, presence of reproductive problems as risk factors	1248
	São Paulo, Rio Grande do Sul	596	Abattoir, meat	353	59.2	IFAT	1:25	Higher in females (63.0%) than rams (53.8%). Extensity of breeding as risk factor	1547
China	Tocantins	182	8 farms	25	13.7	IFAT	1:40		816
	Qinghai	600	Farms in 8 regions	62	10.3	ELISA	IDEXX	Herd size, hygiene. Also tested for T. gondii	1197
Czech Republic	Central Bohemian, U'stí nad Labem	547	9 farms	63	12.0	cELISA	VMRD	Mixed infection with T. gondii in 10%	156
Greece	NS	458	50 farms, dairy	77	16.8	ELISA	IH	Also tested for T. gondii	490
Grenada, West Indies	Grenada and Carriacou	138	7 parishes	18	13.0	ELISA	IDvet	Higher in males (11.1%), than females (3.2%). No statistically significant difference	1839
Iran	Hamedan	358	Aborted ewes	8	2.2	ELISA	CHEKIT	–	761
Italy	Orobie Alps	1010	3 valleys	22	2.2	ELISA	CHEKIT	Also tested for other abortifacients, including T. gondii	716

(Continued)

Table 6.1 (Continued) Prevalence of Antibodies to N. caninum in Sheep

Country	Region[a]	No. Examined	Type of Animals	No. Positive	% Positive	Assay	Cut-Off Titer IFAT or Test[b]	Remarks	Reference
Italy	Lombardy	428	Bergamo, Milan	83	19.3	ELISA, confirmed by WB	In-house	Lower prevalence in sheep from higher altitudes	**744**
Jordan	North	339	62 flocks	213	63.0	ELISA	iELISA-Bio-X	Dogs as risk factor	**5**
	South, Tafelah, Ma'an	320	38 flocks	14	4.3	ELISA	CHEKIT	Dogs and flock size as risk factors	**26**
New Zealand	Southern North Island	67	3 farms	19	28.0	IFAT	1:100	Abortions on the farms	**945**
	Throughout	504	21 farms	–	–	IFAT	1:100	Also used ELISA (IDEXX) and PCR (ITS1). Using blood samples, N. caninum DNA was detected in 6.9% aborting/nonpregnant, 5.0% of pregnant, 2.1% of high fertility ewes	**946**
		220 aborting or not ewes		80	36.4				
		188 pregnant ewes		86	45.7				
		96 ewes with high fertility		20	20.8				
	NS	640 rams	64 farms	4	0.6	IFAT, ELISA	1:50, IDEXX	–	**1692**
	NS	26 pregnant, 16 aborting, 5 fetuses	Flock A	2 pregnant; 9 aborting, 4 fetuses	8.0 pregnant, 56.0 aborting, 80.0 fetuses	IFAT	1:100	–	**2083**
		43 pregnant, 41 aborting	Flock B	2 pregnant, 17 aborting	3.0 pregnant, 41.0 aborting				
Pakistan	Punjab, Azad Kashmir	128	Different origins	35	27.7	cELISA	VMRD	Higher in females (30.8%) than males (23.2%). No statistically significant difference.	**1453**
Philippines	Luzon	38	2 farms	10	26.3	ELISA	In-house	Abortions in cattle present in same farms	**1089**
Slovakia	Košice and Prešov	382	Aborting, dairy	14	3.7	ELISA	IDvet	Also tested for other abortifacients, including *T. gondii*	**1890, 1891**

(Continued)

Table 6.1 (Continued) Prevalence of Antibodies to N. caninum in Sheep

Country	Region[a]	No. Examined	Type of Animals	No. Positive	% Positive	Assay	Cut-Off Titer IFAT or Test[b]	Remarks	Reference
Spain	Galicia	177	Several farms, meat crossbreeds	18	10.1	cELISA	VMRD	Also sampled for *T. gondii*. Lower in mountain area (6.3%) than coast (11.1%)	1550
	Galicia	2400	44 farms, meat	132	5.5	ELISA	IDvet	Seropositivity associated with age. Other abortifacients surveyed, including *T. gondii*	491
	Castilla-La Mancha	180	Dairy farms	7	3.9	cELISA	VMRD	–	1751
	Extremadura, Andalusia	209	Culled ewes, 12 farms, crossbred	4	1.9	ELISA	IDvet	Other abortifacients surveyed including *T. gondii*	97
Switzerland	Zurich	117	Flock with abortions	12	10.3	IFAT	1:160	Persistent abortion problems. Also tested for *T. gondii*. *N. caninum* detected by PCR in 4 out of 20 aborted fetuses	861
Turkey	Kars province	376	5 flocks	8	2.13	ELISA	CHEKIT	Highest prevalence in Tuj breed	772
	Karaman, Konya, Zonguldak	610	NS	13	2.1	ELISA	IH-rNCSAG1	–	2180
UK	England and Wales	660 (abortion)	Sent by practicing veterinary surgeons	3	0.45	IFAT	1:50	Aborted sheep. Also tested for *T. gondii*	875

[a] NS = not stated.

[b] ELISA = Enzyme-linked immunosorbent assay; iELISA-Bio-X = BIO K 192 NEOSPORA CANINUM ELISA antibody KIT, BIO-X Diagnostics, Belgium; IDvet = Neospora caninum Indirect Multispecies, ID.vet, France; CHEKIT = CHEKIT Neospora, indirect ELISA, detergent lysate of tachyzoites, IDEXX Laboratories, The Netherlands; IDEXX = IDEXX HerdChek Neospora caninum antibody, indirect ELISA, sonicate lysate of tachyzoites, IDEXX Laboratories, USA; VMRD = Neospora caninum cELISA Competitive ELISA GP65 surface antigen of tachyzoites VMRD, USA.

Table 6.2 Clinical Neosporosis in Sheep

Country[a]	History	Evidence	Reference
Australia	1 adult ewe with neurological signs suspected to be TSE.	*N. caninum* detected by TEM. Histologically, meningoencephalitis, particularly in the mid brain. *N. caninum* was demonstrable by IHC and the parasite DNA was detected by PCR from the paraffin block	203
England	Lamb born alive with human assistance but died a week later in spite of hand rearing. At necropsy, the thoracic spinal cord narrowed for 2 cm. Lesions were confined to brain and the spinal cord and consisted of nonsuppurative encephalomyelitis. Intact and degenerating tissue cysts were present; tachyzoites were not seen	IHC and TEM examinations	545, 854
England and Wales	281 aborted fetuses tested for abortifacients	*N. caninum* was not identified in tissues of any fetus by IHC, and antibodies to *N. caninum* not found in fetal fluids of 330 fetuses and lambs	1539
England	Brains of 74 aborted fetuses	*N. caninum* DNA found by PCR in 14 fetuses. Histological examination not reported	953
Iran	109 aborted fetal brains tested for *Neospora*	*N. caninum* DNA identified in 1 brain by semi-nested PCR	1782
Iran	Brain, liver, gastric contents, placenta of 70 aborted fetuses	*N. caninum* DNA detected in tissues of 6 aborted fetuses	91
Italy	292 fetuses tested by PCR	*N. caninum* detected in tissues of 6 fetuses	1293
Japan	Surgically delivered twin fetuses at 119 days of gestation. Dam was used in a surgical experiment and died of metritis	*N. caninum* tissue cysts and focal gliosis with mononuclear cells cuffing were detected in the ewe and in both fetuses	1083
New Zealand	a. 61 brains from aborted fetuses + 77 fetal brains from uteri of aborted ewes = total 118 b. 26 fetal brains examined by histology	a. *N. caninum* DNA detected in 23 fetal brains b. 11 fetuses had inflammatory lesions but *Neospora* was not found	945, 946
New Zealand	Higher seropositivity in aborting ewes than with normal pregnancy	*N. caninum* antibodies in fetal fluid of 1 of 7 fetuses	2083
Spain	Lambing rate lower in seropositive ewes versus seronegative ewes. Necropsy performed on 4 fetuses, 15 stillborn lambs, and 2 lambs born with neurological signs	a. *N. caninum* DNA found in brains of 13 of 14 fetuses from seropositive ewes b. By HE staining lesions seen in 4 fetuses/lambs. Tissue cysts identified in the brain of 1 lamb c. By immunostaining, tachyzoites identified in brains of all 4 PCR-positive brains	793
Spain	Brains from 74 fetuses tested	In summary, lesions and *N. caninum* were identified in brains of 1 fetus and stillborn both by PCR and immunohistochemistry *N. caninum* found by PCR in brains of 5, 3 of which had lesions. IHC not performed	1405
Switzerland[a]	20 aborted fetuses from a flock of 117 ewes tested	*N. caninum* DNA found in brains of 4 fetuses; lesions and *N. caninum* found histologically in the brain of 1 of these fetuses	861

[a] 86 fetuses were investigated by histology and IHC in Switzerland, but none had protozoal-associated lesions.[327]

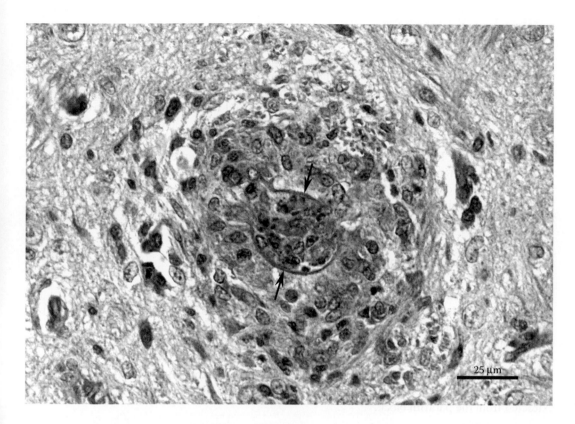

Figure 6.1 Glial nodule around a ruptured *N. caninum* tissue cyst. Inflammatory cells are inside the interior of the cyst. Note tissue cyst wall (arrows). IHC using polyclonal *N. caninum* antibodies. (Adapted from Dubey, J. P. et al. 1990. *J. Parasitol. 76*, 127–130.)

In addition, protozoal-associated lesions were seen in 10.8% of 74 ovine abortions in which previously *N. caninum* DNA had been previously detected.[1405]

6.1.2.2 Detection of N. caninum DNA in Aborted Fetuses

N. caninum DNA has been detected in the tissues of aborted fetuses (Table 6.2) but a causal relationship cannot be established without histological examination and a thorough exclusion of other causes of abortion.[327,1293] Available data are summarized in Table 6.2.

6.1.2.3 Clinical and Subclinical Neosporosis in Adult Sheep

6.1.2.3.1 Isolation of Viable N. caninum from a Clinically Normal Adult Ewe

A dog fed a brain homogenate of a 4-month-old sheep in Brazil excreted *N. caninum* oocysts 10–16 days later.[1579] After sporulation, these oocysts were orally infective to gerbils (*M. unguiculatus*). Thick-walled tissue cysts were found in both gerbils euthanized in good health 2 months PI. *N. caninum* DNA was detected in the brain homogenate of the donor ewe, the dog, as well as from oocysts shed by the dog. The isolate was not named.

6.1.2.3.2 Isolation of Viable N. caninum from a Pregnant Ewe

N. caninum was demonstrated histologically and isolated from the brain of a 5-year-old pregnant Suffolk ewe in Japan.[1083,1097] The ewe was pregnant with twin fetuses and had been used in an unrelated experiment involving fetal hypertension. Laparotomy had been performed at day 118 of gestation and both fetuses were removed and fixed for histological examination. Both fetuses had encephalitis, and *N. caninum* tissue cysts were demonstrable in the brain of 1 fetus.

The ewe had an IFAT *N. caninum* titer of 1:256. She became depressed and died of metritis 1 month after laparotomy. Histologically, the ewe had encephalitis and tissue cysts were identified; parasite DNA was detected by PCR.[1083] Viable *N. caninum* was isolated in immunodeficient mice inoculated with brain homogenate of the ewe; the isolate was further propagated successfully in cell culture.[1097] The isolate was not named.

6.1.2.3.3 Clinical Neosporosis in an Adult Sheep

The cases are summarized in Table 6.2.

A 3-year-old Merino ewe was found to have neurological signs (limb paddling, recumbency).[203] The ewe was euthanized and the tissues were fixed in formalin because of a suspicion of transmissible spongiform encephalopathy (TSE). There was no evidence of TSE by TEM and histological examinations. However, an apicomplexan parasite was detected by TEM; and no other details were provided in the paper. Histologically, there was meningoencephalitis, particularly in the mid brain. *N. caninum* was demonstrated by IHC and the parasite DNA was detected by PCR in DNA extracted from the paraffin block.[203]

6.2 EXPERIMENTAL INFECTIONS

Sheep can be experimentally infected using different routes, strains, and doses of *N. caninum* (Tables 6.3 and 6.4), and they represent an excellent ruminant model for bovine neosporosis. Additionally, rams were successfully infected with *N. caninum*[1929]; ewes mated with infected rams did not seroconvert to *N. caninum*.

The following are the important conclusions from the studies summarized in Table 6.3.

1. The main clinical signs are in fetuses and newborns.
2. The dose can affect the outcome of pregnancy.
3. Clinical effects are mainly related to the gestational period; infection of ewes during early pregnancy results in fetal death, abortion, resorption, and mummification. Inoculation during late gestation results in the birth of weak lambs or normal lambs.
4. Lambs inoculated after birth remained asymptomatic.

6.3 IMMUNITY, PROTECTION, AND VACCINATION

Sheep develop good humoral and cellular responses to *N. caninum* in experimental and natural infections using a variety of tests and antigens (Tables 6.1 through 6.4; and References **73, 88, 259, 260, 1608, 1708,** and **1927.**)

Similar to cattle, chronically infected sheep can abort from neosporosis due to exogenous infection or due to endogenous transplacental transmission of *N. caninum* to the fetus during second

Table 6.3 Experimental Neosporosis in Sheep

Experiment no.	Type	Strain, Stage[a], Dose, Route[a]	No.	Euth. DPI	Main Findings	Reference
1	1-week old	NC1, 1.5×10^6, IM or IV or SC	3	49	Seroconverted 15 DPI, remained healthy	568
2	Pregnant, 3 months	NC1, 1.5×10^7, IV or IM	2[b]	31	Both aborted 2 lambs each, 25 and 26 DPI. Tachyzoites found in lesions of encephalitis of all 4 fetuses. Myositis and placentitis with demonstrable tachyzoites	538, 568
3	Pregnant, day 65	NC2 and Nc-Liv,	12	ND	All 12 aborted, 36–69 DPI, N. caninum in brain of 7 fetuses	1308
	Pregnant, day 90	1.7×10^5 or 10^6, IV	12	ND	8 aborted, 36–61 DPI; 4 gave birth to weak lambs, N. caninum in brain of 5 fetuses	
	Pregnant, day 120		12	ND	15 clinically normal lambs born, N. caninum in brain of 6 lambs	
4	Nonpregnant	NC1, 2.5×10^6, SC	6	ND	3 ewes developed fever, humoral and cellular responses defined	1708
5	Pregnant, day 90[b]	NcNZ1-3, 10^8, IV	10	ND	All 10 aborted, CNS lesions, 7 PCR-positive	2084, 2086
		NcNZ1-3, 10^6, IV	10	ND	All 10 aborted, CNS lesions, 6 PCR-positive	
		NcNZ1-3, 5×10^3, IV	10	ND	5 aborted, 2 premature, 3 live lambs CNS lesions in 7, 8 PCR-positive	
		NcNZ1-3, 50, IV	10	ND	All delivered live lambs, 1 CNS lesion, all PCR-negative	
6	Pregnant, day 40	Nc-Spain7, 10^6, IV	6	ND	All 11 fetuses died, 19–21 DPI, N. caninum detected by PCR in brain of all 11 fetuses, lesions placenta, and fetal livers and brains	87, 88
	Pregnant, day 90		7	ND	All 12 fetuses died, 34–48 DPI, N. caninum detected by PCR in brain of 10 of 12 fetuses, lesions placenta, and fetal livers and brains	
	Pregnant, day 120		7	ND	3 live lambs premature, recumbent lambs born on day 145 of gestation, 6 lambs born clinically normal N. caninum detected by PCR in brains of 6 of 9 lambs	
7	Pregnant, day 45	NC1, 5×10^5, SC	8	ND	Total 10 fetuses, 6 resorbed, 4 aborted	259
	Pregnant, day 65		8	ND	Total 11 fetuses, 1 resorbed, 9 aborted, 1 born alive	
	Pregnant, day 90		8	ND	Total 11 fetuses, 7 aborted, 2 stillborn, 2 born alive, PCR-positive brain in 3 lambs	
8	Pregnant, day 90	Nc-Liv, 10^6, SC	4	25	5 fetuses live, CNS lesions in all	257
			4	40	1 mummified, 4 live, CNS lesions in 4	
			4	53	6 live fetuses, CNS lesions in all	
	Castrated male	Nc-Liv, 10^8, SC	6	ND	Fever 4–9 DPI	
		Nc-Liv, 10^6, SC	5	ND	Fever 6–9 DPI	
		Nc-Liv, 10^4, SC	5	ND	Fever 9 DPI	
9	Nonpregnant ewes	NC2, 10^4 oocysts, orally[c]	6	7 weeks	Serologic responses evaluated. N. caninum DNA detected in blood of ewes as early as 6 DPI and continued until 49 DPI. DNA, but no lesions, found in brains of all ewes	1493

Note: ND = not done.

a In all other experiments tachyzoites were inoculated parenterally.

b Complete study included 6 ewes infected with 1.5×10^7 NC1 tachyzoites IM or IV; all animals seroconverted and presented high titers at 31 DPI.[568]

c Vaccinated with commercial *T. gondii* live vaccine.

Table 6.4 Protective Immunity to Neosporosis in Sheep

Experiment No.	Exposure	Time	No. of Ewes	Challenge	Outcome — Ewes	No. of Fetuses/Lambs — Lesions	No. of Fetuses/Lambs — Tachyzoites	No. of Fetuses/Lambs — Other Assays	Reference
1	Experimentally infected before pregnancy	1 year before present pregnancy	4	None	2 ewes aborted, 1 weak lamb that died, 1 normal lamb	4	3	–	1021
	Experimentally infected before pregnancy	1 year before present pregnancy	5	Day 67 of gestation, 1.7×10^5 tachyzoites, Nc-Liv and NC2, IV	3 ewes aborted, 2 produced normal lambs	5	2	–	
	Not infected	–	2	Day 67 of gestation, 1.7×10^5 tachyzoites, Nc-Liv and NC2, IV	Both gave birth to weak lambs that died	2	0	–	
2	Experimentally infected with 5×10^6 tachyzoites 2 months before pregnancy	1 year earlier	9	10^7 tachyzoites of NcNZ1 strain, day 120 of gestation on year 2	13 infected lambs, 1 died after birth	6	–	*N. caninum* DNA in 0	1928
			7	None	7 of 11 infected lambs	2	–	*N. caninum* DNA in 0	
3	Experimentally infected with 10^7 tachyzoites	3 weeks before pregnancy	12	10^7 tachyzoites NC1, day 90 of gestation, SC	9 dead, 16 live lambs	6	–	*N. caninum* DNA in 2	260
	Not vaccinated		12	None	1 dead by dystocia, 16 live lambs	Not stated	–	*N. caninum* DNA in 1	
		–	12	10^7 tachyzoites NC1, day 90 of gestation, SC	All 14 dead	10	–	*N. caninum* DNA in 5	
4	Experimentally infected with 10^5 tachyzoites	Twice on 21 and 9 months before pregnancy	14	None	All 23 live lambs				
	Vaccinated with killed antigen before pregnancy	Twice (before and after pregnancy-mating), last dose 14 days before challenge	20	Day 140 of gestation, 5×10^6 tachyzoites of Illinois strain, SC	19 ewes gave birth to 21 live lambs, 1 aborted, 5 stillborn, 4 weak that died	6	4	All 33 lambs were seropositive. *N caninum* DNA in 0	1494
	Not vaccinated	–	20		20 ewes gave birth to 35 live lambs, 3 stillborn, 7 weak that died	5	3	18 out of 24 lambs were seropositive. *N caninum* DNA in 9	
5	Vaccinated with killed antigen[a]	–	14	Day 90 of gestation, 5×10^6 tachyzoites Illinois strain, SC	21 live lambs from 12 ewes	–	–	15 of 21 lambs infected[b]	1006
	Not vaccinated	–	11		7 live lambs from 5 ewes	–	–	5 of 7 lambs infected[b]	

a Killed antigen of NC1 strain in adjuvant.
b Presuckling antibodies.

pregnancy (Table 6.4). However, the frequency of repeat abortion through natural infections is unknown.

In one report immunization of ewes with killed *N. caninum* antigens, mixed with adjuvants, elicited protection in ewes against fetal loss but not against congenital infection.[1006] It is also known that immunization with *T. gondii* (using commercial vaccine) does not confer protection against abortion after a challenge with 10^7 NC1 *N. caninum* tachyzoites.[977]

BIBLIOGRAPHY

Serologic Prevalence: **5, 19, 26, 68, 73, 81a, 97, 156, 203, 490, 491, 661, 683, 716, 744, 761, 772, 816, 861, 865, 875, 945, 946, 1089, 1117, 1197, 1248, 1394, 1453, 1547, 1550, 1607, 1608, 1692, 1734, 1746, 1751, 1761, 1839, 1869, 1890, 1891, 1950, 2006, 2048, 2083, 2180.**

Parasite Isolation and Clinical Infections: **91, 203, 327, 545, 793, 854, 861, 945, 946, 953, 1083, 1097, 1293, 1405, 1539, 1579, 1782, 1891, 2083.**

Experimental Infections: **87, 88, 257, 259, 260, 538, 568, 1006, 1021, 1308, 1493, 1494, 1708, 1928, 1929, 2084, 2086.**

Immunity, Protection, and Vaccination: **73, 88, 259, 260, 977, 1006, 1608, 1708, 1927.**

Additional References: **63, 188, 258, 649, 944, 258.**

CHAPTER 7

Neosporosis in Goats

7.1 NATURAL INFECTIONS

7.1.1 DNA Serologic Prevalence

Compared to cattle and sheep, the prevalence in goats is lower (Table 7.1). Seropositivity varied from 0% to 26.6%. Part of these variations maybe related to the duration of the infection, and the serological tests used. Dynamics of antibodies in 13 pregnant goats were followed; titers fluctuated 4–8 fold (decline or rise).[1347] In addition to serosurveys, a study using molecular methods was carried out in Brazil[1860]; tongues, hearts, and brains were collected from 102 adult sheep in an authorized abattoir, all tissues resulted positive for *N. caninum* in 2 animals (1.9%). They used nested PCR and sequencing of amplicons of ITS1 and 18S fragments that also allowed detection of *H. heydorni* and *T. gondii* in 3.9% and 7.8% of animals.

7.1.2 Clinical Disease

Abortion, stillbirth, and birth of weak kids have been reported in goats from several countries (Tables 7.2 and 7.3). Gross abnormalities included hydrocephalus, hypocerebellum, and porencephaly (Figure 7.1a,b). The prominent features of neosporosis-associated encephalitis in goats were (i) glial nodules with numerous tissue cysts (Figure 7.1d) and (ii) the presence of few tachyzoites. The tissue cysts observed were predominantly smaller than 20 μm in diameter.

All reports listed in Tables 7.2 and 7.3 were sporadic. The data reported in Reference **1347** provided evidence for transplacental transmission of *N. caninum* in goats naturally infected prior to pregnancy. Authors screened four 10-year-old goats from several farms in Brazil and selected 13 of them for a longitudinal follow-up. Thirteen goats were seropositive (IFAT, 1:50 or higher); all were seronegative for *T. gondii*. Goats were bred and blood samples were obtained on 0, 30, 60, 90, and 120 days after mating, and 1, 2, and 3 months after parturition. One of these goats aborted 4 fetuses, and another had 2 stillborn kids at 148 days of pregnancy; *Neospora* was found in these animals (see Table 7.2). The goat that aborted had an IFAT titer of 1:800 on day 0 and 1:6400 at the time of abortion. The goat that had stillbirth had an IFAT titer of 1:100 on the day of mating, a titer of 1:3200 at the time of parturition, and at 3 months after parturition the titer declined to 1:400. Of the remaining 11 goats, 8 gave birth to congenitally infected kids, as evidenced by seropositivity in presuckling sera. Thus, endogenous transplacental transmission was evidenced in 9 of 13 (69.2%) goats. At least in 2 congenitally infected kids the IFAT titer had declined from 1:1600 to <1:50 at 120 days of age. These data are of epidemiological significance. Although parasite DNA was detected in tissues of adult goats, *N. caninum* could not be demonstrated histologically.[1448]

Table 7.1 Prevalence of Antibodies to *N. caninum* in Goats

Country	Region	Type	No. Tested	No. Positive	% Positive	Assay	Cut-Off Titer IFAT or Test[a]	Remarks	Reference
Argentina	La Rioja	Not stated. 94 herds, 30 animals per herd	1594	106	6.6	IFAT	1:50	Higher in males 10.3%) than females (6.1%). No association with age	1388
Brazil	Bahia	9 herds	384	58	15.1	IFAT	1:100	Maximum titer at 1:3200. 29% coinfected with *T. gondii*	2013
	Maranhão	5 farms	46	8	17.4	IFAT	1:25		1394
	Minas Gerais	90 herds, several breeds	667	71	10.7	IFAT	1:50	Maximum titer at 1:3200	69
	Paraíba	Abattoir	306	10	3.3	IFAT	1:50	Higher in males (5.0%) than in females (2.2%)	660
	Pernambuco	23 farms	319	85	26.6	IFAT	1:50	–	1950
	Pernambuco	174	Rural villages	5	2.9	IFAT	1:50	*T. gondii* antibodies analyzed	85a
	Piauí	202	Rural villages	4	2.0	IFAT	1:50	*T. gondii* antibodies analyzed	85a
	Rio Grande do Norte	14 farms	381	4	1.0	IFAT	1:50		443
	Santa Catarina	57 municipalities	654	30	4.6	IFAT	1:50	Age and abortion risk factors. Maximum titer at 1:6400	1980
	São Paulo	17 farms	923	161	17.4	NAT	1:25	No correlation with gender	1368
	São Paulo	19 farms	394	25	6.4	IFAT	1:50	Maximum titer at 1:12,800	682
	São Paulo	17 farms	923	161	19.8	NAT	1:25	No association with concurrent *T. gondii* and CAEV infections	409, 1898
China	Qinghai	Cashmere	207	16	7.7	ELISA	IH-NcSAG1	–	1232, 1233
	Qinghai	12 intensive farms	650	47	7.2	ELISA	IDEXX	Confirmed by IFAT (1:50)	1197
	Qinghai	Cashmere	22	4	18.1	ELISA	IH-NcSAG1	–	1243
Costa Rica	Heredia	Dairy goats from 1 farm. Several breeds	81	6	6.1	IFAT	1:100	Confirmed abortion from this herd	566
Czech Republic	8 regions	15 farms	251	15	6.0	cELISA	VMRD	Confirmation by IFAT (1:40)	159
Greece	–	50 farms	375	26	6.9	ELISA	In-house	WB in some random samples	490
Grenada	6 parishes	–	138	8	5.8	ELISA	IDvet	–	1839

(Continued)

Table 7.1 (Continued) Prevalence of Antibodies to *N. caninum* in Goats

Country	Region	Type	No. Tested	No. Positive	% Positive	Assay	Cut-Off Titer IFAT or Test[a]	Remarks	Reference
Iraq	5 regions	Not stated	106	6	5.6	ELISA	IDVet	–	765
Iran	Hamden		450	28	6.2	ELISA	IDVet	–	433
Italy	Lombardy, Bergamo, Milan	Small farms	314	24	5.7	ELISA, confirmed by IB	IH	Highest prevalence in Alpine	744
Jordan	North	27 herds	302	6	2.0	ELISA	iELISA-Bio-X	Risk factors analyzed	5
Mexico	South	24 herds	300	17	5.7	ELISA	CHEKIT	Risk factors analyzed	26
	Veracruz	26 farms	182	7	3.8	ELISA	IDEXX	–	952
Pakistan	Punjab	5 farms. Beetal and crossbred	142	13	8.6	cELISA	VMRD	–	1453
Philippines	Luzon	4 farms	89	21	23.6	ELISA	IH	Other pathogens investigated	1089
Poland	–	Breeding adults, nationwide, 49 herds	1060	5	0.5	cELISA	VMRD, CHEKIT, confirmed by IFAT	–	408
Romania	4 regions	Dairy goats	512	12	2.3	ELISA	IDEXX	–	985
Slovakia	Košice	–	18	3	16.6	ELISA	IDvet	–	1890
	East	1 farm, white shorthaired breed	116	18	15.5	cELISA	VMRD	Positive result by PCR (*Nc-5* gene) in blood in 14 out of 18 seropositive animals	355
Sri Lanka	–	–	486	3	0.6	IFAT	1:160	IH ELISA, confirmed in 3 by WB	1446
Spain	North-western	50 farms	638	45	6.0	cELISA	VMRD	*T. gondii* coinfections in 5	491a
Taiwan	Taichung county	6 farms	24	0	0.0	IFAT	1:50	–	1518
Turkey	Karaman, Konya	Not stated	249	8	3.2	ELISA	rNcSAG1, IH	–	2180
USA	Massachusetts	1 herd, 8-year monitoring period	7086	134	1.9	cELISA, IH-ELISA	VMRD	Negative pathology. Reduction of prevalence close to eradication from 8.4% to 0.1%	51

[a] ELISA = Enzyme-linked immunosorbent assay; IDEXX = IDEXX HerdChek Neospora caninum antibody, indirect ELISA, sonicate lysate of tachyzoites, IDEXX Laboratories, USA; VMRD = Neospora caninum cELISA Competitive ELISA GP65 surface antigen of tachyzoites, VMRD, USA; IDvet = Neospora caninum Indirect Multispecies, ID.vet, France; iELISA-Bio-X = BIO K 192 NEOSPORA CANINUM ELISA antibody KIT, BIO-X Diagnostics, Belgium.

Table 7.2 Fatal Neosporosis in Goats

Country	Region	Case No.	History	Breed	Gross	Serology	Histology	Immuno	PCR	Reference[a,b]
							Diagnosis			
Brazil	Minas Gerais	1-day old	Born weak, unable to nurse	Saanen	Hydrocephalus, porencephaly	IFAT doe 1:400	White matter absent, mild necrosis, perivasculitis, only tissue cysts 9.8–20.5 μm in diameter	Positive	Not done	2032
		Day of birth	Newborn kid, late term	Saanen	Not done	IFAT doe 1:800, presuckling kid 1:400	No lesions in placenta	Negative	Positive in placenta	
		Abortion	Chronically infected, aborted 4 fetuses, 87 days after mating	Mixed breed	None	IFAT doe 1:6400 at abortion day	Positive in 1	Brain positive in 1 fetus	PCR positive brain in the first and heart in the second fetus	1347, 1448
		2 stillborn	Stillborn on 148 days after mating of a chronically infected doe	Pardo-Alpina	None	IFAT doe 1:3200 at parturition	Positive in 2	Positive in both	Positive in CNS of both	
	Rio Grande do Sul	1-day old	Born weak, unable to nurse, ataxic, euthanized day 3	Saanen	None	No data	Encephalitis, more severe in mid brain, many intact and degenerating tissue cysts of 12.4–32.2 μm in diameter	Positive	Not done	382
Costa Rica	Heredia	Aborted fetus	3.5-month gestation	Saanen	Hydrocephalus, smaller cerebellum	IFAT doe 1:800	Meningoencephalitis, tissue cysts 6–20 μm long, degenerating tissue cysts	Positive	Not done	566
Italy	–	Aborted fetus	3–4 month gestation	Not stated	None	Not done	Nonsuppurative encephalitis, myocarditis. 10–20 μm diameter tissue cysts in brain	Not done	Positive brain	635
USA	California	Aborted fetus (Farm A)	130 day gestation	Pygmy	None, autolyzed	Not done	Gliosis, necrosis in brain, numerous tissue cysts 10–32 μm in diameter	Positive	Not done	133
		Aborted fetus (Farm B)	–	Pygmy	None, partially autolyzed	Not done	Encephalitis, 1 tissue cyst 15 μm in diameter	Positive	Not done	
	Pennsylvania	Stillborn	Near-term male	Pygmy	No	Not done	Glial nodules, tissue cysts, lesions only in the brain, and more severe in mid brain	Positive	Not done	551

[a] 144 fetuses were investigated by histology and IHC in Switzerland, but none presented protozoal-associated lesions.[327]
[b] 18 fetal brains and 10 placentae were subjected to bioassay in gerbils in Argentina, none resulted in isolation of *N. caninum*.[2012]

Table 7.3   Detection of *Neospora* DNA in Aborted Goat Fetuses

Country	Region	No. Tested	No. Positive (%)	Positive Tissue	Method	Remarks	Reference
Brazil	Minas Gerais	8 fetuses	6 (75.0)	Brain	Sequence within the XII chromosome, direct PCR	Parasites detected in 4 by IHC	**391**
Italy	Sardinia	23 fetuses	2 (8.6)	Brain, liver, spleen, muscle, abomasum	ITS1, nested PCR	Also found infected with *T. gondii* in 13% of fetuses and 25% of placentae	**1293**
		8 placentae	1 (12.5)	Placenta			
	–	1 fetus	1 (100.0)	Brain	18S, direct PCR	Other pathogens investigated. Lesions in brain	**635**
Spain	–	26 fetuses	3 (11.5)	Brain	Nested PCR, ITS1	Lesions by histology in 15.4%. 2 fetuses positive by PCR presented protozoal-associated lesions	**1405**

Protozoal-associated lesions (mononuclear inflammatory infiltrates and necrotic foci in paren-chyma) were found and properly described in brain, lung, kidney, and heart of abortions caused by *N. caninum* from Spain.[1405] The same was done in the CNS of adult goats and their fetuses from Brazil,[391] but this study was completed with lectin histochemistry and immunohistochemistry of cellular responses. In this later study, neurological lesions were described as multifocal necrosis foci surrounded by astrocytes, perivascular cuffing, and gliosis.

7.2 EXPERIMENTAL INFECTIONS

Results of 2 experiments[1173,1635] reported in pregnant pygmy goats inoculated with the NC1 strain and in Boer goats inoculated with NcSpain7 are summarized in Tables 7.4 and 7.5. The following conclusions were drawn:

1. The main clinical signs were abortion, stillbirth, and birth of weak kids.
2. The stage of pregnancy had a marked effect on the outcome of pregnancy; effects were more severe in does inoculated early in pregnancy.
3. The earliest evidence of placental invasion was day 10 PI.
4. Fetal infection was demonstrable by day 14 PI.
5. Abortions occurring as early as 10–11 DPI were probably due to pyrexia in doe.
6. IgG and IgM antibodies were demonstrable in does as early as 7 DPI.
7. Fetuses developed antibodies as early as 18 DPI when their dams were inoculated at 120 days of gestation.

Two additional studies comprising experimental infections were carried out. In the first one,[568] 2 Nubian goats were IM and SC inoculated with the NC1 isolate (2.55×10^5 and 5.5×10^6 tachyzo-ites), and 4 pygmy goats (used in a previous experiment, see Reference **1173**) were inoculated SC with 10^7 tachyzoites. All goats were subjected to studies on their serological responses. Nubian goats developed antibodies from 2 weeks onwards, peaking 1:3200 (goat 1) and 1:6400 (goat 2) at 24 and 35 DPI, respectively. The titers for all 4 pygmy goats remained <1:50 after 35 DPI.

In a second experiment,[2157] within a study aiming to validate NcMIC10 (*N. caninum* micro-neme protein 10[2157]) as a new diagnostic marker, ten 2–4-years-old Boer goats were IV inoculated

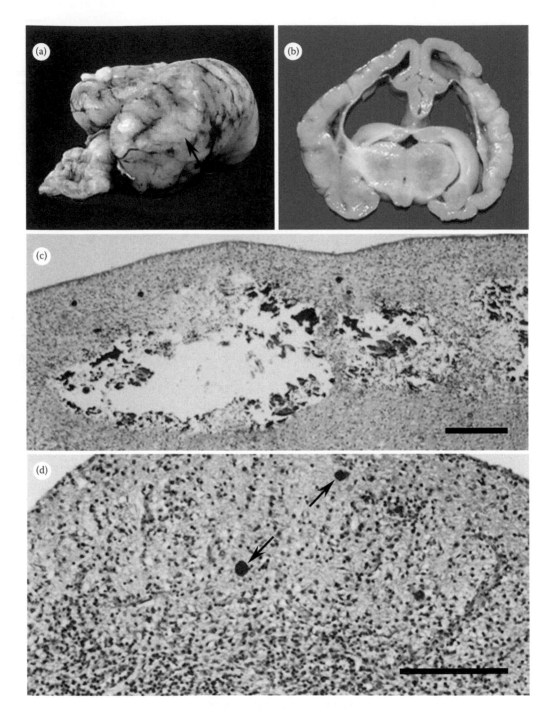

Figure 7.1 Lesions in neonatal goats. (a) Hydrocephalus in a 3.5 month gestational age fetus. The cerebral architecture is collapsed (arrow) because of enlarged lateral ventricles. The cerebral hemispheres were atrophied and the cerebellum was smaller than normal. Unstained. (b) Hydrocephalus in a kid that born weak and died the day of birth. Transverse section of the brain. Note atrophy of cerebral cortex due to dilatation of ventricles. Unstained. (c) Mineralized area with focal cavitations. HE stain. (d) Severe nonsuppurative inflammation. Arrows point to tissue cysts. A few tachyzoites and many tissue cysts were present. IHC with polyclonal rabbit *N. caninum* antibodies. (a, c, and d) Courtesy of authors in Reference **566**; (b) Courtesy of authors in Reference **2032**. Bars = 250 μm.

Table 7.4 Experimental Infections in Pregnant Pygmy Goats Inoculated SC with 10⁶ N. caninum Tachyzoites of the NC1 Strain[a]

		Doe						Progeny		
		Serology (IFAT)								
		IgG		IgM						
ID	Gestation day	Conversion Day (Titer)	Peak Titer (Day)	Titer	Day	Outcome	Histology	IHC	Bioassay (Cell Culture)[b]
53	Early (51 day)	7 (100)	25,600 (28–49)	100	7–14	Aborted 2, autolyzed fetuses day 30	Lesions brain, heart, spinal cord	Positive	N. caninum isolated from placenta
58		7 (100)	25,600 (35–49)	25	7–14	Died in utero, reabsorbed	–	–	–
57	Mid (85 day)	7 (50)	12,800 (77)	25	7	2 apparently healthy kids euthanized day 1 or 28	No lesions	Negative	N. caninum isolated from placenta
59		400 (14)	25,600 (42)	200	14–21	1 live, clinically normal	No lesions	Negative	N. caninum isolated from placenta
55	Late (127 day)	50 (7)	12,800 (35)	100	14	1 stillborn kid, autolyzed	Encephalitis	Positive	Not done
						Weak kid	Negative	Negative	N. caninum isolated from placenta of doe
56		1600 (14)	12,800 (35)	200	14	Weak kid died after birth	Negative	Negative	Not done

Source: Adapted from Lindsay, D. S. et al. 1995. Am. J. Vet. Res. 56, 1176–1180.

[a] All 6 N. caninum inoculated does became pregnant a second time when returned to the herd and rebred. Abortions or stillbirth were not observed during the second pregnancy of these does. Also, N. caninum was not isolated from any placenta or kid born to these does.

[b] Isolates were not named.

Table 7.5 Experimental Infections in Pregnant Boer Goats Inoculated Intravenously with 10^6 *N. caninum* Tachyzoites of the Nc-Spain7 Strain

| | Doe | | | Progeny | |
| | | | | PCR[a] | |
No.	Day Necropsied	Gestation Day	Outcome	Brain	Liver
		Inoculated Day 40			
1	10	50	Dead fetus	−	−
2	10	50	Dead fetus	−	−
3	11	51	Dead fetus	−	−
4	11	51	Dead fetus	−	−
NA	NA	NA	Dead fetus	NA	NA
NA	NA	NA	Dead fetus	NA	NA
5	17	57	Dead fetus	+	+
6	21	61	Dead fetus	+	+
7	21	61	Dead fetus	+	+
		Inoculated Day 90			
8	27	117	Dead fetus	+	NA
9	29	119	Dead fetus	+	NA
10	29	119	Dead fetus	+	−
11	35	125	Dead fetus	+	+
NA	NA	NA	Dead fetus	+	+
12	53	143	Live kid	+	+
13	53	143	Stillborn	−	−
14	55	145	Live kid	+	−
		Inoculated Day 120			
15	12	132	Live kid	−	−
16	14	134	Weak kid	+	+
17	15	135	Live kid	+	+
18	16	136	Weak kid	+	+
19	18	138	Weak kid	+	+
NA	NA	NA	Weak kid	+	+
20	21	141	Weak kid	+	+
21	22	142	Weak kid	+	+

Source: Modified from Porto, W. J. N. et al. 2016. *Vet. Res.* 47, 29.
[a] PCR results: + = positive, − = negative; NA = not available.

with the Nc-Illinois isolate; they were divided into 2 groups (n = 5) that received either 10^4 or 10^6 tachyzoites 90 days after mating. All goats aborted between 24 and 36 DPI.

BIBLIOGRAPHY

Serologic Prevalence: **5, 26, 51, 69, 159, 355, 408, 409, 443, 490, 566, 660, 682, 744, 765, 952, 985, 1089, 1197, 1232, 1233, 1243, 1347, 1368, 1388, 1394, 1446, 1453, 1518, 1839, 1860, 1890, 1898, 1950, 1980, 2013, 2180**.
Clinical Disease: **133, 327, 382, 391, 551, 566, 635, 1293, 1347, 1405, 1448, 2012, 2032**.
Experimental Infections: **568, 1173, 1635, 2157**.
Additional References: **51, 258, 1978**.

CHAPTER 8

Neosporosis in Water Buffaloes (*Bubalus bubalis*)

8.1 NATURAL INFECTIONS

8.1.1 Serologic Prevalence

Reported seroprevalences vary from 1.5% to 100% (Table 8.1) but most of these differences are probably related to the serologic methods and, more importantly, the cut-off values. For example, a very low (1.5%) prevalence in buffaloes was reported in Vietnam,[955] but the results are based on a very high cut-off (1:640) by IFAT.

The management systems could also affect the prevalence. In Asia, most water buffaloes are stall-fed, and most farmers have only a few animals, minimizing exposure to wild canids.[1048,1824]

Compared to cattle, little is known about the epidemiology of *N. caninum* infection in buffaloes. In an epidemiological study of beef cattle and buffaloes grazed together on 4 ranches in Argentina, seroprevalences (IFAT, 1:100) were much higher in buffaloes (43.3%, 584 of 1350) than in cattle (28.6%, 252 of 880).[1393] The age of the surveyed animals may have influenced seroprevalence rates. Higher rates were found in older animals in Argentina; seroprevalences were 57% in buffaloes <3 years old, 63% in 3–5 years old, and 67% in buffaloes older than 5 years.[277] A similar trend was reported in a subsequent study from Argentina[1393] and Thailand.[1048] In 1 study from Brazil seropositivity for *N. caninum* in water buffaloes was higher in extensive production, in dairy production and related to this on farms with a milk colling system.[241] In addition, purchases of animals and presence of other farm animals (pigs, but also small ruminants) seemed to represent a risk factor.[241] In the absence of a longitudinal study, the proportion rates for congenital versus post-natal transmission of *N. caninum* infection in buffaloes are unknown. In Italy, seroprevalences were 24.5% in 1–2-year-old buffaloes, 32.2% in 3–4-year-old buffaloes, 38.6% in 5–6-year-old buffaloes, and 43.1% in buffaloes older than 6 years.[813] Seasonal conditions may also affect the seroprevalence. The highest seroprevalence was recorded in buffaloes sampled in summer (70.5%) versus winter (39.6%).[1451]

8.1.2 Parasitologic Prevalence

Viable *N. caninum* was isolated from buffalo brains.[1722] Brains from 6 asymptomatic culled male buffaloes from São Paulo, Brazil were used to attempt isolation of *N. caninum* using cell culture and bioassay (Table 8.2). For *in vitro* cultivation, 10 g of brain of each of the 6 buffaloes were digested in trypsin, washed, and homogenates seeded onto bovine monocytes. Again in Brazil, viable *N. caninum* was isolated from 3 seropositive water buffaloes by feeding dogs; oocysts recovered were used for further oral infections in dogs.[116]

Table 8.1   Serological Prevalence of *N. caninum* in Water Buffaloes

Country	Region	Type	Number Tested	Number Positive	% Positive (Range)	Serologic Test[b], Cut-Off	Reference
Argentina	Corrientes	4 farms	449	287	63.9 (56.5–75.5)	IFAT, 1:100	277
	NEA[a]	5 farms	500	211	42.2 (26–65)	IFAT, 1:100	1092
	NEA	4 farms	1350	584	43.3 (37.5–65.5)	IFAT, 1:100	1393
	NEA	5 farms	1693	694	40.1	IFAT, 1:100	1090
Australia	Northern Territory	–	480	424	88.3 (33.3–100)	ELISA, IDEXX	1459
Brazil	Bahia	4 farms	117	42	35.9	IFAT, 1:200	777, 787
	Northern Brazil, 13 provinces	Farms	4796	2665	55.5	ELISA	412
				2345	48.8	IFAT, 1:40	
	Pará	3 farms	196	139	70.9	IFAT, 1:25	747
	Pará	4 farms	212	187	88.2	ELISA, IDEXX	2042
	Paraíba	14 farms	136	26	19.1	IFAT, 1:200	241
	Rio Grande do Sul	Farms	164	24	14.6	ELISA, CHEKIT	2048
	São Paulo	5 farms	192	169	88.0	IFAT, 1:50	349
	São Paulo	12 farms	411	230	56.0	IFAT, 1:200	467
	São Paulo	11 farms	222	117	53.0	NAT, 1:40	713, 714
				142	64.0	IFAT, 1:25	
China	Jiangsu	–	40	0	0.0	ELISA, CIVTEST	2163, 2164
Egypt	Cairo	Abattoir	75	51	68.0	NAT, 1:20	572
India	Punjab	2 farms	32	16	50.0	cELISA, VMRD	1337
	Andhra Pradesh, Karnataka	Various sources	341	34	9.9	cELISA, VMRD	1824
Iran	Ahvaz	Abattoir	181	67	37.0	ELISA, IDEXX	830
Iran	Ahvaz	–	122	76	62.3	ELISA, IDEXX	842
Iran							
	Hamedan	Farms	450	28	6.2	ELISA, IDvet	433
Italy	Campania	50 herds	1377	477	34.6	IFAT, 1:200	813
	Campania	2 farms	908	463	51.0	ELISA, IDvet	103
Pakistan	Punjab, Lahore	Farms	300	159	54.7	cELISA, VMRD	1451
	Punjab	10 farms	312	134	42.8	cELISA, VMRD	1453
Philippines	Central Luzon	7 farms	105	4	3.8	ELISA	1089
Thailand	6 provinces	238 farms	628	57	9.1 (8.6–16.7)	IFAT, 1:100	1048
Vietnam	Ho Chi Minh City	Abattoirs	200	3	1.5	ELISA, IFAT, 1:640	955

[a] North East Areas: Formosa, Chaco, Misiones, and Corrientes provinces.
[b] IDEXX = IDEXX HerdChek Neospora caninum antibody, indirect ELISA, sonicate lysate of tachyzoites, IDEXX Laboratories, USA; CHEKIT = CHEKIT Neospora, indirect ELISA, detergent lysate of tachyzoites, IDEXX Laboratories, The Netherlands; CIVTEST = CIVTEST BOVIS NEOSPORA, indirect ELISA, sonicate lysate of tachyzoites, Laboratorios Hipra S.A., Spain; VMRD = Neospora caninum cELISA Competitive ELISA GP65 surface antigen of tachyzoites VMVVMRD, USA.

8.1.3 Clinical Disease

Currently, there is no definitive evidence of clinical neosporosis in buffaloes. Reports of congenital infection and neosporosis-associated clinical signs are summarized here. *N. caninum* DNA was detected in 1 out of 9 fetuses tested from buffaloes slaughtered at an abattoir in Brazil.[347] These fetuses were between 2 and 5 months of gestational age. Antibodies to *N. caninum* were not found in their thoracic fluid, but bioassay and histological examinations were not performed.[347]

Table 8.2 Isolation of Viable *N. caninum* from Brains of Naturally Infected Water Buffaloes from Brazil

No.	Direct Cell Culture from Brain	Bioassay in Gerbils (*Meriones unguiculatus*)	Oocyst Shedding by Dogs	Isolate Designation[a]
1	–	–	+	NCBrBuf-1
2	+	–	+	NCBrBuf-2
4	–	+	–	NCBrBuf-3
5	+	+	+	NCBrBuf-4
6	–	+	–	NCBrBuf-5

Source: Adapted from Rodrigues, A. A. R. et al. 2004. *Vet. Parasitol.* 124, 139–150.
[a] Isolates not archived.

In a heterogeneous population of buffaloes from various sources in Pakistan aborting buffaloes had a higher seroprevalence (78.9%) versus non-aborting buffaloes (59.8%).[1451]

Suggestive evidence of neonatal neosporosis in buffaloes has been reported from Italy.[103,813] In the first report, encephalitis and myocarditis were found in 2 of 4 fetuses examined histologically,[813] but the parasite identity was not confirmed. Additional evidence was provided in the second report.[103] Three fetuses from a herd with seropositive buffaloes were examined at necropsy. *N. caninum* DNA was found in the brains of all 3 fetuses by pNc5 gene-based PCR. Histologically, all 3 fetuses had encephalitis and myocarditis, but *N. caninum* was not demonstrable.[103]

8.2 EXPERIMENTAL INFECTIONS

8.2.1 Infection in Nonpregnant Animals

Six buffalo calves were each inoculated SC with 5×10^6 live culture-derived tachyzoites of the cattle Illinois strain of *N. caninum*, and 2 calves were kept as uninoculated controls. Antibody titers were monitored weekly for 8 weeks and then monthly until 1-year PI using IFAT. All inoculated calves developed IFA titers of 1:100 or higher between 7 and 11 days PI, and peaked after 3 weeks to 1:1600 in 1 buffalo, to 1:800 in 3 other animals, and to 1:400 in the remaining 2 calves. Titers remained elevated until 7 weeks PI, and then declined to 1:25 or 1:50 in all the 6 animals by 12 months PI. All calves remained asymptomatic.[1723]

To follow the dynamics of *N. caninum* antibodies, sera from 29 buffaloes and their calves were collected during 1 year and assayed for *N. caninum* antibodies; 23 of 29 calves were seropositive (IFAT of 1:100 or more) at 1–2 days of age. Of these 23 calves, 17 remained seropositive during the study, while 6 became seronegative at 4 (2 calves), 6 (1 calf), 7 (2 calves), and 8 (1 calf) months of age. These findings indicate a neonatal transmission rate of *N. caninum* in buffaloes of 73%.[1723]

8.2.2 Infection in Pregnant Animals

Transplacental transmission of *N. caninum* has been induced in buffaloes by IV inoculation of dams with 10^8 or 5×10^8 tachyzoites. The results varied with the dose, parasite strain, and the gestational age (Table 8.3). As expected consequences were more severe when dams were inoculated at day 70 versus day 90 of gestation and when the dose was 5 times higher. Fetal death occurred by day 28 in 1 of 3 fetuses necropsied at day 28 PI. Parasite DNA was demonstrable in several fetal tissues. Placentitis was the most prominent lesion, and was found in 90% of the inoculated dams. Inflammatory lesions were found in several organs but *N. caninum* was not demonstrable in tissue sections. Placental inflammation was characterized by the infiltration of CD3+ and CD4+ T cells and T cells exposing γδ T-cell receptor, similar to what was previously found in cattle in early gestations but with milder lesions.[294] There are no data on the birth of infected asymptomatic buffalo calves.

Table 8.3 Experimental *N. caninum* Infection in Pregnant Water Buffaloes Inoculated Intravenously with Tachyzoites

Country	Dam[a]					Progeny				Reference
	No.	Gest. Day	*N. caninum* Strain, Dose	Sero-conversion	Day Killed	Clinical	PCR	Lesions (Histology)	IHC	
Argentina	3	70	NC1, 1×10^8	Day 7	Day 28	Alive	P, B	P, H, B, Lu, Li	B	1091
						Alive	P	H, Lu, Li	ND	
						Dead	P, H, Lu, Li	P, H, Lu, Li	H, Li, Lu, K, Sk	
	3	90			Day 28	Alive	H, Lu	P, B, H, Lu, Li	Ne	
							P, B, H, Lu, Li	P, H, Lu	Ne	
							P, B	P, B, H, Lu, Li	Ne	
	4	90			Day 42		P, B, Lu	P, H, Lu	Ne	
							P, B, H, Lu,	P, Lu	Ne	
							P, B, H, Lu, Li	P, H, Lu, Li	Ne	
							H, Li	P, B, H, Lu	ND	
Brazil	3	70	NC-Bahia, 5×10^8	Day 14	—	All 3 fetuses alive on day 42	B, H, Lu, Li	P 0/3, B 1/2, H 3/3, Li 1/3, Lu 0/3, K 2/3	B of 1	348, 349
	3	70	NC1, 5×10^8	Day 14	—	Fetal death; in all 3 detected on day 35 by ultrasound	P, B, H, Lu	P 3/3, B3/3, H 3/3,Li 2/3, Lu 1/3,K 3/3	P of 1	

Note: B = CNS, K = kidneys, Li = liver, Lu = lung, P = placenta, Sk = skeletal muscle, and Ne = negative results. ND: not done.
a Dams did not develop clinical signs.

BIBLIOGRAPHY

Serologic Prevalence: **103, 241, 277, 349, 412, 433, 467, 572, 713, 714, 747, 777, 783, 813, 830, 842, 955, 1048, 1089, 1090, 1092, 1337, 1393, 1451, 1453, 1459, 1824, 2042, 2048, 2163, 2164**.
Parasite Isolation and Clinical Disease: **103, 116, 347, 813, 1451, 1722**.
Experimental Infections: **294, 348, 349, 1091, 1723**.
Additional References: **777, 1454, 1696**.

CHAPTER 9

Neosporosis in Pigs

9.1 NATURAL INFECTIONS

9.1.1 DNA Serologic Prevalence

Information on serological prevalence in domestic pigs (*Sus scrofa*) and other feral-wild swine is summarized in Tables 9.1 and 9.2. *N. caninum* antibodies were found in 9.6% of 166 subadults versus 19.6% of 301 adult feral swine in USA, suggesting post-natal transmission.[198] Additionally, no positive were detected by PCR in pig tissues in Switzerland.[2133]

9.1.2 Clinical Infections

Currently there is no credible evidence that *N. caninum* causes clinical disease in pigs.

9.2 EXPERIMENTAL INFECTIONS

In 1 study, 2 adult sows inoculated IM with 2.5×10^6 tachyzoites of the NC1 isolate remained clinically normal but developed *N. caninum* antibodies within 2 weeks PI.[568] Peak IFAT titer was 1:6400 and there was no reactivity with *T. gondii*, indicating development of specific antibodies.

An important experiment in pregnant sows was performed in Denmark.[1011] Six sows were inoculated IM with 2.5×10^6 tachyzoites of the Swedish isolate (Nc-SweB1) of *N. caninum* at day 48–101 of gestation and the gilts were euthanized at near the parturition time (Table 9.3). Antibodies were assayed in the dams and their fetuses by IFAT. Tissues of all gilts and most fetuses were studied histologically, by IHC, and by bioassays in cell culture. The following conclusions were drawn:

1. Gilts became seropositive as early as 9 DPI.
2. All gilts remained asymptomatic, but lesions were demonstrable in the liver of 3 gilts, tachyzoites were found in livers of 2 gilts (nos. 5 and 6) and in the endometrium of gilt no. 1.
3. *N. caninum* infection was proven in 3 fetuses from gilt no. 1. Viable *N. caninum* was recovered in cell culture inoculated with tissues of 1 fetus, demonstrated by IHC in the liver of 1, and brains of 2 fetuses.
4. Tissue cysts were not found in tissues of fetuses or gilts.

Table 9.1 Serologic Prevalence of *N. caninum* Antibodies in Domestic Pigs (*S. scrofa*)

Country	Region	No. Examined	Type	No. Positive	% Positive	Assay	Cut-Off Titer or Test[a]	Remarks	Reference
Brazil	Paraiba state	130	Abattoir	4	3.1	IFAT	1:50	Also tested for *T. gondii* antibodies	**438**
	Paraiba state	190	Abattoir	6	3.2	IFAT	1:50	Also tested for *T. gondii* antibodies	**664**
Czech Republic	8 districts	551	Abattoirs	16	3.0	cELISA	VMRD	Clinically healthy animals. Also tested for *T. gondii* antibodies	**158**
Germany	Hesse	2041	94 farms, sows	67	3.3	ELISA	IH	Also tested for *T. gondii* antibodies	**418**
Grenada	–	185	Farms	0	0.0	ELISA	IDvet	Also tested for *T. gondii* antibodies	**1840**
Senegal	Kaolack	60	Wandering sows	35	58.3	cELISA	VMRD	Reported association between serology and stillbirths	**1026**
UK	England, Wales	454	Had aborted/ infertile	40	8.8	ELISA	MASTAZYME	None positive by both IFAT and ELISA	**875**
				0	0.0	IFAT	1:50		

[a] VMRD = Neospora caninum cELISA Competitive ELISA GP65 surface antigen of tachyzoites, VMRD, USA; IDvet = Neospora caninum Indirect Multispecies, ID.vet, France; MASTAZYME = MASTAZYME NEOSPORA, indirect ELISA, formaldehyde-fixed whole tachyzoites, MAST GROUP, UK.

Table 9.2 Serologic Prevalence of *N. caninum* Antibodies in Feral Pigs

Host[a]	Country	Region	No. Examined	No. Positive	% Positive	Assay	Cut-Off Titer or Test[b]	Remarks	Reference
Wild boar (*Sus scrofa ferus*)[c]	Czech Republic	7 regions	565	102	18.1	cELISA	VMRD	58 of 102 (56.9%) confirmed with IFAT at 1:40	154
	Greece	—	94	1	1.1	IFAT	1:40	Other 9 pathogens investigated	1984
	Spain	Central, South, and North-East	298	1	0.3	cELISA	Pourquier	Not confirmed with IFAT	39
	Slovakia	Southern	113	38	33.6	cELISA	VMRD	45 (39.8%) had *T. gondii* antibodies	1703
	Turkey	Erzurum	12	0	0.0	cELISA	VMRD	*T. gondii* also tested	113
Feral pig (*Sus scrofa*)	USA	New Mexico, Oklahoma, Texas	467	74	15.8	cELISA	VMRD	Higher prevalence in adults (19.3%) versus subadults (9.6%)	198
	USA	29 states	1059	159	15.0	NAT	1:25	Higher seroprevalence in adults	319
	Brazil	Mato Grosso do Sul	83	9	10.8	IFAT	1:50	Higher seroprevalence in females	1870
Warthog (*Phacochoerus aethiopicus*)	Kenya	Maasai-Mara Reserve, and other parts	6	4	66.7	NAT	1:40	4 positive at 1:40, 2 at 1:80	677

[a] The absence of anti-*N. caninum* antibodies was detected in 1 babirusa (*Babyrousa babyrussa*), 5 red river hogs (*Potamochoerus porcus pictus*), and 2 wild boars kept in Czech and Slovakian zoological gardens.[1820]

[b] VMRD = Neospora caninum cELISA Competitive ELISA GP65 surface antigen of tachyzoites, VMRD, USA; Pourquier = Competitive ELISA, Institut Pourquier, France.

[c] PCR to detect *N. caninum* DNA resulted negative in brain tissues of a wild boar from Germany.[380]

Table 9.3 Experimental Neosporosis in Pregnant Sows

Sows						Fetuses					
			Antibodies								
Designation of Gilts	Gestation Day	Euth. PID	IgG	IgM	N. caninum IHC	No.	No. Dead	No. Tested	Antibodies	Lesions	IHC
1	48	59	1280	1280	Positive	9	1	7	2	3	3[a]
2	48	59	20480	5120	Negative	10	0	10	–	–	–
3	76	30	5120	1280	Negative	11	2	11	–	–	–
4	72	30	5120	5120	Negative	13	2	13	–	–	–
5	101	9	40	80	Positive	11	0	11	–	–	–
6	101	10	2560	1280	Positive	10	0	10	–	–	–

Source: Data from Jensen, L. et al. 1998. *Acta Pathol. Microbiol. Immunol. Scand*. 106, 475–482.
[a] Parasite isolated in cell culture from tissues of 1 fetus.

BIBLIOGRAPHY

Natural Infections in Domestic Pigs: **158, 418, 438, 664, 875, 1026, 1840, 2133**.
Natural Infections in Wild Swine: **39, 113, 154, 198, 319, 380, 677, 1703, 1820, 1870, 1984**.
Experimental Infections in Domestic Pigs: **568, 1011**.
Additional References: **45**.

CHAPTER 10

Neosporosis in Camels and South American Camelids

10.1 NATURAL INFECTIONS

10.1.1 One Humped Camel (*Camelus dromedarius*)

Antibodies to *N. caninum* were detected in camels from Africa and Middle East (Table 10.1). There is no report of clinical neosporosis in camels.

10.1.2 South American Camelids

10.1.2.1 Serologic Prevalence

Available information for llamas (*Lama glama*), alpacas (*Vicugna pacos*), and vicuñas (*Vicugna vicugna*) is summarized in Table 10.2.

10.1.2.2 Clinical Infections

Neosporosis associated abortion was detected in 7 of 18 llamas and 12 of 32 alpacas (Table 10.3).

BIBLIOGRAPHY

Camels: **25, 841, 909, 935, 964, 965, 1754, 1820, 2082**.
South American Camelids: **312, 329, 331, 358, 1071, 1402, 1417, 1716, 1820, 1828, 1829, 2108**.

Table 10.1   Seroprevalence of *N. caninum* in Camels (*Camelus dromedarius*)

Country[a]	Region	Source	No. Tested	No. Positive	% Positive	Assay	Cut-Off Titer or Test[b]	Remarks	Reference
Egypt	Cairo	Abattoir	161	6	3.7	NAT	1:20	Titer of 1:1280 in 1 camel	**909**
Iran	Mashhad	Abattoir	120	7	5.8	IFAT	1:20	Titer of 1:20 in 3, and 1:40 in 4	**1754**
	Yazd	Abattoir	254	10	3.9	NAT	1:10	Titer of 1:20 in 6, and 1:40 in 4	**841**
	Isfahan	Abattoir	310	10	3.22	IFAT	1:50	Titer of 1:100 in 3	**935**
Saudi Arabia	Riyadh	Farm	412	23	5.6	IFAT	1:20	Titer of 1:20 in 16, and 1:40 in 7	**25**
Sudan	Khartoum	Farm	61	6	9.8	ELISA	VMRD	96.9% of herds were positive	**964, 965**
United Arab Emirates	Dubai	Farms— dams	578	76	13.1	ELISA	VMRD	Samples tested for 7 other infectious agents	**2082**
		Calves	541	77	14.2				

[a] No anti-*N. caninum* antibodies were detected in 10 Bactrian camels (*Camelus bactrianus*) kept in Czech and Slovakian zoos.[1820]
[b] VMRD = Neospora caninum cELISA Competitive ELISA GP65 surface antigen of tachyzoites, VMRD, USA.

Table 10.2 Seroprevalence of *N. caninum* in Farmed South American Camelids

Host[a,b]	Country	Region	No. Tested	No. Positive	% Positive	Assay	Cut-Off Titer or Test[c]	Remarks	Reference
Llama (*Lama glama*)	Argentina	Jujuy	308	14	4.6	IFAT	1:25	11 samples positive at 1:25, 3 positive at 1:50	1402
	Germany	Hesse	20	0	0.0	IB, IFAT	1:20	–	2108
	Peru	Central, South	212	39	18.4	IFAT	1:50	–	329
	Peru	Melgar, Puno	275	46	16.7	IFAT	1:50	–	1417
	Peru	–	73	23	31.5	IFAT	1:50	6 had titer of 1:800. Confirmed by IB when IFAT titer was equal or higher than 1:100	331
	Peru	Puno	81	1	1.2	IB, IFAT	1:20	–	2108
	Peru	Central, South	1845	153	8.3	IFAT	1:100	16 had titer of 1:800. Confirmed by IB	331
Alpaca (*Vicugna pacos*)	Peru	Sierra Central	175	5	2.9	IFAT	1:100	Breeding female	312
	Australia	Alpaca	182	0	0.0	ELISA	IDEXX	–	358
	Germany	Hesse	12	0	0.0	IB, IFAT	1:20	–	2108
	Australia		100	3	3.0	cELISA	VMRD	–	1071
	Peru	Central, South	92	39	42.4	IFAT	1:50	–	329
	Peru	–	78	28	35.9	IFAT	1:50	8 had titer of 1:800. Confirmed by IB when IFAT titer was equal or higher than 1:100	331
	Peru	Puno	675	17	2.6	IB, IFAT	1:20	–	2108
	Peru	Central, South	2874	425	14.8	IFAT	1:100	7 had titer of 1:1600. Confirmed by IB	331
Vicuña (*Vicugna vicugna*)	Peru	Junin	11	0	0.0	IB, IFAT	1:20	–	2108
	Peru	Arequipa	207 (81 free ranging)	2	1.0	ELISA	IDEXX	Other 7 pathogens investigated. Confirmed by IB	1716

[a] 614 additional camelids (571 alpacas and 43 llamas) were surveyed in Arequipa (Peru), seroprevalence was 2.4%; only 2 alpacas were confirmed to have *N. caninum* antibodies by IB.[1716]

[b] The absence of anti-*N. caninum* antibodies was detected in 2 guanacos (*Lama guanicoe*), 3 llamas, 3 alpacas, and 1 vicugna from Czech and Slovakian zoos.[1820]

[c] VMRD = Neospora caninum cELISA Competitive ELISA GP65 surface antigen of tachyzoites; VMRD, USA; IDEXX = IDEXX HerdChek Neospora caninum antibody, indirect ELISA, sonicate lysate of tachyzoites, IDEXX Laboratories, USA.

Table 10.3   Neosporosis Abortion in Llamas and Alpacas from Peruvian Highlands

| Host | No of Fetuses | No. Positive (%) | Diagnosis | | | Reference |
			Histology[a]	IHC	PCR	
Llama	9	2 (22.2)	2	1	1 in brain	**1828**
(*Lama glama*)	18	7 (38.8)	5	4	1 brain and heart, 1 brain	**1829**
Alpaca	6	2 (33.3)	2	1	2 in brain	**1828**
(*Vicugna pacos*)	32	12 (37.5)	8	8	2 in brain and heart, 4 in brain, 1 in heart	**1829**

[a] Indicative of protozoal infection.

CHAPTER 11

Neosporosis in Felids

11.1 NATURAL INFECTIONS

11.1.1 Serologic Prevalence

Information on domestic cats (*Felis catus*) and on other wild Felidae is summarized in Tables 11.1 and 11.2.

11.1.2 Clinical Infections

There is no evidence for clinical neosporosis in felids.

11.2 EXPERIMENTAL INFECTIONS

Cats are susceptible to infection. Information is summarized in Table 11.3. Additionally, information on serological responses of cats to *N. caninum* infection was compiled previously.[568] Briefly, 3 trials were performed. In trial no. 1, three 3-day-old kittens were SC inoculated with tachyzoites, 2 died on days 14–17 PI, and the kitten that survived was seropositive. In trial no. 2, an adult cat seroconverted (>1:800 on day 29 PI) after oral infection with *N. caninum* tissue cysts. In trial no. 3, three 3-month-old cats elicited a pronounced humoral immune response after IM inoculation with NC1 tachyzoites.

The following conclusions can be drawn from Table 11.3:

1. Cats can be infected with *N. caninum* transplacentally and postnatally.
2. Congenital infection can occur due to reactivation of infection in cats infected before pregnancy.
3. Disseminated neosporosis can develop in inoculated cats.
4. Parasite can encyst in cat tissues.
5. Cats can develop antibodies to *N. caninum*.

BIBLIOGRAPHY

Natural Infection in Domestic Cats: **90, 244, 359, 456, 465, 588, 663, 678, 838, 839, 932, 1821, 1855**.
Natural Infection in Other Felids: **70, 333, 677, 1027, 1350, 1517, 1820, 1872, 1888**.
Experimental Infections in Cats: **532, 534, 544, 568, 1310**.
Additional References: **45, 506, 508, 786, 1819**.

Table 11.1 Serologic Prevalence of *N. caninum* Antibodies in Domestic Cats (*Felis catus*)

Country	Region	Type	No. Examined	No. Positive	% Positive	Assay	Cut-Off Titer or Test[a]	Remarks	Reference
Albania	Tirana	Free roaming	146	15	10.3	IFAT	1:100	1 cat had IFAT titer of 1:400	1855
Brazil	Maranhão	Domestic	54	27	50.0	IFAT	1:25	1 cat had IFAT titer of 1:400	240a
	Araçatuba, São Paulo State	Domestic	400	98	24.5	IFAT	1:16	4 cats had IFAT titer of 1:256	244
	Andradina, São Paulo State	Domestic	70	0	0.0	IFAT	1:16	—	359
	Campo Grande, Mato Grosso do Sul	Free roaming and domiciled	151	10	6.6	IFAT	1:50	3 cats had IFAT titer of 1:200	465
	Patos, Paraíba	Free roaming and domiciled	201	0	0.0	IFAT	1:50	—	663
	Pernambuco	Rural villages	32	2	6.2	IFAT	1:50	*T. gondii* antibodies surveyed	85a
	Piauí	Rural villages	3	0		IFAT	1:50	*T. gondii* antibodies surveyed	85a
	Salvador, Bahia	Indoor, outdoor	272	8	2.9	IFAT	1:50	—	456
	São Paulo State	Free roaming and domiciled	502	60	11.9	NAT	1:40	1 cat had an IFAT titer of 1:800; 8 cats were also positive by IB	588
Czech Republic	—	Domestic	414	137	33.0	cELISA	VMRD	5 cats had IFAT titers of 1:100	1821
				16	3.8	IFAT	1:50		
Hungary	—	Rural and urban	330	2	0.6	IFAT	1:40	—	932
Iran	Ahvaz	Stray	100	19	19.0	NAT	1:40	1 cat had a titer of 1:1280	838, 839
Italy	Turin	Stray	282	90	31.9	NAT	1:40	15 cats had titers of 1:320	678
Thailand	Western regions	Dairy farms	36	0	0.0	cELISA	VMRD	—	90
				0	0.0	IFAT	Not stated	—	

[a] VMRD = Neospora caninum cELISA Competitive ELISA GP65 surface antigen of tachyzoites, VMRD, USA.

Table 11.2 Serologic Prevalence of *N. caninum* Antibodies in Wild Felidae

Host Species[a]	Country	No. Examined	Type	No. Positive	% Positive	Assay	Cut-Off, Test[c]	Reference
Amur leopard (*Panthera pardus orientalis*)	USA	1	Zoo	1	100	IFAT	1:50	1888
Feral cat (*Felis silvestris catus*)	Spain	59	Wild	4	6.8	cELISA	VMRD, confirmed with IFAT 1:50	1350
Eurasian wild cat (*Felis silvestris silvestris*)	Spain	6	Wild	1	16.7	cELISA	VMRD, confirmed with IFAT (1:20) and NAT (1:100)	1872
Iberian lynx (*Lynx pardinus*)	Spain	25	Wild	5 (3 by IFAT)	12.0	cELISA	VMRD, confirmed with IFAT (1:20) and NAT (1:100)	1872
Eurasian lynx (*Lynx lynx*)	Czech Republic	2	Zoo	1	50.0	IFAT	1:40	1820
Cheetah (*Acinonyx jubatus*)	Kenya	5	Wild	3	60.0	NAT	1:40	677
	Southern Africa[b]	23	Wild	1	4.3	IFAT	1:50	333
	Czech Republic	15	Zoo	2	13.3	IFAT	1:40	1820
	Southern Africa[b]	41	Wild	3	7.3	IFAT	1:50	333
Lion (*Panthera leo*)	Kenya	20	Wild	11	55.0	NAT	1:40	677
	Senegal	7	Zoo	7	100.0	cELISA	VMRD	1027
	Brazil	9	Zoo	1	11.1	IFAT	1:25	70
	USA	10	Zoo	2	20.0	IFAT	1:50	1888
Indian lion (*Panthera leo goojratensis*)	Czech Republic	2	Zoo	1	50.0	IFAT	1:40	1820
Geoffroy's cat (*Oncifelis geoffroyi*)	USA	1	Zoo	1	100.0	IFAT	1:50	1888
Jaguar (*Panthera onca*)	Brazil	11	Wild	7	63.6	IFAT	1:25	1517
	Brazil	13	Zoo	8	61.5	IFAT	1:25	70
Jaguarundi (*Puma yagouaroundi* syn. *Herpailurus yagouaroundi*)	Czech Republic	1	Zoo	1	100.0	IFAT	1:40	1820
	Brazil	25	Zoo	5	20.0	IFAT	1:25	70
Little-spotted cat (*Leopardus tigrinus*)	Brazil	35	Zoo	11	31.4	IFAT	1:25	70
Ocelot (*Leopardus pardalis*)	Brazil	42	Zoo	30	71.4	IFAT	1:25	70

(Continued)

Table 11.2 (Continued)   Serologic Prevalence of *N. caninum* Antibodies in Wild Felidae

Host Species[a]	Country	No. Examined	Type	No. Positive	% Positive	Assay	Cut-Off, Test[c]	Reference
Puma-cougar (*Puma concolor*)	Brazil	18	Zoo	5	27.8	IFAT	1:25	70
	USA	47	Zoo (42 free ranging)	1	2.2	IFAT	1:50	1888
Tiger (*Panthera tigris*)	Brazil	6	Zoo	4	66.7	IFAT	1:25	70
	USA	11	Zoo	1	9.0	IFAT	1:50	1888
Pampas cat (*Oncifelis colocolo*)	Brazil	3	Zoo	3	100.0	IFAT	1:25	70
Caracal (*Caracal caracal*)	Brazil	1	Zoo	1	100.0	IFAT	1:25	70
Serval (*Letailurus serval*)	Brazil	1	Zoo	1	100.0	IFAT	1:25	70
Fishing cat (*Prionailurus viverrinus*)	Brazil	1	Zoo	1	100.0	IFAT	1:25	70
Clouded leopard (*Neofelis nebulosa*)	USA	2	Zoo	0	0.0	IFAT	1:50	1888
Snow leopard (*Panthera uncia*)	USA	2	Zoo	1	50.0	IFAT	1:50	1888

a Other studies included species of Felidae from which no antibodies against *Neospora* were detected: Reference **1888** (3 zoo cheetahs, *A. jubatus*; 2 zoo Iranian leopards, *Panthera pardus dathei*; 2 zoo jaguars, *P. onca*; 2 zoo jaguarundis, *P. yagouaroundi*; 1 zoo jungle cat, *Felis chaus*; 1 zoo ocelot, *L. pardalis*; 1 zoo leopard, *P. pardus*; 1 zoo lynx, *Lynx canadensis*; 1 zoo caracal, *Caracal caracal*; 3 zoo serval, *L. serval*; 1 zoo fishing cat, *P. viverrinus*), Reference **70** (4 zoo margay, *Leopardus wiedii*; 1 zoo leopard, *P. pardus*), Reference **1820** (1 zoo ocelot, *L. pardalis*; 1 zoo little-spotted cat, *L. tigrinus*; 1 zoo serval, *L. serval*; 2 zoo Pallas's cats, *Otocolobus manul*; 1 zoo clouded leopard, *N. nebulosa*; 2 zoo African lion, *P. leo*; 1 zoo jaguar, *P. onca*; 3 zoo Amur leopard, *P. pardus orientalis*; 2 zoo Siberian tigers, *Panthera tigris altaica*; 6 zoo Sumatran tigers, *Panthera tigris sumatrae*; 1 zoo Bengal tiger, *Panthera tigris tigris*), Reference **677** (2 wild leopards, *P. pardus*; 1 wild serval, *L. serval*), Reference **333** (4 wild leopards, *P. pardus*).

b Southern Africa (Botswana, Namibia, South Africa).

c VMRD = *Neospora caninum* cELISA Competitive ELISA GP65 surface antigen of tachyzoites, VMRD, USA.

Table 11.3 Experimental Neosporosis in Domestic Cats (F. catus)

No. of Cats	Age	Inoculum	Post-Inoculation Day (PID)	Clinical Outcome	Lesions[a]	Reference
2	3 days	NC1 2.5–7.5 × 10⁵ tachyzoites, PO, SC	17	Died	Hepatitis, myositis, encephalomyelitis, tachyzoites seen in kidneys, muscle, and brain	534
1			29	Euthanized	Myositis, encephalitis with tachyzoites. Tissue cysts in brain	
1	5-year old, pregnant	Inoculated SC 2 × 10⁶ NC1 tachyzoites at 47 day gestation	17 (gave birth)	Depressed. Gave birth to a kitten that died 2 days after birth Queen euthanized 3 days after parturition	Disseminated neosporosis, metritis, hepatitis, nephritis with demonstrable tachyzoites. Also in kittens. Macerated fetus found in uterus	532
1	7-month, pregnant	Fed tissue cysts, oral. Mated 111 days later	3 kittens born 174 DPI of the queen	1 healthy kitten euthanized day 2	N. caninum-associated mild lesions in lungs, liver, heart, and skeletal muscle	
				2 kittens euthanized 22 and 30 days of age	No lesions, no N. caninum	
6	84-day old	NC1 tachyzoites, 1 × 10⁶ IM, 5 × 10⁵ PO	3 cats (Nos. 1–3) given MPA[a]	No. 1 died 8 DPI, No. 3 died 16 DPI, No. 3 euthanized 21 DPI	Bacterial septicemia (No.1), Disseminated neosporosis (Nos. 2, 2)	544
			3 cats (Nos. 4–6) not given MPA	Cats Nos. 4–6 euthanized 55 DPI. Clinically normal	Mild myositis, encephalitis (Nos. 4–6). Tissue cysts in brain (No. 6).	
5	Weaned kittens	950–>5000 tissue cysts of 3 NC strains, PO	4–6 weeks	No clinical signs, no seroconversion	Not observed	1310
1	Weaned kittens	1 × 10⁶ tachyzoites parenterally	15	Lethargic	Disseminated neosporosis, IFAT 1:400	

a Methylprednisolone acetate.

CHAPTER 12

Neosporosis in Avian Species

Currently there is no definitive evidence that birds are natural hosts of *N. caninum*. Available information is summarized here.

12.1 NATURAL INFECTIONS

Although *N. caninum* antibodies and DNA were found in several species (Tables 12.1 and 12.2), intact parasites have neither been demonstrated nor isolated by bioassays. Additionally, antibody responses to *N. caninum* were not persistent in experimentally infected birds (Table 12.2).

12.2 EXPERIMENTAL INFECTIONS

None of the birds inoculated with tachyzoites developed clinical signs (Table 12.2). It seems that chickens can be successfully infected with tachyzoites but the host eliminates the infection. Of 7-day-old chickens inoculated with tachyzoites, parasites were found in tissues of chickens killed at 15 DPI but not in chickens killed at 60 DPI.[715] Eggs laid by inoculated hens had no evidence of *N. caninum* infection.[715]

Compared to newborn chickens, embryonated eggs were susceptible to *N. caninum* infection (Table 12.2). Parasites multiplied in several tissues of embryos, and 2 dogs fed infected tissues excreted oocysts.[715] However, oocysts were not seen in the feces of 2 dogs fed infected embryo tissues in another investigation.[1432]

Other experiments were aimed to evaluate the potential role of carnivorous birds as definitive hosts for *N. caninum*; 2 red-tailed hawks (*Buteo jamaicensis*), 2 turkey vultures (*Cathartes aura*), 2 barn owls (*Tyto alba*), and 3 American crows (*Corvus brachyrhynchos*) were fed with rodent tissues (mice and rats infected with 10^5 culture-derived tachyzoites of *N. caninum* 1–6 months before). Oocysts were observed in their feces but it was concluded that their identity was different from *N. caninum* after failure of infection in BALB/c mice.[112]

BIBLIOGRAPHY

Natural Infections: **2, 390, 426, 520, 665, 773, 789, 968, 1288, 1363, 1371, 1440, 1721a, 1762, 1787a, 2117**.
Experimental Infections: **110, 112, 459, 715, 1054, 1274, 1275, 1328, 1360, 1432**.
Additional References: **1295, 1781**.

Table 12.1 Detection of *N. caninum* Infection in Avian Species

Host	Country	No. Tested	No. Positive (%)	DNA or Antibodies	Method (Test, Antibody Titer, PCR, Sequencing)	Reference
Chickens (*Gallus domesticus*)	Americas[a]	1324	30 (39.5)	Antibodies	IFAT, 1:25	1288
	Brazil	200 outdoor	47 (23.5)	Antibodies	IFAT, 1:50	390
	Brazil	100 farm chickens	Antibodies in 17, DNA in 6	Antibodies, DNA, bioassay in KO mice negative	IFAT, 1:50; PCR for pNc5	773
	Brazil	200 indoor	3 (1.5)	Antibodies	IFAT, 1:50	
	Brazil	10 seropositive chickens	6 (60.0)	DNA	3/4 of brain. PCR for pNc5	
	China	700 free range	162 (18.86)	25 hearts digested, negative by PCR	IFAT, 1:25	665
	Egypt	361 free range	56 (15.5)	Antibodies	in-house NcSAG1-ELISA	968
	Iran	150	26 (17.3)	Antibodies	NAT, 1:5	1787
Crow (*Corvus* sp.)	Israel	183	30 (16.4)	Antibodies	NAT,1:100 IFAT, 1:50	1762
		183	2 (1.09)[b]	DNA	Nested PCR of pNc5 and sequencing	
Common raven (*Corvus corax*)	Spain	67 wild	24 (35.8)	Antibodies	IFAT, 1:50	1371
Magpie (*Pica pica*)	Spain[c]	33	2 (6)	DNA	Half-brain homogenized, 0.5 g tested. Nested PCR of pNc5 and sequencing	426
Common buzzard (*Buteo buteo*)	Spain[c]	17	1 (0.58)	DNA	Half-brain homogenized, 0.5 g tested. Nested PCR of pNc5 and sequencing	
Pigeon (species not provided)	China	210 farmed pigeons	63 (30.0)	DNA	0.5 g brain homogenized. Nested PCR of pNc5 and sequencing. Also, sequencing of ITS1	520

(Continued)

Table 12.1 (Continued) Detection of *N. caninum* Infection in Avian Species

Host	Country	No. Tested	No. Positive (%)	DNA or Antibodies	Method (Test, Antibody Titer, PCR, Sequencing)	Reference
Sparrow (*Passer domesticus*)	Iran	217	8 (3.68)	DNA	Whole brains homogenized. Nested PCR of pNc5 and sequencing	2
	Brazil	40	3 (7.5)	DNA	Heart and brain. Nested PCR of pNc5 and sequencing of ITS1	789
Several species[d]	Brazil	294	0	Antibodies	IFAT, 1:50	1363
Several species[e]	Turkey	101-wild	14 (14.0)	DNA	Not clear	1440
Common teal (*Anas crecca*)	Italy	17	6 (42.8)	Antibodies	IFAT, 1:50	1721a
			4 (23.5)	DNA	PCR for pNc5	
Mallard (*Anas platyrhynchos*)	Italy	8	3 (37.5)	Antibodies	IFAT, 1:50	1721a
			2 (25.0)	DNA	PCR for pNc5	
Wigeon (*Anas penelope*)	Italy	3	2 (66.0)	Antibodies	IFAT, 1:50	1721a
			2 (66.0)	DNA	PCR for pNc5	
Northern pintail (*Anas acuta*)	Italy	1	1 (100)	Antibodies	IFAT, 1:50	1721a
			1 (100)	DNA	PCR for pNc5	
Lapwing (*Vanellus vanellus*)	Italy	1	0	Antibodies	IFAT, 1:50	1721a
			1 (100)	DNA	PCR for pNc5	

[a] Antibodies to *N. caninum* were detected in 18.5% of 97 chickens from Mexico, 7.2% of 97 chickens from the USA, 39.5% of 144 chickens from Costa Rica, 71.5% of 102 chickens from Grenada, 44% of 50 chickens from Guatemala, 83.6% of 98 chickens from Nicaragua, 58.1% of 55 chickens from Argentina, 34.3% of 358 chickens from Brazil, 62.3% of 85 chickens from Chile, 11.2% of 62 chickens from Colombia, 38.7% of 80 chickens from Guyana, 18% of 50 chickens from Peru, and 21.7% of 46 chickens from Venezuela.

[b] *Corvus monedula, Corvus cornix.*

[c] DNA not found in 23 Eurasian jays (*Garrulus glandarius*), 105 Griffon vultures (*Gyps fulvus*), and 3 black kites (*Milvus migrans*).

[d] No anti-*N. caninum* antibodies in 4 *Coragyps atratus*, 50 *Columba livia*, 25 *Zenaida auriculata*, 8 *Caracara plancus*, 6 *Oryzoborus maximiliani*, 20 *Serinus canaria*, 2 *Ramphastos toco*, 29 *Amazona aestiva*, 4 *Anordorhynchus hyacinthinus*, 2 *Anordorhynchus leari*, 1 *Ara ararauna*, 2 *Ara chloropterus*, 64 *Melopsittacus undulatus*, 2 *Asio clamator*, 1 *T. alba*, 37 *Rhea americana*, 37 *Struthio camelus.*

[e] *N. caninum* DNA in the brains of 3 of 17 *Larus genei*, 1 of 21 *Corvus corone*, 2 of 3 *Melanitta fusca*, 3 of *Anas clyptea*, 1 of 14 *Perdix perdix*, 1 of 3 *Aquila heliacal*, 3 of 5 *Buteo buteo.*

Table 12.2 Experimental Infection of Avian Species with *N. caninum*[a]

Host	Age	No.	Stage and No. Inoculated	Observation Period	Clinical	Serology	Histopathology	Reference
Pigeon (*Columba livia*)	Breeders	3	NC2 + Nc-Liv, 10^4, 10^5, 10^6	6 weeks	Not ill	IFAT 1:800 in 1, 1:400 in 1, and <1:50 in 1	Nonsuppurative encephalitis but no organisms. 3 positive by PCR and tissue culture	1328
	Feral	4	NC1, 10^7 tachyzoites, IP	45 days	1 pigeon died 25 DPI	IFAT, (1:50), 10–20 DPI, peaked to 1:640 in 1, all became seronegative 25 DPI	Disseminated infection in the pigeon that died. No parasites found in pigeons that survived	1360
	Embryonated eggs, day 8 of incubation	25	NC1 tachyzoites, 10–10^5, 5 per group	Until 8 day of incubation	All embryos inoculated with 10^4 and 10^5 and 2 of 5 inoculated with 10 tachyzoites died	ND	Necrosis in liver, myocardium in 5, cerebral edema in 3, and edema in chorioallantoic membranes in 4. PCR resulted positive in above tissues from 7 except for brain	110
Zebra finches (*Taeniopygia guttata*)	Breeders	3	NC2 + Nc-Liv, 10^4, 10^5, 10^6	6 weeks	Not ill	ND	No lesions. 3 negative by PCR and tissue culture	1328
Quail (*Coturnix japonica*)	20-day old, breeders	48	NC-Bahia, 3.5×10^5 or 5.0×10^6	60 days	Not ill	IFAT (1:10); declined after 14 DPI	No parasites, negative by dog bioassay. PCR and IHC negative at end of experiment	459
Chicken (*Gallus domesticus*)	40-week old	3	NC1, 10^8 tachyzoites, IP	60 days	Not ill	IFAT (1:400) at 15 days and negative by 60 DPI	Parasites not found in eggs laid by these 3 hens	715
	7-day old	40	NC1, 10^3,10^4,10^5, 10^6 tachyzoites, IP, 10 chickens per dose	60 days	Not ill	IFAT (1:400) at 15 days and negative by 60 DPI	Tachyzoites demonstrable in tissues of 12 chickens (3 from each dose) euthanized at 15 DPI but not in 22 chickens euthanized at 60 DPI	
	Embryonated eggs	Several groups of 5	NC1, 10^3,10^4,10^5, 10^6 tachyzoites, intra-allantoic	Incubation period to day of hatch	Mortality	ND	Tachyzoites present in many organs, chickens hatched from inoculated eggs had neurological signs. One dog shed oocysts after feeding highly parasitized embryo chorioallantoic membranes	

(Continued)

Table 12.2 (Continued) Experimental Infection of Avian Species with N. caninum[a]

Host	Age	No.	Stage and No. Inoculated	Observation Period	Clinical	Serology	Histopathology	Reference
Chickens, quail, partridge[b]	Embryonated eggs, day 8 of incubation	60	NC1, 10, 10, 10^2, 10^3, 10^4, 10^5, 10^6 tachyzoites	Incubation period and hatch	Variable mortality before and after hatch	ND	Mortality, arthritis, neurologic signs. 100% mortality rates in all doses for partridges suggesting high susceptibility	1275
Chicken (Gallus domesticus)	Embryonated eggs, day 8 of incubation	60	NC1, 10, 10, 10^2, 10^3, 10^4, 10^5, 10^6 tachyzoites	Incubation period and hatch	Variable mortality before and after hatch	ND	Mortality dose related. One hatched chicken had neurological signs. Arthritis frequently observed[c]	1054, 1274
	90-day old	20	NC1, 3×10^6 tachyzoites, SC	70 days	Not ill	IFAT (1:40) detected 7 DPI, all chickens became seronegative by 70 DPI	N. caninum was not demonstrated in chicken tissues by histopathology, PCR or by bioassay in dogs[d]	1432
	Embryonated eggs, day 10	5	NC1, 1×10^2, intra-allantoic	Day of hatch	Not ill. 2 died 72 h after inoculation	ND	No lesions	
	Embryonated eggs, day 8 of incubation	25	NC1 tachyzoites, 10 to 10^5, 5 per group	Until day 8 of incubation	All embryos inoculated with 10^4 and 10^5 and 3 of 5 inoculated with 10 tachyzoites died	ND	Necrosis in liver in 3, necrosis in myocardium in 5, cerebral edema in 1, and edema in chorioallantoic membranes in 1. PCR resulted positive in above tissues from 7 except for brain	110

Note: ND = not done.

a Two red-tailed hawks (Buteo jamaicensis), 3 American crows (Corvus brachyrhynchos), 2 barn owls (Tyo alba), and 2 turkey vultures (Cathartes aura) remained healthy and did not excrete N. caninum oocysts after ingesting tissues of infected rodents.[112]

b Embryonated eggs of quail, partridge, and boiler chickens were inoculated with 10 to 1 million tachyzoites of the NC1 isolate and mortality rates were calculated.

c IHC lesions in heart, liver, and brain and PCR positive for all tissues when using 10^3 and 10^4 doses. No antigens detected in tissues when using doses 10^5 and 10^6 but remarkable hemorrhage and necrosis.[1054]

d Embryonated eggs were inoculated with 100 NC1 tachyzoites. The chickens born from these eggs were euthanized on day 30 PI. Parasite DNA was found in spleen of 1 chicken. The dogs fed tissues from these chickens did not excrete oocysts.

CHAPTER 13

Neosporosis in Humans and Primates

13.1 NEOSPOROSIS IN HUMANS

Low levels of antibodies to *N. caninum* have been reported in human sera (Table 13.1). The significance of these findings is uncertain because neither parasite DNA nor the parasite has been confirmed in human tissues. Nevertheless, *N. caninum* has been successfully cultured in human cell lines such as MCF-7 human breast carcinoma,[1237] human brain microvascular endothelial cells (HBMECs),[648] human trophoblast (BeWo) and uterine cervical (HeLa),[310] human foreskin fibroblasts (HFFs), human astrocytoma cell line (86HG39), human epithelial lung cells (A549 cells)[1885] and also, the same level of growth was observed when human serum instead of fetal bovine serum was used for tissue culture.[1512]

13.2 EXPERIMENTAL NEOSPOROSIS IN RHESUS MONKEYS

Transplacental neosporosis was demonstrated in *Macaca mulatta* in 2 experiments performed in California, USA.[135] In the first experiment, 2 *M. mulatta* 65-day gestational fetuses were inoculated *in utero* each with 10^6 *N. caninum* tachyzoites; these fetuses were removed by hysterotomy 13 and 22 days later. In the second experiment, 2 *M. mulatta* were inoculated IM and IV with 1.6×10^7 tachyzoites at gestational day 43; fetuses were removed by hysterotomy 67 and 70 days later. The fetuses were found alive but infected with *N. caninum*. In all 4 fetuses *N. caninum*-associated lesions, and live parasites were demonstrated by IHC and bioassay in cell culture. The predominant lesion was encephalitis with demonstrable tachyzoites.[135] Later, using the same specimens, *N. caninum* DNA was detected in the brain and heart of the dam, placenta, and several fetal tissues.[913]

BIBLIOGRAPHY

135, 192, 310, 648, 805, 913, 967, 1198, 1237, 1318, 1449, 1512, 1531, 1601, 1720, 1885, 1987, 1993.
Additional References: **139, 2035**.

Table 13.1   Seroprevalence of *N. caninum* in Humans

Country	Source of Sample	No. of Sera	No. Positive	% Positive	Test (Cut-Off)[a]	Remarks	Reference
Brazil	HIV-infected patients	61	23	37.7	IFAT[a] (1:50), ELISA IH, IB using 29-kDa protein	5 samples were positive by IB	1198
	Neurological disorders	50	9	18.0		8 samples were positive by IB	
	Newborns	91	5	5.5		4 samples were positive by IB	
	Control	54	3	5.6		2 samples were positive by IB	
	HIV-infected patients	342	91	26.6	IFAT (1:50)	1 patient with 1:400	1531
	Healthy farmers	67	7	10.5	IFAT (1:200)	All 7 were positive by IB	192
Denmark	Repeated miscarriage	76	0	0.0	ELISA (ISCOM), IFAT (1:640), IB	No evidence of infection by *N. caninum*	1601
Egypt	Pregnant	101	8	7.9	ELISA IH, NcSAG1t	Results confirmed by IFAT, 1:100	967
France	Healthy women	500	0	0.0	IFAT (1:80)	No evidence of infection by *N. caninum*	1720
	HIV patients	400	4	1.0	IFAT (1:80)	3 positive at 1:80, and 1 positive at 1:160	
Korea	Blood donors	172 *Toxoplasma*-positive sera	12	6.9	ELISA IH, IFAT (1:100)	Some sera reacted to *N. caninum* in IFAT and WB	1449
		110 *Toxoplasma*-negative sera	1	0.9			
UK	Blood donors	199-general population	0	0.0	IFAT (1:160)	No evidence of infection by *N. caninum*	805
		48-farm workers	0	0.0			
	Farm workers and women with miscarriage	400	0	0.0	IFAT (1:400)	Titer <1:200 in 2 individuals. No evidence of infection by *N. caninum*	1993
	Submitted to public health laboratories	3232 general population	0	0.0	ELISA IH, IFAT (1:50)	No evidence of infection by *N. caninum*	1318
		518 farm workers	0	0.0			
USA	Blood donors	1029	69	6.7	IFAT, (1:100), 35 kDa fraction in IB	All samples were negative at 1:200 IFAT. 16 IFAT positive sera reacted in IB. 27.5% of positive cases were also positive for *T. gondii* infection	1987

[a] ISCOM = Detergent extracted tachyzoite antigen incorporated in immune stimulating complex particles.

CHAPTER 14

Neosporosis in Cervids and Other Wild Herbivores

14.1 NATURAL INFECTIONS IN CERVIDS

14.1.1 White-Tailed Deer (*Odocoileus virginianus*)

White-tailed deer (WTD) are considered as one of the most important reservoirs of *N. caninum* infection in the USA. Hundreds of thousands of WTD are harvested by hunters and killed in traffic accidents yearly. The carcasses and viscera are a potential source of infection for wild carnivores, specially wolves, coyotes, but also for dogs, which in turn can spread the infection by excreting environmentally resistant oocysts in the environment.

14.1.1.1 Prevalence

Serologic prevalence is summarized in Table 14.1.

Up to 88% of WTD were seropositive. The lack of correlation with age in 1 study[612] suggested that congenital infection may be important in the epidemiology of neosporosis in WTD (Table 14.1).

Viable *N. caninum* was isolated from brains of 4 adult and 2 WTD fetuses (Table 14.2). Additionally, 1 isolate (NC-deer1) was obtained by feeding a dog the brain of a naturally infected deer from Illinois, USA. The dog excreted *N. caninum*-like oocysts 7–14 days after ingesting the deer tissue. PCR testing and sequencing of ITS1 confirmed the identity of the oocysts as *N. caninum*. A calf fed 2500 oocysts of this isolate developed *N. caninum* antibodies.[783]

14.2 OTHER CERVIDS

14.2.1 Serologic Prevalence

Available information is summarized in Table 14.1.

14.2.2 Clinical Infections

Clinical neosporosis has been reported in 4 species of deer (Table 14.3). A comprehensive report described an outbreak/cluster of cases in a zoo in Argentina.[170] A 2-week-old fawn and 4 neonates were found dead. The fawn had congenital defects (megacolon and anal dilation) and microscopic lesions were detected in histologic sections of several organs. A tissue cyst was seen in a fresh smear of its brain, *N. caninum* DNA was detected by PCR, and viable *N. caninum* was isolated from the

Table 14.1 Seroprevalence of *N. caninum* Antibodies in Cervidae/Cervids

Host[a]	Country	Region	Type	No. Tested	No. Positive	% Positive	Assay	Cut-Off Titer, Antigen, Test[b]	Remarks	Reference
WTD (*Odocoileus virginianus*)	Canada	Northwest	Wild	20	15	75.0	IFAT	1:100	Other pathogens investigated	824
	Mexico	Coahuila, Nuevo León, Tamaulipas	Wild	368	31	8.4	ELISA	IDEXX	Live-captured deer	1509
	USA	Illinois	Wild	400	162	40.5	NAT	1:40	22 samples had titers of 1:1600 or more. No correlation with age or sex	581
		Illinois	Wild	43	20	46.5	IFAT	1:100	6 samples had titer of 1:3200	783
		Iowa	Wild	170	150	88.2	NAT, IFAT, NcGRA6-ELISA, IB	1:25 (NAT)	99 seropositives by WB, 135 by ELISA, 106 by IFA, and 118 by NAT. 37 fawns, 4 yearlings, and 47 adults were seropositive	612
		Minnesota	Wild	485	80	16.5	NAT	1:100	Isolation of *N. caninum* from fetuses of seropositive dams	618
		Minnesota	Wild	62	44	71.0	NAT, IFAT, NcGRA6-ELISA, IB	1:25 (NAT)	10 seropositives by WB, 35 by ELISA, 13 by IFA, and 24 by NAT	612
		Minnesota	Wild	150	30	20.0	IFAT	1:100	6 samples had titer of 1:3200	783
		Missouri	Wild	23	11	47.8	IB	35 kDa, 25 kDa bands	IB was more sensitive than Ncp-29 ELISA for autolyzed samples	67
	USA	Virginia	Wild	110	58	52.7	NAT	1:25	In 3 that had titers of 1:200, *N. caninum* was isolated in mice	2043
		14 states	Wild	305	145	48.0	NAT	1:25	32 had titers of 1:500 or more	1190
		Wisconsin	Wild	147	30	20.4	IB	35 kDa, 25 kDa bands	WB was more sensitive than Ncp-29 ELISA for autolyzed samples	67
		Ohio	Wild	30	11	45.9	cELISA	VMRD	—	1406

(Continued)

Table 14.1 (Continued) Seroprevalence of *N. caninum* Antibodies in Cervidae/Cervids

Host[a]	Country	Region	Type	No. Tested	No. Positive	% Positive	Assay	Cut-Off Titer, Antigen, Test[b]	Remarks	Reference
Mule deer (*Odocoileus hemionus hemionus*)	Canada	Northwest	Wild	1	0	0.0	IFAT	1:100	Other pathogens investigated	824
	USA	Washington state	Wild	42	7	16.6	NAT	1:25	1 had titer of 1:1600	611
	USA	Washington state	Wild	63	3	4.8	ELISA	Not stated	Other 8 pathogens investigated	1441
Black-tailed deer (*Odocoileus hemionus columbianus*)	USA	Alaska	Wild	54	0	0.0	IFAT	1:100	Other pathogens investigated	1910
	USA	Washington state	Wild	43	8	18.6	NAT	1:25	1 had titer of 1:1600	611
Fallow deer (*Dama dama*)	Belgium	–	Wild	4	0	0.0	ELISA	IDvet	RT-PCR negative	439
	Czech Republic	11 regions	Wild and game reserves	143 (79 wild)	2	1.4	cELISA	VMRD, confirmation by IFAT (1:50)	2 were positive by IFAT, 6 were positive by cELISA	155
	Czech and Slovak Republics	–	Zoo	3	0	0.0	IFAT	1:40	–	1820
	Mexico		Wild	19	2	10.5	ELISA	IDEXX		442
	Poland	Kosewo-Górne	Farmed	335	10	2.9	ELISA	IDEXX	Results confirmed by WB	201
	Spain	2 regions	Wild	79	0	0.0	cELISA	Pourquier	Results confirmed by IFAT (1:50)	39
Pampas deer (*Ozotoceros bezoarticus*)	Brazil	Goiás, Mato Grosso	Wild	39	15	38.5	IFAT	1:50	3 deer had titers of 1:12,800	1970
Axis deer (*Axis axis*)	Argentina	La Plata	Zoo	13	12	92.3	IFAT	1:25	8 had titers of >1:800. Clinical cases studied	170
	Mexico		Wild	18	2	11.1	ELISA	IDEXX		442
Père David's deer (*Elaphurus davidianus*)	China	Beijing	Wild	49	13	27	IFAT, IB	1:100	–	846
	Czech and Slovakian Republics	–	Zoo	28	7	25.0	IFAT	1:40	Maximum titer of 1:2560	1820
	USA	Ohio	Semi freedom	103	73	70.9	cELISA	VMRD	10 animals were retested yearly from 2005 to 2011	117
	USA	Ohio	Wild	38	26	40.0	cELISA	VMRD	–	1406

(Continued)

Table 14.1 (Continued) Seroprevalence of N. caninum Antibodies in Cervidae/Cervids

Host[a]	Country	Region	Type	No. Tested	No. Positive	% Positive	Assay	Cut-Off Titer, Antigen, Test[b]	Remarks	Reference
Marsh deer (Blastocerus dichotomus)	Brazil	Foz do Iguaçú	Captive	6	1	16.7	ELISA	IDEXX	–	2182
Pygmy brocket deer (Mazama nana)	Brazil	Foz do Iguaçú	Captive	22	1	4.5	ELISA	IDEXX	–	2182
Red brocket deer (Mazama nana)	Brazil	16 states	Captive and zoo	40	7	17.5	IFAT	1:50	1 deer had titer of 51,200	1971
Red brocket deer (Mazama americana)	Brazil	Foz do Iguaçú	Captive	4	0	0.0	ELISA	IDEXX	–	2182
	Brazil	16 states	Captive and zoo	29	18	62.0	IFAT	1:50	1 deer had titer of 51,200	1971
Gray brocket (Mazama gouazoubira)	Brazil	16 states	Captive and zoo	66	29	43.9	IFAT	1:50	1 deer had titer of 51,200	1971
Rodon (Mazama rondoni)	Brazil	16 states	Captive and zoo	8	3	37.5	IFAT	1:50	1 deer had titer of 51,200	1971
Small red brocket (Mazama bororo)	Brazil	16 states	Captive and zoo	3	2	66.6	IFAT	1:50	1 deer had titer of 51,200	1971
Red deer (Cervus elaphus)	Belgium	–	Farmed	7	0	0.0	ELISA	IDvet	RT-PCR negative	439
	Czech Republic	11 regions	Farmed and game reserves (1 wild)	377	24	6.4	cELISA	VMRD, confirmation by IFAT (1:50)	24 were positive by IFAT, 24 were positive by cELISA	155
	Italy	Western Alps	Wild	102	13	12.7	IFAT	1:40	–	674
	Italy	Trentino	Wild	125	4	3.2	cELISA	VMRD	–	243
	Poland	–	Wild	47	6	12.8	ELISA, IB	ISCOM-ELISA	Results confirmed by WB	802
	Poland		Farmed	106	12	11.3				
	Spain	3 regions	Wild	237	28	11.8	cELISA	Pourquier	Results confirmed by IFAT (1:50)	39
	Spain	North and Central	Wild	131	2	1.5	IFAT	1:50	–	1769
Tarim red deer (Cervus elaphus yarkandensis)	China	Xinjiang	Farmed	218	17	8.0	cELISA	VMRD	2.7% coinfected with T. gondii	1344

(Continued)

Table 14.1 (Continued) Seroprevalence of N. caninum Antibodies in Cervidae/Cervids

Host[a]	Country	Region	Type	No. Tested	No. Positive	% Positive	Assay	Cut-Off Titer, Antigen, Test[b]	Remarks	Reference
Thorold's deer (Cervus albirostris)	Czech and Slovakian Republics	–	Zoo	7	4	57.1	IFAT	1:40	Maximum titer of 1:320	1820
Sika deer (Cervus nippon)	China	3 provinces	Farmed	1800	164	14.6	ELISA	IDEXX	Herds with a history of miscarriages (41.9%) had significantly higher seroprevalence than those without (12.6%)	1343
	Czech Republic	11 regions	Game reserves	14	2	14.3	cELISA	VMRD, confirmation by IFAT (1:50)	2 were positive by IFAT, 2 were positive by cELISA	155
	Japan	Hokkaido	Wild	No DNA in 120 brains by PCR	0	0.0	PCR	pNc5 gene	Also those analyzed for serology, were negative	1513
Vietnam sika deer (Cervus nippon pseudaxis)	Czech and Slovakian Republics	–	Zoo	3	1	33.3	IFAT	1:40	Titer at 1:160	1820
Roe deer (Capreolus capreolus)	Belgium	–	Wild	73	2	2.7	ELISA	IDvet	2 out of 20 positive by RT-PCR	439
	Belgium	Flanders	Wild	168	8	4.8	ELISA	IDvet	Other 12 pathogens investigated	1943
	Czech Republic	11 regions	Wild	79 (10 from game reserves)	11	13.9	cELISA	VMRD, confirmation by IFAT (1:50)	11 were positive by IFAT, 12 were positive by cELISA	155
	Czech and Slovakian Republics	–	Zoo	4	0	0.0	IFAT	1:40	–	1820
	Italy	Bergamo	Wild	117	4	3.4	IFAT	Not specified	–	716
	Italy	Western Alps	Wild	43	16	37.2	IFAT	1:40	–	674
	Italy	Trentino	Wild	66	5	7.6	cELISA	VMRD	–	243
	Spain	Galicia	Wild	160	11	6.8	cELISA	VMRD	5 coinfected with T. gondii	1550
	Spain	2 regions	Wild	33	2	6.1	cELISA	Pourquier	Confirmed by IFAT (1:50)	39
	Spain	North and Central	Wild	228	2	0.9	IFAT	1:50	–	1769

(Continued)

Table 14.1 (Continued) Seroprevalence of *N. caninum* Antibodies in Cervidae/Cervids

Host[a]	Country	Region	Type	No. Tested	No. Positive	% Positive	Assay	Cut-Off Titer, Antigen, Test[b]	Remarks	Reference
Reindeer or caribou (*Rangifer tarandus*)	Spain	–	Wild	72	0	0.0	ELISA	IDvet	–	1409
	Sweden	–	Wild	199	2	1.0	ELISA,IB	ISCOM-ELISA	–	1267
	Canada	Northwest	Wild	20	1	5.0	IFAT	1:100	Other pathogens investigated	824
	Czech Republic	–	Farmed	2	0	0.0	cELISA	VMRD, confirmed by IFAT (1:50)	–	155
	Czech and Slovakian Republics	–	Zoo	9	0	0.0	IFAT	1:40	–	1820
	USA	Alaska	Wild	390	45	11.5	IFAT	1:100	Other pathogens investigated	1910
	USA	Alaska	Wild	160	5	3.1	NAT	1:40	Maximum titer at 1:80	601
	USA	Alaska	Wild	390	45	11.5	IFAT	1:50	8 with titers >1:800	1910
Eastern elk (*Cervus elaphus canadensis*)	Canada	Western Alberta	Wild	278	13	4.7	cELISA	VMRD	3 other pathogens investigated. 5 had IFAT titers	1639, 1641
	Canada	Northwest	Wild	20	5	25.0	IFAT	1:100	Other pathogens investigated	824
	Czech and Slovakian Republics	–	Zoo	1	1	100.0	IFAT	1:40	Titer 1:1280	1820
Moose (*Alces alces*)	Canada	Northwest	Wild	20	2	10	IFAT	1:100	Other pathogens investigated	824
	Czech and Slovakian Republics	–	Zoo	13	0	0.0	IFAT	1:40	–	1820
	Poland	–	Wild	7	3	42.8	ELISA	IDEXX	1 positive by NAT (1:20) Confirmed by IB	1414
					1	14.2	cELISA	VMRD		
	Sweden	–	Wild	417	0	0	ELISA	ISCOM-ELISA	Confirmed by IB	1267
	USA	Alaska	Wild	162	4	2.5	NAT	1:40	Higher titer at 1:80	601
	USA	Alaska	Wild	202	1	0.5	IFAT	1:100	Higher titer >1:800	1910
	USA	Minnesota	Wild	61	8	13.1	IFAT	1:100		783

a Other Cervidae were examined in the Czech and Slovakian Republics zoological gardens but no seropositives were detected[1820]: 2 Chinese water deer (*Hydropotes inermis*), 6 Reeve's muntjac (*Muntiacus reevesi*), 3 Prince Alfred's spotted deer (*Cervus alfredi*), 7 Eld's deer (*R. eldii*), 4 Timor deer (*Cervus timorensis*), 1 Persian fallow deer (*Dama mesopotamica*), 5 Southern pudu (*Pudu pudu*), and 14 Manitoban elk (*Cervus elaphus manitobensis*).

b Pourquier = Competitive ELISA, Institut Pourquier, France; VMRD = Neospora caninum cELISA Competitive ELISA GP65 surface antigen of tachyzoites, VMRD, USA; IDEXX = IDEXX HerdChek Neospora caninum antibody, indirect ELISA, sonicate lysate of tachyzoites, IDEXX Laboratories, USA; IDvet = Neospora caninum Indirect Multispecies, ID.vet, France; ISCOM = Detergent extracted tachyzoite antigen incorporated in immune stimulating complex particles.

Table 14.2 Isolation of Viable *N. caninum* from Wild Ruminant Hosts

Host	Source	Year	Serology (Dams)	Tissues	Bioassay			Isolate Designation	Reference
					Mice	Cell Culture	Dog		
WTD (*Odocoileus virginianus*)	Virginia, USA	2003	NAT, 1:200	Brain of adult	Positive	Positive	ND	NC-WTDVA-1 NC-WTDVA-2 NC-WTDVA-3	2043
	Minnesota, USA	2008 2009	NAT, 1:100	Fetal brain	Positive	Positive	ND	NcWTDMn1 NcWTDMn1	618
	Illinois, USA	2002	IFAT, 1:800		ND	ND	Positive (oocysts were later infective to calf)	NC-deer1	783
Axis deer (*Axis axis*)	La Plata, Argentina	2014	IFAT, 1:25	Brain	Positive (in gerbils)	Positive	ND	NC-Axis	170
European bison (*Bison bonasus*)	North East Poland	2010	ELISA, IDEXX	White blood cells	ND	Positive	ND	NC-PolBb1 NC-PolBb2	200

Note: ND = not done.

Table 14.3 Clinical Neosporosis in Cervids and Antelopes

Host[a]	Country	Source	Age	Lesions	Serology	IHC	PCR	Bioassay	Remarks	Reference
					Diagnosis					
Axis deer (*Axis axis*)	Argentina, La Plata	Zoo	1 fawn born ill, dilated anal sphincter, ataxia first week, died at 14 days	Brain, heart, intestine, lungs	IFAT 1:6400	ND	Positive, pNc5 gene. New microsatellite pattern found	Positive gerbils, KO mouse, cell culture. Isolate named NC-Axis. See Table 14.2	Tissue cysts seen in fresh brain smear of the deer	170
			4 fawns died neonatally	No data	IFAT 1:6400, 1:6400, 1:3200, 1:25	No data	Positive in 1 of 4 fawns. pNc5 gene	ND	–	
Black-tailed deer (*Odocoileus hemionus columbianus*)	California, USA	Wild	2-month-old fawn	Lungs, liver, kidneys, brain not examined. Tachyzoites in lesions	ND	Positive	ND	ND	Most likely cause of death	2113
Eld's deer (*Rucervus eldii*)	France, Paris	Zoo	Full-term stillborn	Encephalitis. Thick-walled tissue cysts in brain. Tachyzoites not detected	No data	Positive	ND	ND		567
Fallow deer (*Dama dama*)	Switzerland, Bern	Unknown	3-week-old female	Necrotizing granulo-matous meningoen-cephalomyelitis. Lesions most severe in the spinal cord. Tachyzoites and tissue cysts found	No data	Positive	Positive, pNc5 gene. Sequencing	ND	Whether transmission was vertical or horizontal remains unclear	1875
Lesser kudu (*Tragelaphus imberbis*)	Germany, Hannover	Zoo	Calf 1 (full term)	Meningoencephalitis, mild myocarditis	IFAT 1:16 (pleural fluid)	Negative	ND	ND	Birth from dam 1 that had IFAT titer at 1:128 at 52 and 165 days post-stillbirth	1600
			Calf 2 (full term)	Meningoencephalitis	IFAT 1:32 (pleural fluid)	Negative	Positive CNS, lungs	ND		
			Calf 3 (stillborn)	No lesions	IFAT <1:2 (pleural fluid)	Negative	Positive CNS, lungs, heart, liver, spleen	ND	Birth from dam 2. No serology data are available	

Note: ND = not done.

[a] 20 dead roe deer (*C. capreolus*) suspected of rabies were subjected to *Neospora* detection by PCR in Belgium, 2 resulted positive but its association to cause of death is not possible.[439]

Table 14.4 Seroprevalence of *N. caninum* Antibodies North American, European, and Asian in Wild Bovidae

Host	Country	Region	Type	No. Tested	No. Positive	% Positive	Assay	Cut-off Titer or Test[b]	Remarks	Reference
Spanish ibex (*Capra pyrenaica hispanica*)	Spain	South	Wild	3	0	0.0	cELISA	Pourquier	Confirmed by IFAT (1:50)	39
	Spain	South	Wild	531	30	5.6	cELISA	VMRD	27 samples were confirmed positive by IFAT (1:50)	721
Mouflon (*Ovis orientalis musimon*)	Czech Republic	11 regions	Wild	105 (29 from game reserves)	3	2.9	cELISA	VMRD, confirmation by IFAT (1:50)	3 were positive by IFAT, 4 were positive by ELISA	155
	Czech and Slovakian Republics	–	Zoo	1	0	0.0	IFAT	1:40	–	1820
Barbary sheep (*Ammotragus lervia*)	Spain	2 regions	Wild	27	0	0.0	cELISA	Pourquier	Confirmed by IFAT (1:50)	39
	Brazil	Curitiba	Zoo	17	4	23.5	IFAT	1:50	–	1407
	Czech and Slovakian Republics	–	Zoo	20	0	0.0	IFAT	1:40	–	1820
European bison (*Bison bonasus*)	Spain	2 regions	Wild	13	1	7.7	cELISA	Pourquier	Confirmed by IFAT (1:50)	39
	Czech and Slovakian Republics	–	Zoo	4	1	25.0	IFAT	1:40	Titer at 1:80	1820
	Poland	–	Wild	320 (48 farmed)	23	7.2	ELISA	IDEXX	Confirmed by IB	264
	Poland	–	Wild	23	3	13.0	ELISA, IB	IDEXX	Viable *N. caninum* isolated from peripheral blood of 1 bison	200
American bison (*Bison bison*)	Canada	Northwest	Wild	3	0	0	IFAT	1:100	Other pathogens investigated	824
	Czech and Slovakian Republics	–	Zoo	9	0	0.0	IFAT	1:40	–	1820
	USA	Alaska	Wild	219	1	0.5	NAT	1:40	Titer at 1:80	601
	USA	Iowa	Wild	30	4	13.3	NAT	1:40	Maximum titer, 4 at 1:320	601
	USA	Ohio	Wild	81	1	1.3	cELISA	VMRD	–	1406

(Continued)

Table 14.4 (Continued) Seroprevalence of *N. caninum* Antibodies North American, European, and Asian in Wild Bovidae.

Host	Country	Region	Type	No. Tested	No. Positive	% Positive	Assay	Cut-off Titer or Test[b]	Remarks	Reference
Musk ox (*Ovibos moschatus*)	Canada	Northwest	Wild	18	0	0.0	IFAT	1:100	Other pathogens investigated	824
	Czech and Slovakian Republics	–	Zoo	1	0	0.0	IFAT	1:40	–	1820
Chamois (*Rupicapra rupicapra*)	USA	Alaska	Wild	224	1	0.4	NAT	1:40	–	601
	Italy	Western Alps	Wild	119	35	29.4	IFAT	1:40	–	674
	Italy	Central Alps	Wild	67	14	21.0	IFAT	Not specified	–	716
	Italy	Trentino	Wild	503	7	1.4	cELISA	VMRD	–	243
	Spain	2 regions	Wild	40	0	0.0	cELISA	Pourquier	Results confirmed by IFAT (1:50)	39
	Spain	North and Central	Wild	149	2	1.34	IFAT	1:50	–	1769
Yak (*Bos grunniens*)	China	Tibet, Sichuan	Wild	749	47	6.3	ELISA			1143
	China	Qinghai	–	181	21	11.6	ELISA	NcSAG1, in-house	–	1242
Alpine ibex (*Capra ibex*)	Czech and Slovakian Republics	–	Zoo	1	0	0.0	IFAT	1:40	–	1820
	Italy	Western Alps	Wild	75	11	14.7	DAT	1:40	8 with titer at 1:80	675

[a] In Czech and Slovakian zoological gardens by IFAT (1:40),[1820] no anti-*N. caninum* antibodies were detected in: 5 West Caucasian tur (*Capra caucasica*), 3 East Caucasian tur (*Capra cylindricornis*), 1 markhor (*Capra falconeri*), 3 Siberian ibex (*Capra sibirica*), 1 long-tailed goral or Amur goral (*Nemorhaedus caudatus*), 1 Japanese serow (*Nemorhaedus crispus*), 3 Rocky Mountain goat (*Oreamnos americanus*), 13 Himalayan tahr (*Hemitragus jemlahicus*), 3 Californian bighorn sheep (*Ovis canadensis californiana*), 5 bharal (*Pseudois nayaur*), 2 golden takin (*Budorcas taxicolor bedfordi*), 3 mishmi takin (*Budorcas taxicolor taxicolor*), 1 gayal (*Bos gaurus* f. *frontalis*), 10 yak (*Bos mutus* f. *grumiens*), 3 watusi cattle (*Bos primigenius* f. *taurus*), and 2 nilgai (*Boselaphus tragocamelus*).

[b] Pourquier = Competitive ELISA, Institut Pourquier, France; VMRD = Neospora caninum cELISA Competitive ELISA GP65 surface antigen of tachyzoites, VMRD, USA; IDEXX = IDEXX HerdChek Neospora caninum antibody, indirect ELISA, sonicate lysate of tachyzoites, IDEXX Laboratories, USA.

Table 14.5 Seroprevalence of *N. caninum* Antibodies in African Wild Bovidae

Host[a]	Country	Region	Type	No. Tested	No. Positive	% Positive	Assay	Cut-Off Titer or Test[b]	Remarks	Reference
Thompson gazelle (*Gazella thomsonii*)	Kenya	–	Wild	26	7	26.9	NAT	1:40	1 at 1:320	677
Eland (*Taurotragus oryx*)	Czech and Slovakian Republics	–	Zoo	12	1	8.3	IFAT	1:40	Titer at 1:80	1820
	Kenya	–	Wild	13	12	92.3	NAT	1:40	5 with titer at 1:320	677
	Senegal	–	Wild	8	0	0.0	cELISA	VMRD	–	506
African buffalo (*Syncerus caffer*)	Czech and Slovakian Republics	–	Zoo	5	1		IFAT	1:40	–	1820
	Kenya	–	Wild	4	2	50.0	NAT	1:40	2 with titer at 1:80	677
	Senegal	–	Wild	4	2	50	cELISA	VMRD	–	506
Sitatunga (*Tragelaphus spekii gratus*)	Czech and Slovakian Republics	–	Zoo	7	1	14.3	IFAT	1:40	Titer at 1:160	1820
Impala (*Aepyceros melampus*)	Kenya	–	Wild	14	2	14.3	NAT	1:40	Titer <1:80	677
	Czech and Slovakian Republics	–	Zoo	4	0	00	IFAT	1:40	–	1820
Lesser kudu (*Tragelaphus imberbis*)	Germany	Hannover	Zoo	4	4	100.0	IFAT	1:2	Pleural fluids of 3 calves and serum sample of 1 dam	1600
Lechwe (*Kobus leche*)	Czech and Slovakian Republics	–	Zoo	4	1	25.0	IFAT	1:40	Titer at 1:80	1820
Blackbuck (*Antilope cervicapra*)	Czech and Slovakian Republics	–	Zoo	9	2	22.2	IFAT	1:40	Maximum titer, 1 at 1:1280	1820

[a] In Czech and Slovakian zoological gardens by IFAT (1:40),[1820] no anti-*N. caninum* antibodies were detected in: 11 blesbok (*Damaliscus pygargus phillipsi*), 3 white-tailed gnu (*Connochaetes gnou*), 2 blue wildebeest (*Connochaetes taurinus*), 17 roan antelope (*Hippotragus equinus*), 17 sable antelope (*Hippotragus niger*), 9 scimitar-horned oryx (*Oryx dammah*), 6 gemsbok (*Oryx gazella*), 8 addax (*Addax nasomaculatus*), 5 mountain reedbuck (*Redunca fulvorufula*), 7 Nile lechwe (*Kobus megaceros*), 2 springbok (*Antidorcas marsupialis*), 1 slender-horned gazelle (*Gazella leptoceros*), 11 Saiga antelopes (*Saiga tatarica*), 2 dwarf forest buffaloes (*Syncerus caffer nanus*), 10 nyalas (*Tragelaphus angasii*), 4 bongo (*Tragelaphus eurycerus*), 2 lesser kudu (*Tragelaphus imberbis*), and 1 greater kudu (*Tragelaphus strepsiceros*). In addition, no antibodies were detected by competitive ELISA in 2 scimitar-horned oryx (*Oryx dammah*), and 1 hippotrague (*Hippotragus equinus koba*) from a wild environment in Senegal.[506]

[b] VMRD = Neospora caninum cELISA Competitive ELISA GP65 surface antigen of tachyzoites, VMRD, USA.

Table 14.6 Seroprevalence of *N. caninum* Antibodies in Other Herbivorous Animals

Host	Country	Region	Type	No. Tested	No. Positive	% Positive	Assay	Cur-Off Titer	Remarks	Reference
Family Ailuridae										
Red Panda (*Ailurus fulgens*)	China	Sichuan	Captive	8	0	0.0	IFAT	1:5, 1:50	8 other pathogens studied	1205
	China	7 locations	Captive	73	3	4.1	IFAT	1:20	8 other pathogens studied	1650
Family Macropodidae										
Western gray kangaroo (*Macropus fuliginosus*)	Australia	Perth	Wild, reserves	102	18	17.6	IFAT	1:50	No association between positive results and reproductive performance	1297
Family Elephantidae										
Indian elephant (*Elephas maximus indicus*)	Thailand	Kanchanaburi	Captive	115	38	33.0	cELISA	VMRD	6.1% coinfection with *T. gondii*	2090

brain of the fawn. *Neospora* DNA was also detected by PCR from 1 of the 4 fetuses (Table 14.3). Microsatellite typing suggested that the infection in the fawn and the fetus was from the same source.

14.3 OTHER HERBIVORES

14.3.1 Serologic Prevalence

Available information is summarized in Table 14.4 (North American, European, and Asian wild Bovidae), Table 14.5 (African wild Bovidae), and Table 14.6 (other non-ruminant herbivorous species).

14.3.2 Clinical Neosporosis

A clinical case in an East African antelope (also called the lesser kudu) (*Tragelaphus imberbis*; Family Bovidae) was reported from neonates in a zoo in Germany (Table 14.3). Evidence for congenital infection was based on the detection of *N. caninum* antibodies in presuckling sera and the detection of parasite DNA in brains of 2 calves. The cause of mortality was not established.

BIBLIOGRAPHY

White-tailed Deer: **67, 170, 581, 612, 618, 783, 824, 1190, 1406, 1509, 2043**.
Other North American and European Cervidae: **39, 155, 201, 243, 439, 442, 601, 611, 674, 716, 783, 802, 824, 1267, 1409, 1414, 1441, 1550, 1639, 1641, 1769, 1820, 1875, 1910, 1943, 2113**.
Other Cervids: **117, 155, 170, 442, 567, 846, 1343, 1344, 1406, 1513, 1820, 1970, 1971, 2182**.
Other Herbivores: **39, 155, 200, 243, 264, 506, 601, 674, 675, 677, 716, 721, 824, 1143, 1205, 1242, 1297, 1406, 1407, 1600, 1650, 1769, 1820, 2090**.
Additional References: **45, 405, 406, 508, 1747, 1819, 1851**.

CHAPTER 15

Neosporosis in White Rhinoceros (*Ceratotherium simum*)

The white rhinoceros is related to wild Equidae (Perissodactyla; odd-toed ungulates). There is no report on serologic prevalence of *N. caninum* in rhinos. However, 3 deaths due to overwhelming neosporosis have been reported (Table 15.1). One rhinoceros from South Africa died of massive myocardial necrosis (Figures 15.1 and 15.2), and the one from Thailand of severe hepatitis with 60% of liver parenchyma necrotic. An aborted rhinoceros in Australia also had extensive hepatic necrosis. Numerous tachyzoites were found in lesions in all 3 animals.

Table 15.1   Clinical Neosporosis in White Rhinoceros (*C. simum*)

Country	Age	History	Clinical	Diagnosis				Reference
				Histology	IHC	PCR	TEM	
Australia	Fetus-7 month gestation	Aborted in captivity in zoo	–	Necrosis in liver with tachyzoites. Tachyzoites seen in cerebellum without any lesions	Tachyzoites in liver[b]	Positive liver using Np6 and Np21 and sequencing. Distinct genotype with MS10	Not done	**1773**
South Africa	16 days old	Free range herd. Born in captivity.	Died suddenly of heart failure	Severe myocarditis	Not in original paper[a]	Not done	Yes	**2102**
Thailand	16 years	Captive in zoo for 15 years	Died suddenly, no signs previous day	Grossly visible necrotic areas in liver, involving 60% of liver. Necrosis in adrenal cortex, and kidneys. Tachyzoites in lesions. No tissue cysts	Positive in liver, kidneys, adrenal, intestine, and bile duct epithelium. No lesions or parasites in brain. Tissue cysts and tachyzoites in heart	Positive liver using pNc5 PCR	Not done	**1876**

[a]   Confirmed by J. P. Dubey (Figure 15.2).
[b]   Histology and IHC illustrations of this case were provided by Reference **508**.

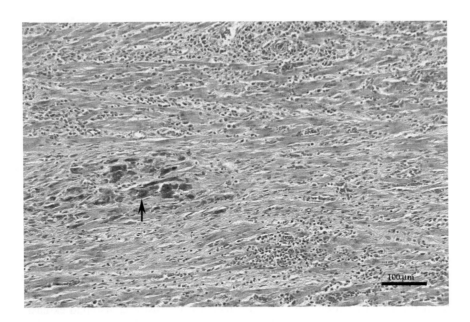

Figure 15.1 Severe myocarditis in a rhinoceros from South Africa from Reference **2102**. Note mononuclear cell infiltration and an area of mineralization (arrow). HE stain. (Courtesy of Dr. June Williams.)

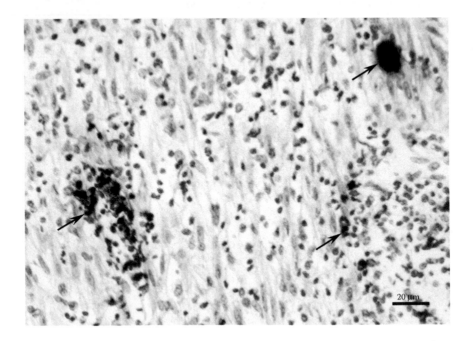

Figure 15.2 Foci of myocarditis from the rhinoceros in Figure 15.1. IHC using anti-*N. caninum* antibodies. All black areas are *N. caninum* tachyzoites (arrows).

BIBLIOGRAPHY

45, 508, 786, 1773, 1820, 1876, 2102.

CHAPTER 16

Neosporosis in Wild Canids and Other Carnivores

16.1 WILD CANIDS

16.1.1 Australian Dingo (*Canis lupus dingo*)

In 2 Australian surveys, antibodies against *N. caninum* were found in 0.9% and 26.9% of dingoes (Table 16.1). Dingoes are considered important in the epidemiology of neosporosis in Australia,[1067,1068] because in the wild, dogs in aboriginal communities are most likely dingo/wild dog hybrids. The evidence for *Neospora*-like oocysts in these dogs is reviewed. Three pure 12-week-old dingo pups were fed beef experimentally infected with *N. caninum*; 1 of these 3 pups excreted few oocysts 12–14 days later.[1067] The oocysts were morphologically similar to *N. caninum* oocysts and their identity was confirmed by PCR and sequencing of pNc5 gene and ITS1 rDNA gene.[1067] *N. caninum*-like oocysts were found in the feces of 2 of 132 aboriginal community dogs; these were PCR-positive for *N. caninum*.[1070] Oocysts from 1 of these dogs were bioassayed in dunnarts (*S. crassicaudata*), that are highly susceptible to *N. caninum* infection.[1069] Oocysts (unknown number) were fed to 2 dunnarts and the experiment was terminated 46 days later. Both dunnarts remained asymptomatic and did not develop antibodies to *N. caninum* as assessed by a cELISA (VMRD). However, tissue cysts were identified in the spleen of 1 dunnart by IHC but not by PCR. *N. caninum* DNA was found in the brain of the second dunnart by using primers targeting the pNc5 gene and this was confirmed by amplification of *N. caninum* MS10 microsatellite DNA using nested PCR.[1069] Although there is strong evidence that dingoes represent another definitive host of *N. caninum*, the *in vitro* isolation of viable *N. caninum* from dingo feces would provide the unambiguous proof that dingoes are definitive hosts for *N. caninum*.

16.1.2 Coyote (*C. latrans*)

Serologic information is summarized in Table 16.1.

Coyotes are considered important in the epidemiology of neosporosis in North America.[125] *N. caninum*-like oocysts were found in the feces of 2 of 185 coyotes in Canada, based on PCR. However, viable *N. caninum* oocysts have not yet been identified in the feces of naturally infected coyotes.[2072]

Experimentally, 1 of 4 coyotes experimentally fed infected beef excreted few *N. caninum*-like oocysts 8–10 days later. Results were confirmed by PCR using Np6/Np21 primers. Bioassay of coyote oocysts was not performed. All 4 coyotes developed IFAT titers of 1:800 or 1:1600, 28 DPI.

In another study, oocysts were not found in the feces of 3 coyote pups fed *N. caninum*-infected tissues[1176]; the pups developed IFAT titers of 1:800 or 1:1600 but *N. caninum* stages were not identified in the tissues of the pup necropsied 33 DPI.

Table 16.1 Seroprevalence of *N. caninum* Antibodies in Wild Canids

Host[a,b]	Country	Region	Type	No. Tested	No. Positive	% Positive	Assay	Cut-Off Titer or Test[c]	Remarks	Reference
Australian dingo (*Canis familiaris dingo*)	Australia	Queensland	Feral	52	14	26.9	IFAT	1:50	Highest titer at 1:3200	123
	Czech Republic and Slovakia	New South Wales	Feral	117	1	0.9	IFAT	1:50	–	1820
		–	Zoo	2	0	0.0	IFAT	1:40	–	
Azara's fox (*Lycalopex gymnocercus*)	Brazil	Rio Grande do Sul	Wild	12	5	41.6	IFAT, NAT	1:50, 1:40	Highest titers 1:1600 IFAT in 1, and NAT 1:640 in 5	291
		Rio Grande do Sul	Captivity	5	1	20.0	IFAT	1:50	–	1296
Crab-eating fox (*Cerdocyon thous*)		São Paulo and Paraná	Wild	15	4	26.6	IFAT, NAT	1:50, 1:40	Highest titers 1:100 IFAT in 3, and NAT 1:40 in 1	291
		Minas Gerais	Captivity	2	0	0.0	IFAT	1:50	–	453
		–	Captivity	25	9	36.0	IFAT	1:50	–	1296
		–	Captivity	7	0	0.0	IFAT	1:25	–	1637
Blue fox (*Alopex lagopus*)	China	Hebei	Farmed	103	28	27.2	cELISA	VMRD	*Neospora*-like cysts were seen in brain and kidney	2165
Chiloé fox (*Pseudalopex fulvipes*)	Chile	IX Region	Zoo	2	2	100.0	NAT	1:20	Highest titer 1:320	1570
Coyote (*Canis latrans*)	Canada	Prince Edward Island	Wild	181	6	3.3	IFAT, NAT, IB, ELISA	1:25, 1:25, Biovet	Positive in all tests	2075
	USA	Colorado	Wild	28	5	17.9	IFAT	1:50	–	783
		Illinois	Wild	40	6	15.0	IFAT	1:50	–	783
		Utah	Wild	45	1	2.2	IFAT	1:50	–	783
		Texas	Wild	52	5	9.6	IFAT	1:25	IFAT titers of 1:25 in 3, 1:50 in 1, and 1:100 in 1	1176
		Texas, Oklahoma, New Mexico	Wild	394	71	18.0	ELISA	Biovet	Samples on filter paper	198
		Alaska	Wild	12	2	16.7	IFAT	1:100	Highest titer 1:200	1910

(Continued)

Table 16.1 (Continued) Seroprevalence of *N. caninum* Antibodies in Wild Canids

Host[a,b]	Country	Region	Type	No. Tested	No. Positive	% Positive	Assay	Cut-Off Titer or Test[c]	Remarks	Reference
Eurasian wolf (*Canis lupus lupus*)	Czech Republic and Slovakia	–	Zoo	10	2	20.0	IFAT	1:40	Highest titer at 1:320	1820
Fennec (*Vulpes zerda*)	Czech Republic and Slovakia	–	Zoo	2	2	100.0	IFAT	1:40	Highest titer in 2 at 1:320	1820
Gray wolf (*Canis lupus*)	Israel	–	Wild	9	1	11.1	IFAT	1:50	Titer of 1:400	1905
	Sweden and Norway	–	Wild	109	4	3.7	IB, ELISA	ISCOM	Samples positive in both tests	225
	Spain	–	Wild	28	7 in cELISA	25.0	cELISA, NAT, IFAT	VMRD	6 were confirmed by NAT	1872
	USA	Alaska	Wild	324	29	9.0	IFAT	1:100	Highest titer in 3 at 1:800	1910
		Alaska	Wild	122	4	3.3	NAT	1:40	–	601
		Alaska	Wild	28	0	0.0	IFAT	Not stated	–	2076
		Yellowstone National Park (ID, MT, WY)	Wild	202	100	49.5	IFAT	1:50	Increased exposure with age	35
		Minnesota	Wild	164	64	39.0	IFAT	1:50	–	783
		Minnesota	Wild	232 adults	153	66.3	IFAT, NAT	1:50, 1:25	–	309
				55 pups	19	35.8				
Golden jackal (*Canis aureus*)	Israel	–	Wild	114	2	1.7	IFAT	1:50	Highest titer at 1:50	1905
Gray fox (*Urocyon cinereoargenteus*)	USA	South Carolina	Wild	26	4	15.4	NAT	1:25	IFAT titers 1:25 in 3 and 1:50 in 1 fox	1188
South American gray fox (*Dusicyon griseus*)	Argentina	Santa Cruz	Wild	56	20	35.7	IFAT	1:25	8 other pathogens investigated	1286
Culpeo fox (*Dusicyon culpaeus*)	Argentina	Santa Cruz	Wild	28	17	60.7	IFAT	1:25	8 other pathogens investigated	1286

(Continued)

Table 16.1 (Continued) Seroprevalence of _N. caninum_ Antibodies in Wild Canids

Host[a,b]	Country	Region	Type	No. Tested	No. Positive	% Positive	Assay	Cut-Off Titer or Test[c]	Remarks	Reference
Maned wolf (_Chrysocyon brachyurus_)	Brazil	Minas Gerais, Distrito Federal, São Paulo	Zoo and preserves	59	5	8.5	IFAT	1:25	Also infected with _T. gondii_ in 74.6%	**1857, 2047**
	Brazil	Minas Gerais, Distrito Federal, Paraná, Rio de Janeiro	Zoo and wild (4)	48	0	0.0	IFAT	1:50	–	**453**
	Brazil	–	Captivity	14	4	26.6	IFAT	1:50	–	**1296**
	Brazil	–	Wild	3	0	0.0	IFAT	1:25	–	**1637**
	Czech Republic and Slovakia	–	Zoo	6	1	16.6	IFAT	1:40	Highest titer at 1:640	**1820**
Raccoon dog (_Nyctereutes procyonoides koreensis_)	Korea	–	Wild	26	6	23.0	NAT	1:50	–	**1063**
Bush dog (_Speothos venaticus_)	Brazil	–	Captivity	6	0	0.0	IFAT	1:50	–	**1296**
Red fox (_Vulpes vulpes_)	Austria	Burgenland	Wild	94	0	0.0	IFAT	1:50	35% infected with _T. gondii_	**2071**
	Belgium	–	Wild	123	96	78.0	IFAT	1:64	21 sera had titers of 1:256–1:8192	**255**
	Czech Republic	8 regions	Wild	80	3	3.8	IFAT	1:50	All had _T. gondii_ titers	**162**
	Canada	Prince Edward Island	Wild	263	3	1.1	IFAT 1:25, ELISA, IB, NAT 1:25	ELISA, Biovet	Positive in all tests	**2075**
	Germany	–	Farmed	122	3	2.5	IB	17, 29, 30, 33, and 37 kDa	Evidence for vertical transmission	**1796**
	Hungary	16 counties	Wild	337	5	1.5	ELISA	IH-ISCOM	4 out of 5 positive were coinfected with _T. gondii_	**992**

(Continued)

Table 16.1 (Continued) Seroprevalence of *N. caninum* Antibodies in Wild Canids

Host[a,b]	Country	Region	Type	No. Tested	No. Positive	% Positive	Assay	Cut-Off Titer or Test[c]	Remarks	Reference
	Ireland	–	Wild	220	6	2.7	IFAT	1:50	3 sera had titers of 1:800, and 3 had titers of 1:6400. Thoracic fluids tested	1439
	Ireland	–	Wild	101	1	1.0	IFAT	1:40	6% positive by PCR in brain	1918
	Ireland	Dublin	Wild	70	1	1.4	IFAT	1:20	Highest titer at 1:1600	2110
	Israel	–	Wild	24	1	4.2	IFAT	1:50	Highest titer at 1:400	1905
	Poland	Lower Silesia	Wild	45	15	33.3	IFAT	1:20	–	1868
	Slovakia		Periurban	177	69	38.9	cELISA	VMRD	–	1703a
			Mountain	126	11	8.7				
	Spain	–	Wild	95	11 in ELISA	11.6	cELISA, NAT, IFAT	VMRD	2 were confirmed by IFAT and 2 by NAT	1872
	Spain	Pyrenees	Wild	53	37	69.8	NAT	1:40	Highest titer of 1:160 in 3	1276
	Sweden	–	Wild	221	0	0.0	ELISA	IH-ISCOM	*T. gondii* also investigated	989
	UK	–	Wild	546	5	0.9	IFAT	1:256	IFAT titers of 1:16 to 1:1:64 were found in additional 8 foxes. Lung fluid, not blood serum, tested	843
	UK	–	Wild	54	1	1.8	IFAT	1:50	Titer at 1:200	123
	UK	Cornwall	Wild	16	1	6.2	IFAT	1:50	–	1861
	USA	Alaska	Wild	9	0	0.0	IFAT	1:100	–	1910
Silver fox (species not specified)	Poland	Lower Silesia	Farmed	60	22	36.7	IFAT	1:20	–	1868

a 30 Hoary foxes (*D. vetulus*) tested negative for *N. caninum* antibodies in Paraiba, Brazil.[291]

b 9 African wild dogs (*L. pictus*) and 3 bush dogs (*S. venaticus*) kept in captivity in Czech and Slovakian zoological gardens tested negative for *N. caninum* antibodies.[1820]

c VMRD = Neospora caninum cELISA Competitive ELISA GP65 surface antigen of tachyzoites, VMRD, USA; Biovet = BIOVET-Neospora caninum, indirect ELISA, sonicate lysate of tachyzoites, BIOVET Laboratories, Canada; ISCOM = Detergent extracted tachyzoite antigen incorporated in immune stimulating complex particles.

16.1.3 Red Fox (*Vulpes vulpes*)

Several serological surveys have been reported in red foxes (Table 16.1) because they were suspected as definitive hosts of *N. caninum*. Indeed, *N. caninum*-like oocysts were not found in the feces of naturally exposed foxes in Spain,[36] Ireland,[2110] and Germany.[380] However, *N. caninum*-like oocysts were reported in the feces of 2 of 271 foxes in Canada, based on PCR testing.[2072] Several studies demonstrated that they are important intermediate hosts.[36,1796] However, foxes that were experimentally fed *N. caninum*-infected tissues did not excrete oocysts.[1799]

There is 1 histologically confirmed report of clinical neosporosis in a fox. A 3-month-old fox developed granulomatous dermatitis during rehabilitation following an automobile accident.[619] The fox had a low (NAT, 1:25) titer to *N. caninum*. Numerous *N. caninum* tachyzoites were found in the lesions by IHC. There was concurrent asymptomatic *T. gondii* infection and viable *T. gondii,* but *N. caninum* was not isolated by bioassay. The fox had been treated with clindamycin that had probably killed the *N. caninum* tachyzoites.[619]

16.1.4 Blue Fox (*Alopex lagopus*)

Antibodies to *N. caninum* were found in blue foxes in China (Table 16.1). *N. caninum* tissue cysts were also reported in the histological sections of kidneys and brain of 5 foxes with an undiagnosed illness.[2165] This report needs confirmation because tissue cysts are rare in renal tissues and in IHC performed by one of us, JPD, the results reported in Reference **2165** could not be confirmed.

A blue fox inoculated IM with *N. caninum*-infected dog brain developed encephalitis and tachyzoites were found in lesions.[205]

16.1.5 Gray Wolf (*C. lupus*)

Wolves are considered to be very important hosts in the epidemiology of neosporosis in wildlife and free range cattle in the USA. *N. caninum*-like oocysts were found in the feces of 3 of 73 wolves.[615] Oocysts from 1 of these 3 samples were proven to be *N. caninum* by bioassays in KO mice, further propagation in cell culture, and by DNA characterization. This isolate was designated NcWolfUS1. Additionally, viable *N. caninum* was isolated by bioassay in KO mice and cell culture from the brain and heart of 2 of 109 wolves from the USA.[620] These isolates were designated NcWolfMn1 and NcWolfMn2.

16.1.6 Other Wild Canids

Serological surveys have been performed in other wild canid species (Table 16.1) such as Azara's fox (*Lycalopex gymnocercus*), crab-eating fox (*Cerdocyon thous*), Chiloé fox (*Pseudalopex fulvipes*), fennec (*Vulpes zerda*), golden jackal (*Canis aureus*), gray fox (*Urocyon cinereoargenteus*), South American gray fox (*Dusicyon griseus*), Culpeo fox (*Dusicyon culpaeus*), hoary fox (*Dusicyon vetulus*), maned wolf (*Chrysocyon brachyurus*), Korean raccoon dog (*Nyctereutes procyonoides koreensis*), African wild dog (*Lycaon pictus*), and bush dog (*Speothos venaticus*).

16.2 OTHER WILD CARNIVORES

16.2.1 Raccoons (*Procyon lotor*)

Serologic prevalence data is summarized in Table 16.2.

Table 16.2 Seroprevalence of *N. caninum* Antibodies in Other Wildlife Carnivores

Host	Country	Region	Type	No. Tested	No. Positive	% Positive	Assay	Cut-Off Titer or Test[d]	Remarks	Reference
Family Didelphidae										
South American opossum (*Didelphis marsupialis*)	Brazil	São Paulo city	Feral	396	84	21.2	IFAT	1:25	Highest titer of 1:400 in 1	2138
Virginia opossum (*Didelphis virginiana*)	USA	Louisiana	Wild	30	0	0.0	IFAT	1:100	–	938
Family Herpestidae										
Egyptian mongoose (*Herpestes ichneumon*)	Spain	Several parts	Wild	23	0	0.0	cELISA, NAT, IFAT	VMRD	–	1872
Family Hyaenidae[a]										
Spotted hyena (*Crocuta crocuta*)	Kenya	–	Wild	3	1	33.3	NAT	1:40	Titer at 1:80	677
	Senegal	–	Wild	2	0	0.0	cELISA	VMRD	–	506
Family Phalangeridae[b]										
Common brushtail possum (*Trichosurus vulpecula*)	Australia	Sydney city	Feral	142	0	0.0	NAT	1:25	–	658
Family Procyonidae										
Raccoon (*Procyon lotor*)	USA	Massachusetts. Florida, Pennsylvania, New Jersey	Wild	99	10	10.1	NAT	1:50	IFAT titer of 1:50 in 9, and 1:100 in 1	1189
Family Ursidae										
Black bear (*Ursus americanus*)	USA	North Carolina	Wild	64	0	0.0	NAT	1:40	–	601
	USA	Alaska	Wild	133	0	0.0	NAT	1:40	–	
	Czech Republic and Slovakia	–	Zoo	3	0	0.0	IFAT	1:40	–	1820
Polar bear (*Ursus maritimus*)	Czech Republic and Slovakia	–	Zoo	1	0	0.0	IFAT	1:40	–	1820
Family Viverridae[c]										
Common genet (*Genetta genetta*)	Spain	Several parts	Wild	19	0	0.0	cELISA, NAT, IFAT	VMRD	–	1872

[a] 3 brown hyenas (*Parahyena brunnea*) kept in Czech and Slovakian zoological gardens tested negative for *N. caninum* antibodies. [1820]

[b] Mainly folivore, but known to eat rodents.

[c] 2 fossas (*Cryptoprocta ferox*) kept in Czech and Slovakian zoological gardens tested negative for *N. caninum* antibodies. [1820]

[d] VMRD = Neospora caninum cELISA Competitive ELISA GP65 surface antigen of tachyzoites; VMRD, USA.

Table 16.3   Seroprevalence of *N. caninum* Antibodies in Wild Mustelids

Host[a]	Country	Region	Type	No. Tested	No. Positive	% Positive	Assay	Cut-Off Titer or Test[b]	Remarks	Reference
American mink (*Mustela vison*)	Ireland	–	Wild	114	10	0.9	IFAT	1:40	Negative by PCR	1918
Eurasian badger (*Meles meles*)	Spain	Several areas	Wild	31	2	6.4	cELISA, NAT, IFAT	VMRD	2 were confirmed by NAT	1872
Fisher (*Martes pennanti*)	Ireland	–	Wild	51	0	0.0	IFAT	1:40	Negative by PCR	1918
	Czech Republic and Slovakia	–	Zoo	2	1	50.0	IFAT	1:40	–	1820
Pole cat (*Mustela putorius*)	Spain	Several areas	Wild	2	1	50.0	cELISA, NAT, IFAT	VMRD	1 was confirmed by NAT	1872
Pine martin (*Martes martes*)	Spain	Several areas	Wild	3	2	66.7	cELISA, NAT, IFAT	VMRD	2 were confirmed by NAT	1872
	Ireland	–	Wild	9	0	0.0	IFAT	1:40	Negative by PCR	1918
Stone marten (*Martes foina*)	Spain	Several areas	Wild	14	3	21.4	cELISA, NAT, IFAT	VMRD	3 were confirmed by NAT	1872
North American river otter (*Lontra canadensis*)	USA	Alaska, Washington state	Wild	40						

[a] 34 wild stoats (*M. erminea*) and 4 feral ferrets (*M. putorius*) from Ireland tested negative for *N. caninum* antibodies and also for PCR.[1918] In the same survey, antibodies were not found in 30 Eurasian otters (*Lutra lutra*) but 4.2% of them tested positive for PCR. In addition, 5 Eurasian otters tested seronegative in Spain.[1872] The absence of antibodies anti-*N. caninum* was reported in a survey on 40 wild North American river otter (*L. canadensis*) from Alaska and Washington state in the USA by IFAT.[743]

[b] VMRD = Neospora caninum cELISA Competitive ELISA GP65 surface antigen of tachyzoites, VMRD, USA.

Table 16.4   Detection of *N. caninum* DNA in Wild Carnivores

Host	Country	No. Tested	No. Positive	%	PCR Method, Locus	Remarks	Reference
Family Delphinidae							
Killer whale (*Orcinus orca*)	Japan	8	0	0.0	Conventional PCR[2141], pNc5	DNA extraction from 0.5 g of brain, liver, diaphragm, testis, heart, and kidney tested	**1515**
Family Canidae							
Blue fox (*Alopex lagopus*)	China	5	4	80.0	Seminested PCR[2141] and sequencing, pNc5	NS	**2165**
Hoary fox (*Lycalopex vetulus*)	Brazil	49	6	12.2	Conventional PCR[2141] and sequencing, pNc5	DNA extraction from 25 mg of brain following commercial kit protocol	**1450**
Red fox (*Vulpes vulpes*)	Czech Republic	152	7	4.6	Conventional PCR[1426], pNc5	DNA extraction from 40 mg of frozen brain	**958**
	Great Britain	83	4	4.8	Nested PCR[259] and sequencing, ITS1	DNA extraction from 1 g of brain tested	**152**
	Ireland	33	0	0.0	Conventional PCR[2141], pNc5	DNA extraction from frozen brain with protozoal-associated lesions by histology	**1439**
	Ireland	151	9	6.0	Nested PCR[259] and sequencing, ITS1	DNA extraction from 200 mg of brain tested	**1918**
	Romania	182	1	0.5	Conventional PCR[2141], pNc5	DNA extraction from 40 mg of brain tested	**1925**
	Spain	122	13	10.7	Conventional PCR[1152], pNc5	DNA extraction from 0.5 g of tissue	**36**
Family Mustelidae							
Eurasian badger (*Meles meles*)	Great Britain	64	7	10.9	Nested PCR[259] and sequencing, ITS1	DNA extraction from 1 g of tissue (brain, muscle, heart, lung, liver, kidney, spleen, spinal cord, blood, lymph node)	**152**
	Ireland	50	0	0.0	Nested PCR[259] and sequencing, ITS1	DNA extraction from 200 mg of brain tested	**1918**

(Continued)

Table 16.4 (Continued) Detection of *N. caninum* DNA in Wild Carnivores

Host	Country	No. Tested	No. Positive	%	PCR Method, Locus	Remarks	Reference
Eurasian otter (*Lutra lutra*)	Ireland	24	0	0.0	Nested PCR[259] and sequencing, ITS1	DNA extraction from 200 mg of brain tested	1918
Ferret (*Mustela furo*)	Great Britain	99	10	10.1	Nested PCR[259] and sequencing, ITS1	DNA extraction from 1 g of brain tested	152
Mink (*Neovison vison*)	Great Britain	65	3	4.6	Nested PCR[259] and sequencing, ITS1	DNA extraction from 1 g of brain and muscle tested	152
	Ireland	197	6	3.1	Nested PCR[259] and sequencing, ITS1	DNA extraction from 200 mg of brain tested	1918
Pine marten (*Martes martes*)	Ireland	8	0	0.0	Nested PCR[259] and sequencing, ITS1	DNA extraction from 200 mg of brain	1918
Polecat (*Mustela putorius*)	Great Britain	70	13	18.6	Nested PCR[259] and sequencing, ITS1	DNA extraction from 1 g of brain	152
Stoat (*Mustela erminea*)	Great Britain	9	0	0.0	Nested PCR[259] and sequencing, ITS1	DNA extraction from 1 g of tissue (brain, muscle, heart, lung, liver, kidney, spleen, spinal cord, blood, lymph node)	152
	Ireland	33	0	0.0	Nested PCR[259] and sequencing, ITS1	DNA extraction from 200 mg of brain	1918
Family Ursidae							
Brown bear (*Ursus arctos*)	Slovakia	45	11	24.4	Nested PCR[259] and sequencing, ITS1 and conventional PCR[2141], pNc5	DNA extraction from 25 mg of muscle, liver, or spleen using commercial kit protocol	354

An *N. caninum* tissue cyst was identified histologically in the brain of an encephalitic raccoon in the USA[1140]; diagnosis was confirmed by PCR. This juvenile raccoon had a clinical canine distemper virus infection, which is known to be immunosuppressive. Thus, it is not clear whether *N. caninum* was associated with encephalitis.

Experimentally, 2 raccoons fed *N. caninum*-infected tissues remained clinically normal, became seropositive, but did not excrete oocysts.[559]

16.2.2 Mustelids

Serologic information is summarized in Table 16.3.

Experimentally, 4 ferrets (*Mustela putorius*), 4 short-tailed weasels (*Mustela erminea*), and 5 long-tailed weasels (*Mustela frenata*) fed *N. caninum* tissue cysts acquired infection, but stayed clinically healthy and did not excrete oocysts.[1306]

16.2.3 DNA Detection in Wild Carnivores

Information is summarized in Table 16.4.

BIBLIOGRAPHY

Wild Canids: **35, 36, 123, 125, 162, 198, 205, 225, 255, 291, 309, 380, 453, 601, 615, 619, 620, 783, 843, 989, 992, 1063, 1067, 1070, 1176, 1188, 1276, 1286, 1296, 1439, 1570, 1637, 1796, 1799, 1820, 1857, 1861, 1868, 1872, 1905, 1910, 1918, 2047, 2071, 2072, 2075, 2076, 2110, 2165**.
Other Wild Carnivores: **36, 152, 354, 506, 559, 601, 658, 677, 743, 938, 958, 1140, 1189, 1306, 1439, 1450, 1515, 1820, 1872, 1918, 1925, 2138, 2165**.
Additional References: **508, 564**.

CHAPTER 17

Neosporosis in Miscellaneous Animals

In this chapter information on neosporosis in all animal species, not included previously, is summarized.

17.1 GIANT PANDA (*AILUROPODA MELANOLEUCA*)

Antibodies to *N. caninum* were not detected in serum samples of 19 Giant pandas from China; sera were tested by IFAT (cut-off 1:50).[1205]

17.2 PARMA WALLABY (*MACROPUS PARMA*)

An 8-year-old captive parma wallaby was found dead in a zoo in Austria. Grossly, the left side of the heart was hypertrophied. Necrosis and mononuclear cell infiltrations were found in the myocardium. The presence of *N. caninum* tachyzoites was confirmed by IHC and PCR testing.[399]

17.3 MARINE MAMMALS

We are not aware of clinical neosporosis in these animals. Serologic prevalence is summarized in Table 17.1.

17.4 SMALL MAMMALS

Serologic prevalence is summarized in Table 17.2.

17.4.1 Rodents and Lagomorphs

Serologic information is summarized in Table 17.2.

Information on the presence of DNA in tissues is summarized in Table 2.2 (see Chapter 2).

An IHC study examined sections of livers, hearts, lung, and brains of 14 squirrels (*Spermophilus variegatus*), 6 rats (*Rattus norvegicus*), and 13 house mice (*Mus musculus*) from Mexico.[1336] *N. caninum*-like parasites were found in the brain and liver of 1 mouse, and in the livers of 2 rats and 6 squirrels. The putative presence of *N. caninum* in the livers of 6 of 6 asymptomatic rats is an unusual finding, and these results need confirmation.

Table 17.1 Seroprevalence of *N. caninum* Antibodies in Marine Mammals

Host	Country	Region	Type	No. Tested	No. Positive	% Positive	Assay[a]	Cut-Off Titer, Antigen or Test	Remarks	Reference
Kuril harbor seal (*Phoca vitulina stejnegeri*)	Japan	Hokkaido	Wild	234–322	14	5.9	ELISA, tNcSAG1	IH	—	712
Spotted seal (*Phoca largha*)	Japan	Hokkaido	Wild	13–46	2	15.3	ELISA tNcSAG1	IH	—	712
Sea otter (*Enhydra lutris neresis*)	USA	California	Wild (found dead)	16	10	62.5	IFAT	1:40	4 with titer >1:320	1354
	USA	California, Washington state	Wild (found dead)	115	17	14.8	NAT	1:40	2 with titer at 1:320	593
	USA	Washington state	Wild (live)	30	11	36.7	NAT	1:40	3 with titer at 1:160	593
Walrus (*Odobenus rosmarus*)	USA	Alaska	Wild	53	3	5.6	NAT	1:40	1 with titer at 1:320	593
Sea lion (*Zalophus californianus*)	USA	Alaska	Wild	27	1	3.7	NAT	1:40	1 with titer at 1:40	593
Harbor seal (*Phoca vitulina*)	USA	Alaska	Wild	331	11	3.5	NAT	1:40	1 with titer at 1:320	593
	USA	Alaska	Wild	34	0		IFAT	1:40		178a
Ringed seal (*Phoca hispida*)	USA	Alaska	Wild	32	4	12.5	NAT	1:40	1 with titer at 1:80	593
Spotted seal (*Phoca largha*)	USA	Alaska	Wild	9	0	0.0	NAT	1:40	—	593
Ribbon seal (*Phoca fasciata*)	USA	Alaska	Wild	14	0	0.0	NAT	1:40	—	593
Bottlenose dolphins (*Tursiops truncatus*)	USA	Florida	Wild	43	47	91.4	NAT	1:40	2 with titer at 1:320	593
Killer whale (*Orcinus orca*)	Japan	Hokkaido	Wild	8	1	12.5	IB	18 kDa	All tested negative by PCR	1515
Family Otariidae										
South American sea lion (*Otaria flavescens*)	Czech Republic and Slovakia	—	Zoo	2	0	0.0	IFAT	1:40	—	1820
Ribbon seal (*P. fasciata*)	Japan	Hokkaido	Wild	4	0	0.0	ELISA, tNcSAG1	IH	—	712
Bearded seal (*Erignathus barbatus*)	Japan	Hokkaido	Wild	1	0	0.0	ELISA tNcSAG1	IH	—	712
	USA	Alaska	Wild	8	1	12.5	NAT	1:40	1 with titer at 1:80	593

Table 17.2 Seroprevalence of *N. caninum* Antibodies in Lagomorphs and Rodents

Host	Country	Region	Type	No. Tested	No. Positive	% Positive	Assay	Cut-Off Titer, Antigen or Test[a]	Remarks	Reference
Family Leporidae										
Iberian hare (*Lepus granatensis*)	Spain	Several parts	Wild	53	1	1.8	cELISA	Pourquier	Not confirmed by IFAT (1:50)	39
European hare (*Lepus europaeus*)	Austria	–	Wild	383	143	37.3	cELISA	VMRD	*T. gondii* prevalence stated	157a
	Czech Republic	–	Wild	333	129	38.7	cELISA	VMRD	*T. gondii* prevalence stated	157a
	Hungary	–	Wild	93	8	8.6	NAT	1:40	Higher titer at 1:320	676
	Italy	Central	Wild	81	9	11.1	IFAT	Not stated	–	632
	Slovakia	–	Wild	44	3	6.8	NAT	1:40	Higher titer at 1:320.	676
	Slovakia	–	Wild	209	8	3.8	cELISA	VMRD	*T. gondii* prevalence stated	146
Wild rabbit (*Oryctolagus cuniculus*)	Italy	–	Domestic	260	3	1.2	IFAT	1:50	–	1247
	Japan	–	Domestic	337	0	0.0	IFAT	1:100	–	1768
	Spain	Several parts	Wild	251	0	0.0	cELISA	Pourquier	Not confirmed by IFAT (1:50)	39
Family Muridae										
Mouse (*Mus musculus*)	USA	Delaware	Wild	79	4	5.0	NAT	1:20	–	1008
Yellow necked mouse (*Apodemus flavicollis*)	Czech Republic	4 localities	Wild	240	1	0.4	cELISA	VMRD	–	1247
Rat (*Rattus norvegicus*)	Grenada	6 parishes	Wild	242	11	4.5	NAT	1:20	Positive by PCR of ITS1 (21.9%) and pNc5 (9.5%) fragments	1008
Capybaras (*Hydrochaeris hydrochaeris*)	Brazil	São Paulo, 11 counties	Wild	213	20	9.4	IFAT	1:25	–	2139
	Brazil	São Paulo, 6 counties	Wild	63	1	1.54	IFAT	1:25	–	2018
	Brazil	São Paulo, Itu	Wild	170	0	0.0	IFAT	1:50	10% seropositive for *T. gondii*	436

a Pourquier = Competitive ELISA, Institut Pourquier, France; VMRD = Neospora caninum cELISA Competitive ELISA GP65 surface antigen of tachyzoites, VMRD, USA.

N. caninum antibodies were found in 9 of 55 *R. norvegicus* from cattle farms in Taiwan.[949] Two of these seropositive rats had detectable *N. caninum* DNA in brains, but the parasite was not demonstrated histologically in their tissues. Brains of all rats were bioassayed in outbred mice immunosuppressed with prednisolone. When the outbred mice died, their brains were homogenized and inoculated IP into a nude mouse. Six months after inoculation with the brain homogenate of 1 seropositive rat, the brain homogenate of 1 outbred mouse (mouse no. A) was inoculated into a nude mouse. The nude mouse was euthanized (PI day not stated) and tested for *N. caninum* infection. A tissue cyst was found in the brain of the nude mouse, and stained positively with *N. caninum* antibody. DNA was isolated from the brain of the nude mouse and the presence of *N. caninum* confirmed by PCR. Results of this report need confirmation.

BIBLIOGRAPHY

39, 146, 178a, 399, 436, 439, 593, 632, 676, 712, 743, 949, 1008, 1205, 1246, 1247, 1336, 1354, 1515, 1768, 1820, 2018, 2139, 2167.

N. hughesi and Neosporosis in Horses and Other Equids

18.1 INTRODUCTION

In 1998, a new species of *Neospora*, *N. hughesi* was described from a horse in California, USA.[1284] To date, *N. hughesi* has been reported only from horses. It is uncertain at present whether there are 1 or 2 species of *Neospora* that infect horses. Therefore, available information on *N. hughesi* and neosporosis in horses is summarized in this chapter.

18.2 DIFFERENCES BETWEEN *N. HUGHESI* AND *N. CANINUM*

18.2.1 Morphological

18.2.1.1 Tachyzoites

In the original description of *N. hughesi,* tachyzoites in smears were 4.0–7.0 × 1.8–3.0 (n = 10) µm and 4.9–5.3 (n = 3) × 1.4–2.5 (n = 18) µm in sections of the spinal cord of the horse.[1284] However, any appreciable visual differences between tachyzoites of another isolate (Oregon isolate) of *N. hughesi* (Figure 18.1) and the NC1 isolate of *N. caninum* were not found when viewed by microscopy of cell culture derived tachyzoites.[585]

18.2.1.2 Tissue Cysts

In the original description of *N. hughesi,* tissue cysts (n = 6) in the spinal cord of the naturally affected horse were reported to be 6.9–16.0 × 10.7–19.3 µm in size, with a thin cyst wall (0.15–1.0 µm).[1284] The thickness of the tissue cyst wall was proposed as one of the distinguishing features of *N. hughesi,* because tissue cysts of *N. caninum* are generally thick-walled (1–3 µm). However, thick-walled tissue cysts (1.0–3.0 µm) have been reported in horses,[414,585,1175] (Figure 18.2) and thin-walled (<1.0 µm) tissue cysts were reported in dogs naturally infected with *N. caninum.*[573] Thus, there are no clear morphological differences between *N. caninum* and *N. hughesi* tissue cyst wall morphology.

18.2.1.3 Antigenic and Molecular Differences

Antigenic and molecular differences between *N. hughesi* and *N. caninum* have been reported.[585,1284,1285,1887] There is a 6%–9% difference in the amino acid sequences of 2 surface antigens (SAG1 and SRS2) between these 2 species.[585,1284,1285] Additionally, their dense granules

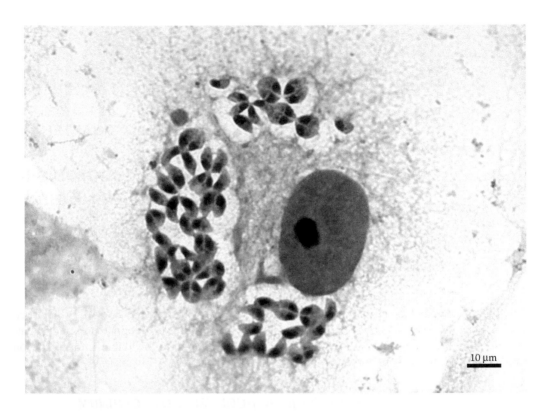

Figure 18.1   Individual and dividing tachyzoites in CV1 cells inoculated with 5×10^5 tachyzoites of the Alabama strain of *N. hughesi*. (Courtesy of Dr. David Lindsay).

proteins (GRA6 and GRA7) are also different.[2066] The nucleotide sequence of the ITS-1 region of these 2 parasites has a 7 bases difference.[585,1284]

18.2.1.4 Biological Differences

Rodents are more susceptible to experimental infections with *N. caninum* than *N. hughesi*.[585,2065] The KO mice inoculated with *N. caninum* died sooner than after infection with *N. hughesi*, and lesions observed by histology were different. Necrosis of the myocardium (Figure 18.3), even visible macroscopically, was the predominant feature of *N. hughesi* infection whereas lesions in liver, lungs, and brain predominated in *N. caninum* infections. Gerbils (*Meriones unguiculatus*) inoculated with *N. caninum* often developed fatal infections whereas they remained without symptoms after inoculation with *N. hughesi*. Whether these differences are due to species differences or represent strain-specific features remains to be elucidated in future experiments with a larger number of representative isolates using identical procedures.

18.3 NATURAL INFECTIONS IN HORSES

18.3.1 Serologic Prevalence

Information on horses is summarized in Table 18.1, and data regarding other equids is compiled in Table 18.2. In most of these surveys *N. caninum* tachyzoites were used as antigens. Only a few surveys employed *N. hughesi* tachyzoites and these studies are indicated in Table 18.1.

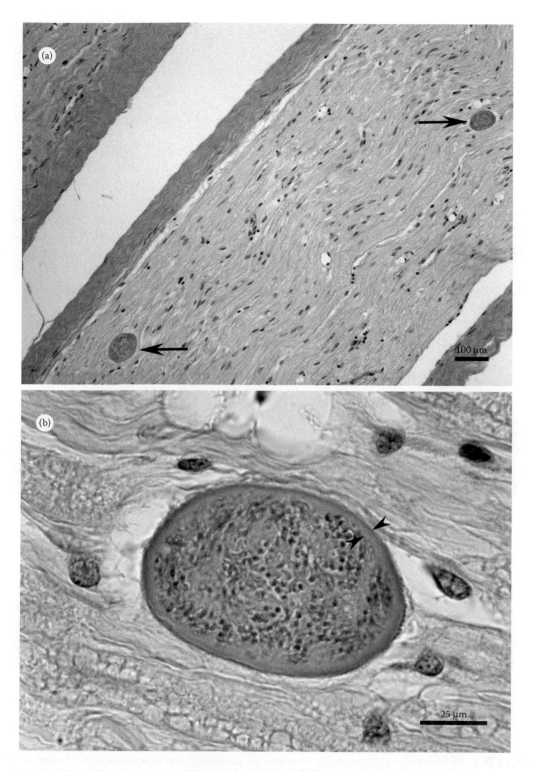

Figure 18.2 (a,b). Tissue cysts (arrows) in glossopharyngeal nerve of a horse reported in Reference **414**. HE stain. Note the absence of inflammation and thick cyst wall (arrowhead) of tissue cysts. (Sample courtesy of Dr. Barbara Daft).

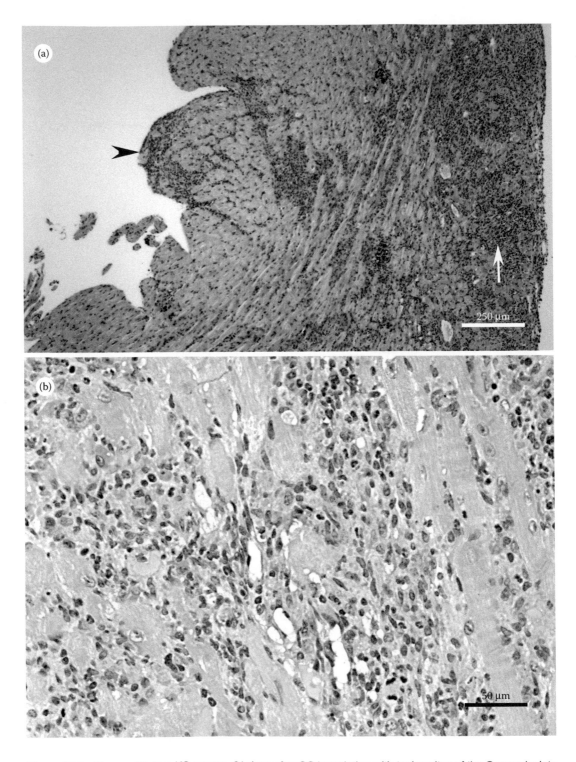

Figure 18.3 Myocarditis in a KO mouse, 21 days after SC inoculation with tachyzoites of the Oregon isolate of *N. hughesi*. HE stain. (a) Pan-myocarditis. Arrow points to submyocardium, and arrowhead points to endocardium. (b) Necrosis and inflammation. Infiltration mainly by mononuclear cells.

Table 18.1 Serologic Prevalence of *N. caninum* Antibodies in Horses

Country	Region	No. Tested	Type	No. Positive	% Positive	Assay	Cut-Off Titer or Test	Remarks	Reference
Argentina	Chaco	76	Draft horses	0	0.0	NAT	1:50	Also tested for other pathogens	579
Brazil	Mato Grosso	200	Healthy	30	15.0	IFAT	1:50	Highest titer 1:400 in 1 horse	1122
	Minas Gerais	506	Healthy	105	23.9	IFAT	1:64	Also tested for *S. neurona* and *T. gondii*	1710
	Pará	411	Healthy horses	28	6.8	IFAT	1:50	No risk factors detected	1488
	Paraná, Curitiba city	97	Healthy horses	14	14.4	IFAT	1:50	Highest titer 1:200 in 2 horses	2046
	Paraná, Curitiba city	14	Pregnant mares	12	85.7	IFAT	1:50	Highest titer 1:400	919
	Paraná, 3 locations	72	Mares	28	38.8	IFAT	1:50	2 foals had precolostral antibodies	1203
	Paraná and Santa Catarina	112	Mares from 5 breeding farms	14	12.5	IFAT	1:50	25.7% (9/35) prevalence in mares with reproductive problem versus 6.4% (5/77) without problems Highest titer only 1:50	6
	Rio Grande do Sul	181	Pregnant mares	39	21.5	ELISA	IH	9.3% of their paired foals had precolostral anti-*Neospora* antibodies	1628
	Rio Grande do Sul	241	Cart horses and Crioulo breed	34	15.9	IFAT	1:50	–	1982
	Rio Grande do Sul, São Paulo, Rio de Janeiro	101	Race horses	0	0.0	NAT	1:25	–	578
	Santa Catarina	615	Healthy	25	4.1	IFAT	1:50	72 with history of neurological and reproductive problems	458
	Santa Catarina	174	NS	84	48.2	IFAT	1:50	Age, higher rates in older horses	317
	São Paulo	325	Healthy	19	5.8	IFAT	1:50	Highest titer 1:400	2045
		483	Diseased	73	15.1				
	São Paulo	26	History of ataxia	15	57.6	IFAT	1:2	26 CSF negative	1906
	South	203	Mares	129	63.3	IFAT	1:50	Of 129, 34.8% gave birth to seropositive foals	79
	10 states	961	Old horses from abattoirs	24	2.5	ELISA	rNhSAG1 IH	–	915
Chile	IX and VII regions	145	Healthy	47	32.0	NAT	1:40	Highest titer 1:320 in 5 horses	1571

(Continued)

Table 18.1 (Continued) Serologic Prevalence of *N. caninum* Antibodies in Horses

Country	Region	No. Tested	Type	No. Positive	% Positive	Assay	Cut-Off Titer or Test	Remarks	Reference
China	Xinjiang	430	–	45	10.4	rELISA	–	–	1195
Costa Rica	7 provinces	315	Healthy horses	11	3.5	ELISA	rNhSAG1 IH	Only 1 sample was confirmed positive by IB	420[a]
Czech Republic	9 regions	552	Healthy horses from farms	131	24.0	cELISA	VMRD	2 of 7 samples ELISA positive were positive by IFAT at 1:50 dilution	157
France	Normandy	67	Investigated for rhinovirus pneumonia	37	55.2	NAT	1:40	4 horses with titers at 1:320	1638
	Normandy	175	Various sources	54	30.8	NAT	1:40	Higher prevalence in aborting mares (50%) versus 22.4% in general population	1625
	Normandy	50	2 farms with neurological signs	3	6.0	NAT	1:100	Association with *S. neurona*	1626
	Normandy	434	Submission to clinical labs	99	23.0	NAT	1:80	Highest titer 1:800 in 3 horses	1623
Iraq	Nineveh	81	Healthy	3	3.7	ELISA	rNhSAG1 IH	–	27
		9	Clinical signs	7	77.8			Ataxia, muscular atrophy	2137
Iran	East-Azerbaijan	100	–	28	28.0	NAT	1:80		1399
	Fars	200	Urban and work horses	64	32.0	NAT	1:80		762
	Hamedan	120	Rural and race horses	49	40.8	NAT	1:80		763
	Khuzestan	235	Arabian horses	47	20.0	NAT	1:40	3 horses with titer at 1:160	1942
	Razavi Khorasan	150	Healthy	45	30.0	NAT	1:80		934
Israel	48 farms	800		95	11.9	IFAT	1:50	Higher prevalence in horses with neurological signs (21.2%), and aborted mares (37.5%)	1078
Italy	3 Southern provinces	150	Healthy horses	42	28.0	IFAT	1:50	Highest titer 1:800 in one horse	350
	South, Center, and North	297	Healthy horses	28	9.4	IFAT	1:50	Highest titer 1:200 in 3 horses	1605
	51 municipalities in South	643	Healthy horses	70	10.9	cELISA	VMRD	No significant risk factors	161
				15	2.3	IFAT	1:50		

(Continued)

Table 18.1 (*Continued*) Serologic Prevalence of *N. caninum* Antibodies in Horses

Country	Region	No. Tested	Type	No. Positive	% Positive	Assay	Cut-Off Titer or Test	Remarks	Reference
Jordan	5 regions	227	Healthy horses	7	3.0	ELISA	IDEXX	Risk factors studied	1932
Mexico	Durango	495	Healthy horses	15	3.0	ELISA	rNhSAG1 IH	Confirmed by IB with *N. hughesi* antigen	2154
New Zealand	—	21	Healthy horses	1	5.0	IFAT	1:100	—	2034
Peru	Lima	163	Race horses	19	11.6	IFAT	1:100	—	1236
Saudi Arabia	6 provinces	229	Healthy	23	10.0	cELISA	VMRD	—	25a
South Korea	Jeju Island	191	Healthy thoroughbred	4	2.0	IFAT, WB	1:50	*N. hughesi* used as antigen Samples were negative by IB	823
Sweden	Linköping	414	Abattoir	39	9.0	ISCOM-ELISA	—	4 sera confirmed positive by IB	991
Turkey	Ankara	75	Healthy race horses	7	9.3	cELISA	VMRD	—	1056
	Niğde	125	Healthy	30	24.0	IFAT	1:50	—	1035
	6 regions	616	Draft	2	0.3	ELISA	rNcSAG1	—	2180
USA	Alabama	536	Horses with unknown history used for infectious anemia testing	62	11.5	IFAT	1:50	Highest titer 1:1600 in 1 horse	334
	25 states	296	Abattoir	69	23.3	NAT	1:40	17 horses had titer of 1:800	580
	Washington state	140	Abortion	18	13.0	IFAT	1:50	Not statistically different between 2 groups	1319
		160	Healthy	14	8.0				
	Wyoming	276	Wild horses	86	31.1	NAT	1:25	Highest titer 1:12,800 in 1 horse	592
	49 states	3,123	Samples submitted to a diagnostic facility in California	38	1.2	IFAT	1:320	Seropositive horses from 21 states Highest prevalence in Quarter horse (63.1%) than other breeds	1647a
	4 states	208	Healthy (only 6 with neurological signs)	36	17.3	IFAT	1:100	*N. hughesi* used as antigen-2 had titer of 1:5,120. 4 of 36 IFAT positive were confirmed positive by IB	2034a
	18 states	5250	Healthy	1785	34.0	IFAT	Not stated	Increasing with age, warmblood breed and geography (South)	993a

[a] *N. hughesi* used as antigen.

Table 18.2 Serologic Prevalence of *N. caninum* Antibodies in Other Equids

Host[a,b]	Country	Region	No. Tested	No. Positive	% Positive	Assay	Cut-Off Titer or Test	Remarks	Reference
Donkey (*Equus asinus*)	Brazil[c]	Bahia state	500	2	0.4	IFAT	1:100	Confirmed by IB (37 kDa antigen)	720
		Alagoas, Paraíba, Pernambuco, Piauí, Rio Grande do Norte	333	7	2.1	IFAT	1:40	21% seroprevalence of *S. neurona*	749
	Colombia	Sucre	56	11	19.7	Dot-ELISA	1:200	–	227
	Iran	Hamedan province	100	52	52.0	NAT	1:80	Higher prevalence in females (56.4) than in males (36.4)	762, 763
	Italy	South	238 of 5 breeds	28	11.8	cELISA	VMRD	–	1245
Zebra (*Equus zebra*)	Kenya	–	41	24	58.5	IFAT	1:40	–	677

[a] In a survey carried out in zoological collections from Czech Republic and Slovakia, no antibodies against *Neospora* sp. were detected in 1 Somali wild ass (*Equus africanus somaliensis*), 4 Boehm's zebras (*Equus burchelli boehmi*), 2 Chapman's zebras (*Equus burchelli chapmanni*), 3 Grevy's zebras (*Equus grevyi*), 1 Kulan (*Equus hemionus kulan*), 8 Tibetan wild asses (*Equus kiang*), 13 Przewalski's horses (*Equus przewalskii*), and 14 Hartmann's mountain zebras (*Equus zebra hartmannae*).[1279,1820]

[b] A clinical case of EPM due to *N. hughesi* was diagnosed in a 23-year-old mule from USA; bilateral ocular abnormalities and an abnormal pelvic limb gait associated to 1:160 IFAT titer in CSF were found.[688]

[c] Absence of antibodies against *N. caninum* was found in 6 donkeys and 9 mules from Pará state, Brazil.[1488]

18.3.2 Transplacental Infection

The presence of antibodies in presuckling foal serum is indicative of transplacental infection because the placental barrier limits, but does not completely preclude, the passage of maternal antibodies to fetuses in horses.[521,522] There is documentation that *Neospora* can be transmitted congenitally in horses but it seems to be a rare event. Antibodies to *Neospora* were detected in 2 foals in Brazil,[1203] but the information is not definitive because of the low titer (1:50) and the use of *N. caninum* as antigen (Table 18.1). In an extensive study of mares from 4 farms in California, USA over a 3-year period, antibodies to *Neospora* were not detected in pre-suckling sera of 366 foals from 261 mares,[522] and 69 of 324 (21.2%) mares were seropositive to *N. hughesi* at the time of parturition. Additionally, *Neospora* was not found in fetal tissues or placenta from 22 abortions by histology on these farms.[522]

In another study, antibodies to *N. hughesi* were detected in 17 of 74 (22.9%) mares.[1645] Presuckling sera from 74 foals and their 58 dams were tested for *N. hughesi* antibodies. Only 3 foals were seropositive in pre-suckling sera (IFAT 1:1280, 1:2560, and 1:20,480). These seropositive foals were from 2 mares (A, B) with IFAT titers of 1:640 and 1:10,240. Two congenitally infected foals were born to the same mare, and *N. hughesi* DNA was detected in its placenta.

N. hughesi can be transmitted in subsequent pregnancies. Mares A and B from the above study[1645] gave birth to an additional 4 congenitally infected foals in subsequent pregnancies.[1646] All 7 congenitally infected foals in these 2 studies were healthy. Colostrally acquired antibodies to *N. hughesi* disappeared or declined after 3 months post-suckling.[1645]

Contrary to the findings from California, USA, a Brazilian study reported a high prevalence of *Neospora* antibodies in the sera of mares as well as their foals prior to suckling.[79] Sera from 203 thoroughbred mares and their foals were screened at 1:50 dilution by IFAT using *N. caninum* as antigen; 129 of 203 (63.3%) were seropositive. Of these 129 seropositive mares, a very high (34.8%) percentage of mares gave birth to seropositive foals. Additionally, 6 of 74 seronegative mares gave birth to seropositive foals. Although mare sera were titrated further to 1:200 dilution and the seropositivity decreased to 33%, foal sera were not titrated. As discussed earlier, low levels of maternal IgG can cross the placenta in mares.[521,522] Therefore, further studies are needed to confirm results of this investigation from Brazil. In another study from Brazil, *Neospora* antibodies were found in 21.5% (39 of 181) mares and in 9.3% (17 of 181) of their foals in pre-suckling sera.[1628] In this study, antibodies were assayed using an in-house ELISA.

18.3.3 Clinical Infections

18.3.3.1 Histologically Confirmed Cases

Confirmed fatal neosporosis has been reported in very young or old horses (Table 18.3). The earliest reported case was a late-term fetus. The fetus and associated placenta were autolyzed and it was suspected that the fetus died *in utero* several days before abortion. Only the lungs and placenta were submitted for diagnosis.[524,540] Numerous tachyzoites were seen histologically in lungs that had pneumonitis.

The second case was in a 1-month-old foal. The foal had circled its mare to the right from birth, indicative of a neurological disorder. The foal was blind in 1 eye and had minimal vision in the second eye. Histologically, it had granulomatous encephalitis, marked by gliosis especially in the mid brain and optic nerve. Thick-walled *Neospora* tissue cysts, but no tachyzoites, were detected in the brain.[1175]

Another 6 cases of neosporosis were diagnosed in aged horses, 3 of them had other non-protozoal complications. Among these, the clinical presentation in the horse from Georgia was unusual.[807] The horse had anemia, was icteric, had abdominal lymphadenopathy, and foci of thrombosis in

Table 18.3   Clinical Neosporosis in horses in USA and Canada Confirmed by Necropsy

No.	Location	Age	Sex	Breed	Clinical Signs	Histologic	Diagnosis			Reference
							IHC	PCR/ Molecular	Isolation	
1	North Carolina	Aborted fetus	Not stated	Appaloosa x Quarter Horse	2 months before term	Pneumonia, tachyzoites in lungs	Yes	ND	ND	524, 540
2	Wisconsin	1 month	Mare	Quarter Horse	Neurological	Meningoencephalitis, 1–3 μm thick-walled tissue cysts in brain	Yes	ND	ND	1175
3	Georgia	10 years	Mare	Appaloosa	Weight loss, anemia	Mesenteric lymphadenitis, enteritis, tachyzoites in lesions, no encephalitis	Yes	ND	ND	807
4	California	11 years	Gelding	Quarter Horse	Neurological	Encephalomyelitis, tachyzoites and thin-walled (2 μm) tissue cysts	Yes	Yes	Yes-NE1 (ATCC-209622)	1283, 1284
5	California	19 years	Mare	Pinto	Neurological	Encephalomyelitis, tachyzoites, and tissue cysts, nerves and nerve roots, tachyzoites in myocardiocytes	Yes	ND	ND	414
6	Oregon	20 years	Gelding	Morgan x Quarter Horse	Neurological, ataxia	Encephalomyelitis, tachyzoites, but no tissue cysts	Yes	Yes	Yes-Oregon isolate	585, 844
7	Alabama	13 years	Mare	Palomino Quarter Horse	Neurological	Inflammatory lesions in spinal cord; protozoa not seen	No	ND	Yes-NA1	334
8	Saskatchewan, Canada	10 years	Gelding	Arabian x Quarter Horse	Neurological	Encephalomyelitis; tachyzoites seen	Yes	Yes	ND	2107

Note: ND = not done.

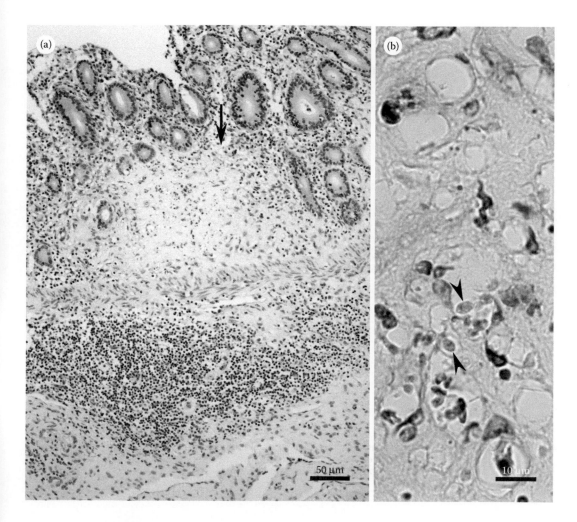

Figure 18.4   Histological section of colon of the mare. HE stain. (a) Note a focus of necrosis in the lamina propria adjacent to a necrotic crypt of Lieberkühn (arrow). (b) Higher magnification of the lesion showing many tachyzoites (arrowheads) among necrotic cells. (From Gray, M. L. et al. 1996. *J. Vet. Diagn. Invest.* 8, 130–133.)

lymph nodes and lungs. Numerous *Neospora* tachyzoites were present in the small intestines and in parinodal connective tissue (Figure 18.4).

A 19-year-old Pinto mare had Cushing's disease; manifested by adrenal cortical hyperplasia, pituitary adenoma, and mycobacterial pulmonary granuloma.[414] It addition to CNS lesions, there were inflammatory lesions in hypoglossal, pharyngeal, and facial nerves, associated with *Neospora* tissue cysts (Figure 18.2) and tachyzoites.

In a case report on a 20-year-old neosporosis horse from Oregon, thyroid adenoma was also found.[844] The horse had encephalomyelitis and tachyzoites were demonstrated, but no tissue cysts were found (Figure 18.5).

18.3.3.2 *Antemortem Diagnosed Cases*

Clinical neosporosis was diagnosed in 4 equids, all from California (Table 18.4). All of these were initially presented with signs reminiscent of *Sarcocystis neurona*-induced encephalomyelitis, including

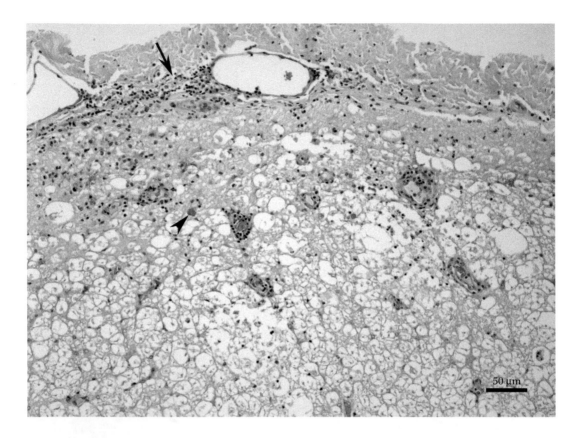

Figure 18.5 Demyelination and inflammation (arrow) in submeningeal area of the thoracic spinal cord of a horse. HE stain. Arrowhead points to a group of tachyzoites. (From case reported in Hamir, A. N. et al. 1998. *Vet. Parasitol.* 79, 269–274.)

Table 18.4 Clinical Neosporosis in Equids in California, USA, Diagnosed *Antemortem*

No.	Age	Sex/Condition	Breed	Clinical Signs	Diagnosis Serology	Treatment	Outcome	Reference
1	24 years	Gelding	Appaloosa	Rear limb incoordination, ataxia	CSF-*N. hughesi* IFAT titer 1:2560	Ponazuril (5 mg/kg, PO, for 60 days)	Improved clinically	**686**
2	16 years	Gelding	Quarter Horse	Ataxia of all 4 limbs	Serum titer 1:1280, CSF titer 1:10	Ponazuril (5 mg/kg, PO, for 45 days)	Improved clinically	
3	4 months	Filly	Percheron	Ataxia	Serum titer 1:1280, CSF titer 1:20	Ponazuril (5 mg/kg, PO, for 30 days)	Improved clinically	
4	23 years	Female	Mule	Ataxia, facial nerve paralysis	CSF titer 1:40, *N. hughesi*	Ponazuril (5 mg/kg, PO, for 60 days)	No clinical improvement	**688**

Note: PO = per os.

neurological symptoms, muscle atrophy, and facial nerve involvement. Detailed laboratory investigations excluded *S. neurona* infection. The treatment with ponazuril was proven to be promising.

18.3.3.3 Abortion

Neospora-associated abortion in horses is rare.[522] Three types of evidences have been presented. There is only 1 histologically confirmed case in an equine fetus (Table 18.3). *Neospora* DNA was detected by PCR in equine fetal tissues (Table 18.5) but *Neospora* organisms were not demonstrated. Thus, the direct cause of the effect was not conclusively established.

Higher antibody prevalences were reported in sera of mares with abortion than in healthy mares (Table 1; e.g., References **6, 1319, 1625,** and **2045**) but the results are not conclusive with respect to correlation between seropositivity and abortion.

18.3.4 Diagnosis

Clinically, neurological disease in horses (equine protozoal myeloencephalitis, EPM) caused by *N. hughesi* is similar to the EPM caused by *S. neurona*. Parasite morphology, serology, molecular differences, and IHC can be useful in differential diagnosis.[411,586,621,687,788,914,1282,1644,1672]

18.3.4.1 Serological Diagnosis

An ELISA test was developed which showed differences in the levels of antibodies against NcSAG1 and NhSAG1, depending on whether an animal was infected or immunized with *N. caninum* or *N. hughesi*, respectively.[914,1705] The NhSAG1 test had a higher sensitivity to detect infections with *N. hughesi* than infections with *N. caninum*. However, an unambiguous differentiation between an antibody response against *N. caninum* and *N. hughesi* remains impossible with the currently available tools, neither by Nh-IFAT[1546] nor NhSAG1-ELISA.[914] ELISA or IFAT titer ratios between serum and CFT are important in the diagnosis of EPM.[1705] High CSF relative to ELISA titers indicates that the parasite under investigation (*S. neurona* or *N. hughesi*) is involved in EPM.

Table 18.5 Detection of *Neospora* DNA in Aborted Equine Fetuses

Country	No. Tested	No. Positive (%)	Positive Tissue	Method	Remarks	Reference
France	12	1 (8.3)	Brain	PCR, pNc5 gene[a]	–	**1638**
	26	6 (23.1)	Brain	PCR, pNc5 gene	Positive: 3 of 12 stillbirth, 3 of 14 fetuses, histologically negative	**2041**
	407	3 (0.7)	Brain, heart, placenta	RT-PCR, pNc5 gene	1 only infected by *Neospora*, 1 coinfected with EHV-1 (equine herpes virus), 1 coinfected with *Streptococcus*	**1141**
	91	3 (3.3)	Brain, heart, placenta	PCR, pNc5 gene	3 of 91 brains, 2 of 77 hearts, and 1 of 1 placenta were positive	**1625**
USA	74	1 (1.3)	Placenta	PCR sequencing, ITS-1 gene	From a mare that gave birth to 2 congenitally infected foals, positive for *N. hughesi*	**1645**

[a] Currently there is no evidence for the presence of pNc5 gene in *N. hughesi*.

18.3.4.2 Polymerase Chain Reaction

There are distinct uncertainties in the literature because most diagnostic *Neospora*-PCRs were developed prior to the description of *N. hughesi*.[1284] Subsequently, ITS1 and 28S rDNA sequences of *N. hughesi* were compared with the published primer sequences for *N. caninum*. Using ITS1 or 28S rDNA as a target, at least 1 primer has minimal sequence differences.[643,644] Thus, it is possible that these ITS1 and 28S rDNA-based PCRs for *N. caninum* either do not detect *N. hughesi* DNA, or do so with reduced sensitivity. Prior to the use of these primer pairs to detect *N. hughesi* DNA, an evaluation of their sensitivity and specificity is necessary. Fragments of the pNc5 gene of *N. hughesi* were not amplified by a PCR using Np6/Np21 as a primer pair.[164,1887] Therefore, it is likely that also the PCR using the modified primers Np6plus and Np21plus[1426] are *N. caninum*-specific and do not amplify *N. hughesi* DNA. However, other primer pairs also developed to amplify pNc5 gene fragments[174,196,364,1043,2141] have not yet been tested for the amplification of *N. hughesi* DNA. An *N. hughesi*-specific pNc5 gene PCR is not available. *N. hughesi* can be distinguished from *N. caninum* by DNA-based techniques using specific gene[2066] or microsatellite sequences.[29]

18.4 EXPERIMENTAL INFECTIONS

A 15-year-old horse was inoculated with 6×10^7 (IM) and 7×10^7 (IV) tachyzoites of BPA1 isolate of *N. caninum*.[2034] No neurological signs associated to the infection were detected, but the animal became febrile 32-h PI. *N. hughesi* IFAT titers were <1:40 the day of inoculation, peaked to 1:5120 at 12 DPI, and dropped to 1:640 at 7 weeks PI. No protozoal inflammatory lesions or parasites were detected in gross or microscopic examination of spinal cord, brain stem, lung, or kidney. Finding of high antibody titers using *N. hughesi* as antigen in the horse inoculated with *N. caninum* indicates cross reactivity between these 2 parasites.

In a second, more comprehensive study, 7 adult horses were inoculated IM with 5×10^7 tachyzoites of the EN-1 isolate of *N. hughesi*.[1546] Pre-inoculation IFAT titers of these horses on day 0 were <1:80 in 3, 1:80 in 2, and 1:160 in 2. All of the inoculated horses developed IFAT titers of 1:1280 or higher 28 DPI. Antibody titers of 7 uninoculated control horses were 1:320 or lower. None of the horses developed confirmed EPM. The authors concluded that a titer of 1:640 discriminated between infected and uninfected horses. The NAT gave inconsistent results because of spontaneous agglutination when using *N. hughesi* tachyzoites in combination with horse serum. ELISAs based on whole parasite antigens and recombinant proteins also gave inconsistent results. These findings should be considered while interpreting results of *Neospora* serological studies in horses.

BIBLIOGRAPHY

Biology and Speciation: **414, 573, 585, 1175, 1284, 1285, 1887, 2065, 2066.**
Serologic Prevalence: **6, 25a, 27, 79, 157, 161, 227, 317, 334, 350, 420, 458, 578, 580, 592, 677, 688, 720, 749, 762, 763, 823, 915, 919, 934, 991, 993, 1035, 1056, 1078, 1122, 1195, 1203, 1236, 1245, 1279, 1319, 1399, 1488, 1571, 1605, 1623, 1625, 1626, 1628, 1638, 1647, 1710, 1820, 1906, 1932, 1942, 1982, 2034, 2045, 2046, 2137, 2154, 2180.**
Transplacental Infection: **79, 521, 522, 1203, 1628, 1645, 1646.**
Clinical Infections: **6, 334, 414, 522, 524, 540, 585, 686, 688, 807, 844, 1141, 1175, 1283, 1284, 1319, 1625, 1638, 1645, 2041, 2045, 2107.**
Diagnosis: **29, 164, 174, 196, 364, 411, 586, 621, 643, 644, 687, 788, 914, 1043, 1282, 1284, 1426, 1644, 1672, 1705, 1887, 2066, 2141.**
Experimental Infections: **1546, 2034.**
Additional References: **1186.**

References

1. Abbitt, B., Craig, T. M., Jones, L. P., Huey, R. L., Eugster, A. K. 1993. Protozoal abortion in a herd of cattle concurrently infected with *Hammondia pardalis*. *J. Am. Vet. Med. Assoc.* 203, 444–448.
2. Abdoli, A., Arbabi, M., Dalimi, A., Pirestani, M. 2015. Molecular detection of *Neospora caninum* in house sparrows (*Passer domesticus*) in Iran. *Avian Pathol.* 44, 319–322.
3. Abe, C., Tanaka, S., Ihara, F., Nishikawa, Y. 2014. Macrophage depletion prior to *Neospora caninum* infection results in severe neosporosis in mice. *Clin. Vaccine Immunol.* 21, 1185–1188.
4. Abe, C., Tanaka, S., Nishimura, M., Ihara, F., Xuan, X., Nishikawa, Y. 2015. Role of the chemokine receptor CCR5-dependent host defense system in *Neospora caninum* infections. *Parasit. Vectors* 8, e5.
5. Abo-Shehada, M. N., Abu-Halaweh, M. M. 2010. Flock-level seroprevalence of, and risk factors for, *Neospora caninum* among sheep and goats in northern Jordan. *Prev. Vet. Med.* 93, 25–32.
6. Abreu, R. A., Weiss, R. R., Thomaz-Soccol, V., Locatelli-Dittrich, R., Laskoski, L. M., Bertol, M. A. F., Koch, M. O., Alban, S. M., Green, K. T. 2014. Association of antibodies against *Neospora caninum* in mares with reproductive problems and presence of seropositive dogs as a risk factor. *Vet. Parasitol.* 202, 128–131.
6a. Acosta, I. C. L., Centoducatte, L. D., Soares, H. S., Marcili, A., Gondim, M. F. N., Rossi, J. L., Gennari, S. M. 2017. Occurrence of *Neospora caninum* and *Toxoplasma gondii* antibodies in dogs from rural properties surrounding a biological reserve, Espírito Santo, Brazil. *Braz. J. Vet. Parasitol.* 25, 536–539. DOI: 10.1590/S1984-29612016075.
7. Adomako-Ankomah, Y., Wier, G. M., Borges, A. L., Wand, H. E., Boyle, J. P. 2014. Differential locus expansion distinguishes *Toxoplasmatinae* species and closely related strains of *Toxoplasma gondii*. *mBio* 5, e01003–e01013.
8. Adomako-Ankomah, Y., English, E. D., Danielson, J. J., Pernas, L. F., Parker, M. L., Boulanger, M. J., Dubey, J. P., Boyle, J. P. 2016. Host mitochondrial association in the human parasite *Toxoplasma gondii* via neofunctionalization of a gene duplicate. *PLoS Genet.* 203, 283–298.
9. Afonso, E., Poulle, M. L., Lemoine, M., Villena, I., Aubert, D., Gilot-Fromont, E. 2007. Prevalence of *Toxoplasma gondii* in small mammals from the Ardennes Region, France. *Folia Parasitol. (Praha)* 54, 313–314.
10. Agerholm, J. S., Barr, B. C. 1994. Bovine abortions associated with *Neospora* in Denmark. *Acta Vet. Scand.* 35, 461–464.
11. Agerholm, J. S., Willadsen, C. M., Nielsen, T. K., Giese, S. B., Holm, E., Jensen, L., Agger, J. F. 1997. Diagnostic studies of abortion in Danish dairy herds. *J. Vet. Med. Assoc.* 44, 551–558.
12. Agrawal, P., Sharma, J. 1999. Recent trend in the laboratory diagnosis of *Neospora caninum*: A review. *Int. J. Anim. Sci.* 14, 157–161.
13. Aguado-Martínez, A., Álvarez-García, G., Arnaiz-Seco, I., Innes, E., Ortega-Mora, L. M. 2005. Use of avidity enzyme-linked immunosorbent assay and avidity Western blot to discriminate between acute and chronic *Neospora caninum* infection in cattle. *J. Vet. Diagn. Invest.* 17, 442–450.
14. Aguado-Martínez, A., Álvarez-García, G., Fernández-García, A., Risco-Castillo, V., Arnaiz-Seco, I., Rebordosa-Trigueros, X., Navarro-Lozano, V., Ortega-Mora, L. M. 2008. Usefulness of rNcGRA7- and rNcSAG4-based ELISA tests for distinguishing primo-infection, recrudescence, and chronic bovine neosporosis. *Vet. Parasitol.* 157, 182–195.
15. Aguado-Martínez, A., Álvarez-García, G., Fernández-García, A., Risco-Castillo, V., Marugán-Hernández, V., Ortega-Mora, L. M. 2009. Failure of a vaccine using immunogenic recombinant proteins rNcSAG4 and rNcGRA7 against neosporosis in mice. *Vaccine* 27, 7331–7338.
16. Aguado-Martínez, A., Ortega-Mora, L. M., Álvarez-García, G., Rodríguez-Marco, S., Risco-Castillo, V., Marugán-Hernández, V., Fernández-García, A. 2009. Stage-specific expression of NcSAG4 as a marker of chronic *Neospora caninum* infection in a mouse model. *Parasitology* 136, 757–764.
17. Aguado-Martínez, A., Álvarez García, G., Schares, G., Risco-Castillo, V., Fernández-García, A., Marugán-Hernández, V., Ortega-Mora, L. M. 2010. Characterisation of NcGRA7 and NcSAG4 proteins: Immunolocalization and their role in the host cell invasion by *Neospora caninum* tachyzoites. *Acta Parasitol.* 55, 304–312.
18. Aguado-Martínez, A., Basto, A. P., Müller, J., Balmer, V., Manser, V., Leitão, A., Hemphill, A. 2016. N-terminal fusion of a toll-like receptor 2-ligand to a *Neospora caninum* chimeric antigen efficiently modifies the properties of the specific immune response. *Parasitology* 143, 606–616.

19. Aguiar, D. M., Chiebao, D. P., Rodrigues, A. A. R., Cavalcante, G. T., Labruna, M. B., Gennari, S. M. 2004. Prevalência de anticorpos anti-*Neospora caninum* em ovinos do município de Monte Negro, RO, Amazônia Ocidental Brasileira. *Arq. Inst. Biol. (São Paulo)* 71(Suppl), 616–618.

20. Aguiar, D. M., Cavalcante, G. T., Rodrigues, A. A. R., Labruna, M. B., Camargo, L. M. A., Camargo, E. P., Gennari, S. M. 2006. Prevalence of anti-*Neospora caninum* antibodies in cattle and dogs from Western Amazon, Brazil, in association with some possible risk factors. *Vet. Parasitol.* 142, 71–77.

21. Ahn, H. J., Kim, S., Kim, D. Y., Nam, H. W. 2003. ELISA detection of IgG antibody against a recombinant major surface antigen (Nc-p43) fragment of *Neospora caninum* in bovine sera. *Korean J. Parasitol.* 41, 175–177.

22. Akca, A., Gokce, H. I., Guy, C. S., McGarry, J. W., Williams, D. J. L. 2005. Prevalence of antibodies to *Neospora caninum* in local and imported cattle breeds in the Kars province of Turkey. *Res. Vet. Sci.* 78, 123–126.

23. Aktas, F., Vural, G., Sezen, I. Y. 2007. Serological survey of *Neospora caninum* infection in dairy cattle. *Indian Vet. J.* 84, 419–420.

24. Aktas, M., Saki, C. E., Altay, K., Simsek, S., Utuk, A. E., Koroglu, E., Dumanli, N. 2005. Survey of *Neospora caninum* in cattle in some provinces in the Eastern Anatolian region. *Türkiye Parazitol. Derg.* 29, 22–25.

25. Al-Anazi, A. D. 2011. Prevalence of *Neospora caninum* and *Toxoplasma gondii* antibodies in sera from camels (*Camelus dromedarius*) in Riyadh Province, Saudi Arabia. *J. Egypt. Soc. Parasitol.* 41, 245–250.

25a. Al-Anazi, A. D., Mohamed, S. A. 2013. Seroprevalence of *Neospora* spp. in horses from Central Province of Saudi Arabia. *Afr. J. Biotechnol.* 12, 982–985.

26. Al-Majali, A. M., Jawasreh, K. I., Talafha, H. A., Talafha, A. Q. 2008. Neosporosis in sheep and different breeds of goats from southern Jordan: Prevalence and risk factors analysis. *Am. J. Anim. Vet. Sci.* 3, 47–52.

27. Al-Obaidii, W. A., Al-Kennany, E. R. 2014. Investigation of *Neospora hughesi* antibodies by using ELISA in horses in Nineveh Province. *Assiut Vet. Med. J.* 60, 167–170.

28. Al-Qassab, S., Reichel, M. P., Ivens, A., Ellis, J. T. 2009. Genetic diversity amongst isolates of *Neospora caninum*, and the development of a multiplex assay for the detection of distinct strains. *Mol. Cell. Probes* 23, 132–139.

29. Al-Qassab, S., Reichel, M. P., Ellis, J. 2010. A second generation multiplex PCR for typing strains of *Neospora caninum* using six DNA targets. *Mol. Cell. Probes* 24, 20–26.

30. Al-Qassab, S. E., Reichel, M. P., Ellis, J. T. 2010. On the biological and genetic diversity in *Neospora caninum*. *Diversity* 2, 411–438.

31. Alaeddine, F., Keller, N., Leepin, A., Hemphill, A. 2005. Reduced infection and protection from clinical signs of cerebral neosporosis in C57BL/6 mice vaccinated with recombinant microneme antigen NcMIC1. *J. Parasitol.* 91, 657–665.

32. Alaeddine, F., Hemphill, A., Debache, K., Guionaud, C. 2013. Molecular cloning and characterization of NcROP2Fam-1, a member of the ROP2 family of rhoptry proteins in *Neospora caninum* that is targeted by antibodies neutralizing host cell invasion *in vitro*. *Parasitology* 140, 1033–1050.

33. Aleksic, N., Perovic, M., Bakrac, T. 1995. Neosporosis—A newly-discovered parasitic infection of domestic animals. *Vet. Glas.* 49, 437–442.

34. Alkurashi, M., Eastick, F. A., Kuchipudi, S. V., Rauch, C., Madouasse, A., Zhu, X. Q., Elsheikha, H. M. 2011. Influence of culture medium pH on internalization, growth and phenotypic plasticity of *Neospora caninum*. *Vet. Parasitol.* 177, 267–274.

35. Almberg, E. S., Mech, L. D., Smith, D. W., Sheldon, J. W., Crabtree, R. L. 2009. A serological survey of infectious disease in Yellowstone National Park's canid community. *PLoS ONE* 4, e7042.

36. Almería, S., Ferrer, D., Pabón, M., Castellà, J., Mañas, S. 2002. Red foxes (*Vulpes vulpes*) are a natural intermediate host of *Neospora caninum*. *Vet. Parasitol.* 107, 287–294.

37. Almería, S., de Marez, T., Dawson, H., Araujo, R., Padilla, T., Dubey, J. P., Gasbarre, L. C. 2002. Immune responses in dam and fetus after experimental infection of pregnant cows with *Neospora caninum*. *Proceedings of the 10th International Congress of Parasitology*. August 4–9, Vancouver, Canada. Monduzzi, S. p. A. Medimond, Paris. 413–418.

38. Almería, S., de Marez, T., Dawson, H., Araujo, R., Dubey, J. P., Gasbarre, L. C. 2003. Cytokine gene expression in dams and foetuses after experimental *Neospora caninum* infection of heifers at 110 days of gestation. *Parasite Immunol.* 25, 383–392.

39. Almería, S., Vidal, D., Ferrer, D., Pabón, M., Fernández-de-Mera, M. I. G., Ruiz-Fons, F., Alzaga, V., Marco, I., Calvete, C., Lavin, S., Gortazar, C., López-Gatius, F., Dubey, J. P. 2007. Seroprevalence of *Neospora caninum* in non-carnivorous wildlife from Spain. *Vet. Parasitol.* 143, 21–28.

40. Almería, S., Nogareda, C., Santolaria, P., Garcia-Ispierto, I., Yániz, J. L., López-Gatius, F. 2009. Specific anti-*Neospora caninum* IgG1 and IgG2 antibody responses during gestation in naturally infected cattle and their relationship with gamma interferon production. *Vet. Immunol. Immunopathol.* 130, 35–42.

41. Almería, S., López-Gatius, F., García-Ispierto, I., Nogareda, C., Bech-Sàbat, G., Serrano, B., Santolaria, P., Yániz, J. L. 2009. Effects of crossbreed pregnancies on the abortion risk of *Neospora caninum*-infected dairy cows. *Vet. Parasitol.* 163, 323–329.

42. Almería, S., Araujo, R., Tuo, W., López-Gatius, F., Dubey, J. P., Gasbarre, L. C. 2010. Fetal death in cows experimentally infected with *Neospora caninum* at 110 days of gestation. *Vet. Parasitol.* 169, 304–311.

43. Almería, S., Araujo, R. N., Darwich, L., Dubey, J. P., Gasbarre, L. C. 2011. Cytokine gene expression at the materno-foetal interface after experimental *Neospora caninum* infection of heifers at 110 days of gestation. *Parasite Immunol.* 33, 517–523.

44. Almería, S., Serrano, B., Yàniz, J. L., Darwich, L., López-Gatius, F. 2012. Cytokine gene expression profiles in peripheral blood mononuclear cells from *Neospora caninum* naturally infected dams throughout gestation. *Vet. Parasitol.* 183, 237–243.

45. Almería, S. 2013. *Neospora caninum* and wildlife. *ISRN Parasitol.* Article ID 947347, 1–23.

46. Almería, S., López-Gatius, F. 2013. Bovine neosporosis: Clinical and practical aspects. *Res. Vet. Sci.* 95, 303–309.

47. Almería, S., Serrano-Pérez, B., Darwich, L., Araujo, R. N., López-Gatius, F., Dubey, J. P., Gasbarre, L. C. 2014. Maternal and fetal immune response patterns in heifers experimentally infected with *Neospora caninum* in the second trimester of pregnancy—A descriptive study. *Vet. Parasitol.* 204, 146–152.

48. Almería, S., López-Gatlus, F. 2015. Markers related to the diagnosis and to the risk of abortion in bovine neosporosis. *Res. Vet. Sci.* 100, 169–175.

49. Almeria, S., Serrano-Pérez, B., Darwich, L., Mur-Novales, R., Garcia-Ispierto, I., Cabezón, O., López Gatius, F. 2016. Cytokine gene expression in aborting and non-aborting dams and in their foetuses after experimental infection with *Neospora caninum* at 110 days of gestation. *Vet. Parasitol.* 227, 138–142.

50. Almería, S., Serrano-Perez, B., Darwich, L., Domingo, M., Mur-Novales, R., Regidor-Cerrillo, J., Cabezón, O., Pérez-Maillo, M., Lopez-Helguera, I., Fernández-Aguilar, X., Puig-Ribas, M., Ortega-Mora, L. M., García-Ispierto, I., Dubey, J. P., López-Gatius, F. 2016. Foetal death in naive heifers inoculated with *Neospora caninum* isolate Nc-Spain7 at 110 days of pregnancy. *Exp. Parasitol.* 168, 62–69.

51. Altbuch, J. A., Schofield, M. J., Porter, C. A., Gavin, W. G. 2012. *Neospora caninum*: A successful testing and eradication program in a dairy goat herd. *Small Ruminant Res.* 105, 341–344.

52. Álvarez García, G., López Pérez, I., Innes, E., Collantes Fernandez, E., Fernandez Garcia, A., Gomez Bautista, M., Ortega Mora, L. M. 2006. Use of an immunodominant p17 antigenic fraction of *Neospora caninum* in detection of antibody response in cattle. *Mem. Inst. Oswaldo Cruz* 101, 529–534.

53. Álvarez-García, G., Pereira-Bueno, J., Gómez-Bautista, M., Ortega-Mora, L. M. 2002. Pattern of recognition of *Neospora caninum* tachyzoite antigens by naturally infected pregnant cattle and aborted foetuses. *Vet. Parasitol.* 107, 15–27.

54. Álvarez-García, G., Collantes-Fernández, E., Costas, E., Rebordosa, X., Ortega-Mora, L. M. 2003. Influence of age and purpose for testing on the cut-off selection of serological methods in bovine neosporosis. *Vet. Res.* 34, 341–352.

55. Álvarez-García, G., Pitarch, A., Zaballos, A., Fernández-García, A., Gil, C., Gómez-Bautista, M., Aguado-Martínez, A., Ortega-Mora, L. M. 2007. The NcGRA7 gene encodes the immunodominant 17 kDa antigen of *Neospora caninum*. *Parasitology* 137, 41–50.

56. Alvarez-García, G., García-Culebras, A., Gutiérrez-Expósito, D., Navarro-Lozano, V., Pastor-Fernández, I., Ortega-Mora, L. M. 2013. Serological diagnosis of bovine neosporosis: A comparative study of commercially available ELISA tests. *Vet. Parasitol.* 198, 85–95.

57. Alves, D., McEwen, B., Hazlett, M., Maxie, G., Anderson, N. 1996. Trends in bovine abortions submitted to the Ontario Ministry of Agriculture, Food and Rural Affairs, 1993–1995. *Can. Vet. J.* 37, 287–288.

58. Ammann, P., Waldvogel, A., Breyer, I., Esposito, M., Müller, N., Gottstein, B. 2004. The role of B- and T-cell immunity in toltrauril-treated C57BL/6 WT, μMT and nude mice experimentally infected with *Neospora caninum*. *Parasitol. Res.* 93, 178–187.

59. Anderson, M., Picanso, J., Thurmond, M., Blanchard, P., Layton, B., Palmer, C., Case, J., Barr, B., Dubey, J. P., Conrad, P. 1992. Epidemiological investigations of bovine protozoal abortion in California dairy herds. *Proceedings of the XVII World Buiatrics Congress.* 25th American Association of Bovine Practitioners. August 31–September 4, St Paul, Minnesota. Williums, E. I. (Editor). Frontier Printers, Inc., Stillwater, Oklahoma. Volume 2, 74–78.

60. Anderson, M. 1999. Neosporosis in cattle. *Bovine Proc.* 32, 161–166.

61. Anderson, M. L., Barr, B., Conrad, P., Thurmond, M., Picanso, J., Dubey, J. P. 1991. Bovine protozoal abortions in California. *Bovine Pract.* 26, 102–104.

62. Anderson, M. L., Blanchard, P. C., Barr, B. C., Dubey, J. P., Hoffman, R. L., Conrad, P. A. 1991. *Neospora*-like protozoan infection as a major cause of abortion in California dairy cattle. *J. Am. Vet. Med. Assoc.* 198, 241–244.

63. Anderson, M. L., Barr, B. C., Conrad, P. A. 1994. Protozoal causes of reproductive failure in domestic ruminants. *Vet. Clin. N. Am. Food Amin. Pract.* 10, 439–461.

64. Anderson, M. L., Palmer, C. W., Thurmond, M. C., Picanso, J. P., Blanchard, P. C., Breitmeyer, R. E., Layton, A. W., McAllister, M., Daft, B., Kinde, H., Read, D. H., Dubey, J. P., Conrad, P. A., Barr, B. C. 1995. Evaluation of abortions in cattle attributable to neosporosis in selected dairy herds in California. *J. Am. Vet. Med. Assoc.* 207, 1206–1210.

65. Anderson, M. L., Reynolds, J. P., Rowe, J. D., Sverlow, K. W., Packham, A. E., Barr, B. C., Conrad, P. A. 1997. Evidence of vertical transmission of *Neospora* sp. infection in dairy cattle. *J. Am. Vet. Med. Assoc.* 210, 1169–1172.

66. Anderson, M. L., Andrianarivo, A. G., Conrad, P. A. 2000. Neosporosis in cattle. *Anim. Reprod. Sci.* 60–61, 417–431.

67. Anderson, T., Dejardin, A., Howe, D. K., Dubey, J. P., Michalski, M. L. 2007. *Neospora caninum* antibodies detected in Midwestern white-tailed deer (*Odocoileus virginianus*) by Western blot and ELISA. *Vet. Parasitol.* 145, 152–155.

68. Andrade, G. S., Bruhn, F. R. P., Rocha, C. M. B. M., Guimarães, A. S., Gouveia, A. M. G., Guimarães, A. M. 2012. Seroprevalence and risk factors for *Neospora caninum* in sheep in the state Minas Gerais, southeastern Brazil. *Vet. Parasitol.* 188, 168–171.

69. Andrade, G. S., Bruhn, F. R. P., Rocha, C. M. B. M., Guimarães, A. S., Gouveia, A. M. G., Guimarães, A. M. 2013. Seroprevalence for *Neospora caninum* in goats of Minas Gerais state, Brazil. *Res. Vet. Sci.* 94, 584–586.

70. André, M. R., Adania, C. H., Teixeira, R. H. F., Silva, K. F., Jusi, M. M. G., Machado, S. T. Z., de Bortolli, C. P., Falcade, M., Sousa, L., Alegretti, S. M., Felippe, P. A. N., Machado, R. Z. 2010. Antibodies to *Toxoplasma gondii* and *Neospora caninum* in captive neotropical and exotic wild canids and felids. *J. Parasitol.* 96, 1007–1009.

71. Andreotti, R., Pinckney, R. D., Pires, P. P., Silva, E. A. E. 2004. Evidence of *Neospora caninum* in beef cattle and dogs in the state of Mato Grosso do Sul, center-western region, Brazil. *Braz. J. Vet. Parasitol.* 13, 129–131.

72. Andreotti, R., Oliveira, J. M., Araujo e Silva, E., Oshiro, L. M., Matos, M. F. C. 2006. Occurrence of *Neospora caninum* in dogs and its correlation with visceral leishmaniasis in the urban area of Campo Grande, Mato Grosso do Sul, Brazil. *Vet. Parasitol.* 135, 375–379.

73. Andreotti, R., Matos, M. F. C., Gonçalves, K. N., Oshiro, L. M., Lima-Junior, M. S. C., Paiva, F., Leite, F. L. 2009. Comparison of indirect ELISA based on recombinant protein NcSRS2 and IFAT for detection of *Neospora caninum* antibodies in sheep. *Rev. Bras. Parasitol. Vet.* 18, 19–22.

74. Andreotti, R., Barros, J. C., Pereira, A. R., Oshiro, L. M., Cunha, R. C., Neto, L. F. F. 2010. Association between seropositivity for *Neospora caninum* and reproductive performance of beef heifers in the Pantanal of Mato Grosso do Sul, Brazil. *Rev. Bras. Parasitol. Vet.* 19, 119–123.

75. Andrianarivo, A. G., Choromanski, L., McDonough, S. P., Packham, A. E., Conrad, P. A. 1999. Immunogenicity of a killed whole *Neospora caninum* tachyzoite preparation formulated with different adjuvants. *Int. J. Parasitol.* 29, 1613–1625.

76. Andrianarivo, A. G., Rowe, J. D., Barr, B. C., Anderson, M. L., Packham, A. E., Sverlow, K. W., Choromanski, L., Loui, C., Grace, A., Conrad, P. A. 2000. A Polygen-adjuvanted killed *Neospora caninum* tachyzoite preparation failed to prevent foetal infection in pregnant cattle following i.v./i.m. experimental tachyzoite challenge. *Int. J. Parasitol.* 30, 985–990.

77. Andrianarivo, A. G., Barr, B. C., Anderson, M. L., Rowe, J. D., Packham, A. E., Sverlow, K. W., Conrad, P. A. 2001. Immune responses in pregnant cattle and bovine fetuses following experimental infection with *Neospora caninum*. *Parasitol. Res.* 87, 817–825.

78. Andrianarivo, A. G., Anderson, M. L., Rowe, J. D., Gardner, I. A., Reynolds, J. P., Choromanski, L., Conrad, P. A. 2005. Immune responses during pregnancy in heifers naturally infected with *Neospora caninum* with and without immunization. *Parasitol. Res.* 96, 24–31.

79. Antonello, A. M., Pivoto, F. L., Camillo, G., Braunig, P., Sangioni, L. A., Pompermayer, E., Vogel, F. S. F. 2012. The importance of vertical transmission of *Neospora* sp. in naturally infected horses. *Vet. Parasitol.* 187, 367–370.

79a. Antonello, A. M., Camillo, G., Braunig, P., Sangioni, L. A., Vogel, F. S. F. 2016. Diâmica sorológica de anticorpos contra *Neospora caninum* durante a gestação de vacas naturalmente infectadas. *Ciênc. Anim. Bras.* 16, 553–559.

80. Antony, A., Williamson, N. B. 2001. Recent advances in understanding the epidemiology of *Neospora caninum* in cattle. *N. Z. Vet. J.* 49, 42–47.

81. Antony, A., Williamson, N. B. 2003. Prevalence of antibodies to *Neospora caninum* in dogs of rural or urban origin in central New Zealand. *N. Z. Vet. J.* 51, 232–237.

81a. Arbabi, M., Abdoli, A., Dalimi, A., Pirestani, M. 2016. Identification of latent neosporosis in sheep in Tehran, Iran by polymerase chain reaction using primers specific for the *Nc-5* gene. *Onderstepoort J. Vet. Res.* 83, Article No. 1058e.

82. Araujo, F. G., Slifer, T. 2003. Different strains of *Toxoplasma gondii* induce different cytokine responses in CBA/Ca mice. *Infect. Immun.* 71, 4171–4174.

83. Armengol, R., Pabón, M., Adelantado, C., López-Gatius, F., Almería, S. 2006. First report of *Neospora caninum* abortion in a beef cow-calf herd from Andorra, Europe. *J. Parasitol.* 92, 1361–1362.

84. Armengol, R., Pabón, M., Santolaria, P., Cabezón, O., Adelantado, C., Yániz, J., López-Gatius, F., Almería, S. 2007. Low seroprevalence of *Neospora caninum* infection associated with the limousin breed in cow-calf herds in Andorra, Europe. *J. Parasitol.* 93, 1029–1032.

85. Arnold, L. M. 2013. Investigation of an abortion epidemic due to *Neospora caninum* in a beef herd on pasture. *Bovine Pract.* 47, 1–6.

85a. Arraes-Santos, A. I., Araújo, A. C., Guimarães, M. F., Santos, J. R., Pena, H. F. J., Gennari, S. M., Azevedo, S. S., Labruna, M. B., Horta, M. C. 2016. Seroprevalence of anti-*Toxoplasma gondii* and anti-*Neospora caninum* antibodies in domestic mammals from two distinct regions in the semi-arid region of Northeastern Brazil. *Vet. Parasitol. Reg. Stud. Rep.* 5, 14–18.

86. Arranz-Solís, D., Aguado-Martínez, A., Müller, J., Regidor-Cerrillo, J., Ortega-Mora, L. M., Hemphill, A. 2015. Dose-dependent effects of experimental infection with the virulent *Neospora caninum* Nc-Spain7 isolate in a pregnant mouse model. *Vet. Parasitol.* 211, 133–140.

87. Arranz-Solís, D., Benavides, J., Regidor-Cerrillo, J., Fuertes, M., Ferre, I., Ferreras, M. C., Collantes-Fernández, E., Hemphill, A., Pérez, V., Ortega-Mora, L. M. 2015. Influence of the gestational stage on the clinical course, lesional development and parasite distribution in experimental ovine neosporosis. *Vet. Res.* 46, e19.

88. Arranz-Solís, D., Benavides, J., Regidor-Cerrillo, J., Horcajo, P., Castaño, P., Ferreras, M. C., Jiménez-Pelayo, L., Collantes-Fernández, E., Ferre, I., Hemphill, A., Pérez, V., Ortega-Mora, L. M. 2016. Systemic and local immune responses in sheep after *Neospora caninum* experimental infection at early, mid and late gestation. *Vet. Res.* 47, 2.

89. Arreola-Camberos, S., Garcia-Marquez, L. J., Macedo-Barragan, R., Morales-Salinas, E., Figueroa-Chavez, D. 2012. Risk factors and seroprevalence against *Neospora caninum* in dual-purpose and beef cattle in Colima, Mexico. *J. Anim. Vet. Adv.* 11, 2440–2444.

90. Arunvipas, P., Inpankaew, T., Jittapalapong, S. 2012. Risk factors of *Neospora caninum* infection in dogs and cats in dairy farms in Western Thailand. *Trop. Anim. Health Prod.* 44, 1117–1121.

91. Asadpour, R., Jafari-Joozani, R., Salehi, N. 2013. Detection of *Neospora caninum* in ovine abortion in Iran. *J. Parasit. Dis.* 37, 105–109.

92. Asai, T., Howe, D. K., Nakajima, K., Nozaki, T., Takeuchi, T., Sibley, L. D. 1998. *Neospora caninum*: Tachyzoites express a potent type-I nucleoside triphosphate hydrolase, but lack nucleoside diphosphate hydrolase activity. *Exp. Parasitol.* 90, 277–285.

92a. Ash, A., Elliot, A., Thompson, R.C.A., 2015. The detection and characaterisation of *Neospora/Hammondia*-like oocysts from naturally infected dogs within the same urban region of Australia. *Vet. Parasitol. Reg. Stud. Rep.* 1–2, 47–50.

93. Asmare, K., Regassa, F., Robertson, L. J., Skijerve, E. 2013. Seroprevalence of *Neospora caninum* and associated risk factors in intensive or semi-intensively managed dairy and breeding cattle of Ethiopia. *Vet. Parasitol.* 193, 85–94.

94. Asmare, K., Regassa, F., Robertson, L. J., Martin, A. D., Skjerve, E. 2013. Reproductive disorders in relation to *Neospora caninum, Brucella* spp. and bovine viral diarrhoea virus serostatus in breeding and dairy farms of central and southern Ethiopia. *Epidemiol. Infect.* 141, 1772–1780.

95. Asmare, K. 2014. *Neospora caninum* versus *Brucella* spp. exposure among dairy cattle in Ethiopia: A case control study. *Trop. Anim. Health Prod.* 46, 961–966.

96. Asmare, K., Skjerve, E., Bekele, J., Sheferaw, D., Stachurska-Hagen, T., Robertson, L. J. 2014. Molecular identification of *Neospora caninum* from calf/foetal brain tissue and among oocysts recovered from faeces of naturally infected dogs in southern Ethiopia. *Acta Trop.* 130, 88–93.

97. Astorga, R. J., Reguillo, L., Hernández, M., Cardoso-Toset, F., Tarradas, C., Maldonado, A., Gómez-Laguna, J. 2014. Serosurvey on Schmallenberg virus and selected ovine reproductive pathogens in culled ewes from southern Spain. *Transbound. Emerg. Dis.* 61, 4–11.

98. Atkinson, R., Harper, P. A. W., Ryce, C., Morrison, D. A., Ellis, J. T. 1999. Comparison of the biological characteristics of two isolates of *Neospora caninum*. *Parasitology* 118, 363–370.

99. Atkinson, R., Harper, P. A. W., Reichel, M. P., Ellis, J. T. 2000. Progress in the serodiagnosis of *Neospora caninum* infections of cattle. *Parasitol. Today* 16, 110–114.

100. Atkinson, R. A., Cook, R. W., Reddacliff, L. A., Rothwell, J., Broady, K. W., Harper, P. A. W., Ellis, J. T. 2000. Seroprevalence of *Neospora caninum* infection following an abortion outbreak in a dairy cattle herd. *Aust. Vet. J.* 78, 262–266.

101. Atkinson, R. A., Ryce, C., Miller, C. M. D., Balu, S., Harper, P. A. W., Ellis, J. T. 2001. Isolation of *Neospora caninum* genes detected during a chronic murine infection. *Int. J. Parasitol.* 31, 67–71.

102. Augustine, P. C., Jenkins, M. C., Dubey, J. P. 1999. Effect of polyclonal antisera developed against dense granule-associated *Neospora caninum* proteins on cell invasion and development *in vitro* by *N. caninum* tachyzoites. *Parasitology* 119, 441–445.

103. Auriemma, C., Lucibelli, M. G., Borriello, G., de Carlo, E., Martucciello, A., Schiavo, L., Gallo, A., Bove, F., Corrado, F., Girardi, S., Amoroso, M. G., Degli Uberti, B., Galiero, G. 2014. PCR detection of *Neospora caninum* in water buffalo foetal tissues. *Acta Parasitol.* 59, 1–4.

104. Ayinmode, A. B., Adediran, O. A., Schares, G. 2016. Seroprevalence of *Toxoplasma gondii* and *Neospora caninum* in urban and rural dogs from southwestern Nigeria. *Afr. J. Infect. Dis.* 10, 25–28.

105. de Azevedo, S. S., Batista, C. S. A., Vasconcellos, S. A., Aguiar, D. M., Ragozo, A. M. A., Rodrigues, A. A. R., Alves, C. J., Gennari, S. M. 2005. Seroepidemiology of *Toxoplasma gondii* and *Neospora caninum* in dogs from the state of Paraíba, northeast region of Brazil. *Res. Vet. Sci.* 79, 51–56.

106. Bacigalupe, D., Basso, W., Caspe, S. G., Moré, G., Lischinsky, L., Gos, M. L., Leunda, M., Campero, L., Moore, D. P., Schares, G., Campero, C. M., Venturini, M. C. 2013. *Neospora caninum* NC-6 Argentina induces fetopathy in both serologically positive and negative experimentally inoculated pregnant dams. *Parasitol. Res.* 112, 2585–2592.

107. Bacsadi, A., Bajmócy, E., Matiz, K., Kiss, I. 2001. Bovine abortion associated with *Neospora caninum* in Hungary. *Acta Vet. Hung.* 49, 185–189.

108. Bae, J. S., Kim, D. Y., Hwang, W. S., Kim, J. H., Lee, N. S., Nam, H. W. 2000. Detection of IgG antibody against *Neospora caninum* in cattle in Korea. *Korean J. Parasitol.* 38, 245–249.

109. Bae, J. S., Kim, J. H., Hur, K., Kim, K. S., Hwang, W. S., Choi, Y. K., Hyun, B. H., Kim, D. Y. 2000. Experimental infection of Korean *Neospora caninum* isolates in mice. *Korean J. Vet. Res.* 40, 138–144.

110. Bahrami, S., Rezaie, A., Boroomand, Z., Namavari, M., Ghavami, S. 2016. Embryonated pigeon eggs as a model to investigate *Neospora caninum* infection. *Lab. Anim.* Doi: 10.1177/0023677216652373.

111. Baillargeon, P., Fecteau, G., Paré, J., Lamothe, P., Sauvé, R. 2001. Evaluation of the embryo transfer procedure proposed by the International Embryo Transfer Society as a method of controlling vertical transmission of *Neospora caninum* in cattle. *J. Am. Vet. Med. Assoc.* 218, 1803–1806.

112. Baker, D. G., Morishita, T. Y., Brooks, D. L., Shen, S. K., Lindsay, D. S., Dubey, J. P. 1995. Experimental oral inoculations in birds to evaluate potential definitive host of *Neospora caninum*. *J. Parasitol.* 81, 783–785.

113. Balkaya, I., Utuk, A. E., Babur, C., Beyhan, Y. E., Piskin, F. C., Sozdutmaz, I. 2015. Detection of *Toxoplasma gondii* and *Neospora caninum* antibodies in wild boars (*Sus scrofa*) in eastern Turkey. *Israel J. Vet. Med.* 70, 28–31.

114. Balthazar, L. M. C., Leal, P. D. S., Teixeira Filho, W. L., Lopes, C. W. G. 2013. Cães sororreagentes a *Neospora caninum* (Apicomplexa: Toxoplasmatinae) atendidos em uma clínica veterinária na cidade do Rio de Janeiro, RJ. *Rev. Bras. Med. Vet.* 35(Suppl 2), 48–51.

115. Bañales, P., Fernandez, L., Repiso, M. V., Gil, A., Dargatz, D. A., Osawa, T. 2006. A nationwide survey on seroprevalence of *Neospora caninum* infection in beef cattle in Uruguay. *Vet. Parasitol.* 139, 15–20.

116. Bandini, L. A., Neto, A. F. A., Pena, H. F. J., Cavalcante, G. T., Schares, G., Nishi, S. M., Gennari, S. M. 2011. Experimental infection of dogs (*Canis familiaris*) with sporulated oocysts of *Neospora caninum*. *Vet. Parasitol.* 176, 151–156.

117. Bapodra, P., Wolfe, B. A. 2015. Investigation of *Neospora caninum* seroprevalence and potential impact on reproductive success in semi-free-ranging Père David's deer (*Elaphurus davidianus*). *Vet. Rec. Open* 2, e000123.

118. Barber, J., Trees, A. J., Owen, M., Tennant, B. 1993. Isolation of *Neospora caninum* from a British dog. *Vet. Rec.* 133, 531–532.

119. Barber, J. S., Holmdahl, O. J. M., Owen, M. R., Guy, F., Uggla, A., Trees, A. J. 1995. Characterization of the first European isolate of *Neospora caninum* (Dubey, Carpenter, Speer, Topper and Uggla). *Parasitology* 111, 563–568.

120. Barber, J. S., Trees, A. J. 1996. Clinical aspects of 27 cases of neosporosis in dogs. *Vet. Rec.* 139, 439–443.

121. Barber, J. S., Payne-Johnson, C. E., Trees, A. J. 1996. Distribution of *Neospora caninum* within the central nervous system and other tissues of six dogs with clinical neosporosis. *J. Small Anim. Pract.* 37, 568–574.

122. Barber, J. S., van Ham, L., Polis, I., Trees, A. J. 1997. Seroprevalence of antibodies to *Neospora caninum* in Belgian dogs. *J. Small Anim. Pract.* 38, 15–16.

123. Barber, J. S., Gasser, R. B., Ellis, J., Reichel, M. P., McMillan, D., Trees, A. J. 1997. Prevalence of antibodies to *Neospora caninum* in different canid populations. *J. Parasitol.* 83, 1056–1058.

124. Barber, J. S., Trees, A. J. 1998. Naturally occurring vertical transmission of *Neospora caninum* in dogs. *Int. J. Parasitol.* 28, 57–64.

125. Barling, K. S., Sherman, M., Peterson, M. J., Thompson, J. A., McNeill, J. W., Craig, T. M., Adams, L. G. 2000. Spatial associations among density of cattle, abundance of wild canids, and seroprevalence to *Neospora caninum* in a population of beef calves. *J. Am. Vet. Med. Assoc.* 217, 1361–1365.

126. Barling, K. S., McNeill, J. W., Thompson, J. A., Paschal, J. C., McCollum, F. T., Craig, T. M., Adams, L. G. 2000. Association of serologic status for *Neospora caninum* with postweaning weight gain and carcass measurements in beef calves. *J. Am. Vet. Med. Assoc.* 217, 1356–1360.

127. Barling, K. S., McNeill, J. W., Paschal, J. C., McCollum, F. T. III, Craig, T. M., Adams, L. G., Thompson, J. A. 2001. Ranch-management factors associated with antibody seropositivity for *Neospora caninum* in consignments of beef calves in Texas, USA. *Prev. Vet. Med.* 52, 53–61.

127a. Barling, K. S., Lunt, D. K., Snowden, K. F., Thompson, J. A. 2001. Association of serologic status for *Neospora caninum* and postweaning feed efficiency in beef steers. *J. Am. Vet. Med. Assoc.* 219, 1259–1262.

128. Barling, K. S., Lunt, D. K., Graham, S. L., Choromanski, L. J. 2003. Evaluation of an inactivated *Neospora caninum* vaccine in beef feedlot steers. *J. Am. Vet. Med. Assoc.* 222, 624–627.

129. Barna, F., Debache, K., Vock, C. A., Küster, T., Hemphill, A. 2013. *In vitro* effects of novel ruthenium complexes in *Neospora caninum* and *Toxoplasma gondii* tachyzoites. *Antimicrob. Agents Chemother.* 57, 5747–5754.

130. Barr, B. C., Anderson, M. L., Blanchard, P. C., Daft, B. M., Kinde, H., Conrad, P. A. 1990. Bovine fetal encephalitis and myocarditis associated with protozoal infections. *Vet. Pathol.* 27, 354–361.

131. Barr, B. C., Anderson, M. L., Dubey, J. P., Conrad, P. A. 1991. *Neospora*-like protozoal infections associated with bovine abortions. *Vet. Pathol.* 28, 110–116.

132. Barr, B. C., Conrad, P. A., Dubey, J. P., Anderson, M. L. 1991. *Neospora*-like encephalomyelitis in a calf: Pathology, ultrastructure, and immunoreactivity. *J. Vet. Diagn. Invest.* 3, 39–46.

133. Barr, B. C., Anderson, M. L., Woods, L. W., Dubey, J. P., Conrad, P. A. 1992. *Neospora*-like protozoal infections associated with abortion in goats. *J. Vet. Diagn. Invest.* 4, 365–367.

134. Barr, B. C., Conrad, P. A., Breitmeyer, R., Sverlow, K., Anderson, M. L., Reynolds, J., Chauvet, A. E., Dubey, J. P., Ardans, A. A. 1993. Congenital *Neospora* infection in calves born from cows that had previously aborted *Neospora*-infected fetuses: Four cases (1990–1992). *J. Am. Vet. Med. Assoc.* 202, 113–117.

135. Barr, B. C., Conrad, P. A., Sverlow, K. W., Tarantal, A. F., Hendrickx, A. G. 1994. Experimental fetal and transplacental *Neospora* infection in the nonhuman primate. *Lab. Invest.* 71, 236–242.

136. Barr, B. C., Rowe, J. D., Sverlow, K. W., BonDurant, R. H., Ardans, A. A., Oliver, M. N., Conrad, P. A. 1994. Experimental reproduction of bovine fetal *Neospora* infection and death with a bovine *Neospora* isolate. *J. Vet. Diagn. Invest.* 6, 207–215.

137. Barr, B. C., Anderson, M. L., Sverlow, K. W., Conrad, P. A. 1995. Diagnosis of bovine fetal *Neospora* infection with an indirect fluorescent antibody test. *Vet. Rec.* 137, 611–613.

138. Barratt, J., Al Qassab, S., Reichel, M. P., Ellis, J. T. 2008. The development and evaluation of a nested PCR assay for detection of *Neospora caninum* and *Hammondia heydorni* in feral mouse tissues. *Mol. Cell. Probes* 22, 228–233.

139. Barratt, J. L. N., Harkness, J., Marriott, D., Ellis, J. T., Stark, D. 2010. Importance of nonenteric protozoan infections in immunocompromised people. *Clin. Microbiol. Rev.* 23, 795–836.

140. Barta, J. R., Dubey, J. P. 1992. Characterization of anti-*Neospora caninum* hyperimmune rabbit serum by Western blot analysis and immunoelectron microscopy. *Parasitol. Res.* 78, 689–694.

141. Bartels, C. J. M., Wouda, W., Schukken, Y. H. 1999. Risk factors for *Neospora caninum*-associated abortion storms in dairy herds in the Netherlands (1995–1997). *Theriogenology* 52, 247–257.

142. Bartels, C. J. M., van Maanen, C., van der Meulen, A. M., Dijkstra, T., Wouda, W. 2005. Evaluation of three enzyme-linked immunosorbent assays for detection of antibodies to *Neospora caninum* in bulk milk. *Vet. Parasitol.* 131, 235–246.

143. Bartels, C. J. M., van Schaik, G., Veldhuisen, J. P., van den Borne, B. H. P., Wouda, W., Dijkstra, T. 2006. Effect of *Neospora caninum* serostatus on culling, reproductive performance and milk production in Dutch dairy herds with and without a history of *Neospora caninum* associated abortion storms. *Prev. Vet. Med.* 77, 186–198.

144. Bartels, C. J. M., Arnaiz-Seco, J. I., Ruiz-Santa-Quitera, A., Björkman, C., Frössling, J., von Blumröder, D., Conraths, F. J., Schares, G., van Maanen, C., Wouda, W., Ortega-Mora, L. M. 2006. Supranational comparison of *Neospora caninum* seroprevalences in cattle in Germany, The Netherlands, Spain and Sweden. *Vet. Parasitol.* 137, 17–27.

145. Bartels, C. J. M., van Schaik, G., van Maanen, K., Wouda, W., Dijkstra, T. 2007. Factors associated with variation in *Neospora caninum* bulk-milk S/P ratios in initially bulk-milk negative testing Dutch dairy herds. *Prev. Vet. Med.* 81, 265–273.

146. Bartels, C. J. M., Huinink, I., Beiboer, M. L., van Schaik, G., Wouda, W., Dijkstra, T., Stegeman, A. 2007. Quantification of vertical and horizontal transmission of *Neospora caninum* infection in Dutch dairy herds. *Vet. Parasitol.* 148, 83–92.

147. Bartley, P. M., Kirvar, E., Wright, S., Swales, C., Esteban-Redondo, I., Buxton, D., Maley, S. W., Schock, A., Rae, A. G., Hamilton, C., Innes, E. A. 2004. Maternal and fetal immune responses of cattle inoculated with *Neospora caninum* at mid-gestation. *J. Comp. Pathol.* 130, 81–91.

148. Bartley, P. M., Wright, S., Sales, J., Chianini, F., Buxton, D., Innes, E. A. 2006. Long-term passage of tachyzoites in tissue culture can attenuate virulence of *Neospora caninum in vivo*. *Parasitology* 133, 421–432.

149. Bartley, P. M., Wright, S., Chianini, F., Buxton, D., Innes, E. A. 2007. Inoculation of Balb/c mice with live attenuated tachyzoites protects against a lethal challenge of *Neospora caninum*. *Parasitology* 135, 13–21.

150. Bartley, P. M., Wright, S. E., Maley, S. W., Buxton, D., Nath, M., Innes, E. A. 2009. The development of immune responses in Balb/c mice following inoculation with attenuated or virulent *Neospora caninum* tachyzoites. *Parasite Immunol.* 31, 392–401.

151. Bartley, P. M., Wright, S. E., Maley, S. W., Macaldowie, C. N., Nath, M., Hamilton, C. M., Katzer, F., Buxton, D., Innes, E. A. 2012. Maternal and foetal immune responses of cattle following an experimental challenge with *Neospora caninum* at day 70 of gestation. *Vet. Res.* 43, 38.

152. Bartley, P. M., Wright, S. E., Zimmer, I. A., Roy, S., Kitchener, A. C., Meredith, A., Innes, E. A., Katzer, F. 2013. Detection of *Neospora caninum* in wild carnivorans in Great Britain. *Vet. Parasitol.* 192, 279–283.

153. Bartley, P. M., Wright, S. E., Zimmer, I. A., Roy, S., Kitchener, A. C., Meredith, A., Innes, E. A., Katzer, F. 2013. Development of maternal and foetal immune responses in cattle following experimental challenge with *Neospora caninum* at day 210 of gestation. *Vet. Res.* 44, e91.

154. Bártová, E., Sedlák, K., Literák, I. 2006. Prevalence of *Toxoplasma gondii* and *Neospora caninum* antibodies in wild boars in the Czech Republic. *Vet. Parasitol.* 142, 150–153.

155. Bártová, E., Sedlák, K., Pavlik, I., Literák, I. 2007. Prevalence of *Neospora caninum* and *Toxoplasma gondii* antibodies in wild ruminants from the countryside or captivity in the Czech Republic. *J. Parasitol.* 93, 1216–1218.

156. Bártová, E., Sedlák, K., Literák, I. 2009. *Toxoplasma gondii* and *Neospora caninum* antibodies in sheep in the Czech Republic. *Vet. Parasitol.* 161, 131–132.

157. Bártová, E., Sedlák, K., Syrová, M., Literák, I. 2010. *Neospora* spp. and *Toxoplasma gondii* antibodies in horses in the Czech Republic. *Parasitol. Res.* 107, 783–785.

157a. Bártová, E., Sedlák, K., Treml, F., Holko, I., Literák, I. 2010. *Neospora caninum* and *Toxoplasma gondii* antibodies in European brown hares in the Czech Republic, Slovakia, and Austria. *Vet. Parasitol.* 171, 155–158.

158. Bártová, E., Sedlák, K. 2011. Seroprevalence of *Toxoplasma gondii* and *Neospora caninum* in slaughtered pigs in the Czech Republic. *Parasitology* 138, 1369–1371.

159. Bartova, E., Sedlak, K. 2012. *Toxoplasma gondii* and *Neospora caninum* antibodies in goats in the Czech Republic. *Vet. Med.* 57, 111–114.

160. Bártová, E., Sedlák, K., Budíková, M. 2015. A study of *Neospora caninum* and *Toxoplasma gondii* antibody seroprevalence in healthy cattle in the Czech Republic. *Ann. Agric. Environ. Med.* 22, 32–34.

161. Bártová, E., Machacová, T., Sedlák, K., Budíková, M., Mariani, U., Veneziano, V. 2015. Seroprevalence of antibodies of *Neospora* spp. and *Toxoplasma gondii* in horses from southern Italy. *Folia Parasitol. (Praha)* 62, 043.

162. Bártová, E., Slezáková, R., Nágl, I., Sedlák, K. 2016. *Neospora caninum* and *Toxoplasma gondii* antibodies in red foxes (*Vulpes vulpes*) in the Czech Republic. *Ann. Agric. Environ. Med.* 23, 84–86.

163. Barutzki, D., Schaper, R. 2003. Endoparasites in dogs and cats in Germany 1999–2002. *Parasitol. Res.* 90, S148–S150.

164. Basso, W., Venturini, L., Venturini, M. C., Hill, D. E., Kwok, O. C. H., Shen, S. K., Dubey, J. P. 2001. First isolation of *Neospora caninum* from the feces of a naturally infected dog. *J. Parasitol.* 87, 612–618.

165. Basso, W., Venturini, L., Venturini, M. C., Moore, P., Rambeau, M., Unzaga, J. M., Campero, C., Bacigalupe, D., Dubey, J. P. 2001. Prevalence of *Neospora caninum* infection in dogs from beef-cattle farms, dairy farms, and from urban areas of Argentina. *J. Parasitol.* 87, 906–907.

166. Basso, W., Venturini, M. C., Bacigalupe, D., Kienast, M., Unzaga, J. M., Larsen, A., Machuca, M., Venturini, L. 2005. Confirmed clinical *Neospora caninum* infection in a boxer puppy from Argentina. *Vet. Parasitol.* 131, 299–303.

167. Basso, W., Herrmann, D. C., Conraths, F. J., Pantchev, N., Vrhovec, M. G., Schares, G. 2009. First isolation of *Neospora caninum* from the faeces of a dog from Portugal. *Vet. Parasitol.* 159, 162–166.

168. Basso, W., Schares, S., Bärwald, A., Herrmann, D. C., Conraths, F. J., Pantchev, N., Vrhovec, M. G., Schares, G. 2009. Molecular comparison of *Neospora caninum* oocyst isolates from naturally infected dogs with cell culture-derived tachyzoites of the same isolates using nested polymerase chain reaction to amplify microsatellite markers. *Vet. Parasitol.* 160, 43–50.

169. Basso, W., Schares, S., Minke, L., Bärwald, A., Maksimov, A., Peters, M., Schulze, C., Müller, M., Conraths, F. J., Schares, G. 2010. Microsatellite typing and avidity analysis suggest a common source of infection in herds with epidemic *Neospora caninum*-associated bovine abortion. *Vet. Parasitol.* 173, 24–31.

170. Basso, W., Moré, G., Quiroga, M. A., Balducchi, D., Schares, G., Venturini, M. C. 2014. *Neospora caninum* is a cause of perinatal mortality in axis deer (*Axis axis*). *Vet. Parasitol.* 199, 255–258.

171. Basto, A. P., Leitão, A. 2014. Targeting TLR2 for vaccine development. *J. Immunol. Res.* 2014, 1–22. Article ID 619410.

172. Baszler, T. V., Knowles, D. P., Dubey, J. P., Gay, J. M., Mathison, B. A., McElwain, T. F. 1996. Serological diagnosis of bovine neosporosis by *Neospora caninum* monoclonal antibody-based competitive inhibition enzyme-linked immunosorbent assay. *J. Clin. Microbiol.* 34, 1423–1428.

173. Baszler, T. V., Long, M. T., McElwain, T. F., Mathison, B. A. 1999. Interferon-γ and interleukin-12 mediate protection to acute *Neospora caninum* infection in BALB/c mice. *Int. J. Parasitol.* 29, 1635–1646.

174. Baszler, T. V., Gay, L. J. C., Long, M. T., Mathison, B. A. 1999. Detection by PCR of *Neospora caninum* in fetal tissues from spontaneous bovine abortions. *J. Clin. Microbiol.* 37, 4059–4064.

175. Baszler, T. V., McElwain, T. F., Mathison, B. A. 2000. Immunization of BALB/c mice with killed *Neospora caninum* tachyzoite antigen induces a type 2 immune response and exacerbates encephalitis and neurological disease. *Clin. Diagn. Lab. Immunol.* 7, 893–898.

176. Baszler, T. V., Adams, S., Vander-Schalie, J., Mathison, B. A., Kostovic, M. 2001. Validation of a commercially available monoclonal antibody-based competitive-inhibition enzyme-linked immunosorbent assay for detection of serum antibodies to *Neospora caninum* in cattle. *J. Clin. Microbiol.* 39, 3851–3857.

177. Baszler, T. V., Shkap, V., Mwangi, W., Davies, C. J., Mathison, B. A., Mazuz, M., Resnikov, D., Fish, L., Leibovitch, B., Staska, L. M., Savitsky, I. 2008. Bovine immune response to inoculation with *Neospora caninum* surface antigen SRS2 lipopeptides mimics immune response to infection with live parasites. *Clin. Vaccine Immunol.* 15, 659–667.

178. Batmaz, H., Senturk, S., Aydin, L. 2004. Clinical neosporosis in a dog in Turkey. *Aust. Vet. Pract.* 34, 108–110.

178a. Bauer, K. L., Goertz, C. E. C., Belovarac, J. A., Walton, R. W., Dunn, J. L., Tuomi, P. 2016. Infectious disease and toxicological monitoring of stranded Pacific harbor seals (*Phoca vitulina richardsi*) in Cook Inlet as surrogates for monitoring endangered belugas (*Delphinapterus leucas*). *J. Zoo Wildl. Med.* 47, 770–780.

179. Bech-Sàbat, G., Serrano, B., García-Ispierto, I., Santolaria, P., Yániz, J. L., Almería, S., López-Gatius, F. 2006. Effect of progesterone supplementation during early foetal period in *Neospora caninum* seropositive dairy cows. *Reprod. Dom. Anim.* 41, 104.

180. Bech-Sàbat, G., López-Gatius, F., Santolaria, P., García-Ispierto, I., Pabón, M., Nogareda, C., Yániz, J. L., Almería, S. 2007. Progesterone supplementation during mid-gestation increases the risk of abortion in *Neospora*-infected dairy cows with high antibody titres. *Vet. Parasitol.* 145, 164–167.

181. Beck, R., Marinculić, A., Mihaljević, Ž., Benić, M., Martinković, F. 2010. Seroprevalence and potential risk factors of *Neospora caninum* infection in dairy cattle in Croatia. *Vet. Arhiv* 80, 163–171.

182. Beckers, C. J. M., Wakefield, T., Joiner, K. A. 1997. The expression of *Toxoplasma* proteins in *Neospora caninum* and the identification of a gene encoding a novel rhoptry protein. *Mol. Biochem. Parasitol.* 89, 209–223.

183. Behnke, M. S., Khan, A., Lauron, E. J., Jimah, J. R., Wang, Q., Tolia, N. H., Sibley, L. D. 2015. Rhoptry proteins ROP5 and ROP18 are major murine virulence factors in genetically divergent South American strains of *Toxoplasma gondii*. *PLoS Genet.* 11, e1005434.

184. Behrendt, J. H., Taubert, A., Zahner, H., Hermosilla, C. 2008. Studies on synchronous egress of coccidian parasites (*Neospora caninum, Toxoplasma gondii, Eimeria bovis*) from bovine endothelial host cells mediated by calcium ionophore A23187. *Vet. Res. Commun.* 32, 325–332.

185. Beiting, D. P., Peixoto, L., Akopyants, N. S., Beverley, S. M., Wherry, E. J., Christian, D. A., Hunter, C. A., Brodsky, I. E., Roos, D. S. 2014. Differential induction of TLR3-dependent innate immune signaling by closely related parasite species. *PLoS ONE* 9, e88398.

186. Beiting, D. P., Hidano, S., Baggs, J. E., Geskes, J. M., Fang, Q., Wherry, E. J., Hunter, C. A., Roos, D. S., Cherry, S. 2015. The orphan nuclear receptor TLX is an enhancer of STAT1-mediated transcription and immunity to *Toxoplasma gondii*. *PLoS Biol.* 13, e1002200.

187. Bell, A. S., Ranford-Cartwright, L. C. 2002. Real-time quantitative PCR in parasitology. *Trends Parasitol.* 18, 337–342.

188. Benavides, J., Maley, S., Pang, Y., Palarea, J., Eaton, S., Katzer, F., Innes, E. A., Buxton, D., Chianini, F. 2011. Development of lesions and tissue distribution of parasite in lambs orally infected with sporulated oocysts of *Toxoplasma gondii*. *Vet. Parasitol.* 179, 209–215.

189. Benavides, J., Katzer, F., Maley, S. W., Bartley, P. M., Cantón, G., Palarea-Albaladejo, J., Purslow, C. A., Pang, Y., Rocchi, M. S., Chianini, F., Buxton, D., Innes, E. A. 2012. High rate of transplacental infection and transmission of *Neospora caninum* following experimental challenge of cattle at day 210 of gestation. *Vet. Res.* 43, e83.

190. Benavides, J., Collantes-Fernández, E., Ferre, I., Pérez, V., Campero, C., Mota, R., Innes, E., Ortega-Mora, L. M. 2014. Experimental ruminant models for bovine neosporosis: What is known and what is needed. *Parasitology* 141, 1471–1488.

191. Benetti, A. H., Toniollo, G. H., dos Santos, T. R., Gennari, S. M., da Costa, A. J., Dias, R. A. 2008. Ocorrência de anticorpos anti-*Neospora caninum* em cães no município de Cuiabá, Mato Grosso. *Ciênc. Anim. Bras.* 9, 177–180.

192. Benetti, A. H., Schein, F. B., dos Santos, T. R., Toniollo, G. H., da Costa, A. J., Mineo, J. R., Lobato, J., de Oliveira Silva, D. A., Gennari, S. M. 2009. Pesquisa de anticorpos anti-*Neospora caninum* em bovinos leiteiros, cães e trabalhadores rurais da região Sudoeste do Estado de Mato Grosso. *Rev. Bras. Parasitol. Vet.* 18(Suppl 1), 29–33.

193. Berger-Schoch, A. E., Herrmann, D. C., Schares, G., Müller, N., Bernet, D., Gottstein, B., Frey, C. F. 2011. Prevalence and genotypes of *Toxoplasma gondii* in feline faeces (oocysts) and meat from sheep, cattle and pigs in Switzerland. *Vet. Parasitol.* 177, 290–297.

194. Bergeron, N., Fecteau, G., Paré, J., Martineau, R., Villeneuve, A. 2000. Vertical and horizontal transmission of *Neospora caninum* in dairy herds in Québec. *Can. Vet. J.* 41, 464–467.

195. Bergeron, N., Fecteau, G., Villeneuve, A., Girard, C., Paré, J. 2001. Failure of dogs to shed oocysts after being fed bovine fetuses naturally infected by *Neospora caninum*. *Vet. Parasitol.* 97, 145–152.

196. Bergeron, N., Girard, C., Paré, J., Fecteau, G., Robinson, J., Baillargeon, P. 2001. Rare detection of *Neospora caninum* in placentas from seropositive dams giving birth to full-term calves. *J. Vet. Diagn. Invest.* 13, 173–175.

197. Besteiro, S., Dubremetz, M. F., Lebrun, M. 2011. The moving junction of apicomplexan parasites: A key structure for invasion. *Cell. Microbiol.* 13, 797–805.

198. Bevins, S., Blizzard, E., Bazan, L., Whitley, P. 2013. *Neospora caninum* exposure in overlapping populations of coyotes (*Canis latrans*) and feral swine (*Sus scrofa*). *J. Wildl. Dis.* 49, 1028–1032.

199. Bielanski, A., Robinson, J., Phipps-Todd, B. 2002. Effect of *Neospora caninum* on *in vitro* development of preimplantation stage bovine embryos and adherence to the zona pellucida. *Vet. Rec.* 150, 316–318.

200. Bieé, J., Moskwa, B., Cabaj, W. 2010. *In vitro* isolation and identification of the first *Neospora caninum* isolate from European bison (*Bison bonasus bonasus* L.). *Vet. Parasitol.* 173, 200–205.

201. Bieé, J., Moskwa, B., Bogdaszewski, M., Cabaj, W. 2012. Detection of specific antibodies anti-*Neospora caninum* in the fallow deer (*Dama dama*). *Res. Vet. Sci.* 92, 96–98.

202. Bildfell, R., Davidson, J., Dubey, J. P. 1994. *Neospora*-induced protozoal bovine abortion in Prince Edward Island. *Can. Vet. J.* 35, 122.

203. Bishop, S., King, J., Windsor, P., Reichel, M. P., Ellis, J., Šlapeta, J. 2010. The first report of ovine cerebral neosporosis and evaluation of *Neospora caninum* prevalence in sheep in New South Wales. *Vet. Parasitol.* 170, 137–142.

204. Biyikoglu, G., Oncel, T., Bagci, O. 2005. Serological survey of *Neospora caninum* infection. *Indian Vet. J.* 82, 345–346.

205. Bjerkås, I., Mohn, S. F., Presthus, J. 1984. Unidentified cyst-forming sporozoon causing encephalomyelitis and myositis in dogs. *Z. Parasitenkd.* 70, 271–274.

206. Bjerkås, I., Landsverk, T. 1986. Identification of *Toxoplasma gondii* and *Encephalitozoon cuniculi* by immunoperoxidase techniques and electron microscopy, in stored, formalin-fixed, paraffin-embedded tissue. *Acta Vet. Scand.* 27, 11–22.

207. Bjerkås, I., Presthus, J. 1988. Immuno-histochemical and ultrastructural characteristics of a cyst-forming sporozoon associated with encephalomyelitis and myositis in dogs. *Acta Pathol. Microbiol. Immunol. Scand.* 96, 445–454.

208. Bjerkås, I., Presthus, J. 1989. The neuropathology in toxoplasmosis-like infection caused by a newly recognized cyst-forming sporozoon in dogs. *Acta Pathol. Microbiol. Immunol. Scand.* 97, 459–468.

209. Bjerkås, I., Dubey, J. P. 1991. Evidence that *Neospora caninum* is identical to the *Toxoplasma*-like parasite of Norwegian dogs. *Acta Vet. Scand.* 32, 407–410.

210. Bjerkås, I. 1992. Infections with *Neospora caninum* and Neospora-like parasites in dogs, with special emphasis on infections in Norway. *Proceedings of the 12th International Symposium on New and Emerging Infectious Diseases*. September 8, Davis, California. World Association of Veterinary Microbiologists and Specialists in Infectious Diseases. 275–279.

211. Bjerkås, I., Jenkins, M. C., Dubey, J. P. 1994. Identification and characterization of *Neospora caninum* tachyzoite antigens useful for diagnosis of neosporosis. *Clin. Diagn. Lab. Immunol.* 1, 214–221.

212. Björkman, C., Lundén, A., Uggla, A. 1994. Prevalence of antibodies to *Neospora caninum* and *Toxoplasma gondii* in Swedish dogs. *Acta Vet. Scand.* 35, 445–447.

213. Björkman, C., Lundén, A., Holmdahl, J., Barber, J., Trees, A. J., Uggla, A. 1994. *Neospora caninum* in dogs: Detection of antibodies by ELISA using an iscom antigen. *Parasite Immunol.* 16, 643–648.

214. Björkman, C., Gustafsson, K., Holmdahl, J., Kindahl, H., Lundén, A., Magnusson, U., Stenlund, S., Uggla, A. 1994. *Neospora caninum*, en nyupptäckt patogen hos nötkreatur och hund i Sverige. *Svensk Veterinärtidning* 46, 433–435.

215. Björkman, C., Johansson, O., Stenlund, S., Holmdahl, O. J. M., Uggla, A. 1996. *Neospora* species infection in a herd of dairy cattle. *J. Am. Vet. Med. Assoc.* 208, 1441–1444.

216. Björkman, C., Holmdahl, O. J. M., Uggla, A. 1997. An indirect enzyme-linked immunoassay (ELISA) for demonstration of antibodies to *Neospora caninum* in serum and milk of cattle. *Vet. Parasitol.* 68, 251–260.

217. Björkman, C., Hemphill, A. 1998. Characterization of *Neospora caninum* iscom antigens using monoclonal antibodies. *Parasite Immunol.* 20, 73–80.

218. Björkman, C., Lundén, A. 1998. Application of iscom antigen preparations in ELISAs for diagnosis of *Neospora* and *Toxoplasma* infections. *Int. J. Parasitol.* 28, 187–193.

219. Björkman, C., Näslund, K., Stenlund, S., Maley, S. W., Buxton, D., Uggla, A. 1999. An IgG avidity ELISA to discriminate between recent and chronic *Neospora caninum* infection. *J. Vet. Diagn. Invest.* 11, 41–44.

219a. Björkman, C., Uggla, A. 1999. Serological diagnosis of *Neospora caninum* infection. *Int. J. Parasitol.* 29, 1497–1507.

220. Björkman, C., Alenius, S., Emanuelsson, U., Uggla, A. 2000. *Neospora caninum* and bovine virus diarrhoea virus infections in Swedish dairy cows in relation to abortion. *Vet. J.* 159, 201–206.

221. Björkman, C., McAllister, M. M., Frössling, J., Näslund, K., Leung, F., Uggla, A. 2003. Application of the *Neospora caninum* IgG avidity ELISA in assessment of chronic reproductive losses after an outbreak of neosporosis in a herd of beef cattle. *J. Vet. Diagn. Invest.* 15, 3–7.

222. Björkman, C., Gondim, L. F. P., Näslund, K., Trees, A. J., McAllister, M. M. 2005. IgG avidity pattern in cattle after ingestion of *Neospora caninum* oocysts. *Vet. Parasitol.* 128, 195–200.

223. Björkman, C., Álvarez-García, G., Conraths, F. J., Mattsson, J. G., Ortega-Mora, L. M., Sager, H., Schares, G. 2006. *Neospora caninum* IgG avidity tests: An interlaboratory comparison. *Vet. Parasitol.* 140, 273–280.

224. Björkman, C., Sager, H., Schares, G. 2007. Serology in neosporosis. In *Protozoal Abortion in Farm Animals. Guidelines for Diagnosis and Control*. Ortega-Mora, L. M., Gottestin, B., Conraths, F. J., Buxton, D. (Editors). Atheneaeum Press, Gateshead, UK, CAB International. 63–75.

225. Björkman, C., Jakubek, E. B., Arnemo, J. M., Malmsten, J. 2010. Seroprevalence of *Neospora caninum* in gray wolves in Scandinavia. *Vet. Parasitol.* 173, 139–142.

226. Blader, I. J., Koshy, A. A. 2014. *Toxoplasma gondii* development of its replicative niche: In its host cell and beyond. *Eukaryot. Cell* 13, 965–976.

227. Blanco, R. D., Patarroyo, J. H., Vargas, M. I., Cardona, J. A., Araújo, L. S., Gomez, V. E. 2014. Ocorrência de anticorpos anti-*Neospora* spp. em jumentos (*Equus asinus*) no estado de Sucre—Colômbia. *Arq. Bras. Med. Vet. Zootec.* 66, 450–454.

228. Boas, R. V., Pacheco, T. A., Melo, A. L. T., de Oliveira, A. C. S., de Aguiar, D. M., Pacheco, R. C. 2015. Infection by *Neospora caninum* in dairy cattle belonging to family farmers in the northern region of Brazil. *Braz. J. Vet. Parasitol.* 24, 204–208.

229. Boaventura, C. M., de Oliveira, V. S. F., Melo, D. P. G., Borges, L. M. F., da Silva, A. C. 2008. Prevalência de *Neospora caninum* em cães de Goiânia. *Rev. Patol. Trop.* 37, 15–22.

230. Boger, L. A., Hattel, A. L. 2003. Additional evaluation of undiagnosed bovine abortion cases may reveal fetal neosporosis. *Vet. Parasitol.* 113, 1–6.

231. Borel, N., Frey, C. F., Gottstein, B., Hilbe, M., Pospischil, A., Franzoso, F. D., Waldvogel, A. 2014. Laboratory diagnosis of ruminant abortion in Europe. *Vet. J.* 200, 218–229.

232. Borsuk, S., Andreotti, R., Leite, F. P. L., Pinto, L. S., Simionatto, S., Hartleben, C. P., Goetze, M., Oshiro, L. M., Matos, M. F. C., Berne, M. E. A. 2010. Development of an indirect ELISA-NcSRS2 for detection of *Neospora caninum* antibodies in cattle. *Vet. Parasitol.* 177, 33–38.

233. Botelho, A. S., Teixeira, L., Correia da Costa, J. M., Faustino, A. M. R., Castro, A. G., Vilanova, M. 2007. *Neospora caninum*: High susceptibility to the parasite in C57BL/10ScCr mice. *Exp. Parasitol.* 115, 68–75.

234. Bottari, N. B., Tonin, A. A., Fighera, R., Flores, M. M., França, R. T., Camillo, G., Toscan, G., Vogel, F. S. F., Sangoi, M. B., Bochi, G. V., Moresco, R. N., Lopes, S. T. A., Da Silva, A. S. 2014. *Neospora caninum* and *Toxoplasma gondii*: Relationship between hepatic lesions, cytological and biochemical analysis of the cavitary liquid during the acute phase of the diseases in experimental models. *Exp. Parasitol.* 136, 68–73.

235. Boulton, J. G., Gill, P. A., Cook, R. W., Fraser, G. C., Harper, P. A. W., Dubey, J. P. 1995. Bovine *Neospora* abortion in north-eastern New South Wales. *Aust. Vet. J.* 72, 119–120.

236. Bourdoiseau, G. 1993. La néosporose des carnivores. *Rec. Méd. Vét.* 169, 473–475.
237. Bourdoiseau, G. 2000. Connaissances actuelles sur l'epidemiologie de la neosporose en France. *Proceedings of the Société Française de Buiatrie.* November 15–17, Paris. 116–118.
238. Boyd, S. P., Barr, P. A., Brooks, H. W., Orr, J. P. 2005. Neosporosis in a young dog presenting with dermatitis and neuromuscular signs. *J. Small Anim. Pract.* 46, 85–88.
239. Boydell, P., Brogan, N. 2000. Horner's syndrome associated with *Neospora* infection. *J. Small Anim. Pract.* 41, 571–572.
240. Boysen, P., Klevar, S., Olsen, I., Storset, A. K. 2006. The protozoan *Neospora caninum* directly triggers bovine NK cells to produce gamma interferon and to kill infected fibroblasts. *Infect. Immun.* 74, 953–960.
240a. Braga, M. S. C. O., Andre, M. R., Jusi, M. M. G., Freschi, C. R., Teixeira, M. C. A., Machado, R. Z. 2012. Occurrence of anti-*Toxoplasma gondii* and anti-*Neospora caninum* antibodies in cats with outdoor access in São Luís, Maranhão, Brazil. *Rev. Bras. Parasitol. Vet.* 21, 107–111.
241. Brasil, A. W. L., Parentoni, R. N., Feitosa, T. F., Bezerra, C. S., Vilela, V. L. R., Pena, H. F. J., de Azevedo, S. S. 2015. Risk factors for *Toxoplasma gondii* and *Neospora caninum* seropositivity in buffaloes in Paraiba State, Brazil. *Braz. J. Vet. Parasitol.* 24, 459–463.
242. Braund, K. G., Blagburn, B. L., Toivio-Kinnucan, M., Amling, K. A., Pidgeon, G. L. 1988. *Toxoplasma* polymyositis/polyneuropathy—A new clinical variant in two mature dogs. *J. Am. Anim. Hosp. Assoc.* 24, 93–97.
243. Bregoli, M., Gioia, C., Stefano, N., Mariapia, C., Claudio, P. 2006. Serological survey of *Neospora caninum* in free-ranging wild ruminants. *Vet. Arhiv* 76(Suppl), S111–S115.
244. Bresciani, K. D. S., Gennari, S. M., Serrano, A. C. M., Rodrigues, A. A. R., Ueno, T., Franco, L. G., Perri, S. H. V., Amarante, A. F. T. 2007. Antibodies to *Neospora caninum* and *Toxoplasma gondii* in domestic cats from Brazil. *Parasitol. Res.* 100, 281–285.
245. Bresciani, K. D. S., Costa, A. J., Nunes, C. M., Serrano, A. C. M., Moura, A. B., Stobbe, N. S., Perri, S. H. V., Dias, R. A., Gennari, S. M. 2007. Ocorrência de anticorpos contra *Neospora caninum* e *Toxoplasma gondii* e estudo de fatores de risco em cães de Araçatuba—SP. *Ars Vet.* 23, 40–46.
246. Brickell, J. S., McGowan, M. M., Wathes, D. C. 2010. Association between *Neospora caninum* seropositivity and perinatal mortality in dairy heifers at first calving. *Vet. Rec.* 167, 82–85.
247. Brom, P. R. F., Regidor-Cerrillo, J., Collantes-Fernández, E., Ortega-Mora, L. M., Guimarães, M. S., da Silva, A. C. 2014. Genetic characterisation of *Neospora caninum* strains from clinical samples of zebuine foetuses obtained in abattoirs in Goiás, Brazil. *Vet. Parasitol.* 204, 381–387.
248. Bruhn, F. R. P., Figueiredo, V. C., Andrade, G. S., Costa-Júnior, L. M., da Rocha, C. M. B. M., Guimarães, A. M. 2012. Occurrence of anti-*Neospora caninum* antibodies in dogs in rural areas in Minas Gerais, Brazil. *Rev. Bras. Parasitol. Vet.* 21, 161–164.
249. Bruhn, F. R. P., Daher, D. O., Lopes, E., Barbieri, J. M., da Rocha, C. M. B. M., Guimarães, A. M. 2013. Factors associated with seroprevalence of *Neospora caninum* in dairy cattle in southeastern Brazil. *Trop. Anim. Health Prod.* 45, 1093–1098.
250. Bruno, S., Duschak, V. G., Ledesma, B., Ferella, M., Andersson, B., Guarnera, E. A., Angel, S. O. 2004. Identification and characterization of serine proteinase inhibitors from *Neospora caninum. Mol. Biochem. Parasitol.* 136, 101–107.
251. Bryan, L. A., Gajadhar, A. A., Dubey, J. P., Haines, D. M. 1994. Bovine neonatal encephalomyelitis associated with a *Neospora* sp. protozoan. *Can. Vet. J.* 35, 111–113.
252. Bukau, B., Horwich, A. L. 1998. The Hsp70 and Hsp60 chaperone machines. *Cell* 92, 351–366.
253. Burkhardt, E., Dubey, J. P., Korte, G., Bauer, C. 1992. Zwei Erkrankungen infolge einer Infektion mit *Neospora caninum* bei Hundewelpen in Deutschland. *Kleintierpraxis* 37, 701–706.
254. Butler, J. E. 1983. Bovine immunoglobulins: An augmented review. *Vet. Immunol. Immunopathol.* 4, 43–152.
255. Buxton, D., Maley, S. W., Pastoret, P. P., Brochier, B., Innes, E. A. 1997. Examination of red foxes (*Vulpes vulpes*) from Belgium for antibody to *Neospora caninum* and *Toxoplasma gondii. Vet. Rec.* 141, 308–309.
256. Buxton, D., Caldow, G. L., Maley, S. W., Marks, J., Innes, E. A. 1997. Neosporosis and bovine abortion in Scotland. *Vet. Rec.* 141, 649–651.
257. Buxton, D., Maley, S. W., Thomson, K. M., Trees, A. J., Innes, E. A. 1997. Experimental infection of non-pregnant and pregnant sheep with *Neospora caninum. J. Comp. Pathol.* 117, 1–16.
258. Buxton, D. 1998. Protozoan infections (*Toxoplasma gondii, Neospora caninum* and *Sarcocystis* spp.) in sheep and goats: Recent advances. *Vet. Res.* 29, 289–310.

259. Buxton, D., Maley, S. W., Wright, S., Thomson, K. M., Rae, A. G., Innes, E. A. 1998. The pathogenesis of experimental neosporosis in pregnant sheep. *J. Comp. Pathol.* 118, 267–279.

260. Buxton, D., Wright, S., Maley, S. W., Rae, A. G., Lundén, A., Innes, E. A. 2001. Immunity to experimental neosporosis in pregnant sheep. *Parasite Immunol.* 23, 85–91.

261. Buxton, D., McAllister, M. M., Dubey, J. P. 2002. The comparative pathogenesis of neosporosis. *Trends Parasitol.* 18, 546–552.

262. Byrem, T. M., Bartlett, P. C., Donohue, H., Voisinet, B. D., Houseman, J. T. 2012. Performance of a commercial serum ELISA for the detection of antibodies to *Neospora caninum* in whole and skim milk samples. *Vet. Parasitol.* 190, 249–253.

263. Cabaj, W., Choromanski, L., Rodgers, S., Moskwa, B. E., Malczewski, A. 2000. *Neospora caninum* infections in aborting dairy cows in Poland. *Acta Parasitol.* 45, 113–114.

264. Cabaj, W., Moskwa, B., Pastusiak, K., Gill, J. 2005. Antibodies to *Neospora caninum* in the blood of European bison (*Bison bonasus bonasus* L.) living in Poland. *Vet. Parasitol.* 128, 163–168.

265. Cabezón, O., Millán, J., Gomis, M., Dubey, J. P., Ferroglio, E., Almería, S. 2010. Kennel dogs as sentinels of *Leishmainia infantum*, *Toxoplasma gondii*, and *Neospora caninum* in Majorca Island, Spain. *Parasitol. Res.* 107, 1505–1508.

266. Cabral, A. D., Camargo, C. N., Galleti, N. T. C., Okuda, L. H., Pituco, E. M., Del Fava, C. 2009. Diagnosis of *Neospora caninum* in bovine fetuses by histology, immunohistochemistry, and nested-PCR. *Rev. Bras. Parasitol. Vet.* 18, 14–19.

267. Cadore, G. C., Vogel, F. S. F., Flores, E. F., Sangioni, L. A., Camillo, G. 2009. Suscetibilidade de linhagens celulares e cultivos primários ao *Neospora caninum*. *Ciência Rural* 39, 1581–1585.

268. Caetano-da-Silva, A., Ferre, I., Collantes-Fernández, E., Navarro, V., Aduriz, G., Ugarte-Garagalza, C., Ortega-Mora, L. M. 2004. Occasional detection of *Neospora caninum* DNA in frozen extended semen from naturally infected bulls. *Theriogenology* 62, 1329–1336.

269. Caetano-da-Silva, A., Ferre, I., Aduriz, G., Álvarez-García, G., del-Pozo, I., Atxaerandio, R., Regidor-Cerrillo, J., Ugarte-Garagalza, C., Ortega-Mora, L. M. 2004. *Neospora caninum* infection in breeder bulls: Seroprevalence and comparison of serological methods used for diagnosis. *Vet. Parasitol.* 124, 19–24.

270. Cai, G. L., Lu, C., Gao, C. S., Zhang, X. P., Wu, D. 2006. [Epidemiological survey of cattle neosporosis in Jilin Province, China]. *J. Agr. Sci. Yanbian Univ.* 28, 110–114 (in Chinese).

271. Calandra, P. M., Di Matía, J. M., Cano, D. B., Odriozola, E. A., García, J. A., Späth, E. A. J., Odeón, A. C., Paolicchi, F. A., Morrell, E. L., Campero, C. M., Moore, D. P. 2014. Neosporosis epidémica y endémica: Descripcción de dos eventos en bovinos para cría. *Rev. Argent. Microbiol.* 46, 315–319.

272. Caldow, G. L., Buxton, D., Spence, J. A., Holisz, J. 1996. Diagnoses of bovine abortion in Scotland. *Proceedings of the XIX World Buiatrics Conference.* July 8–12, Edinburgh, Scotland, Volume 1, 191–194.

273. Caldow, G. L. 1998. Bovine abortion outbreak associated with *Neospora* and other infectious agents. *Vet. Rec.* 142, 118–119.

274. Campero, C. M., Anderson, M. L., Conosciuto, G., Odriozola, H., Bretschneider, G., Poso, M. A. 1998. *Neospora caninum*-associated abortion in a dairy herd in Argentina. *Vet. Rec.* 143, 228–229.

275. Campero, C. M., Moore, D. P., Odeón, A. C., Cipolla, A. L., Odriozola, E. 2003. Aetiology of bovine abortion in Argentina. *Vet. Res. Commun.* 27, 359–369.

276. Campero, C. M., Moore, D. P., Lagomarsino, H., Odeón, A. C., Castro, M., Visca, H. 2003. Serological status and abortion rate in progeny obtained by natural service or embryo transfer from *Neospora caninum*-seropositive cows. *J. Vet. Med. B* 50, 458–460.

277. Campero, C. M., Pérez, A., Moore, D. P., Crudeli, G., Benitez, D., Draghi, M. G., Cano, D., Konrad, J. L., Odeón, A. C. 2007. Occurrence of antibodies against *Neospora caninum* in water buffaloes (*Bubalus bubalis*) on four ranches in Corrientes province, Argentina. *Vet. Parasitol.* 150, 155–158.

278. Campero, L. M., Minke, L., Moré, G., Rambeaud, M., Bacigalupe, D., Moore, D. P., Hecker, Y., Campero, C. M., Schares, G., Venturini, M. C. 2015. Evaluation and comparison of serological methods for the detection of bovine neosporosis in Argentina. *Rev. Argent. Microbiol.* 47, 295–301.

279. Campero, L. M., Venturini, M. C., Moore, D. P., Massola, L., Lagomarsino, H., García, B., Bacigalupe, D., Rambeaud, M., Pardini, L., Leunda, M. R., Schares, G., Campero, C. M. 2015. Isolation and molecular characterization of a new *Neospora caninum* isolate from cattle in Argentina. *Exp. Parasitol.* 155, 8–12.

280. Canada, N., Meireles, C. S., Rocha, A., da Costa, J. M. C., Erickson, M. W., Dubey, J. P. 2002. Isolation of viable *Toxoplasma gondii* from naturally infected aborted bovine fetuses. *J. Parasitol.* 88, 1247–1248.

281. Canada, N., Meireles, C. S., Rocha, A., Sousa, S., Thompson, G., Dubey, J. P., Romand, S., Thulliez, P., Correia da Costa, J. M. 2002. First Portuguese isolate of *Neospora caninum* from an aborted fetus from a dairy herd with endemic neosporosis. *Vet. Parasitol.* 110, 11–15.

282. Canada, N., Carvalheira, J., Meireles, C. S., Correia da Costa, J. M., Rocha, A. 2004. Prevalence of *Neospora caninum* infection in dairy cows and its consequences for reproductive management. *Theriogenology* 62, 1229–1235.

283. Canada, N., Meireles, C. S., Carvalheira, J., Rocha, A., Sousa, S., Correia da Costa, J. M. 2004. Determination of an optimized cut-off value for the *Neospora* agglutination test for serodiagnosis in cattle. *Vet. Parasitol.* 121, 225–231.

284. Canada, N., Meireles, C. S., Mezo, M., González-Warleta, M., Correia da Costa, J. M., Sreekumar, C., Hill, D. E., Miska, K. B., Dubey, J. P. 2004. First isolation of *Neospora caninum* from an aborted bovine fetus in Spain. *J. Parasitol.* 90, 863–864.

285. Canada, N., Meireles, C. S., Ferreira, P., da Costa, J. M. C., Rocha, A. 2006. Artificial insemination of cows with semen *in vitro* contaminated with *Neospora caninum* tachyzoites failed to induce neosporosis. *Vet. Parasitol.* 139, 109–114.

286. Canatan, H. E., Polat, I. M., Bayramoglu, R., Kuplulu, S., Vural, M. R., Aktug, E. 2014. Effects of *Neospora caninum* on reproductive performance and the efficacy of treatment with a combination of sulphadiazine-trimethoprim and toltrazuil: A longitudinal field study. *Vet. Med.* 59, 22–28.

287. Cannas, A., Naguleswaran, A., Müller, N., Gottstein, B., Hemphill, A. 2003. Reduced cerebral infection of *Neospora caninum*-infected mice after vaccination with recombinant microneme protein NcMIC3 and RIBI adjuvant. *J. Parasitol.* 89, 44–50.

288. Cannas, A., Naguleswaran, A., Müller, N., Eperon, S., Gottstein, B., Hemphill, A. 2003. Vaccination of mice against experimental *Neospora caninum* infections using NcSAG1- and NcSRS2-based recombinant antigens and DNA vaccines. *Parasitology* 126, 303–312.

289. Cañón-Franco, W. A., Bergamaschi, D. P., Camargo, L. M. A., Paula, V. S. O., Souza, S. L. P., Gennari, S. M. 2003. Comparison between direct agglutination test and indirect fluorescent antibody test for the detection of *Neospora caninum* antibodies in naturally exposed dogs. *Rev. Bras. Parasitol. Vet.* 12, 4–6.

290. Cañón-Franco, W. A., Bergamaschi, D. P., Labruna, M. B., Camargo, L. M. A., Souza, S. L. P., Silva, J. C. R., Pinter, A., Dubey, J. P., Gennari, S. M. 2003. Prevalence of antibodies to *Neospora caninum* in dogs from Amazon, Brazil. *Vet. Parasitol.* 115, 71–74.

291. Cañón-Franco, W. A., Yai, L. E. O., Souza, S. L. P., Santos, L. C., Farias, N. A. R., Ruas, J., Rossi, F. W., Gomes, A. A. B., Dubey, J. P., Gennari, S. M. 2004. Detection of antibodies to *Neospora caninum* in two species of wild canids, *Lycalopex gymnocercus* and *Cerdocyon thous* from Brazil. *Vet. Parasitol.* 123, 275–277.

292. Cantile, C., Arispici, M. 2002. Necrotizing cerebellitis due to *Neospora caninum* infection in an old dog. *J. Vet. Med. A* 49, 47–50.

293. Cantón, G. J., Katzer, F., Benavides-Silván, J., Maley, S. W., Palarea-Albaladejo, J., Pang, Y., Smith, S., Bartley, P. M., Rocchi, M., Innes, E. A., Chianini, F. 2013. Phenotypic characterisation of the cellular immune infiltrate in placentas of cattle following experimental inoculation with *Neospora caninum* in late gestation. *Vet. Res.* 44, e60.

294. Cantón, G. J., Konrad, J. L., Moore, D. P., Caspe, S. G., Palarea-Albaladejo, J., Campero, C. M., Chianini, F. 2014. Characterization of immune cell infiltration in the placentome of water buffaloes (*Bubalus bubalis*) infected with *Neospora caninum* during pregnancy. *J. Comp. Pathol.* 150, 463–468.

295. Cantón, G. J., Katzer, F., Maley, S. W., Bartley, P. M., Benavides-Silván, J., Palarea-Albaladejo, J., Pang, Y., Smith, S. H., Rocchi, M. S., Buxton, D., Innes, E. A., Chianini, F. 2014. Inflammatory infiltration into placentas of *Neospora caninum* challenged cattle correlates with clinical outcome of pregnancy. *Vet. Res.* 45, e11.

296. Cantón, G. J., Katzer, F., Maley, S. W., Bartley, P. M., Benavides-Silván, J., Palarea-Albaladejo, J., Pang, Y., Smith, S. H., Rocchi, M., Buxton, D., Innes, E. A., Chianini, F. 2014. Cytokine expression in the placenta of pregnant cattle after inoculation with *Neospora caninum*. *Vet. Immunol. Immunopathol.* 161, 77–89.

297. Cao, L., Zhang, X., Tuo, W. 2011. Tunicamycins, a class of nucleoside antibiotics similar to corynetoxins of the *Rathayibacter toxicus*, increase susceptibility of mice to *Neospora caninum*. *Vet. Parasitol.* 177, 13–19.

298. Capelli, G., Nardelli, S., di Regalbono, A. F., Scala, A., Pietrobelli, M. 2004. Sero-epidemiological survey of *Neospora caninum* infection in dogs in north-eastern Italy. *Vet. Parasitol.* 123, 143–148.

299. Capelli, G., Natale, A., Nardelli, S., di Regalbono, A. F., Pietrobelli, M. 2006. Validation of a commercially available cELISA test for canine neosporosis against an indirect fluorescent antibody test (IFAT). *Prev. Vet. Med.* 73, 315–320.

299a. Cardona, J. A., Martínez, Y., Betancur, C. A. 2015. Seroepidemiología de hembras bovinas naturalmente infectadas por *Neospora caninum* en Córdoba, Colombia. *Revista U.D.C.A Actualidad & Divulgación Científica*. 18, 401–408.

300. Cardoso, J. M. S., Funada, M. R., Soares, R. M., Gennari, S. M. 2008. Perfil sorológico dos anticorpos colostrais para *Neospora caninum* em bezerros livres da infecção. *Braz. J. Vet. Res. Anim. Sci.* 45, 379–384.

301. Cardoso, J. M. S., Nishi, S. M., Funada, M. R., Amaku, M., Guimarães, J. S., Gennari, S. M. 2009. Antibody dynamics during gestation in cows naturally infected with *Neospora caninum* from four dairy herds in Brazil. *Braz. J. Vet. Res. Anim. Sci.* 46, 395–399.

302. Cardoso, J. M. S., Amaku, M., Araújo, A. J. S., Gennari, S. M. 2012. A longitudinal study of *Neospora caninum* infection on three dairy farms in Brazil. *Vet. Parasitol.* 187, 553–557.

303. Cardoso, J. M. S., Amaku, M., Arav, A. J. U., Araújo, A. J. U. S., Gennari, S. M. 2012. *Neospora caninum*: Analysis of reproductive parameters in dairy herds in Brazil. *Braz. J. Vet. Res. Anim. Sci.* 49, 459–464.

304. Cardoso, M. R. D., Mota, C. M., Ribeiro, D. P., Noleto, P. G., Andrade, W. B. F., Souza, M. A., Silva, N. M., Mineo, T. W. P., Mineo, J. R., Silva, D. A. O. 2012. Adjuvant and immunostimulatory effects of a D-galactose-binding lectin from *Synadenium carinatum* latex (ScLL) in the mouse model of vaccination against neosporosis. *Vet. Res.* 43, 76.

305. Cardoso, R., Nolasco, S., Goncalves, J., Cortes, H. C., Leitao, A., Soares, H. 2014. *Besnoitia besnoiti* and *Toxoplasma gondii*: Two apicomplexan strategies to manipulate the host cell centrosome and Golgi apparatus. *Parasitology* 141, 1436–1454.

306. Cardoso, R., Soares, H., Hemphill, A., Leitão, A. 2016. Apicomplexans pulling the strings: Manipulation of the host cell cytoskeleton dynamics. *Parasitology* 143, 957–970.

307. Carreno, R. A., Schnitzler, B. E., Jeffries, A. C., Tenter, A. M., Johnson, A. M., Barta, J. R. 1998. Phylogenetic analysis of coccidia based on 18S rDNA sequence comparison indicates that *Isospora* is most closely related to *Toxoplasma* and *Neospora*. *J. Eukaryot. Microbiol.* 45, 184–188.

308. Carruthers, V. B., Blackman, M. J. 2005. A new release on life: Emerging concepts in proteolysis and parasite invasion. *Mol. Microbiol.* 55, 1617–1630.

309. Carstensen, M., Giudice, J. H., Hildebrand, E. C., Dubey, J. P., Erb, J., Stark, D., Hart, J., Barber-Meyer, S., Mech, L. D., Windels, S. K., Edwards, A. J. 2016. Serological survey of diseases of free-ranging gray wolves (*Canis lupus*) in Minnesota. *J. Wildl. Dis.* DOI: 10.7589/2016-06-140.

310. Carvalho, J. V., Alves, C. M. O. S., Cardoso, M. R. D., Mota, C. M., Barbosa, B. F., Ferro, E. A. V., Silva, N. M., Mineo, T. W. P., Mineo, J. R., Silva, D. A. O. 2010. Differential susceptibility of human trophoblastic (BeWo) and uterine cervical (HeLa) cells to *Neospora caninum* infection. *Int. J. Parasitol.* 40, 1629–1637.

311. Carvalho-Patrício, M. A., Richartz, R. R. T. B., Gasino-Joineau, M. E., Zonta-de-Carvalho, R. C., Meirelles, A. C., Locatelli-Dittrich, R. 2013. *Neospora*-DNA prevalence in rabies-negative cattle with neurological disorders. *Vet. Rec.* 172, 238.

312. Casas V. G., Chávez V. A., Casas A. E., Leyva V. V., Alvarado S. A., Serrano M. E., Ticona S. D., Puray Ch. N. 2006. Presencia de *Neospora caninum* en llamas de una empresa ganadera de la Sierra Central. *Rev. Inv. Vet. Perú* 17, 8–13.

313. Caspe, S. G., Moore, D. P., Leunda, M. R., Cano, D. B., Lischinsky, L., Regidor-Cerrillo, J., Álvarez García, G., Echaide, I. G., Bacigalupe, D., Ortega Mora, L. M., Odeón, A. C., Campero, C. M. 2012. The *Neospora caninum*-Spain 7 isolate induces placental damage, fetal death and abortion in cattle when inoculated in early gestation. *Vet. Parasitol.* 189, 171–181.

314. Cavalcante, G. T., Monteiro, R. M., Soares, R. M., Nishi, S. M., Alves Neto, A. F., Esmerini, P. O., Sercundes, M. K., Martins, J., Gennari, S. M. 2011. Shedding of *Neospora caninum* oocysts by dogs fed different tissues from naturally infected cattle. *Vet. Parasitol.* 179, 220–223.

315. Cavalcante, G. T., Soares, R. M., Nishi, S. M., Hagen, S. C. F., Vannucchi, C. I., Maiorka, P. C., Paixão, A. S., Gennari, S. M. 2012. Experimental infection with *Neospora caninum* in pregnant bitches. *Rev. Bras. Parasitol. Vet.* 21, 232–236.

316. Cavirani, S., Cabassi, C. S., Taddei, S., Donofrio, G., Bottarelli, E. 2005. Association between *Neospora caninum* antibodies and blue tongue vaccination in dairy cows. *Vet. Res. Commun.* 29(Suppl 2), 233–236.

317. Cazarotto, C. J., Balzan, A., Grosskopf, R. K., Boito, J. P., Portella, L. P., Vogel, F. F., Fávero, J. F., Cucco, D. C., Biazus, A. H., Machado, G., Da Silva, A. S. 2016. Horses seropositive for *Toxoplasma gondii*, *Sarcocystis* spp. and *Neospora* spp.: Possible risk factors for infection in Brazil. *Microb. Pathog.* 99, 30–35.

318. Cedillo, C. J. R., Martínez, M. J. J., Santacruz, A. M., Banda, R. V. M., Morales, S. E. 2008. Models for experimental infection of dogs fed with tissue from fetuses and neonatal cattle naturally infected with *Neospora caninum*. *Vet. Parasitol.* 154, 151–155.

319. Cerqueira-Cézar, C. K., Pedersen, K., Calero-Bernal, R., Kwok, O. C., Villena, I., Dubey, J. P. 2016. Seroprevalence of *Neospora caninum* in feral swine (*Sus scrofa*) in the United States. *Vet. Parasitol.* 226, 35–37.

320. Chahan, B., Gaturaga, I., Huang, X., Liao, M., Fukumoto, S., Hirata, H., Nishikawa, Y., Suzuki, H., Sugimoto, C., Nagasawa, H., Fujisaki, K., Igarashi, I., Mikami, T., Xuan, X. 2003. Serodiagnosis of *Neospora caninum* infection in cattle by enzyme-linked immunosorbent assay with recombinant truncated NcSAG1. *Vet. Parasitol.* 118, 177–185.

321. Chanlun, A. 2002. *Neospora caninum* infection in cattle: The use of bulk milk for detection of infection in dairy herds in Thailand. *Masters thesis*. Report no. 28. Swedish University of Agricultural Sciences, Uppsala, Sweden. 1–52.

322. Chanlun, A., Näslund, K., Aiumlamai, S., Björkman, C. 2002. Use of bulk milk for detection of *Neospora caninum* infection in dairy herds in Thailand. *Vet. Parasitol.* 110, 35–44.

323. Chanlun, A. 2006. Epidemiology of *Neospora caninum* infection in dairy cattle in Thailand. *Doctoral thesis*. No. 2006:86. Swedish University of Agricultural Sciences, Uppsala, Sweden. 1–250.

324. Chanlun, A., Emanuelson, U., Aiumlamai, S., Björkman, C. 2006. Variations of *Neospora caninum* antibody levels in milk during lactation in dairy cows. *Vet. Parasitol.* 141, 349–355.

325. Chanlun, A., Emanuelson, U., Chanlun, S., Aiumlamai, S., Björkman, C. 2006. Application of repeated bulk milk testing for identification of infection dynamics of *Neospora caninum* in Thai dairy herds. *Vet. Parasitol.* 136, 243–250.

326. Chanlun, A., Emanuelson, U., Frössling, J., Aiumlamai, S., Björkman, C. 2007. A longitudinal study of seroprevalence and seroconversion of *Neospora caninum* infection in dairy cattle in northeast Thailand. *Vet. Parasitol.* 146, 242–248.

327. Chanton-Greutmann, H., Thoma, R., Corboz, L., Borel, N., Pospischil, A. 2002. Aborte beim kleinen Wiederkäuer in der Schweiz: Untersuchungen während zwei Ablammperioden (1996–1998) unter besonderer Beachtung des Chlamydienabortes. *Schweiz. Arch. Tierheilkd.* 144, 483–492.

328. Chao, W. D., Ma, L. Q., Li, W. C., Zhang, Z. X., Li, D. Q., Han, W. G. 2005. Serological investigation of *Neospora caninum* infection in dairy cows in Geermu County of Qinghai Province. *Chin. Vet. Sci.* 12, 87–89. (also published in *Chin. J. Vet. Sci. Tech.* 35, 1012–1014) (in Chinese).

329. Chávez, V. A., Serrano, M. E., Casas, A. E., Ortega, M. L. M. 2002. *Neospora caninum* en camélidos sudamericanos peruanos. *Rev. Inv. Vet. Perú* 13, 92–93.

330. Chávez-Velásquez, A., Álvarez-García, G., Collantes-Fernández, E., Casas-Astos, E., Rosadio-Alcántara, R., Serrano-Martínez, E., Ortega-Mora, L. M. 2004. First report of *Neospora caninum* infection in adult alpacas (*Vicugna pacos*) and llamas (*Lama glama*). *J. Parasitol.* 90, 864–866.

331. Chávez-Velásquez, A., Aguado-Martínez, A., Ortega-Mora, L. M., Casas-Astos, E., Serrano-Martínez, E., Casas-Velásquez, G., Ruiz-Santa-Quitera, J. A., Álvarez-García, G. 2014. *Toxoplasma gondii* and *Neospora caninum* seroprevalences in domestic South American camelids of the Peruvian Andes. *Trop. Anim. Health Prod.* 46, 1141–1147.

332. Cheadle, M. A., Lindsay, D. S., Blagburn, B. L. 1999. Prevalence of antibodies to *Neospora caninum* in dogs. *Vet. Parasitol.* 85, 325–330.

333. Cheadle, M. A., Spencer, J. A., Blagburn, B. L. 1999. Seroprevalences of *Neospora caninum* and *Toxoplasma gondii* in nondomestic felids from southern Africa. *J. Zoo Wildl. Med.* 30, 248–251.

334. Cheadle, M. A., Lindsay, D. S., Rowe, S., Dykstra, C. C., Williams, M. A., Spencer, J. A., Toivio-Kinnucan, M. A., Lenz, S. D., Newton, J. C., Rolsma, M. D., Blagburn, B. L. 1999. Prevalence of antibodies to *Neospora* sp. in horses from Alabama and characterisation of an isolate recovered from a naturally infected horse. *Int. J. Parasitol.* 29, 1537–1543.

335. Cheah, T. S., Mattsson, J. G., Zaini, M., Sani, R. A., Jakubek, E. B., Uggla, A., Chandrawathani, P. 2004. Isolation of *Neospora caninum* from a calf in Malaysia. *Vet. Parasitol.* 126, 263–269.

336. Chi, J., VanLeeuwen, J. A., Weersink, A., Keefe, G. P. 2002. Direct production losses and treatment costs from bovine viral diarrhoea virus, bovine leukosis virus, *Mycobacterium avium* subspecies *paratuberculosis*, and *Neospora caninum*. *Prev. Vet. Med.* 55, 137–153.

337. Chi, J., VanLeeuwen, J. A., Weersink, A., Keefe, G. P. 2002. Management factors related to seroprevalences to bovine viral-diarrhoea virus, bovine-leukosis virus, *Mycobacterium avium* subspecies *paratuberculosis*, and *Neospora caninum* in dairy herds in the Canadian Maritimes. *Prev. Vet. Med.* 55, 57–68.

338. Chi, J., Weersink, A., VanLeeuwen, J. A., Keefe, G. P. 2003. The economics of controlling infectious diseases on dairy farms. *Can. J. Agr. Econ.* 50, 237–256.

339. Chiebao, D. P., Valadas, S. Y. O. B., Minervino, A. H. H., Castro, V., Romaldini, A. H. C. N., Calhau, A. S., De Souza, R. A. B., Gennari, S. M., Keid, L. B., Soares, R. M. 2015. Variables associated with infections of cattle by *Brucella abortus*, *Leptospira* spp. and *Neospora* spp. in Amazon Region in Brazil. *Transbound. Emerg. Dis.* 62, e30–e36.

340. Chigerwe, M., Tyler, J. W., Nagy, D. W., Middleton, J. R. 2008. Frequency of detectable serum IgG concentrations in precolostral calves. *Am. J. Vet. Res.* 69, 791–795.

341. Chin Lee, S. H., Jefferies, R., Watt, P., Hopkins, R., Sotzik, F., Reid, S., Armson, A., Boxell, A., Ryan, U. 2008. *In vitro* analysis of the TAT protein transduction domain as a drug delivery vehicle in protozoan parasites. *Exp. Parasitol.* 118, 303–307.

342. Cho, J. H., Chung, W. S., Song, K. J., Na, B. K., Kang, S. W., Song, C. Y., Kim, T. S. 2005. Protective efficacy of vaccination with *Neospora caninum* multiple recombinant antigens against experimental *Neospora caninum* infection. *Korean J. Parasitol.* 43, 19–25.

343. Cho, M. H., Na, B. K., Song, K. J., Cho, J. H., Kang, S. W., Lee, K. H., Song, C. Y., Kim, T. S. 2004. Cloning, expression, and characterization of iron-containing superoxide dismutase from *Neospora caninum*. *J. Parasitol.* 90, 278–285.

344. Choi, W. Y., Nam, H. W., Youn, J. H., Kim, D. J., Kong, Y., Kang, S. Y., Cho, S. Y. 1992. Detection of antibodies in serum and cerebrospinal fluid to *Toxoplasma gondii* by indirect latex agglutination test and enzyme-linked immunosorbent assay. *Korean J. Parasitol.* 30, 83–90.

345. Choi, Y. K., Johnson, W. O., Thurmond, M. C. 2006. Diagnosis using predictive probabilities without cut-offs. *Stat. Med.* 25, 699–717.

346. Choromanski, L., Block, W. 2000. Humoral immune responses and safety of experimental formulations of inactivated *Neospora* vaccines. *Parasitol. Res.* 86, 851–853.

347. Chryssafidis, A. L., Soares, R. M., Rodrigues, A. A. R., Carvalho, N. A. T., Gennari, S. M. 2011. Evidence of congenital transmission of *Neospora caninum* in naturally infected water buffalo (*Bubalus bubalis*) fetus from Brazil. *Parasitol. Res.* 108, 741–743.

348. Chryssafidis, A. L., Cantón, G., Chianini, F., Innes, E. A., Madureira, E. H., Gennari, S. M. 2014. Pathogenicity of Nc-Bahia and Nc-1 strains of *Neospora caninum* in experimentally infected cows and buffaloes in early pregnancy. *Parasitol. Res.* 113, 1521–1528.

349. Chryssafidis, A. L., Cantón, G., Chianini, F., Innes, E. A., Madureira, E. H., Soares, R. M., Gennari, S. M. 2015. Abortion and foetal lesions induced by *Neospora caninum* in experimentally infected water buffalos (*Bubalus bubalis*). *Parasitol. Res.* 114, 193–199.

350. Ciaramella, P., Corona, M., Cortese, L., Piantedosi, D., Santoro, D., di Loria, A., Rigato, R. 2004. Seroprevalence of *Neospora* spp. in asymptomatic horses in Italy. *Vet. Parasitol.* 123, 11–15.

351. Clancy, D., French, N. P. 2001. A stochastic model for disease transmission in a managed herd, motivated by *Neospora caninum* amongst dairy cattle. *Math. Bio.* 170, 113–132.

352. Clemente, M., de Miguel, N., Lia, V. V., Matrajt, M., Angel, S. O. 2004. Structure analysis of two *Toxoplasma gondii* and *Neospora caninum* satellite DNA families and evolution of their common monomeric sequence. *J. Mol. Evol.* 58, 557–567.

353. Clothier, K., Anderson, M. 2015. Evaluation of bovine abortion cases and tissue suitability for identification of infectious agents in California diagnostic laboratory cases from 2007 to 2012. *Theriogenology* 85, 933–938.

354. Čobádiová, A., Víchova, B., Majláthová, V., Reiterová, K. 2013. First molecular detection of *Neospora caninum* in European brown bear (*Ursus arctos*). *Vet. Parasitol.* 197, 346–349.

355. Čobadiová, A., Reiterová, K., Derdáková, M., Spilovská, S., Turčeková, L., Hviščova, I., Hisira, V. 2013. *Toxoplasma gondii*, *Neospora caninum* and tick-transmitted bacterium *Anaplasma phagocytophilum* infections in one selected goat farm in Slovakia. *Acta Parasitol.* 58, 541–546.

356. Coceres, V. M., Alonso, A. M., Alomar, M. L., Corvi, M. M. 2012. Rabbit antibodies against *Toxoplasma* Hsp20 are able to reduce parasite invasion and gliding motility in *Toxoplasma gondii* and parasite invasion in *Neospora caninum*. *Exp. Parasitol.* 132, 274–281.

357. Cochrane, S. M., Dubey, J. P. 1993. Neosporosis in a golden retriever dog from Ontario. *Can. Vet. J.* 34, 232–233.

358. Cockcroft, P. D., Dornin, L. E., Lambert, R. C., Perry, J., Reichel, M. P. 2015. Serological survey for antibodies against bovine viral diarrhoea virus and *Neospora caninum* in a population of South Australian alpacas (*Vicugna pacos*). *Aust. Vet. J.* 93, 476–478.

359. Coelho, W. M. D., do Amarante, A. F. T., Apolinário, J. C., Coelho, N. M. D., de Lima, V. M. F., Perri, S. H. V., Bresciani, K. D. S. 2011. Seroepidemiology of *Toxoplasma gondii*, *Neospora caninum*, and *Leishmania* spp. infections and risk factors for cats from Brazil. *Parasitol. Res.* 109, 1009–1013.

360. Cole, R. A., Lindsay, D. S., Dubey, J. P., Blagburn, B. L. 1993. Detection of *Neospora caninum* in tissue sections using a murine monoclonal antibody. *J. Vet. Diagn. Invest.* 5, 579–584.

361. Cole, R. A., Lindsay, D. S., Dubey, J. P., Toivio-Kinnucan, M. A., Blagburn, B. L. 1994. Characterization of a murine monoclonal antibody generated against *Neospora caninum* by western blot analysis and immunoelectron microscopy. *Am. J. Vet. Res.* 55, 1717–1722.

362. Cole, R. A., Lindsay, D. S., Blagburn, B. L., Dubey, J. P. 1995. Vertical transmission of *Neospora caninum* in mice. *J. Parasitol.* 81, 730–732.

363. Cole, R. A., Lindsay, D. S., Blagburn, B. L., Sorjonen, D. C., Dubey, J. P. 1995. Vertical transmission of *Neospora caninum* in dogs. *J. Parasitol.* 81, 208–211.

364. Collantes-Fernández, E., Zaballos, Á., Álvarez-García, G., Ortega-Mora, L. M. 2002. Quantitative detection of *Neospora caninum* in bovine aborted fetuses and experimentally infected mice by real-time PCR. *J. Clin. Microbiol.* 40, 1194–1198.

365. Collantes-Fernández, E., Álvarez-García, G., Pérez-Pérez, V., Pereira-Bueno, J., Ortega-Mora, L. M. 2004. Characterization of pathology and parasite load in outbred and inbred mouse models of chronic *Neospora caninum* infection. *J. Parasitol.* 90, 579–583.

366. Collantes-Fernández, E., López-Pérez, I., Álvarez-García, G., Ortega-Mora, L. M. 2006. Temporal distribution and parasite load kinetics in blood and tissues during *Neospora caninum* infection in mice. *Infect. Immun.* 74, 2491–2494.

367. Collantes-Fernández, E., Rodríguez-Bertos, A., Arnáiz-Seco, I., Moreno, B., Aduriz, G., Ortega-Mora, L. M. 2006. Influence of the stage of pregnancy on *Neospora caninum* distribution, parasite loads and lesions in aborted bovine foetuses. *Theriogenology* 65, 629–641.

368. Collantes-Fernández, E., Arnáiz-Seco, I., Moreno Burgos, B., Rodriguez-Bertos, A., Aduriz, G., Fernández-García, A., Ortega-Mora, L. M. 2006. Comparison of *Neospora caninum* distribution, parasite loads and lesion between epidemic and endemic bovine abortion cases. *Vet. Parasitol.* 142, 187–191.

369. Collantes-Fernández, E., Gómez-Bautista, M., Miró, G., Álvarez-García, G., Pereira-Bueno, J., Frisuelos, C., Ortega-Mora, L. M. 2008. Seroprevalence and risk factors associated with *Neospora caninum* infection in different dog populations in Spain. *Vet. Parasitol.* 152, 148–151.

370. Collantes-Fernandez, E., Arrighi, R. B. G., Álvarez-García, G., Weidner, J. M., Regidor-Cerrillo, J., Boothroyd, J. C., Ortega-Mora, L. M., Barragan, A. 2012. Infected dendritic cells facilitate systemic dissemination and transplacental passage of the obligate intracellular parasite *Neospora caninum* in mice. *PLoS ONE* 7, e32123.

371. Collery, P. M. 1995. *Neospora* abortion in cattle in Ireland. *Vet. Rec.* 136, 595.

372. Conrad, P., Barr, B., Anderson, M., Sverlow, K., Rowe, J., Thurmond, M., Breitmeyer, R., Picanso, J., Dubey, J. P., Palmer, C., Reynolds, J., Ardans, A. 1992. A newly recognized protozoan causing bovine abortion. *Proceedings of the 12th International Symposium on New and Emerging Infectious Diseases.* September 8, Davis, California. Osborn, B., Casturcci, G. Schore, C. (Editors). World Association of Microbiologists, Immunologists and Specialists of Infectious Diseases. Sponsored by School of Veterinary Medicine, University of California, Davis. 281–286.

373. Conrad, P. A., Barr, B. C., Sverlow, K. W., Anderson, M., Daft, B., Kinde, H., Dubey, J. P., Munson, L., Ardans, A. 1993. *In vitro* isolation and characterization of a *Neospora* sp. from aborted bovine foetuses. *Parasitology* 106, 239–249.

374. Conrad, P. A., Sverlow, K., Anderson, M., Rowe, J., BonDurant, R., Tuter, G., Breitmeyer, R., Palmer, C., Thurmond, M., Ardans, A., Dubey, J. P., Duhamel, G., Barr, B. 1993. Detection of serum antibody responses in cattle with natural or experimental *Neospora* infections. *J. Vet. Diagn. Invest.* 5, 572–578.

375. Conraths, F. J., Bauer, C., Becker, W. 1996. Nachweis von Antikörpern gegen *Neospora caninum* bei Kühen in hessischen Betrieben mit Abort- und Fruchtbarkeitsproblemen. *Dtsch. Tierärztl. Wochenschr.* 103, 221–224.

376. Conraths, F. J., Schares, G. 1999. Diagnostik und Epidemiologie *Neospora-caninum*-assoziierter Aborte beim Rind. *Tierärztl. Prax.* 27, 145–153.

377. Conraths, F. J., Schares, G., Tchernychova, G., Bessonov, O. A. S. 2000. Seroepidemiological evidence for bovine neosporosis and *Neospora caninum*-associated abortions in the Russian Federation. *Int. J. Parasitol.* 30, 890–891.

378. Conraths, F. J., Schares, G. 2006. Validation of molecular-diagnostic techniques in the parasitological laboratory. *Vet. Parasitol.* 136, 91–98.

379. Conraths, F. J., Schares, G., Ortega-Mora, L. M., Gottstein, B. 2007. *Control Measures: Neosporosis.* CAB International, Oxfordshire, UK. 276–284.

379a. Constantino, C., Pellizzaro, M., de Paula, E. F. E., Vieira, T. S. W. J., Brandão, A. P. D., Ferreira, F., Vieira, R. F. C., Langoni, H., Biondo, A. W. 2016. Serosurvey for *Leishmania* spp., *Toxoplasma gondii*, *Trypanosoma cruzi* and *Neospora caninum* in neighborhood dogs in Curitiba-Paraná, Brazil. *Braz. J. Vet. Parasitol.* 25, 504–510. DOI: 10.1590/S1984-29612016062.

380. Constantin, E. M., Schares, G., Groβmann, E., Sauter, K., Romig, T., Hartmann, S. 2011. Untersuchungen zur Rolle des Rotfuchses (*Vulpes vulpes*) als möglicher Endwirt von *Neospora caninum*. *Berl. Münch. Tierärztl. Wochenschr.* 124, 148–153.

381. Corbellini, L. G., Driemeier, D., Cruz, C., Dias, M. M. 2000. Aborto bovino por *Neospora caninum* no Rio Grande do Sul. *Ciência Rural, Santa Maria* 30, 863–868.

382. Corbellini, L. G., Colodel, E. M., Driemeier, D. 2001. Granulomatous encephalitis in a neurologically impaired goat kid associated with degeneration of *Neospora caninum* tissue cysts. *J. Vet. Diagn. Invest.* 13, 416–419.

383. Corbellini, L. G., Driemeier, D., Cruz, C. F. E., Gondim, L. F. P., Wald, V. 2002. Neosporosis as a cause of abortion in dairy cattle in Rio Grande do Sul, southern Brazil. *Vet. Parasitol.* 103, 195–202.

384. Corbellini, L. G., Pescador, C. A., Frantz, F., Wunder, E., Steffen, D., Smith, D. R., Driemeier, D. 2006. Diagnostic survey of bovine abortion with special reference to *Neospora caninum* infection: Importance, repeated abortion and concurrent infection in aborted fetuses in Southern Brazil. *Vet. J.* 172, 114–120.

385. Corbellini, L. G., Smith, D. R., Pescador, C. A., Schmitz, M., Correa, A., Steffen, D. J., Driemeier, D. 2006. Herd-level risk factors for *Neospora caninum* seroprevalence in dairy farms in southern Brazil. *Prev. Vet. Med.* 74, 130–141.

386. Correia, A., Ferreirinha, P., Costa, A. A., Dias, J., Melo, J., Costa, R., Ribeiro, A., Faustino, A., Teixeira, L., Rocha, A., Vilanova, M. 2013. Mucosal and systemic T cell response in mice intragastrically infected with *Neospora caninum* tachyzoites. *Vet. Res.* 44, 69.

387. Coskun, S. Z., Aydyn, L., Bauer, C. 2000. Seroprevalence of *Neospora caninum* infection in domestic dogs in Turkey. *Vet. Rec.* 146, 649.

388. Cosoroaba, I., Chitimia, L. 2006. Neosporoza bovinelor. *Sci. Parasitol.* 1–2, 55–66.

389. Costa, G. H. N., Cabral, D. D., Varandas, N. P., Sobral, E. A., Borges, F. A., Castagnolli, K. C. 2001. Freqüência de anticorpos anti-*Neospora caninum* e anti-*Toxoplasma gondii* em soros de bovinos pertencentes aos estados de São Paulo e de Minas Gerais. *Semina Ci. Agrárias Londrina* 22, 61–66.

390. Costa, K. S., Santos, S. L., Uzêda, R. S., Pinheiro, A. M., Almeida, M. A. O., Araújo, F. R., McAllister, M. M., Gondim, L. F. P. 2008. Chickens (*Gallus domesticus*) are natural intermediate hosts of *Neospora caninum*. *Int. J. Parasitol.* 38, 157–159.

391. Costa, R. C., Orlando, D. R., Abreu, C. C., Nakagaki, K. Y. R., Mesquita, L. P., Nascimento, L. C., Silva, A. C., Maiorka, P. C., Peconick, A. P., Raymundo, D. L., Varaschin, M. S. 2014. Histological and immunohistochemical characterization of the inflammatory and glial cells in the central nervous system of goat fetuses and adult male goats naturally infected with *Neospora caninum*. *BMC Vet. Res.* 10, 291.

392. Couzinet, S., Dubremetz, J. F., Buzoni-Gatel, D., Jeminet, G., Prensier, G. 2000. *In vitro* activity of the polyether inophorous antibiotic monensin against the cyst form of *Toxoplasma gondii*. *Parasitology* 121, 359–365.

393. Cox, B. T., Reichel, M. P., Griffiths, L. M. 1998. Serology of a *Neospora* abortion outbreak on a dairy farm in New Zealand: A case study. *N. Z. Vet. J.* 46, 28–31.

394. Cramer, G., Kelton, D., Duffield, T. F., Hobson, J. C., Lissemore, K., Hietala, S. K., Peregrine, A. S. 2002. *Neospora caninum* serostatus and culling of Holstein cattle. *J. Am. Vet. Med. Assoc.* 221, 1165–1168.

395. Crawshaw, W. M., Brocklehurst, S. 2003. Abortion epidemic in a dairy herd associated with horizontally transmitted *Neospora caninum* infection. *Vet. Rec.* 152, 201–206.

396. Cringoli, G., Capuano, F., Veneziano, V., Romano, L., Solimene, R., Barber, J. S., Trees, A. J. 1996. Prevalence of antibodies against *Neospora caninum* in dog sera. *Parassitologia* 38, 282.

397. Cringoli, G., Rinaldi, L., Capuano, F., Baldi, L., Veneziano, V., Capelli, G. 2002. Serological survey of *Neospora caninum* and *Leishmania infantum* co-infection in dogs. *Vet. Parasitol.* 106, 307–313.

398. Cringoli, G., Capuano, F., Landolfi, M. C., Veneziano, V., Barber, J. S., Trees, A. J. 2016. Anticorpi verso *Neospora caninum* in cani della Campania. *Acta Med. Vet.* 42, 197–202.

399. Cronstedt-Fell, A., Richter, B., Voracek, T., Kübber-Heiss, A. 2012. Neosporosis in a captive Parma wallaby (*Macropus parma*). *J. Comp. Pathol.* 146, 274–277.

400. Crookshanks, J. L., Taylor, S. M., Haines, D. M., Shelton, G. D. 2007. Treatment of canine pediatric *Neospora caninum* myositis following immunohistochemical identification of tachyzoites in muscle biopsies. *Can. Vet. J.* 48, 506–508.

401. Cruz-Vázquez, C., Medina-Esparza, L., Marentes, A., Morales-Salinas, E., Garcia-Vázquez, Z. 2008. Seroepidemiological study of *Neospora caninum* infection in dogs found in dairy farms and urban areas of Aguascalientes, Mexico. *Vet. Parasitol.* 157, 139–143.

402. Cuddon, P., Lin, D. S., Bowman, D. D., Lindsay, D. S., Miller, T. K., Duncan, I. D., DeLahunta, A., Cummings, J., Suter, M., Cooper, B., King, J. M., Dubey, J. P. 1992. *Neospora caninum* infection in English Springer spaniel littermates: Diagnostic evaluation and organism isolation. *J. Vet. Intern. Med.* 6, 325–332.

403. Cui, X., Lei, T., Yang, D. Y., Hao, P., Liu, Q. 2012. Identification and characterization of a novel *Neospora caninum* immune mapped protein 1. *Parasitology* 139, 998–1004.

404. Cummings, J. F., de Lahunta, A., Suter, M. M., Jacobson, R. H. 1988. Canine protozoan polyradiculoneuritis. *Acta Neuropathol.* 76, 46–54.

405. Curry, P. S., Ribble, C., Sears, W. C., Hutchins, W., Orsel, K., Godson, D., Lindsay, R., Dibernardo, A., Kutz, S. J. 2014. Blood collected on filter paper for wildlife serology: Detecting antibodies to *Neospora caninum*, West Nile virus, and five bovine viruses in reindeer. *J. Wildl. Dis.* 50, 297–307.

406. Curry, P. S., Ribble, C., Sears, W. C., Orsel, K., Hutchins, W., Godson, D., Lindsay, R., Dibernardo, A., Campbell, M., Kutz, S. J. 2014. Blood collected on filter paper for wildlife serology: Evaluating storage and temperature challenges of field collections. *J. Wildl. Dis.* 50, 308–321.

407. Cuteri, V., Nisoli, L., Preziuso, S., Attili, A. R., Guerra, C., Lulla, D., Traldi, G. 2005. Application of a new therapeutic protocol against *Neospora caninum*-induced abortion in cattle: A field study. *J. Anim. Vet. Adv.* 4, 510–514.

408. Czopowicz, M., Kaba, J., Szalus-Jordanow, O., Nowicki, M., Witkowski, L., Frymus, T. 2011. Seroprevalence of *Toxoplasma gondii* and *Neospora caninum* infections in goats in Poland. *Vet. Parasitol.* 178, 339–341.

409. da Costa, H. F., Stachissini, A. V. M., Langoni, H., Padovani, C. R., Gennari, S. M., Modolo, J. R. 2012. Reproductive failures associated with antibodies against caprine arthritis-encephalitis virus, *Toxoplasma gondii* and *Neospora caninum* in goats in the state of São Paulo, Brazil. *Braz. J. Vet. Res. Anim. Sci.* 49, 67–72.

410. da Cunha, N. A., Lucas, A. S., Pappen, F. G., Ragozo, A. M. A., Gennari, S. M., Lucia, T., Farias, N. A. R. 2008. Fatores de risco e prevalência de anticorpos anti-*Neospora caninum* em cães urbanos e rurais do Rio Grande do Sul, Brasil. *Rev. Bras. Parasitol. Vet.* 17(Suppl 1), 301–306.

411. da Silva, A. S., Radavelli, W. M., Moura, A. B., Duarte, T., Duarte, M. M. M. F., Stefani, L. M. 2014. Horses seropositive for *Neospora* spp.: Immunoglobulins, cytokines, and C-reactive protein levels. *J. Equine Vet. Sci.* 34, 1240–1243.

412. da Silva, J. B., dos Santos, P. N., Castro, G. N. S., da Fonseca, A. H., Barbosa, J. D. 2014. Prevalence survey of selected bovine pathogens in water buffaloes in the north region of Brazil. *J. Parasitol. Res.* 2014, e603484.

413. Dabritz, H. A., Miller, M. A., Atwill, E. R., Gardner, I. A., Leutenegger, C. M., Melli, A. C., Conrad, P. A. 2007. Detection of *Toxoplasma gondii*-like oocysts in cat feces and estimates of the environmental oocyst burden. *J. Am. Vet. Med. Assoc.* 231, 1676–1684.

414. Daft, B. M., Barr, B. C., Collins, N., Sverlow, K. 1996. *Neospora* encephalomyelitis and polyradiculoneuritis in an aged mare with Cushing's disease. *Equine Vet. J.* 28, 240–243.

415. Daguer, H., Vicente, R. T., da Costa, T., Virmond, M. P., Hamann, W., Amendoeira, M. R. R. 2004. Soroprevalência de anticorpos anti-*Toxoplasma gondii* em bovinos e funcionários de matadouros da microrregião de Pato Branco, Paraná, Brasil. *Ciênc. Rural, Santa Maria* 34, 1133–1137.

416. Dalimi, A., Sabevarinejad, G., Ghafarifar, F., Forouzandeh-Moghadam, M. 2014. Molecular detection of *Neospora caninum* from naturally infected dogs in Lorestan province, West of Iran. *Arch. Razi Inst.* 69, 185–190.

417. Daly, P., Drudy, D., Chalmers, W. S. K., Baxendale, W., Fanning, S., Callanan, J. J. 2006. Greyhound meningoencephalitis: PCR-based detection methods highlight an absence of the most likely primary inducing agents. *Vet. Microbiol.* 118, 189–200.

418. Damriyasa, I. M., Bauer, C., Edelhofer, R., Failing, K., Lind, P., Petersen, E., Schares, G., Tenter, A. M., Volmer, R., Zahner, H. 2004. Cross-sectional survey in pig breeding farms in Hesse, Germany: Seroprevalence and risk factors of infections with *Toxoplasma gondii*, *Sarcocystis* spp. and *Neospora caninum* in sows. *Vet. Parasitol.* 126, 271–286.

419. Damriyasa, I. M., Schares, G., Bauer, C. 2010. Seroprevalence of antibodies to *Neospora caninum* in *Bos javanicus* ("Bali cattle") from Indonesia. *Trop. Anim. Health Prod.* 42, 95–98.

420. Dangoudoubiyam, S., Oliveira, J. B., Víquez, C., Gómez-García, A., González, O., Romero, J. J., Kwok, O. C. H., Dubey, J. P., Howe, D. K. 2011. Detection of antibodies against *Sarcocystis neurona*, *Neospora* spp., and *Toxoplasma gondii* in horses from Costa Rica. *J. Parasitol.* 97, 522–524.

421. Dannatt, L., Guy, F., Trees, A. J. 1995. Abortion due to *Neospora* species in a dairy herd. *Vet. Rec.* 137, 566–567.

422. Dannatt, L. 1997. *Neospora caninum* antibody levels in an endemically-infected dairy herd. *Cattle Pract.* 5, 335–337.

423. Darius, A. K., Mehlhorn, H., Heydorn, A. O. 2004. Effects of toltrazuril and ponazuril on *Hammondia heydorni* (syn. *Neospora caninum*) infections in mice. *Parasitol. Res.* 92, 520–522.

424. Darius, A. K., Mehlhorn, H., Heydorn, A. O. 2004. Effects of toltrazuril and ponazuril on the fine structure and multiplication of tachyzoites of the NC-1 strain of *Neospora caninum* (a synonym of *Hammondia heydorni*) in cell cultures. *Parasitol. Res.* 92, 453–458.

425. Darkin-Rattray, S. J., Gurnett, A. M., Myers, R. W., Dulski, P. M., Crumley, T. M., Allocco, J. J., Cannova, C., Meinke, P. T., Colletti, S. L., Bednarek, M. A., Singh, S. B., Goetz, M. A., Dombrowski, A. W., Polishook, J. D., Schmatz, D. M. 1996. Apicidin: A novel antiprotozoal agent that inhibits parasite histone deacetylase. *Proc. Natl. Acad. Sci. USA* 93, 13143–13147.

426. Darwich, L., Cabezón, O., Echeverria, I., Pabón, M., Marco, I., Molina-López, R., Alarcia-Alejos, O., López-Gatius, F., Lavín, S., Almería, S. 2011. Presence of *Toxoplasma gondii* and *Neospora caninum* DNA in the brain of wild birds. *Vet. Parasitol.* 183, 377–381.

427. Darwich, L., Li, Y., Serrano-Pérez, B., Mur-Novales, R., Garcia-Ispierto, I., Cabezón, O., López-Gatius, F., Almería, S. 2016. Maternal and foetal cytokine production in dams naturally and experimentally infected with *Neospora caninum* on gestation day 110. *Res. Vet. Sci.* 107, 55–61.

428. Davis, S. W., Dubey, J. P. 1995. Mediation of immunity to *Toxoplasma gondii* oocyst shedding in cats. *J. Parasitol.* 81, 882–886.

429. Davison, H. C., French, N. P., Trees, A. J. 1999. Herd-specific and age-specific seroprevalence of *Neospora caninum* in 14 British dairy herds. *Vet. Rec.* 144, 547–550.

430. Davison, H. C., Otter, A., Trees, A. J. 1999. Estimation of vertical and horizontal transmission parameters of *Neospora caninum* infections in dairy cattle. *Int. J. Parasitol.* 29, 1683–1689.

431. Davison, H. C., Otter, A., Trees, A. J. 1999. Significance of *Neospora caninum* in British dairy cattle determined by estimation of seroprevalence in normally calving cattle and aborting cattle. *Int. J. Parasitol.* 29, 1189–1194.

432. Davison, H. C., Guy, F., Trees, A. J., Ryce, C., Ellis, J. T., Otter, A., Jeffrey, M., Simpson, V. R., Holt, J. J. 1999. *In vitro* isolation of *Neospora caninum* from a stillborn calf in the UK. *Res. Vet. Sci.* 67, 103–105.

433. Gharekhani, J., Esmaeilnejad, B., Rezaei, H., Yakhali, M., Heidari, H., Azhari, M. 2016. Prevalence of anti-*Neospora caninum* antibodies in Iranian goats. *Ann. Parasitol. In (Polish).* 62, 111–114.

434. Davison, H. C., Guy, C. S., McGarry, J. W., Guy, F., Williams, D. J. L., Kelly, D. F., Trees, A. J. 2001. Experimental studies on the transmission of *Neospora caninum* between cattle. *Res. Vet. Sci.* 70, 163–168.

435. Davoli-Ferreira, M., Fonseca, D. M., Mota, C. M., Dias, M. S., Lima-Junior, D. S., da Silva, M. V., Quirino, G. F. S., Zamboni, D. S., Silva, J. S., Mineo, T. W. P. 2016. Nucleotide-binding oligomerization domain-containing protein 2 prompts potent inflammatory stimuli during *Neospora caninum* infection. *Sci. Rep.* 6, 29289.

436. de Abreu, J. A. P., Krawczak, F. S., Nunes, F. P., Labruna, M. B., Pena, H. F. J. 2016. Anti-*Toxoplasma gondii* and anti-*Neospora caninum* antibodies in capybaras (*Hydrochoerus hydrochaeris*) from Itu Municipality, São Paulo. *Rev. Bras. Parasitol. Vet.* 25, 116–118.

437. de Almeida, M. A. O. 2004. Epidemiologia de *Neospora caninum*. *Proceedings of the XIII Congresso Brasileiro de Parasitologia Veterinária & I Simpósio Latino-Americano de Ricketisioses*, Ouro Preto, MG, 2004. *Rev. Bras. Parasitol. Vet.* 13(Suppl 1), 37–40.

438. de Azevedo, S. S., Pena, H. F. J., Alves, C. J., Guimarães, A. A. M., Oliveira, R. M., Maksimov, P., Schares, G., Gennari, S. M. 2010. Prevalence of anti-*Toxoplasma gondii* and anti-*Neospora caninum* antibodies in swine from Northeastern Brazil. *Rev. Bras. Parasitol. Vet.* 19, 80–84.

439. de Craeye, S., Speybroeck, N., Ajzenberg, D., Dardé, M. L., Collinet, F., Tavernier, P., Van Gucht, S., Dorny, P., Dierick, K. 2011. *Toxoplasma gondii* and *Neospora caninum* in wildlife: Common parasites in Belgian foxes and Cervidae? *Vet. Parasitol.* 178, 64–69.

440. de Jesus, E. E. V., Santos, P. O. M., Barbosa, M. V. F., Pinheiro, A. M., Gondim, L. F. P., Guimarães, J. E., Almeida, M. Â. O. 2006. Frequência de anticorpos anti-*Neospora caninum* em cães nos municípios de Salvador e Lauro de Freitas, Estado da Bahia—Brasil. *Braz. J. Vet. Res. Anim. Sci.* 43, 5–10.

441. de Jesus, E. E. V., dos Santos, A. B., Ribeiro, C. S. O., Pinheiro, A. M., Freire, S. M., El-Bachá, R. S., Costa, S. L., de Fatima Dias Costa, M. 2014. Role of IFN-γ and LPS on neuron/glial co-cultures infected by *Neospora caninum*. *Front. Cell. Neurosci.* 8, 340.

442. de La Torre, J. R., Bautista-Piña, C., Ortega-S., J. A., Cantu-Covarruvias, A., Alvarez-Ojeda, M. G., Romero-Salas, D., Henke, S. E., Hilton, C. D., Hewitt, D. G., de Young, R. W., Campbell, T. A., Bryant, F. C. 2016. *Neospora caninum* in axis and fallow deer in northern Mexico. *J. Wildl. Dis.* 53, 186–187.

443. de Lima, J. T. R., Ahid, S. M. M., Barrêto, R. A., Pena, H. F. J., Dias, R. A., Gennari, S. M. 2008. Prevalência de anticorpos anti-*Toxoplasma gondii* e anti-*Neospora caninum* em rebanhos caprinos do município de Mossoró, Rio Grande do Norte. *Braz. J. Vet. Res. Anim. Sci.* 45, 81–86.

444. de Magalhães, V. C. S., de Olivelra, U. V., Costa, S. C. L., Sántos, I. A., Pereira, M. J. S., Munhoz, A. D. 2014. Transmission paths of *Neospora caninum* in a dairy herd of crossbred cattle in the northeast of Brazil. *Vet. Parasitol.* 202, 257–264.

445. de Marez, T., Liddell, S., Dubey, J. P., Jenkins, M. C., Gasbarre, L. 1999. Oral infection of calves with *Neospora caninum* oocysts from dogs: Humoral and cellular immune responses. *Int. J. Parasitol.* 29, 1647–1657.

446. de Matos, R. B., Braga-de-Souza, S., Pitanga, B. P. S., da Silva, V. D. A., de Jesus, E. E. V., Pinheiro, A. M., Costa, M. F. D., El-Bacha, R. S., Ribeiro, C. S. O., Costa, S. L. 2014. Flavonoids modulate the proliferation of *Neospora caninum* in glial ell primary cultures. *Korean J. Parasitol.* 52, 613–619.

447. De Meerschman, F., Losson, B. 1998. *Neospora caninum* et la néosporose: Biologie et description de la maladie chez le chien. *Ann. Méd. Vét.* 142, 247–253.

448. De Meerschman, F., Focant, C., Boreux, R., Leclipteux, T., Losson, B. 2000. Cattle neosporosis in Belgium: A case-control study in dairy and beef cattle. *Int. J. Parasitol.* 30, 887–890.

449. De Meerschman, F., Rettigner, C., Focant, C., Boreux, R., Pinset, C., Leclipteux, T., Losson, B. 2002. Use of a serum-free medium to produce *in vitro Neospora caninum* and *Toxoplasma gondii* tachyzoites on Vero cells. *Vet. Res.* 33, 159–168.

450. De Meerschman, F., Speybroeck, N., Berkvens, D., Rettigner, C., Focant, C., Leclipteux, T., Cassart, D., Losson, B. 2002. Fetal infection with *Neospora caninum* in dairy and beef cattle in Belgium. *Theriogenology* 58, 933–945.

451. De Meerschman, F., Focant, C., Detry, J., Rettigner, C., Cassart, D., Losson, B. 2005. Clinical, pathological and diagnostic aspects of congenital neosporosis in a series of naturally infected calves. *Vet. Rec.* 157, 115–118.

452. de Melo, C. B., Leite, R. C., de Souza, G. N., Leite, R. C. 2001. Freqüência de infecção por *Neospora caninum* em dois diferentes sistemas de produção de leite e fatores predisponentes à infecção em bovinos em Minas Gerais. *Rev. Bras. Parasitol. Vet.* 10, 67–74.

453. Melo, C. B., Leite, R. C., Leite, F. S. C., Leite, R. C. 2002. Serological surveillance on South American wild canids for *Neospora caninum*. *Arq. Bras. Med. Vet. Zootec.* 54, 444–447.

454. de Melo, C. B., Leite, R. C., Leite, R. C. 2004. Neosporose em bovinos. *Rev. Bras. Reprod. Anim.* 28, 13–22.

455. de Melo, C. B., Leite, R. C., Lobato, Z. I. P., Leite, R. C. 2004. Infection by *Neospora caninum* associated with bovine herpesvirus 1 and bovine viral diarrhea virus in cattle from Minas Gerais State, Brazil. *Vet. Parasitol.* 119, 97–105.

456. de Meneses, I. D., Andrade, M. R., Uzêda, R. S., Bittencourt, M. V., Lindsay, D. S., Gondim, L. F. P. 2014. Frequency of antibodies against *Sarcocystis neurona* and *Neospora caninum* in domestic cats in the state of Bahia, Brazil. *Rev. Bras. Parasitol. Vet.* 23, 526–529.

457. de Moraes, C. C. G., Megid, J., Pituco, E. M., Okuda, L. H., Del Fava, C., de Stefano, E., Crocci, A. J. 2008. Ocorrência de anticorpos anti-*Neospora caninum* em cães da microrregião da Serra de Botucatu, Estado de São Paulo, Brasil. *Rev. Bras. Parasitol. Vet.* 17, 1–6.

458. de Moura, A. B., da Silva, M. O., Farias, J. A., Vieira-Nero, A., de Souza, A. P., Sartor, A. A., Fonteque, J. H., Bunn, S. 2013. *Neospora* spp. antibodies in horses from two geographical regions of the state of Santa Catarina, Brazil. *Rev. Bras. Parasitol. Vet.* 22, 597–601.

458a. de Moura, A. B., Osaki, S. C., Zulpo, D. L., Garcia, J. L., Teixeira, E. B. 2012. Occurrence of anti-antibodies in beef cattle of microregion of Guarapuava, Paraná State, Brazil. *Arq. Inst. Biol. (São Paulo)* 79, 419–422.

459. de Ol iveira, U. V., de Magalhães, V. C. S., Almeida, C. P., Santos, I. A., Mota, D. A., Macêdo, L. S., Silva, F. L., Carvalho, F. S., Wenceslau, A. A., Munhoz, A. D. 2013. Quails are resistant to infection with *Neospora caninum* tachyzoites. *Vet. Parasitol.* 198, 209–213.

460. de Oliveira, J. M., Matos, M. F. C., Oshiro, L. M., Andreotti, R. 2004. Prevalence of anti-*Neospora caninum* antibodies in dogs in the urban area of Campo Grande, MS, Brazil. *Rev. Bras. Parasitol. Vet.* 13, 155–158.

461. de Oliveira, V. S. F., Álvarez-Garcia, G., Ortega-Mora, L. M., Borges, L. M. F., da Silva, A. C. 2010. Abortions in bovines and *Neospora caninum* transmission in an embryo transfer center. *Vet. Parasitol.* 173, 206–210.

462. de Sá, G. L., Pacheco, D. B., Monte, L. G., Sinnott, F. A., Xavier, M. A., Rizzi, C., Borsuk, S., Berne, M. E. A., Andreotti, R., Hartleben, C. P. 2014. Diagnostic potential of anti-rNcp-43 polyclonal antibodies for the detection of *Neospora caninum*. *Curr. Microbiol.* 68, 472–476.

463. de Schuyter, T., De Cock, H. E. V., Lemmens, P. 2013. Cutaneous neosporosis in an adult dog in Belgium. *Vlaams Diergeneesk. Tijdsch.* 82, 59–62.

464. de Seabra, N. M., Pereira, V. F., Kuwassaki, M. V., Benassi, J. C., Oliveira, T. M. F. S. 2015. *Toxoplasma gondii*, *Neospora caninum*, *Leishmania* spp. serology and *Leishmania* spp. PCR in dogs from Pirassununga, SP. *Braz. J. Vet. Parasitol.* 24, 454–458.

465. de Sousa, K. C. M., Herrera, H. M., Domingos, I. H., Campos, J. B. V., dos Santos, I. M. C., Neves, H. H., Machado, R. Z., André, M. R. 2014. Serological detection of *Toxoplasma gondii*, *Leishmania infantum* and *Neospora caninum* in cats from an area endemic for leishmaniasis in Brazil. *Rev. Bras. Parasitol. Vet.* 23, 449–455.

466. de Sousa, M. E., Porto, W. J. N., de Albuquerque, P. P. F., Neto, O. L. S., Pinheiro, J. W., Mota, R. A. 2012. Seroprevalence of antibodies to *Neospora caninum* in dogs in the state of Alagoas, Brazil. *Rev. Bras. Parasitol. Vet.* 21, 287–290.

467. de Souza, L. M., do Nascimento, A. A., Furuta, P. I., Basso, L. M. S., da Silveira, D. M., da Costa, A. J. 2001. Detecção de anticorpos contra *Neospora caninum* e *Toxoplasma gondii* em soros de bubalinos (*Bubalus bulalis*) no Estado de São Paulo, Brasil. *Semina Ci. Agrárias* 22, 39–48.

468. de Souza, S. L. P., Guimarães, J. S., Ferreira, F., Dubey, J. P., Gennari, S. M. 2002. Prevalence of *Neospora caninum* antibodies in dogs from dairy cattle farms in Parana, Brazil. *J. Parasitol.* 88, 408–409.

469. de Souza, W., Attias, M. 2015. New views of the *Toxoplasma gondii* parasitophorous vacuole as revealed by Helium Ion Microscopy (HIM). *J. Struct. Biol.* 191, 76–85.

470. de Yaniz, M. G., Moore, D. P., Odeón, A. C., Cano, A., Cano, D. B., Leunda, M. R., Campero, C. M. 2007. Humoral immune response in pregnant heifers inoculated with *Neospora caninum* tachyzoites by conjunctival route. *Vet. Parasitol.* 148, 213–218.

471. Debache, K., Guionaud, C., Alaeddine, F., Mevissen, M., Hemphill, A. 2008. Vaccination of mice with recombinant NcROP2 antigen reduces mortality and cerebral infection in mice infected with *Neospora caninum* tachyzoites. *Int. J. Parasitol.* 38, 1455–1463.

472. Debache, K., Alaeddine, F., Guionaud, C., Monney, T., Müller, J., Strohbusch, M., Leib, S. L., Grandgirard, D., Hemphill, A. 2009. Vaccination with recombinant NcROP2 combined with recombinant NcMIC1 and NcMIC3 reduces cerebral infection and vertical transmission in mice experimentally infected with *Neospora caninum* tachyzoites. *Int. J. Parasitol.* 39, 1373–1384.

473. Debache, K., Guionaud, C., Alaeddine, F., Hemphill, A. 2010. Intraperitoneal and intra-nasal vaccination of mice with three distinct recombinant *Neospora caninum* antigens results in differential effects with regard to protection against experimental challenge with *Neospora caninum* tachyzoites. *Parasitology* 137, 229–240.

474. Debache, K., Guionaud, C., Kropf, C., Boykin, D., Stephens, C. E., Hemphill, A. 2011. Experimental treatment of *Neospora caninum*-infected mice with the arylimidamide DB750 and the thiazolide nitazoxanide. *Exp. Parasitol.* 129, 95–100.

475. Debache, K., Kropf, C., Schütz, C. A., Harwood, L. J., Käuper, P., Monney, T., Rossi, N., Laue, C., McCullough, K. C., Hemphill, A. 2011. Vaccination of mice with chitosan nanogel-associated recombinant NcPDI against challenge infection with *Neospora caninum* tachyzoites. *Parasite Immunol.* 33, 81–94.

476. Debache, K., Hemphill, A. 2012. Effects of miltefosine treatment in fibroblast cell cultures and in mice experimentally infected with *Neospora caninum* tachyzoites. *Parasitology* 139, 934–944.

477. Debache, K., Hemphill, A. 2012. Intra-cisternal vaccination induces high-level protection against *Neospora caninum* infection in mice. *Vaccine* 30, 4209–4215.

478. Debache, K., Hemphill, A. 2013. Differential effects of intranasal vaccination with recombinant NcPDI in different mouse models of *Neospora caninum* infection. *Parasite Immunol.* 35, 11–20.

479. del Campos, S. J., Chávez, V. A., Delgado, C. A., Falcón, P. N., Ornelas, A. Â., Casas, A. E., Serrano, M. E. 2003. Frecuencia de *Neospora caninum* en perros de establos lecheros del valle de Lima. *Rev. Inv. Vet. Perú* 14, 145–149.

480. Del Fava, C., Pituco, E. M., Okuda, L. H., de Stefano, E., Ferrari, C. I. L., Demarchi, J. J. A. A., Marini, A., de Oliveira, F. P., Fonzar, J. F., Gasparelli, A. G. 2007. Coeficientes reprodutivos e características de desempenho em matrizes bovinas de corte soro reagentes para *Neospora caninum*. *B. Indústr. Anim.* 64, 191–196.

481. Dellarupe, A., Regidor-Cerrillo, J., Jiménez-Ruiz, E., Schares, G., Unzaga, J. M., Venturini, M. C., Ortega-Mora, L. M. 2014. Comparison of host cell invasion and proliferation among *Neospora caninum* isolates obtained from oocysts and from clinical cases of naturally infected dogs. *Exp. Parasitol.* 145, 22–28.

482. Dellarupe, A., Regidor-Cerrillo, J., Jiménez-Ruiz, E., Schares, G., Unzaga, J. M., Venturini, M. C., Ortega-Mora, L. M. 2014. Clinical outcome and vertical transmission variability among canine *Neospora caninum* isolates in a pregnant mouse model of infection. *Parasitology* 141, 356–366.

483. Dempster, R. P. 1983. Comparison of techniques for the purification of *Toxoplasma gondii* tachyzoites from mouse peritoneal exudates (meeting abstract). *N. Z. J. Zool.* 10, 128.

484. Dempster, R. P. 1984. *Toxoplasma gondii*: Purification of zoites from peritoneal exudates by eight methods. *Exp. Parasitol.* 57, 195–207.

485. Deng, C., Zhang, W., Liu, Q., Liu, J., Ding, J. 2009. [Balb/c mouse model of neosporosis and the development process with infection of *Neospora caninum*]. *Scientia Agricultura Sinica* 42, 1123–1128 (in Chinese).

486. Denkers, E. Y., Bzik, D. J., Fox, B. A., Butcher, B. A. 2012. An inside job: Hacking into Janus kinase/signal transducer and activator of transcription signaling cascades by the intracellular protozoan *Toxoplasma gondii*. *Infect. Immun.* 80, 476–482.

487. Deo, V. K., Yoshimatsu, K., Otsuki, T., Dong, J., Kato, T., Park, E. Y. 2013. Display of *Neospora caninum* surface protein related sequence 2 on *Rous sarcoma* virus-derived gag protein virus-like particles. *J. Biotechnol.* 165, 69–75.

488. Desmonts, G., Remington, J. S. 1980. Direct agglutination test for diagnosis of *Toxoplasma* infection: Method for increasing sensitivity and specificity. *J. Clin. Microbiol.* 11, 562–568.

489. Di Lorenzo, C., Venturini, C., Castellano, C., Venturini, L., Unzaga, J. M., Bacigalupe, D. 1997. Detección de anticuerpos anti-*Neospora caninum* y anti-*Toxoplasma gondii* en perros de área urbana. *Rev. Med. Vet.* 78, 325–326.

490. Diakou, A., Papadopoulos, E., Panousis, N., Karatzias, C., Giadinis, N. 2013. *Toxoplasma gondii* and *Neospora caninum* seroprevalence in dairy sheep and goats mixed stock farming. *Vet. Parasitol.* 198, 387–390.

491. Díaz, J. M., Fernández, G., Prieto, A., Valverde, S., Lago, N., Díaz, P., Panadero, R., López, C., Morrondo, P., Díez-Baños, P. 2014. Epidemiology of reproductive pathogens in semi-intensive lamb-producing flocks in North-West Spain: A comparative serological study. *Vet. J.* 200, 335–338.

492. Dijkstra, T., Barkema, H. W., Eysker, M., Wouda, W. 2001. Evidence of post-natal transmission of *Neospora caninum* in Dutch dairy herds. *Int. J. Parasitol.* 31, 209–215.

493. Dijkstra, T., Eysker, M., Schares, G., Conraths, F. J., Wouda, W., Barkema, H. W. 2001. Dogs shed *Neospora caninum* oocysts after ingestion of naturally infected bovine placenta but not after ingestion of colostrum spiked with *Neospora caninum* tachyzoites. *Int. J. Parasitol.* 31, 747–752.

494. Dijkstra, T. 2002. Horizontal and vertical transmission of *Neospora caninum*. *Doctoral thesis.* Universiteit Utrecht, Utrecht, The Netherlands. 1–140.

495. Dijkstra, T., Barkema, H. W., Hesselink, J. W., Wouda, W. 2002. Point source exposure of cattle to *Neospora caninum* consistent with periods of common housing and feeding and related to the introduction of a dog. *Vet. Parasitol.* 105, 89–98.

496. Dijkstra, T., Barkema, H. W., Björkman, C., Wouda, W. 2002. A high rate of seroconversion for *Neospora caninum* in a dairy herd without an obvious increased incidence of abortions. *Vet. Parasitol.* 109, 203–211.

497. Dijkstra, T., Barkema, H. W., Eysker, M., Hesselink, J. W., Wouda, W. 2002. Natural transmission routes of *Neospora caninum* between farm dogs and cattle. *Vet. Parasitol.* 105, 99–104.

498. Dijkstra, T., Barkema, H. W., Eysker, M., Beiboer, M. L., Wouda, W. 2003. Evaluation of a single serological screening of dairy herds for *Neospora caninum* antibodies. *Vet. Parasitol.* 110, 161–169.

499. Dijkstra, T., Bartels, C. J. M., Wouda, W. 2005. Control of bovine neosporosis: Experiences from the Netherlands. The 20th International Conference of the World Association for the Advancement of Veterinary Parasitology 2005, Session M. Diagnosis and Control of Protozoan-Associated Abortion in Ruminants. October 16–20, Christchurch, New Zealand.

500. Dijkstra, T., Lam, T. J. G. M., Bartels, C. J. M., Eysker, M., Wouda, W. 2008. Natural postnatal *Neospora caninum* infection in cattle can persist and lead to endogenous transplacental infection. *Vet. Parasitol.* 152, 220–225.

501. Ding, D., Yu, L., Jia, L., Zhang, S. 2006. [Epidemiological survey of *Neospora caninum* infection in beef cattle in Jilin Province, China]. *Anim. Husbandry Vet. Med.* 38, 34–36 (in Chinese).

502. Dion, S., Germon, S., Guiton, R., Ducournau, C., Dimier-Poisson, I. 2011. Functional activation of T cells by dendritic cells and macrophages exposed to the intracellular parasite *Neospora caninum*. *Int. J. Parasitol.* 41, 685–695.

503. Dirikolu, L., Yohn, R., Garrett, E. F., Chakkath, T., Ferguson, D. C. 2009. Detection, quantifications and pharmacokinetics of toltrazuril sulfone (Ponazuril®) in cattle. *J. Vet. Pharmacol. Therap.* 32, 280–288.

504. do Nascimento, E. E., Sammi, A. S., Dos Santos, J. R., Nino, B. S. L., Bogado, A. L. G., Taroda, A., Vidotto, O., Garcia, J. L. 2014. Anti-*Neospora caninum* antibody detection and vertical transmission rate in pregnant zebu beef cows (*Bos indicus*): *Neospora caninum* in pregnant beef cows (*Bos indicus*). *Comp. Immunol. Microbiol. Infect. Dis.* 37, 267–270.

505. Dogga, S. K., Bartošová-Sojková, P., Lukeš, J., Soldati-Favre, D. 2015. Phylogeny, morphology, and metabolic and invasive capabilities of epicellular fish coccidium *Goussia janae*. *Protist* 166, 659–676.

506. Dombou, E., Kamga Waladjo, A. R., Gbati, O. B., Bakou, S. N., Mukakanamugire, A., Chatagnon, G., Akakpo, J. A., Diop, P. E. H., Pangui, L. J., Tainturier, D. 2008. Neosporosis in Senegalese wildlife. *Proceedings of the World Buiatrics Congress.* July 6–10, Budapest, Hungary. Magyar Allatorvosk Lapja 130. Suppl II. 275.

507. Domingues, L. M., Machado, R. Z., Tinucci Costa, M., Carvalho, C. S., Costa, A. J., Malheiros, E. B. 1998. Canine toxoplasmosis: A comparative evaluation of the detection of anti-*Toxoplasma gondii* antibodies by the indirect immunoenzymatic assay (ELISA) and the indirect immunofluorescence reaction (IIF). *Rev. Bras. Parasitol. Vet.* 7, 79–85.

508. Donahoe, S. L., Lindsay, S. A., Krockenberger, M., Phalen, D., Slapeta, J. 2015. A review of neosporosis and pathologic findings of *Neospora caninum* infection in wildlife. *Int. J. Parasitol. Parasites Wildl.* 4, 216–238.

509. Dong, J., Otsuki, T., Kato, T., Park, E. Y. 2012. Development of a diagnostic method for neosporosis in cattle using recombinant *Neospora caninum* proteins. *BMC Biotechnol.* 12, 19.

510. Dong, J., Otsuki, T., Kato, T., Kohsaka, T., Ike, K., Park, E. Y. 2013. Development of two murine antibodies against *Neospora caninum* using phage display technology and application on the detection of *N. caninum*. *PLoS ONE* 8, e53264.

511. Dong, J., Otsuki, T., Kato, T., Park, E. Y. 2014. Tracking *Neospora caninum* parasites using chimera monoclonal antibodies against its surface antigen-related sequences (rNcSRS2). *J. Biosci. Bioeng.* 117, 351–357.

512. Dorchies, P., Dubey, J. P. 1996. *Neospora caninum*: Une nouvelle menace parasitaire? *Proceedings of the Protozooses Bovines Actualités.* Societe Francaise de Buiatrie. October 3, Annecy, France. Navetat, H., Schelcher, F. (Editors). Societe Francaise de Buiatrie, Ecole Nationale Veterinaire. 14–18.

513. dos Santos, A. P. M. E., Navarro, I. T., Freire, R. L., Vidotto, O., Bracarense, A. P. F. R. L. 2005. *Neospora caninum* in dairy cattle in Paraná State, Brazil: Histological and immunohistochemical analysis in fetuses. *Semina Ci. Agrárias* 26, 559–562.

514. dos Santos, A. P. M. E., Navarro, I. T., Vidotto, O., Bracarense, A. P. F. R. L. 2006. Encefalomielite congênita em bezerro associada ao *Neospora caninum* no estado do Paraná, Brasil. *Semina Ci. Agrárias* 27, 111–114.

515. dos Santos, D. S., Andrade, M. P., Varaschin, M. S. 2011. *Neospora caninum* in bovine fetuses of Minas Gerais, Brazil: Genetic characteristics of rDNA. *Rev. Bras. Parasitol. Vet.* 20, 281–288.

516. Dowse, T. J., Soldati, D. 2005. Rhomboid-like proteins in Apicomplexa: Phylogeny and nomenclature. *Trends Parasitol.* 21, 254–258.

517. Dreier, K. J., Stewarter, L. W., Kerlin, R. L., Ritter, D. M., Brake, D. A. 1999. Phenotypic characterisation of a *Neospora caninum* temperature-sensitive strain in normal and immunodeficient mice. *Int. J. Parasitol.* 29, 1627–1634.

518. Drewry, L. L., Sibley, L. D. 2015. *Toxoplasma* actin is required for efficient host cell invasion. *mBio* 6, e00557–15.

519. Dryden, M. W., Payne, P. A., Ridley, R., Smith, V. 2005. Comparison of common fecal flotation techniques for the recovery of parasite eggs and oocysts. *Vet. Therapeut.* 6, 15–28.

520. Du, L., Yang, D., Zhai, T., Gong, P., Zhang, X., Li, J. 2015. Detection of *Neospora caninum*-DNA in brain tissues from pigeons in Changchun, Jilin (China). *Vet. Parasitol.* 214, 171–173.

521. Duarte, P. C., Conrad, P. A., Wilson, W. D., Ferraro, G. L., Packham, A. E., Bowers-Lepore, J., Carpenter, T. E., Gardner, I. A. 2004. Risk of postnatal exposure to *Sarcocystis neurona* and *Neospora hughesi* in horses. *Am. J. Vet. Res.* 65, 1047–1052.

522. Duarte, P. C., Conrad, P. A., Barr, B. C., Wilson, W. D., Ferraro, G. L., Packham, A. E., Carpenter, T. E., Gardner, I. A. 2004. Risk of transplacental transmission of *Sarcocystis neurona* and *Neospora hughesi* in California horses. *J. Parasitol.* 90, 1345–1351.

523. Dubey, J. P., Frenkel, J. K. 1976. Feline toxoplasmosis from acutely infected mice and the development of *Toxoplasma* cysts. *J. Protozool.* 23, 537–546.

524. Dubey, J. P., Porterfield, M. L. 1986. *Toxoplasma*-like sporozoa in an aborted equine fetus. *J. Am. Vet. Med. Assoc.* 188, 1312–1313.

525. Dubey, J. P., Desmonts, G. 1987. Serological responses of equids fed *Toxoplasma gondii* oocysts. *Equine Vet. J.* 19, 337–339.

526. Dubey, J. P., Hughes, H. P. A., Lillehoj, H. S., Gamble, H. R., Munday, B. L. 1987. Placental transfer of specific antibodies during ovine congenital toxoplasmosis. *Am. J. Vet. Res.* 48, 474–476.

527. Dubey, J. P., Beattie, C. P. 1988. *Toxoplasmosis of Animals and Man.* CRC Press, Boca Raton, Florida. 1–220.

528. Dubey, J. P., Hattel, A. L., Lindsay, D. S., Topper, M. J. 1988. Neonatal *Neospora caninum* infection in dogs: Isolation of the causative agent and experimental transmission. *J. Am. Vet. Med. Assoc.* 193, 1259–1263.

529. Dubey, J. P., Carpenter, J. L., Speer, C. A., Topper, M. J., Uggla, A. 1988. Newly recognized fatal protozoan disease of dogs. *J. Am. Vet. Med. Assoc.* 192, 1269–1285.

530. Dubey, J. P. 1989. Congenital neosporosis in a calf. *Vet. Rec.* 125, 486.

531. Dubey, J. P. 1989. *Neospora caninum*, a recently recognized cyst-forming coccidian. Coccidia and Intestinal Coccidiomorphs, Vth International Coccidiosis Conference. October 17–20, Tours, France, INRA Publ. 19–24.

532. Dubey, J. P., Lindsay, D. S. 1989. Transplacental *Neospora caninum* infection in cats. *J. Parasitol.* 75, 765–771.

533. Dubey, J. P., Lindsay, D. S. 1989. Transplacental *Neospora caninum* infection in dogs. *Am. J. Vet. Res.* 50, 1578–1579.

534. Dubey, J. P., Lindsay, D. S. 1989. Fatal *Neospora caninum* infection in kittens. *J. Parasitol.* 75, 148–151.

535. Dubey, J. P., Leathers, C. W., Lindsay, D. S. 1989. *Neospora caninum*-like protozoon associated with fatal myelitis in newborn calves. *J. Parasitol.* 75, 146–148.

536. Dubey, J. P., Carpenter, J. L., Topper, M. J., Uggla, A. 1989. Fatal toxoplasmosis in dogs. *J. Am. Anim. Hosp. Assoc.* 25, 659–664.

537. Dubey, J. P. 1990. *Neospora caninum*: A look at a new *Toxoplasma*-like parasite of dogs and other animals. *Comp. Cont. Edu. Pract. Vet.* 12, 653–663.

538. Dubey, J. P., Lindsay, D. S. 1990. *Neospora caninum* induced abortion in sheep. *J. Vet. Diagn. Invest.* 2, 230–233.

539. Dubey, J. P., Lindsay, D. S. 1990. Neosporosis in dogs. *Vet. Parasitol.* 36, 147–151.

540. Dubey, J. P., Porterfield, M. L. 1990. *Neospora caninum* (Apicomplexa) in an aborted equine fetus. *J. Parasitol.* 76, 732–734.

541. Dubey, J. P., Koestner, A., Piper, R. C. 1990. Repeated transplacental transmission of *Neospora caninum* in dogs. *J. Am. Vet. Med. Assoc.* 197, 857–860.

542. Dubey, J. P., Hartley, W. J., Lindsay, D. S. 1990. Congenital *Neospora caninum* infection in a calf with spinal cord anomaly. *J. Am. Vet. Med. Assoc.* 197, 1043–1044.

543. Dubey, J. P., Greene, C. E., Lappin, M. R. 1990. *Toxoplasmosis and Neosporosis.* W. B. Saunders, Philadelphia, Pennsylvania. 818–834.

544. Dubey, J. P., Lindsay, D. S., Lipscomb, T. P. 1990. Neosporosis in cats. *Vet. Pathol.* 27, 335–339.

545. Dubey, J. P., Hartley, W. J., Lindsay, D. S., Topper, M. J. 1990. Fatal congenital *Neospora caninum* infection in a lamb. *J. Parasitol.* 76, 127–130.

546. Dubey, J. P., Miller, S., Lindsay, D. S., Topper, M. J. 1990. *Neospora caninum*-associated myocarditis and encephalitis in an aborted calf. *J. Vet. Diagn. Invest.* 2, 66–69.

547. Dubey, J. P., Higgins, R. J., Smith, J. H., O'Toole, T. D. 1990. *Neospora caninum* encephalomyelitis in a British dog. *Vet. Rec.* 126, 193–194.

548. Dubey, J. P. 1992. *Neospora caninum* infections. In *Kirk's Current Veterinary Therapy. XI Small Animal Practice.* Kirk, R. W., Bongura, J. D. (Editors). W. B. Saunder's Co., Philadelphia, Pennsylvania. 263–266.

549. Dubey, J. P., Sréter, T. 1992. A neosporosis: A háziállatok nemrég felfedezett protozoan okozta betegsége Szemlecikk. *Magy. Allatorvosok Lapja* 47, 533–536.

550. Dubey, J. P., Janovitz, E. B., Skowronek, A. J. 1992. Clinical neosporosis in a 4-week-old Hereford calf. *Vet. Parasitol.* 43, 137–141.

551. Dubey, J. P., Acland, H. M., Hamir, A. N. 1992. *Neospora caninum* (Apicomplexa) in a stillborn goat. *J. Parasitol.* 78, 532–534.

552. Dubey, J. P., Lindsay, D. S., Anderson, M. L., Davis, S. W., Shen, S. K. 1992. Induced transplacental transmission of *Neospora caninum* in cattle. *J. Am. Vet. Med. Assoc.* 201, 709–713.

553. Dubey, J. P. 1993. Neosporosis. *Proceedings of the 11th American College of Veterinary Internal Medicine Forum.* May 1993, Washington, DC. 710–712.

554. Dubey, J. P. 1993. Protozoal abortion in cattle. *J. Am. Vet. Med. Assoc.* 203, 1250–1251.

555. Dubey, J. P. 1993. Recent advances in neosporosis, toxoplasmosis and sarcocystosis with special reference to abortions in livestock. *Proceedings of the VIth International Coccidiosis Conference.* June 21–25, Guelph, Ontario, Canada. 27–35.

556. Dubey, J. P. 1993. *Toxoplasma, Neospora, Sarcocystis,* and other tissue cyst-forming coccidia of humans and animals. In *Parasitic Protozoa.* Kreir, J. P. (Editor). Academic Press, New York. Volume 6, 1–158.

557. Dubey, J. P., de Lahunta, A. 1993. Neosporosis associated congenital limb deformities in a calf. *Appl. Parasitol.* 34, 229–233.

558. Dubey, J. P., Lindsay, D. S. 1993. Neosporosis. *Parasitol. Today* 9, 452–458.

559. Dubey, J. P., Hamir, A. N., Shen, S. K., Thulliez, P., Rupprecht, C. E. 1993. Experimental *Toxoplasma gondii* infection in raccoons (*Procyon lotor*). *J. Parasitol.* 79, 548–552.

560. Dubey, J. P. 1994. Neosporosis. *Agri-Practice* 15, 16–17.

561. Dubey, J. P. 1994. Neosporosis—A newly recognized parasitic disease. *Akbash Dogs Int.* 29, 6–7.

562. Dubey, J. P., Metzger, F. L. Jr., Hattel, A. L., Lindsay, D. S., Fritz, D. L. 1995. Canine cutaneous neosporosis: Clinical improvement with clindamycin. *Vet. Dermatol.* 6, 37–43.

563. Dubey, J. P. 1996. WAAP and Pfizer Award for excellence in veterinary parasitology research pursuing life cycles and transmission of cyst-forming coccidia of animals and humans. *Vet. Parasitol.* 64, 13–20.

564. Dubey, J. P., Lindsay, D. S. 1996. A review of *Neospora caninum* and neosporosis. *Vet. Parasitol.* 67, 1–59.

565. Dubey, J. P., Lindsay, D. S. 1996. Neosporosis—A newly recognized protozoan disease. *J. Vet. Parasitol.* 10, 99–145.

566. Dubey, J. P., Morales, J. A., Villalobos, P., Lindsay, D. S., Blagburn, B. L., Topper, M. J. 1996. Neosporosis-associated abortion in a dairy goat. *J. Am. Vet. Med. Assoc.* 208, 263–265.

567. Dubey, J. P., Rigoulet, J., Lagourette, P., George, C., Longeart, L., LeNet, J. L. 1996. Fatal transplacental neosporosis in a deer (*Cervus eldi siamensis*). *J. Parasitol.* 82, 338–339.

568. Dubey, J. P., Lindsay, D. S., Adams, D. S., Gay, J. M., Baszler, T. V., Blagburn, B. L., Thulliez, P. 1996. Serologic responses of cattle and other animals infected with *Neospora caninum*. *Am. J. Vet. Res.* 57, 329–336.

569. Dubey, J. P., Jenkins, M. C., Adams, D. S., McAllister, M. M., Anderson-Sprecher, R., Baszler, T. V., Kwok, O. C. H., Lally, N. C., Björkman, C., Uggla, A. 1997. Antibody responses of cows during an outbreak of neosporosis evaluated by indirect fluorescent antibody test and different enzyme-linked immunosorbent assays. *J. Parasitol.* 83, 1063–1069.

570. Dubey, J. P., Lindsay, D. S., Speer, C. A. 1998. Structure of *Toxoplasma gondii* tachyzoites, bradyzoites and sporozoites, and biology and development of tissue cysts. *Clin. Microbiol. Rev.* 11, 267–299.

571. Dubey, J. P., Abbitt, B., Topper, M. J., Edwards, J. F. 1998. Hydrocephalus associated with *Neospora caninum*-infection in an aborted bovine fetus. *J. Comp. Pathol.* 118, 169–173.

572. Dubey, J. P., Romand, S., Hilali, M., Kwok, O. C. H., Thulliez, P. 1998. Seroprevalence of antibodies to *Neospora caninum* and *Toxoplasma gondii* in water buffaloes (*Bubalus bubalis*) from Egypt. *Int. J. Parasitol.* 28, 527–529.

573. Dubey, J. P., Dorough, K. R., Jenkins, M. C., Liddell, S., Speer, C. A., Kwok, O. C. H., Shen, S. K. 1998. Canine neosporosis: Clinical signs, diagnosis, treatment and isolation of *Neospora caninum* in mice and cell culture. *Int. J. Parasitol.* 28, 1293–1304.

574. Dubey, J. P. 1999. Neosporosis—The first decade of research. *Int. J. Parasitol.* 29, 1485–1488.

575. Dubey, J. P. 1999. Recent advances in *Neospora* and neosporosis. *Vet. Parasitol.* 84, 349–367.

576. Dubey, J. P. 1999. Neosporosis: A newly recognized protozoan disease similar to toxoplasmosis. *Infect. Dis. Rev.* 1, 129–130.

577. Dubey, J. P. 1999. Neosporosis in cattle: Biology and economic impact. *J. Am. Vet. Med. Assoc.* 214, 1160–1163.

578. Dubey, J. P., Kerber, C. E., Granstrom, D. E. 1999. Serologic prevalence of *Sarcocystis neurona*, *Toxoplasma gondii*, and *Neospora caninum* in horses in Brazil. *J. Am. Vet. Med. Assoc.* 215, 970–972.

579. Dubey, J. P., Venturini, M. C., Venturini, L., McKinney, J., Pecoraro, M. 1999. Prevalence of antibodies to *Sarcocystis neurona*, *Toxoplasma gondii*, and *Neospora caninum* in horses from Argentina. *Vet. Parasitol.* 86, 59–62.

580. Dubey, J. P., Romand, S., Thulliez, P., Kwok, O. C. H., Shen, S. K., Gamble, H. R. 1999. Prevalence of antibodies to *Neospora caninum* in horses in North America. *J. Parasitol.* 85, 968–969.

581. Dubey, J. P., Hollis, K., Romand, S., Thulliez, P., Kwok, O. C. H., Hungerford, L., Anchor, C., Etter, D. 1999. High prevalence of antibodies to *Neospora caninum* in white-tailed deer (*Odocoileus virginianus*). *Int. J. Parasitol.* 29, 1709–1711.

582. Dubey, J. P. 2000. *Neospora caninum*: Un agent abortif majeeur des bovins. *Bulletin des GTV* 7, 85.

583. Dubey, J. P., Lindsay, D. S. 2000. Gerbils (*Meriones unguiculatus*) are highly susceptible to oral infection with *Neospora caninum* oocysts. *Parasitol. Res.* 86, 165–168.

584. Dubey, J. P., Odening, K. 2001. Toxoplasmosis and related infections. In *Parasitic Diseases of wild Mammals*. Samuel, W. M., Pybus, M. J., Kocan, A. A. (Editors). Iowa State University Press, Ames, Iowa. 478–519.

585. Dubey, J. P., Liddell, S., Mattson, D., Speer, C. A., Howe, D. K., Jenkins, M. C. 2001. Characterization of the Oregon isolate of *Neospora hughesi* from a horse. *J. Parasitol.* 87, 345–353.

586. Dubey, J. P., Lindsay, D. S., Saville, W. J. A., Reed, S. M., Granstrom, D. E., Speer, C. A. 2001. A review of *Sarcocystis neurona* and equine protozoal myeloencephalitis (EPM). *Vet. Parasitol.* 95, 89–131.

587. Dubey, J. P., Hill, D. E., Lindsay, D. S., Jenkins, M. C., Uggla, A., Speer, C. A. 2002. *Neospora caninum* and *Hammondia heydorni* are separate species. *Trends Parasitol.* 18, 66–69.

588. Dubey, J. P., Lindsay, D. S., Hill, D., Romand, S., Thulliez, P., Kwok, O. C. H., Silva, J. C. R., Oliveira-Camargo, M. C., Gennari, S. M. 2002. Prevalence of antibodies to *Neospora caninum* and *Sarcocystis neurona* in sera of domestic cats from Brazil. *J. Parasitol.* 88, 1251–1252.

589. Dubey, J. P., Barr, B. C., Barta, J. R., Bjerkås, I., Björkman, C., Blagburn, B. L., Bowman, D. D., Buxton, D., Ellis, J. T., Gottstein, B., Hemphill, A., Hill, D. E., Howe, D. K., Jenkins, M. C., Kobayashi, Y., Koudela, B., Marsh, A. E., Mattsson, J. G., McAllister, M. M., Modrý, D., Omata, Y., Sibley, L. D., Speer, C. A., Trees, A. J., Uggla, A., Upton, S. J., Williams, D. J. L., Lindsay, D. S. 2002. Redescription of *Neospora caninum* and its differentiation from related coccidia. *Int. J. Parasitol.* 32, 929–946.

590. Dubey, J. P. 2003. Review of *Neospora caninum* and neosporosis in animals. *Korean J. Parasitol.* 41, 1–16.

591. Dubey, J. P. 2003. Neosporosis in cattle. *J. Parasitol.* 89(Suppl), S42–S56.

592. Dubey, J. P., Mitchell, S. M., Morrow, J. K., Rhyan, J. C., Stewart, L. M., Granstrom, D. E., Romand, S., Thulliez, P., Saville, W. J., Lindsay, D. S. 2003. Prevalence of antibodies to *Neospora caninum, Sarcocystis neurona,* and *Toxoplasma gondii* in wild horses from central Wyoming. *J. Parasitol.* 89, 716–720.

593. Dubey, J. P., Zarnke, R., Thomas, N. J., Wong, S. K., Van Bonn, W., Briggs, M., Davis, J. W., Ewing, R., Mense, M., Kwok, O. C. H., Romand, S., Thulliez, P. 2003. *Toxoplasma gondii, Neospora caninum, Sarcocystis neurona,* and *Sarcocystis canis*-like infections in marine mammals. *Vet. Parasitol.* 116, 275–296.

594. Dubey, J. P. 2004. Toxoplasmosis—A waterborne zoonosis. *Vet. Parasitol.* 126, 57–72.

595. Dubey, J. P. 2004. Neosporosis in cattle. *Vet. Parasitol.* 125, 138–140.

596. Dubey, J. P. 2004. Neosporosis. In *Infectious Diseases of Livestock with Special Reference to Southern Africa.* Coetzer, J. A. W., Thomson, G. R., Tustin, R. C., Kirk, N. P. J. (Editors). Oxford University Press, Ni City, South Africa. 382–393.

597. Dubey, J. P., Morales, E. S., Lehmann, T. 2004. Isolation and genotyping of *Toxoplasma gondii* from free-ranging chickens from Mexico. *J. Parasitol.* 90, 411–413.

598. Dubey, J. P., Sreekumar, C., Knickman, E., Miska, K. B., Vianna, M. C. B., Kwok, O. C. H., Hill, D. E., Jenkins, M. C., Lindsay, D. S., Greene, C. E. 2004. Biologic, morphologic, and molecular characterisation of *Neospora caninum* isolates from littermate dogs. *Int. J. Parasitol.* 34, 1157–1167.

599. Dubey, J. P. 2005. Unexpected oocyst shedding by cats fed *Toxoplasma gondii* tachyzoites: *In vivo* stage conversion and strain variation. *Vet. Parasitol.* 133, 289–298.

600. Dubey, J. P. 2005. Neosporosis in cattle. *Vet. Clin. Food Anim.* 21, 473–483.

601. Dubey, J. P., Thulliez, P. 2005. Prevalence of antibodies to *Neospora caninum* in wild animals. *J. Parasitol.* 91, 1217–1218.

602. Dubey, J. P., Knickman, E., Greene, C. E. 2005. Neonatal *Neospora caninum* infections in dogs. *Acta Parasitol.* 50, 176–179.

603. Dubey, J. P., Schares, G. 2006. Diagnosis of bovine neosporosis. *Vet. Parasitol.* 140, 1–34.

604. Dubey, J. P., Lindsay, D. S. 2006. Neosporosis, toxoplasmosis, and sarcocystosis in ruminants. *Vet. Clin. Food Anim.* 22, 645–671.

605. Dubey, J. P., Buxton, D., Wouda, W. 2006. Pathogenesis of bovine neosporosis. *J. Comp. Pathol.* 134, 267–289.

606. Dubey, J. P., Chapman, J. L., Rosenthal, B. M., Mense, M., Schueler, R. L. 2006. Clinical *Sarcocystis neurona, Sarcocystis canis, Toxoplasma gondii,* and *Neospora caninum* infections in dogs. *Vet. Parasitol.* 137, 36–49.

607. Dubey, J. P., Schares, G., Ortega-Mora, L. M. 2007. Epidemiology and control of neosporosis and *Neospora caninum. Clin. Microbiol. Rev.* 20, 323–367.

608. Dubey, J. P., Vianna, M. C. B., Kwok, O. C. H., Hill, D. E., Miska, K. B., Tuo, W., Velmurugan, G. V., Conors, M., Jenkins, M. C. 2007. Neosporosis in Beagle dogs: Clinical signs, diagnosis, treatment, isolation and genetic characterization of *Neospora caninum. Vet. Parasitol.* 149, 158–166.

609. Dubey, J. P., Alvarado-Esquivel, C., Liesenfeld, O., Herrera-Flores, R. G., Ramírez-Sánchez, B. E., González-Herrera, A., Martínez-García, S. A., Bandini, L. A., Kwok, O. C. H. 2007. *Neospora caninum* and *Toxoplasma gondii* antibodies in dogs from Durango City, Mexico. *J. Parasitol.* 93, 1033–1035.

610. Dubey, J. P., Stone, D., Kwok, O. C. H., Sharma, R. N. 2008. *Toxoplasma gondii* and *Neospora caninum* antibodies in dogs from Grenada, West Indies. *J. Parasitol.* 94, 750–751.

611. Dubey, J. P., Mansfield, K., Hall, B., Kwok, O. C. H., Thulliez, P. 2008. Seroprevalence of *Neospora caninum* and *Toxoplasma gondii* in black tailed deer (*Odocoileus hemionus columbianus*) and mule deer (*Odocoileus hemionus hemionus*). *Vet. Parasitol.* 156, 310–313.

612. Dubey, J. P., Jenkins, M. C., Kwok, O. C. H., Zink, R. L., Michalski, M. L., Ulrich, V., Gill, J., Carstensen, M., Thulliez, P. 2009. Seroprevalence of *Neospora caninum* and *Toxoplasma gondii* antibodies in white-tailed deer (*Odocoileus virginianus*) from Iowa and Minnesota using four serologic tests. *Vet. Parasitol.* 161, 330–334.

613. Dubey, J. P. 2010. *Toxoplasmosis of Animals and Humans.* 2nd edition. CRC Press, Boca Raton, Florida. 1–313.

614. Dubey, J. P., Schares, G. 2011. Neosporosis in animals—The last five years. *Vet. Parasitol.* 180, 90–108.

615. Dubey, J. P., Jenkins, M. C., Rajendran, C., Miska, K., Ferreira, L. R., Martins, J., Kwok, O. C. H., Choudhary, S. 2011. Gray wolf (*Canis lupus*) is a natural definitive host for *Neospora caninum*. *Vet. Parasitol.* 181, 382–387.

616. Dubey, J. P., Lappin, M. R. 2012. Toxoplasmosis and neosporosis. In *Infectious Diseases of the Dog and Cat.* Greene, C. (Editor). 4th edition. Elsevier, St. Louis, Missouri. 806–827.

617. Dubey, J. P. 2013. Neosporosis in dogs. *Commonwealth Agric. Bureau Rev.* 8, 055.

618. Dubey, J. P., Jenkins, M. C., Kwok, O. C. H., Ferreira, L. R., Choudhary, S., Verma, S. K., Villena, I., Butler, E., Cartensen, M. 2013. Congenital transmission of *Neospora caninum* in white-tailed deer (*Odocoileus virginianus*). *Vet. Parasitol.* 196, 519–522.

619. Dubey, J. P., Whitesell, L. E., Culp, W. E., Daye, S. 2014. Diagnosis and treatment of *Neospora caninum*-associated dermatitis in a red fox (*Vulpes vulpes*) with concurrent *Toxoplasma gondii* infection. *J. Zoo Wildl. Med.* 45, 454–457.

620. Dubey, J. P., Jenkins, M. C., Ferreira, L. R., Choudhary, S., Verma, S. K., Kwok, O. C. H., Fetterer, R., Butler, E., Carstensen, M. 2014. Isolation of viable *Neospora caninum* from brains of wild gray wolves (*Canis lupus*). *Vet. Parasitol.* 201, 150–153.

621. Dubey, J. P., Howe, D. K., Furr, M., Saville, W. J., Marsh, A. E., Reed, S. M., Grigg, M. E. 2015. An update on *Sarcocystis neurona* infections in animals and equine protozoal myeloencephalitis (EPM). *Vet. Parasitol.* 209, 1–42.

622. Dubey, J. P., Calero-Bernal, R., Rosenthal, B. M., Speer, C. A., Fayer, R. 2016. *Sarcocystosis of Animals and Humans.* 2nd edition. CRC Press, Boca Raton, Florida. 1–481.

623. Dubinsky, P., Reiterova, K., Moskwa, B., Bobakova, M., Durecko, R., Cabaj, W. 2006. *Neospora caninum* as a potential cause of abortions in dairy cows. *Slovensky Veterinarsky Casopis* 31, 175–177.

624. Duff, J. P., Otter, A. 1994. *Neospora*-associated abortions in cattle. *Vet. Rec.* 135, 415.

625. Duffield, T. F., Peregrine, A. S., McEwen, B. J., Hietala, S. K., Bagg, R., Dick, P. 2001. Seroprevalence of *Neospora caninum* infection in 25 Ontario dairy herds and its association with periparturient health and production. *Bovine Pract.* 35, 8–12.

626. Duivenvoorden, J., Lusis, P. 1995. *Neospora* abortions in eastern Ontario dairy herds. *Can. Vet. J.* 36, 623.

627. Dumètre, A., Dardé, M. L. 2003. How to detect *Toxoplasma gondii* oocysts in environmental samples? *FEMS Microbiol. Rev.* 27, 651–661.

628. Dumètre, A., Dardé, M. L. 2005. Immunomagnetic separation of *Toxoplasma gondii* oocysts using a monoclonal antibody directed against the oocyst wall. *J. Microbiol. Methods* 61, 209–217.

629. Duong, M. C., Alenius, S., Huong, L. T. T., Björkman, C. 2008. Prevalence of *Neospora caninum* and bovine viral diarrhoea virus in dairy cows in Southern Vietnam. *Vet. J.* 175, 390–394.

630. Dyer, R. M., Jenkins, M. C., Kwok, O. C. H., Douglas, L. W., Dubey, J. P. 2000. Serologic survey of *Neospora caninum* infection in a closed dairy cattle herd in Maryland: Risk of serologic reactivity by production groups. *Vet. Parasitol.* 90, 171–181.

631. Eastick, F. A., Elsheikha, H. M. 2010. Stress-driven stage transformation of *Neospora caninum*. *Parasitol. Res.* 106, 1009–1014.

632. Ebani, V. V., Poli, A., Rocchigiani, G., Bertelloni, F., Nardoni, S., Papini, A., Mancianti, F. 2016. Serological survey on some pathogens in wild brown hares (*Lepus europaeus*) in Central Italy. *Asian Pac. J. Trop. Med.* 9, 465–469.

633. Edelhofer, R., Loeschenberger, K., Peschke, R., Sager, H., Nowotny, N., Kolodziejek, J., Tews, A., Doneus, G., Prosl, H. 2003. First PCR-confirmed report of a *Neospora caninum*-associated bovine abortion in Austria. *Vet. Rec.* 152, 471–473.

634. Eiras, C., Arnaiz, I., Álvarez-García, G., Ortega-Mora, L. M., Sanjuánl, M. L., Yus, E., Diéguez, F. J. 2011. *Neospora caninum* seroprevalence in dairy and beef cattle from the northwest region of Spain, Galicia. *Prev. Vet. Med.* 98, 128–132.

635. Eleni, C., Crotti, S., Manuali, E., Costarelli, S., Filippini, G., Moscati, L., Magnino, S. 2004. Detection of *Neospora caninum* in an aborted goat foetus. *Vet. Parasitol.* 123, 271–274.

636. Ellis, J., Luton, K., Baverstock, P. R., Brindley, P. J., Nimmo, K. A., Johnson, A. M. 1994. The phylogeny of *Neospora caninum*. *Mol. Biochem. Parasitol.* 64, 303–311.

637. Ellis, J., Miller, C., Quinn, H., Ryce, C., Reichel, M. P. 2008. Evaluation of recombinant proteins of *Neospora caninum* as vaccine candidates (in a mouse model). *Vaccine* 26, 5989–5996.

638. Ellis, J., Sinclair, D., Morrison, D., Al-Qassab, S., Springett, K., Ivens, A. 2010. Microarray analyses of mouse responses to infection by *Neospora caninum* identifies disease associated cellular pathways in the host response. *Mol. Biochem. Parasitol.* 174, 117–127.

639. Ellis, J., Goodswen, S., Kennedy, P. J., Bush, S. 2012. The core mouse response to infection by *Neospora caninum* defined by gene set enrichment analyses. *Bioinformatics and Biology Insights* 6, 187–202.

640. Ellis, J. T. 1998. Polymerase chain reaction approaches for the detection of *Neospora caninum* and *Toxoplasma gondii*. *Int. J. Parasitol.* 28, 1053–1060.

641. Ellis, J. T., Morrison, D. A., Jeffries, A. C. 1998. The phylum apicomplexa: An update on the molecular phylogeny. In *Evolutionary Relationships among Protozoa*. Coombs, G. H., Vickerman, K., Sleigh, M. A., Warren, A. (*Editors*). Chapman & Hall, London. 255–274.

642. Ellis, J. T., Amoyal, G., Ryce, C., Harper, P. A. W., Clough, K. A., Homan, W. L., Brindley, P. J. 1998. Comparison of the large subunit ribosomal DNA of *Neospora* and *Toxoplasma* and development of a new genetic marker for their differentiation based on the D2 domain. *Mol. Cell. Probes* 12, 1–13.

643. Ellis, J. T., McMillan, D., Ryce, C., Payne, S., Atkinson, R., Harper, P. A. W. 1999. Development of a single tube nested polymerase chain reaction assay for the detection of *Neospora caninum* DNA. *Int. J. Parasitol.* 29, 1589–1596.

644. Ellis, J. T., Morrison, D. A., Liddell, S., Jenkins, M. C., Mohammed, O. B., Ryce, C., Dubey, J. P. 1999. The genus *Hammondia* is paraphyletic. *Parasitology* 118, 357–362.

645. Ellis, J. T., Ryce, C., Atkinson, R., Balu, S., Holmdahl, O. J. M. 2000. *Neospora caninum*: Gene discovery through analysis of expressed sequence tags. *Int. J. Parasitol.* 30, 909–913.

646. Ellis, J. T., Ryce, C., Atkinson, R., Balu, S., Jones, P., Harper, P. A. W. 2000. Isolation, characterization and expression of a GRA2 homologue from *Neospora caninum*. *Parasitology* 120, 383–390.

646a. Ellis, J. T., Holmdahl, O. J., Ryce, C., Njenga, J. M., Harper, P. A., Morrison, D. A. 2000. Molecular phylogeny of *Besnoitia* and the genetic relationships among *Besnoitia* of cattle, wildebeest and goats. *Protist* 151, 329–336.

647. Elsheikha, H. M., McKinlay, C. L., Elsaied, N. A., Smith, P. A. 2013. Effects of *Neospora caninum* infection on brain microvascular endothelial cells bioenergetics. *Parasit. Vectors* 6, 24.

648. Elsheikha, H. M., Alkurashi, M., Kong, K., Zhu, X. Q. 2014. Metabolic footprinting of extracellular metabolites of brain endothelium infected with *Neospora caninum in vitro*. *BMC Res. Notes* 7, 406

649. Elsinghorst, T. A. M. 2003. First cases of animal diseases published since 2000. 5. Sheep. *Vet. Q.* 25, 165–169.

650. Enachescu, V., Ionita, M., Mitrea, I. L. 2014. Comparative study for the detection of antibodies to *Neospora caninum* in milk and sera in dairy cattle in southern Romania. *Acta Parasitol.* 59, 5–10.

651. English, E. D., Adomako-Ankomah, Y., Boyle, J. P. 2015. Secreted effectors in *Toxoplasma gondii* and related species: Determinants of host range and pathogenesis? *Parasite Immunol.* 37, 127–140.

652. Eperon, S., Brönnimann, K., Hemphill, A., Gottstein, B. 1999. Susceptibility of B-cell deficient C57BL/6 (μMT) mice to *Neospora caninum* infection. *Parasite Immunol.* 21, 225–236.

653. Esposito, M., Stettler, R., Moores, S. L., Pidathala, C., Müller, N., Stachulski, A., Berry, N. G., Rossignol, J. F., Hemphill, A. 2005. *In vitro* efficacies of nitazoxanide and other thiazolides against *Neospora caninum* tachyzoites reveal antiparasitic activity independent of the nitro group. *Antimicrob. Agents Chemother.* 49, 3715–3723.

654. Esposito, M., Müller, N., Hemphill, A. 2007. Structure-activity relationships from *in vitro* efficacies of the thiazolide series against the intracellular apicomplexan protozoan *Neospora caninum*. *Int. J. Parasitol.* 37, 183–190.

655. Esposito, M., Moores, S., Naguleswaran, A., Müller, J., Hemphill, A. 2007. Induction of tachyzoite egress from cells infected with the protozoan *Neospora caninum* by nitro- and bromo-thiazolides, a class of broad-spectrum anti-parasitic drugs. *Int. J. Parasitol.* 37, 1143–1152.

656. Estes, D. M., Brown, W. C. 2002. Type 1 and type 2 responses in regulation of Ig isotype expression in cattle. *Vet. Immunol. Immunopathol.* 90, 1–10.

657. Evans, J., Levesque, D., Shelton, G. D. 2004. Canine inflammatory myopathies: A clinicopathologic review of 200 cases. *J. Vet. Intern. Med.* 18, 679–691.

658. Eymann, J., Herbert, C. A., Cooper, D. W., Dubey, J. P. 2006. Serologic survey for *Toxoplasma gondii* and *Neospora caninum* in the common brushtail possum (*Trichosurus vulpecula*) from urban Sydney, Australia. *J. Parasitol.* 92, 267–272.

659. Fard, S. R. N., Khalili, M., Aminzadeh, A. 2008. Prevalence of antibodies to *Neospora caninum* in cattle in Kerman province, south east Iran. *Vet. Arhiv* 78, 253–259.

660. Faria, E. B., Gennari, S. M., Pena, H. F. J., Athayde, A. C. R., Silva, M. L. C. R., Azevedo, S. S. 2007. Prevalence of anti-*Toxoplasma gondii* and anti-*Neospora caninum* antibodies in goats slaughtered in the public slaughterhouse of Patos city, Paraíba State, Northeast region of Brazil. *Vet. Parasitol.* 149, 126–129.

661. Faria, E. B., Cavalcanti, E. F. T. S. F., Medeiros, E. S., Pinheiro-Júnior, J. W., Azevedo, S. S., Athayde, A. C. R., Mota, R. A. 2010. Risk factors associated with *Neospora caninum* seropositivity in sheep from the State of Alagoas, in the northeast region of Brazil. *J. Parasitol.* 96, 197–199.

662. Fazaeli, A., Carter, P. E., Pennington, T. H. 2000. Intergenic spacer (IGS) polymorphism: A new genetic marker for differentiating *Toxoplasma gondii* strains and *Neospora caninum*. *J. Parasitol.* 86, 716–723.

663. Feitosa, T. F., Vilela, V. L. R., Dantas, E. S., Souto, D. V. O., Pena, H. F. J., Athayde, A. C. R., Azevêdo, S. S. 2014. *Toxoplasma gondii* and *Neospora caninum* in domestic cats from the Brazilian semi-arid: Seroprevalence and risk factors. *Arq. Bras. Med. Vet. Zootec.* 66, 1060–1066.

664. Feitosa, T. F., Vilela, V. L. R., de Melo, L. R. B., de Almeida Neto, J. L., Souto, D. V. O., de Morais, D. F., Athayde, A. C. R., Azevedo, S. S., Pena, H. F. J. 2014. *Toxoplasma gondii* and *Neospora caninum* in slaughtered pigs from Northeast, Brazil. *Vet. Parasitol.* 202, 305–309.

664a. Feng, X., Zhang, N., Tuo, W. 2010. *Neospora caninum* tachyzoite- and antigen-stimulated cytokine production by bone marrow-derived dendritic cells and spleen cells of naïve BALB/c mice. *J. Parasitol.* 96, 717–723.

665. Feng, Y., Lu, Y., Wang, Y., Liu, J., Zhang, L., Yang, Y. 2016. *Toxoplasma gondii* and *Neospora caninum* in free-range chickens in Henan Province of China. *Biomed Res. Int.* 2016, e8290536.

666. Fereig, R. M., AbouLaila, M. R., Mohamed, S. G. A., Mahmoud, H. Y. A. H., Ali, A. O., Ali, A. F., Hilali, M., Zaid, A., Mohamed, A. E. A., Nishikawa, Y. 2016. Serological detection and epidemiology of *Neospora caninum* and *Cryptosporidium parvum* antibodies in cattle in southern Egypt. *Acta Trop.* 162, 206–211.

667. Fernandes, B. C. T. M., Gennari, S. M., Souza, S. L. P., Carvalho, J. M., Oliveira, W. G., Cury, M. C. 2004. Prevalence of anti-*Neospora caninum* antibodies in dogs from urban, periurban and rural areas of the city of Uberlândia, Minas Gerais—Brazil. *Vet. Parasitol.* 123, 33–40.

668. Fernández-García, A., Risco-Castillo, V., Zaballos, A., Álvarez-García, G., Ortega-Mora, L. M. 2006. Identification and molecular cloning of the *Neospora caninum SAG4* gene specifically expressed at bradyzoite stage. *Mol. Biochem. Parasitol.* 146, 89–97.

669. Ferrari, A., Trees, A. J., Dellepiane, M., Carbonaro, M. 1996. Prime segnalazioni di *Neospora caninum* negli allevamenti bovini italiani. *Proceedings of the IV Mediterranean Federation of Health and Reproduction of Ruminants*. October 28–30, Murcia, Spain.

670. Ferrari, A., Donofrio, G., Dellepiane, M., Cabassi, C. S., Bigliardi, E., Cavirani, S. 1997. Anticorpi verso *Neospora caninum* in bovine da latte con aborto a carattere enzootico. *Att. Soc. Ital. Buiatria* 29, 223–227.

671. Ferre, I., Aduriz, G., del-Pozo, I., Regidor-Cerrillo, J., Atxaerandio, R., Collantes-Fernández, E., Hurtado, A., Ugarte-Garagalza, C., Ortega-Mora, L. M. 2005. Detection of *Neospora caninum* in the semen and blood of naturally infected bulls. *Theriogenology* 63, 1504–1518.

672. Ferre, I., Serrano-Martínez, E., Martínez, A., Osoro, K., Mateos-Sanz, A., del-Pozo, I., Aduriz, G., Tamargo, C., Hidalgo, C. O., Ortega-Mora, L. M. 2008. Effects of re-infection with *Neospora caninum* in bulls on parasite detection in semen and blood and immunological responses. *Theriogenology* 69, 905–911.

672a. Ferreira, M. S. T., Vogel, F. S. F., Sangioni, L. A., Weber, A., Bräunig, P., Vaz, M. A. B., Cezar, A. S. 2016. Oral infection of neonate gerbils by *Neospora caninum* tachyzoites. *Ciência Rural, Santa Maria.* 46, 654–649.

673. Ferreirinha, P., Dias, J., Correia, A., Pérez-Cabezas, B., Santos, C., Teixeira, L., Ribeiro, A., Rocha, A., Vilanova, M. 2014. Protective effect of intranasal immunization with *Neospora caninum* membrane antigens against murine neosporosis established through the gastrointestinal tract. *Immunology* 141, 256–267.

673a. Ferreirinha, P., Correia, A., Teixeira-Carvalho, M., Osório, H., Teixeira, L., Rocha, A., Vilanova, M. 2016. Mucosal immunization confers long-term protection against intragastrically established *Neospora caninum* infection. *Vaccine.* 34, 6250–6258.

674. Ferroglio, E., Rossi, L. 2001. Prevalence of *Neospora caninum* antibodies in wild ruminants from the Italian Alps. *Vet. Rec.* 148, 754–755.

675. Ferroglio, E., Bassano, B., Trisciuoglio, A., Rossi, L. 2001. Antibodies to *Neospora caninum* in Alpine ibex from the Italian Alps. *Z. Jagdwiss.* 47, 226–228.

676. Ferroglio, E., Trisciuoglio, A. 2003. Antibodies to *Neospora caninum* in European brown hare (*Lepus europaeus*). *Vet. Parasitol.* 115, 75–78.

677. Ferroglio, E., Wambwa, E., Castiello, M., Trisciuoglio, A., Prouteau, A., Pradere, E., Ndungu, S., De Meneghi, D. 2003. Antibodies to *Neospora caninum* in wild animals from Kenya, East Africa. *Vet. Parasitol.* 118, 43–49.

678. Ferroglio, E., Guiso, P., Pasino, M., Accossato, A., Trisciuoglio, A. 2005. Antibodies to *Neospora caninum* in stray cats from north Italy. *Vet. Parasitol.* 131, 31–34.

679. Ferroglio, E., Pasino, M., Ronco, F., Benà, A., Trisciuoglio, A. 2007. Seroprevalence of antibodies to *Neospora caninum* in urban and rural dogs in north-west Italy. *Zoonoses Public Health* 54, 135–139.

680. Ferroglio, E., Pasino, M., Romano, A., Grande, D., Pregel, P., Trisciuoglio, A. 2007. Evidence of *Neospora caninum* DNA in wild rodents. *Vet. Parasitol.* 148, 346–349.

681. Fidan, A. F., Cingi, C. C., Karafakioglu, Y. S., Utuk, A. E., Pekkaya, S., Piskin, F. C. 2010. The levels of antioxidant activity, malondialdehyde and nitric oxide in cows naturally infected with *Neospora caninum*. *J. Anim. Vet. Adv.* 9, 1707–1711.

682. Figliuolo, L. P. C., Rodrigues, A. A. R., Viana, R. B., Aguiar, D. M., Kasai, N., Gennari, S. M. 2004. Prevalence of anti-*Toxoplasma gondii* and anti-*Neospora caninum* antibodies in goat from São Paulo State, Brazil. *Small Ruminant Res.* 55, 29–32.

683. Figliuolo, L. P. C., Kasai, N., Ragozo, A. M. A., de Paula, V. S. O., Dias, R. A., Souza, S. L. P., Gennari, S. M. 2004. Prevalence of anti-*Toxoplasma gondii* and anti-*Neospora caninum* antibodies in ovine from São Paulo State, Brazil. *Vet. Parasitol.* 123, 161–166.

684. Figueredo, L. A., Dantas-Torres, F., de Faria, E. B., Gondim, L. F. P., Simões-Mattos, L., Brandão-Filho, S. P., Mota, R. A. 2008. Occurrence of antibodies to *Neospora caninum* and *Toxoplasma gondii* in dogs from Pernambuco, Northeast Brazil. *Vet. Parasitol.* 157, 9–13.

685. Figueroa, V. J., Morales, S. E., Martínez, M. J. J. 2012. Risk factors associated with infection by *Neospora caninum* in dual-purpose cattle in the central region of Veracruz, Mexico. *Inernet J. Vet. Med.* 9(1), 1–9.

686. Finno, C. J., Aleman, M., Pusterla, N. 2007. Equine protozoal myeloencephalitis associated with neosporosis in 3 horses. *J. Vet. Intern. Med.* 21, 1405–1408.

687. Finno, C. J., Packham, A. E., Wilson, W. D., Gardner, I. A., Conrad, P. A., Pusterla, N. 2007. Effects of blood contamination of cerebrospinal fluid on results of indirect fluorescent antibody tests for detection of antibodies against *Sarcocystis neurona* and *Neospora hughesi*. *J. Vet. Diagn. Invest.* 19, 286–289.

688. Finno, C. J., Eaton, J. S., Aleman, M., Hollingsworth, S. R. 2010. Equine protozoal myeloencephalitis due to *Neospora hughesi* and equine motor neuron disease in a mule. *Vet. Ophthalmol.* 13, 259–265.

689. Fioretti, D. P., Rosignoli, L., Ricci, G., Moretti, A., Pasquali, P., Polidori, G. A. 2000. *Neospora caninum* infection in a clinically healthy calf: Parasitological study and serological follow-up. *J. Vet. Med. B* 47, 47–53.

690. Fioretti, D. P., Pasquai, P., Diaferia, M., Mangili, V., Rosignoli, L. 2003. *Neospora caninum* infection and congenital transmission: Serological and parasitological study of cows up to the fourth gestation. *J. Vet. Med. B* 50, 399–404.

691. Fischer, I., Furrer, K., Audigé, L., Fritsche, A., Giger, T., Gottstein, B., Sager, H. 2003. Von der Bedeutung der bovinen Neosporose beim Abortgeschehen in der Schweiz. *Schweiz. Arch. Tierheilkd.* 145, 114–123.

692. Fish, L., Mazuz, M., Molad, T., Savitsky, I., Shkap, V. 2007. Isolation of *Neospora caninum* from dairy zero grazing cattle in Israel. *Vet. Parasitol.* 149, 167–171.

693. Flagstad, A., Jensen, H. E., Bjerkås, I., Rasmussen, K. 1995. *Neospora caninum* infection in a litter of Labrador retriever dogs in Denmark. *Acta Vet. Scand.* 36, 387–391.

694. Flynn, R. J., Marshall, E. S. 2011. Parasite limiting macrophages promote IL-17 secretion in naive bovine CD4+ T-cells during *Neospora caninum* infection. *Vet. Immunol. Immunopathol.* 144, 423–429.

695. Fondevila, D., Añor, S., Pumarola, M., Dubey, J. P. 1998. *Neospora caninum* identification in an aborted bovine fetus in Spain. *Vet. Parasitol.* 77, 187–189.

696. Fort, M. 2003. *Neospora caninum*: Estudio seroepidemiológico en bovinos de la provincia de La Pampa. *Publicacion Tecnica—INTA* 52 [*Neospora caninum*: Seroepidemiologic study in cattle from La Pampa province.], 1–44.

697. Franco, M., Shastri, A. J., Boothroyd, J. C. 2014. Infection by *Toxoplasma gondii* specifically induces host c-Myc and the genes this pivotal transcription factor regulates. *Eukaryot. Cell* 13, 483–493.

698. French, N. P., Davison, H. C., Clancy, D., Begon, M., Trees, A. J. 1998. Modelling of *Neospora* species infection in dairy cattle: The importance of horizontal and vertical transmission and differential culling. *Proceedings of the Society for Veterinary Epidemiology and Preventive Medicine Meeting, West County Hotel.* March 25–27, Ennis, Co., Clare. 113–122.

699. French, N. P., Clancy, D., Davison, H. C., Trees, A. J. 1999. Mathematical models of *Neospora caninum* infection in dairy cattle: Transmission and options for control. *Int. J. Parasitol.* 29, 1691–1704.

700. Fridlund-Plugge, N., Montiani-Ferreira, F., Richartz, R. R. T. B., Dal Pizzol, J., Machado, P. C., Patrício, L. F. L., Rosinelli, A. S., Locatelli-Dittrich, R. 2008. Frequency of antibodies against *Neospora caninum* in stray and domiciled dogs from urban, periurban and rural areas from Paraná State, Southern Brazil. *Rev. Bras. Parasitol. Vet.* 17, 222–226.

701. Friedrich, N., Santos, J. M., Liu, Y., Palma, A. S., Leon, E., Saouros, S., Kiso, M., Blackman, M. J., Matthews, S., Feizi, T., Soldati-Favre, D. 2010. Members of a novel protein family containing microneme adhesive repeat domains act as sialic acid-binding lectins during host cell invasion by apicomplexan parasites. *J. Biol. Chem.* 285, 2064–2076.

702. Fritz, D., George, C., Dubey, J. P., Trees, A. J., Barber, J. S., Hopfner, C. L., Mehaut, S., Le Net, J. L., Longeart, L. 1997. *Neospora caninum*: Associated nodular dermatitis in a middle-aged dog. *Canine Pract.* 22, 21–24.

703. Frössling, J., Bonnett, B., Lindberg, A., Björkman, C. 2003. Validation of a *Neospora caninum* iscom ELISA without a gold standard. *Prev. Vet. Med.* 57, 141–153.

704. Frössling, J. 2004. Epidemiology of *Neospora caninum* infection in cattle: Evaluation of diagnostic tests and herd studies. *Doctoral thesis.* Swedish University of Agricultural Sciences, Uppsala. 1–153.

705. Frössling, J., Uggla, A., Björkman, C. 2005. Prevalence and transmission of *Neospora caninum* within infected Swedish dairy herds. *Vet. Parasitol.* 128, 209–218.

706. Frössling, J., Lindberg, A., Björkman, C. 2006. Evaluation of an iscom ELISA used for detection of antibodies to *Neospora caninum* in bulk milk. *Prev. Vet. Med.* 74, 120–129.

707. Frössling, J., Nødtvedt, A., Lindberg, A., Björkman, C. 2008. Spatial analysis of *Neospora caninum* distribution in dairy cattle from Sweden. *Geospat. Health* 3, 39–45.

708. Fry, D. R., McSporran, K. D., Ellis, J. T., Harvey, C. 2009. Protozoal hepatitis associated with immunosuppressive therapy in a dog. *J. Vet. Intern. Med.* 23, 366–368.

709. Fuchs, N., Sonda, S., Gottstein, B., Hemphill, A. 1998. Differential expression of cell surface- and dense granule-associated *Neospora caninum* proteins in tachyzoites and bradyzoites. *J. Parasitol.* 84, 753–758.

710. Fuchs, N., Ingold, K., Sonda, S., Bütikofer, P., Hemphill, A. 1999. Detection of surface-associated and intracellular glycoconjugates and glycoproteins in *Neospora caninum* tachyzoites. *Int. J. Parasitol.* 29, 1597–1611.

711. Fuehrer, H. P., Blöschl, I., Siehs, C., Hassl, A. 2010. Detection of *Toxoplasma gondii, Neospora caninum*, and *Encephalitozoon cuniculi* in the brains of common voles (*Microtus arvalis*) and water voles (*Arvicola terrestris*) by gene amplification techniques in western Austria (Vorarlberg). *Parasitol. Res.* 107, 469–473.

712. Fujii, K., Kakumoto, C., Kobayashi, M., Saito, S., Kariya, T., Watanabe, Y., Xuan, X., Igarashi, I., Suzuki, M. 2007. Seroepidemiology of *Toxoplasma gondii* and *Neospora caninum* in seals around Hokkaido, Japan. *J. Vet. Med. Sci.* 69, 393–398.

713. Fujii, T. U., Kasai, N., Nishi, S. M., Dubey, J. P., Gennari, S. M. 2001. Seroprevalence of *Neospora caninum* in female water buffaloes (*Bubalus bubalis*) from the southeastern region of Brazil. *Vet. Parasitol.* 99, 331–334.

714. Fujii, T. U., Kasai, N., Vasconcellos, S. A., Richtzenhain, L. J., Cortez, A., Souza, S. L. P., Baruselli, P. S., Nishi, S. M., Ferreira, F., Gennari, S. M. 2001. Anticorpos anti-*Neospora caninum* e contra outros agentes de abortamentos em búfalas da região do Vale do Ribeira, São Paulo, Brasil. *Arq. Inst. Biol. (São Paulo)* 68, 5–9.

715. Furuta, P. I., Mineo, T. W. P., Carrasco, A. O. T., Godoy, G. S., Pinto, A. A., Machado, R. Z. 2007. *Neospora caninum* infection in birds: Experimental infections in chicken and embryonated eggs. *Parasitology* 134, 1931–1939.

716. Gaffuri, A., Giacometti, M., Tranquillo, V. M., Magnino, S., Cordioli, P., Lanfranchi, P. 2006. Serosurvey of roe deer, chamois, and domestic sheep in the central Italian Alps. *J. Wildl. Dis.* 42, 685–690.

717. Gagnon, C. A., Allam, O., Drolet, R., Tremblay, D. 2010. Cross Canada Disease Report, Quebec: Detection of bovine lymphotropic herpesvirus DNA in tissues of a bovine aborted fetus. *Can. Vet. J.* 51, 1021–1022.

718. Gaitero, L., Añor, S., Montoliu, P., Zamora, Á., Pumarola, M. 2006. Detection of *Neospora caninum* tachyzoites in canine cerebrospinal fluid. *J. Vet. Intern. Med.* 20, 410–414.

719. Galgut, B. I., Janardhan, K. S., Grondin, T. M., Harkin, K. R., Wight-Carter, M. T. 2010. Detection of *Neospora caninum* tachyzoites in cerebrospinal fluid of a dog following prednisone and cyclosporine therapy. *Vet. Clin. Pathol.* 39, 386–390.

720. Galvão, C. M. M. Q., Rezende-Gondim, M. M., Chaves, A. C. R., Schares, G., Ribas, J. R. L., Gondim, L. F. P. 2015. Brazilian donkeys (*Equus asinus*) have a low exposure to *Neospora* spp. *Braz. J. Vet. Parasitol.* 24, 340–344.

721. García-Bocanegra, I., Cabezón, O., Pabón, M., Gómez-Guillamón, F., Arenas, A., Alcaide, E., Salas-Vega, R., Dubey, J. P., Almería, S. 2012. Prevalence of *Toxoplasma gondii* and *Neospora caninum* antibodies in Spanish ibex (*Capra pyrenaica hispanica*). *Vet. J.* 191, 257–260.

722. García-Chaparro, J. F., Moreno-Figueredo, G., Cruz-Carrillo, A. C. 2014. Prevalencia de *Neospora caninum* y DVB en una finca con problemas reproductivos en Sopó (Cundinamarca). *Ciencia Agric.* 11, 9–16

723. García-Ispierto, I., López-Gatius, F., Santolaria, P., Yániz, J. L., Nogareda, C., López-Béjar, M. 2007. Factors affecting the fertility of high producing dairy herds in northeastern Spain. *Theriogenology* 67, 632–638.

724. García-Ispierto, I., López-Gatius, F., Almería, S., Yániz, J., Santolaria, P., Serrano, B., Bech-Sàbat, G., Nogareda, C., Sulon, J., de Sousa, N. M., Beckers, J. F. 2009. Factors affecting plasma prolactin concentrations throughout gestation in high producing dairy cows. *Domest. Anim. Endocrinol.* 36, 57–66.

725. García-Ispierto, I., Nogareda, C., Yániz, J. L., Almería, S., Martínez-Bello, D., de Sousa, N. M., Beckers, J. F., López-Gatius, F. 2010. *Neospora caninum* and *Coxiella burnetii* seropositivity are related to endocrine pattern changes during gestation in lactating dairy cows. *Theriogenology* 74, 212–220.

726. García-Ispierto, I., Almería, S., Serrano, B., de Sousa, N. M., Beckers, J. F., López-Gatius, F. 2013. Plasma concentrations of pregnancy-associated glycoproteins measured using anti-bovine PAG-2 antibodies on Day 120 of gestation predict abortion in dairy cows naturally infected with *Neospora caninum*. *Reprod. Dom. Anim.* 48, 613–618.

727. García-Ispierto, I., Serrano-Pérez, B., Almería, S., Martínez-Bello, D., Tchimbou, A. F., de Sousa, N. M., Beckers, J. F., López-Gatius, F. 2015. Effects of crossbreeding on endocrine patterns determined in pregnant beef/dairy cows naturally infected with *Neospora caninum*. *Theriogenology* 83, 491–496.

728. García-Lunar, P., Moré, G., Campero, L., Ortega-Mora, L. M., Álvarez-García, G. 2015. Anti-*Neospora caninum* and anti-*Sarcocystis* spp. specific antibodies cross-react with *Besnoitia besnoiti* and influence the serological diagnosis of bovine besnoitiosis. *Vet. Parasitol.* 214, 49–54.

729. García-Melo, D. P., Regidor-Cerrillo, J., Ortega-Mora, L. M., Collantes-Fernández, E., de Oliveira, V. S. F., de Oliveira, M. A. P., da Silva, A. C. 2009. Isolation and biological characterisation of a new isolate of *Neospora caninum* from an asymptomatic calf in Brazil. *Acta Parasitol.* 54, 180–185.

730. García-Melo, D. P., Regidor-Cerrillo, J., Collantes-Fernández, E., Aguado-Martínez, A., Del Pozo, I., Minguijón, E., Gómez-Bautista, M., Aduriz, G., Ortega-Mora, L. M. 2010. Pathogenic characterization in mice of *Neospora caninum* isolates obtained from asymptomatic calves. *Parasitology* 137, 1057–1068.

731. García-Vázquez, Z., Cruz-Vázquez, C., Medina-Espinoza, L., García-Tapia, D., Chavarria-Martinez, B. 2002. Serological survey of *Neospora caninum* infection in dairy cattle herds in Aguascalientes, Mexico. *Vet. Parasitol.* 106, 115–120.

732. García-Vazquez, Z., Rosario-Cruz, R., Ramos-Aragon, A., Cruz-Vazquez, C., Mapes-Sanchez, G. 2005. *Neospora caninum* seropositivity and association with abortions in dairy cows in Mexico. *Vet. Parasitol.* 134, 61–65.

733. García-Vazquez, Z., Rosario-Cruz, R., Mejia-Estrada, F., Rodriguez-Vivas, I., Romero-Salas, D., Fernandez-Ruvalcaba, M., Cruz-Vazquez, C. 2009. Seroprevalence of *Neospora caninum* antibodies in beef cattle in three southern states of Mexico. *Trop. Anim. Health Prod.* 41, 749–753.

734. Gargala, G., Baishanbo, A., Favennec, L., François, A., Ballet, J. J., Rossignol, J. F. 2005. Inhibitory activities of epidermal growth factor receptor tyrosine kinase-targeted dihydroxyisoflavone and trihydroxydeoxybenzoin derivatives on *Sarcocystis neurona*, *Neospora caninum*, and *Cryptosporidium parvum* development. *Antimicrob. Agents Chemother.* 49, 4628–4634.

735. Garosi, L., Dawson, A., Couturier, J., Matiasek, L., de Stefani, A., Davies, E., Jeffery, N., Smith, P. 2010. Necrotizing cerebellitis and cerebellar atrophy caused by *Neospora caninum* infection: Magnetic resonance imaging and clinicopathologic findings in seven dogs. *J. Vet. Intern. Med.* 24, 571–578.

736. Gasser, R. B., Edwards, G., Cole, R. A. 1993. Neosporosis in a dog. *Aust. Vet. Pract.* 23, 190–193.

737. Gaturaga, I., Chahan, B., Xuan, X., Huang, X., Liao, M., Fukumoto, S., Hirata, H., Nishikawa, Y., Takashima, Y., Suzuki, H., Fujisaki, K., Sugimoto, C. 2005. Detection of antibodies to *Neospora caninum* in cattle by enzyme-linked immunosorbent assay with truncated NcSR2 expressed in *Escherichia coli*. *J. Parasitol.* 91, 191–192.

738. Gavrea, R., Mircean, V., Pastiu, A., Cozma, V. 2012. Epidemiological survey of *Neospora caninum* infection in dogs from Romania. *Vet. Parasitol.* 188, 382–385.

739. Gavrea, R. R., Iovu, A., Losson, B., Cozma, V. 2011. Seroprevalence of *Neospora caninum* in dairy cattle from north-west and centre of Romania. *Parasite* 18, 349–351.

740. Gavrea, R. R., Cozma, V. 2010. Seroprevalence of *Neospora caninum* in cows with reproductive failure in center and northwest of Romania. *Sci. Parasitol.* 11, 67–70.

741. Gavrilović, P., Živulj, A., Todorović, I., Jovanović, M., Parunović, J. 2013. Investigation of importance of *Neospora caninum* in aetiology of abortion in dairy cows in Serbia. *Revue Méd. Vét.* 164, 100–104.

742. Gay, J. M. 2006. Neosporosis in dairy cattle: An update from an epidemiological perspective. *Theriogenology* 66, 629–632.

743. Gaydos, J. K., Conrad, P. A., Gilardi, K. V. K., Blundell, G. M., Ben-David, M. 2007. Does human proximity affect antibody prevalence in marine-foraging river otters (*Lontra canadensis*)? *J. Wildl. Dis.* 43, 116–123.

744. Gazzonis, A. L., Alvarez Garcia, G., Zanzani, S. A., Ortega Mora, L. M., Invernizzi, A., Manfredi, M. T. 2016. *Neospora caninum* infection in sheep and goats from north-eastern Italy and associated risk factors. *Small Ruminant Res.* 140, 7–12.

745. Gennari, S. M., Yai, L. E. O., D'Áuria, S. N. R., Cardoso, S. M. S., Kwok, O. C. H., Jenkins, M. C., Dubey, J. P. 2002. Occurrence of *Neospora caninum* antibodies in sera from dogs of the city of São Paulo, Brazil. *Vet. Parasitol.* 106, 177–179.

746. Gennari, S. M. 2004. *Neospora caninum* no Brasil: Situação atual da pesquisa. XIII Congresso Brasileiro de Parasitologia Veterinária & I Simpósio Latino-Americano de Ricketisioses, Ouro Preto, MG, 2004. *Rev. Bras. Parasitol. Vet.* 13(Suppl I), 23–28.

747. Gennari, S. M., Rodrigues, A. A. R., Viana, R. B., Cardoso, E. C. 2005. Occurrence of anti-*Neospora caninum* antibodies in water buffaloes (*Bubalus bubalis*) from the northern region of Brazil. *Vet. Parasitol.* 134, 169–171.

748. Gennari, S. M., Cañón-Franco, W. A., Feitosa, M. M., Ikeda, F. A., de Lima, V. M. F., Amaku, M. 2006. Presence of anti-*Neospora caninum* and *Toxoplasma gondii* antibodies in dogs with visceral leishmaniosis from the region of Araçatuba, São Paulo, Brazil. *Braz. J. Vet. Res. Anim. Sci.* 43, 613–615.

749. Gennari, S. M., Pena, H. F. J., Lindsay, D. S., Lopes, M. G., Soares, H. S., Cabral, A. D., Vitaliano, S. N., Amaku, M. 2016. Prevalence of antibodies against *Neospora* spp. and *Sarcocystis neurona* in donkeys from northeastern Brazil. *Rev. Bras. Parasitol. Vet.* 25, 109–111.

750. Georgieva, D. A., Prelezov, P. N., Koinarski, V. T. 2006. *Neospora caninum* and neosporosis in animals—A review. *Bulg. J. Vet. Med.* 9, 1–26.

751. Geurden, T., Somers, R., Thanh, N. T. G., Vien, L. V., Nga, V. T., Giang, H. H., Dorny, P., Giao, H. K., Vercruysse, J. 2008. Parasitic infections in dairy cattle around Hanoi, northern Vietnam. *Vet. Parasitol.* 153, 384–388.

752. Ghalmi, F., China, B., Losson, B. 2007. Diagnostic et surveillance épidémiologique de *Neospora caninum*. *Ann. Méd. Vét.* 151, 123–149.

753. Ghalmi, F., China, B., Kaidi, R., Daube, G., Losson, B. 2008. Detection of *Neospora caninum* in dog organs using real time PCR systems. *Vet. Parasitol.* 155, 161–167.

754. Ghalmi, F., Dramchini, N., China, B. 2009. Risk factors for abortion in cattle herds in Algeria. *Vet. Rec.* 165, 475–476.

755. Ghalmi, F., China, B., Kaidi, R., Losson, B. 2009. Evaluation of a SRS2 sandwich commercial enzyme-linked immunosorbent assay for the detection of anti-*Neospora caninum* antibodies in bovine and canine sera. *J. Vet. Diagn. Invest.* 21, 108–111.

756. Ghalmi, F., China, B., Kaidi, R., Losson, B. 2009. First epidemiological study on exposure to *Neospora caninum* in different canine populations in the Algiers District (Algeria). *Parasitol. Int.* 58, 444–450.

757. Ghalmi, F., China, B., Kaidi, R., Losson, B. 2011. *Neospora caninum* is associated with abortion in Algerian cattle. *J. Parasitol.* 97, 1121–1124.

758. Ghalmi, F., China, B., Ghalmi, A., Hammitouche, D., Losson, B. 2012. Study of the risk factors associated with *Neospora caninum* seroprevalence in Algerian cattle populations. *Res. Vet. Sci.* 93, 655–661.

759. Ghalmi, F., China, B., Jenkins, M., Azzag, N., Losson, B. 2014. Comparison of different serological methods to detect antibodies specific to *Neospora caninum* in bovine and canine sera. *J. Vet. Diagn. Invest.* 26, 136–140.

760. Ghanem, M. E., Suzuki, T., Akita, M., Nishibori, M. 2009. *Neospora caninum* and complex vertebral malformation as possible causes of bovine fetal mummification. *Can. Vet. J.* 50, 389–392.

761. Gharekhani, J., Tavoosidana, G. R., Zandieh, M. 2013. Seroprevalence of *Neospora caninum* in sheep from Western Iran. *Vet. World* 6, 709–710.

762. Gharekhani, J., Tavoosidana, G. R., Naderisefat, G. R. 2013. Seroprevalence of *Neospora* infection in horses and donkeys in Hamedan province, Western Iran. *Vet. World* 6, 620–622.

763. Gharekhani, J., Heidari, H. 2014. Serology based comprehensive study of *Neospora* infection in domestic animals in Hamedan province, Iran. *J. Adv. Vet. Anim. Res.* 1, 119–124.

764. Gharekhani, J., Haddadzadeh, H., Bahonar, A. 2015. Prevalence of immunoglobulin G (IgG) antibody to *Neospora caninum* in dairy cattle of Hamedan province, west of Iran. *Vet. Res. Forum* 5, 149–152.

765. Ghattof, H. H., Faraj, A. A. 2015. Seroprevalence of *Neospora caninum* in goats in Wasit Province—Iraq. *Int. J. Curr. Microbiol. App. Sci.* 4, 182–191.

766. Gibney, E. H., Kipar, A., Rosbottom, A., Guy, C. S., Smith, R. F., Hetzel, U., Trees, A. J., Williams, D. J. L. 2008. The extent of parasite-associated necrosis in the placenta and foetal tissues of cattle following *Neospora caninum* infection in early and late gestation correlates with foetal death. *Int. J. Parasitol.* 38, 579–588.

767. Gibson, A. K., Raverty, S., Lambourn, D. M., Huggins, J., Magargal, S. L., Grigg, M. E. 2011. Polyparasitism is associated with increased disease severity in *Toxoplasma gondii*-infected marine sentinel species. *PLoS Negl. Trop. Dis.* 5(5), e1142.

768. Giraldi, J. H., Bracarense, A. P. F. R. L., Vidotto, O., Tudury, E. A., Navarro, I. T., Batista, T. N. 2002. Sorologia e histopatologia de *Toxoplasma gondii* e *Neospora caninum* em cães portadores de distúrbios neurológicos. *Semina Ci. Agrárias* 23, 9–14.

769. Givens, M. D. 2006. A clinical, evidence-based approach to infectious causes of infertility in beef cattle. *Theriogenology* 66, 648–654.

770. Gjerde, B. 2013. Characterisation of full-length mitochondrial copies and partial nuclear copies (numts) of the cytochrome b and cytochrome c oxidase subunit I genes of *Toxoplasma gondii*, *Neospora caninum*, *Hammondia heydorni* and *Hammondia triffittae* (Apicomplexa: Sarcocystidae). *Parasitol. Res.* 112, 1493–1511.

771. Gleeson, M. T., Johnson, A. M. 1999. Physical characterisation of the plastid DNA in *Neospora caninum*. *Int. J. Parasitol.* 29, 1563–1573.

772. Gökçe, G., Mor, N., Kirmizigül, A. H., Bozukluhan, K., Erkiliç, E. E. 2015. The first report of seropositivity for *Neospora caninum* in sheep from Turkey. *Israel J. Vet. Med.* 70, 40–44.

772a. Gomez, F., Massone, A., Mouras, F., Basso, W., Schares, G., del Amo, A. 2011. Gastric nodules with *Neospora caninum* zooites in a confirmed clinical *Neospora caninum* infection in a dog. *Analecta. Vet.* 31, 25–28.

773. Gonçalves, I. N., Uzêda, R. S., Lacerda, G. A., Moreira, R. R. N., Araújo, F. R., Oliveira, R. H. M., Corbellini, L. G., Gondim, L. F. P. 2012. Molecular frequency and isolation of cyst-forming coccidia from free ranging chickens in Bahia State, Brazil. *Vet. Parasitol.* 190, 74–79.

774. Gondim, L. F. P., Wolf, A., Vrhovec, M. G., Pantchev, N., Bauer, C., Langenmayer, M. C., Bohne, W., Teifke, J. P., Dubey, J. P., Conraths, F. J., Schares, G. 2016. Characterization of an IgG monoclonal antibody targeted to both tissue cyst and sporocyst walls of *Toxoplasma gondii*. *Exp. Parasitol.* 163, 46–56.

775. Gondim, L. F. P., Sartor, I. F., Monteiro, L. A., Haritani, M. 1999. *Neospora caninum* infection in an aborted bovine foetus in Brazil. *N. Z. Vet. J.* 47, 35.

776. Gondim, L. F. P., Sartor, I. F., Hasegawa, M., Yamane, I. 1999. Seroprevalence of *Neospora caninum* in dairy cattle in Bahia, Brazil. *Vet. Parasitol.* 86, 71–75.

777. Gondim, L. F. P., Souza, R. M., Guimaraes, J. E., Almeida, M. Â. O. 1999. Frequência de anticorpos contra *Neospora caninum* em búfalos criados no Estado da Bahia. XI Seminário Brasileiro de Parasitolgia Veterinária. Anais/Colégio Brasileiro de Parasitologia Veterinária. Salvador, Brazil. 227–227.

778. Gondim, L. F. P., Saeki, H., Onaga, H., Haritani, M., Yamane, I. 1999. Maintenance of *Neospora caninum* tachyzoites using Mongolian gerbils (*Meriones unguiculatus*). *N. Z. Vet. J.* 47, 36.

779. Gondim, L. F. P., Pinheiro, A. M., Santos, P. O. M., Jesus, E. E. V., Ribeiro, M. B., Fernandes, H. S., Almeida, M. A. O., Freire, S. M., Meyer, R., McAllister, M. M. 2001. Isolation of *Neospora caninum* from the brain of a naturally infected dog, and production of encysted bradyzoites in gerbils. *Vet. Parasitol.* 101, 1–7.

780. Gondim, L. F. P., Gao, L., McAllister, M. M. 2002. Improved production of *Neospora caninum* oocysts, cyclical oral transmission between dogs and cattle, and *in vitro* isolation from oocysts. *J. Parasitol.* 88, 1159–1163.

781. Gondim, L. F. P., McAllister, M. M., Pitt, W. C., Zemlicka, D. E. 2004. Coyotes (*Canis latrans*) are definitive hosts of *Neospora caninum*. *Int. J. Parasitol.* 34, 159–161.

782. Gondim, L. F. P., Laski, P., Gao, L., McAllister, M. M. 2004. Variation of the internal transcribed spacer 1 sequence within individual strains and among different strains of *Neospora caninum*. *J. Parasitol.* 90, 119–122.

783. Gondim, L. F. P., McAllister, M. M., Mateus-Pinilla, N. E., Pitt, W. C., Mech, L. D., Nelson, M. E. 2004. Transmission of *Neospora caninum* between wild and domestic animals. *J. Parasitol.* 90, 1361–1365.

784. Gondim, L. F. P., McAllister, M. M., Anderson-Sprecher, R. C., Björkman, C., Lock, T. F., Firkins, L. D., Gao, L., Fischer, W. R. 2004. Transplacental transmission and abortion in cows administered *Neospora caninum* oocysts. *J. Parasitol.* 90, 1394–1400.

785. Gondim, L. F. P., McAllister, M. M., Gao, L. 2005. Effects of host maturity and prior exposure history on the production of *Neospora caninum* oocysts by dogs. *Vet. Parasitol.* 134, 33–39.

786. Gondim, L. F. P. 2006. *Neospora caninum* in wildlife. *Trends Parasitol.* 22, 247–252.

787. Gondim, L. F. P., Pinheiro, A. M., Almeida, M. A. O. 2007. Frequência de anticorpos anti-*Neospora caninum* em búfalos (*Bubalus bubalis*) criados no estado da Bahia. *Rev. Bras. Saúde Prod. Anim.* 8, 92–96.

788. Gondim, L. F. P., Lindsay, D. S., McAllister, M. M. 2009. Canine and bovine *Neospora caninum* control sera examined for cross-reactivity using *Neospora caninum* and *Neospora hughesi* indirect fluorescent antibody tests. *J. Parasitol.* 95, 86–88.

789. Gondim, L. S. Q., Abe-Sandes, K., Uzêda, R. S., Silva, M. S. A., Santos, S. L., Mota, R. A., Vilela, S. M. O., Gondim, L. F. P. 2010. *Toxoplasma gondii* and *Neospora caninum* in sparrows (*Passer domesticus*) in the Northeast of Brazil. *Vet. Parasitol.* 168, 121–124.

790. González, L., Buxton, D., Atxaerandio, R., Aduriz, G., Maley, S., Marco, J. C., Cuervo, L. A. 1999. Bovine abortion associated with *Neospora caninum* in northern Spain. *Vet. Rec.* 144, 145–150.

791. González-Warleta, M., Castro-Hermida, J. A., Carro-Corral, C., Cortizo-Mella, J., Mezo, M. 2008. Epidemiology of neosporosis in dairy cattle in Galicia (NW Spain). *Parasitol. Res.* 102, 243–249.

792. González-Warleta, M., Castro-Hermida, J. A., Carro-Corral, C., Mezo, M. 2011. Anti-*Neospora caninum* antibodies in milk in relation to production losses in dairy cattle. *Prev. Vet. Med.* 101, 58–64.

793. González-Warleta, M., Castro-Hermida, J. A., Regidor-Cerrillo, J., Benavides, J., Álvarez-García, G., Fuertes, M., Ortega-Mora, L. M., Mezo, M. 2014. *Neospora caninum* infection as a cause of reproductive failure in a sheep flock. *Vet. Res.* 45, 88.

794. Goodswen, S. J., Kennedy, P. J., Ellis, J. T. 2013. A review of the infection, genetics, and evolution of *Neospora caninum*: From the past to the present. *Infect. Genet. Evol.* 13, 133–150.

795. Goodswen, S. J., Kennedy, P. J., Ellis, J. T. 2014. Discovering a vaccine against neosporosis using computers: Is it feasible? *Trends Parasitol.* 30, 401–411.

796. Goodswen, S. J., Barratt, J. L. N., Kennedy, P. J., Ellis, J. T. 2015. Improving the gene structure annotation of the apicomplexan parasite *Neospora caninum* fulfils a vital requirement towards an in silico-derived vaccine. *Int. J. Parasitol.* 45, 305–318.

797. Gottstein, B., Hentrich, B., Wyss, R., Thür, B., Busato, A., Stärk, K. D. C., Müller, N. 1998. Molecular and immunodiagnostic investigations on bovine neosporosis in Switzerland. *Int. J. Parasitol.* 28, 679–691.

798. Gottstein, B., Hentrich, B., Wyss, R., Thür, B., Bruckner, L., Müller, N., Kaufmann, H., Waldvogel, A. 1999. Molekular- und immundiagnostische Untersuchungen zur bovinen Neosporose in der Schweiz. *Schweiz. Arch. Tierheilkd.* 141, 59–68.

799. Gottstein, B., Eperon, S., Dai, W. J., Cannas, A., Hemphill, A., Greif, G. 2001. Efficacy of toltrazuril and ponazuril against experimental *Neospora caninum* infection in mice. *Parasitol. Res.* 87, 43–48.

800. Gottstein, B., Razmi, G. R., Ammann, P., Sager, H., Müller, N. 2005. Toltrazuril treatment to control diaplacental *Neospora caninum* transmission in experimentally infected pregnant mice. *Parasitology* 130, 41–48.

801. Goździk, K., Cabaj, W. 2007. Characterization of the first Polish isolate of *Neospora caninum* from cattle. *Acta Parasitol.* 52, 295–297.

802. Goździk, K., Jakubek, E. B., Björkman, C., Bie , J., Moskwa, B., Cabaj, W. 2010. Seroprevalence of *Neospora caninum* in free living and farmed red deer (*Cervus elaphus*) in Poland. *Pol. J. Vet. Sci.* 13, 117–120.

803. Goździk, K., Wrzesień, R., Wielgosz-Ostolska, A., Bien, J., Kozak-Ljunggren, M., Cabaj, W. 2011. Prevalence of antibodies against *Neospora caninum* in dogs from urban areas in Central Poland. *Parasitol. Res.* 108, 991–996.

804. Graham, D. A., Smyth, J. A., McLaren, I. E., Ellis, W. A. 1996. Stillbirth/perinatal weak calf syndrome: Serological examination for evidence of *Neospora caninum* infection. *Vet. Rec.* 139, 523–524.

805. Graham, D. A., Calvert, V., Whyte, M., Marks, J. 1999. Absence of serological evidence for human *Neospora caninum* infection. *Vet. Rec.* 144, 672–673.

806. Grandi, G., Genchi, C., Bazzocchi, C., Mortarino, M., Borghetti, P., De Angelis, E., Kramer, L. H. 2008. Generation and infection of bovine PBMC-derived dendritic cells with *Neospora caninum*. *Vet. Res. Commun.* 32(Suppl 1), S207–S209.

807. Gray, M. L., Harmon, B. G., Sales, L., Dubey, J. P. 1996. Visceral neosporosis in a 10-year-old horse. *J. Vet. Diagn. Invest.* 8, 130–133.

808. Greca, H., Silva, A. V., Langoni, H. 2010. Associação entre a presença de anticorpos anti-*Leishmania* sp. e anti-*Neospora caninum* cães de Bauru, SP. *Arq. Bras. Med. Vet. Zootec.* 62, 224–227.

809. Greene, C. E., Cook, J. R., Mahaffey, E. A. 1985. Clindamycin for treatment of *Toxoplasma* polymyositis in a dog. *J. Am. Vet. Med. Assoc.* 187, 631–634.

810. Greif, G., Harder, A., Haberkorn, A. 2001. Chemotherapeutic approaches to protozoa: Coccidiae–current level of knowledge and outlook. *Parasitol. Res.* 87, 973–975.

811. Greig, B., Rossow, K. D., Collins, J. E., Dubey, J. P. 1995. *Neospora caninum* pneumonia in an adult dog. *J. Am. Vet. Med. Assoc.* 206, 1000–1001.

812. Greiner, M., Gardner, I. A. 2000. Application of diagnostic tests in veterinary epidemiologic studies. *Prev. Vet. Med.* 45, 43–59.

813. Guarino, A., Fusco, G., Savini, G., Di Francesco, G., Cringoli, G. 2000. Neosporosis in water buffalo (*Bubalus bubalis*) in southern Italy. *Vet. Parasitol.* 91, 15–21.

814. Guedes, M. H. P., Guimarães, A. M., Rocha, C. M. B. M., Hirsch, C. 2008. Frequência de anticorpos anti-*Neospora caninum* em vacas e fetos provenientes de municípios do sul de Minas Gerais. *Rev. Bras. Parasitol. Vet.* 17, 189–194.

815. Guido, S., Katzer, F., Nanjiani, I., Milne, E., Innes, E. A. 2016. Serology-based diagnostics for the control of bovine neosporosis. *Trends Parasitol.* 32, 131–143.

816. Guimarães, A., Raimundo, J. M., Moraes, L. M. B., Silva, A. T., Santos, H. A., Pires, M. S., Machado, R. Z., Baldani, C. D. 2015. Occurrences of anti-*Toxoplasma gondii* and anti-*Neospora caninum* antibodies in sheep from four districts of Tocantins state, Brazilian Legal Amazon Region. *Pesq. Vet. Bras.* 35, 110–114.

817. Guimarães, A. M., Rocha, C. M. B. M., Oliveira, T. M. F. S., Rosado, I. R., Morais, L. G., Santos, R. R. D. 2009. Fatores associados à soropositividade para *Babesia*, *Toxoplasma*, *Neospora* e *Leishmania* em cães atendidos em nove clínicas veterinárias do município de Lavras, MG. *Rev. Bras. Parasitol. Vet.* 18(Suppl 1), 49–53.

818. Guimarães, J. S., Souza, S. L. P., Bergamaschi, D. P., Gennari, S. M. 2004. Prevalence of *Neospora caninum* antibodies and factors associated with their presence in dairy cattle of the north of Paraná state, Brazil. *Vet. Parasitol.* 124, 1–8.

819. Guionaud, C., Hemphill, A., Mevissen, M., Alaeddine, F. 2010. Molecular characterization of *Neospora caninum* MAG1, a dense granule protein secreted into the parasitophorous vacuole, and associated with the cyst wall and the cyst matrix. *Parasitology* 137, 1605–1619.

820. Gunning, R. F., Gumbrell, R. C., Jeffrey, M. 1994. *Neospora* infection and congenital ataxia in calves. *Vet. Rec.* 134, 558.

821. Guo, Z. G., Johnson, A. M. 1995. Genetic comparison of *Neospora caninum* with *Toxoplasma* and *Sarcocystis* by random amplified polymorphic DNA-polymerase chain reaction. *Parasitol. Res.* 81, 365–370.

822. Gupta, A., Stroup, S., Dedeaux, A., Bauer, R. W., Gaunt, S. D. 2011. What is your diagnosis? Fine-needle aspirate of ulcerative skin lesions in a dog. *Vet. Clin. Pathol.* 40, 401–402.

823. Gupta, G. D., Lakritz, J., Kim, J. H., Kim, D. Y., Kim, J. K., Marsh, A. E. 2002. Seroprevalence of *Neospora, Toxoplasma gondii*, and *Sarcocystis neurona* antibodies in horses from Jeju island, South Korea. *Vet. Parasitol.* 106, 193–201.

824. Gutiérrez-Expósito, D., Ortega-Mora, L. M., Gajadhar, A. A., García-Lunar, P., Dubey, J. P., Álvarez García, G. 2012. Serological evidence of *Besnoitia* spp. infection in Canadian wild ruminants and strong cross-reaction between *Besnoitia besnoiti* and *Besnoitia tarandi*. *Vet. Parasitol.* 190, 19–28.

825. Guy, C. S., Williams, D. J. L., Kelly, D. F., McGarry, J. W., Guy, F., Björkman, C., Smith, R. F., Trees, A. J. 2001. *Neospora caninum* in persistently infected, pregnant cows: Spontaneous transplacental infection is associated with an acute increase in maternal antibody. *Vet. Rec.* 149, 443–449.

826. Haddad, J. P. A., Dohoo, I. R., VanLeewen, J. A. 2005. A review of *Neospora caninum* in dairy and beef cattle—A Canadian perspective. *Can. Vet. J.* 46, 230–243.

827. Haddadzadeh, H. R., Sadrebazzaz, A., Malmasi, A., Ardakani, H. T., Nia, P. K., Sadreshirazi, N. 2007. Seroprevalence of *Neospora caninum* infection in dogs from rural and urban environments in Tehran, Iran. *Parasitol. Res.* 101, 1563–1565.

828. Haerdi, C., Haessig, M., Sager, H., Greif, G., Staubli, D., Gottstein, B. 2006. Humoral immune reaction of newborn calves congenitally infected with *Neospora caninum* and experimentally treated with toltrazuril. *Parasitol. Res.* 99, 534–540.

829. Hain, A. U., Miller, A. S., Levitskaya, J., Bosch, J. 2016. Virtual screening and experimental validation identify novel inhibitors of the *Plasmodium falciparium* Atg8–Atg3 protein–protein interaction. *Chem. Med. Chem.*

830. Hajikolaei, M. R. H., Goraninejad, S., Hamidinejat, H., Ghorbanpour, M., Paryab, R. 2007. Occurrence of *Neospora caninum* antibodies in water buffaloes (*Bubalus bulalis*) from the south-western region of Iran. *Bull. Vet. Inst. Pulawy* 51, 233–235.

831. Haldorson, G. J., Mathison, B. A., Wenberg, K., Conrad, P. A., Dubey, J. P., Trees, A. J., Yamane, I., Baszler, T. V. 2005. Immunization with native surface protein NcSRS2 induces a Th2 immune response and reduces congenital *Neospora caninum* transmission in mice. *Int. J. Parasitol.* 35, 1407–1415.

832. Haldorson, G. J., Stanton, J. B., Mathison, B. A., Suarez, C. E., Baszler, T. V. 2006. *Neospora caninum*: Antibodies directed against tachyzoite surface protein NcSRS2 inhibit parasite attachment and invasion of placental trophoblasts *in vitro*. *Exp. Parasitol.* 112, 172–178.

833. Hall, C. A., Reichel, M. P., Ellis, J. T. 2005. *Neospora* abortions in dairy cattle: Diagnosis, mode of transmission and control. *Vet. Parasitol.* 128, 231–241.

834. Hall, C. A., Reichel, M. P., Ellis, J. T. 2006. Performance characteristics and optimisation of cut-off values of two enzyme-linked immunosorbent assays for the detection of antibodies to *Neospora caninum* in the serum of cattle. *Vet. Parasitol.* 140, 61–68.

835. Hall, C. A., Reichel, M. P., Ellis, J. T. 2006. Prevalence of *Neospora caninum* infection in Australian (NSW) dairy cattle estimated by a newly validated ELISA for milk. *Vet. Parasitol.* 142, 173–178.

836. Hamada, H., Petrino, M. G., Kakunaga, T. 1982. A novel repeated element with Z-DNA-forming potential is widely found in evolutionarily diverse eukaryotic genomes. *Proc. Natl. Acad. Sci. USA* 79, 6465–6469.

837. Hamel, D., Shukullari, E., Rapti, D., Silaghi, C., Pfister, K., Rehbein, S. 2016. Parasites and vector-borne pathogens in client-owned dogs in Albania. Blood pathogens and seroprevalences of parasitic and other infectious agents. *Parasitol. Res.* 115, 489–499.

838. Hamidinejat, H., Mosalanejad, B., Jalali, M. H. R., Avizeh, R., Ghorbanpour, M. 2011. Prevalence of *Neospora caninum* in domestic cats from Ahvaz, Iran. Animal hygiene and sustainable livestock production. *Proceedings of the XVth International Congress of the International Society for Animal Hygiene.* July 3–7, Vienna, Austria, Volume 2, 899–900.

839. Hamidinejat, H., Mosalanejad, B., Avizeh, R., Jalali, M. H. R., Ghorbanpour, M., Namavari, M. 2011. *Neospora caninum* and *Toxoplasma gondii* antibody prevalence in Ahvaz feral cats, Iran. *Jundishapur J. Microbiol.* 4, 217–222.

840. Hamidinejat, H., Haji Hajikolaei, M. R., Ghorbanpoor, M., Namavari, M., Gol, S. M. A. 2013. Development and standardization of Dot-ELISA for detection of *Neospora caninum* antibodies in cattle and comparison with standard indirect ELISA and direct aggultination test (DAT). *Iranian J. Parasitol.* 8, 634–640.

841. Hamidinejat, H., Ghorbanpour, M., Rasooli, A., Nouri, M., Hekmatimoghaddam, S., Namavari, M. M., Pourmehdi-Borojeni, M., Sazmand, A. 2013. Occurrence of anti-*Toxoplasma gondii* and *Neospora caninum* antibodies in camels (*Camelus dromediarius*) in the center of Iran. *Turk. J. Vet. Anim. Sci.* 37, 277–281.

842. Hamidinejat, H., Seifi Abad Shapouri, M. R., Namavari, M. M., Shayan, P., Kefayat, M. 2015. Development of an indirect ELISA using different fragments of recombinant Ncgra7 for detection of *Neospora caninum* infection in cattle and water buffalo. *Iranian J. Parasitol.* 10, 69–77.

843. Hamilton, C. M., Gray, R., Wright, S. E., Gangadharan, B., Laurenson, K., Innes, E. A. 2005. Prevalence of antibodies to *Toxoplasma gondii* and *Neospora caninum* in red foxes (*Vulpes vulpes*) from around the UK. *Vet. Parasitol.* 130, 169–173.

844. Hamir, A. N., Tornquist, S. J., Gerros, T. C., Topper, M. J., Dubey, J. P. 1998. *Neospora caninum*-associated equine protozoal myeloencephalitis. *Vet. Parasitol.* 79, 269–274.

845. Han, F., Liu, J., Nan, H., Liu, Q. 2015. [Serologic investigation of neosporosis in a dairy farm in Beijing]. *Chin. J. Vet. Med.* 51, 50–52 (in Chinese).

846. Han, F., Fu, Y., Zhang, C., Liu, Q., Xu, J., Liu, J. 2016. Prevalence of antibodies against *Neospora caninum* in Père David's deer (*Elaphurus davidianus*) in Beijing, China. *J. Wildl. Dis.* 52, 387–390.

847. Han, J. I., Chang, D. W., Na, K. J. 2015. A multiplex quantitative real-time polymerase chain reaction panel for detecting neurologic pathogens in dogs with meningoencephalitis. *J. Vet. Sci.* 16, 341–347.

848. Hancock, K., Tsang, V. C. W. 1983. India ink staining of proteins on nitrocellulose paper. *Anal. Biochem.* 133, 157–162.

849. Hao, P., Yang, N., Cui, X., Liu, J., Yang, D., Liu, Q. 2014. First isolation of *Neospora caninum* blood of a naturally infected adult dairy cow in Beijing, China. *J. Parasitol.* 100, 812–816.

850. Hara, O. A., Liao, M., Baticados, W., Bannai, H., Zhang, G., Zhang, S., Lee, E., Nishikawa, Y., Claveria, F., Igarashi, M., Nagasawa, H., Xuan, X. 2006. Expression of recombinant dense granule protein 7 of *Neospora caninum* and evaluation of its diagnostic potential for canine neosporosis. *J. Protozool. Res.* 16, 34–41.

851. Haritani, M. 1996. Research on bovine *Neospora* infection. *J. Vet. Med.* 49, 857–859.

852. Harkins, D., Clements, D. N., Maley, S., Marks, J., Wright, S., Esteban, I., Innes, E. A., Buxton, D. 1998. Western blot analysis of the IgG responses of ruminants infected with *Neospora caninum* and with *Toxoplasma gondii*. *J. Comp. Pathol.* 119, 45–55.

853. Harmelin, A., Perl, S., Nyska, A., Yakobson, B., Shpigel, N., Orgad, U., Dubey, J. P. 1995. Neosporosis-associated bovine abortion in Israel. *Vet. Rec.* 136, 80.

854. Hartley, W. J., Bridge, P. S. 1975. A case of suspected congenital *Toxoplasma* encephalomyelitis in a lamb associated with a spinal cord anomaly. *Br. Vet. J.* 131, 380–384.

855. Hasegawa, M. Y., Sartor, I. F., Canavessi, A. M. O., Pinckney, R. D. 2004. Occurrence of *Neospora caninum* antibodies in beef cattle and in farm dogs from Avaré Region of São Paulo, Brazil [in Portuguese]. *Semina Ci. Agrárias* 25, 370–384.

856. Häsler, B., Stärk, K. D. C., Sager, H., Gottstein, B., Reist, M. 2006. Simulating the impact of four control strategies on the population dynamics of *Neospora caninum* infection in Swiss dairy cattle. *Prev. Vet. Med.* 77, 254–283.

857. Häsler, B., Regula, G., Stärk, K. D. C., Sager, H., Gottstein, B., Reist, M. 2006. Financial analysis of various strategies for the control of *Neospora caninum* in dairy cattle in Switzerland. *Prev. Vet. Med.* 77, 230–253.

858. Häsler, B., Hernandez, J. A., Reist, M., Sager, H., Steiner-Moret, C., Staubli, D., Stärk, K. D. C., Gottstein, B. 2006. *Neospora caninum*: Serological follow-up in dairy cows during pregnancy. *Vet. Parasitol.* 137, 222–230.

859. Häsler, B., Stärk, K., Gottstein, B., Reist, M. 2008. Epidemiologische und finanzielle Entscheidungsgrundlagen zur Kontrolle von *Neospora caninum* auf Schweizer Milchviehbetrieben. *Schweiz. Arch. Tierheilkd.* 150, 273–280.

860. Hässig, M., Gottstein, B. 2002. Epidemiological investigations of abortions due to *Neospora caninum* on Swiss dairy farms. *Vet. Rec.* 150, 538–542.

861. Hässig, M., Sager, H., Reitt, K., Ziegler, D., Strabel, D., Gottstein, B. 2003. *Neospora caninum* in sheep: A herd case report. *Vet. Parasitol.* 117, 213–220.

862. Hattel, A. L., Castro, M. D., Gummo, J. D., Weinstock, D., Reed, J. A., Dubey, J. P. 1998. Neosporosis-associated bovine abortion in Pennsylvania. *Vet. Parasitol.* 74, 307–313.

863. Hay, W. H., Shell, L. G., Lindsay, D. S., Dubey, J. P. 1990. Diagnosis and treatment of *Neospora caninum* infection in a dog. *J. Am. Vet. Med. Assoc.* 197, 87–89.

864. He, P., Li, J., Gong, P., Liu, C., Zhang, G., Yang, J., Tuo, W., Yang, B., Zhang, X. 2013. *Neospora caninum* surface antigen (p40) is a potential diagnostic marker for cattle neosporosis. *Parasitol. Res.* 112, 2117–2120.

865. Hecker, Y. P., Moore, D. P., Manazza, J. A., Unzaga, J. M., Späth, E. J. A., Pardini, L. L., Venturini, M. C., Roberi, J. L., Campero, C. M. 2013. First report of seroprevalence of *Toxoplasma gondii* and *Neospora caninum* in dairy sheep from Humid Pampa, Argentina. *Trop. Anim. Health Prod.* 45, 1645–1647.

866. Hecker, Y. P., Moore, D. P., Quattrocchi, V., Regidor-Cerrillo, J., Verna, A., Leunda, M. R., Morrell, E., Ortega-Mora, L. M., Zamorano, P., Venturini, M. C., Campero, C. M. 2013. Immune response and protection provided by live tachyzoites and native antigens from the NC-6 Argentina strain of *Neospora caninum* in pregnant heifers. *Vet. Parasitol.* 197, 436–446.

867. Hecker, Y. P., Cóceres, V., Wilkowsky, S. E., Ortiz, J. M. J., Morrell, E. L., Verna, A. E., Ganuza, A., Cano, D. B., Lischinsky, L., Ángel, S. O., Zamorano, P., Odeón, A. C., Leunda, M. R., Campero, C. M., Morein, B., Moore, D. P. 2014. A *Neospora caninum* vaccine using recombinant proteins fails to prevent foetal infection in pregnant cattle after experimental intravenous challenge. *Vet. Immunol. Immunopathol.* 162, 142–153.

868. Hecker, Y. P., Cantón, G., Regidor-Cerrillo, J., Chianini, F., Morrell, E., Lischinsky, L., Ortega-Mora, L. M., Innes, E. A., Odeón, A., Campero, C. M., Moore, D. P. 2015. Cell mediated immune responses in the placenta following challenge of vaccinated pregnant heifers with *Neospora caninum*. *Vet. Parasitol.* 214, 247–254.

869. Heckeroth, A. R., Tenter, A. M., Hemphill, A., Innes, E. A., Buxton, D. 2000. Evaluation of cross-reactivity between tachyzoites of *Neospora caninum* and *Toxoplasma gondii* by two-dimensional gel electrophoresis. *Int. J. Parasitol.* 30, 913–919.

870. Heckeroth, A. R., Tenter, A. M. 2007. Immunoanalysis of three litters born to a Doberman bitch infected with *Neospora caninum*. *Parasitol. Res.* 100, 837–846.

871. Heidari, H., Mohammadzadeh, A., Gharekhani, J. 2014. Seroprevalence of *Neospora caninum* in slaughtered native cattle in Kurdistan province, Iran. *Vet. Res. Forum* 5, 69–72.

872. Hein, H. E., Machado, G., Miranda, I. C. S., Costa, E. F., Pellegrini, D. C. P., Driemeier, D., Corbellini, L. G. 2012. Neosporose bovina: Avaliação da transmissão vertical e fração atribuível de aborto em uma população de bovinos no Estado do Rio Grande do Sul. *Pesq. Vet. Bras.* 32, 396–400.

873. Heintzelman, M. B., Schwartzman, J. D. 2001. Myosin diversity in Apicomplexa. *J. Parasitol.* 87, 429–432.

874. Helman, R. G., Stair, E. L., Lehenbauer, T. W., Rodgers, S., Saliki, J. T. 1998. Neosporal abortion in Oklahoma cattle with emphasis on the distribution of brain lesions in aborted fetuses. *J. Vet. Diagn. Invest.* 10, 292–295.

875. Helmick, B., Otter, A., McGarry, J., Buxton, D. 2002. Serological investigation of aborted sheep and pigs for infection by *Neospora caninum*. *Res. Vet. Sci.* 73, 187–189.

876. Hemphill, A. 1996. Subcellular localization and functional characterization of Nc-p43, a major *Neospora caninum* tachyzoite surface protein. *Infect. Immun.* 64, 4279–4287.

877. Hemphill, A., Gottstein, B. 1996. Identification of a major surface protein on *Neospora caninum* tachyzoites. *Parasitol. Res.* 82, 497–504.

878. Hemphill, A., Gottstein, B., Kaufmann, H. 1996. Adhesion and invasion of bovine endothelial cells by *Neospora caninum*. *Parasitology* 112, 183–197.

879. Hemphill, A., Fuchs, N., Sonda, S., Gottstein, B., Hentrich, B. 1997. Identification and partial characterization of a 36 kDa surface protein on *Neospora caninum* tachyzoites. *Parasitology* 115, 371–380.

880. Hemphill, A., Felleisen, R., Connolly, B., Gottstein, B., Hentrich, B., Müller, N. 1997. Characterization of a cDNA-clone encoding Nc-p43, a major *Neospora caninum* tachyzoite surface protein. *Parasitology* 115, 581–590.

881. Hemphill, A., Gajendran, N., Sonda, S., Fuchs, N., Gottstein, B., Hentrich, B., Jenkins, M. 1998. Identification and characterisation of a dense granule-associated protein in *Neospora caninum* tachyzoites. *Int. J. Parasitol.* 28, 429–438.

882. Hemphill, A. 1999. The host-parasite relationship in neosporosis. *Adv. Parasitol.* 43, 47–104.

883. Hemphill, A., Fuchs, N., Sonda, S., Hehl, A. 1999. The antigenic composition of *Neospora caninum*. *Int. J. Parasitol.* 29, 1175–1188.

884. Hemphill, A., Müller, N., Sager, H., Gottstein, B. 2000. [*Neospora caninum* and neosporosis—Basic science at the Institute of Parasitology and possible implications] (in German). *Schweiz. Arch. Tierheilkd.* 142, 257–261.

885. Hemphill, A., Vonlaufen, N., Naguleswaran, A., Keller, N., Riesen, M., Guetg, N., Srinivasan, S., Alaeddine, F. 2004. Tissue culture and explant approaches to studying and visualizing *Neospora caninum* and its interactions with the host cell. *Microsc. Microanal.* 10, 602–620.

886. Hemphill, A., Gottstein, B. 2006. *Neospora caninum* and neosporosis—Recent achievements in host and parasite cell biology and treatment. *Acta Parasitol.* 51, 15–25.

887. Hemphill, A., Vonlaufen, N., Naguleswaran, A. 2006. Cellular and immunological basis of the host-parasite relationship during infection with *Neospora caninum*. *Parasitology* 133, 261–278.

888. Hemphill, A. 2007. Generation of parasite cysts in cultured cells instead of living animals. *ALTEX* 24 Spec No, 29–31.

889. Hemphill, A., Gottstein, B., Conraths, F. J., de Meerschman, F., Ellis, J. T., Innes, E. A., McAllister, M. M., Ortega-Mora, L. M., Tenter, A. M., trees, A. J., Williums, D. J. L., Wouda, W. 1999. A European perspective on *Neospora caninum*. *Int. J. Parasitol.* 30, 877–924.

890. Hemphill, A., Vonlaufen, N., Golaz, J. L., Burgener, I. A. 2009. Infection of primary canine duodenal epithelial cell cultures with *Neospora caninum*. *J. Parasitol.* 95, 372–380.

891. Hemphill, A., Debache, K., Monney, T., Schorer, M., Guionaud, C., Alaeddine, F., Mueller, N., Mueller, J. 2013. Proteins mediating the *Neospora caninum*-host cell interaction as targets for vaccination. *Front. Biosci.* 5, 23–36.

892. Hemphill, A., Aguado-Martínez, A., Müller, J. 2016. Approaches for the vaccination and treatment of *Neospora caninum* infections in mice and ruminant models. *Parasitology* 143, 245–259.

893. Henning, K., Schares, G., Granzow, H., Polster, U., Hartmann, M., Hotzel, H., Sachse, K., Peters, M., Rauser, M. 2002. *Neospora caninum* and *Waddlia chondrophila* strain 2032/99 in a septic stillborn calf. *Vet. Microbiol.* 85, 285–292.

894. Henriquez, F. L., Nickdel, M. B., McLeod, R., Lyons, R. E., Lyons, K., Dubremetz, J. F., Grigg, M. E., Samuel, B. U., Roberts, C. W. 2005. *Toxoplasma gondii* dense granule protein 3 (GRA3) is a type I transmembrane protein that possesses a cytoplasmic dilysine (KKXX) endoplasmic reticulum (ER) retrieval motif. *Parasitology* 131, 169–179.

895. Herman, R. K., Molestina, R. E., Sinai, A. P., Howe, D. K. 2007. The apicomplexan pathogen *Neospora caninum* inhibits host cell apoptosis in the absence of discernible NF-κB activation. *Infect. Immun.* 75, 4255–4262.

896. Hermosilla, C., Silva, L. M. R., Kleinertz, S., Prieto, R., Silva, M. A., Taubert, A. 2016. Endoparasite survey of free-swimming baleen whales (*Balaenoptera musculus*, *B. physalus*, *B. borealis*) and sperm whales (*Physeter macrocephalus*) using non/minimally invasive methods. *Parasitol. Res.* 115, 889–896.

897. Hernandez, J., Risco, C., Donovan, A. 2001. Association between exposure to *Neospora caninum* and milk production in dairy cows. *J. Am. Vet. Med. Assoc.* 219, 632–635.

898. Hernandez, J., Risco, C., Donovan, A. 2002. Risk of abortion associated with *Neospora caninum* during different lactations and evidence of congenital transmission in dairy cows. *J. Am. Vet. Med. Assoc.* 221, 1742–1746.

899. Herrmann, D. C., Pantchev, N., Vrhovec, M. G., Barutzki, D., Wilking, H., Fröhlich, A., Lüder, C. G., Conraths, F. J., Schares, G. 2010. Atypical *Toxoplasma gondii* genotypes identified in oocysts shed by cats in Germany. *Int. J. Parasitol.* 40, 285–292.

900. Herrmann, D. C., Maksimov, A., Pantchev, N., Vrhovec, M. G., Conraths, F. J., Schares, G. 2011. Comparison of different commercial DNA extraction kits to detect *Toxoplasma gondii* oocysts in cat faeces. *Berl. Münch. Tierärztl. Wochenschr.* 124, 497–502.

901. Heuer, C., Nicholson, C., Russel, D., Weston, J. 2004. Field study in dairy cattle from New Zealand. In *Intervet Symposium: Bovine Neosporosis*. Schetters, T. (Editor). *Vet. Parasitol.* 125, 137–146.

902. Heydorn, A. O., Mehlhorn, H. 2002. *Neospora caninum* is an invalid species name: An evaluation of facts and statements. *Parasitol. Res.* 88, 175–184.

903. Heydorn, A. O., Mehlhorn, H. 2002. A re-evaluation of *Neospora* and *Hammondia* spp. *Trends Parasitol.* 18, 246.

904. Hiasa, J., Kohara, J., Nishimura, M., Xuan, X., Tokimitsu, H., Nishikawa, Y. 2012. ELISAs based on rNcGRA7 and rNcSAG1 antigens as an indicator of *Neospora caninum* activation. *Vet. Parasitol.* 187, 379–385.

905. Hiasa, J., Nishimura, M., Itamoto, K., Xuan, X., Inokuma, H., Nishikawa, Y. 2012. Enzyme-linked immunosorbent assays based on *Neospora caninum* dense granule protein 7 and profilin for estimating the stage of neosporosis. *Clin. Vaccine Immunol.* 19, 411–417.

906. Hietala, S. K., Thurmond, M. C. 1999. Postnatal *Neospora caninum* transmission and transient sero-logic responses in two dairies. *Int. J. Parasitol.* 29, 1669–1676.

907. Higa, A. C., Machado, R. Z., Tinucci-Costa, M., Domingues, L. M., Malheiros, E. B. 2000. Evaluation of cross-reactivity of *Toxoplasma gondii* and *Neospora caninum* antigens in dogs sera. *Rev. Bras. Parasitol. Vet.* 9, 91–95.

908. Hilali, M., Lindberg, R., Waller, T., Wallin, B. 1986. Enigmatic cyst-forming sporozoon in the spinal cord of a dog. *Acta Vet. Scand.* 27, 623–625.

909. Hilali, M., Romand, S., Thulliez, P., Kwok, O. C. H., Dubey, J. P. 1998. Prevalence of *Neospora caninum* and *Toxoplasma gondii* antibodies in sera from camels from Egypt. *Vet. Parasitol.* 75, 269–271.

910. Hill, D. E., Liddell, S., Jenkins, M. C., Dubey, J. P. 2001. Specific detection of *Neospora caninum* oocyst in fecal samples from experimentally-infected dogs using the polymerase chain reaction. *J. Parasitol.* 87, 395–398.

911. Ho, M. S. Y., Barr, B. C., Marsh, A. E., Anderson, M. L., Rowe, J. D., Tarantal, A. F., Hendrickx, A. G., Sverlow, K., Dubey, J. P., Conrad, P. A. 1996. Identification of bovine *Neospora* parasites by PCR amplification and specific small subunit rRNA sequence probe hybridization. *J. Clin. Microbiol.* 34, 1203–1208.

912. Ho, M. S. Y., Barr, B. C., Rowe, J. D., Anderson, M. L., Sverlow, K. W., Packham, A., Marsh, A. E., Conrad, P. A. 1997. Detection of *Neospora* sp. from infected bovine tissues by PCR and probe hybrid-ization. *J. Parasitol.* 83, 508–514.

913. Ho, M. S. Y., Barr, B. C., Tarantal, A. F., Lai, L. T. Y., Hendrickx, A. G., Marsh, A. E., Sverlow, K. W., Packham, A. E., Conrad, P. A. 1997. Detection of *Neospora* from tissues of experimentally infected rhesus macaques by PCR and specific DNA probe hybridization. *J. Clin. Microbiol.* 35, 1740–1745.

914. Hoane, J. S., Yeargan, M. R., Stamper, S., Saville, W. J., Morrow, J. K., Lindsay, D. S., Howe, D. K. 2005. Recombinant NhSAG1 ELISA: A sensitive and specific assay for detecting antibodies against *Neospora hughesi* in equine serum. *J. Parasitol.* 91, 446–452.

915. Hoane, J. S., Gennari, S. M., Dubey, J. P., Ribeiro, M. G., Borges, A. S., Yai, L. E. O., Aguiar, D. M., Cavalcante, G. T., Bonesi, G. L., Howe, D. K. 2006. Prevalence of *Sarcocystis neurona* and *Neospora* spp. infection in horses from Brazil based on presence of serum antibodies to parasite surface antigen. *Vet. Parasitol.* 136, 155–159.

915a. Hoar, B. R., Ribble, C. S., Spitzer, C. C., Spitzer, P. G., Janzen, E. D. 1996. Investigation of pregnancy losses in beef cattle herds associated with *Neospora* sp. infection. *Can. Vet. J.* 37, 364–366.

916. Hoar, B. R., McQuarry, A. C., Hietala, S. K. 2007. Prevalence of *Neospora caninum* and persistent infection with bovine viral diarrhea virus in dairy-breed steers in a feedlot. *J. Am. Vet. Med. Assoc.* 230, 1038–1043.

917. Hobson, J. C., Duffield, T. F., Kelton, D., Lissemore, K., Hietala, S. K., Leslie, K. E., McEwen, B., Cramer, G., Peregrine, A. S. 2002. *Neospora caninum* serostatus and milk production of Holstein cat-tle. *J. Am. Vet. Med. Assoc.* 221, 1160–1164.

918. Hobson, J. C., Duffield, T. F., Kelton, D., Lissemore, K., Hietala, S. K., Leslie, K. E., McEwen, B., Peregrine, A. S. 2005. Risk factors associated with *Neospora caninum* abortion in Ontario Holstein dairy herds. *Vet. Parasitol.* 127, 177–188.

919. Hoffmann Kormann, D. C. S., Locatelli-Dittrich, R., Richartz, R. R. T. B., Antunes, J., Dittrich, J. R., Patrício, L. F. L. 2008. Soroprevalência e cinética mensal de anticorpos anti-*Neospora* sp. em éguas gestantes. *Rev. Bras. Parasitol. Vet.* 17(Suppl 1), 335–338.

920. Holmberg, T. A., Vernau, W., Melli, A. C., Conrad, P. A. 2006. *Neospora caninum* associated with septic peritonitis in an adult dog. *Vet. Clin. Pathol.* 35, 235–238.

921. Holmdahl, J., Björkman, C., Stenlund, S., Uggla, A., Dubey, J. P. 1997. Bovine *Neospora* and *Neospora caninum*: One and the same. *Parasitol. Today* 13, 40–41.

922. Holmdahl, O. J. M., Mattsson, J. G., Uggla, A., Johansson, K. E. 1994. The phylogeny of *Neospora caninum* and *Toxoplasma gondii* based on ribosomal RNA sequences. *FEMS Microbiol. Lett.* 119, 187–192.

923. Holmdahl, O. J. M., Björkman, C., Uggla, A. 1995. A case of *Neospora* associated bovine abortion in Sweden. *Acta Vet. Scand.* 36, 279–281.

924. Holmdahl, O. J. M., Mattsson, J. G. 1996. Rapid and sensitive identification of *Neospora caninum* by *in vitro* amplification of the internal transcribed spacer 1. *Parasitology* 112, 177–182.

925. Holt, N., Murray, M., Cuddon, P. A., Lappin, M. R. 2011. Seroprevalence of various infectious agents in dogs with suspected acute canine polyradiculoneuritis. *J. Vet. Intern. Med.* 25, 261–266.

926. Homan, W. L., Limper, L., Verlaan, M., Borst, A., Vercammen, M., van Knapen, F. 1997. Comparison of the internal transcribed spacer, ITS 1, from *Toxoplasma gondii* isolates and *Neospora caninum*. *Parasitol. Res.* 83, 285–289.

927. Hoon-Hanks, L. L., Regan, D., Dubey, J. P., Porter, M. C., Duncan, C. G. 2013. Hepatic neosporosis in a dog treated for pemphigus foliaceus. *J. Vet. Diagn. Invest.* 25, 807–810.

928. Horcajo, P., Regidor-Cerrillo, J., Aguado-Martínez, A., Hemphill, A., Ortega-Mora, L. M. 2016. Vaccine for bovine neosporosis: Current status and key aspects for development. *Parasitol. Immunol.* Doi: 10.1111pim.12342.

929. Hornok, S., Näslund, K., Hajtós, I., Tanyi, J., Tekes, L., Varga, I., Uggla, A., Björkman, C. 1998. Detection of antibodies to *Neospora caninum* in bovine postabortion blood samples from Hungary. *Acta Vet. Hung.* 46, 431–436.

930. Hornok, S., Edelhofer, R., Hajtós, I. 2006. Seroprevalence of neosporosis in beef and dairy cattle breeds in northeast Hungary. *Acta Vet. Hung.* 54, 485–491.

931. Hornok, S., Edelhofer, R., Fok, É., Berta, K., Fejes, P., Répási, A., Farkas, R. 2006. Canine neosporosis in Hungary: Screening for seroconversion of household, herding and stray dogs. *Vet. Parasitol.* 137, 197–201.

932. Hornok, S., Edelhofer, R., Joachim, A., Farkas, R., Berta, K., Répási, A., Lakatos, B. 2008. Seroprevalence of *Toxoplasma gondii* and *Neospora caninum* infection of cats in Hungary. *Acta Vet. Hung.* 56, 81–88.

933. Hoskins, J. D., Bunge, M. M., Dubey, J. P., Duncan, D. E. 1991. Disseminated infection with *Neospora caninum* in a ten-year-old dog. *Cornell Vet.* 81, 329–334.

934. Hosseini, M. H., Moraveji, M., Tahamtan, Y., Rahimian, A., Mohammadi, Gh., Namavari, M. M. 2011. Seroprevalence of *Neospora* spp. in horses in north east of Iran. *Iranian J. Parasitol.* 6, 64–68.

935. Hosseininejad, M., Pirali-Kheirabadi, K., Hosseini, F. 2009. Seroprevalence of *Neospora caninum* infection in camels (*Camelus dromedarius*) in Isfahan Province, Center of Iran. *Iranian J. Parasitol.* 4, 61–64.

936. Hosseininejad, M., Hosseini, F., Mosharraf, M., Shahbaz, S., Mahzounieh, M., Schares, G. 2010. Development of an indirect ELISA test using an affinity purified surface antigen (P38) for sero-diagnosis of canine *Neospora caninum* infection. *Vet. Parasitol.* 171, 337–342.

937. Hosseininejad, M., Hosseini, F. 2011. Seroprevalence of *Neospora caninum* and *Toxoplasma gondii* infection in dogs from west and central parts of Iran using two indirect ELISA tests and assessment of associate risk factors. *Iranian J. Vet. Res.* 12, 46–51.

938. Houk, A. E., Goodwin, D. G., Zajac, A. M., Barr, S. C., Dubey, J. P., Lindsay, D. S. 2010. Prevalence of antibodies to *Trypanosoma cruzi*, *Toxoplasma gondii*, *Encephalitozoon cuniculi*, *Sarcocystis neurona*, *Besnoitia darlingi*, and *Neospora caninum* in North American opossums, *Didelphis virginiana*, from Southern Louisiana. *J. Parasitol.* 96, 1119–1122.

939. Howe, D. K., Sibley, L. D. 1997. Development of molecular genetics for *Neospora caninum*: A complementary system to *Toxoplasma gondii*. *Methods* 13, 123–133.

940. Howe, D. K., Mercier, C., Messina, M., Sibley, L. D. 1997. Expression of *Toxoplasma gondii* genes in the closely-related apicomplexan parasite *Neospora caninum*. *Mol. Biochem. Parasitol.* 86, 29–36.

941. Howe, D. K., Crawford, A. C., Lindsay, D., Sibley, L. D. 1998. The p29 and p35 immunodominant antigens of *Neospora caninum* tachyzoites are homologous to the family of surface antigens of *Toxoplasma gondii*. *Infect. Immun.* 66, 5322–5328.

942. Howe, D. K., Sibley, L. D. 1999. Comparison of the major antigens of *Neospora caninum* and *Toxoplasma gondii*. *Int. J. Parasitol.* 29, 1489–1496.

943. Howe, D. K., Tang, K., Conrad, P. A., Sverlow, K., Dubey, J. P., Sibley, L. D. 2002. Sensitive and specific identification of *Neospora caninum* infection of cattle based on detection of serum antibodies to recombinant Ncp29. *Clin. Diagn. Lab. Immunol.* 9, 611–615.

944. Howe, L., West, D., Pomroy, B., Collett, M., Kenyon, P., Morris, S., Pattison, R. 2007. Investigations into the involvement of *Neospora caninum* in unexplained sheep abortions. *Proceedings of the Society of Sheep & Beef Cattle Veterinarians of the NZVA*, Palmerston North, New Zealand. 129–134.

945. Howe, L., West, D. M., Collett, M. G., Tattersfield, G., Pattison, R. S., Pomroy, W. E., Kenyon, P. R., Morris, S. T., Williamson, N. B. 2008. The role of *Neospora caninum* in three cases of unexplained ewe abortions in the southern North Island of New Zealand. *Small Ruminant Res.* 75, 115–122.

946. Howe, L., Collett, M. G., Pattison, R. S., Marshall, J., West, D. M., Pomroy, W. E. 2012. Potential involvement of *Neospora caninum* in naturally occurring ovine abortions in New Zealand. *Vet. Parasitol.* 185, 64–71.

947. Hu, J., Ferroglio, E., Trisciuoglio, A. 2011. Immunoblot diagnosis of infection with *Neospora caninum* in cattle based on recombinant NcSAG4 antigen. *Parasitol. Res.* 108, 1055–1058.

948. Huang, C. C., Ting, L. J., Shiau, J. R., Chen, M. C., Ooi, H. K. 2004. An abortion storm in cattle associated with neosporosis in Taiwan. *J. Vet. Med. Sci.* 66, 465–467.

949. Huang, C. C., Yang, C. H., Watanabe, Y., Liao, Y. K., Ooi, H. K. 2004. Finding of *Neospora caninum* in the wild brown rat (*Rattus norvegicus*). *Vet. Res.* 35, 283–290.

950. Huang, P., Liao, M., Zhang, H., Lee, E. G., Nishikawa, Y., Xuan, X. 2007. Dense-granule protein NcGRA7, a new marker for the serodiagnosis of *Neospora caninum* infection in aborting cows. *Clin. Vaccine Immunol.* 14, 1640–1643.

951. Hudson, A., Ellis, J. T. 2005. Culture of *Neospora caninum* in the presence of a mycoplasma removal agent results in the selection of a mutant population of tachyzoites. *Parasitology* 130, 607–610.

952. Huerta-Peña, J. C., Martínez-Herrera, D. I., Peniche-Cardeña, Á. E. J., Villanueva-Valencia, M., Hernández-Ruiz, S. G., Villagómez-Cortés, J. A., Barradas-Piña, F. T., Morales-Álvarez, J. F., Flores-Castro, R. 2011. Seroprevalence and risk factors associated with *Neospora caninum* in goats from municipalities of the central region of Veracruz. *Trop. Subtrop. Agroecosystem* 13, 445–454.

953. Hughes, J. M., Williams, R. H., Morley, E. K., Cook, D. A. N., Terry, R. S., Murphy, R. G., Smith, J. E., Hide, G. 2006. The prevalence of *Neospora caninum* and co-infection with *Toxoplasma gondii* by PCR analysis in naturally occurring mammal populations. *Parasitology* 132, 29–36.

954. Hughes, J. M., Thomasson, D., Craig, P. S., Georgin, S., Pickles, A., Hide, G. 2008. *Neospora caninum*: Detection in wild rabbits and investigation of co-infection with *Toxoplasma gondii* by PCR analysis. *Exp. Parasitol.* 120, 255–260.

955. Huong, L. T. T., Ljungström, B. L., Uggla, A., Björkman, C. 1998. Prevalence of antibodies to *Neospora caninum* and *Toxoplasma gondii* in cattle and water buffaloes in southern Vietnam. *Vet. Parasitol.* 75, 53–57.

956. Hur, K., Kim, J. H., Hwang, W. S., Hwang, E. K., Jean, Y. H., Lee, B. C., Bae, J. S., Kang, Y. B., Yamane, I., Kim, D. Y. 1998. Seroepidemiological study of *Neospora caninum* in Korean dairy cattle by indirect immunofluorescent antibody assay. *Korean J. Vet. Res.* 38, 859–866.

957. Hurkova, L., Halova, D., Modry, D. 2005. The prevalence of *Neospora caninum* antibodies in bulk milk of dairy herds in the Czech Republic: A case report. *Vet. Med. Czech* 50, 549–552.

958. Hurková, L., Modrý, D. 2006. PCR detection of *Neospora caninum*, *Toxoplasma gondii* and *Encephalitozoon cuniculi* in brains of wild carnivores. *Vet. Parasitol.* 137, 150–154.

959. Hurková-Hofmannová, L., Václavek, P., Škoric, M., Fictum, P., Modrý, D. 2007. Multimammate rat (*Mastomys natalensis*), Tristram's jird (*Meriones tristrami*) and Wagner's gerbil (*Gerbillus dasyurus*) as laboratory models of acute neosporosis. *Res. Vet. Sci.* 82, 377–381.

960. Hurková-Hofmannová, L., Qablan, M. A., Juránková, J., Modrý, D., Piálek, J. 2014. A survey of *Toxoplasma gondii* and *Neospora caninum* infecting house mice from a hybrid zone. *J. Parasitol.* 100, 139–141.

961. Hussien, M. O., Elfahal, A. M., Enan, K. A., Mohammed, M. S., Ibrahim, A. M., Taha, K. M., El-Hussein, A. M. 2012. Seroprevalence of *Neospora caninum* in cattle in Sudan. *Vet. World* 5, 465–468.

962. Huynh, M. H., Boulanger, M. J., Carruthers, V. B. 2014. A conserved apicomplexan microneme protein contributes to *Toxoplasma gondii* invasion and virulence. *Infect. Immun.* 82, 4358–4368.

963. Hyun, C., Gupta, G. D., Marsh, A. E. 2003. Sequence comparison of *Sarcocystis neurona* surface antigen from multiple isolates. *Vet. Parasitol.* 112, 11–20.

964. Ibrahim, A. M., Ismail, A. A., Angara, T. E. E., Osman, O. M. 2014. Seroprevalence of *Neospora caninum* in dairy cattle and the co-herded camels, sheep and goats in dairy farms in the Khartoum State, Sudan. *J. Appl. Ind. Sci.* 2, 206–212.

965. Ibrahim, A. M., Ismail, A. A., Angara, T. E. E., Osman, O. M. 2015. Detection of antibodies against Toxoplasma gondii and *Neospora caninum* in dairy camels from the Khartoum State, Sudan. *The Regional Conference of Camel Management and Production under Open range System (RCCMPR)*. March 2–4, Khartoum-Sudan. 71–72.

966. Ibrahim, A. M. E., Elfahal, A. M., El Hussein, A. R. M. 2012. First report of *Neospora caninum* infection in cattle in Sudan. *Trop. Anim. Health Prod.* 44, 769–772.

967. Ibrahim, H. M., Huang, P., Salem, T. A., Talaat, R. M., Nasr, M. I., Xuan, X., Nishikawa, Y. 2009. Short report: Prevalence of *Neospora caninum* and *Toxoplasma gondii* antibodies in northern Egypt. *Am. J. Trop. Med. Hyg.* 80, 263–267.

968. Ibrahim, H. M. 2013. Seroprevalence of *Neospora caninum* antibodies in chicken samples from Delta Egypt using a recombinant NcSAG1 protein-based ELISA. *Egypt. J. Immunol.* 20, 29–37.

969. Iça, A., Yildirim, A., Düzlü, Ö., Inci, A. 2006. [Seroprevalence of *Neospora caninum* in cattle in the region of Kayseri]. *Turkiye Parazitol. Derg.* 30, 92–94 (in Turkish).

970. Ichikawa-Seki, M., Guswanto, A., Allamanda, P., Mariamah, E. S., Wibowo, P. E., Nishikawa, Y. 2016. Seroprevalence of antibody to NcSAG1 antigen of *Neospora caninum* in cattle from Western Java, Indonesia. *J. Vet. Med. Sci.* 78, 121–123.

971. Ihara, F., Nishimura, M., Muroi, Y., Furuoka, H., Yokoyama, N., Nishikawa, Y. 2016. Changes in neurotransmitter levels and expression of immediate early genes in brain of mice infected with *Neospora caninum*. *Sci. Rep.* 6, 23052.

972. Imre, K., Morariu, S., Ilie, M. S., Imre, M., Ferrari, N., Genchi, C., Darabus, G. 2012. Serological survey of *Neospora caninum* infection in cattle herds from western Romania. *J. Parasitol.* 98, 683–685.

973. Innes, E. A., Panton, W. R. M., Marks, J., Trees, A. J., Holmdahl, J., Buxton, D. 1995. Interferon gamma inhibits the intracellular multiplication of *Neospora caninum* as shown by incorporation of ^{3H} uracil. *J. Comp. Pathol.* 113, 95–100.

974. Innes, E. A., Buxton, D., Eperon, S., Gottstein, B. 2000. Immunology of *Neospora caninum* infection in cattle and mice. *Int. J. Parasitol.* 30, 896–900.

975. Innes, E. A., Buxton, D., Maley, S., Wright, S., Marks, J., Esteban, I., Rae, A., Schock, A., Wastling, J. 2000. Neosporosis: Aspects of epidemiology and host immune response. *Ann. N. Y. Acad. Sci.* 916, 93–101.

976. Innes, E. A., Wright, S. E., Maley, S., Rae, A., Schock, A., Kirvar, E., Bartley, P., Hamilton, C., Carey, I. M., Buxton, D. 2001. Protection against vertical transmission in bovine neosporosis. *Int. J. Parasitol.* 31, 1523–1534.

977. Innes, E. A., Lundén, A., Esteban, I., Marks, J., Maley, S., Wright, S., Rae, A., Harkins, D., Vermeulen, A., McKendrick, I. J., Buxton, D. 2001. A previous infection with *Toxoplasma gondii* does not protect against a challenge with *Neospora caninum* in pregnant sheep. *Parasite Immunol.* 23, 121–132.

978. Innes, E. A., Andrianarivo, A. G., Björkman, C., Williams, D. J. L., Conrad, P. A. 2002. Immune responses to *Neospora caninum* and prospects for vaccination. *Trends Parasit.* 18, 497–504.

979. Innes, E. A., Wright, S., Bartley, P., Maley, S., Macaldowie, C., Esteban-Redondo, I., Buxton, D. 2005. The host-parasite relationship in bovine neosporosis. *Vet. Immunol. Immunopathol.* 108, 29–36.

980. Innes, E. A., Vermeulen, A. N. 2006. Vaccination as a control strategy against the coccidial parasites, *Eimeria*, *Toxoplasma* and *Neospora*. *Parasitology* 133, S145–S168.

981. Innes, E. A. 2007. The host-parasite relationship in pregnant cattle infected with *Neospora caninum*. *Parasitology* 134, 1903–1910.

982. Innes, E. A., Mattsson, J. G. 2007. *Neospora caninum* emerges from the shadow of *Toxoplasma gondii*. *Trends Parasitol.* 23, 43–44.

983. Innes, E. A., Bartley, P. M., Maley, S. W., Wright, S. E., Buxton, D. 2007. Comparative host-parasite relationships in ovine toxoplasmosis and bovine neosporosis and strategies for vaccination. *Vaccine* 25, 5495–5503.

984. Inpankaew, T., Jittapalapong, S., Mitchell, T. J., Sununta, C., Igarashi, I., Xuan, X. 2014. Seroprevalence of *Neospora caninum* infection in dairy cows in Northern provinces, Thailand. *Acta Parasit.* 59, 305–309.

985. Iovu, A., Györke, A., Mircean, V., Gavrea, R., Cozma, V. 2012. Seroprevalence of *Toxoplasma gondii* and *Neospora caninum* in dairy goats from Romania. *Vet. Parasitol.* 186, 470–474.

986. Ishigaki, K., Noya, M., Kagawa, Y., Ike, K., Orima, H., Imai, S. 2012. Detection of *Neospora caninum*-specific DNA from cerebrospinal fluid by polymerase chain reaction in a dog with confirmed neosporosis. *J. Vet. Med. Sci.* 74, 1051–1055.

987. Jackson, W., de Lahunta, A., Adaska, J., Cooper, B., Dubey, J. P. 1995. *Neospora caninum* in an adult dog with progressive cerebellar signs. *Progr. Vet. Neurol.* 6, 124–127.

988. Jacobson, L. S., Jardine, J. E. 1993. *Neospora caninum* infection in three Labrador littermates. *J. S. Afr. Vet. Assoc.* 64, 47–51.

989. Jakubek, E. B., Bröjer, C., Regnersen, C., Uggla, A., Schares, G., Björkman, C. 2001. Seroprevalences of *Toxoplasma gondii* and *Neospora caninum* in Swedish red foxes (*Vulpes vulpes*). *Vet. Parasitol.* 102, 167–172.

990. Jakubek, E. B., Uggla, A. 2005. Persistence of *Neospora caninum*-specific immunoglobulin G antibodies in bovine blood and lung tissue stored at room temperature. *J. Vet. Diagn. Invest.* 17, 458–460.

991. Jakubek, E. B., Lundén, A., Uggla, A. 2006. Seroprevalences in *Toxoplasma gondii* and *Neospora* sp. infections in Swedish horses. *Vet. Parasitol.* 138, 194–199.

992. Jakubek, E. B., Farkas, R., Pálfi, V., Mattsson, J. G. 2007. Prevalence of antibodies against *Toxoplasma gondii* and *Neospora caninum* in Hungarian red foxes (*Vulpes vulpes*). *Vet. Parasitol.* 144, 39–44.

993. James, K. E., Smith, W. A., Conrad, P. A., Packham, A. E., Guerrero, L., Ng, M., Pusterla, N. 2015. Seroprevalence of *Sarcocystis neurona* and *Neospora hughesi* among healthy horses in the United States. *Proc. Am. Assoc. Equine Pract.* 61, 524.

994. Jardine, J. E., Dubey, J. P. 1992. Canine neosporosis in South Africa. *Vet. Parasitol.* 44, 291–294.

995. Jardine, J. E., Last, R. D. 1993. *Neospora caninum* in aborted twin calves. *J. S. Afr. Vet. Assoc.* 64, 101–102.

996. Jardine, J. E., Last, R. D. 1995. The prevalence of neosporosis in aborted bovine foetuses submitted to the Allerton Regional Veterinary Laboratory. *Onderstepoort J. Vet. Res.* 62, 207–209.

997. Jardine, J. E., Wells, B. H. 1995. Bovine neosporosis in Zimbabwe. *Vet. Rec.* 137, 223.

998. Jardine, J. E. 1996. The ultrastructure of bradyzoites and tissue cysts of *Neospora caninum* in dogs: Absence of distinguishing morphological features between parasites of canine and bovine origin. *Vet. Parasitol.* 62, 231–240.

999. Jefferies, R., Ryan, U. M., Irwin, P. J. 2007. PCR-RFLP for the detection and differentiation of the canine piroplasm species and its use with filter paper-based technologies. *Vet. Parasitol.* 144, 20–27.

1000. Jenkins, M., Baszler, T., Björkman, C., Schares, G., Williams, D. 2002. Diagnosis and seroepidemiology of *Neospora caninum*-associated bovine abortion. *Int. J. Parasitol.* 32, 631–636.

1001. Jenkins, M., Parker, C., Tuo, W., Vinyard, B., Dubey, J. P. 2004. Inclusion of CpG adjuvant with plasmid DNA coding for NcGRA7 improves protection against congenital neosporosis. *Infect. Immun.* 72, 1817–1819.

1002. Jenkins, M., Soares, R., Murphy, C., Hemphill, A., O'Handley, R., Dubey, J. P. 2004. Localization of a 56-kDa antigen that is present in multiple developmental stages of *Neospora caninum*. *J. Parasitol.* 90, 660–663.

1003. Jenkins, M. C., Wouda, W., Dubey, J. P. 1997. Serological response over time to recombinant *Neospora caninum* antigens in cattle after a neosporosis-induced abortion. *Clin. Diagn. Lab. Immunol.* 4, 270–274.

1004. Jenkins, M. C., Caver, J. A., Björkman, C., Anderson, T. C., Romand, S., Vinyard, B., Uggla, A., Thulliez, P., Dubey, J. P. 2000. Serological investigation of an outbreak of *Neospora caninum*-associated abortion in a dairy herd in southeastern United States. *Vet. Parasitol.* 94, 17–26.

1005. Jenkins, M. C. 2001. Advances and prospects for subunit vaccines against protozoa of veterinary importance. *Vet. Parasitol.* 101, 291–310.

1006. Jenkins, M. C., Tuo, W. B., Dubey, J. P. 2004. Evaluation of vaccination with *Neospora caninum* protein for prevention of fetal loss associated with experimentally induced neosporosis in sheep. *Am. J. Vet. Res.* 65, 1404–1408.

1007. Jenkins, M. C., Fetterer, R., Schares, G., Björkman, C., Wapenaar, W., McAllister, M., Dubey, J. P. 2005. HPLC purification of recombinant NcGRA6 antigen improves enzyme-linked immunosorbent assay for serodiagnosis of bovine neosporosis. *Vet. Parasitol.* 131, 227–234.

1008. Jenkins, M. C., Parker, C., Hill, D., Pinckney, R. D., Dyer, R., Dubey, J. P. 2007. *Neospora caninum* detected in feral rodents. *Vet. Parasitol.* 143, 161–165.

1009. Jenkins, M. C., Tuo, W., Feng, X., Cao, L., Murphy, C., Fetterer, R. 2010. *Neospora caninum*: Cloning and expression of a gene coding for cytokine-inducing profilin. *Exp. Parasitol.* 125, 357–362.

1010. Jensen, A. M., Björkman, C., Kjeldsen, A. M., Wedderkopp, A., Willadsen, C., Uggla, A., Lind, P. 1999. Associations of *Neospora caninum* seropositivity with gestation number and pregnancy outcome in Danish dairy herds. *Prev. Vet. Med.* 40, 151–163.

1011. Jensen, L., Jensen, T. K., Lind, P., Henriksen, S. A., Uggla, A., Bille-Hansen, V. 1998. Experimental porcine neosporosis. *Acta Pathol. Microbiol. Immunol. Scand.* 106, 475–482.

1012. Jesus, E. E. V., Almeida, M. A. O., Atta, A. M. 2007. Anti-neosporal IgG and IgE antibodies in canine neosporosis. *Zoonoses Public Health* 54, 387–392.

1013. Jesus, E. E. V., Pinheiro, A. M., Santos, A. B., Freire, S. M., Tardy, M. B., El-Bachá, R. S., Costa, S. L., Costa, M. F. D. 2013. Effects of IFN-γ, TNF-α, IL-10 and TGF-β on *Neospora caninum* infection in rat glial cells. *Exp. Parasitol.* 133, 269–274.

1014. Jia, J., Zhang, S. F., Liu, M. M., Qian, N. C., Guo, H. P. 2014. Isolation, identification, and pathogenicity of *Neospora caninum* China Yanbian strain. *Iranian J. Parasitol.* 9, 394–401.

1015. Jiménez-Ruiz, E., Álvarez-García, G., Aguado-Martínez, A., Salman, H., Irache, J. M., Marugán-Hernández, V., Ortega-Mora, L. M. 2012. Low efficacy of NcGRA7, NcSAG4, NcBSR4 and NcSRS9 formulated in poly-ε-caprolactone against *Neospora caninum* infection in mice. *Vaccine* 30, 4983–4992.

1016. Jiménez-Ruiz, E., Álvarez-García, G., Aguado-Martínez, A., Ortega-Mora, L. M. 2013. Low rates of *Neospora caninum* infection reactivation during gestation are observed in both chronically and congenitally infected mice. *Parasitology* 140, 220–228.

1017. Jiménez-Ruiz, E., Álvarez-García, G., Aguado-Martínez, A., Ortega-Mora, L. M. 2013. Mice congenitally infected with low-to-moderate virulence *Neospora caninum* isolates exhibited clinical reactivation during the mating period without transmission to the next generation. *Exp. Parasitol.* 134, 244–248.

1018. Jiménez-Ruiz, E., Bech-Sàbat, G., Álvarez-García, G., Regidor-Cerrillo, J., Hinojal-Campaña, L., Ortega-Mora, L. M. 2013. Specific antibody responses against *Neospora caninum* recombinant rNcGRA7, rNc-SAG4, rNcBSR4 and rNcSRS9 proteins are correlated with virulence in mice. *Parasitology* 140, 569–579.

1019. Jimenez-Ruiz, E., Morlon-Guyot, J., Daher, W., Meissner, M. 2016. Vacuolar protein sorting mechanisms in apicomplexan parasites. *Mol. Bioch. Parasitol.* 209, 18–25. Doi: 10.1016/j.milbiopara2016.01.007.

1020. Jin, C., Yu, L., Wang, Y., Hu, S., Zhang, S. 2015. Evaluation of *Neospora caninum* truncated dense granule protein 2 for serodiagnosis by enzyme-linked immunosorbent assay in dogs. *Exp. Parasitol.* 157, 88–91.

1021. Jolley, W. R., McAllister, M. M., McGuire, A. M., Wills, R. A. 1999. Repetitive abortion in *Neospora*-infected ewes. *Vet. Parasitol.* 82, 251–257.

1022. Joly, A. 2000. Néosporose bovine: Observation dans 162 élevages et suivi de 35 élevages contaminés. *Bulletin des GTV* 7, 35–40.

1023. Jonker, F. H. 2004. Fetal death: Comparative aspects in large domestic animals. *Anim. Reprod. Sci.* 82–83, 415–430.

1024. Journel, C., Tainturier, D., Pitel, P. H., Chatagnon, G. 1999. *Neospora caninum*: étude d'un élevage contaminé, quelques hypothèses de transmission. *Point Vét.* 30, 397–404.

1025. Kameyama, K., Nishimura, M., Punsantsogvoo, M., Ibrahim, H. M., Xuan, X., Furuoka, H., Nishikawa, Y. 2012. Immunological characterization of *Neospora caninum* cyclophilin. *Parasitology* 139, 294–301.

1026. Kamga-Waladjo, A. R., Chatagnon, G., Bakou, S. N., Boly, H., Diop, P. E. H., Tainturier, D. 2009. *Neospora caninum* antibodies and its consequences for reproductive characteristics in wandering sows from Senegal, West Africa. *Asian J. Anim. Vet. Adv.* 4, 263–266.

1027. Kamga-Waladjo, A. R., Gbati, O. B., Kone, P., Lapo, R. A., Dombou, E., Chatagnon, G., Bakou, S. N., Diop, P. E. H., Pangui, L. J., Tainturier, D., Akakpo, J. A. 2009. *Neospora caninum* and *Toxoplasma gondii* in lion (*Panthera leo*) from Senegal, West Africa. *Asian J. Anim. Vet. Adv.* 4, 346–349.

1028. Kamga-Waladjo, A. R., Gbati, O. B., Kone, P., Lapo, R. A., Chatagnon, G., Bakou, S. N., Pangui, L. J., Diop, P. E. H., Akakpo, J. A., Tainturier, D. 2010. Seroprevalence of *Neospora caninum* antibodies and its consequences for reproductive parameters in dairy cows from Dakar-Senegal, West Africa. *Trop. Anim. Health Prod.* 42, 953–959.

1029. Kamga-Waladjo, A. R., Allanonto, V., Gbati, O. B., Kone, P. S., Koffi, J. F. A., Coulibaly, F., Ndour, A. P. N., Efoua-Tomo, N., Kante, S., Syll, M., Mime, P. I., Bakou, S. N., Diop, B. M., Diop, P. E. H., Pangui, L. J. 2013. Seroprevalence of *Neospora caninum* and *Toxoplasma gondii* in dogs and risk of infection of dogs and women by the city Saint Louis, Senegal. *Sci. Parasitol.* 14, 129–137.

1030. Kang, S. W., Lee, E. H., Jean, Y. H., Choe, S. E., Quyen, D. V., Lee, M. S. 2008. The differential protein expression profiles and immunogenicity of tachyzoites and bradyzoites of *in vitro* cultured *Neospora caninum*. *Parasitol. Res.* 103, 905–913.

1031. Kang, S. W., Kweon, C. H., Lee, E. H., Choe, S. E., Jung, S. C., Quyen, D. V. 2008. The differentiation of transcription between tachyzoites and bradyzoites of *in vitro* cultured *Neospora caninum*. *Parasitol. Res.* 103, 1011–1018.

1032. Kang, S. W., Park, S. S., Choe, S. E., Jean, Y. H., Jung, S. C., Kim, K., Quyen, D. V. 2009. Characterization of tissue distribution and histopathological lesions in *Neospora caninum* experimentally infected gerbils. *Parasitol. Res.* 104, 1261–1268.

1033. Kano, R., Masukata, Y., Omata, Y., Kobayashi, Y., Maeda, R., Saito, A. 2005. Relationship between type 1/type 2 immune responses and occurrence of vertical transmission in BALB/c mice infected with *Neospora caninum*. *Vet. Parasitol.* 129, 159–164.

1034. Kano, R., Kudo, A., Kamiya, H., Kobayashi, Y., Maeda, R., Omata, Y. 2007. C57BL/6 mice infected with *Neospora caninum* during administration of progesterone show bias toward type 2 immune response. *J. Vet. Med. Sci.* 69, 1095–1097.

1035. Karatepe, M., Karatepe, B. 2012. [Investigation of seroprevalence of *Neospora* spp. in horses in Nigde Province (Turkey)]. *Kafkas Univ. Vet. Fak. Derg.* 18, A39–A42 (in Turkish).

1035a. Karatepe, B., Karatepe, M. 2016. Seroprevalence of *Neospora caninum* in cattle in Nigde Province, Turkey. *Israel J. Vet. Med.* 71, 39–42.

1036. Kargar, M., Mojaver, S., Namavari, M., Sayari, M., Rahimian, A. 2013. Suspension culture of *Neospora caninum* by *Theileria annulata*-infected cell line. *Trop. Biomed.* 30, 349–354.

1037. Kasari, T. R., Barling, K., McGrann, J. M. 1999. Estimated production and economic losses from *Neospora caninum* infection in Texas beef herds. *Bovine Pract.* 33, 113–120.

1038. Kashiwazaki, Y., Pholpark, S., Charoenchai, A., Polsar, C., Teeverapanya, S., Pholpark, M. 2001. Postnatal neosporosis in dairy cattle in northeast Thailand. *Vet. Parasitol.* 94, 217–220.

1039. Kashiwazaki, Y., Gianneechini, R. E., Lust, M., Gil, J. 2004. Seroepidemiology of neosporosis in dairy cattle in Uruguay. *Vet. Parasitol.* 120, 139–144.

1040. Kasper, L. H., Khan, I. A. 1998. Antigen-specific CD8+ T cells protect against lethal toxoplasmosis in mice infected with *Neospora caninum*. *Infect. Immun.* 66, 1554–1560.

1041. Katkiewicz, M., Wierzchoń, M. 2002. [Occurrence of *Neospora caninum* in domestic animals and its role in the etiology of reproduction disturbances]. *Medycyna Wet.* 58, 332–336 (in Polish).

1042. Kato, T., Otsuki, T., Yoshimoto, M., Itagaki, K., Kohsaka, T., Matsumoto, Y., Ike, K., Park, E. Y. 2015. *Bombyx mori* nucleopolyhedrovirus displaying *Neospora caninum* antigens as a vaccine candidate against *N. caninum* infection in mice. *Mol. Biotechnol.* 57, 145–154.

1043. Kaufmann, H., Yamage, M., Roditi, I., Dobbelaere, D., Dubey, J. P., Holmdahl, O. J. M., Trees, A., Gottstein, B. 1996. Discrimination of *Neospora caninum* from *Toxoplasma gondii* and other apicomplexan parasites by hybridization and PCR. *Mol. Cell. Probes* 10, 289–297.

1044. Keefe, G. P., VanLeeuwen, J. A. 2000. *Neospora* then and now: Prevalence of *Neospora caninum* in maritime Canada in 1979, 1989, and 1998. *Can. Vet. J.* 41, 864–866.

1045. Kefayat, M., Hamidinejat, H., Seifiabadshapoori, M. R., Namavari, M. M., Shayan, P., Gooraninejad, S. 2014. Cloning and expression of *Neospora caninum* dense-granule 7 in *E. coli*. *J. Parasit. Dis.* 38, 196–200.

1046. Keller, N., Naguleswaran, A., Cannas, A., Vonlaufen, N., Bienz, M., Björkman, C., Bohne, W., Hemphill, A. 2002. Identification of a *Neospora caninum* microneme protein (NcMIC1) which interacts with sulfated host cell surface glycosaminoglycans. *Infect. Immun.* 70, 3187–3198.

1047. Keller, N., Riesen, M., Naguleswaran, A., Vonlaufen, N., Stettler, R., Leepin, A., Wastling, J. M., Hemphill, A. 2004. Identification and characterization of a *Neospora caninum* microneme-associated protein (NcMIC4) that exhibits unique lactose-binding properties. *Infect. Immun.* 72, 4791–4800.

1048. Kengradomkij, C., Inpankaew, T., Kamyingkird, K., Wongpanit, K., Wongnakphet, S., Mitchell, T. J., Xuan, X., Igarashi, I., Jittapalapong, S., Stich, R. W. 2015. Seroprevalence and risk factors associated with exposure of water buffalo (*Bubalus bubalis*) to *Neospora caninum* in northeast Thailand. *Vet. Parasitol.* 207, 156–160.

1049. Keyloun, K. R., Reid, M. C., Choi, R., Song, R., Fox, A. M. W., Hillesland, H. K., Zhang, Z., Vidadala, R., Merritt, E. A., Lau, A. O. T., Maly, D. J., Fan, E., Barrett, L. K., Van Voorhis, W. C., Ojo, K. K. 2014. The gatekeeper residue and beyond: Homologous calcium dependent protein kinases as drug development targets for veterinarian Apicomplexa parasites. *Parasitology* 141, 1499–1509.

1050. Khaitsa, M. L., Barigye, R., Dyer, N. W., Doetkott, D. M., Foster, J. R. 2006. Serologic and other diagnostic evidence of *Neospora caninum* presence in North Dakota beef herds. *Bovine Pract.* 40, 51–56.

1051. Khan, I. A., Schwartzman, J. D., Fonseka, S., Kasper, L. H. 1997. *Neospora caninum*: Role for immune cytokines in host immunity. *Exp. Parasitol.* 85, 24–34.

1052. Khanmohammadi, M., Fallah, E. 2011. Prevalence of *Neospora caninum* antibodies in shepherd dogs in Sarab district, East Azerbaijan Province, Iran. *African Journal of Microbiology Reserch* 5, 5062–5066.

1053. Khodakaram-Tafti, A., Ikede, B. O. 2005. A retrospective study of sporadic bovine abortions, stillbirths, and neonatal abnormalities in Atlantic Canada, from 1990 to 2001. *Can. Vet. J.* 46, 635–637.

1054. Khodakaram-Tafti, A., Mansourian, M., Namavari, M., Hosseini, A. 2012. Immunohistochemical and polymerase chain reaction studies in *Neospora caninum* experimentally infected broiler chicken embryonated eggs. *Vet. Parasitol.* 188, 10–13.

1055. Khordadmehr, M., Namavari, M., Khodakaram-Tafti, A., Mansourian, M., Rahimian, A., Daneshbod, Y. 2013. Comparison of use of Vero cell line and suspension culture of murine macrophage to attenuation of *Neospora caninum*. *Res. Vet. Sci.* 95, 515–521.

1056. Kilbaş, Z. G., Adanir, R., Avcioglu, H. 2008. Seroprevalence of *Neospora caninum* in racehorses in Ankara, Turkey. *Acta Parasitol.* 53, 315–316.

1057. Kim, D. Y., Hwang, W. S., Kim, J. H., Hur, K., Lee, B. C., Jean, Y. H., Rhee, J. C., Choi, S. H. 1997. Bovine abortion associated with *Neospora* in Korea. *Korean J. Vet. Res.* 37, 607–612.

1058. Kim, J. H., Hwang, E. K., Sohn, H. J., Jean, Y. H., Yoon, S. S., Kim, D. Y. 1998. [Repeated bovine abortion associated with *Neospora caninum* in Korea]. *Korean J. Vet. Res.* 38, 853–858 (in Korean).

1059. Kim, J. H., Sohn, H. J., Hwang, E. K., Hwang, W. S., Hur, K., Jean, Y. H., Lee, B. C., Rhee, J. C., Kang, Y. B., Yamane, I., Kim, D. J. 1998. *In vitro* isolation of a bovine *Neospora* in Korea. *Korean J. Vet. Res.* 38, 139–145.

1060. Kim, J. H., Sohn, H. J., Hwang, W. S., Hwang, E. K., Jean, Y. H., Yamane, I., Kim, D. Y. 2000. *In vitro* isolation and characterization of bovine *Neospora caninum* in Korea. *Vet. Parasitol.* 90, 147–154.

1061. Kim, J. H., Lee, J. K., Hwang, E. K., Kim, D. Y. 2002. Prevalence of antibodies to *Neospora caninum* in Korean native beef cattle. *J. Vet. Med. Sci.* 64, 941–943.

1062. Kim, J. H., Lee, J. K., Lee, B. C., Park, B. K., Yoo, H. S., Hwang, W. S., Shin, N. R., Kang, M. S., Jean, Y. H., Yoon, H. J., Kang, S. K., Kim, D. Y. 2002. Diagnostic survey of bovine abortion in Korea: With special emphasis on *Neospora caninum*. *J. Vet. Med. Sci.* 64, 1123–1127.

1063. Kim, J. H., Kang, M. S., Lee, B. C., Hwang, W. S., Lee, C. W., So, B. J., Dubey, J. P., Kim, D. Y. 2003. Seroprevalence of antibodies to *Neospora caninum* in dogs and raccoon dogs in Korea. *Korean J. Parasitol.* 41, 243–245.

1064. Kim, J. T., Park, J. Y., Seo, H. S., Oh, H. G., Noh, J. W., Kim, J. H., Kim, D. Y., Youn, H. J. 2002. *In vitro* antiprotozoal effects of artemisinin on *Neospora caninum*. *Vet. Parasitol.* 103, 53–63.

1065. Kim, S. K., Fouts, A. E., Boothroyd, J. C. 2007. *Toxoplasma gondii* dysregulates IFN-γ-inducible gene expression in human fibroblasts: Insights from a genome-wide transcriptional profiling. *J. Immunol.* 178, 5154–5165.

1066. Kindahl, H., Björkman, C., Stenlund, S., Uggla, A. 2000. *Neospora* abortion in cattle—Background and the present Swedish situation. *Reprod. Dom. Anim.* 6(Suppl), 73–75.

1067. King, J. S., Šlapeta, J., Jenkins, D. J., Al-Qassab, S. E., Ellis, J. T., Windsor, P. A. 2010. Australian dingoes are definitive host of *Neospora caninum*. *Int. J. Parasitol.* 40, 945–950.

1068. King, J. S., Jenkins, D. J., Ellis, J. T., Fleming, P., Windsor, P. A., Šlapeta, J. 2011. Implications of wild dog ecology on the sylvatic and domestic life cycle of *Neospora caninum* in Australia. *Vet. J.* 188, 24–33.

1069. King, J. S., McAllan, B., Spielman, D. S., Lindsay, S. A., Hurková-Hofmannová, L., Hartigan, A., Al-Qassab, S. E., Ellis, J. T., Šlapeta, J. 2011. Extensive production of *Neospora caninum* tissue cysts in a carnivorous marsupial succumbing to experimental neosporosis. *Vet. Res.* 42, 75.

1070. King, J. S., Brown, G. K., Jenkins, D. J., Ellis, J. T., Fleming, P. J. S., Windsor, P. A., Šlapeta, J. 2012. Oocysts and high seroprevalence of *Neospora caninum* in dogs living in remote Aboriginal communities and wild dogs in Australia. *Vet. Parasitol.* 187, 85–92.

1071. King, J. S., Vaughan, J. L., Windsor, P. A. 2015. Serological evidence of *Neospora caninum* in alpacas from eastern Australia. *Aust. Vet. J.* 93, 259–261.

1072. Klauck, V., Machado, G., Pazinato, R., Radavelli, W. M., Santos, D. S., Berwaguer, J. C., Braunig, P., Vogel, F. F., Da Silva, A. S. 2016. Relation between *Neospora caninum* and abortion in dairy cows: Risk factors and pathogenesis of disease. *Microb. Pathog.* 92, 46–49.

1073. Klein, B. U., Müller, E. 2001. Seroprävalenz von Antikörpern gegen *Neospora caninum* bei Hunden mit und ohne klinischem Neosporoseverdacht in Deutschland. *Praktische Tierarzt.* 82, 437–440.

1074. Klein, F., Hietala, S. K., Berthet, H., Very, P., Gradinaru, D. 1997. *Neospora caninum*: enquête sérologique sur les avortements des bovins normands et charolais. *Point Vét.* 28, 1283–1286.

1075. Klein, F., Ould-Amrouche, A., Osdoit, C., Touratier, A., Sanaa, M. 2000. *Neospora caninum*: une enquête séroépidémiologique dans l'Orne. *Bulletin des GTV* 7, 41–45.

1076. Klevar, S., Kulberg, S., Boysen, P., Storset, A. K., Moldal, T., Björkman, C., Olsen, I. 2007. Natural killer cells act as early responders in an experimental infection with *Neospora caninum* in calves. *Int. J. Parasitol.* 37, 329–339.

1077. Klevar, S., Norström, M., Tharaldsen, J., Clausen, T., Björkman, C. 2010. The prevalence and spatial clustering of *Neospora caninum* in dairy herds in Norway. *Vet. Parasitol.* 170, 153–157.

1078. Kligler, E. B., Shkap, V., Baneth, G., Mildenberg, Z., Steinman, A. 2007. Seroprevalence of *Neospora* spp. among asymptomatic horses, aborted mares and horses demonstrating neurological signs in Israel. *Vet. Parasitol.* 148, 109–113.

1079. Klopfleisch, R., von Deetzen, M., Weiss, A. T., Weigner, J., Weigner, F., Plendl, J., Gruber, A. D. 2016. Weigners fixative—An alternative to formalin fixation for histology with improved preservation of nucleic acids. *Vet. Pathol.* 50, 191–199.

1080. Knowler, C., Wheeler, S. J. 1995. *Neospora caninum* infection in three dogs. *J. Small Anim. Pract.* 36, 172–177.

1081. Kobayashi, A., Katagiri, S., Kimura, T., Ochiai, K., Umemura, T. 2002. Steroid hormones do not reactivate *Neospora caninum* in ovariectomized mice. *J. Vet. Med. Sci.* 64, 773–777.

1082. Kobayashi, T., Narabu, S., Yanai, Y., Hatano, Y., Ito, A., Imai, S., Ike, K. 2013. Gene cloning and characterization of the protein encoded by the *Neospora caninum* bradyzoite-specific antigen gene *BAG1*. *J. Parasitol.* 99, 453–458.

1083. Kobayashi, Y., Yamada, M., Omata, Y., Koyama, T., Saito, A., Matsuda, T., Okuyama, K., Fujimoto, S., Furuoka, H., Matsui, T. 2001. Naturally-occurring *Neospora caninum* infection in an adult sheep and her twin fetuses. *J. Parasitol.* 87, 434–436.

1084. Koestner, A., Cole, C. R. 1960. Neuropathology of canine toxoplasmosis. *Am. J. Vet. Res.* 21, 831–844.

1085. Koiwai, M., Hamaoka, T., Haritani, M., Shimizu, S., Kimura, K. 2005. Proportion of abortions due to neosporosis among dairy cattle in Japan. *J. Vet. Med. Sci.* 67, 1173–1175.

1086. Koiwai, M., Hamaoka, T., Haritani, M., Shimizu, S., Tsutsui, T., Eto, M., Yamane, I. 2005. Seroprevalence of *Neospora caninum* in dairy and beef cattle with reproductive disorders in Japan. *Vet. Parasitol.* 130, 15–18.

1087. Koiwai, M., Hamaoka, T., Haritani, M., Shimizu, S., Zeniya, Y., Eto, M., Yokoyama, R., Tsutsui, T., Kimura, K., Yamane, I. 2006. Nationwide seroprevalence of *Neospora caninum* among dairy cattle in Japan. *Vet. Parasitol.* 135, 175–179.

1088. Kong, K., Rowlands, C. J., Elsheikha, H., Notingher, I. 2012. Label-free molecular analysis of live *Neospora caninum* tachyzoites in host cells by selective scanning Raman micro-spectroscopy. *Analyst* 137, 4119–4122.

1089. Konnai, S., Mingala, C. N., Sato, M., Abes, N. S., Venturina, F. A., Gutierrez, C. A., Sano, T., Omata, Y., Cruz, L. C., Onuma, M., Ohashi, K. 2008. A survey of abortifacient infectious agents in livestock in Luzon, the Philippines, with emphasis on the situation in a cattle herd with abortion problems. *Acta Tropica* 105, 269–273.

1090. Konrad, J. L., Crudeli, G. Á., Olazarri, M. J., Cano, D. B., Leunda, M. R., Odeón, A. C., Moore, D. P., Campero, C. M. 2009. Seroprevalence of *Neospora caninum* in buffaloes (*Bubalus bubalis*) in five ranches of Northeast Argentina. *Pakistan J. Zool. Suppl. Ser.* 9, 789–791.

1091. Konrad, J. L., Moore, D. P., Crudeli, G., Caspe, S. G., Cano, D. B., Leunda, M. R., Lischinsky, L., Regidor-Cerrillo, J., Odeón, A. C., Ortega-Mora, L. M., Echaide, I., Campero, C. M. 2012. Experimental inoculation of *Neospora caninum* in pregnant water buffalo. *Vet. Parasitol.* 187, 72–78.

1092. Konrad, J. L., Campero, L. M., Caspe, G. S., Brihuega, B., Draghi, G., Moore, D. P., Crudeli, G. A., Venturini, M. C., Campero, C. M. 2013. Detection of antibodies against *Brucella abortus*, *Leptospira* spp., and Apicomplexa protozoa in water buffaloes in the Northeast of Argentina. *Trop. Anim. Health Prod.* 45, 1751–1756.

1093. Kornberg, M., Kosfeld, H. U. 1997. *Neospora caninum* bei einem Hund. [*Neospora caninum* in a dog]. *Kleintierpraxis* 42, 235–240.

1094. Koudela, B., Doležel, R. 1997. Kokcidie *Neospora caninum*—nový puvodce infekcního zmetání skotu. [Coccidia *Neospora caninum*—new producer of cattle infectious abortion]. *Veterinárství* 6, 240–242.

1095. Koudela, B., Svoboda, M., Björkman, C., Uggla, A. 1998. Neosporosis in dogs: The first case report in the Czech Republic. *Vet. Med. Czech* 43, 51–54.

1096. Kowalczyk, S. J., Czopowicz, M., Weber, C. N., Müller, E., Witkowski, L., Kaba, J. 2016. Herd-level seroprevalence of *Neospora caninum* infection in dairy catle in central and northeastern Poland. *Acta Parasitol.* 61, 63–65.

1097. Koyama, T., Kobayashi, Y., Omata, Y., Yamada, M., Furuoka, H., Maeda, R., Matsui, T., Saito, A., Mikami, T. 2001. Isolation of *Neospora caninum* from the brain of a pregnant sheep. *J. Parasitol.* 87, 1486–1488.

1098. Kramer, A. M., Wouda, W., Kooistra, H. S. 2000. [Clinical neosporosis in the dog: A review]. *Tijdschr. Diergeneeskd.* 125, 609–613.

1099. Kramer, L., de Risio, L., Tranquillo, V. M., Magnino, S., Genchi, C. 2004. Analysis of risk factors associated with seropositivity to *Neospora caninum* in dogs. *Vet. Rec.* 154, 692–693.

1100. Kramsky, J. A., Manning, E. J. B., Collins, M. T. 2003. Protein G to enriched serum immunoglobulin from nondomestic hoofstock species. *J. Vet. Diagn. Invest.* 15, 253–261.

1101. Krishna, R., Xia, D., Sanderson, S., Shanmugasundram, A., Vermont, S., Bernal, A., Daniel-Naguib, G., Ghali, F., Brunk, B. P., Roos, D. S., Wastling, J. M., Jones, A. R. 2015. A large-scale proteogenomics study of apicomplexan pathogens-*Toxoplasma gondii* and *Neospora caninum*. *Proteomics* 15, 2618–2628.

1102. Kritzner, S., Sager, H., Blum, J., Krebber, R., Greif, G., Gottstein, B. 2002. An explorative study to assess the efficacy of Toltrazuril-sulfone (Ponazuril) in calves experimentally infected with *Neospora caninum*. *Ann. Clin. Microbiol. Antimicrob.* 1, 4.

1103. Kubota, N., Sakata, Y., Miyazaki, N., Itamoto, K., Bannai, H., Nishikawa, Y., Xuan, X., Inokuma, H. 2008. Serological survey of *Neospora caninum* infection among dogs in Japan through species-specific ELISA. *J. Vet. Med. Sci.* 70, 869–872.

1104. Kul, O., Kabakci, N., Yildiz, K., Öcal, N., Kalender, H., Ilkme, N. A. 2009. *Neospora caninum* associated with epidemic abortions in dairy cattle: The first clinical neosporosis report in Turkey. *Vet. Parasitol.* 159, 69–72.

1105. Kul, O., Atmaca, H. T., Anteplioglu, T., Ocal, N., Canpolat, S. 2015. *Neospora caninum*: The first demonstration of the enteroepithelial stages in the intestines of a naturally infected dog. *J. Comp. Pathol.* 153, 9–13.

1106. Kurtdede, A., Küplülü, S., Ural, K., Cingi, C. Ç., Güzel, M., Karakurum, M. C., Haydardedeoğlu, A. E. 2006. Serodiagnosis of bovine neosporosis with immunocomb assay in Ankara region. *Ankara Üniv. Vet. Fak. Derg.* 53, 207–209.

1107. Kuruca, L., Spasojević-Kosić, L., Simin, S., Savović, M., Laus, S., Lalosević, V. 2013. *Neospora caninum* antibodies in dairy cows and domesic dogs from Vojvodina, Serbia. *Parasite* 20, 40.

1108. Kwon, H. J., Kim, J. H., Kim, M., Lee, J. K., Hwang, W. S., Kim, D. Y. 2003. Anti-parasitic activity of depudecin on *Neospora caninum* via the inhibition of histone deacetylase. *Vet. Parasitol.* 112, 269–276.

1109. Kyaw, T., Virakul, P., Muangyai, M., Suwimonteerabutr, J. 2004. *Neospora caninum* seroprevalence in dairy cattle in central Thailand. *Vet. Parasitol.* 121, 255–263.

1110. Kyaw, T., Suwimonteerabutr, J., Virakul, P., Lohachit, C., Kalpravidh, W. 2005. Seronegative conversion in four *Neospora caninum*-infected cows, with a low rate of transplacental transmission. *Vet. Parasitol.* 131, 145–150.

1111. La Perle, K. M. D., Del Piero, F., Carr, R. F., Harris, C., Stromberg, P. C. 2001. Cutaneous neosporosis in two adult dogs on chronic immunosuppressive therapy. *J. Vet. Diagn. Invest.* 13, 252–255.

1112. Lally, N., Jenkins, M., Liddell, S., Dubey, J. P. 1997. A dense granule protein (NCDG1) gene from *Neospora caninum*. *Mol. Biochem. Parasitol.* 87, 239–243.

1113. Lally, N. C., Jenkins, M. C., Dubey, J. P. 1996. Evaluation of two *Neospora caninum* recombinant antigens for use in an enzyme-linked immunosorbent assay for the diagnosis of bovine neosporosis. *Clin. Diagn. Lab. Immunol.* 3, 275–279.

1114. Lally, N. C., Jenkins, M. C., Dubey, J. P. 1996. Development of a polymerase chain reaction assay for the diagnosis of neosporosis using the *Neospora caninum* 14-3-3 gene. *Mol. Biochem. Parasitol.* 75, 169–178.

1115. Landmann, J. K., Jillella, D., O'Donoghue, P. J., McGowan, M. R. 2002. Confirmation of the prevention of vertical transmission of *Neospora caninum* in cattle by the use of embryo transfer. *Aust. Vet. J.* 80, 502–503.

1116. Landmann, J. K., Gunn, A. A., O'Donoghue, P. J., Tranter, W. P., McGowan, M. R. 2011. Epidemiology and impact of *Neospora caninum* infection in three Queensland tropical dairy herds. *Reprod. Dom. Anim.* 46, 734–737.

1117. Langoni, H., Greca, H., Guimarães, F. F., Ullmann, L. S., Gaio, F. C., Uehara, R. S., Rosa, E. P., Amorim, R. M., da Silva, R. C. 2011. Serological profile of *Toxoplasma gondii* and *Neospora caninum* infection in commercial sheep from São Paulo State, Brazil. *Vet. Parasitol.* 177, 50–54.

1118. Langoni, H., Matteucci, G., Medici, B., Camossi, L. G., Richini-Pereira, V. B., da Silva, R. C. 2012. Detection and molecular analysis of *Toxoplasma gondii* and *Neospora caninum* from dogs with neurological disorders. *Rev. Soc. Bras. Med. Trop.* 45, 365–368.

1119. Langoni, H., Fornazari, F., da Silva, R. C., Monti, E. T., Villa, F. B. 2013. Prevalence of antibodies against *Toxoplasma gondii* and *Neospora caninum* in dogs. *Braz. J. Microbiol.* 44, 1327–1330.

1120. Larson, R. L., Hardin, D. K. 2003. Review: *Neospora caninum*-induced abortion in cattle. *Bovine Practitioner* 37, 121–126.

1121. Larson, R. L., Hardin, D. K., Pierce, V. L. 2004. Economic considerations for diagnostic and control options for *Neospora caninum*-induced abortions in endemically infected herds of beef cattle. *J. Am. Vet. Med. Assoc.* 224, 1597–1604.

1122. Laskoski, L. M., Muraro, L. S., Dittrich, R. L., Abreu, R. A., Koch, M. O., Silva, F. T., Hagi, R. H. 2015. Occurrence of anti-*Neospora caninum* and anti-*Toxoplasma gondii* antibodies in horses in the Pantanal of Mato Grosso, Brazil. *Semi. Ci. Agrár.* 36, 895–900.

1123. Lasri, S., De Meerschman, F., Rettigner, C., Focant, C., Losson, B. 2004. Comparison of three techniques for the serological diagnosis of *Neospora caninum* in the dog and their use for epidemiological studies. *Vet. Parasitol.* 123, 25–32.

1124. Lassen, B., Viltrop, A., Raaperi, K., Järvis, T. 2008. *Neospora caninum* antibodies in bulk milk and serum from Estonian dairy farms. *Proceedings of the International Scientific Conference Animals Health, Food Hygiene*. November 14, Jelgava, Latvia, 105–107.

1125. Lassen, B., Orro, T., Aleksejev, A., Raaperi, K., Järvis, T., Viltrop, A. 2012. *Neospora caninum* in Estonian dairy herds in relation to herd size, reproduction parameters, bovine virus diarrhoea virus, and bovine herpes virus 1. *Vet. Parasitol.* 190, 43–50.

1126. Lathe, C. L. 1994. *Neospora caninum* in British dogs. *Vet. Rec.* 134, 532.

1127. Latif, B. M., Jakubek, E. B. 2008. Determination of the specificities of monoclonal and polyclonal antibodies to *Neospora*, *Toxoplasma* and *Cryptosporidium* by fluorescent antibody test (FAT). *Trop. Biomed.* 25, 225–231.

1128. Lee, E. G., Kim, J. H., Shin, Y. S., Shin, G. W., Suh, M. D., Kim, D. Y., Kim, Y. H., Kim, G. S., Jung, T. S. 2003. Establishment of a two-dimensional electrophoresis map for *Neospora caninum* tachyzoites by proteomics. *Proteomics* 3, 2339–2350.

1129. Lee, E. G., Kim, J. H., Shin, Y. S., Shin, G. W., Kim, Y. H., Kim, G. S., Kim, D. Y., Jung, T. S., Suh, M. D. 2004. Two-dimensional gel electrophoresis and immunoblot analysis of *Neospora caninum* tachyzoites. *J. Vet. Sci.* 5, 139–145.

1130. Lee, E. G., Kim, J. H., Shin, Y. S., Shin, G. W., Kim, Y. R., Palaksha, K. J., Kim, D. Y., Yamane, I., Kim, Y. H., Kim, G. S., Suh, M. D., Jung, T. S. 2005. Application of proteomics for comparison of proteome of *Neospora caninum* and *Toxoplasma gondii* tachyzoites. *J. Chromatogr. B* 815, 305–314.

1131. Lee, J. K., Kim, J. H., Kim, J. H., Lee, B. C., Hwang, W. S., Youn, H. J., Nam, H. W., Jean, Y. W., Kim, D. Y. 2001. [Establishment of diagnostic method for bovine neosporosis by PCR using paraffin block]. *Korean J. Vet. Res.* 41, 381–385 (in Korean).

1132. Leepin, A., Stüdli, A., Brun, R., Stephens, C. E., Boykin, D. W., Hemphill, A. 2008. Host cells participate in the *in vitro* effects of novel diamidine analogues against tachyzoites of the intracellular apicomplexan parasites *Neospora caninum* and *Toxoplasma gondii*. *Antimicrob. Agents Chemother.* 52, 1999–2008.

1133. Legnani, S., Pantchev, N., Forlani, A., Zini, E., Schares, G., Balzer, J., Roccabianca, P., Ferri, F., Zanna, G. 2015. Emergence of cutaneous neosporosis in a dog receiving immunosuppressive therapy: Molecular identification and management. *Vet. Dermatol.* 27, 49–e14.

1134. Lehenbauer, T. W., Rodgers, S. J., Helman, R. G., Saliki, J. T. 1998. Epidemiology of *Neospora caninum* infection in Oklahoma beef and dairy cattle. *Proceedings of the 31st Annual Convention of the American Association of Bovine Practioners*. September 24–28, Spokane, WA, 31, 225.

1135. Lei, T., Wang, H., Liu, J., Nan, H., Liu, Q. 2014. ROP18 is a key factor responsible for virulence difference between *Toxoplasma gondii* and *Neospora caninum*. *PLoS ONE* 9, e99744.

1136. Lei, Y., Davey, M., Ellis, J. T. 2005. Attachment and invasion of *Toxoplasma gondii* and *Neospora caninum* to epithelial and fibroblast cell lines *in vitro*. *Parasitology* 131, 583–590.

1137. Lei, Y., Davey, M., Ellis, J. 2005. Autofluorescence of *Toxoplasma gondii* and *Neospora caninum* cysts *in vitro*. *J. Parasitol.* 91, 202–208.

1138. Lei, Y., Birch, D., Davey, M., Ellis, J. T. 2005. Subcellular fractionation and molecular characterization of the pellicle and plasmalemma of *Neospora caninum*. *Parasitology* 131, 467–475.

1139. Lélu, M., Gilot-Fromont, E., Aubert, D., Richaume, A., Afonso, E., Dupuis, E., Gotteland, C., Marnef, F., Poulle, M. L., Dumètre, A., Thulliez, P., Dardé, M. L., Villena, I. 2011. Development of a sensitive method for *Toxoplasma gondii* oocyst extraction in soil. *Vet. Parasitol.* 183, 59–67.

1140. Lemberger, K. Y., Gondim, L. F. P., Pessier, A. P., McAllister, M. M., Kinsel, M. J. 2005. *Neospora caninum* infection in a free-ranging raccoon (*Procyon lotor*) with concurrent canine distemper virus infection. *J. Parasitol.* 91, 960–961.

1141. Leon, A., Richard, E., Fortier, C., Laugier, C., Fortier, G., Pronost, S. 2012. Molecular detection of *Coxiella burnetii* and *Neospora caninum* in equine aborted foetuses and neonates. *Prev. Vet. Med.* 104, 179–183.

1142. Li, J., He, P., Yu, Y., Du, L., Gong, P., Zhang, G., Zhang, X. 2014. Detection of *Neospora caninum*-DNA in feces collected from dogs in Shenyang (China) and ITS1 phylogenetic analysis. *Vet. Parasitol.* 205, 361–364.

1143. Li, K., Han, Z., Shahzad, M., Li, J. 2015. Seroepidemiology of *Neospora caninum* infection in yaks (*Bos grunniens*) in Tibet and Hongyuan of Sichuan, China. *Cattle Pract.* 23, 142–143.

1144. Li, L., Crabtree, J., Fischer, S., Pinney, D., Stoeckert, C. J., Sibley, L. D., Roos, D. S. 2004. ApiEST-DB: Analyzing clustered EST data of the apicomplexan parasites. *Nucleic Acids Res.* 32, D326–D328.

1145. Li, R. W., Tuo, W. 2011. *Neospora caninum*: Comparative gene expseression profiling of *Neospora caninum* wild type and a temperature sensitive clone. *Exp. Parasitol.* 129, 346–354.

1146. Li, W., Liu, J., Wang, J., Fu, Y., Nan, H., Liu, Q. 2015. Identification and characaterization of a microneme protein (NcMIC6) in *Neospora caninum*. *Parasitol. Res.* 114, 2893–2902.

1147. Li, W. C., Liu, D. Y., Zhou, L., Cao, J. T., Mei, N., Zhang, H., Chen, C., Chen, T., Han, M. M., Zhang, W., Fan, Z. L., Gu, Y. F. 2015. [Prevalence of *Neospora caninum* infection in the intestine of pet dogs in some areas of Anhui and Zhejiang]. *Chin. J. Parasitol. Parasit. Dis.* 33, 283–286 (in Chinese).

1148. Liao, M., Zhang, S., Xuan, X., Zhang, G., Huang, X., Igarashi, I., Fujisaki, K. 2005. Development of rapid immunochromatographic test with recombinant NcSAG1 for detection of antibodies to *Neospora caninum* in cattle. *Clin. Diagn. Lab. Immunol.* 12, 885–887.

1149. Liao, M., Xuan, X., Huang, X., Shirafuji, H., Fukumoto, S., Hirata, H., Suzuki, H., Fujisaki, K. 2005. Identification and characterization of cross-reactive antigens from *Neospora caninum* and *Toxoplasma gondii*. *Parasitology* 130, 481–488.

1150. Liao, M., Ma, L., Bannai, H., Lee, E. G., Xie, Z., Tang, X., Zhang, H., Xuan, X., Fujisaki, K. 2006. Identification of a protein disulfide isomerase of *Neospora caninum* in excretory-secretory products and its IgA binding and enzymatic activities. *Vet. Parasitol.* 139, 47–56.

1151. Liddell, S., Lally, N. C., Jenkins, M. C., Dubey, J. P. 1998. Isolation of the cDNA encoding a dense granule associated antigen (NCDG2) of *Neospora caninum*. *Mol. Biochem. Parasitol.* 93, 153–158.

1152. Liddell, S., Jenkins, M. C., Dubey, J. P. 1999. A competitive PCR assay for quantitative detection of *Neospora caninum*. *Int. J. Parasitol.* 29, 1583–1587.

1153. Liddell, S., Jenkins, M. C., Dubey, J. P. 1999. Vertical transmission of *Neospora caninum* in BALB/c mice determined by polymerase chain reaction detection. *J. Parasitol.* 85, 550–555.

1154. Liddell, S., Jenkins, M. C., Collica, C. M., Dubey, J. P. 1999. Prevention of vertical transfer of *Neospora caninum* in BALB/c mice by vaccination. *J. Parasitol.* 85, 1072–1075.

1155. Liddell, S., Parker, C., Vinyard, B., Jenkins, M., Dubey, J. P. 2003. Immunization of mice with plasmid DNA coding for NcGRA7 or NcsHSP33 confers partial protection against vertical transmission of *Neospora caninum*. *J. Parasitol.* 89, 496–500.

1156. Lima Junior, M. S. C., Andreotti, R., Caetano, A. R., Paiva, F., Matos, M. F. C. 2007. Cloning and expression of an antigenic domain of a major surface protein (NC-P43) of *Neospora caninum*. *Rev. Bras. Parasitol. Vet.* 16, 61–66.

1157. Lindsay, D. S., Dubey, J. P. 1989. Immunohistochemical diagnosis of *Neospora caninum* in tissue sections. *Am. J. Vet. Res.* 50, 1981–1983.

1158. Lindsay, D. S., Dubey, J. P. 1989. *In vitro* development of *Neospora caninum* (Protozoa: Apicomplexa) from dogs. *J. Parasitol.* 75, 163–165.

1159. Lindsay, D. S., Dubey, J. P. 1989. *Neospora caninum* (Protozoa: Apicomplexa) infections in mice. *J. Parasitol.* 75, 772–779.

1160. Lindsay, D. S., Dubey, J. P. 1989. Evaluation of anti-coccidial drugs' inhibition of *Neospora caninum* development in cell cultures. *J. Parasitol.* 75, 990–992.

1161. Lindsay, D. S., Dubey, J. P. 1990. *Neospora caninum* (Protozoa: Apicomplexa) infections in rats. *Can. J. Zool.* 68, 1595–1599.

1162. Lindsay, D. S., Dubey, J. P. 1990. Effects of sulfadiazine and amprolium on *Neospora caninum* (Protozoa: Apicomplexa) infections in mice. *J. Parasitol.* 76, 177–179.

1163. Lindsay, D. S., Dubey, J. P. 1990. Infections in mice with tachyzoites and bradyzoites of *Neospora caninum* (Protozoa: Apicomplexa). *J. Parasitol.* 76, 410–413.

1164. Lindsay, D. S., Blagburn, B. L., Dubey, J. P. 1990. Infection of mice with *Neospora caninum* (Protozoa: Apicomplexa) does not protect against challenge with *Toxoplasma gondii*. *Infect. Immun.* 58, 2699–2700.

1165. Lindsay, D. S., Dubey, J. P., Upton, S. J., Ridley, R. K. 1990. Serological prevalence of *Neospora caninum* and *Toxoplasma gondii* in dogs from Kansas. *J. Helminthol. Soc. Wash.* 57, 86–88.

1166. Lindsay, D. S., Dubey, J. P., Blagburn, B. L. 1991. Characterization of a *Neospora caninum* (Protozoa: Apicomplexa) isolate (NC-3) in mice. *J. Alabama Acad. Sci.* 62, 1–8.

1167. Lindsay, D. S., Blagburn, B. L., Dubey, J. P. 1992. Factors affecting the survival of *Neospora caninum* bradyzoites in murine tissues. *J. Parasitol.* 78, 70–72.

1168. Lindsay, D. S., Speer, C. A., Toivio-Kinnucan, M. A., Dubey, J. P., Blagburn, B. L. 1993. Use of infected cultured cells to compare ultrastructural features of *Neospora caninum* from dogs and *Toxoplasma gondii*. *Am. J. Vet. Res.* 54, 103–106.

1169. Lindsay, D. S., Dubey, J. P., Cole, R. A., Nuehring, L. P., Blagburn, B. L. 1993. *Neospora*-induced protozoal abortions in cattle. *Comp. Cont. Edu. Pract. Vet.* 15, 882–889.

1170. Lindsay, D. S., Blagburn, B. L., Dubey, J. P. 1994. Neosporosis: A recently recognized cause of protozoal abortion and neonatal mortality in cattle and other animals. *Curr. Top. Vet. Res.* 1, 53–62.

1171. Lindsay, D. S., Rippey, N. S., Cole, R. A., Parsons, L. C., Dubey, J. P., Tidwell, R. R., Blagburn, B. L. 1994. Examination of the activities of 43 chemotherapeutic agents against *Neospora caninum* tachyzoites in cultured cells. *Am. J. Vet. Res.* 55, 976–981.

1172. Lindsay, D. S., Lenz, S. D., Cole, R. A., Dubey, J. P., Blagburn, B. L. 1995. Mouse model for central nervous system *Neospora caninum* infections. *J. Parasitol.* 81, 313–315.

1173. Lindsay, D. S., Rippey, N. S., Powe, T. A., Sartin, E. A., Dubey, J. P., Blagburn, B. L. 1995. Abortions, fetal death, and stillbirths in pregnant pygmy goats inoculated with tachyzoites of *Neospora caninum*. *Am. J. Vet. Res.* 56, 1176–1180.

1174. Lindsay, D. S., Butler, J. M., Rippey, N. S., Blagburn, B. L. 1996. Demonstration of synergistic effects of sulfonamides and dihydrofolate reductase/thymidylate synthase inhibitors against *Neospora caninum* tachyzoites in cultured cells, and characterization of mutants resistant to pyrimethamine. *Am. J. Vet. Res.* 57, 68–72.

1175. Lindsay, D. S., Steinberg, H., Dubielzig, R. R., Semrad, S. D., Konkle, D. M., Miller, P. E., Blagburn, B. L. 1996. Central nervous system neosporosis in a foal. *J. Vet. Diagn. Invest.* 8, 507–510.

1176. Lindsay, D. S., Kelly, E. J., McKown, R. D., Stein, F. J., Plozer, J., Herman, J., Blagburn, B. L., Dubey, J. P. 1996. Prevalence of *Neospora caninum* and *Toxoplasma gondii* antibodies in coyotes (*Canis latrans*) and experimental infections of coyotes with *Neospora caninum*. *J. Parasitol.* 82, 657–659.

1177. Lindsay, D. S., Butler, J. M., Blagburn, B. L. 1997. Efficacy of decoquinate against *Neospora caninum* tachyzoites in cell cultures. *Vet. Parasitol.* 68, 35–40.

1178. Lindsay, D. S., Lenz, S. D., Dykstra, C. C., Blagburn, B. L., Dubey, J. P. 1998. Vaccination of mice with *Neospora caninum*: Response to oral challenge with *Toxoplasma gondii* oocysts. *J. Parasitol.* 84, 311–315.

1179. Lindsay, D. S. 1999. *Neospora caninum* and neosporosis. *Int. J. Parasitol.* 29, 1483.

1180. Lindsay, D. S., Dubey, J. P., Duncan, R. B. 1999. Confirmation that the dog is a definitive host for *Neospora caninum*. *Vet. Parasitol.* 82, 327–333.

1181. Lindsay, D. S., Upton, S. J., Dubey, J. P. 1999. A structural study of the *Neospora caninum* oocyst. *Int. J. Parasitol.* 29, 1521–1523.

1182. Lindsay, D. S., Dubey, J. P., McAllister, M. E. 1999. *Neospora caninum* and the potential for parasite transmission. *Comp. Cont. Edu. Pract. Vet.* 21, 317–321.

1183. Lindsay, D. S., Lenz, S. D., Blagburn, B. L., Brake, D. A. 1999. Characterization of temperature-sensitive strains of *Neospora caninum* in mice. *J. Parasitol.* 85, 64–67.

1184. Lindsay, D. S., Dubey, J. P. 2000. Canine neosporosis. *J. Vet. Parasitol.* 14, 1–11.

1185. Lindsay, D. S., Dykstra, C. C., Williams, A., Spencer, J. A., Lenz, S. D., Palma, K., Dubey, J. P., Blagburn, B. L. 2000. Inoculation of *Sarcocystis neurona* merozoites into the central nervous system of horses. *Vet. Parasitol.* 92, 157–163.

1186. Lindsay, D. S. 2001. Neosporosis: An emerging protozoal disease of horses. *Equine Vet. J.* 33, 116–118.

1187. Lindsay, D. S., Ritter, D. M., Brake, D. 2001. Oocyst excretion in dogs fed mouse brains containing tissue cysts of a cloned line of *Neospora caninum*. *J. Parasitol.* 87, 909–911.

1188. Lindsay, D. S., Weston, J. L., Little, S. E. 2001. Prevalence of antibodies to *Neospora caninum* and *Toxoplasma gondii* in gray foxes (*Urocyon cinereoargenteus*) from South Carolina. *Vet. Parasitol.* 97, 159–164.

1189. Lindsay, D. S., Spencer, J., Rupprecht, C. E., Blagburn, B. L. 2001. Prevalence of agglutinating antibodies to *Neospora caninum* in raccoons, *Procyon lotor. J. Parasitol.* 87, 1197–1198.

1190. Lindsay, D. S., Little, S. E., Davidson, W. R. 2002. Prevalence of antibodies to *Neospora caninum* in white-tailed deer, *Odocoileus virginianus*, from the Southeastern United States. *J. Parasitol.* 88, 415–417.

1191. Lista-Alves, D., Palomares-Naveda, R., Garcia, F., Obando, C., Arrieta, D., Hoet, A. E. 2006. Serological evidence of *Neospora caninum* in dual-purpose cattle herds in Venezuela. *Vet. Parasitol.* 136, 347–349.

1192. Little, P. B. 1996. Central nervous system rendez-vous: Canine progressive posterior paresis. *Can. Vet. J.* 37, 55–56.

1193. Liu, J., Yu, J., Wang, M., Liu, Q., Zhang, W., Deng, C., Ding, J. 2007. Serodiagnosis of *Neospora caninum* infection in cattle using a recombinant tNcSRS2 protein-based ELISA. *Vet. Parasitol.* 143, 358–363.

1194. Liu, J., Cai, J. Z., Zhang, W., Liu, Q., Chen, D., Han, J. P., Liu, Q. R. 2008. Seroepidemiology of *Neospora caninum* and *Toxoplasma gondii* infection in yaks (*Bos grunniens*) in Qinghai, China. *Vet. Parasitol.* 152, 330–332.

1195. Liu, M., Chen, Q., Jiergele, Xu, Z., Bayinchahan. 2016. [Initial report on seroprevalence of neosporosis in horses in northern areas of Xinjiang]. *Heilongjiang Anim. Husb. Vet. Med.* 2016(01), 91–93 (in Chinese).

1196. Liu, Q., Li, B., Chai, C., Zhu, Y. 2003. [Serological diagnosis of *Neospora caninum* infection in dairy cattle: Preliminary report]. *Chin. J. Vet. Med.* 39, 8–9 (in Chinese).

1197. Liu, Z. K., Li, J. Y., Pan, H. 2015. Seroprevalence and risk factors of *Toxoplasma gondii* and *Neospora caninum* infections in small ruminants in China. *Prev. Vet. Med.* 118, 488–492.

1198. Lobato, J., Silva, D. A. O., Mineo, T. W. P., Amaral, J. D. H. F., Segundo, G. R. S., Costa-Cruz, J. M., Ferreira, M. S., Borges, A. S., Mineo, J. R. 2006. Detection of immunoglobulin G antibodies to *Neospora caninum* in humans: High seropositivity rates in patients who are infected by human immunodeficiency virus or have neurological disorders. *Clin. Vaccine Immunol.* 13, 84–89.

1199. Locatelli-Dittrich, R., Soccol, V. T., Richartz, R. R. T. B., Gasino-Joineau, M. E., Vinne, R., Pinckney, R. D. 2001. Serological diagnosis of neosporosis in a herd of dairy cattle in southern Brazil. *J. Parasitol.* 87, 1493–1494.

1200. Locatelli-Dittrich, R., Thomaz Soccol, V., Richartz, R. R. T. B., Gasino-Joineau, M. E., van der Vinne, R., Silva, R., Leite, L. C., Pinckney, R. 2001. Detecção de anticorpos contra *Neospora caninum* em vacas leiteiras e bezerros no estado do Paraná. [Detection of antibodies against *Neospora caninum* in dairy cattle and calf from Paraná state]. *Arch. Vet. Sci.* 6, 37–41.

1201. Locatelli-Dittrich, R., Richartz, R. R. T. B., Gasino-Joineau, M. E., Pinckney, R. D., de Sousa, R. S., Leite, L. C., Thomaz-Soccol, V. 2003. Isolation of *Neospora caninum* from a blind calf in Paraná, southern Brazil. *Vet. Rec.* 153, 366–367.

1202. Locatelli-Dittrich, R., Thomaz-Soccol, V., Richartz, R. R. T. B., Gasino-Joineau, M. E., van der Vinne, R., Pinckney, R. D. 2004. Isolamento de *Neospora caninum* de feto bovino de rebanho leiteiro no Paraná. *Rev. Bras. Parasitol. Vet.* 13, 103–109.

1203. Locatelli-Dittrich, R., Dittrich, J. R., Richartz, R. R. T. B., Gasino Joineau, M. E., Antunes, J., Pinckney, R. D., Deconto, I., Hoffmann, D. C. S., Thomaz-Soccol, V. 2006. Investigation of *Neospora* sp. and *Toxoplasma gondii* antibodies in mares and in precolostral foals from Parana State, Southern Brazil. *Vet. Parasitol.* 135, 215–221.

1204. Locatelli-Dittrich, R., Machado, P. C., Fridlund-Plugge, N., Richartz, R. R. T. B., Montiani-Ferreira, F., Patrício, L. F. L., Patrício, M. A. C., Joineau, M. G., Pieppe, M. 2008. Determinação e correlação de anticorpos anti-*Neospora caninum* em bovinos e cães do Paraná, Brasil. [Determination and correlation of anti-*Neospora caninum* antibodies in cattle and dogs from Paraná, Brazil]. *Rev. Bras. Parasitol. Vet.* 17 (Suppl 1), 191–196.

1205. Loeffler, I. K., Howard, J., Montali, R. J., Hayek, L. A., Dubovi, E., Zhang, Z., Yan, Q., Guo, W., Wildt, D. E. 2007. Serosurvey of ex situ giant pandas (*Ailuropoda melanoleuca*) and red pandas (*Ailurus fulgens*) in China with implications for species conservation. *J. Zoo Wildl. Med.* 38, 559–566.

1206. Long, M. T., Baszler, T. V. 1996. Fetal loss in BALB/c mice infected with *Neospora caninum. J. Parasitol.* 82, 608–611.

1207. Long, M. T., Baszler, T. V., Mathison, B. A. 1998. Comparison of intracerebral parasite load, lesion development, and systemic cytokines in mouse strains infected with *Neospora caninum. J. Parasitol.* 84, 316–320.

1208. Long, M. T., Baszler, T. V. 2000. Neutralization of maternal IL-4 modulates congenital protozoal transmission: Comparison of innate versus acquired immune responses. *J. Immunol.* 164, 4768–4774.

1209. Longshore, R. C. 1996. What is your neurologic diagnosis? *J. Am. Vet. Med. Assoc.* 208, 667–668.

1209a. Loo, C. S. N., Lam, N. S. K., Yu, D., Su, X.-Z., Lu, F. 2017. Artemisinin and its derivatives in treating protozoan infections beyond malaria. *Pharmacological Research.* 117, 192–217.

1210. Loobuyck, M., Frössling, J., Lindberg, A., Björkman, C. 2009. Seroprevalence and spatial distribution of *Neospora caninum* in a population of beef cattle. *Prev. Vet. Med.* 92, 116–122.

1211. Lopes, M. G., Mendonça, I. L., Fortes, K. P., Amaku, M., Pena, H. F. J., Gennari, S. M. 2011. Presence of antibodies against *Toxoplasma gondii, Neospora caninum* and *Leishmania infantum* in dogs from Piauí. *Rev. Bras. Parasitol. Vet.* 20, 111–114.

1212. Lopes, M. G., Hernandez, M., de Lima, J. T. R., Grisi Filho, J. J. H., Gennari, S. M. 2015. Occurrence of antibodies anti-*Toxoplasma gondii* and anti-*Neospora caninum* in dogs from Natal, RN, Brazil. *Braz. J. Vet. Res. Anim. Sci.* 52, 120–124.

1213. López-Gatius, F. 2003. Is fertility declining in dairy cattle? A retrospective study in northeastern Spain. *Theriogenology* 60, 89–99.

1214. López-Gatius, F., Pabón, M., Almería, S. 2004. *Neospora caninum* infection does not affect early pregnancy in dairy cattle. *Theriogenology* 62, 606–613.

1215. López-Gatius, F., López-Béjar, M., Murugavel, K., Pabón, M., Ferrer, D., Almería, S. 2004. *Neospora*-associated abortion episode over a 1-year period in a dairy herd in North-east Spain. *J. Vet. Med. B* 51, 348–352.

1216. López-Gatius, F., Santolaria, P., Almería, S. 2005. *Neospora caninum* infection does not affect the fertility of dairy cows in herds with high incidence of *Neospora*-associated abortions. *J. Vet. Med. B* 52, 51–53.

1217. López-Gatius, F., Santolaria, P., Yániz, J. L., Garbayo, J. M., Almería, S. 2005. The use of beef bull semen reduced the risk of abortion in *Neospora*-seropositive dairy cows. *J. Vet. Med. B* 52, 88–92.

1218. López-Gatius, F., García-Ispierto, I., Santolaria, P., Yániz, J. L., López-Béjar, M., Norgareda, C., Almería, S. 2005. Relationship between rainfall and *Neospora caninum*-associated abortion in two dairy herds in a dry environment. *J. Vet. Med. B* 52, 147–152.

1219. López-Gatius, F., Garbayo, J. M., Santolaria, P., Yániz, J. L., Almería, S., Ayad, A., de Sousa, N. M., Beckers, J. F. 2007. Plasma pregnancy-associated glycoprotein-1 (PAG-1) concentrations during gestation in *Neospora*-infected dairy cows. *Theriogenology* 67, 502–508.

1220. López-Gatius, F., Almería, S., Donofrio, G., Nogareda, C., García-Ispierto, I., Bech-Sàbat, G., Santolaria, P., Yániz, J. L., Pabón, M., de Sousa, N. M., Beckers, J. F. 2007. Protection against abortion linked to gamma interferon production in pregnant dairy cows naturally infected with *Neospora caninum*. *Theriogenology* 68, 1067–1073.

1221. López-Pérez, I. C., Risco-Castillo, V., Collantes-Fernández, E., Ortega-Mora, L. M. 2006. Comparative effect of *Neospora caninum* infection in BALB/c mice at three different gestation periods. *J. Parasitol.* 92, 1286–1291.

1222. López-Pérez, I. C., Collantes-Fernández, E., Aguado-Martínez, A., Rodríguez-Bertos, A., Ortega-Mora, L. M. 2008. Influence of *Neospora caninum* infection in BALB/c mice during pregnancy in post-natal development. *Vet. Parasitol.* 155, 175–183.

1223. López-Pérez, I. C., Collantes-Fernández, E., Rojo-Montejo, S., Navarro-Lozano, V., Risco-Castillo, V., Pérez-Pérez, V., Pereira-Bueno, J., Ortega-Mora, L. M. 2010. Effects of *Neospora caninum* infection at mid-gestation on placenta in a pregnant mouse model. *J. Parasitol.* 96, 1017–1020.

1224. Lorenzo, V., Pumarola, M., Siso, S. 2002. Neosporosis with cerebellar involvement in an adult dog. *J. Small Anim. Pract.* 43, 76–79.

1225. Löschenberger, K., Rössel, C., Edelhofer, R., Suchy, A., Prosl, H. 2000. Diagnose von *Neospora caninum* beim Hund anhand eines Fallbeispiels. [Diagnosis of *Neospora caninum* in a dog—a case report]. *Tierärztl. Prax.* 28, 390–394.

1226. Löschenberger, K., Szölgyényi, W., Peschke, R., Prosl, H. 2004. Detection of the protozoan *Neospora caninum* using *in situ* polymerase chain reaction. *Biotechnic. Histochem.* 79, 101–105.

1227. Losson, B., Bourdoiseau, G. 2000. *Neospora caninum* un nouvel agent abortif chez les bovins. [*Neospora caninum*, a new abortive agent in cattle]. *Bull. des GTV* 7, 107–114.

1228. Louie, K., Sverlow, K. W., Barr, B. C., Anderson, M. L., Conrad, P. A. 1997. Cloning and characterization of two recombinant *Neospora* protein fragments and their use in serodiagnosis of bovine neosporosis. *Clin. Diagn. Lab. Immunol.* 4, 692–699.

1229. Louie, K., Conrad, P. A. 1999. Characterization of a cDNA encoding a subtilisin-like serine protease (NC-p65) of *Neospora caninum*. *Mol. Biochem. Parasitol.* 103, 211–223.

1230. Louie, K., Nordhausen, R., Robinson, T. W., Barr, B. C., Conrad, P. A. 2002. Characterization of *Neospora caninum* protease, NcSUB1 (NC-p65), with rabbit anti-N54. *J. Parasitol.* 88, 1113–1119.

1231. Lovett, J. L., Howe, D. K., Sibley, L. D. 2000. Molecular characterization of a thrombospondin-related anonymous protein homologue in *Neospora caninum*. *Mol. Biochem. Parasitol.* 107, 33–43.

1232. Lu, Y., Wang, G., Ma, L. 2006. [Serological investigation of neosporosis in cashmere goats]. *Chin. Anim. Prot.* 23, 36–37 (in Chinese).

1233. Lu, Y., Wang, G., Ma, L. 2007. [Serodiagnosis of *Neospora caninum* infection in cashmere goat by enzyme-linked immunosorbent assay with recombinant truncated NcSAG1t]. *Chin. J. Anim. Husb. Vet. Med.* 34, 109–111 (in Chinese).

1234. Lundén, A., Marks, J., Maley, S. W., Innes, E. A. 1998. Cellular immune responses in cattle experimentally infected with *Neospora caninum*. *Parasite Immunol.* 20, 519–526.

1235. Lundén, A., Wright, S., Allen, J. E., Buxton, D. 2002. Immunization of mice against neosporosis. *Int. J. Parasitol.* 32, 867–876.

1236. Luza, M., Serrano-Martínez, E., Tantaleán, M., Quispe, M., Casas, G. 2013. Primer reporte de *Neospora caninum*, en caballos de carrera de Lima, Perú. *Salud Tecnol. Vet.* 1, 40–45.

1237. Lv, Q., Li, J., Gong, P., Xing, S., Zhang, X. 2010. *Neospora caninum*: In vitro culture of tachyzoites in MCF-7 human breast carcinoma cells. *Exp. Parasitol.* 126, 536–539.

1238. Lv, Q., Xing, S., Gong, P., Chang, L., Bian, Z., Wang, L., Zhang, X., Li, J. 2015. A 78 kDa host cell invasion protein of *Neospora caninum* as a potential vaccine candidate. *Exp. Parasitol.* 148, 56–65.

1239. Lyon, C. 2010. Update on the diagnosis and management of *Neospora caninum* infections in dogs. *Top. Comp. Anim. Med.* 25, 170–175.

1240. Ma, L. 2006. Serological diagnosis of *Neospora caninum* infection in dairy cattle. *Heilongjiang Anim. Sci. Vet. Med.* 2006, 61–62.

1241. Ma, L. 2006. Diagnosis of neosporosis in beef cattle in Qinghai, China. *Animal Husb. Veter. Med.* 38, 46–47.

1242. Ma, L., Shen, Y. 2006. Serological diagnosis of *Neospora caninum* infection in yaks in Qinghai, China. *Chin. J. Vet. Med.* 42, 33–34.

1243. Ma, S., Ma, L. 2006. [Serological investigation of neosporosis in cashmere goats]. *Chin. J. Vet. Med.* 9, 25–26 (in Chinese).

1244. Macaldowie, C., Maley, S. W., Wright, S., Bartley, P., Esteban-Redondo, I., Buxton, D., Innes, E. 2004. Placental pathology associated with fetal death in cattle inoculated with *Neospora caninum* by two different routes in early pregnancy. *J. Comp. Pathol.* 131, 142–156.

1245. Machacová, T., Bártová, E., di Loria, A., Sedlák, K., Guccione, J., Fulgione, D., Veneziano, V. 2013. Seroprevalence and risk factore of *Neospora* spp. in donkeys from Southern Italy. *Vet. Parasitol.* 198, 201–204.

1246. Machacova, T., Bartova, E., Sedlak, K., Budikova, M., Piccirillo, A. 2015. Risk factors involved in transmission of *Toxoplasma gondii* and *Neospora caninum* infection in rabbit farms in Northern Italy. *Ann. Agric. Environ. Med.* 22, 677–679.

1247. Machacová, T., Ajzenberg, D., Žákovská, A., Sedlák, K., Bártová, E. 2016. *Toxoplasma gondii* and *Neospora caninum* in wild small mammals: Seroprevalence, DNA detection and genotyping. *Vet. Parasitol.* 223, 88–90.

1248. Machado, G. P., Kikuti, M., Langoni, H., Paes, A. C. 2011. Seroprevalence and risk factors associated with neosporosis in sheep and dogs from farms. *Vet. Parasitol.* 182, 356–358.

1249. Machado, R. Z., Mineo, T. W. P., Landim, L. P., Carvalho, A. F., Gennari, S. M., Miglino, M. A. 2007. Possible role of bovine trophoblast giant cells in transplacental transmission of *Neospora caninum* in cattle. *Rev. Bras. Parasitol. Vet.* 16, 21–25.

1250. Magaña, A., Sánchez, F., Villa, K., Rivera, L., Morales, E. 2015. Systemic neosporosis in a dog treated for immune-mediated thrombocytopenia and hemolytic anemia. *Vet. Clin. Pathol.* 44, 592–596.

1251. Magnino, S., Vigo, P. G., Bandi, C., Colombo, M., de Giuli, L., Fabbi, M., Genchi, C. 1998. PCR diagnosis for *Neospora caninum* infection in aborted bovine fetuses and for *Toxoplasma gondii* infection in hares and goats in Italy. *Proceedings of the IX International Congress of Parasitology, Makuhari Messe.* August 24–28, Chiba, Japan, 1269–1272.

1252. Magnino, S., Vigo, P. G., Bandi, C., Colombo, M., de Giuli, L., Fabbi, M., Genchi, C. 1998. A proposito di neosporosi in Italia. [*Neospora caninum*: Italian situation]. *Summa* 5, 25–27.

1253. Magnino, S., Vigo, P. G., Fabbi, M., Colombo, M., Bandi, C., Genchi, C. 1999. Isolation of a bovine *Neospora* from a newborn calf in Italy. *Vet. Rec.* 144, 456.

1254. Magnino, S., Vigo, P. G., Bazzocchi, C., Genchi, C., Bandi, C., Fabbi, M. 1999. Bovine neosporosis in Italy: Serological examinations, PCR and isolation in cell culture. Poster, 17th International Conference of the World Association for the Advancement of Veterinary Parasitology. August 15–19, Copenhagen, Denmark.

1255. Magnino, S., Vigo, P. G., Bandi, C., Bazzocchi, C., Fabbi, M., Genchi, C. 2000. Small-subunit rDNA sequencing of the Italian bovine *Neospora caninum* isolate (NC-PV1 strain). *Parassitologia* 42, 191–192.

1256. Magnino, S., Bandi, C., Vigo, P. G., Fabbi, M., Colombo, M., Colombo, N., Genchi, C. 2000. Diagnostica della neosporosi bovina nel nord Italia. [Diagnosis of bovine neosporosis in northern Italy]. *Large Anim. Rev.* 6, 25–29.

1257. Magnino, S., Vigo, P. G., Bandi, C., Rosignoli, C., Boldini, M., Vezzoli, F., Alborali, L., Cammi, G., Foni, E., Colombo, N., Colombo, M., Bergami, C., Mellini, A., Fabbi, M., Genchi, C. 2000. Neosporosi bovina in Italia: un biennio di attività diagnostica. [Bovine neosporosis in Italy: Two years of diagnostics]. *La Selezione Veterinaria* (Suppl), S15–S23.

1258. Magno, R. C., Lemgruber, L., Vommaro, R. C., de Souza, W., Attias, M. 2005. Intravacuolar network may act as a mechanical support for *Toxoplasma gondii* inside the parasitophorous vacuole. *Microsc. Res. Tech.* 67, 45–52.

1259. Maia, C., Cortes, H., Brancal, H., Lopes, A. P., Pimenta, P., Campino, L., Cardoso, L. 2014. Prevalence and correlates of antibodies to *Neospora caninum* in dogs in Portugal. *Parasite* 21, 29.

1260. Mainar-Jaime, R. C., Thurmond, M. C., Berzal-Herranz, B., Hietala, S. K. 1999. Seroprevalence of *Neospora caninum* and abortion in dairy cows in northern Spain. *Vet. Rec.* 145, 72–75.

1261. Mainar-Jaime, R. C., Berzal-Herranz, B., Arias, P., Rojo-Vázquez, F. A. 2001. Epidemiological pattern and risk factors associated with bovine viral-diarrhoea virus (BVDV) infection in a non-vaccinated dairy-cattle population from the Asturias region of Spain. *Prev. Vet. Med.* 52, 63–73.

1262. Malaguti, J. M., Cabral, A. D., Abdalla, R. P., Salgueiro, Y. O., Galleti, N. T., Okuda, L. H., Cunha, E. M. S., Pituco, E. M., Del Fava, C. 2012. *Neospora caninum* as causative agent of bovine encephalitis in Brazil. *Rev. Bras. Parasitol. Vet.* 21, 48–54.

1263. Maley, S. W., Buxton, D., Thomson, K. M., Schriefer, C. E. S., Innes, E. A. 2001. Serological analysis of calves experimentally infected with *Neospora caninum*: A 1-year study. *Vet. Parasitol.* 96, 1–9.

1264. Maley, S. W., Buxton, D., Rae, A. G., Wright, S. E., Schock, A., Bartley, P. M., Esteban-Redondo, I., Swales, C., Hamilton, C. M., Sales, J., Innes, E. A. 2003. The pathogenesis of neosporosis in pregnant cattle: Inoculation at mid-gestation. *J. Comp. Pathol.* 129, 186–195.

1265. Maley, S. W., Buxton, D., Macaldowie, C. N., Anderson, I. E., Wright, S. E., Bartley, P. M., Esteban-Redondo, I., Hamilton, C. M., Storset, A. K., Innes, E. A. 2006. Characterization of the immune response in the placenta of cattle experimentally infected with *Neospora caninum* in early gestation. *J. Comp. Pathol.* 135, 130–141.

1266. Malmasi, A., Hosseininejad, M., Haddadzadeh, H., Badii, A., Bahonar, A. 2007. Serologic study of anti-*Neospora caninum* antibodies in household dogs and dogs living in dairy and beef cattle farms in Tehran, Iran. *Parasitol. Res.* 100, 1143–1145.

1267. Malmsten, J., Jakubek, E. B., Björkman, C. 2011. Prevalence of antibodies against *Toxoplasma gondii* and *Neospora caninum* in moose (*Alces alces*) and roe deer (*Capreolus capreolus*) in Sweden. *Vet. Parasitol.* 177, 275–280.

1268. Mann, T. R., Cadore, G. C., Camillo, G., Vogel, F. S. F., Schmidt, C., Andrade, C. M. 2016. Canine cutaneous neosporosis in Brazil. *Vet. Dermatol.* 27, 195–197. Doi: 10.1111/vde.12294.

1269. Mansilla, F. C., Franco-Mahecha, O. L., Lavoria, M. A., Moore, D. P., Giraldez, A. N., Iglesias, M. E., Wilda, M., Capozzo, A. V. 2012. The immune enhancement of a novel soy lecithin/beta-glucans based adjuvant on native *Neospora caninum* tachyzoite extract vaccine in mice. *Vaccine* 30, 1124–1131.

1270. Mansilla, F. C., Czepluch, W., Malacari, D. A., Hecker, Y. P., Bucafusco, D., Franco-Mahecha, O. L., Moore, D. P., Capozzo, A. V. 2013. Dose-dependent immunogenicity of a soluble *Neospora caninum* tachyzoite-extract vaccine formulated with a soy lecithin/β-glucan adjuvant in cattle. *Vet. Parasitol.* 197, 13–21.

1271. Mansilla, F. C., Moore, D. P., Quintana, M. E., Cardoso, N., Hecker, Y. P., Gual, I., Czepluch, W., Odeón, A. C., Capozzo, A. V. 2015. Safety and immunogenicity of a soluble native *Neospora caninum* tachyzoite-extract vaccine formulated with a soy lecithin/beta-glucan adjuvant in pregnant cattle. *Vet. Immunol. Immunopathol.* 165, 75–80.

1272. Mansilla, F. C., Quintana, M. E., Cardoso, N. P., Capozzo, A. V. 2016.. Fusion of foreign T-cell epitopes and addition of TLR agonists enhance immunity against *Neospora caninum* profilin in cattle. *Parasite Immunol.* 38, 663–669.

1273. Mansilla, F. C., Quintana, M. E., Langellotti, C., Wilda, M., Martinez, A., Fonzo, A., Moore, D. P., Cardoso, N., Capozzo, A. V. 2016. Immunization with *Neospora caninum* profilin induces limited protection and a regulatory T-cell response in mice. *Exp. Parasitol.* 160, 1–10.

1274. Mansourian, M., Khodakaram-Tafti, A., Namavari, M. 2009. Histopathological and clinical investigations in *Neospora caninum* experimentally infected broiler chicken embryonated eggs. *Vet. Parasitol.* 166, 185–190.

1275. Mansourian, M., Namavari, M., Khodakaram-Tafti, A., Rahimian, A. 2015. Experimental *Neospora caninum* infection in domestic bird's embryonated eggs. *J. Parasit. Dis.* 39, 241–244.

1276. Marco, I., Ferroglio, E., López-Olvera, J. R., Montané, J., Lavín, S. 2008. High seroprevalence of *Neospora caninum* in the red fox (*Vulpes vulpes*) in the Pyrenees (NE Spain). *Vet. Parasitol.* 152, 321–324.

1277. Marin, R. E. 2004. Fallas reproductivas en vacas lecheras en la Provincia de Jujuy. *Vet. Arg.* 21, 90–100.

1278. Marks, J., Lundén, A., Harkins, D., Innes, E. 1998. Identification of *Neospora* antigens recognized by CD4+ T cells and immune sera from experimentally infected cattle. *Parasite Immunol.* 20, 303–309.

1279. Markus, M. B., Daly, T. J. M., Biggs, H. C. 1983. Domestic dog as a final host of *Sarcocystis* of the mountain zebra *Equus zebra hartmannae*. *S. Afr. J. Sci.* 79, 471.

1280. Marques, F. A. C., Headley, A. S., Figueredo-Pereira, V., Taroda, A., Barros, L. D., Cunha, I. A. L., Munhoz, K., Bugni, F. M., Zulpo, D. L., Igarashi, M., Vidotto, O., Guimarães, J. S., Garcia, J. L. 2011. *Neospora caninum*: Evaluation of vertical transmission in slaughtered beef cows (*Bos indicus*). *Parasitol. Res.* 108, 1015–1019.

1281. Marsh, A. E., Barr, B. C., Sverlow, K., Ho, M., Dubey, J. P., Conrad, P. A. 1995. Sequence analysis and comparison of ribosomal DNA from bovine *Neospora* to similar coccidial parasites. *J. Parasitol.* 81, 530–535.

1282. Marsh, A. E., Barr, B. C., Madigan, J. E., Conrad, P. A. 1996. *In vitro* cultivation & characterization of a Neospora isolate obtained from a horse with protozoal myeloencephalitis. Program Guide & Abstracts for the Joint Meeting of the American Society of Parasitologists & the Society of Protozoologists. June 11–15, Tucson, Arizona, 108–109.

1283. Marsh, A. E., Barr, B. C., Madigan, J., Lakritz, J., Nordhausen, R., Conrad, P. A. 1996. Neosporosis as a cause of equine protozoal myeloencephalitis. *J. Am. Vet. Med. Assoc.* 209, 1907–1913.

1284. Marsh, A. E., Barr, B. C., Packham, A. E., Conrad, P. A. 1998. Description of a new *Neospora* species (Protozoa: Apicomplexa: Sarcocystidae). *J. Parasitol.* 84, 983–991.

1285. Marsh, A. E., Howe, D. K., Wang, G., Barr, B. C., Cannon, N., Conrad, P. A. 1999. Differentiation of *Neospora hughesi* from *Neospora caninum* based on their immunodominant surface antigen, SAG1 and SRS2. *Int. J. Parasitol.* 29, 1575–1582.

1286. Martino, P. E., Montenegro, J. L., Preziosi, J. A., Venturini, C., Bacigalupe, D., Stanchi, N. O., Bautista, E. L. 2004. Serological survey of selected pathogens of free-ranging foxes in southern Argentina, 1998–2001. *Rev. Sci. Tech.* 23, 801–806.

1287. Martins, A. A., de Zamprogna, T. O., Lucas, T. M., da Cunha, I. A. L., Garcia, J. L., da Silva, A. V. 2012. Frequency and risk factors for infection by *Neospora caninum* in dairy farms of Umuarama, PR, Brazil. *Arq. Ciên. Vet. Zool.* 15, 137–142.

1288. Martins, J., Kwok, O. C. H., Dubey, J. P. 2011. Seroprevalence of *Neospora caninum* in free-range chickens (*Gallus domesticus*) from the Americas. *Vet. Parasitol.* 182, 349–351.

1289. Marugán-Hernández, V., Álvarez-García, G., Risco-Castillo, V., Regidor-Cerrillo, J., Ortega-Mora, L. M. 2010. Identification of *Neospora caninum* proteins regulated during the differentiation process from tachyzoite to bradyzoite stage by DIGE. *Proteomics* 10, 1740–1750.

1289a. Martins, N. É. X., Freschi, C. R., Baptista, F., Machado, R. Z., Freitas, F. L. C., Almeida, K. S. 2011. Ocorrência de anticorpos anti-*Neospora caninum* em vacas lactanties do município de Araguaína, estado do Tocantins, Brasil. [Occurrence of anti *Neospora caninum* antibodies in lactating cows from the municipality of Araguaiana, Tocantins state, Brazil]. *Rev. Patol. Trop.* 40, 231–238.

1290. Marugán-Hernández, V., Ortega-Mora, L. M., Aguado-Martínez, A., Alvarez-García, G. 2011. Genetic manipulation of *Neospora caninum* to express the bradyzoite-specific protein NcSAG4 in tachyzoites. *Parasitology* 138, 472–480.

1291. Marugán-Hernández, V., Ortega-Mora, L. M., Aguado-Martínez, A., Jiménez-Ruíz, E., Álvarez-García, G. 2011. Transgenic *Neospora caninum* strains constitutively expressing the bradyzoite NcSAG4 protein proved to be safe and conferred significant levels of protection against vertical transmission when used as live vaccines in mice. *Vaccine* 29, 7867–7874.

1292. Marugán-Hernández, V., Álvarez-García, G., Tomley, F., Hemphill, A., Regidor-Cerrillo, J., Ortega-Mora, L. M. 2011. Identification of novel rhoptry proteins in *Neospora caninum* by LC/MS-MS analysis of subcellular fractions. *J. Paroteomics* 74, 629–642.

1293. Masala, G., Porcu, R., Daga, C., Denti, S., Canu, G., Patta, C., Tola, S. 2007. Detection of pathogens in ovine and caprine abortion samples from Sardinia, Italy, by PCR. *J. Vet. Diagn. Invest.* 19, 96–98.

1293a. Masuda, T., Kobayashi, Y., Maeda, R., Omata, Y. 2007. Possibility of *Neospora caninum* infection by venereal transmission in CB-17 *scid* mice. *Vet. Parasitol.* 149, 130–133.

1294. Matoba, K., Shiba, T., Takeuchi, T., Sibley, L. D., Seiki, M., Kikyo, F., Horiuchi, T., Asai, T., Harada, S. 2010. Crystallization and preliminary X-ray structural analysis of nucleoside triphosphate hydrolases from *Neospora caninum* and *Toxoplasma gondii*. *Acta Cryst.* F66, 1445–1448.

1295. Matsubayashi, M., Kimata, I., Iseki, M., Lillehoj, H. S., Matsuda, H., Nakanishi, T., Tani, H., Sasai, K., Baba, E. 2005. Cross-reactivities with *Cryptosporidium* spp. by chicken monoclonal antibodies that recognize avian Eimeria spp. *Vet. Parasitol.* 128, 47–57.

1296. Mattos, B. C., Patrício, L. F. L., Plugge, N. F., Lange, R. R., Richartz, R. R. T. B., Dittrich, R. L. 2008. Soroprevalência de anticorpos anti-*Neospora caninum* e anti-*Toxoplasma gondii* em canídeos selvagens cativos. [Seroprevalence of anti-*Neospora caninum* and anti-*Toxoplasma gondii* antibodies in captive wild canids]. *Rev. Bras. Parasitol. Vet.* 17 (Supl 1), 267–272.

1297. Mayberry, C., Maloney, S. K., Mitchell, J., Mawson, P. R., Bencini, R. 2014. Reproductive implications of exposure to *Toxoplasma gondii* and *Neospora caninum* in western grey kangaroos (*Macropus fuliginosus ocydromus*). *J. Wildl. Dis.* 50, 364–368.

1298. Mayhew, I. G., Smith, K. C., Dubey, J. P., Gatward, L. K., McGlennon, N. J. 1991. Treatment of encephalomyelitis due to *Neospora caninum* in a litter of puppies. *J. Small Anim. Pract.* 32, 609–612.

1299. Mazuz, L. M., Fish, L., Molad, T., Savitsky, I., Wolkomirsky, R., Leibovitz, B., Shkap, V. 2011. *Neospora caninum* as causative-pathogen of abortion in cattle. *Israel J. Vet. Med.* 66, 14–18.

1300. Mazuz, M. L., Haynes, R., Shkap, V., Fish, L., Wollkomirsky, R., Leibovich, B., Molad, T., Savitsky, I., Golenser, J. 2012. *Neospora caninum*: *In vivo* and *in vitro* treatment with artemisone. *Vet. Parasitol.* 187, 99–104.

1301. Mazuz, M. L., Fish, L., Reznikov, D., Wolkomirsky, R., Leibovitz, B., Savitzky, I., Golenser, J., Shkap, V. 2014. Neosporosis in naturally infected pregnant dairy cattle. *Vet. Parasitol.* 205, 85–91.

1302. Mazuz, M. L., Fish, L., Wolkomirsky, R., Leibovich, B., Reznikov, D., Savitsky, I., Golenser, J., Shkap, V. 2015. The effect of a live *Neospora caninum* tachyzoite vaccine in naturally infected pregnant dairy cows. *Prev. Vet. Med.* 120, 232–235.

1303. Mazuz, M. L., Shkap, V., Wollkomirsky, R., Leibovich, B., Savitsky, I., Fleiderovitz, L., Noam, S., Elena, B., Molad, T., Golenser, J. 2016. *Neospora caninum*: Chronic and congenital infection in consecutive pregnancies of mice. *Vet. Parasitol.* 219, 66–70.

1304. McAllister, D., Latham, S. 2002. *Neospora* 2001. *Trends Parasitol.* 18, 4–5.

1305. McAllister, M., Huffman, E. M., Hietala, S. K., Conrad, P. A., Anderson, M. L., Salman, M. D. 1996. Evidence suggesting a point source exposure in an outbreak of bovine abortion due to neosporosis. *J. Vet. Diagn. Invest.* 8, 355–357.

1306. McAllister, M., Wills, R. A., McGuire, A. M., Jolley, W. R., Tranas, J. D., Williams, E. S., Lindsay, D. S., Björkman, C., Belden, E. L. 1999. Ingestion of *Neospora caninum* tissue cysts by *Mustela* species. *Int. J. Parasitol.* 29, 1531–1536.

1307. McAllister, M. M., Parmley, S. F., Weiss, L. M., Welch, V. J., McGuire, A. M. 1996. An immunohistochemical method for detecting bradyzoite antigen (BAG5) in *Toxoplasma gondii*-infected tissues cross-reacts with a *Neospora caninum* bradyzoite antigen. *J. Parasitol.* 82, 354–355.

1308. McAllister, M. M., McGuire, A. M., Jolley, W. R., Lindsay, D. S., Trees, A. J., Stobart, R. H. 1996. Experimental neosporosis in pregnant ewes and their offspring. *Vet. Pathol.* 33, 647–655.

1309. McAllister, M. M., Dubey, J. P., Lindsay, D. S., Jolley, W. R., Wills, R. A., McGuire, A. M. 1998. Dogs are definitive hosts of *Neospora caninum*. *Int. J. Parasitol.* 28, 1473–1478.

1310. McAllister, M. M., Jolley, W. R., Wills, R. A., Lindsay, D. S., McGuire, A. M., Tranas, J. D. 1998. Oral inoculation of cats with tissue cysts of *Neospora caninum*. *Am. J. Vet. Res.* 59, 441–444.

1311. McAllister, M. M. 1999. Uncovering the biology and epidemiology of *Neospora caninum*. *Parasitol. Today* 15, 216–217.

1312. McAllister, M. M. 2000. *Neospora caninum*: Its oocysts and its identity: An opinion. *Parasitol. Res.* 86, 860.

1313. McAllister, M. M. 2000. A European perspective on *Neospora caninum*: Summary of the COST820 1999 annual meeting in Interlaken, Switzerland. *Int. J. Parasitol.* 30, 877–879.

1314. McAllister, M. M., Björkman, C., Anderson-Sprecher, R., Rogers, D. G. 2000. Evidence of point-source exposure to *Neospora caninum* and protective immunity in a herd of beef cows. *J. Am. Vet. Med. Assoc.* 217, 881–887.

1314a. McAllister, M. M. 2001. Do cows protect fetuses from *Neospora caninum* transmission?. *Trends Parasitol.* 17, 6.

1315. McAllister, M. M., Wallace, R. L., Björkman, C., Gao, L., Firkins, L. D. 2005. A probable source of *Neospora caninum* infection in an abortion outbreak in dairy cows. *Bovine Practitioner* 39, 69–74.

1316. McAllister, M. M. 2016. Diagnosis and control of bovine neosporosis. *Vet. Clin. Food Anim.* 32, 443–463.

1316a. McAllister, M. M., Funnell, O., Donahoe, S. L., Šlapeta, J. 2016. Unusual presentation of neosporosis in a neonatal puppy from a litter of bulldogs. *Aust. Vet. J.* 94, 411–414.

1317. McCann, C. M., McAllister, M. M., Gondim, L. F. P., Smith, R. F., Cripps, P. J., Kipar, A., Williams, D. J. L., Trees, A. J. 2007. *Neospora caninum* in cattle: Experimental infection with oocysts can result in exogenous transplacental infection, but not endogenous transplacental infection in the subsequent pregnancy. *Int. J. Parasitol.* 37, 1631–1639.

1318. McCann, C. M., Vyse, A. J., Salmon, R. L., Thomas, D., Williams, D. J. L., McGarry, J. W., Pebody, R., Trees, A. J. 2008. Lack of serologic evidence of *Neospora caninum* in humans, England. *Emerg. Infect. Dis.* 14, 978–980.

1319. McDole, M. G., Gay, J. M. 2002. Seroprevalence of antibodies against *Neospora caninum* in diagnostic equine serum samples and their possible association with fetal loss. *Vet. Parasitol.* 105, 257–260.

1320. McDonald, S. 1999. *Neospora caninum*, canine neosporosis. *Doberman Q.* 8–99.

1321. McDonald, S. 1999. *Neospora caninum*, the common denominator in canine health problems. *Doberman Digest.* 9–99.

1322. McFadden, G. I., Waller, R. F., Reith, M. E., Lang-Unnasch, N. 1997. Plastids in apicomplexan parasites. *Pl. Syst. Evol.* 11, 261–287.

1323. McGarry, J. W., Guy, F., Trees, A. J., Williams, D. J. L., Davisen, H. C., Bjorkman, C. 2000. Validation and application of an inhibition ELISA to detect serum antibodies to *Neospora caninum* in different host species. *Int. J. Parasitol.* 30, 880–884.

1324. McGarry, J. W., Stockton, C. M., Williams, D. J. L., Trees, A. J. 2003. Protracted shedding of oocysts of *Neospora caninum* by a naturally infected foxhound. *J. Parasitol.* 89, 628–630.

1325. McGlennon, N. J., Jefferies, A. R., Casas, C. 1990. Polyradiculoneuritis and polymyositis due to a *Toxoplasma*-like protozoan: Diagnosis and treatment. *J. Small Anim. Pract.* 31, 102–104.

1326. McGuire, A. M., McAllister, M. M., Jolley, W. R. 1997. Separation and cryopreservation of *Neospora caninum* tissue cysts from murine brain. *J. Parasitol.* 83, 319–321.

1327. McGuire, A. M., McAllister, M. M., Jolley, W. R., Anderson-Sprecher, R. C. 1997. A protocol for the production of *Neospora caninum* tissue cysts in mice. *J. Parasitol.* 83, 647–651.

1328. McGuire, A. M., McAllister, M., Wills, R. A., Tranas, J. D. 1999. Experimental inoculation of domestic pigeons (*Columbia livia*) and zebra finches (*Poephila guttata*) with *Neospora caninum* tachyzoites. *Int. J. Parasitol.* 29, 1525–1529.

1329. McInnes, L. M., Irwin, P., Palmer, D. G., Ryan, U.M. 2006. *In vitro* isolation and characterisation of the first canine *Neospora caninum* isolate in Australia. *Vet. Parasitol.* 137, 355–363.

1330. McInnes, L. M., Ryan, U. M., O'Handley, R., Sager, H., Forshaw, D., Palmer, D. G. 2006. Diagnostic significance of *Neospora caninum* DNA detected by PCR in cattle serum. *Vet. Parasitol.* 142, 207–213.

1331. McIntosh, D. W., Haines, D. M. 1994. *Neospora* infection in an aborted fetus in British Columbia. *Can. Vet. J.* 35, 114–115.

1332. McIntyre, K. M., Setzkorn, C., Wardeh, M., Hepworth, P. J., Radford, A. D., Baylis, M. 2014. Using open-access taxonomic and spatial information to create a comprehensive database for the study of mammalian and avian liverstock and pet infections. *Prev. Vet. Med.* 116, 325–335.

1333. McNamee, P. T., Jeffrey, M. 1994. *Neospora*-associated bovine abortion in Northern Ireland. *Vet. Rec.* 134, 48.

1334. McNamee, P. T., Trees, A. J., Guy, F., Moffett, D., Kilpatrick, D. 1996. Diagnosis and prevalence of neosporosis in cattle in Northern Ireland. *Vet. Rec.* 138, 419–420.

1335. Medina, L., Cruz-Vázquez, C., Quezada, T., Morales, E., García-Vázquez, Z. 2006. Survey of *Neospora caninum* infection by nested PCR in aborted fetuses from dairy farms in Aguascalientes, Mexico. *Vet. Parasitol.* 136, 187–191.

1336. Medina-Esparza, L., Macías, L., Ramos-Parra, M., Morales-Salinas, E., Quezada, T., Cruz-Vázquez, C. 2013. Frequency of infection by *Neospora caninum* in wild rodents associated with dairy farms in Aguascalientes, Mexico. *Vet. Parasitol.* 191, 11–14.

1336a. Medina-Esparza, L., Regidor-Cerrillo, J., García-Ramos, D., Álvarez-García, G., Benavides, J., Ortega-Mora, L.M., Cruz-Vázquez, C. 2016. Genetic characterization of *Neospora caninum* from aborted bovine foetuses in Aguascalientes, Mexico. *Vet. Parasitol.* 228, 183–187.

1337. Meenakshi, Sandhu, K. S., Ball, M. S., Kumar, H., Sharma, S., Sidhu, P. K., Sreekumar, C., Dubey, J. P. 2007. Seroprevalence of *Neospora caninum* antibodies in cattle and water buffaloes in India. *J. Parasitol.* 93, 1374–1377.

1338. Meerburg, B. G., de Craeye, S., Dierick, K., Kijlstra, A. 2012. *Neospora caninum* and *Toxoplasma gondii* in brain tissue of feral rodents and insectivores caught on farms in the Netherlands. *Vet. Parasitol.* 184, 317–320.

1339. Mehlhorn, H., Heydorn, A. O. 2000. *Neospora caninum*: Is it really different from *Hammondia heydorni* or is it a strain of *Toxoplasma gondii*? An opinion. *Parasitol. Res.* 86, 169–178.

1339a. Mejia, R. W. 2008. Identificacion de *Neospora* sp. en felidos y canidos silvestres en cautiverio en el zoologico jamie duque. *Thesis.* Inversidad de la Salle, Bogota, Colombia. 1–78.

1340. Meléndez, J. A. S., García, J. J. M., Ramos, J. J. Z., Valdés, V. M. R., Vidal, G. H., Aranda, G. D., Romero, R. R., Alejo, L. C. G., Ramírez, R. Á. 2005. Frecuencia de anticuerpos contra *Neospora caninum* en ganado bovino del noreste de México. [Frequency of *Neospora caninum* antibodies in cattle from northeastern Mexico]. *Vet. Méx.* 36, 303–311.

1341. Mello, R. C., Andreotti, R., Barros, J. C., Tomich, R. G. P., Mello, A. K. M., Campolim, A. I., Pellegrin, A. O. 2008. Levantamento epidemiológico de *Neospora caninum* em bovinos de assentamentos rurais em Corubá, MS. [Epidemiologic survey of *Neospora caninum* in bovines of rural settlements in Corubá, MS]. *Rev. Bras. Parasitol. Vet.* 17(Suppl 1), 311–316.

1342. Melo, D. P. G., da Silva, A. C., Ortega-Mora, L. M., Bastos, S. A., Boaventura, C. M. 2006. Prevalência de anticorpos anti-*Neospora caninum* em bovinos das microrregiões de Goiânia e Anápolis, Goiás, Brasil. [Prevalence of antibodies anti-*Neospora caninum* in bovines from Anápolis and Goiânia micro regions, Goiás, Brazil]. *Rev. Bras. Parasitol. Vet.* 15, 105–109.

1343. Meng, Q. F., Li, Y., Zhou, Y., Bai, Y. D., Wang, W. L., Wang, W. L., Cong, W. 2015. Seroprevalence of *Neospora caninum* infection in farmed sika deer (*Cervus nippon*) in China. *Vet. Parasitol.* 211, 289–292.

1344. Meng, Q. L., Qiao, J., Wang, W. S., Chen, C. F., Zhang, Z. C., Cai, K. J., Cai, X. P., Tian, G. F., Tian, Z. Z., Yang, L. H. 2012. Seroprevalence of *Toxoplasma gondii* and *Neospora caninum* in Tarim red deer (*Cervus elaphus yarkandensis*) from Xinjiang Province, Northwest China. *J. Anim. Vet. Adv.* 11, 912–915.

1345. Mercier, C., Cesbron-Delauw, M. F. 2015. *Toxoplasma* secretory granules: One population or more? *Trends Parasitol.* 31, 60–71.

1346. Meseck, E. K., Njaa, B. L., Haley, N. J., Park, E. H., Barr, S. C. 2005. Use of a multiplex polymerase chain reaction to rapidly differentiate *Neospora caninum* from *Toxoplasma gondii* in an adult dog with necrotizing myocarditis and myocardial infarct. *J. Vet. Diagn. Invest.* 17, 565–568.

1347. Mesquita, L. P., Nogueira, C. I., Costa, R. C., Orlando, D. R., Bruhn, F. R., Lopes, P. F. R., Nakagaki, K. Y. R., Peconick, A. P., Seixas, J. N., Bezerra Júnior, P. S., Raymundo, D. L., Varaschin, M. S. 2013. Antibody kinetics in goats and conceptuses naturally infected with *Neospora caninum*. *Vet. Parasitol.* 196, 327–333.

1348. Metcalfe, R. V., Bettelheim, K. A., Berry, M. E., Hobbs, K. M., Thompson, A. L., Cole, S. P. 1979. Studies on antibody levels to Brucella abortus, *Toxoplasma gondii* and Leptospira serogroups in sera collected by the National Serum Bank during 1974–1976. *Zentralbl. Bakteriol. Mikrobiol. Hyg. A* 245, 520–526.

1349. Micheloud, J. F., Moore, D. P., Canal, A. M., Lischinsky, L., Hecker, Y. P., Canton, G. J., Odriozola, E., Odeon, A. C., Campero, C. M. 2015. First report of congenital *Neospora caninum* encephalomyelitis in two newborn calves in the Argentinean Pampas. *J. Vet. Sci. Technol.* 6, 251.

1350. Millán, J., Cabezón, O., Pabón, M., Dubey, J. P., Almería, S. 2009. Seroprevalence of *Toxoplasma gondii* and *Neospora caninum* in feral cats (*Felis silvestris catus*) in Majorca, Balearic Islands, Spain. *Vet. Parasitol.* 165, 323–326.

1351. Millán, J., Candela, M. G., Palomares, F., Cubero, M. J., Rodríguez, A., Barral, M., de la Fuente, J., Almería, S., León-Vizcaíno, L. 2009. Disease threats to the endangered Iberian lynx (*Lynx pardinus*). *Vet. J.* 182, 114–124.

1352. Miller, C., Quinn, H., Ryce, C., Reichel, M. P., Ellis, J. T. 2005. Reduction in transplacental transmission of *Neospora caninum* in outbred mice by vaccination. *Int. J. Parasitol.* 35, 821–828.

1353. Miller, C. M. D., Quinn, H. E., Windsor, P. A., Ellis, J. T. 2002. Characterization of the first Australian isolate of *Neospora caninum* form cattle. *Aust. Vet. J.* 80, 620–625.

1354. Miller, M. A., Conrad, P. A., Harris, M., Hatfield, B., Langlois, G., Jessup, D. A., Magargal, S. L., Packham, A. E., Toy-Choutka, S., Melli, A. C., Murray, M. A., Gulland, F. M., Grigg, M. E. 2010. A protozoal-associated epizootic impacting marine wildlife: Mass-mortality of southern sea otters (*Enhydra lutris nereis*) due to *Sarcocystis neurona* infection. *Vet. Parasitol.* 172, 183–194.

1355. Milne, E., Crawshaw, M., Brocklehurst, S., Wright, S., Maley, S., Innes, E. 2006. Associations between *Neospora caninum* specific antibodies in serum and milk in two dairy herds in Scotland. *Prev. Vet. Med.* 77, 31–47.

1356. Mineo, T. W. P., Silva, D. A. O., Costa, G. H. N., von Ancken, A. C. B., Kasper, L. H., Souza, M. A., Cabral, D. D., Costa, A. J., Mineo, J. R. 2001. Detection of IgG antibodies to *Neospora caninum* and *Toxoplasma gondii* in dogs examined in veterinary hospital from Brazil. *Vet. Parasitol.* 98, 239–245.

1357. Mineo, T. W. P., Silva, D. A. O., Naslund, K., Björkman, C., Uggla, A., Mineo, J. R. 2004. *Toxoplasma gondii* and *Neospora caninum* serological status of different canine populations from Uberlândia, Minas Gerais. *Arq. Bras. Med. Vet. Zootec.* 56, 414–417.

1358. Mineo, T. W. P., Alenius, S., Näslund, K., Montassier, H. J., Björkman, C. 2006. Distribution of antibodies against *Neospora caninum*, BVDV and BHV-1 among cows in Brazilian dairy herds with reproductive disorders. *Rev. Bras. Parasitol. Vet.* 15, 188–192.

1359. Mineo, T. W. P., Benevides, L., Silva, N. M., Silva, J. S. 2009. Myeloid differentiation factor 88 is required for resistance to *Neospora caninum* infection. *Vet. Res.* 40, 32.

1360. Mineo, T. W. P., Carrasco, A. O. T., Marciano, J. A., Werther, K., Pinto, A. A., Machado, R. Z. 2009. Pigeons (*Columba livia*) are a suitable experimental model for *Neospora caninum* infection in birds. *Vet. Parasitol.* 159, 149–153.

1361. Mineo, T. W. P., Oliveira, C. J. F., Gutierrez, F. R. S., Silva, J. S. 2010. Recognition by toll-like receptor 2 induces antigen-presenting cell activation and Th1 programming during infection by *Neospora caninum*. *Immunol. Cell Biol.* 88, 825–833.

1362. Mineo, T. W. P., Oliveira, C. J. F., Silva, D. A. O., Oliveira, L. L., Abatepaulo, A. R., Ribeiro, D. P., Ferreira, B. R., Mineo, J. R., Silva, J. S. 2010. *Neospora caninum* excreted/secreted antigens trigger CC-chemokine receptor 5-dependent cell migration. *Int. J. Parasitol.* 40, 797–805.

1363. Mineo, T. W. P., Carrasco, A. O. T., Raso, T. F., Werther, K., Pinto, A. A., Machado, R. Z. 2011. Survey for natural *Neospora caninum* infection in wild and captive birds. *Vet. Parasitol.* 182, 352–355.

1364. Minervino, A. H. H., Ragozo, A. M. A., Monteiro, R. M., Ortolani, E. L., Gennari, S. M. 2008. Prevalence of *Neospora caninum* antibodies in cattle from Santarém, Pará, Brazil. *Res. Vet. Sci.* 84, 254–256.

1365. Minervino, A. H. H., Cassinelli, A. B. M., de Lima, J. T. R., Soares, H. S., Malheiros, A. F., Marcili, A., Gennari, S. M. 2012. Prevalence of anti-*Neospora caninum* and anti-*Toxoplasma gondii* antibodies in dogs from two different indigenous communities in the Brazilian Amazon Region. *J. Parasitol.* 98, 1276–1278.

1366. Mitrea, I. L., Enachescu, V., Radulescu, R., Ionita, M. 2012. Seroprevalence of *Neospora caninum* infection on dairy dattle in farms from southern Romania. *J. Parasitol.* 98, 69–72.

1367. Mitrea, I. L., Enachescu, V., Ionita, M. 2013. *Neospora caninum* infection in dogs from southern Romania: Coproparasitological study and serological follow-up. *J. Parasitol.* 99, 365–367.

1368. Modolo, J. R., Stachissini, A. V. M., Gennari, S. M., Dubey, J. P., Langoni, H., Padovani, C. R., Barrozo, L. V., Leite, B. L. S. 2008. Freqüência de anticorpos anti-*Neospora caninum* em soros de caprinos do estado de São Paulo e sua relação com o manejo dos animais. [Frequency of anti-*Neospora caninum* antibodies in sera of goats of the State São Paulo and its relationship with flock management]. *Pesq. Vet. Bras.* 28, 597–600.

1369. Moen, A. R., Wouda, W., van Werven, T. 1995. Clinical and sero-epidemiological follow-up study in four dairy herds with an outbreak of *Neospora* abortion. *Proceedings, Dutch Society for Veterinary Epidemiology and Economics.* December 13, Lelystad, 93–103.

1370. Moen, A. R., Wouda, W., Mul, M. F., Graat, E. A. M., van Werven, T. 1998. Increased risk of abortion following *Neospora caninum* abortion outbreaks: A retrospective and prospective cohort study in four dairy herds. *Theriogenology* 49, 1301–1309.

1371. Molina-López, R., Cabezón, O., Pabón, M., Darwich, L., Obón, E., Lopez-Gatius, F., Dubey, J. P., Almería, S. 2012. High seroprevalence of *Toxoplasma gondii* and *Neospora caninum* in the common raven (*Corvus corax*) in the Northeast of Spain. *Res. Vet. Sci.* 93, 300–302.

1372. Mols-Vorstermans, T., Hemphill, A., Monney, T., Schaap, D., Boerhout, E. 2013. Differential effects on survival, humoral immune responses and brain lesions in inbred BALB/c, CBA/CA, and C57BL/6 mice experimentally infected with *Neospora caninum* tachyzoites. *ISRN Parasitology* 2013, 830980.

1373. Monney, T., Debache, K., Hemphill, A. 2011. Vaccines against a major cause of abortion in cattle, *Neospora caninum* infection. *Animals* 1, 306–325.

1374. Monney, T., Rütti, D., Schorer, M., Debache, K., Grandgirard, D., Leib, S. L., Hemphill, A. 2011. RecNcMIC3-1-R is a microneme- and rhoptry-based chimeric antigen that protects against acute neosporosis and limits cerebral parasite load in the mouse model for *Neospora caninum* infection, *Vaccine* 29, 6967–6975.

1375. Monney, T., Debache, K., Grandgirard, D., Leib, S. L., Hemphill, A. 2012. Vaccination with the recombinant chimeric antigen recNcMIC3-1-R induces a non-prtective Th2-type immune response in the pregnant mouse model for *N. caninum* infection. *Vaccine* 30, 6588–6594.

1376. Monney, T., Grandgirard, D., Leib, S. L., Hemphill, A. 2013. Use of a Th1 stimulator adjuvant for vaccination against *Neospora caninum* infection in the pregnant mouse model. *Pathogens* 2, 193–208.

1376a. Monney, T., Hemphill, A. 2014. Vaccines against neosporosis: What can we learn from the past studies. *Exp. Parasitol.* 140, 52–70.

1377. Monteiro, R. M., Richtzenhain, L. J., Pena, H. F. J., Souza, S. L. P., Funada, M. R., Gennari, S. M., Dubey, J. P., Sreekumar, C., Keid, L. B., Soares, R. M. 2007. Molecular phylogenetic analysis in *Hammondia*-like organisms based on partial Hsp70 coding sequences. *Parasitology* 134, 1195–1203.

1378. Monteiro, R. M., Pena, H. F. J., Gennari, S. M., de Sousa, S. O., Richtzenhain, L. J., Soares, R. M. 2008. Differential diagnosis of oocysts of *Hammondia*-like organisms of dogs and cats by PCR-RFLP analysis of 70-kilodalton heat shock protein (HSP70) gene. *Parasitol. Res.* 103, 235–238.

1379. Moore, D., Reichel, M., Spath, E., Campero, C. 2013. *Neospora caninum* causes severe economic losses in cattle in the humid pampa region of Argentina. *Trop. Anim. Health Prod.* 45, 1237–1241.

1380. Moore, D. P., Campero, C. M., Odeón, A. C., Posso, M. A., Cano, D., Leunda, M. R., Basso, W., Venturini, M. C., Späth, E. 2002. Seroepidemiology of beef and dairy herds and fetal study of *Neospora caninum* in Argentina. *Vet. Parasitol.* 107, 303–316.

1381. Moore, D. P., Campero, C. M., Odeón, A. C., Chayer, R., Bianco, M. A. 2003. Reproductive losses due to *Neospora caninum* in a beef herd in Argentina. *J. Vet. Med. B* 50, 304–308.

1382. Moore, D. P., Campero, C. M., Odeón, A. C., Bardón, J. C., Silva-Paulo, P., Paolicchi, F. A., Cipolla, A. L. 2003. Humoral immune response to infectious agents in aborted bovine fetuses in Argentina. *Rev. Argent. Microbiol.* 35, 143–148.

1383. Moore, D. P., Draghi, M. G., Campero, C. M., Cetrá, B., Odeón, A. C., Alcaraz, E., Späth, E. A. J. 2003. Serological evidence of *Neospora caninum* infections in beef bulls in six counties of the Corrientes province, Argentina. *Vet. Parasitol.* 114, 247–252.

1384. Moore, D. P. 2005. Fate of *Neospora*-seropositive animals: An opinion. *Parasitol. Latinoam.* 60, 192–195.

1385. Moore, D. P. 2005. Neosporosis in South America. *Vet. Parasitol.* 127, 87–97.

1386. Moore, D. P., Odeón, A. C., Venturini, M. C., Campero, C. M. 2005. Neosporosis bovina: Conceptos generales, inmunidad y perspectivas para la vacunación. *Rev. Argent. Microbiol.* 37, 217–228.

1387. Moore, D. P., Leunda, M. R., Zamorano, P. I., Odeón, A. C., Romera, S. A., Cano, A., de Yaniz, G., Venturini, M. C., Campero, C. M. 2005. Immune response to *Neospora caninum* in naturally infected heifers and heifers vaccinated with inactivated antigen during the second trimester of gestation. *Vet. Parasitol.* 130, 29–39.

1388. Moore, D. P., de Yaniz, M. G., Odeón, A. C., Cano, D., Leunda, M. R., Späth, E. A. J., Campero, C. M. 2007. Serological evidence of *Neospora caninum* infections in goats from La Rioja Province, Argentina. *Small Ruminant Res.* 73, 256–258.

1389. Moore, D. P., Regidor-Cerrillo, J., Morrell, E., Poso, M. A., Cano, D. B., Leunda, M. R., Linschinky, L., Odeón, A. C., Odriozola, E., Ortega-Mora, L. M., Campero, C. M. 2008. The role of *Neospora caninum* and *Toxoplasma gondii* in spontaneous bovine abortion in Argentina. *Vet. Parasitol.* 156, 163–167.

1390. Moore, D. P., Pérez, A., Agliano, S., Brace, M., Cantón, G., Cano, D., Leunda, M. R., Odeón, A. C., Odriozola, E., Campero, C. M. 2009. Risk factors associated with *Neospora caninum* infections in cattle in Argentina. *Vet. Parasitol.* 161, 122–125.

1391. Moore, D. P., Echaide, I., Verna, A. E., Leunda, M. R., Cano, A., Pereyra, S., Zamorano, P. I., Odeón, A. C., Campero, C. M. 2011. Immune response to *Neospora caninum* native antigens formulated with immune stimulating complexes in calves. *Vet. Parasitol.* 175, 245–251.

1392. Moore, D. P., Alvarez-García, G., Chiapparrone, M. L., Regidor-Cerrillo, J., Lischinsky, L. H., de Yaniz, M. G., Odeón, A. C., Ortega-Mora, L. M., Campero, C. M. 2014. *Neospora caninum* tachyzoites inoculated by the conjunctival route are not vertically transmitted in pregnant cattle: A descriptive study. *Vet. Parasitol.* 199, 1–7.

1393. Moore, D. P., Konrad, J. L., San Martino, S., Reichel, M. P., Cano, D. B., Méndez, S., Späth, E. J. L., Odeón, A. C., Crudeli, G., Campero, C. M. 2014. *Neospora caninum* serostatus is affected by age and species variables in cohabiting water buffaloes and beef cattle. *Vet. Parasitol.* 203, 259–263.

1394. Moraes, L. M. B., Raimundo, J. M., Guimarães, A., Santos, H. A., Macedo, G. L., Massard, C. L., Machado, R. Z., Baldani, C. D. 2011. Occurrence of anti-*Neospora caninum* and anti-*Toxoplasma gondii* antibodies in goats and sheep in western Maranhão, Brazil. *Rev. Bras. Parasitol. Vet.* 20, 312–317.

1395. Morales, S. E., Ramírez, L. J., Trigo, T. F., Ibarra V. F., Puente C. E., Santa Cruz, M. 1997. Descripción de un caso de aborto bovino asociado a infección por *Neospora* sp en México. [Description of a case of bovine abortion associated with *Neospora* sp. infection in Mexico]. *Vet. Méx.* 28, 353–357.

1396. Morales S. E., Trigo T. F. J., Ibarra V. F., Puente, C. E., Santacruz, M. 2001. Seroprevalence study of bovine neosporosis in Mexico. *J. Vet. Diagn. Invest.* 13, 413–415.

1397. Morales, E., Trigo, F. J., Ibarra, F., Puente, E., Santacruz, M. 2001. Neosporosis in Mexican dairy herds: Lesions and immunohistochemical detection of *Neospora caninum* in fetuses. *J. Comp. Pathol.* 125, 58–63.

1398. Morales, J. A., Dubey, J. P., Rodriguez, F., Esquivel, R. L., Fritz, D. 1995. Neosporosis and toxoplasmosis-associated paralysis in dogs in Costa Rica. *Appl. Parasitol.* 36, 179–184.

1399. Moraveji, M., Hosseini, M. H., Amrabadi, O., Rahimian, A., Namazi, F., Namavari, M. 2011. Seroprevalence of *Neospora* spp. in horses in South of Iran. *Trop. Biomed.* 28, 514–517.

1400. Moraveji, M., Hosseini, A., Moghaddar, N., Namavari, M. M., Eskandari, M. H. 2012. Development of latex agglutination test with recombinant NcSAG1 for the rapid detection of antibodies to *Neospora caninum* in cattle. *Vet. Parasitol.* 189, 211–217.

1401. Moré, G., Basso, W., Bacigalupe, D., Venturini, M. C., Venturini, L. 2008. Diagnosis of *Sarcocystis cruzi*, *Neospora caninum*, and *Toxoplasma gondii* infections in cattle. *Parasitol. Res.* 102, 671–675.

1402. Moré, G., Pardini, L., Basso, W., Marín, R., Bacigalupe, D., Auad, G., Venturini, L., Venturini, M. C. 2008. Seroprevalence of *Neospora caninum*, *Toxoplasma gondii* and *Sarcocystis* sp. in llamas (*Lama glama*) from Jujuy, Argentina. *Vet. Parasitol.* 155, 158–160.

1403. Moré, G., Bacigalupe, D., Basso, W., Rambeaud, M., Beltrame, F., Ramirez, B., Venturini, M. C., Venturini, L. 2009. Frequency of horizontal and vertical transmission for *Sarcocystis cruzi* and *Neospora caninum* in dairy cattle. *Vet. Parasitol.* 160, 51–54.

1404. Moré, G., Bacigalupe, D., Basso, W., Rambeaud, M., Venturini, M. C., Venturini, L. 2010. Serologic profiles for *Sarcocystis* sp. and *Neospora caninum* and productive performance in naturally infected beef calves. *Parasitol. Res.* 106, 689–693.

1405. Moreno, B., Collantes-Fernández, E., Villa, A., Navarro, A., Regidor-Cerrillo, J., Ortega-Mora, L. M. 2012. Occurrence of *Neospora caninum* and *Toxoplasma gondii* infections in ovine and caprine abortions. *Vet. Parasitol.* 187, 312–318.

1406. Moreno-Torres, K., Wolfe, B., Saville, W., Garabed, R. 2016. Estimating *Neospora caninum* prevalence in wildlife populations using Bayesian inference. *Ecol. Evol.* 6, 2216–2225.

1407. Morikawa, V. M., Zimpel, C. K., Paploski, I. A. D., Lara, M. C. C. S. H., Villalobos, E. M. C., Romaldini, A. H. C. N., Okuda, L. H., Biondo, A. W., de Barros Filho, I. R. 2014. Occurrences of anti-*Toxoplasma gondii* and anti-*Neospora caninum* antibodies in Barbary sheep at Curitiba zoo, southern Brazil. *Rev. Bras. Parasitol. Vet.* 23, 255–259.

1408. Morris, M. T., Cheng, W. C., Zhou, X. W., Brydges, S. D., Carruthers, V. B. 2004. *Neospora caninum* expresses an unusual single-domain Kazal protease inhibitor that is discharged into the parasitophorous vacuole. *Int. J. Parasitol.* 34, 693–701.

1409. Morrondo, P., Díaz-Cao, J. M., Prieto, A., Pérez, A., Cabanelas, E., Panadero, R., Díaz, P., Fernández, G., Pajares, G., Díez-Baños, P. 2015. Antibody prevalence of *Toxoplasma gondii* and *Neospora caninum* in Spanish roe deer. XXII International Congress of Mediteranean Federation for Health and Production of Ruminants. Sardina, Italy. 219–223.

1410. Moskwa, B., Cabaj, W. 2003. [*Neospora caninum*: A newly recognized agent causing spontaneous abortion in Polish cattle]. *Medycyna Wet.* 59, 23–26 (in Polish).

1411. Moskwa, B., Cabaj, W., Pastusiak, K., Bien, J. 2003. The suitability of milk in detection of *Neospora caninum* infection in cows. *Acta Parasitol.* 48, 138–141.

1412. Moskwa, B., Pastusiak, K., Bien, J., Cabaj, W. 2007. The first detection of *Neospora caninum* DNA in the colostrum of infected cows. *Parasitol. Res.* 100, 633–636.

1413. Moskwa, B., Gozdzik, K., Bien, J., Cabaj, W. 2008. Studies on *Neospora caninum* DNA detection in the oocytes and embryos collected from infected cows. *Vet. Parasitol.* 158, 370–375.

1414. Moskwa, B., Gozdzik, K., Bien, J., Kornacka, A., Cybulska, A., Reiterová, K., Cabaj, W. 2014. Detection of antibodies to *Neospora caninum* in moose (*Alces alces*): The first report in Europe. *Folia Parasitol. (Praha)* 61, 34–36.

1415. Mota, C. M., Ferreira, M. D., Costa, L. F., Barros, P. S. C., Silva, M. V., Santiago, F. M., Mineo, J. R., Mineo, T. W. P. 2014. Fluorescent ester dye-based assays for the *in vitro* measurement of *Neospora caninum* proliferation. *Vet. Parasitol.* 205, 14–19.

1415a. Mota, C. M., Oliveira, A. C. M., Davoli-Ferreira, M., Silva, M. V., Santiago, F. M., Nadipuram, S. M., Vashisht, A. A., Wohlschlegel, J. A., Bradley, P. J., Silva, J. S., Mineo, J. R., Mineo, T. W. P. 2016. *Neospora caninum* activates p38 MAPK as an evasion mechanism against innate immunity. *Front. Microbiol.* 7, 1456.

1416. Mota, R. A., Ferre, I., Faria, E. B. 2008. Situação da neosporose bovina no Brasil e métodos de diagnóstico. [Situation of Brazilian cattle neosporosis and diagnostic methods]. *Med. Vet.* 2, 38–48.

1417. Moya, F. R., Chávez, V. A., Casas, A. E., Serrano, M. E., Falcón, P. N., Pezo C. D. 2003. Seroprevalencia de *Neospora caninum* en llamas de la Provincia de Melgar, Puno. [Seroprevalence of *Neospora caninum* in llamas from Melgar province, Puno]. *Rev. Inv. Vet. Perú* 14, 155–160.

1418. Mugridge, N. B., Morrison, D. A., Heckeroth, A. R., Johnson, A. M., Tenter, A. M. 1999. Phylogenetic analysis based on full-length large subunit ribosomal RNA gene sequence comparison reveals that *Neospora caninum* is more closely related to *Hammondia heydorni* than to *Toxoplasma gondii*. *Int. J. Parasitol.* 29, 1545–1556.

1419. Müller, J., Naguleswaran, A., Müller, N., Hemphill, A. 2008. *Neospora caninum*: Functional inhibition of protein disulfide isomerase by the broad-spectrum anti-parasitic drug nitazoxanide and other thiazolides. *Exp. Parasitol.* 118, 80–88.

1420. Müller, J., Hemphill, A. 2011. Drug target identification in intracellular and exracellular protozoan parasites. *Curr. Top. Med. Chem.* 11, 2029–2038.

1421. Müller, J., Hemphill, A. 2013. *In vitro* culture systems for the study of apicomplexan parasites in farm animals. *Int. J. Parasitol.* 43, 115–124.

1422. Müller, J., Hemphill, A. 2013. New approaches for the identification of drug targets in protozoan parasites. *Int. Rev. Cell Mol. Biol.* 30, 359–401.

1423. Müller, J., Balmer, V., Winzer, P., Rahman, M., Manser, V., Haynes, R. K., Hemphill, A. 2015. *In vitro* effects of new artemisinin derivatives in *Neospora caninum*-infected human fibroblasts. *Int. J. Antimicrob. Agents* 46, 88–93.

1424. Müller, J., Aguado-Martínez, A., Manser, V., Balmer, V., Winzer, P., Ritler, D., Hostettler, I., Arranz-Solís, D., Ortega-Mora, L., Hemphill, A. 2015. Buparvaquone is active against *Neospora caninum in vitro* and in experimentally infected mice. *Int. J. Parasitol. Drugs Drug Resist.* 5, 16–25.

1425. Müller, J., Aguado-Martínez, A., Manser, V., Wong, H. N., Haynes, R. K., Hemphill, A. 2016. Repurposing of antiparasitic drugs: The hydroxy-naphthoquinone buparvaquone inhibits vertical transmission in the pregnant neosporosis mouse model. *Vet. Res.* 47, 32.

1426. Müller, N., Zimmermann, V., Hentrich, B., Gottstein, B. 1996. Diagnosis of *Neospora caninum* and *Toxoplasma gondii* infection by PCR and DNA hybridization immunoassay. *J. Clin. Microbiol.* 34, 2850–2852.

1427. Müller, N., Sager, H., Hemphill, A., Mehlhorn, H., Heydorn, A. O., Gottstein, B. 2001. Comparative molecular investigation of Nc5-PCR amplicons from *Neospora caninum* NC-1 and *Hammondia heydorni*-Berlin-1996. *Parasitol. Res.* 87, 883–885.

1428. Müller, N., Vonlaufen, N., Gianinazzi, C., Leib, S. L., Hemphill, A. 2002. Application of real-time fluorescent PCR for quantitative assessment of *Neospora caninum* infections in organotypic slice cultures of rat central nervous system tissue. *J. Clin. Microbiol.* 40, 252–255.

1429. Munday, B. L., Dubey, J. P., Mason, R. W. 1990. *Neospora caninum* infection in dogs. *Aust. Vet. J.* 67, 76.

1430. Munhoz, A. D., Flausino, W., da Silva, R. T., de Almeida, C. R. R., Lopes, C. W. G. 2006. Distribuição de anticorpos contra *Neospora caninum* em vacas leiteiras dos municípios de resende e Rio Claro, Estado do Rio de Janeiro, Brasil. [Distribution of anti-*Neospora caninum* antibodies in dairy cows at municipalities of Resende and Rio Claro in the State of Rio de Janeiro, Brazil]. *Rev. Bras. Parasitol. Vet.* 15, 101–104.

1431. Munhoz, A. D., Pereira, M. J. S., Flausino, W., Lopes, C. W. G. 2009. *Neospora caninum* seropositivity in cattle breeds in the South Fluminense Paraíba Valley, state of Rio de Janeiro. *Pesq. Vet. Bras.* 29, 29–32.

1432. Munhoz, A. D., do Amaral, T. F., Gonçalves, L. R., de Moraes, V. M. B., Machado, R. Z. 2014. *Gallus gallus domesticus* are resistant to infection with *Neospora caninum* tachyzoites of the NC-1 strain. *Vet. Parasitol.* 206, 123–128.

1433. Munhoz, A. D., Mineo, T. W. P., Alessi, A. C., Lopes, C. W. G., Machado, R. Z. 2016. Assessment of experimental infection for dogs using *Gallus gallus* chorioallantoic membranes inoculated with *Neospora caninum*. *Rev. Bras. Parasitol. Vet.* 22, 565–570.

1434. Muñoz-Zanzi, C. A., Thurmond, M. C., Hietala, S. K. 2004. Effect of bovine viral diarrhea virus infection on fertility of dairy heifers. *Theriogenology* 61, 1085–1099.

1435. Mur-Novales, R., Serrano-Pérez, B., García Ispierto, I., de Sousa, N. M., Beckers, J. F., Almería, S., López-Gatius, F. 2015. Experimental *Neospora caninum* infection modifies trophoblast cell populations and plasma pregnancy-associated glycoprotien 1 and 2 dynamics in pregnant dairy heifers. *Vet. Parasitol.* 216, 7–12.

1436. Mur-Novales, R., López-Gatius, F., Serrano-Pérez, B., García-Ispierto, I., Darwich, L., Cabezón, O., de Sousa, N. M., Beckers, J. F., Almería, S. 2016. Experimental *Neospora caninum* infecton in pregnant dairy heifers raises concentrations of pregnancy-associated glycoproteins 1 and 2 in foetal fluids. *Reprod. Dom. Anim.* 51, 282–286.

1437. Muradian, V. 2009. *Isolamento e caracterização molecular e biológica de Toxoplasma gondii e pesquisa de Neospora caninum em roedores urbanos da Grande São Paulo (SP).* Faculty of Veterinary Medicine and Animal Science, São Paulo University, São Paulo, Brazil. [Isolation, molecular and biological characterization of *Toxoplasma gondii* and survey for *Neospora caninum* in urban rodents from Great Sao Paulo (SP)]. 1–109.

1438. Muradian, V., Ferreira, L. R., Lopes, E. G., Esmerini, P. O., Pena, H. F. J., Soares, R. M., Gennari, S. M. 2012. A survey of *Neospora caninum* and *Toxoplasma gondii* infection in urban rodents from Brazil. *J. Parasitol.* 98, 128–134.

1439. Murphy, T. M., Walochnik, J., Hassl, A., Moriarty, J., Mooney, J., Toolan, D., Sanchez-Miguel, C., O'Loughlin, A., McAuliffe, A. 2007. Study on the prevalence of *Toxoplasma gondii* and *Neospora caninum* and molecular evidence of *Encephalitozoon cuniculi* and *Encephalitozoon (Septata) intestinalis* infections in red foxes (*Vulpes vulpes*) in rural Ireland. *Vet. Parasitol.* 146, 227–234.

1440. Muz, M. N., Kilinç, Ô. O., Işler, C. T., Altuğ, E., Karakavuk, M. 2015. Bazi Yabani kuslarin beyin dokularinda *Toxoplasma gondii* ve *Neospora caninum*'un moleküler tanisi. *Kafkas Univ. Vet. Fak. Derg.* 21, 173–178.

1441. Myers, W. L., Foreyt, W. J., Talcott, P. A., Evermann, J. F., Chang, W. Y. 2015. Serologic, trace element, and fecal parasite survey of free-ranging, femaled mule deer (*Odocoileus hemionus*) in eastern Washington, USA. *J. Wildl. Dis.* 51, 125–136.

1442. Myhre, E. B., Kronvall, G. 1981. Specific binding of bovine, ovine, caprine and equine IgG subclasses to defined types of immunoglobulin receptors in Gram-positive cocci. *Comp. Immunol. Microbiol. Infect. Dis.* 4, 317–328.

1443. Naguleswaran, A., Cannas, A., Keller, N., Vonlaufen, N., Schares, G., Conraths, F. J., Björkman, C., Hemphill, A. 2001. *Neospora caninum* microneme protein NcMIC3: Secretion, subcellular localization, and functional involvement in host cell interaction. *Infect. Immun.* 69, 6483–6494.

1444. Naguleswaran, A., Cannas, A., Keller, N., Vonlaufen, N., Björkman, C., Hemphill, A. 2002. Vero cell surface proteoglycan interaction with the microneme protein NcMIC$_3$ mediates adhesion of *Neospora caninum* tachyzoites to host cells unlike that in *Toxoplasma gondii*. *Int. J. Parasitol.* 32, 695–704.

1445. Naguleswaran, A., Müller, N., Hemphill, A. 2003. *Neospora caninum* and *Toxoplasma gondii*: A novel adhesion/invasion assay reveals distinct differences in tachyzoite-host cell interactions. *Exp. Parasitol.* 104, 149–158.

1446. Naguleswaran, A., Hemphill, A., Rajapakse, R. P. V. J., Sager, H. 2004. Elaboration of a crude antigen ELISA for serodiagnosis of caprine neosporosis: Validation of the test by detection of *Neospora caninum*-specific antibodies in goats from Sri Lanka. *Vet. Parasitol.* 126, 257–262.

1447. Naguleswaran, A., Alaeddine, F., Guionaud, C., Vonlaufen, N., Sonda, S., Jenoe, P., Mevissen, M., Hemphill, A. 2005. *Neospora caninum* protein disulfide isomerase is involved in tachyzoite-host cell interaction. *Int. J. Parasitol.* 35, 1459–1472.

1448. Nakagaki, K. Y. R., Abreu, C. C., Costa, R. C., Orlando, D. R., Freire, L. R., Bruhn, F. R. P., Peconick, A. P., Wouters, F., Wouters, A. T. B., Raymundo, D. L., Varaschin, M. S. 2016. Lesions and distribution of *Neospora caninum* in tissues of naturally infected female goats. *Small Ruminant Res.* 140, 57–62.

1449. Nam, H. W., Kang, S. W., Choi, W. Y. 1998. Antibody reaction of human anti-*Toxoplasma gondii* positive and negative sera with *Neospora caninum* antigens. *Korean J. Parasitol.* 36, 269–275.

1450. Nascimento, C. O. M., Silva, M. L. C. R., Kim, P. C. P., Gomes, A. A. B., Gomes, A. L. V,, Maia, R. C. C., Almeida, J. C., Mota, R. A. 2015. Occurrence of *Neospora caninum* and *Toxoplasma gondii* DNA in brain tissue from hoary foxes (*Pseudalopex vetulus*) in Brazil. *Acta Trop.* 146, 60–65.

1451. Nasir, A., Ashraf, M., Khan, M. S., Yaqub, T., Javeed, A., Avais, M., Akhtar, F. 2011. Seroprevalence of *Neospora caninum* in dairy buffaloes in Lahore District, Pakistan. *J. Parasitol.* 97, 541–543.

1452. Nasir, A., Lanyon, S. R., Schares, G., Anderson, M. L., Reichel, M. P. 2012. Sero-prevalence of *Neospora caninum* and *Besnoitia besnoiti* in South Australian beef and dairy cattle. *Vet. Parasitol.* 186, 480–485.

1453. Nasir, A., Ashraf, M., Khan, M. S., Javeed, A., Yaqub, T., Avais, M., Reichel, M. P. 2012. Prevalence of *Neospora caninum* antibodies in sheep and goats in Pakistan. *J. Parasitol.* 98, 213–215.

1454. Nasir, A., Ashraf, M., Shakoor, A., Adil, M., Abbas, T., Kashif, M., Younus, M., Reichel, M. P. 2014. Co-infection of water buffaloes in Punjab, Pakistan, with *Neospora caninum and Brucella abortus*. *Turk. J. Vet. Anim. Sci.* 38, 572–576.

1455. Nazir, M. M., Maqbool, A., Khan, M. S., Sajjid, A., Lindsay, D. S. 2013. Effects of age and breed on the prevalence of *Neospora caninum* in commercial dairy cattle from Pakistan. *J. Parasitol.* 99, 368–370.

1456. Nazir, M. M., Maqbool, A., Akhtar, M., Ayaz, M., Ahmad, A. N., Ashraf, K., Ali, A., Alam, M. A., Ali, M. A., Khalid, A. R., Lindsay, D. S. 2014. *Neospora caninum* prevalence in dogs raised under different living conditions. *Vet. Parasitol.* 204, 364–368.

1457. Nematollahi, A., Jaafari, R., Moghaddam, G. 2011. Seroprevalence of *Neospora caninum* infection in dairy cattle in Tabriz, Northwest Iran. *Iranian J. Parasitol.* 6, 95–98.

1458. Neto, A. F. A., Bandini, L. A., Nishi, S. M., Soares, R. M., Driemeier, D., Antoniassi, N. A. B., Schares, G., Gennari, S. M. 2011. Viability of sporulated oocysts of *Neospora caninum* after exposure to different physical and chemical treatments. *J. Parasitol.* 97, 135–139.

1459. Neverauskas, C. E., Nasir, A., Reichel, M. P. 2015. Prevalence and distribution of *Neospora caninum* in water buffalo (*Bubalus bubalis*) and cattle in the Northern Territory of Australia. *Parasitol. Int.* 64, 392–396.

1460. Nghiem, P. P., Schatzberg, S. J. 2010. Conventional and molecular diagnostic testing for the acute neurologic patient. *J. Vet. Emerg. Crit Care* 20, 46–61.

1461. Nguyen, T. T. D., Choe, S. E., Byun, J. W., Koh, H. B., Lee, H. S., Kang, S. W. 2012. Seroprevalence of *Toxoplasma gondii* and *Neospora caninum* in dogs from Korea. *Acta Parasitol.* 57, 7–12.

1462. Nietfeld, J. C., Dubey, J. P., Anderson, M. L., Libal, M. C., Yaeger, M. J., Neiger, R. D. 1992. *Neospora*-like protozoan infection as a cause of abortion in dairy cattle. *J. Vet. Diagn. Invest.* 4, 223–226.

1463. Nishi, S. M., Viero, L. M., Soares, R. M., Maiorka, P. C., Gennari, S. M. 2009. Emprego da RT-PCR em tempo real para a quantificação da expressão de genes associados à resposta imune em bezerros bovinos experimentalmente infectados por *Neospora caninum. Rev. Bras. Parasitol. Vet.* 18, 8–14.

1464. Nishikawa, Y., Xuan, X., Nagasawa, H., Igarashi, I., Fujisaki, K., Otsuka, H., Mikami, T. 2000. Monoclonal antibody inhibition of *Neospora caninum* tachyzoite invasion into host cells. *Int. J. Parasitol.* 30, 51–58.

1465. Nishikawa, Y., Ikeda, H., Fukumoto, S., Xuan, X., Nagasawa, H., Otsuka, H., Mikami, T. 2000. Immunisation of dogs with a canine herpesvirus vector expressing *Neospora caninum* surface protein, NcSRS2. *Int. J. Parasitol.* 30, 1167–1171.

1466. Nishikawa, Y., Kousaka, Y., Fukumoto, S., Xuan, X., Nagasawa, H., Igarashi, I., Fujisaki, K., Otsuka, H., Mikami, T. 2000. Delivery of *Neospora caninum* surface protein, NcSRS2 (Nc-p43), to mouse using recombinant vaccinia virus. *Parasitol. Res.* 86, 934–939.

1467. Nishikawa, Y., Iwata, A., Nagasawa, H., Fujisaki, K., Otsuka, H., Mikami, T. 2001. Comparison of the growth inhibitory effects of canine IFN-α, -β and -γ on canine cells infected with *Neospora caninum* tachyzoites. *J. Vet. Med. Sci.* 63, 445–448.

1468. Nishikawa, Y., Mishima, M., Nagasawa, H., Igarashi, I., Fujisaki, K., Otsuka, H., Mikami, T. 2001. Interferon-gamma-induced apoptosis in host cells infected with *Neospora caninum. Parasitology* 123, 25–31.

1469. Nishikawa, Y., Tragoolpua, K., Inoue, N., Makala, L., Nagasawa, H., Otsuka, H., Mikami, T. 2001. In the absence of endogenous gamma interferon, mice acutely infected with *Neospora caninum* succumb to a lethal immune response characterized by inactivation of peritoneal macrophages. *Clin. Diagn. Lab. Immunol.* 8, 811–817.

1470. Nishikawa, Y., Xuan, X., Nagasawa, H., Igarashi, I., Fujisaki, K., Otsuka, H., Mikami, T. 2001. Prevention of vertical transmission of *Neospora caninum* in BALB/c mice by recombinant vaccinia virus carrying NcSRS2 gene. *Vaccine* 19, 1710–1716.

1471. Nishikawa, Y., Inoue, N., Xuan, X., Nagasawa, H., Igarashi, I., Fujisaki, K., Otsuka, H., Mikami, T. 2001. Protective efficacy of vaccination by recombinant vaccinia virus against *Neospora caninum* infection. *Vaccine* 19, 1381–1390.

1472. Nishikawa, Y., Kousaka, Y., Tragoolpua, K., Xuan, X., Makala, L., Fujisaki, K., Mikami, T., Nagasawa, H. 2001. Characterization of *Neospora caninum* surface protein NcSRS2 based on baculovirus expression system and its application for serodiagnosis of *Neospora* infection. *J. Clin. Microbiol.* 39, 3987–3991.

1473. Nishikawa, Y., Mikami, T., Nagasawa, H. 2002. Vaccine development against *Neospora caninum* infection. *J. Vet. Med. Sci.* 64, 1–5.

1474. Nishikawa, Y., Claveria, F. G., Fujisaki, K., Nagasawa, H. 2002. Studies on serological cross-reaction of *Neospora caninum* with *Toxoplasma gondii* and *Hammondia heydorni. J. Vet. Med. Sci.* 64, 161–164.

1475. Nishikawa, Y., Makala, L., Otsuka, H., Mikami, T., Nagasawa, H. 2002. Mechanisms of apoptosis in murine fibroblasts by two intracellular protozoan parasites, *Toxoplasma gondii* and *Neospora caninum. Parasite Immunol.* 24, 347–354.

1476. Nishikawa, Y., Tragoolpua, K., Makala, L., Xuan, X., Nagasawa, H. 2002. *Neospora caninum* NcSRS2 is a transmembrane protein that contains a glycosylphosphatidylinositol anchor in insect cells. *Vet. Parasitol.* 109, 191–201.

1477. Nishikawa, Y., Inoue, N., Makala, L., Nagasawa, H. 2003. A role for balance of interferon-gamma and interleukin-4 production in protective immunity against *Neospora caninum* infection. *Vet. Parasitol.* 116, 175–184.

1478. Nishikawa, Y., Zhang, H., Huang, P., Zhang, G., Xuan, X. 2009. Effects of a transferring antibody against *Neospora caninum* infection in a murine model. *Vet. Parasitol.* 160, 60–65.

1479. Nishikawa, Y., Zhang, H., Ikehara, Y., Kojima, N., Xuan, X., Yokoyama, N. 2009. Immunization with oligomannose-coated liposome-entrapped dense granule protein 7 protects dams and offspring from *Neospora caninum* infection in mice. *Clin. Vaccine Immunol.* 16, 792–797.

1480. Nishikawa, Y., Zhang, H., Ibrahim, H. M., Yamada, K., Nagasawa, H., Xuan, X. 2010. Roles of CD122+ cells in resistance against *Neospora caninum* infection in a murine model. *J. Vet. Med. Sci.* 72, 1275–1282.

1481. Nishimura, M., Kohara, J., Hiasa, J., Muroi, Y., Yokoyama, N., Kida, K., Xuan, X., Furuoka, H., Nishikawa, Y. 2013. Tissue distribution of *Neospora caninum* in experimentally infected cattle. *Clin. Vaccine Immunol.* 20, 309–312.

1482. Nishimura, M., Kohara, J., Kuroda, Y., Hiasa, J., Tanaka, S., Muroi, Y., Kojima, N., Furuoka, H., Nishikawa, Y. 2013. Oligomannose-coated liposome-entrapped dense granule protein 7 induces protective immune response to *Neospora caninum* in cattle. *Vaccine* 31, 3528–3535.

1483. Nishimura, M., Tanaka, S., Ihara, F., Muroi, Y., Yamagishi, J., Furuoka, H., Suzuki, Y., Nishikawa, Y. 2015. Transcriptome and histopathological changes in mouse brain infected with *Neospora caninum*. *Sci. Rep.* 5, 7936, 1–11.

1484. Nogareda, C., López-Gatius, F., Santolaria, P., García-Ispierto, I., Bech-Sàbat, G., Pabón, M., Mezo, M., Gonzalez-Warleta, M., Castro-Hermida, J. A., Yániz, J., Almeria, S. 2007. Dynamics of anti-*Neospora caninum* antibodies during gestation in chronically infected dairy cows. *Vet. Parasitol.* 148, 193–199.

1485. Nogareda, C., Jubert, A., Kantzoura, V., Kouam, M. K., Feidas, H., Theodoropoulos, G. 2013. Geographical distribution modelling for *Neospora caninum* and *Coxiella burnetii* infections in dairy cattle farms in northeastern Spain. *Epidemiol. Infect.* 141, 81–90.

1486. Nogueira, C. I., Mesquita, L. P., Abreu, C. C., Nakagaki, K. Y. R., Seixas, J. N., Bezerra, P. S., Rocha, C. M. B. M., Guimaraes, A. M., Peconick, A. P., Varaschin, M. S. 2013. Risk factors associated with seroprevalence of *Neospora caninum* in dogs from urban and rural areas of milk and coffee production in Minas Gerais state, Brazil. *Epidemiol. Infect.* 141, 2286–2293.

1487. Nolan, S. J., Romano, J. D., Luechtefeld, T., Coppens, I. 2015. *Neospora caninum* recruits host cell structures to its parasitophorous vacuole and salvages lipids from organelles. *Eukaryot. Cell* 14, 454–473.

1488. Norlander, E. 2014. Seroprevalence of *Toxoplasma gondii* and *Neospora* spp. in equids from three municipalities in Pará, Brazil. *Doctor of Veterinary Medicine Thesis*. Swedish University of Agricultural Sciences, Uppsala, Sweden. 1–31.

1489. Novoselov, S. V., Lobanov, A. V., Hua, D., Kasaikina, M. V., Hatfield, D. L., Gladyshev, V. N. 2007. A highly efficient form of the selenocysteine insertion sequence element in protozoan parasites and its use in mammalian cells. *Proc. Natl. Acad. Sci. USA* 104, 7857–7862.

1490. O' Doherty, E., Sayers, R., O' Grady, L. 2013. Temporal trends in bulk milk antibodies to *Salmonella*, *Neospora caninum*, and *Leptospira interrogans* serovar hardjo in Irish dairy herds. *Prev. Vet. Med.* 109, 343–348.

1491. O' Doherty, E., Berry, D. P., O' Grady, L., Sayers, R. 2014. Management practices as risk factors for the presence of bulk milk antibodies to *Salmonella*, *Neospora caninum* and *Leptospira interrogans* serovar hardjo in Irish dairy herds. *Animal* 8, 1010–1019.

1492. O' Doherty, E., Sayers, R., O' Grady, L., Shalloo, L. 2015. Effect of exposure to *Neospora caninum*, *Salmonella*, and *Leptospira interrogans* serovar *Hardjo* on the economic performance of Irish dairy herds. *J. Dairy Sci.* 98, 2789–2800.

1493. O'Handley, R., Liddell, S., Parker, C., Jenkins, M. C., Dubey, J. P. 2002. Experimental infection of sheep with *Neospora caninum* oocysts. *J. Parasitol.* 88, 1120–1123.

1494. O'Handley, R. M., Morgan, S. A., Parker, C., Jenkins, M. C., Dubey, J. P. 2003. Vaccination of ewes for prevention of vertical transmission of *Neospora caninum*. *Am. J. Vet. Res.* 64, 449–452.

1495. O'Toole, D., Jeffrey, M. 1987. Congenital sporozoan encephalomyelitis in a calf. *Vet. Rec.* 121, 563–566.

1496. O'Toole, D. 2010. Monitoring and investigating natural disease by veterinary pathologists in diagnostic laboratories. *Vet. Pathol.* 47, 40–44.

1497. Obendorf, D. L., Murray, N., Veldhuis, G., Munday, B. L., Dubey, J. P. 1995. Abortion caused by neosporosis in cattle. *Aust. Vet. J.* 72, 117–118.

1498. Öcal, N., Atmaca, H. T., Albay, M. K., Deniz, A., Kalender, H., Yildiz, K., Kul, O. 2014. A new approach to *Neospora caninum* infection epidemiology: Neosporosis in integrated and rural dairy farms in Turkey. *Turk. J. Vet. Anim. Sci.* 38, 161–168.

1499. Ochsenreither, S., Kuhls, K., Schaar, M., Presber, W., Schönian, G. 2006. Multilocus microsatellite typing as a new tool for discrimination of *Leishmania infantum* MON-1 strains. *J. Clin. Microbiol.* 44, 495–503.

1500. Odin, M., Dubey, J. P. 1993. Sudden death associated with *Neospora caninum* myocarditis in a dog. *J. Am. Vet. Med. Assoc.* 203, 831–833.

1501. Ogawa, L., Freire, R. L., Vidotto, O., Gondim, L. F. P., Navarro, I. T. 2005. Occurrence of antibodies to *Neospora caninum* and *Toxoplasma gondii* in dairy cattle from the northern region of the Paraná State, Brazil. *Arq. Bras. Med. Vet. Zootec.* 57, 312–316.

1502. Ogedengbe, J. D., Hanner, R. H., Barta, J. R. 2011. DNA barcoding identifies Eimeria species and contributes to the phylogenetics of coccidian parasites (Eimeriorina, Apicomplexa, Alveolata). *Int. J. Parasitol.* 41, 843–850.

1502a. Ogedengbe, M. E., Ogedengbe, J. D., Whale, J. C., Elliot, K., Juárez-Estrada, M. A., Arta, J. R. 2016. Molecular phylogenetic analyses of tissue coccidia (sarcocystidae; apicomplexa) based on nuclear 18 s RDNA and mitochondrial COI sequences confirms the paraphyly of the genus *Hammondia*. *Parasitology Open*. 2, e2.

1503. Ogino, H., Watanabe, E., Watanabe, S., Agawa, H., Narita, M., Haritani, M., Kawashima, K. 1992. Neosporosis in the aborted fetus and newborn calf. *J. Comp. Pathol*. 107, 231–237.

1504. Ojo, K. K., Reid, M. C., Siddaramaiah, L. K., Müller, J., Winzer, P., Zhang, Z., Keyloun, K. R., Vidadala, R. S. R., Merritt, E. A., Hol, W. G. J., Maly, D. J., Fan, E., Van Voorhis, W. C., Hemphill, A. 2014. *Neospora caninum* calcium-dependent protein kinase 1 is an effective drug target for neosporosis therapy. *PLoS ONE* 9, e92929.

1505. Okeoma, C. M., Williamson, N. B., Pomroy, W. E., Stowell, K. M. 2004. Recognition patterns of *Neospora caninum* tachyzoite antigens by bovine IgG at different IFAT titres. *Parasite Immunol*. 26, 177–185.

1506. Okeoma, C. M., Williamson, N. B., Pomroy, W. E., Stowell, K. M., Gillespie, L. 2004. The use of PCR to detect *Neospora caninum* DNA in the blood of naturally infected cows. *Vet. Parasitol*. 122, 307–315.

1507. Okeoma, C. M., Williamson, N. B., Pomroy, W. E., Stowell, K. M., Gillespie, L. M. 2004. Isolation and molecular characterisation of *Neospora caninum* in cattle in New Zealand. *N. Z. Vet. J*. 52, 364–370.

1508. Okeoma, C. M., Stowell, K. M., Williamson, N. B., Pomroy, W. E. 2005. *Neospora caninum*: Quantification of DNA in the blood of naturally infected aborted and pregnant cows using real-time PCR. *Exp. Parasitol*. 110, 48–55.

1508a. Okumu, T. A., Munene, J. N., Wabacha, J., Tsuma, V., Van Leeuwen, J. 2016. Seroepidemiological survey of *Neospora caninum* and its risk factors in farm dogs in Nakuru district, Kenya. *Vet. World*. 9, 1162–1166.

1509. Olamendi-Portugal, M., Caballero-Ortega, H., Correa, D., Sánchez-Alemán, M. A., Cruz-Vázquez, C., Medina-Esparza, L., Ortega-SJ. A., Cantu, A., García-Vázquez, Z. 2012. Serosurvey of antibodies against *Toxoplasma gondii* and *Neospora caninum* in white-tailed deer from Northern Mexico. *Vet. Parasitol*. 189, 369–373.

1510. Oliveira, A. C., Lopes, L. B., Melo, C. B., Leite, R. C. 2004. Influência do *Neospora caninum* na eficiência reprodutiva de um rebanho Mestiço-zebu no extremo sul da Bahia. [Influence of *Neospora caninum* on reproductive efficiency in a zebu crossbred herd in Bahia State, Brazil]. *Rev. Bras. Reprod. Anim*. 28, 295–297.

1510a. Oliveira, S., Soares, R. M., Aizawa, J., Soares, H. S., Chiebao, D. P., Ortega-Mora, L. M., Regidor-Cerrillo, J., Silva, N. Q. B., Gennari, S. M., Pena, H. F. J. 2017. Isolation and biological and molecular characterizsation of *Neospora caninum* (NC-SP1) from a naturally infected adult cattle (*Bos taurus*) in the state of São Paulo, Brazil. *Parasitology*. doi: 10.1017/S0031182016002481.

1511. Omata, Y., Nidaira, M., Kano, R., Kobayashi, Y., Koyama, T., Furuoka, H., Maeda, R., Matsui, T., Saito, A. 2004. Vertical transmission of *Neospora caninum* in BALB/c mice in both acute and chronic infection. *Vet. Parasitol*. 121, 323–328.

1512. Omata, Y., Kano, R., Masukata, Y., Kobayashi, Y., Igarashi, I., Maeda, R., Saito, A. 2005. Development of *Neospora caninum* cultured with human serum *in vitro* and *in vivo*. *J. Parasitol*. 91, 222–225.

1513. Omata, Y., Ishiguro, N., Kano, R., Masukata, Y., Kudo, A., Kamiya, H., Fukui, H., Igarashi, M., Maeda, R., Nishimura, M., Saito, A. 2005. Prevalence of *Toxoplasma gondii* and *Neospora caninum* in sika deer from eastern Hokkaido, Japan. *J. Wildl. Dis*. 41, 454–458.

1514. Omata, Y., Kamiya, H., Kano, R., Kobayashi, Y., Maeda, R., Saito, A. 2006. Footpad reaction induced by *Neospora caninum* tachyzoite extract in infected BALB/c mice. *Vet. Parasitol*. 139, 102–108.

1515. Omata, Y., Umeshita, Y., Watarai, M., Tachibana, M., Sasaki, M., Murata, K., Yamada, T. K. 2006. Investigation for presence of *Neospora caninum*, *Toxoplasma gondii* and *Brucella* species infection in killer whales (*Orcinus orca*) mass-stranded on the coast of Shiretoko, Hokkaido, Japan. *J. Vet. Med. Sci*. 68, 523–526.

1516. Öncel, T., Biyikoğlu, G. 2003. [*Neosporosis caninum* in dairy cattle in Sakarya, Turkey]. Uludag Univ. *J. Fac. Vet. Med*. 22, 87–89 (in Turkish).

1517. Onuma, S. S. M., Melo, A. L. T., Kantek, D. L. Z., Crawshaw-Junior, P. G., Morato, R. G., May-Júnior, J. A., Pacheco, T. A., de Aguiar, D. M. 2014. Exposure of free-living jaguars to *Toxoplasma gondii*, *Neospora caninum* and *Sarcocystis neurona* in the Brazilian Pantanal. *Braz. J. Parasitol. Vet*. 23, 547–553.

1518. Ooi, H. K., Huang, C. C., Yang, C. H., Lee, S. H. 2000. Serological survey and first finding of *Neospora caninum* in Taiwan, and the detection of its antibodies in various body fluids of cattle. *Vet. Parasitol.* 90, 47–55.

1519. Opel, K. L., Chung, D., McCord, B. R. 2010. A study of PCR inhibition mechanisms using real time PCR. *J. Forensic Sci.* 55, 25–33.

1520. Opsteegh, M., Teunis, P., Züchner, L., Koets, A., Langelaar, M., van der Giessen, J. 2011. Low predictive value of seroprevalence of *Toxoplasma gondii* in cattle for detection of parasite DNA. *Int. J. Parasitol.* 41, 343–354.

1521. Ordeix, L., Lloret, A., Fondevila, D., Dubey, J. P., Ferrer, L., Fondati, A. 2002. Cutaneous neosporosis during treatment of pemphigus foliaceus in a dog. *J. Am. Anim. Hosp. Assoc.* 38, 415–419.

1522. Orozco, M. A., Morales, E., Salmerón, F. 2013. Characterization of the inflammatory response in the uteri of cows infected naturally by *Neospora caninum*. *J. Comp. Pathol.* 148, 148–156.

1523. Ortega, Y. R., Torres, M. P., Mena, K. D. 2007. Presence of *Neospora caninum* specific antibodies in three dairy farms in Georgia and two in Texas. *Vet. Parasitol.* 144, 353–355.

1524. Ortega-Mora, L. M., Ferre, I., del Pozo, I., Caetano da Silva, A., Collantes-Fernández, E., Regidor-Cerrillo, J., Ugarte-Garagalza, C., Aduriz, G. 2003. Detection of *Neospora caninum* in semen of bulls. *Vet. Parasitol.* 117, 301–308.

1525. Ortega-Mora, L. M., Fernández-García, A., Gómez-Bautista, M. 2006. Diagnosis of bovine neosporosis: Recent advances and perspectives. *Acta Parasitol.* 51, 1–14.

1526. Ortuño, A., Castellà, J., Almería, S. 2002. Seroprevalence of antibodies to *Neospora caninum* in dogs from Spain. *J. Parasitol.* 88, 1263–1266.

1527. Osawa, T., Wastling, J., Maley, S., Buxton, D., Innes, E. A. 1998. A multiple antigen ELISA to detect *Neospora*-specific antibodies in bovine sera, bovine foetal fluids, ovine and caprine sera. *Vet. Parasitol.* 79, 19–34.

1528. Osawa, T., Wastling, J., Acosta, L., Ortellado, C., Ibarra, J., Innes, E. A. 2002. Seroprevalence of *Neospora caninum* infection in dairy and beef cattle in Paraguay. *Vet. Parasitol.* 110, 17–23.

1529. Osburn, B. I., MacLachlan, N. J., Terrell, T. G. 1982. Ontogeny of the immune system. *J. Am. Vet. Med. Assoc.* 181, 1049–1052.

1530. Oshiro, L. M., Matos, M. F. C., de Oliveira, J. M., Monteiro, L. A. R. C., Andreotti, R. 2007. Prevalence of anti-*Neospora caninum* antibodies in cattle from the state of Mato Grosso do Sul, Brazil. *Rev. Bras. Parasitol. Vet.* 16, 133–138.

1531. Oshiro, L. M., Motta-Castro, A. R. C., Freitas, S. Z., Cunha, R. C., Dittrich, R. L., Meirelles, A. C. F., Andreotti, R. 2015. *Neospora caninum* and *Toxoplasma gondii* serodiagnosis in human immunodeficiency virus carriers. *Rev. Soc. Bras. Med. Trop.* 48, 568–572.

1532. Osoro, K., Ortega-Mora, L. M., Martínez, A., Serrano-Martínez, E., Ferre, I. 2009. Natural breeding with bulls experimentally infected with *Neospora caninum* failed to induce seroconversion in dams. *Theriogenology* 71, 639–642.

1533. Otranto, D., Llazari, A., Testini, G., Traversa, D., di Regalbono, A. F., Badan, M., Capelli, G. 2003. Seroprevalence and associated risk factors of neosporosis in beef and dairy cattle in Italy. *Vet. Parasitol.* 118, 7–18.

1534. Otsuki, T., Dong, J., Kato, T., Park, E. Y. 2013. Expression, purification and antigenicity of *Neospora caninum*-antigens using silkworm larvae targeting for subunit vaccines. *Vet. Parasitol.* 192, 284–287.

1535. Otter, A., Griffiths, I. B., Jeffrey, M. 1993. Bovine *Neospora caninum* abortion in the UK. *Vet. Rec.* 133, 375.

1536. Otter, A., Jeffrey, M., Griffiths, I. B., Dubey, J. P. 1995. A survey of the incidence of *Neospora caninum* infection in aborted and stillborn bovine fetuses in England and Wales. *Vet. Rec.* 136, 602–606.

1537. Otter, A. 1997. Neospora and bovine abortion. *Vet. Rec.* 140, 239.

1538. Otter, A., Jeffrey, M., Scholes, S. F. E., Helmick, B., Wilesmith, J. W., Trees, A. J. 1997. Comparison of histology with maternal and fetal serology for the diagnosis of abortion due to bovine neosporosis. *Vet. Rec.* 141, 487–489.

1539. Otter, A., Wilson, B. W., Scholes, S. F. E., Jeffrey, M., Helmick, B., Trees, A. J. 1997. Results of a survey to determine whether *Neospora* is a significant cause of ovine abortion in England and Wales. *Vet. Rec.* 140, 175–177.

1540. Otter, A. 2000. Neosporosis: Diagnosis and future breeding considerations. *Irish Vet. J.* 53, 146–150.

1541. Ould-Amrouche, A., Klein, F., Osdoit, C., Mohamed, H. O., Touratier, A., Sanaa, M., Mialot, J. P. 1999. Estimation of *Neospora caninum* seroprevalence in dairy cattle from Normandy, France. *Vet. Res.* 30, 531–538.

1542. Pabón, M., López-Gatius, F., García-Ispierto, I., Bech-Sàbat, G., Nogareda, C., Almería, S. 2007. Chronic *Neospora caninum* infection and repeat abortion in dairy cows: A 3-year study. *Vet. Parasitol.* 147, 40–46.

1543. Paciejewski, S. 1997. [Neosporosis in cattle]. *Zycie Weterynaryjne* 72, 317–319 (in Polish).

1544. Paciello, O., D'Orazi, A., Borzacchiello, G., Martano, M., Restucci, B., Maiolino, P., Papparella, S. 2004. Expression of major histocompatibility complex class I and II in a case of *Neospora caninum* myositis in a dog. *Acta Myologica* 23, 151–153.

1545. Packham, A. E., Sverlow, K. W., Conrad, P. A., Loomis, E. F., Rowe, J. D., Anderson, M. L., Marsh, A. E., Cray, C., Barr, B. C. 1998. A modified agglutination test for *Neospora caninum*: Development, optimization, and comparison to the indirect fluorescent-antibody test and enzyme-linked immunosorbent assay. *Clin. Diagn. Lab. Immunol.* 5, 467–473.

1546. Packham, A. E., Conrad, P. A., Wilson, W. D., Jeanes, L. V., Sverlow, K. W., Gardner, I. A., Daft, B. M., Marsh, A. E., Blagburn, B. L., Ferraro, G. L., Barr, B. C. 2002. Qualitative evaluation of selective tests for detection of *Neospora hughesi* antibodies in serum and cerebrospinal fluid of experimentally infected horses. *J. Parasitol.* 88, 1239–1246.

1547. Paiz, L. M., Silva, R. C., Menozzi, B. D., Langoni, H. 2015. Antibodies to *Neospora caninum* in sheep from slaughterhouses in the state of São Paulo, Brazil. *Rev. Bras. Parasitol. Vet.* 24, 95–100.

1548. Palavicini, P., Romero, J. J., Dolz, G., Jiménez, A. E., Hill, D. E., Dubey, J. P. 2007. Fecal and serological survey of *Neospora caninum* in farm dogs in Costa Rica. *Vet. Parasitol.* 149, 265–270.

1549. Pan, Y., Jansen, G. B., Duffield, T. F., Hietala, S., Kelton, D., Lin, C. Y., Peregrine, A. S. 2004. Genetic susceptibility to *Neospora caninum* infection in Holstein cattle in Ontario. *J. Dairy Sci.* 87, 3967–3975.

1550. Panadero, R., Painceira, A., López, C., Vázquez, L., Paz, A., Díaz, P., Dacal, V., Cienfuegos, S., Fernández, G., Lago, N., Díez-Baños, P., Morrondo, P. 2010. Seroprevalence of *Toxoplasma gondii* and *Neospora caninum* in wild and domestic ruminants sharing pastures in Galicia (Northwest Spain). *Res. Vet. Sci.* 88, 111–115.

1551. Pannwitz, G. 2001. *Neospora-caninum*-Infektion bei einem adulten Irischen Wolfshund assoziiert mit eosinophiler Pleozytose—Ein Fallbericht. [*Neospora caninum* infection in an adult Irish wolfhound associated with eosinophilic pleocytosis—a case report]. *Tierärztl. Umschau* 56, 244–248.

1552. Paradies, P., Capelli, G., Testini, G., Cantacessi, C., Trees, A. J., Otranto, D. 2007. Risk factors for canine neosporosis in farm and kennel dogs in southern Italy. *Vet. Parasitol.* 145, 240–244.

1553. Paré, J., Thurmond, M. C., Hietala, S. K. 1994. Congenital *Neospora* infection in dairy cattle. *Vet. Rec.* 134, 531–532.

1554. Paré, J. 1995. Mise à jour sur les infections à *Neospora* sp. chez les bovins. [Review of *Neospora* sp. infection in cattle]. *Méd. Vét. Québec* 25, 12–16.

1555. Paré, J., Hietala, S. K., Thurmond, M. C. 1995. An enzyme-linked immunosorbent assay (ELISA) for serological diagnosis of *Neospora* sp. infection in cattle. *J. Vet. Diagn. Invest.* 7, 352–359.

1556. Paré, J., Hietala, S. K., Thurmond, M. C. 1995. Interpretation of an indirect fluorescent antibody test for diagnosis of *Neospora* sp. infection in cattle. *J. Vet. Diagn. Invest.* 7, 273–275.

1557. Paré, J., Thurmond, M. C., Hietala, S. K. 1996. Congenital *Neospora caninum* infection in dairy cattle and associated calfhood mortality. *Can. J. Vet. Res.* 60, 133–139.

1558. Paré, J., Thurmond, M. C., Hietala, S. K. 1997. *Neospora caninum* antibodies in cows during pregnancy as a predictor of congenital infection and abortion. *J. Parasitol.* 83, 82–87.

1559. Paré, J., Fecteau, G., Fortin, M., Marsolais, G. 1998. Seroepidemiologic study of *Neospora caninum* in dairy herds. *J. Am. Vet. Med. Assoc.* 213, 1595–1598.

1560. Parish, S. M., Maag-Miller, L., Besser, T. E., Weidner, J. P., McElwain, T., Knowles, D. P., Leathers, C. W. 1987. Myelitis associated with protozoal infection in newborn calves. *J. Am. Med. Assoc.* 191, 1599–1600.

1561. Park, C. H., Sawada, M., Morita, T., Shimada, A., Ochiai, K., Umemura, T. 2000. *Neospora caninum* infected the alimentary tract of nude mice and was transmitted to other mice by intraperitoneal inoculation with the intestinal contents. *J. Vet. Med. Sci.* 62, 525–527.

1562. Parzefall, B., Driver, C. J., Benigni, L., Davies, E. 2014. Magnetic resonance imaging characteristics in four dogs with central nervous system neosporosis. *Vet. Radiol. Ultrasound* 55, 539–546.

1563. Pasquali, P., Mandara, M. T., Adamo, F., Ricci, G., Polidori, G. A., Dubey, J. P. 1998. Neosporosis in a dog in Italy. *Vet. Parasitol.* 77, 297–299.

1564. Pastor-Fernández, I., Arranz-Solís, D., Regidor-Cerrillo, J., Álvarez-García, G., Hemphill, A., García-Culebras, A., Cuevas-Martín, C., Ortega-Mora, L. M. 2015. A vaccine formulation combining rhoptry proteins NcROP40 and NcROP2 improves pup survival in a pregnant mouse model of neosporosis. *Vet. Parasitol.* 207, 203–215.

1565. Pastor-Fernández, I., Regidor-Cerrillo, J., Jiménez-Ruiz, E., Álvarez-García, G., Marugán-Hernández, V., Hemphill, A., Ortega-Mora, L. M. 2016. Characterization of the *Neospora caninum* NcROP40 and NcROP2Fam-1 rhoptry proteins during the tachyzoite lytic cycle. *Parasitology* 143, 97–113.

1566. Pastor-Fernández, I., Regidor-Cerrillo, J., Álvarez-García, G., Marugán-Hernández, V., García-Lunar, P., Hemphill, A., Ortega-Mora, L. M. 2016. The tandemly repeated NTPase (NTPDase) from *Neospora caninum* is a canonical dense granule protein whose RNA expression, protein secretion and phosphorylation coincides with the tachyzoite egress. *Parasit. Vectors* 9, 352.

1567. Patitucci, A. N., Alley, M. R., Jones, B. R., Charleston, W. A. G. 1997. Protozoal encephalomyelitis of dogs involving *Neospora caninum* and *Toxoplasma gondii* in New Zealand. *N. Z. Vet. J.* 45, 231–235.

1568. Patitucci, A. N., Charleston, W. A. G., Alley, M. R., O'Connor, R. J., Pomroy, W. E. 1999. Serological study of a dairy herd with a recent history of *Neospora* abortion. *N. Z. Vet. J.* 47, 28–30.

1569. Patitucci, A. N., Pérez, M. J., Israel, K. F., Rozas, M. A. 2000. Prevalencia de anticuerpos séricos contra *Neospora caninum* en dos rebaños lecheros de la IX región de Chile. [Prevalence of *Neospora caninum* in two dairy herds of the IX Region of Chile]. *Arch. Med. Vet.* 32, 209–214.

1570. Patitucci, A. N., Phil, M., Pérez, M. J., Rozas, M. A., Israel, K. F. 2001. *Neosporosis canina*: presencia de anticuerpos séricos en poblaciones caninas rurales y urbanas de Chile. [Influence of *Neospora caninum* on reproductive efficiency in a zebu crossbred herd in Bahia state, Brazil]. *Arch. Med. Vet.* 33, 227–232.

1571. Patitucci, A. N., Perez, M. J., Carcamo, C. M., Baeza, L. 2004. Presencia de anticuerpos sericos contra *Neospora caninum* en equinos en Chile [Presence of serum antibodies to *Neospora caninum* in Chilean horses]. *Arch. Med. Vet.* 36, 203–206.

1572. Paula, V. S. O., Rodrigues, A. A. R., Richtzenhain, L. J., Cortez, A., Soares, R. M., Gennari, S. M. 2004. Evaluation of a PCR based on primers to Nc5 gene for the detection of *Neospora caninum* in brain tissues of bovine aborted fetuses. *Vet. Res. Commun.* 28, 581–585.

1573. Paulan, S. C., Lins, A. G. S., Tenório, M. S., da Silva, D. T., Pena, H. F. J., Machado, R. Z., Gennari, S. M., Buzetti, W. A. T. 2013. Seroprevalence rates of antibodies against *Leishmania infantum* and other protozoan and rickettsial parasites in dogs. *Rev. Bras. Parasitol. Vet.* 22, 162–166.

1574. Pavicic, L., Lalosevic, V., Spasojevic-Kosic, L., Laus, S., Simin, S. 2011. Seroprevalence of *Neospora caninum* in dogs. *Contemporary Agriculture* 60, 453–457.

1575. Payne, S., Ellis, J. 1996. Detection of *Neospora caninum* DNA by the polymerase chain reaction. *Int. J. Parasitol.* 26, 347–351.

1576. Paz, G. F., Leite, R. C., Rocha, M. A. 2007. Associação entre sorologia para *Neospora caninum* e taxa de prehez em vacas receptoras de embriões. [Association between seropositivity for *Neospora caninum* and pregnancy rate in bovine receipts submitted to embryo transfer technology]. *Arq. Bras. Med. Vet. Zootec.* 59, 1323–1325.

1577. Peckham, R. K., Brill, R., Foster, D. S., Bowen, A. L., Leigh, J. A., Coffey, T. J., Flynn, R. J. 2014. Two distinct populations of bovine IL-17+ T-cells can be induced and WC1+IL-17+γδ T-cells are effective killers of protozoan parasites. *Sci. Rep.* 4, 5431.

1578. Pedraza-Díaz, S., Marugán-Hernández, V., Collantes-Fernández, E., Regidor-Cerrillo, J., Rojo-Montejo, S., Gómez-Bautista, M., Ortega-Mora, L. M. 2009. Microsatellite markers for the molecular characterization of *Neospora caninum*: Application to clinical samples. *Vet. Parasitol.* 166, 38–46.

1579. Pena, H. F. J., Soares, R. M., Ragozo, A. M. A., Monteiro, R. M., Yai, L. E. O., Nishi, S. M., Gennari, S. M. 2007. Isolation and molecular detection of *Neospora caninum* from naturally infected sheep from Brazil. *Vet. Parasitol.* 147, 61–66.

1580. Penarete-Vargas, D. M., Mévélec, M. N., Dion, S., Sèche, E., Dimier-Poisson, I., Fandeur, T. 2010. Protection against lethal *Neospora caninum* infection in mice induced by heterologous vaccination with a mic1 mic3 knockout *Toxoplasma gondii* strain. *Infect. Immun.* 78, 651–660.

1581. Peng, X., Qu, K., Wang, C., Guo, Y., Wang, T., Qian, W., Yan, W. 2016. [Nested PCR test for diagnosis of *Neospora*-associated abortion in dairy cattle in Luoyang area]. *Heilongjiang Anim. Husb. Veter. Med.* 2016, 146–148. (in Chinese).

1582. Peregrine, A. S., Duffield, T. F., Wideman, G., Kelton, D., Hobson, J., Cramer, G., Hietala, S. K. 2004. Udder health in dairy cattle infected with *Neospora caninum. Prev. Vet. Med.* 64, 101–112.

1583. Peregrine, A. S., Martin, S. W., Hopwood, D. A., Duffield, T. F., McEwen, B., Hobson, J. C., Hietala, S. K. 2006. *Neospora caninum* and *Leptospira* serovar serostatus in dairy cattle in Ontario. *Can. Vet. J.* 47, 467–470.

1584. Pereira, G. R., Vogel, F. S. F., Bohrer, R. C., da Nóbrega, J. E., Ilha, G. F., da Rosa, P. R. A., Glanzner, W. G., Camillo, G., Braunig, P., de Oliveira, J. F. C., Gonçalves, P. B. D. 2014. *Neospora caninum* DNA detection by TaqMan real-time PCR assay in experimentally infected pregnant heifers. *Vet. Parasitol.* 199, 129–135.

1585. Pereira, L. M., Candido-Silva, J. A., De Vries, E., Yatsuda, A. P. 2010. A new thrombospondin-related anonymous protein homologue in *Neospora caninum* (NcMIC2-like1). *Parasitology* 138, 287–297.

1586. Pereira, L. M., Yatsuda, A. P. 2014. The chloramphenicol acetyltransferase vector as a tool for stable tagging of *Neospora caninum. Mol. Biochem. Parasitol.* 196, 75–81.

1587. Pereira, L. M., Yatsuda, A. P. 2014. Comparison of an ELISA assay for the detection of adhesive/invasive *Neospora caninum* tachyzoites. *Rev. Bras. Parasitol. Vet.* 23, 36–43.

1588. Pereira, L. M., Baroni, L., Yatsuda, A. P. 2014. A transgenic *Neospora caninum* strain based on mutations of the dihydrofolate reductase-thymidylate synthase gene. *Exp. Parasitol.* 138, 40–47.

1589. Pereira-Bueno, J., Quintanilla-Gozalo, A., Seijas-Carballedo, A., Costas, E., Ortega-Mora, L. M. 2000. Observational studies in *Neospora caninum* infected dairy cattle: Pattern of transmission and age-related antibody fluctuations. *Int. J. Parasitol.* 30, 906–909.

1590. Pereira-Bueno, J., Quintanilla-Gozalo, A., Pérez-Pérez, V., Espi-Felgueroso, A., Álvarez-García, G., Collantes-Fernández, E., Ortega-Mora, L. M. 2003. Evaluation by different diagnostic techniques of bovine abortion associated with *Neospora caninum* in Spain. *Vet. Parasitol.* 111, 143–152.

1591. Perez, E., Gonzalez, O., Dolz, G., Morales, J. A., Barr, B., Conrad, P. 1998. First report of bovine neosporosis in dairy cattle in Costa Rica. *Vet. Rec.* 142, 520–521.

1592. Pérez-Zaballos, F. J., Ortega-Mora, L. M., Álvarez-García, G., Collantes-Fernández, E., Navarro-Lozano, V., García-Villada, L., Costas, E. 2005. Adaptation of *Neospora caninum* isolates to cell-culture changes: An argument in favor of its clonal population structure. *J. Parasitol.* 91, 507–510.

1593. Perl, S., Harrus, S., Satuchne, C., Yakobson, B., Haines, D. 1998. Cutaneous neosporosis in a dog in Israel. *Vet. Parasitol.* 79, 257–261.

1594. Pernas, L., Adomako-Ankomah, Y., Shastri, A. J., Ewald, S. E., Treeck, M., Boyle, J. P., Boothroyd, J. C. 2014. Toxoplasma effector MAF1 mediates recruitment of host mitochondria and impacts the host response. *PLoS Biol.* 12, e1001845.

1595. Pescador, C. A., Corbellini, L. G., Oliveira, E. C., Raymundo, D. L., Driemeier, D. 2007. Histopathological and immunohistochemical aspects of *Neospora caninum* diagnosis in bovine aborted fetuses. *Vet. Parasitol.* 150, 159–163.

1596. Pessoa, G. A., Martini, A. P., Trentin, J. M., Dalcin, V. C., Leonardi, C. E. P., Vogel, F. S. F., de Sá Filho, M. F., Rubin, M. I. B., Silva, C. A. M. 2016. Impact of spontaneous *Neospora caninum* infection on pregnancy loss and subsequent pregnancy in grazing lactating dairy cows. *Theriogenology* 85, 519–527.

1597. Peter, A. T. 2000. Abortions in dairy cows: New insights and economic impact. *Adv. Dairy Technol.* 12, 233–244.

1598. Peters, M., Wagner, F., Schares, G. 2000. Canine neosporosis: Clinical and pathological findings and first isolation of *Neospora caninum* in Germany. *Parasitol. Res.* 86, 1–7.

1599. Peters, M., Lütkefels, E., Heckeroth, A. R., Schares, G. 2001. Immunohistochemical and ultrastructural evidence for *Neospora caninum* tissue cysts in skeletal muscles of naturally infected dogs and cattle. *Int. J. Parasitol.* 31, 1144–1148.

1600. Peters, M., Wohlsein, P., Knieriem, A., Schares, G. 2001. *Neospora caninum* infection associated with stillbirths in captive antelopes (*Tragelaphus imberbis*). *Vet. Parasitol.* 97, 153–157.

1601. Petersen, E., Lebech, M., Jensen, L., Lind, P., Rask, M., Bagger, P., Björkman, C., Uggla, A. 1999. *Neospora caninum* infection and repeated abortions in humans. *Emerg. Infect. Dis.* 5, 278–280.

1602. Pfeiffer, D. U., Williamson, N. B., Reichel, M. P. 2000. Long-term serological monitoring as a tool for epidemiological investigation of *Neospora caninum* infection in a New Zealand dairy herd. *Proceedings of 9th Symposium of the International Society for Veterinary Epidemiology and Economics*, August 7–11, Breckenridge, Colorado, USA. 616–618.

1603. Pfeiffer, D. U., Williamson, N. B., Reichel, M. P., Wichtel, J. J., Teague, W. R. 2002. A longitudinal study of *Neospora caninum* infection on a dairy farm in New Zealand. *Prev. Vet. Med.* 54, 11–24.

1604. Piagentini, M., Moya-Araujo, C. F., Prestes, N. C., Sartor, I. F. 2012. *Neospora caninum* infection dynamics in dairy cattle. *Parasitol. Res.* 111, 717–721.

1605. Piantedosi, D., Giudice, E., Pietra, M., Luciani, A., Brini, E., Guglielmini, C., Pugliese, A., Ciaramella, P. 2009. Seroprevalence of *Neospora* spp. in asymptomatic horses in Italy. *Ippologia* 20, 3–8.

1606. Piergili-Fioretti, D., Rosignoli, L., Ricci, G., Moretti, A., Pasquali, P., Polidori, G. A. 2000. *Neospora caninum* infection in a clinically healthy calf: Parasitological study serological follow-up. *J. Vet. Med. B Infect. Dis. Vet. Public Health* 47, 47–53.

1607. Pinheiro, A. F., Borsuk, S., Berne, M. E. A., Pinto, L. S., Andreotti, R., Roos, T., Roloff, B. C., Leite, F. P. L. 2013. Expression of *Neospora caninum* NcSRS2 surface protein in *Pichia pastoris* and its application for serodiagnosis of *Neospora* infection. *Pathog. Glob. Health* 107, 116–121.

1608. Pinheiro, A. F., Borsuk, S., Berne, M. E. A., Pinto, L. S., Andreotti, R., Roos, T., Roloff, B. C., Leite, F. P. L. 2015. Use of ELISA based on NcSRS2 of *Neospora caninum* expressed in *Pichia pastoris* for diagnosing neosporosis in sheep and dogs. *Braz. J. Vet. Parasitol.* 24, 148–154.

1609. Pinheiro, A. M., Costa, M. F., Paule, B., Vale, V., Ribeiro, M., Nascimento, I., Schaer, R. E., Almeida, M. A. O., Meyer, R., Freire, S. M. 2005. Serologic immunoreactivity to *Neospora caninum* antigens in dogs determined by indirect immunofluorescence, Western blotting and dot-ELISA. *Vet. Parasitol.* 130, 73–79.

1610. Pinheiro, A. M., Costa, S. L., Freire, S. M., Almeida, M. Â. O., Tardy, M., El Bachá, R., Costa, M. F. D. 2006. Astroglial cells in primary culture: A valid model to study *Neospora caninum* infection in the CNS. *Vet. Immunol. Immunopathol.* 113, 243–247.

1611. Pinheiro, A. M., Costa, S. L., Freire, S. M., Meyer, R., Almeida, M. Â. O., Tardy, M., El Bachá, R., Costa, M. F. D. 2006. *Neospora caninum*: Infection induced IL-10 overexpression in rat astrocytes *in vitro. Exp. Parasitol.* 112, 193–197.

1612. Pinheiro, A. M., Santos, C. V., Costa, M. F. D., Rodrigues, L. E. A. 2007. Host/parasite relationship in the *in vitro* infection of rat gliocytes by *Neospora caninum*: Evaluation of cell respiration. *Res. Vet. Sci.* 83, 27–29.

1613. Pinheiro, A. M., Costa, S. L., Freire, S. M., Ribeiro, C. S. O., Tardy, M., El-Bachá, R. S., Costa, M. F. D. 2010. *Neospora caninum*: Early immune response of rat mixed glial cultures after tachyzoites infection. *Exp. Parasitol.* 124, 442–447.

1614. Pinheiro, A. M., Santos, C. V. C. D., Rodrigues, L. E. A. 2013. *Neospora caninum*: Infection induces high lysosomal activity. *Exp. Parasitol.* 134, 409–412.

1615. Pinitkiatisakul, S., Mattsson, J. G., Wikman, M., Friedman, M., Bengtsson, K. L., Ståhl, S., Lundén, A. 2005. Immunisation of mice against neosporosis with recombinant NcSRS2 iscoms. *Vet. Parasitol.* 129, 25–34.

1616. Pinitkiatisakul, S. 2007. Recombinant subunit vaccines against *Neospora caninum*. *Thesis.* Swedish University of Agricultural Sciences, Uppsala, Sweden. 1–53.

1617. Pinitkiatisakul, S., Friedman, M., Wikman, M., Mattsson, J. G., Lövgren-Bengtsson, K., Ståhl, S., Lundén, A. 2007. Immunogenicity and protective effect against murine cerebral neosporosis of recombinant NcSRS2 in different iscom formulations. *Vaccine* 25, 3658–3668.

1618. Pinitkiatisakul, S., Mattsson, J. G., Lundén, A. 2008. Quantitative analysis of parasite DNA in the blood of immunized and naive mice after infection with *Neospora caninum*. *Parasitology* 135, 175–182.

1619. Pipano, E., Shkap, V., Fish, L., Savitsky, I., Perl, S., Orgad, U. 2002. Susceptibility of *Psammomys obesus* and *Meriones tristrami* to tachyzoites of *Neospora caninum*. *J. Parasitol.* 88, 314–319.

1620. Piper, R. C., Cole, C. R., Shadduck, J. A. 1970. Natural and experimental ocular toxoplasmosis in animals. *Am. J. Ophthalmol.* 69, 662–668.

1621. Pitel, P. H., Pronost, S., Legendre, M. F., Chatagnon, G., Tainturier, D., Fortier, G. 2000. Infection des bovins par *Neospora caninum*: deux années d'observations dans l'Ouest de la France. [*Neospora caninum* infection in cattle: A two year study in Western France]. *Point Vét.* 31, 53–58.

1622. Pitel, P. H., Pronost, S., Chatagnon, G., Tainturier, D., Fortier, G., Ballet, J. J. 2001. Neosporosis in bovine dairy herds from the west of France: Detection of *Neospora caninum* DNA in aborted fetuses, seroepidemiology of *N. caninum* in cattle and dogs. *Vet. Parasitol.* 102, 269–277.

1623. Pitel, P. H., Pronost, S., Romand, S., Thulliez, P., Fortier, G., Ballet, J. J. 2001. Prevalence of antibodies to *Neospora caninum* in horses in France. *Equine Vet. J.* 33, 205–207.

1624. Pitel, P. H., Pronost, S., Gargala, G., Anrioud, D., Toquet, M.-P., Foucher, N., Collobert-Laugier, C., Fortier, G., Ballet, J.-J. 2002. Detection of *Sarcocystis neurona* antibodies in French horses with neurological signs. *Int. J. Parasitol.* 32, 481–485.

1625. Pitel, P. H., Romand, S., Pronost, S., Foucher, N., Gargala, G., Maillard, K., Thulliez, P., Collobert-Laugier, C., Tainturier, D., Fortier, G., Ballet, J. J. 2003. Investigation of *Neospora* sp. antibodies in aborted mares from Normandy, France. *Vet. Parasitol.* 118, 1–6.

1626. Pitel, P. H., Lindsay, D. S., Caure, S., Romand, S., Pronost, S., Gargala, G., Mitchell, S. M., Hary, C., Thulliez, P., Fortier, G., Ballet, J. J. 2003. Reactivity against *Sarcocystis neurona* and *Neospora* by serum antibodies in healthy French horses from two farms with previous equine protozoal myeloencephalitis-like cases. *Vet. Parasitol.* 111, 1–7.

1627. Pitel, P. P. H., Legrand, L., Pronost, S., Maillard, K., Marcillaud-Pitel, C., Richard, E., Fortier, G. 2010. Néosporose bovine: de l'étude du cycle parasitarire à la définition des méthodes de lutte. [Bovine neosporosis: From life cycle to prophylaxis]. *Bull. Acad. Vét. France* 163, 131–142.

1628. Pivoto, F. L., de Macêdo Junior, A. G., da Silva, M. V., Ferreira, F. B., Silva, D. A. O., Pompermayer, E., Sangioni, L. A., Mineo, T. W. P., Vogel, F. S. F. 2014. Serological status of mares in parturition and the levels of antibodies (IgG) against protozoan family Sarcocystidae from their pre colostral foals. *Vet. Parasitol.* 199, 107–111.

1629. Ploneczka, K., Mazurkiewicz, M. 2008. Seroprevalence of *Neospora caninum* in dogs in south-western Poland. *Vet. Parasitol.* 153, 168–171.

1630. Plugge, N. F., Ferreira, F. M., Richartz, R. R. T. B., de Siqueira, A., Dittrich, R. L. 2011. Occurrence of antibodies against *Neospora caninum* and/or *Toxoplasma gondii* in dogs with neurological signs. *Rev. Bras. Parasitol. Vet.* 20, 202–206.

1631. Poli, A., Mancianti, F., Carli, M. A., Stroscio, M. C., Kramer, L. 1998. *Neospora caninum* infection in a Bernese cattle dog from Italy. *Vet. Parasitol.* 78, 79–85.

1632. Pollo-Oliveira, L., Post, H., Acencio, M. L., Lemke, N., van den Toorn, H., Tragante, V., Heck, A. J. R., Altelaar, A. F. M., Yatsuda, A. P. 2013. Unravelling the *Neospora caninum* secretome through the secreted fraction (ESA) and quantification of the discharged tachyzoite using high-resolution mass spectrometry-based proteomics. *Parasit. Vectors* 6, 335.

1633. Poncelet, L., Coignoul, F., Fontaine, J., Balligand, M. 1990. Infestation de chiots par *Neospora canis* en Belgique et en France? [Pups infestation by *Neospora caninum* in Belgium and France?]. *Ann. Méd. Vét.* 134, 167–171.

1634. Poncelet, L., Bjerkås, I., Charlier, G., Coignoul, F., Losson, B., Balligand, M. 1990. Confirmation de la présence de *Neospora caninum* en Belgique. [*Neospora caninum* in Belgium]. *Ann. Méd. Vét.* 134, 501–503.

1635. Porto, W. J. N., Regidor-Cerrillo, J., Kim, P. C. P., Benavides, J., Silva, A. C. S., Horcajo, P., Oliveira, A. A. F., Ferre, I., Mota, R. A., Ortega-Mora, L. M. 2016. Experimental caprine neosporosis: The influence of gestational stage on the outcome of infection. *Vet. Res.* 47, 29.

1636. Prandini da Costa Reis, R., Crisman, R., Roser, M., Malik, R., Šlapeta, J. 2016. Neonatal neosporosis in a 2-week-old Bernese mountain dog infected with multiple *Neospora caninum* strains based on MS10 microsatellite analysis. *Vet. Parasitol.* 221, 134–138.

1637. Proença, L. M., Silva, J. C. R., Galera, P. D., Lion, M. B., Marinho-Filho, J. S., Ragozo, A. M. A., Gennari, S. M., Dubey, J. P., Vasconcellos, S. A., Souza, G. O., Pinheiro Júnior, J. W., Santana, V. L. A., França, G. L., Rodrigues, F. H. G. 2013. Serologic survey of infectious diseases in populations of maned wolf (*Chrysocyon brachyurus*) and crab-eating fox (*Cerdocyon thous*) from Águas Emendadas Ecological Station, Brazil. *J. Zoo Wildl. Med.* 44, 152–155.

1638. Pronost, S., Pitel, P. H., Romand, S., Thulliez, P., Collobert, C., Fortier, G. 1999. *Neospora caninum*: première mise en évidence en France sur un avorton équin, analyse et perspectives. [First PCR detection on an equine aborted fetus in France—Analysis and prospects]. *Prat. Vét. Equine* 31, 111–114.

1639. Pruvot, M., Hutchins, W., Orsel, K. 2014. Statistical evaluation of a commercial *Neospora caninum* competitive ELISA in the absence of a gold standard: Application to wild elk (*Cervus elaphus*) in Alberta. *Parasitol. Res.* 113, 2899–2905.

1640. Pruvot, M., Kutz, S., Barkema, H. W., De Buck, J., Orsel, K. 2014. Occurrence of *Mycobacterium avium* subspecies paratuberculosis and *Neospora caninum* in Alberta cow-calf operations. *Prev. Vet. Med.* 117, 95–102.

1641. Pruvot, M., Kutz, S., van der Meer, F., Musiani, M., Barkema, H. W., Orsel, K. 2014. Pathogens at the livestock-wildlife interface in Western Alberta: Does transmission route matter? *Vet. Res.* 45, 18.

1641a. Pulido-Medellín, M. O., García-Corredor, D. J., Vargas-Abella, J. C. 2016. Seroprevalencia de *Neospora caninum* en un hato lechero de Boyacá, Colombia. *Rev. Inv. Vet. Perú*. 27, 355–362.

1642. Pumarola, M., Añor, S., Ramis, A. J., Borràs, D., Gorraiz, J., Dubey, J. P. 1996. *Neospora caninum* infection in a Napolitan mastiff dog from Spain. *Vet. Parasitol*. 64, 315–317.

1643. Puray, Ch. N., Chávez, V. A., Casas, A. E., Falcón, P. N., Casas, V. G. 2006. Prevalencia de *Neospora caninum* en bovinos de una empresa ganaderde la sierra central del Perú. [Prevalence of *Neospora caninum* in cattle of a livestock company of the central highlands of Peru]. *Rev. Inv. Vet. Perú* 17, 189–194.

1644. Pusterla, N., Wilson, W. D., Conrad, P. A., Barr, B. C., Ferraro, G. L., Daft, B. M., Leutenegger, C. M. 2006. Cytokine gene signatures in neural tissue of horses with equine protozoal myeloencephalitis or equine herpes type 1 myeloencephalopathy. *Vet. Rec*. 159, 341–346.

1645. Pusterla, N., Conrad, P. A., Packham, A. E., Mapes, S. M., Finno, C. J., Gardner, I. A., Barr, B. C., Ferraro, G. L., Wilson, W. D. 2011. Endogenous transplacental transmmission of *Neospora hughesi* in naturally infected horses. *J. Parasitol*. 97, 281–285.

1646. Pusterla, N., Mackie, S., Packham, A., Conrad, P. A. 2014. Serological investigation of transplacental infection with *Neospora hughesi* and *Sarcocystis neurona* in broodmares. *Vet. J*. 202, 649–650.

1647. Pusterla, N., Tamez-Trevino, E., White, A., VanGeem, J., Packham, A., Conrad, P. A., Kass, P. 2014. Comparison of prevalence factors in horses with and without seropositivity to *Neospora hughesi* and/or *Sarcocystis neurona*. *Vet. J*. 200, 332–334.

1648. Qian, W., Wang, H., Shan, D., Li, B., Liu, J., Liu, Q. 2015. Activity of several kinds of drugs against *Neospora caninum*. *Parasitol. Int*. 64, 597–602.

1649. Qian, W., Wang, T., Yan, W., Han, L., Zhai, K., Duan, B., Lv, C. 2016. Occurrence and first multilocus microsatellite genotyping of *Neospora caninum* from naturally infected dogs in dairy farms in Henan, Central China. *Parasitol. Res*. 115, 3267–3273.

1650. Qin, Q., Wei, F., Li, M., Dubovi, E. J., Loeffler, I. K. 2007. Serosurvey of infectious disease agents of carnivores in captive red pandas (*Ailurus fulgens*) in China. *J. Zoo Wildl. Med*. 38, 42–50.

1651. Qu, G., Fetterer, R., Jenkins, M., Leng, L., Shen, Z., Murphy, C., Han, W., Bucala, R., Tuo, W. 2013. Characterization of *Neospora caninum* macrophage migration inhibitory factor. *Exp. Parasitol*. 135, 246–256.

1652. Quinn, H. E., Ellis, J. T., Smith, N. C. 2002. *Neospora caninum*: A cause of immune-mediated failure of pregnancy? *Trends Parasitol*. 18, 391–394.

1653. Quinn, H. E., Miller, C. M. D., Ryce, C., Windsor, P. A., Ellis, J. T. 2002. Characterization of an outbred pregnant mouse model of *Neospora caninum* infection. *J. Parasitol*. 88, 691–696.

1654. Quinn, H. E., Miller, C. M. D., Ellis, J. T. 2004. The cell-mediated immune response to *Neospora caninum* during pregnancy in the mouse is associated with a bias towards production of interleukin-4. *Int. J. Parasitol*. 34, 723–732.

1655. Quinn, H. E., Windsor, P. A., Kirkland, P. D., Ellis, J. T. 2004. An outbreak of abortion in a dairy herd associated with *Neospora caninum* and bovine pestivirus infection. *Aust. Vet. J*. 82, 99–101.

1656. Quintanilla-Gozalo, A., Pereira-Bueno, J., Tabarés, E., Innes, E. A., González-Paniello, R., Ortega-Mora, L. M. 1999. Seroprevalence of *Neospora caninum* infection in dairy and beef cattle in Spain. *Int. J. Parasitol*. 29, 1201–1208.

1657. Quintanilla-Gozalo, A., Pereira-Bueno, J., Seijas-Carballedo, A., Costas, E., Ortega-Mora, L. M. 2000. Observational studies in *Neospora caninum* infected dairy cattle: Relationship infection-abortion and gestational antibody fluctuations. *Int. J. Parasitol*. 30, 900–906.

1658. Ragozo, A. M. A., Paula, V. S. O., Souza, S. L. P., Bergamaschi, D. P., Gennari, S. M. 2003. Ocorrência de anticorpos anti-*Neospora caninum* em soros bovinos procedentes de seis estados brasileiros. [Occurrence of anti-*Neospora caninum* antibodies in bovine sera from six Brazilian states]. *Rev. Bras. Parasitol. Vet*. 12, 33–37.

1659. Raimundo, J. M., Guimarães, A., Moraes, L. M. B., Santos, L. A., Neponuceno, L. L., Barbosa, S. M., Pires, M. S., Santos, H. A., Massard, C. L., Machado, R. Z., Baldani, C. D. 2015. *Toxoplasma gondii* and *Neospora caninum* in dogs from the state of Tocantins: Serology and associated factors. *Braz. J. Vet. Parasitol*. 24, 475–481.

1660. Rajkhowa, S., Rajkhowa, C., Dutta, P. R., Michui, P., Das, R. 2008. Serological evidence of *Neospora caninum* infection in mithun (*Bos frontalis*) from India. *Res. Vet. Sci*. 84, 250–253.

1661. Ramamoorthy, S., Sriranganathan, N., Lindsay, D. S. 2005. Gerbil model of acute neosporosis. *Vet. Parasitol*. 127, 111–114.

1662. Ramamoorthy, S., Lindsay, D. S., Schurig, G. G., Boyle, S. M., Duncan, R. B., Vemulapalli, R., Sriranganathan, N. 2006. Vaccination with γ-irradiated *Neospora caninum* tachyzoites protects mice against acute challenge with *N. caninum*. *J. Eukaryot. Microbiol.* 53, 151–156.

1663. Ramamoorthy, S., Duncan, R., Lindsay, D. S., Sriranganathan, N. 2007. Optimization of the use of C57BL/6 mice as a laboratory animal model for *Neospora caninum* vaccine studies. *Vet. Parasitol.* 145, 253–259.

1664. Ramamoorthy, S., Sanakkayala, N., Vemulapalli, R., Jain, N., Lindsay, D. S., Schurig, G. S., Boyle, S. M., Sriranganathan, N. 2007. Prevention of vertical transmission of *Neospora caninum* in C57BL/6 mice vaccinated with *Brucella abortus* strain RB51 expressing *N. caninum* protective antigens. *Int. J. Parasitol.* 37, 1531–1538.

1665. Ramamoorthy, S., Sanakkayala, N., Vemulapalli, R., Duncan, R. B., Lindsay, D. S., Schurig, G. S., Boyle, S. M., Kasimanickam, R., Sriranganathan, N. 2007. Prevention of lethal experimental infection of C57BL/6 mice by vaccination with *Brucella abortus* strain RB51 expressing *Neospora caninum* antigens. *Int. J. Parasitol.* 37, 1521–1529.

1666. Ramaprasad, A., Mourier, T., Naeem, R., Malas, T. B., Moussa, E., Panigrahi, A., Vermont, S. J., Otto, T. D., Wastling, J., Pain, A. 2015. Comprehensive evaluation of *Toxoplasma gondii* VEG and *Neospora caninum* LIV genomes with tachyzoite stage transcriptome and proteome defines novel transcript features. *PLoS ONE* 10, e0124473.

1666a. Ramos, I. A. S., da Silva, R. J., Maciel, T. A., Afonso da Silva, J. A. B., Fidelis, O. L., Soares, P. C., Machado, R. Z., André, M. R., de Mendonça, C. L. 2016. Assessment of transplacental transmission of *Neospora caninum* in dairy cattle in the Agreste region of Pernambuco. *Rev. Bras. Parasitol. Vet.* 25, 516–522. Doi: 10.1590/S1984-29612016055.

1667. Rasmussen, K., Jensen, A. L. 1996. Some epidemiologic features of canine neosporosis in Denmark. *Vet. Parasitol.* 62, 345–349.

1668. Razmi, G. 2009. Fecal and molecular survey of *Neospora caninum* in farm and household dogs in Mashhad Area, Khorasan Province, Iran. *Korean J. Parasitol.* 47, 417–420.

1669. Razmi, G. R., Mohammadi, G. R., Garrosi, T., Farzaneh, N., Fallah, A. H., Maleki, M. 2006. Seroepidemiology of *Neospora caninum* infection in dairy cattle herds in Mashhad area, Iran. *Vet. Parasitol.* 135, 187–189.

1670. Razmi, G. R., Maleki, M., Farzaneh, N., Talebkhan Garoussi, M., Fallah, A. H. 2007. First report of *Neospora caninum*-associated bovine abortion in Mashhad area, Iran. *Parasitol. Res.* 100, 755–757.

1671. Razmi, G. R., Zarea, H., Naseri, Z. 2010. A survey of *Neospora caninum*-associated bovine abortion in large dairy farms of Mashhad, Iran. *Parasitol. Res.* 106, 1419–1423.

1672. Reed, S. M., Furr, M., Howe, D. K., Johnson, A. L., MacKay, R. J., Morrow, J. K., Pusterla, N., Witonsky, S. 2016. Equine protozoal myeloencephalitis: An updated consensus statement with a focus on parasite biology, diagnosis, treatment, and prevention. *J. Vet. Intern. Med.* 30, 491–502. Doi: 10.1111/jvim.13834.

1673. Regidor-Cerrillo, J., Pedraza-Díaz, S., Gómez-Bautista, M., Ortega-Mora, L. M. 2006. Multilocus microsatellite analysis reveals extensive genetic diversity in *Neospora caninum*. *J. Parasitol.* 92, 517–524.

1674. Regidor-Cerrillo, J., Gómez-Bautista, M., Pereira-Bueno, J., Aduriz, G., Navarro-Lozano, V., Risco-Castillo, V., Férnandez-García, A., Pedraza-Díaz, S., Ortega-Mora, L. M. 2008. Isolation and genetic characterization of *Neospora caninum* from asymptomatic calves in Spain. *Parasitology* 135, 1651–1659.

1675. Regidor-Cerrillo, J., Gómez-Bautista, M., Del Pozo, I., Jiménez-Ruiz, E., Aduriz, G., Ortega-Mora, L. M. 2010. Influence of *Neospora caninum* intra-specific variability in the outcome of infection in a pregnant BALB/c mouse model. *Vet. Res.* 41, 52.

1676. Regidor-Cerrillo, J., Pedraza-Diaz, S., Rojo-Montejo, S., Vazquez-Moreno, E., Arnaiz, I., Gomez-Bautista, M., Jimenez-Palacios, S., Ortega-Mora, L. M., Collantes-Fernandez, E. 2010. *Neospora caninum* infection in stray and farm dogs: Seroepidemiological study and oocyst shedding. *Vet. Parasitol.* 174, 332–335.

1677. Regidor-Cerrillo, J., Gómez-Bautista, M., Sodupe, I., Aduriz, G., Álvarez-García, G., del-Pozo, I., Ortega-Mora, L. M. 2011. *In vitro* invasion efficiency and intracellular proliferation rate comprise virulence-related phenotypic traits of *Neospora caninum*. *Vet. Res.* 42, 41.

1678. Regidor-Cerrillo, J., Álvarez-García, G., Pastor-Fernández, I., Marugán-Hernández, V., Gómez-Bautista, M., Ortega-Mora, L. M. 2012. Proteome expression changes among virulent and attenuated *Neospora caninum* isolates. *J. Proteomics* 75, 2306–2318.

1679. Regidor-Cerrillo, J., Díez-Fuertes, F., García-Culebras, A., Moore, D. P., González-Warleta, M., Cuevas, C., Schares, G., Katzer, F., Pedraza-Díaz, S., Mezo, M., Ortega-Mora, L. M. 2013. Genetic diversity and geographic population structure of bovine *Neospora canium* determined by microsatellite genotyping analysis. *PLoS ONE* 8, e72678.

1680. Regidor-Cerrillo, J., Arranz-Solís, D., Benavides, J., Gómez-Bautista, M., Castro-Hermida, J. A., Mezo, M., Pérez, V., Ortega-Mora, L. M., González-Warleta, M. 2014. *Neospora caninum* infection during early pregnancy in cattle: How the isolate influences infection dynamics, clinical outcome and peripheral and local immune responses. *Vet. Res.* 45, 10.

1681. Regidor-Cerrillo, J., García-Lunar, P., Pastor-Fernández, I., Álvarez-García, G., Collantes-Fernández, E., Gómez-Bautista, M., Ortega-Mora, L. M. 2015. *Neospora caninum* tachyzoite immunome study reveals differences among three biologically different isolates. *Vet. Parasitol.* 212, 92–99.

1682. Reichel, M. P., Drake, J. M. 1996. The diagnosis of *Neospora* abortions in cattle. *N. Z. Vet. J.* 44, 151–154.

1683. Reichel, M. P. 1998. Prevalence of *Neospora* antibodies in New Zealand dairy cattle and dogs. *N. Z. Vet. J.* 46, 38.

1684. Reichel, M. P., Thornton, R. N., Morgan, P. L., Mills, R. J. M., Schares, G. 1998. Neosporosis in a pup. *N. Z. Vet. J.* 46, 106–110.

1685. Reichel, M. P. 2000. *Neospora caninum* infections in Australia and New Zealand. *Aust. Vet. J.* 78, 258–261.

1686. Reichel, M. P., Pfeiffer, D. U. 2002. An analysis of the performance characteristics of serological tests for the diagnosis of *Neospora caninum* infection in cattle. *Vet. Parasitol.* 107, 197–207.

1687. Reichel, M. P., Ellis, J. T. 2002. Control options for *Neospora caninum* infections in cattle—Current state of knowledge. *N. Z. Vet. J.* 50, 86–92.

1688. Reichel, M. P., Pfeiffer, D. U. 2004. An analysis of the performance characteristics of serological tests for the diagnosis of *Neospora caninum* infection in cattle. *N. Z. J. Zool.* 31, 82–83.

1689. Reichel, M. P., Ellis, J. T. 2006. If control of *Neospora caninum* infection is technically feasible does it make economic sense? *Vet. Parasitol.* 142, 23–34.

1690. Reichel, M. P., Ellis, J. T., Dubey, J. P. 2007. Neosporosis and hammondiosis in dogs. *J. Small Anim. Pract.* 48, 308–312.

1691. Reichel, M. P., Ellis, J. T. 2008. Re-valuating the economics of neosporosis control. *Vet. Parasitol.* 156, 361–362.

1692. Reichel, M. P., Ross, G. P., McAllister, M. M. 2008. Evaluation of an enzyme-linked immunosorbent assay for the serological diagnosis of *Neospora caninum* infection in sheep and determination of the apparent prevalence of infection in New Zealand. *Vet. Parasitol.* 151, 323–326.

1693. Reichel, M. P., Ellis, J. T. 2009. *Neospora caninum*—How close are we to development of an efficacious vaccine that prevents abortion in cattle? *Int. J. Parasitol.* 39, 1173–1187.

1694. Reichel, M. P., Ayanegui-Alcérreca, M. A., Gondim, L. F. P., Ellis, J. T. 2013. What is the global economic impact of *Neospora caninum* in cattle—The billion dollar question. *Int. J. Parasitol.* 43, 133–142.

1695. Reichel, M. P., McAllister, M. M., Pomroy, W. E., Campero, C., Ortega-Mora, L. M., Ellis, J. T. 2014. Control options for *Neospora caninum*—Is there anything new or are we going backwards? *Parasitology* 141, 1455–1470.

1696. Reichel, M. P., McAllister, M. M., Nasir, A., Moore, D. P. 2015. A review of *Neospora caninum* in water buffalo (*Bubalus bubalis*). *Vet. Parasitol.* 212, 75–79.

1697. Reichel, M. P., Moore, D. P., Hemphill, A., Ortega-Mora, L. M., Dubey, J. P., Ellis, J. T. 2015. A live vaccine against *Neospora caninum* abortions in cattle. *Vaccine* 33, 1299–1301.

1698. Reid, A. J., Vermont, S. J., Cotton, J. A., Harris, D., Hill-Cawthorne, G. A., Könen-Waisman, S., Latham, S. M., Mourier, T., Norton, R., Quail, M. A., Sanders, M., Shanmugam, D., Sohal, A., Wasmuth, J. D., Brunk, B., Grigg, M. E., Howard, J. C., Parkinson, J., Roos, D. S., Trees, A. J., Berriman, M., Pain, A., Wastling, J. M. 2012. Comparative genomics of the apicomplexan parasites *Toxoplasma gondii* and *Neospora caninum*: Coccida differing in host range and transmission strategy. *PLoS Pathog.* 8, e1002567.

1699. Reisberg, K., Selim, A. M., Gaede, W. 2013. Simultaneous detection of *Chlamydia* spp., *Coxiella burnetii*, and *Neospora caninum* in abortion material of ruminants by multiplex real-time polymerase chain reaction. *J. Vet. Diagn. Invest.* 25, 614–619.

1700. Reiterová, K., Špilovská, S., Antolová, D., Dubinský, P. 2009. *Neospora caninum*, potential cause of abortions in dairy cows: The current serological follow-up in Slovakia. *Vet. Parasitol.* 159, 1–6.

1701. Reiterová, K., Špilovská, S., Pošivák, J., Dubinský, P., Novotný, F., Valocký, I. 2009. *Neospora caninum*—Possible causative agent of abortions in dairy farm in Slovakia. *Folia Vet.* 1(Suppl 53), 88–91.

1702. Reiterová, K., Špilovská, S., Cobádiová, A., Mucha, R. 2011. First *in vitro* isolation of *Neospora caninum* from a naturally infected adult dairy cow in Slovakia. *Acta Parasitol.* 56, 111–115.

1703. Reiterová, K., Špilovská, S., Blanarová, L., Derdáková, M., Cobádiová, A., Hisira, V. 2016. Wild boar (*Sus scrofa*)—Reservoir host of *Toxoplasma gondii*, *Neospora caninum* and *Anaplasma phagocytophilum* in Slovakia. *Acta Parasitol.* 61, 255–260.

1703a. Reiterová, K., Špilovská, S., Cobádiová, A., Hurniková, Z. 2016. Prevalence of *Toxoplasma gondii* and *Neospora caninum* in red foxes in Slovakia. *Acta Parasitol.* 61, 762–768.

1704. Reitt, K., Hilbe, M., Voegtlin, A., Corboz, L., Haessig, M., Pospischil, A. 2007. Aetiology of bovine abortion in Switzerland from 1986 to 1995—A retrospective study with emphasis on detection of *Neospora caninum* and *Toxoplasma gondii* by PCR. *J. Vet. Med. A* 54, 15–22.

1705. Renier, A. C., Morrow, J. K., Graves, A. J., Finno, C. J., Howe, D. K., Owens, S. D., Tamez-Trevino, E., Packham, A. E., Conrad, P. A., Pusterla, N. 2016. Diagnosis of equine protozoal myeloencephalitis using indirect fluorescent antibody testing and enzyme-linked immunosorbent assay titer ratios for *Sarcocystis neurona* and *Neospora hughesi*. *J. Equine Vet. Sci.* 36, 49–51.

1706. Rettigner, C., Leclipteux, T., De Meerschman, F., Focant, C., Losson, B. 2004. Survival, immune responses and tissue cyst production in outbred (Swiss white) and inbred (CBA/Ca) strains of mice experimentally infected with *Neospora caninum* tachyzoites. *Vet. Res.* 35, 225–232.

1707. Rettigner, C., De Meerschman, F., Focant, C., Vanderplasschen, A., Losson, B. 2004. The vertical transmission following the reactivation of a *Neospora caninum* chronic infection does not seem to be due an alteration of the systemic immune response in pregnant CBA/Ca mice. *Parasitology* 128, 149–160.

1708. Rettigner, C., Lasri, S., De Meerschman, F., Focant, C., Beckers, J. F., Losson, B. 2004. Immune response and antigen recognition in non-pregnant ewes experimentally infected with *Neospora caninum* tachyzoites. *Vet. Parasitol.* 122, 261–271.

1709. Ribeiro, D. P., Freitas, M. M. P., Cardoso, M. R. D., Pajuaba, A. C. A. M., Silva, N. M., Mineo, T. W. P., Silva, J. S., Mineo, J. R., Silva, D. A. O. 2009. CpG-ODN combined with *Neospora caninum* lysate, but not with excreted-secreted antigen, enhances protection against infection in mice. *Vaccine* 27, 2570–2579.

1710. Ribeiro, M. J. M., Rosa, M. H. F., Bruhn, F. R. P., Garcia, A. M., da Rocha, C. M. B. M., Guimarães, A. M. 2016. Seroepidemiology of *Sarcocystis neurona*, *Toxoplasma gondii* and *Neospora* spp. among horses in the south of the state of Minas Gerais, Brazil. *Braz. J. Vet. Parasitol.* 25, 142–150.

1711. Rinaldi, L., Fusco, G., Musella, V., Veneziano, V., Guarino, A., Taddei, R., Cringoli, G. 2005. *Neospora caninum* in pastured cattle: Determination of climatic, environmental, farm management and individual animal risk factors using remote sensing and geographical information systems. *Vet. Parasitol.* 128, 219–230.

1712. Rinaldi, L., Pacelli, F., Iovane, G., Pagnini, U., Veneziano, V., Fusco, G., Cringoli, G. 2007. Survey of *Neospora caninum* and bovine herpes virus 1 coinfection in cattle. *Parasitol. Res.* 100, 359–364.

1713. Risco-Castillo, V., Fernández-García, A., Ortega-Mora, L. M. 2004. Comparative analysis of stress agents in a simplified *in vitro* system of *Neospora caninum* bradyzoite production. *J. Parasitol.* 90, 466–470.

1714. Risco-Castillo, V., Fernández-García, A., Zaballos, A., Aguado-Martínez, A., Hemphill, A., Rodríguez-Bertos, A., Álvarez-García, G., Ortega-Mora, L. M. 2007. Molecular characterisation of *BSR4*, a novel bradyzoite-specific gene from *Neospora caninum*. *Int. J. Parasitol.* 37, 887–896.

1715. Risco-Castillo, V., Marugán-Hernández, V., Fernández-García, A., Aguado-Martínez, A., Jiménez-Ruiz, E., Rodríguez-Marco, S., Álvarez-García, G., Ortega-Mora, L. M. 2011. Identification of a gene cluster for cell-surface genes of the SRS superfamily in *Neospora caninum* and characterization of the novel *SRS9* gene. *Parasitology* 138, 1832–1842.

1716. Risco-Castillo, V., Wheeler, J. C., Rosadio, R., García-Peña, F. J., Arnaiz-Seco, I., Hoces, D., Castillo, H., Veliz, Á., Ortega-Mora, L. M. 2014. Health impact evaluation of alternative management systems in vicuña (*Vicugna vicugna mensalis*) populations in Peru. *Trop. Anim. Health Prod.* 46, 641–646.

1717. Ritter, D. M., Kerlin, R., Sibert, G., Brake, D. 2002. Immune factors influencing the course of infection with *Neospora caninum* in the murine host. *J. Parasitol.* 88, 271–280.

1718. Robbe, D., Passarelli, A., Gloria, A., Di Cesare, A., Capelli, G., Iorio, R., Traversa, D. 2016. *Neospora caninum* seropositivity and reproductive risk factors in dogs. *Exp. Parasitol.* 164, 31–35.

1719. Robert, F. 1996. Néosporose: Bilan des connaissances actuelles sur cette parasitose de découverte récente. [Neosporosis: Assessment of the current knowledge on this new discovery parasitosis]. *Rev. Fr. Lab.* 285, 29–34.

1720. Robert-Gangneux, F., Klein, F. 2009. Serologic screening for *Neospora caninum*, France. *Emerg. Infect. Dis.* 15, 987–988.

1721. Rocchi, M. S., Bartley, P. M., Inglis, N. F., Collantes-Fernandez, E., Entrican, G., Katzer, F., Innes, E. A. 2011. Selection of *Neospora caninum* antigens stimulating bovine CD4^{+ve} T cell responses through immunopotency screening and proteomic approaches. *Vet. Res.* 42, 91.

1721a. Rocchigiani, G., Poli, A., Nardoni, S., Roberto, P., Mancianti, F. 2017. *Neospora caninum* in wild waterfowl: Occurrence of parasite DNA and low antibody titers. *J. Parasitol.* 103, 142–145. DOI: 10.1645/16-34.

1722. Rodrigues, A. A. R., Gennari, S. M., Aguiar, D. M., Sreekumar, C., Hill, D. E., Miska, K. B., Vianna, M. C. B., Dubey, J. P. 2004. Shedding of *Neospora caninum* oocysts by dogs fed tissues from naturally infected water buffaloes (*Bubalus bubalis*) from Brazil. *Vet. Parasitol.* 124, 139–150.

1723. Rodrigues, A. A. R., Gennari, S. M., Paula, V. S. O., Aguiar, D. M., Fujii, T. U., Starke-Buzeti, W., Machado, R. Z., Dubey, J. P. 2005. Serological resonses to *Neospora caninum* in experimentally and naturally infected water buffaloes (*Bubalus bubalis*). *Vet. Parasitol.* 129, 21–24.

1724. Rodríguez, A. M., Maresca, S., Cano, D. B., Armendano, J. I., Combessies, G., Lopéz-Valiente, S., Odriozola, E. R., Späth, E. J. L., Odeón, A. C., Campero, C. M., Moore, D. P. 2016. Frequency of *Neospora caninum* infections in beef cow-calf operations under extensive management. *Vet. Parasitol.* 219, 40–43.

1725. Rodriguez, I., Choromanski, L., Rodgers, S. J., Weinstock, D. 2003. Survey of *Neospora caninum* antibodies in dairy and beef cattle from five regions of the United States. *Vet. Therapeut.* 3, 396–401.

1726. Roelandt, S., Van der Stede, Y., Czaplicki, G., Van Loo, H., Van Driessche, E., Dewulf, J., Hooyberghs, J., Faes, C. 2015. Serological diagnosis of bovine neosporosis: A Bayesian evaluation of two antibody ELISA tests for in vivo diagnosis in purchased and abortion cattle. *Vet. Rec.* 176, 598.

1727. Rogers, D. G., Grotelueschen, D. M., Anderson, M. L., McCullough, M. S., Shain, W. S., Dubey, J. P. 1993. Endemic protozoal abortions in a dairy cow herd. *Agri-Practice* 14, 16–21.

1728. Rojo-Montejo, S., Collantes-Fernández, E., Blanco-Murcia, J., Rodríguez-Bertos, A., Risco-Castillo, V., Ortega-Mora, L. M. 2009. Experimental infection with a low virulence isolate of *Neospora caninum* at 70 days gestation in cattle did not result in foetopathy. *Vet. Res.* 40, 49.

1729. Rojo-Montejo, S., Collantes-Fernández, E., Regidor-Cerrillo, J., Álvarez-García, G., Marugan-Hernández, V., Pedraza-Díaz, S., Blanco-Murcia, J., Prenafeta, A., Ortega-Mora, L. M. 2009. Isolation and characterization of a bovine isolate of *Neospora caninum* with low virulence. *Vet. Parasitol.* 159, 7–16.

1730. Rojo-Montejo, S., Collantes-Fernández, E., Regidor-Cerrillo, J., Rodríguez-Bertos, A., Prenafeta, A., Gomez-Bautista, M., Ortega-Mora, L. M. 2011. Influence of adjuvant and antigen dose on protection induced by an inactivated whole vaccine against *Neospora caninum* infection in mice. *Vet. Parasitol.* 175, 220–229.

1731. Rojo-Montejo, S., Collantes-Fernández, E., López-Pérez, I., Risco-Castillo, V., Prenafeta, A., Ortega-Mora, L. M. 2012. Evaluation of the protection conferred by a naturally attenuated *Neospora caninum* isolate against congenital and cerebral neosporosis in mice. *Vet. Res.* 43, 62.

1732. Rojo-Montejo, S., Collantes-Fernández, E., Pérez-Zaballos, F., Rodríguez-Marco, S., Blanco-Murcia, J., Rodríguez-Bertos, A., Prenafeta, A., Ortega-Mora, L. M. 2013. Effect of vaccination of cattle with the low virulence Nc-Spain 1H isolate of *Neospora caninum* against a heterologous challenge in early and mid-gestation. *Vet. Res.* 44, 106.

1733. Romand, S., Thulliez, P., Dubey, J. P. 1998. Direct agglutination test for serologic diagnosis of *Neospora caninum* infection. *Parasitol. Res.* 84, 50–53.

1734. Romanelli, P. R., Freire, R. L., Vidotto, O., Marana, E. R. M., Ogawa, L., de Paula, V. S. O., Garcia, J. L., Navarro, I. T. 2007. Prevalence of *Neospora caninum* and *Toxoplasma gondii* in sheep and dogs from Guarapuava farms, Paraná State, Brazil. *Res. Vet. Sci.* 82, 202–207.

1735. Romano, A., Trisciuoglio, A., Grande, D., Ferroglio, E. 2009. Comparison of two PCR protocols for the detection of *Neospora canium* DNA in rodents. *Vet. Parasitol.* 159, 159–161.

1736. Romero Zúñiga, J. J. 2005. Appraisal of the epidemiology of *Neospora caninum* infection in Costa Rican dairy cattle. *Doctoral thesis*. Wageningen University, Wagenigen, the Netherlands. 1–137.

1737. Romero, J. J., Perez, E., Dolz, G., Frankena, K. 2002. Factors associated with *Neospora caninum* serostatus in cattle of 20 specialised Costa Rican dairy herds. *Prev. Vet. Med.* 53, 263–273.

1738. Romero, J. J., Frankena, K. 2003. The effect of the dam-calf relationship on serostatus to *Neospora caninum* on 20 Costa Rican dairy farms. *Vet. Parasitol.* 114, 159–171.

1739. Romero, J. J., Pérez, E., Frankena, K. 2004. Effect of a killed whole *Neospora caninum* tachyzoite vaccine on the crude abortion rate of Costa Rican dairy cows under field conditions. *Vet. Parasitol.* 123, 149–159.

1740. Romero, J. J., Van Breda, S., Vargas, B., Dolz, G., Frankena, K. 2005. Effect of neosporosis on productive and reproductive performance of dairy cattle in Costa Rica. *Theriogenology* 64, 1928–1939.

1741. Romero-Salas, D., García-Vázquez, Z., Montiel-Palacios, F., Montiel-Peña, T., Aguilar-Domínguez, M., Medina-Esparza, L., Cruz-Vázquez, C. 2010. Seroprevalence of *Neospora caninum* antibodies in cattle in Veracruz, Mexico. *J. Anim. Vet. Adv.* 9, 1445–1451.

1742. Lang-unnasch, N., Aiello, D. P. 1999. Sequence evidence for an altered genetic code in the *Neospora caninum* plastid. *Int. J. parasitol.* 29, 1557–1562.

1743. Rosbottom, A., Guy, C. S., Gibney, E. H., Smith, R. F., Valarcher, J. F., Taylor, G., Williams, D. J. L. 2007. Peripheral immune responses in pregnant cattle following *Neospora caninum* infection. *Parasite Immunol.* 29, 219–228.

1744. Rosbottom, A., Gibney, E. H., Guy, C. S., Kipar, A., Smith, R. F., Kaiser, P., Trees, A. J., Williams, D. J. L. 2008. Upregulation of cytokines is detected in the placentas of cattle infected with *Neospora caninum* and is more marked early in gestation when fetal death in observed. *Infect. Immun.* 76, 2352–2361.

1745. Rosbottom, A., Gibney, H., Kaiser, P., Hartley, C., Smith, R. F., Robinson, R., Kipar, A., Williams, D. J. L. 2011. Up regulation of the maternal immune response in the placenta of cattle naturally infected with *Neospora caninum*. *PLoS ONE* 6, e15799.

1746. Rossi, G. F., Cabral, D. D., Ribeiro, D. P., Pajuaba, A. C. A. M., Corrêa, R. R., Moreira, R. Q., Mineo, T. W. P., Mineo, J. R., Silva, D. A. O. 2011. Evaluation of *Toxoplasma gondii* and *Neospora caninum* infections in sheep from Uberlândia, Minas Gerais State, Brazil, by different serological methods. *Vet. Parasitol.* 175, 252–259.

1747. Rosypal, A. C., Lindsay, D. S. 2005. The sylvatic cycle of *Neospora caninum*: Where do we go from here? *Trends Parasitol.* 21, 439–440.

1748. Rudbäck, E., Mannonen, J., Nikander, S., Henriksson, K. 1991. *Neopsora caninum*—uusi parasiitti Suomessa? [*Neospora caninum*—a new parasite in Finland?]. *Suomen Eläinlääkärilehti* 97, 526–529.

1749. Ruehlmann, D., Podell, M., Oglesbee, M., Dubey, J. P. 1995. Canine neosporosis: A case report and literature review. *J. Am. Anim. Hosp. Assoc.* 31, 174–183.

1750. Ruíz, R. N., Casas, A, E., Suárez, A. F., Díaz, C. D., Fernández, P. V. 2012. Frecuencia de anticuerpos contra *Neospora caninum* y *Toxoplasma gondii* en canes con signos clínicos de afección neuromuscular. [Frequency of antibodies against *Neospora caninum* and *Toxoplasma gondii* in dogs with clinical signs of neuromuscular disease]. *Rev. Inv. Vet. Perú* 23, 441–447.

1751. Ruiz-Fons, F., González-Barrio, D., Aguilar-Ríos, F., Soler, A. J., Garde, J. J., Gortázar, C., Fernández-Santos, M. R. 2014. Infectious pathogens potentially transmitted by semen of the black variety of the Manchega sheep breed: Health constraints for conservation purposes. *Anim. Reprod. Sci.* 149, 152–157.

1752. Ruttkowski, B., Joachim, A., Daugschies, A. 2001. PCR-based differentiation of three porcine *Eimeria* species and *Isospora suis*. *Vet. Parasitol.* 95, 17–23.

1753. Sadrebazzaz, A., Haddadzadeh, H., Esmailnia, K., Habibi, G., Vojgani, M., Hashemifesharaki, R. 2004. Serological prevalence of *Neospora caninum* in healthy and aborted dairy cattle in Mashhad, Iran. *Vet. Parasitol.* 124, 201–204.

1754. Sadrebazzaz, A., Haddadzadeh, H., Shayan, P. 2006. Seroprevalence of *Neospora caninum* and *Toxoplasma gondii* in camels (*Camelus dromedarius*) in Mashhad, Iran. *Parasitol. Res.* 98, 600–601.

1755. Sadrebazzaz, A., Habibi, G., Haddadzadeh, H., Ashrafi, J. 2007. Evaluation of bovine abortion associated with *Neospora caninum* by different diagnostic techniques in Mashhad, Iran. *Parasitol. Res.* 100, 1257–1260.

1756. Saey, V., Martlé, V., van Ham, L., Chiers, K. 2010. Neuritis of the cauda equina in a dog. *J. Small Anim. Pract.* 51, 549–552.

1757. Sager, H., Fischer, I., Furrer, K., Strasser, M., Waldvogel, A., Boerlin, P., Audigé, L., Gottstein, B. 2001. A Swiss case-control study to assess *Neospora caninum*-associated bovine abortions by PCR, histopathology and serology. *Vet. Parasitol.* 102, 1–15.

1758. Sager, H., Gloor, M., Björkman, C., Kritzner, S., Gottstein, B. 2003. Assessment of antibody avidity in aborting cattle by a somatic *Neospora caninum* tachyzoite antigen IgG avidity ELISA. *Vet. Parasitol.* 112, 1–10.

1759. Sager, H., Hüssy, D., Kuffer, A., Schreve, F., Gottstein, B. 2005. Mise en évidence d'un cas de "abortion storm" (transmission transplacentaire exogène de *Neospora caninum*) dans une exploitation de vaches laitières: une première en Suisse. [First documentation of a *Neospora*-induced "abortion storm" (exogenous transplacental transmission of *Neospora caninum*) in a Swiss dairy farm]. *Schweiz. Arch. Tierheilkd.* 147, 113–120.

1760. Sager, H., Steiner-Moret, C., Müller, N., Staubli, D., Esposito, M., Schares, G., Hässig, M., Stärk, K., Gottstein, B. 2006. Incidence of *Neospora caninum* and other intestinal protozoan parasites in populations of Swiss dogs. *Vet. Parasitol.* 139, 84–92.

1761. Salaberry, S. R. S., Okuda, L. H., Nassar, A. F. C., de Castro, J. R., Lima-Ribeiro, A. M. C. 2010. Prevalence of *Neospora caninum* antibodies in sheep flocks of Uberlândia county, MG. *Rev. Bras. Parasitol. Vet.* 19, 148–151.

1762. Salant, H., Mazuz, M. L., Savitsky, I., Nasereddin, A., Blinder, E., Baneth, G. 2015. *Neospora caninum* in crows from Israel. *Vet. Parasitol.* 212, 375–378.

1763. Salb, A. L., Barkema, H. W., Elkin, B. T., Thompson, R. C. A., Whiteside, D. P., Black, S. R., Dubey, J. P., Kutz, S. J. 2008. Dogs as sources and sentinels of parasites in humans and wildlife, Northern Canada. *Emerg. Infect. Dis.* 14, 60–63.

1764. Salehi, N., Haddadzadeh, H., Ashrafihelan, J., Shayan, P., Sadrebazzaz, A. 2009. Molecular and pathological study of bovine aborted fetuses and placenta from *Neospora caninum* infected dairy cattle. *Iranian J. Parasitol.* 4, 40–51.

1765. Salehi, N., Haddadzadeh, H. R., Shayan, P., Vodjgani, M., Bolourchi, M. 2010. Serological study of *Neospora caninum* in pregnant dairy cattle in Tehran, Iran. *Int. J. Vet. Res.* 4, 113–116.

1766. Salehi, N., Haddadzadeh, H. R., Shayan, P., Koohi, M. K. 2012. Isolation of *Neospora caninum* from an aborted fetus of seropositive cattle in Iran. *Veterinarski Arhiv* 82, 545–553.

1767. Salehi, N., Gottstein, B., Haddadzadeh, H. R. 2015. Genetic diversity of bovine *Neospora caninum* determined by microsatellite markers. *Parasitol. Int.* 64, 357–361.

1768. Salman, D., Oohashi, E., Mohamed, A. E. A., Abd El-Mottelib, A. E. R., Okada, T., Igarashi, M. 2014. Seroprevalences of *Toxoplasma gondii* and *Neospora caninum* in pet rabbits in Japan. *J. Vet. Med. Sci.* 76, 855–862.

1769. San Miguel, J. M., Gutiérrez-Expósito, D., Aguado-Martínez, A., González-Zotes, E., Pereira-Bueno, J., Gómez-Bautista, M., Rubio, P., Ortega-Mora, L. M., Collantes-Fernández, E., Álvarez-García, G. 2016. Effect of different ecosystems and management practices on *Toxoplasma gondii* and *Neospora caninum* infections in wild ruminants in Spain. *J. Wildl. Dis.* 52, 293–300.

1770. Sánchez, G. F., Morales, S. E., Martínez, M. J., Trigo, J. F. 2003. Determination and correlation of anti-*Neospora caninum* antibodies in dogs and cattle from Mexico. *Can. J. Vet. Res.* 67, 142–145.

1771. Sánchez, G. F., Banda, R. V. M., Sahagun, R. A., Ledesma, M. N., Morales, S. E. 2009. Comparison between immunohistochemistry and two PCR methods for detection of *Neospora caninum* in formalin-fixed and paraffin-embedded brain tissue of bovine fetuses. *Vet. Parasitol.* 164, 328–332.

1772. Sanderson, M. W., Gay, J. M., Baszler, T. V. 2000. *Neospora caninum* seroprevalence and associated risk factors in beef cattle in the northwestern United States. *Vet. Parasitol.* 90, 15–24.

1773. Sangster, C., Bryant, B., Campbell-Ward, M., King, J. S., Šlapeta, J. 2010. Neosporosis in an aborted southern white rhinoceros (*Ceratotherium simum simum*) fetus. *J. Zoo Wildl. Med.* 41, 725–728.

1774. Sanhueza, J. M., Heuer, C., West, D. 2013. Contribution of *Leptospira*, *Neospora caninum* and bovine viral diarrhea virus to fetal loss of beef cattle in New Zealand. *Prev. Vet. Med.* 112, 90–98.

1775. Santolaria, P., López-Gatius, F., Yániz, J., García Ispierto, I., Nogareda, C., Bech-Sàbat, G., Serrano, B., Almeria, S. 2009. Early postabortion recovery of *Neospora*-infected lactating dairy cows. *Theriogenology* 72, 798–802.

1776. Santolaria, P., Almería, S., Martínez-Bello, D., Nogareda, C., Mezo, M., Gonzalez-Warleta, M., Castro-Hermida, J. A., Pabón, M., Yániz, J. L., López-Gatius, F. 2011. Different humoral mechanisms against *Neospora caninum* infection in purebreed and crossbreed beef/dairy cattle pregnancies. *Vet. Parasitol.* 178, 70–76.

1777. Santos, S. L., de Souza Costa, K., Gondim, L. Q., da Silva, M. S. A., Uzêda, R. S., Abe-Sandes, K., Gondim, L. F. P. 2010. Investigation of *Neospora caninum*, *Hammondia* sp., and *Toxoplasma gondii* in tissues from slaughtered beef cattle in Bahia, Brazil. *Parasitol. Res.* 106, 457–461.

1778. Sari, Y. C. 2013. *Neospora caninum* infection: Pathomorphological studies supervised by Ekowati handharyani. *Assay rerquirement for Bacheolar of Veterinary Medicine at the Faculty of Veterinary Medicine*, Institut Pertanian Bogor, Indonesia. 1–16.

1779. Sartor, I. F., Hasegawa, M. Y., Canavessi, A. M. O., Pinckney, R. D. 2003. Ocorrência de anticorpos de *Neospora caninum* em vacas leiteiras avaliados pelos métodos ELISA e RIFI no município de Avaré, SP. [Occurrence of *Neospora caninum* antibody in dairy cows assayed by ELISA and IFAT from Avare county, SP]. *Semin. Ci. Agrár.* 24, 3–10.

1780. Sartor, I. F., Garcia Filho, A., Vianna, L. C., Pituco, E. M., Dal Pai, V., Sartor, R. 2005. Ocorrência de anticorpos anti-*Neospora caninum* em bovinos leiteiros e de corte da região de Presidente Prudente, SP. [Occurrence of antibodies anti-*Neospora caninum* in dairy and beef cattle in the region of Presidente Prudente, SP, Brazil]. *Arq. Inst. Biol. (São Paulo)* 72, 413–418.

1781. Sasai, K., Lillehoj, H. S., Hemphill, A., Matsuda, H., Hanioka, Y., Fukata, T., Baba, E., Arakawa, A. 1998. A chicken anti-conoid monoclonal antibody identifies a common epitope which is present on motile stages of *Eimeria*, *Neospora*, and *Toxoplasma*. *J. Parasitol.* 84, 654–656.

1782. Sasani, F., Javanbakht, J., Seifori, P., Fathi, S., Hassan, M. A. 2013. *Neospora caninum* as causative agent of ovine encephalitis in Iran. *Pathol. Discov.* 1, 1–e5.

1783. Savović, M., Lalošević, V., Simin, S., Pavičić, L. J., Boboš, S. 2012. Seroprevalence of *Neospora caninum* in dairy cows with reproductive disorders in Vojvodina province, Serbia. *Lucrări Ştiintifice Med. Vet.*, 45, 161–166.

1784. Sawada, M., Park, C. H., Morita, T., Shimada, A., Umemura, T., Haritani, M. 1997. Pathological findings of nude mice inoculated with bovine *Neospora*. *J. Vet. Med. Sci.* 59, 947–948.

1785. Sawada, M., Park, C. H., Kondo, H., Morita, T., Shimada, A., Yamane, I., Umemura, T. 1998. Serological survey of antibody to *Neospora caninum* in Japanese dogs. *J. Vet. Med. Sci.* 60, 853–854.

1786. Sawada, M., Kondo, H., Tomioka, Y., Park, C. H., Morita, T., Shimada, A., Umemura, T. 2000. Isolation of *Neospora caninum* from the brain of a naturally infected adult dairy cow. *Vet. Parasitol.* 90, 247–252.

1787. Sayari, M., Namavari, M., Mojaver, S. 2016. Seroprevalence of *Neospora caninum* infection in free ranging chickens (*Gallus domesticus*). *J. Parasit. Dis.* 40, 845–847.

1788. Schares, G., Peters, M., Wurm, R., Tackmann, K., Henning, K., Conraths, F. J. 1997. *Neospora caninum* verursacht Aborte in einem Rinderbestand in Nordrhein-Westfalen. [*Neospora caninum* causes abortions in a cattle herd in North Rhine-Westphalia]. *Dtsch. Tierärztl. Wochenschr.* 104, 208–212.

1789. Schares, G., Peters, M., Wurm, R., Bärwald, A., Conraths, F. J. 1998. The efficiency of vertical transmission of *Neospora caninum* in dairy cattle analysed by serological techniques. *Vet. Parasitol.* 80, 87–98.

1790. Schares, G., Conraths, F. J., Reichel, M. P. 1999. Bovine neosporosis: Comparison of serological methods using outbreak sera from a dairy herd in New Zealand. *Int. J. Parasitol.* 29, 1659–1667.

1791. Schares, G., Dubremetz, J. F., Dubey, J. P., Bärwald, A., Loyens, A., Conraths, F. J. 1999. *Neospora caninum*: I entification of 19-, 38-, and 40-kDa surface antigens and a 33-kDa dense granule antigen using monoclonal antibodies. *Exp. Parasitol.* 92, 109–119.

1792. Schares, G., Rauser, M., Zimmer, K., Peters, M., Wurm, R., Dubey, J. P., de Graaf, D. C., Edelhofer, R., Mertens, C., Hess, G., Conraths, F. J. 1999. Serological differences in *Neospora caninum*-associated epidemic and endemic abortions. *J. Parasitol.* 85, 688–694.

1793. Schares, G., Zinecker, C. F., Schmidt, J., Azzouz, N., Conraths, F. J., Gerold, P., Schwarz, R. T. 2000. Structural analysis of free and protein-bound glycosyl-phosphatidylinositols of *Neospora caninum*. *Mol. Biochem. Parasitol.* 105, 155–161.

1794. Schares, G., Rauser, M., Söndgen, P., Rehberg, P., Bärwald, A., Dubey, J. P., Edelhofer, R., Conraths, F. J. 2000. Use of purified tachyzoite surface antigen p38 in an ELISA to diagnose bovine neosporosis. *Int. J. Parasitol.* 30, 1123–1130.

1795. Schares, G., Conraths, F. J. 2001. Placentophagia—An alternative way for horizontal transmission of *Neospora caninum* in cattle? *Trends Parasitol.* 17, 574–575.

1796. Schares, G., Wenzel, U., Müller, T., Conraths, F. J. 2001. Serological evidence for naturally occurring transmission of *Neospora caninum* among foxes (*Vulpes vulpes*). *Int. J. Parasitol.* 31, 418–423.

1797. Schares, G., Heydorn, A. O., Cüppers, A., Conraths, F. J., Mehlhorn, H. 2001. Cyclic transmission of *Neospora caninum*: Serological findings in dogs shedding oocysts. *Parasitol. Res.* 87, 873–877.

1798. Schares, G., Heydorn, A. O., Cüppers, A., Conraths, F. J., Mehlhorn, H. 2001. *Hammondia heydorni*-like oocysts shed by a naturally infected dog and *Neospora caninum* NC-1 cannot be distinguished. *Parasitol. Res.* 87, 808–816.

1799. Schares, G., Heydorn, A. O., Cüppers, A., Mehlhorn, H., Geue, L., Peters, M., Conraths, F. J. 2002. In contrast to dogs, red foxes (*Vulpes vulpes*) did not shed *Neospora caninum* upon feeding of intermediate host tissues. *Parasitol. Res.* 88, 44–52.

1800. Schares, G., Bärwald, A., Staubach, C., Söndgen, P., Rauser, M., Schröder, R., Peters, M., Wurm, R., Selhorst, T., Conraths, F. J. 2002. p38-avidity-ELISA: Examination of herds experiencing epidemic or endemic *Neospora caninum*-associated bovine abortion. *Vet. Parasitol.* 106, 293–305.

1801. Schares, G., Bärwald, A., Staubach, C., Ziller, M., Klöss, D., Wurm, R., Rauser, M., Labohm, R., Dräger, K., Fasen, W., Hess, R. G., Conraths, F. J. 2003. Regional distriution of bovine *Neospora caninum* infection in the German state of Rhineland-Palatinate modelled by logistic regression. *Int. J. Parasitol.* 33, 1631–1640.

1802. Schares, G., Bärwald, A., Staubach, C., Wurm, R., Rauser, M., Conraths, F. J., Schroeder, C. 2004. Adaptation of a commercial ELISA for the detection of antibodies against *Neospora caninum* in bovine milk. *Vet. Parasitol.* 120, 55–63.

1803. Schares, G., Bärwald, A., Staubach, C., Ziller, M., Klöss, D., Schroder, R., Labohm, R., Dräger, K., Fasen, W., Hess, R. G., Conraths, F. J. 2004. Potential risk factors for bovine *Neospora caninum* infection in Germany are not under the control of the farmers. *Parasitology* 129, 301–309.

1804. Schares, G., Bärwald, A., Conraths, F. J. 2005. Adaptation of a surface antigen-based ELISA for the detection of antibodies against *Neospora caninum* in bovine milk. *J. Vet. Med. B* 52, 45–48.

1805. Schares, G., Pantchev, N., Barutzki, D., Heydorn, A. O., Bauer, C., Conraths, F. J. 2005. Oocysts of *Neospora caninum*, *Hammondia heydorni*, *Toxoplasma gondii* and *Hammondia hammondi* in faeces collected from dogs in Germany. *Int. J. Parasitol.* 35, 1525–1537.

1806. Schares, G. 2007. Oocyst detection and differentiation. In *Protozoal Abortion in Farm Animals. Guidelines for Diagnosis and control*. Ortega-Mora, L. M., Gottestin, B., Conraths, F. J., Buxton, D. (Editors.) Atheneaeum Press, Gateshead, U.K. 80–88.

1807. Schares, G., Vrhovec, M. G., Pantchev, N., Herrmann, D. C., Conraths, F. J. 2008. Occurrence of *Toxoplasma gondii* and *Hammondia hammondi* oocysts in the faeces of cats from Germany and other European countries. *Vet. Parasitol.* 152, 34–45.

1808. Schares, G., Wilking, H., Bolln, M., Conraths, F. J., Bauer, C. 2009. *Neospora caninum* in dairy herds in Schleswig-Holstein, Germany. *Berl. Munch. Tierarztl. Wochenschr.* 122, 47–50.

1809. Schares, G., Ziller, M., Herrmann, D. C., Globokar, M. V., Pantchev, N., Conraths, F. J. 2016. Seasonality in the proportions of domestic cats shedding *Toxoplasma gondii* or *Hammondia hammondi* oocysts is associated with climatic factors. *Int. J. Parasitol.* 46, 263–273.

1810. Schatzberg, S. J., Haley, N. J., Barr, S. C., de Lahunta, A., Olby, N., Munana, K., Sharp, N. J. H. 2003. Use of a multiplex polymerase chain reaction assay in the antemortem diagnosis of toxoplasmosis and neosporosis in the central nervous system of cats and dogs. *Am. J. Vet. Res.* 64, 1507–1513.

1811. Schetters, T. 2004. Intervet symposium: Bovine neosporosis. *Vet. Parasitol.* 125, 137–146.

1812. Schock, A., Buxton, D., Spence, J. A., Low, J. C., Baird, A. 2000. Histopathological survey of aborted bovine fetuses in Scotland with special reference to *Neospora caninum*. *Vet. Rec.* 147, 687–688.

1813. Schock, A., Innes, E. A., Yamane, I., Latham, S. M., Wastling, J. M. 2001. Genetic and biological diversity among isolates of *Neospora caninum*. *Parasitology* 123, 13–23.

1814. Schönian, G., Mauricio, I., Gramiccia, M., Cañavate, C., Boelaert, M., Dujardin, J. C. 2008. Leishmaniases in the Mediterranean in the era of molecular epidemiology. *Trends Parasitol.* 24, 135–142.

1815. Schorer, M., Debache, K., Barna, F., Monney, T., Müller, J., Boykin, D. W., Stephens, C. E., Hemphill, A. 2012. Di-cationic arylimidamides act against *Neospora caninum* tachyzoites by interference in membrane structure and nucleolar integrity and are active against challenge infection in mice. *Int. J. Parasitol. Drugs Drug Resist.* 2, 109–120.

1816. Schwab, A. E., Geary, T. G., Baillargeon, P., Schwab, A. J., Fecteau, G. 2009. Association of BoLA DRB3 and DQA1 alleles with susceptibly to *Neospora caninum* and reproductive outcome in Quebec Holstein cattle. *Vet. Parasitol.* 165, 136–140.

1817. Scott, H. M., Sorensen, O., Wu, J. T. Y., Chow, E. Y. W., Manninen, K., VanLeeuwen, J. A. 2006. Seroprevalence of *Mycobacterium avium* subspecies *paratuberculosis*, *Neospora caninum*, *Bovine leukemia virus*, and *Bovine viral diarrhea virus* infection among dairy cattle and herds in Alberta and agroecological risk factors associated with seropositivity. *Can. Vet. J.* 47, 981–991.

1818. Scott, H. M., Sorensen, O., Wu, J. T. Y., Chow, E. Y. W., Manninen, K. 2007. Seroprevalence of and agroecological risk factors for *Mycobacterium avium* subspecies *paratuberculosis* and *Neospora caninum* infection among adult beef cattle in cow-calf herds in Alberta, Canada. *Can. Vet. J.* 48, 397–406.

1819. Sedlák, K. 2004. Prevalence of *Neospora caninum* antibodies in zoo animals in the Czech Republic. *J. Eukaryot. Microbiol.* 51, 20A.

1820. Sedlák, K., Bártová, E. 2006. Seroprevalences of antibodies to *Neospora caninum* and *Toxoplasma gondii* in zoo animals. *Vet. Parasitol.* 136, 223–231.

1821. Sedlák, K., Bártová, E., Machacová, T. 2014. Seroprevalence of *Neospora caninum* in cats from the Czech Republic. *Acta Parasitol.* 59, 359–361.

1822. Segura-Correa, J. C., Domínguez-Díaz, D., Alvalos-Ramírez, R., Argaez-Sosa, J. 2010. Intraherd correlation coefficients and design effects for bovine viral diarrhoea, infectious bovine rhinotracheitis, leptospirosis and neosporosis in cow-calf system herds in North-eastern Mexico. *Prev. Vet. Med.* 96, 272–275.

1823. Selahi, F., Namavari, M., Hosseini, M. H., Mansourian, M., Tahamtan, Y. 2013. Development of a disperse dye immunoassay technique for detection of antibodies against *Neospora caninum* in cattle. *Korean J. Parasitol.* 51, 129–132.

1824. Sengupta, P. P., Balumahendiran, M., Raghavendra, A. G., Honnappa, T. G., Gajendragad, M. R., Prabhudas, K. 2013. Prevalence of *Neospora caninum* antibodies in dairy cattle and water buffaloes and associated abortions in the plateau of Southern Peninsular India. *Trop. Anim. Health Prod.* 45, 205–210.

1825. Seo, H. S., Kim, K. H., Kim, D. Y., Park, B. K., Shin, N. S., Kim, J. H., Youn, H. 2013. GC/MS analysis of high-performance liquid chromatography fractions from *Sophora flavescens* and *Torilis japonica* extracts and their *in vitro* anti-neosporal effects on *Neospora caninum*. *J. Vet. Sci.* 14, 241–248.

1825a. Seppänen, V., Syrjälä, P., Skrzypczak, T., Kukowski, D., Korpikallio, A., Taponen, J. 2016. Control of *Neospora* infection in cattle – a review and a case report [In Finnish]. *Suomen Eläinlääkärilehti* 122, 66–73.

1826. Serrano, B., Almería, S., García-Ispierto, I., Yániz, J. L., Abdelfattah-Hassan, A., López-Gatius, F. 2011. Peripheral white blood cell counts throughout pregnancy in non-aborting *Neospora caninum*-seronegative and seropositive high-producing dairy cows in a Holstein Friesian herd. *Res. Vet. Sci.* 90, 457–462.

1827. Serrano, E., Ferre, I., Osoro, K., Aduriz, G., Mateos-Sanz, A., Martínez, A., Atxaerandio, R., Hidalgo, C. O., Ortega-Mora, L. M. 2006. Intrauterine *Neospora caninum* inoculation of heifers. *Vet. Parasitol.* 135, 197–203.

1828. Serrano-Martínez, E., Collantes-Fernández, E., Rodríguez-Bertos, A., Casas-Astos, E., Álvarez-García, G., Chávez-Velásquez, A., Ortega-Mora, L. M. 2004. *Neospora* species-associated abortion in alpacas (*Vicugna pacos*) and llamas (*Llama glama*). *Vet. Rec.* 155, 748–749.

1829. Serrano-Martínez, E., Collantes-Fernández, E., Chávez-Velásquez, A., Rodríguez-Bertos, A., Casas-Astos, E., Risco-Castillo, V., Rosadio-Alcantara, R., Ortega-Mora, L. M. 2007. Evaluation of *Neospora caninum* and *Toxoplasma gondii* infections in alpaca (*Vicugna pacos*) and llama (*Lama glama*) aborted foetuses from Peru. *Vet. Parasitol.* 150, 39–45.

1830. Serrano-Martínez, E., Ferre, I., Osoro, K., Aduriz, G., Mota, R. A., Martínez, A., del-Pozo, I., Hidalgo, C. O., Ortega-Mora, L. M. 2007. Intraurine *Neospora caninum* inoculation of heifers and cows using contaminated semen with different numbers of tachyzoites. *Theriogenology* 67, 729–737.

1831. Serrano-Martínez, E., Ferre, I., Martínez, A., Osoro, K., Mateos-Sanz, A., del-Pozo, I., Aduriz, G., Tamargo, C., Hidalgo, C. O., Ortega-Mora, L. M. 2007. Experimental neosporosis in bulls: Parasite detection in semen and blood and specific antibody and interferon-gamma responses. *Theriogenology* 67, 1175–1184.

1832. Serrano-Pérez, B., Garcia-Ispierto, I., de Sousa, N. M., Beckers, J. F., Almería, S., López-Gatius, F. 2014. Gamma interferon production and plasma concentrations of pregnancy-associated glycoproteins 1 and 2 in gestating dairy cows naturally infected with *Neospora caninum*. *Reprod. Dom. Anim.* 49, 275–280.

1833. Serrano-Pérez, B., Hansen, P. J., Mur-Novales, R., García-Ispierto, I., de Sousa, N. M., Beckers, J. F., Almería, S., López-Gatius, F. 2016. Crosstalk between uterine serpin (SERPINA14) and pregnancy-associated glycoproteins at the fetal-maternal interface in pregnant dairy heifers experimentally infected with *Neospora caninum*. *Theriogenology* 86, 824–830.

1834. Sevgili, M., Altaş, M. G., Keskin, O. 2005. Seroprevalence of *Neospora caninum* in cattle in the province of Sanliurfa. *Turk. J. Vet. Anim. Sci.* 29, 127–130.

1835. Shabbir, M. Z., Nazir, M. M., Maqbool, A., Lateef, M., Shabbir, M. A. B., Ahmad, A., Rabbani, M., Yaqub, T., Sohail, M. U., Ijaz, M. 2011. Seroprevalence of *Neospora caninum* and *Brucella abortus* among dairy cattle herds with high abortion rates. *J. Parasitol.* 97, 740–742.

1836. Shapiro, K., Mazet, J. A., Schriewer, A., Wuertz, S., Fritz, H., Miller, W. A., Largier, J., Conrad, P. A. 2010. Detection of *Toxoplasma gondii* oocysts and surrogate microspheres in water using ultrafiltration and capsule filtration. *Water Res.* 44, 893–903.

1837. Sharma, R., Mcmillan, M., Tiwari, K., Chikweto, A., Thomas, D., Bhaiyat, M. I. 2014. Seroprevalence of *Toxoplasma gondii* and *Neospora caninum* infection in cattle in Grenada, West Indies. *Glob. J. Med. Res.:G Vet. Sci. Vet. Med.* 14, 2.

1838. Sharma, R., Kimmitt, T., Tiwari, K., Chikweto, A., Thomas, D., Lanza Perea, M., Bhaiyat, M. I. 2015. Serological evidence of antibodies to *Neospora caninum* in stray and owned Grenadian dogs. *Trop. Biomed.* 32, 286–290.

1839. Sharma, R. N., Bush, J., Tiwari, K., Chikweto, A., Bhaiyat, M. I. 2015. Seroprevalence of *Neospora caninum* in sheep and goats from Grenada, West Indies. *Open J. Vet. Med.* 5, 219–223.

1840. Sharma, R. N., Tiwari, K., Chikweto, A., DeAllie, C., Bhaiyat, M. I. 2015. Prevalence of antibodies to *Toxoplasma gondii* and *Neospora caninum* in pigs in Grenada, West Indies. *Open J.Veterinary Med.* 5, 138–141.

1841. Sharma, S., Bal, M. S., Meenakshi, Kaur, K., Sandhu, K. S., Dubey, J. P. 2008. Seroprevalence of *Neospora caninum* antibodies in dogs in India. *J. Parasitol.* 94, 303–304.

1842. Sheahan, B. J., Caffrey, J. F., Dubey, J. P., McHenry, D. F. 1993. *Neospora caninum* encephalomyelitis in seven dogs. *Irish Vet. J.* 46, 3–7.

1843. Shen, L. P., Liu, P. H., Xu, F., Xue, X. 2006. [Prevalence of serum antibodies to bovine neosporosis in Shanghai]. *Chin. J. Vet. Parasitol.* 14, 14–16 (in Chinese).

1844. Shibahara, T., Kokuho, T., Eto, M., Haritani, M., Hamaoka, T., Shimura, K., Nakamura, K., Yokomizo, Y., Yamane, I. 1999. Pathological and immunological findings of athymic nude and congenic wild type BALB/c mice experimentally infected with *Neospora caninum*. *Vet. Pathol.* 36, 321–327.

1845. Shiel, R. E., Mooney, C. T., Brennan, S. F., Nolan, C. M., Callanan, J. J. 2010. Clinical and clinicopathological features of non-suppurative meningoencephalitis in young greyhounds in Ireland. *Vet. Rec.* 167, 333–337.

1846. Shin, Y. S., Lee, E. G., Shin, G. W., Kim, Y. R., Lee, E. Y., Kim, J. H., Jang, H., Gershwin, L. J., Kim, D. Y., Kim, Y. H., Kim, G. S., Suh, M. D., Jung, T. S. 2004. Identification of antigenic proteins from *Neospora caninum* recognized by bovine immunoglobulins M, E, A and G using immunoproteomics. *Proteomics* 4, 3600–3609.

1847. Shin, Y. S., Lee, E. G., Jung, T. S. 2005. Exploration of immunoblot profiles of *Neospora caninum* probed with different bovine immunoglobulin classes. *J. Vet. Sci.* 6, 157–160.

1848. Shin, Y. S., Shin, G. W., Kim, Y. R., Lee, E. Y., Yang, H. H., Palaksha, K. J., Youn, H. J., Kim, J. H., Kim, D. Y., Marsh, A. E., Lakritz, J., Jung, T. S. 2005. Comparison of proteome and antigenic proteome between two *Neospora caninum* isolates. *Vet. Parasitol.* 134, 41–52.

1849. Shivaprasad, H. L., Ely, R., Dubey, J. P. 1989. A *Neospora*-like protozoon found in an aborted bovine placenta. *Vet. Parasitol.* 34, 145–148.

1850. Shkap, V., Reske, A., Pipano, E., Fish, L., Baszler, T. 2002. Immunological relationship between *Neospora caninum* and *Besnoitia besnoiti*. *Vet. Parasitol.* 106, 35–43.

1851. Shoemaker, M. E. 2014. The effect of stress on the ecology of *Neospora caninum* in bison (*Bison bison*). *Masters Degree thesis*. Ohio State University, Columbus, Ohio, USA. 1–93.

1852. Sibley, L. D., Niesman, I. R., Parmley, S. F., Cesbron-Delauw, M. F. 1995. Regulated secretion of multi-lamellar vesicles leads to formation of a tubulovesicular network in host-cell vacuoles occupied by *Toxoplasma gondii*. *J. Cell Sci.* 108, 1669–1677.

1853. Sicupira, P. M. L., de Magalhães, V. C. S., Galvão, G. S., Pereira, M. J. S., Gondim, L. F. P., Munhoz, A. D. 2012. Factors associated with infection by *Neospora caninum* in dogs in Brazil. *Vet. Parasitol.* 185, 305–308.

1854. Sierra, R. C., Medina-Esparza, L., Parra, M. R., García-Vázquez, Z., Cruz-Vázquez, C. 2011. Risk factors associated with *Neospora caninum* antibody seroprevalence in dairy cattle in Aguascalientes, Mexico. *Rev. Mex. Cienc. Pecu* 2, 15–24.

1855. Silaghi, C., Knaus, M., Rapti, D., Kusi, I., Shukullari, E., Hamel, D., Pfister, K., Rehbein, S. 2014. Survey of *Toxoplasma gondii* and *Neospora caninum*, haemotropic mycoplasmas and other arthropod-borne pathogens in cats from Albania. *Parasit. Vectors* 7, 62.

1856. Silva, A. F., Rangel, L., Ortiz, C. G., Morales, E., Zanella, E. L., Castillo-Velázquez, U., Gutierrez, C. G. 2012. Increased incidence of DNA amplification in follicular than in uterine and blood samples indicates possible tropism of *Neospora caninum* to the ovarian follicle. *Vet. Parasitol.* 188, 175–178.

1857. Silva, D. A. O., Vitaliano, S. N., Mineo, T. W. P., Ferreira, R. A., Bevilacqua, E., Mineo, J. R. 2005. Evaluation of homologous, heterologous, and affinity conjugates for the serodiagnosis of *Toxoplasma gondii* and *Neospora caninum* in maned wolves (*Chrysocyon brachyurus*). *J. Parasitol.* 91, 1212–1216.

1858. Silva, D. A. O., Lobato, J., Mineo, T. W. P., Mineo, J. R. 2007. Evaluation of serological tests for the diagnosis of *Neospora caninum* infection in dogs: Optimization of cut off titers and inhibition studies of cross-reactivity with *Toxoplasma gondii*. *Vet. Parasitol.* 143, 234–244.

1859. Silva, M. I. S., Almeida, M. Â. O., Mota, R. A., Pinheiro-Junior, J. W., Rabelo, S. S. A. 2008. Fatores de riscos associados à infecção por *Neospora caninum* em matrizes bovinas leiteiras em Pernambuco. [Risk factors associated to *Neospora caninum* infection in dairy cows in Pernambuco]. *Ciência Anim. Brasil.* 9, 455–461.

1860. Silva, M. S. A., Uzêda, R. S., Costa, K. S., Santos, S. L., Macedo, A. C. C., Abe-Sandes, K., Gondim, L. F. P. 2009. Detection of *Hammondia heydorni* and coccidia (*Neospora caninum* and *Toxoplasma gondii*) in goats slaughtered in Bahia, Brazil. *Vet. Parasitol.* 162, 156–159.

1861. Simpson, V. R., Monies, R. J., Riley, P., Cromey, D. S. 1997. Foxes and neosporosis. *Vet. Rec.* 141, 503.

1862. Simsek, S., Utuk, A. E., Koroglu, E., Dumanli, N., Risvanli, A. 2008. Seroprevalence of *Neospora caninum* in repeat breeder dairy cows in Turkey. *Arch. Tierz.* 51, 143–148.

1863. Sinnott, F. A., Monte, L. G., Collares, T. F., De Matos, B. M., Pacheco, D. B., Borsuk, S., Andreotti, R., Hartleben, C. P. 2015. Blocking ELISA using recombinant NcSRS2 protein for diagnosing bovine neosporosis. *Curr. Microbiol.* 70, 429–432.

1864. Siverajah, S., Ryce, C., Morrison, D. A., Ellis, J. T. 2003. Characterization of an alpha tubulin gene sequence from *Neospora caninum* and *Hammondia heydorni*, and their comparison to homologous from Apicomplexa. *Parasitology* 126, 561–569.

1865. Šlapeta, J. R., Modrý, D., Kyselová, I., Horejš, R., Lukeš, J., Koudela, B. 2002. Dog shedding oocysts of *Neospora caninum*: PCR diagnosis and molecular phylogenetic approach. *Vet. Parasitol.* 109, 157–167.

1866. Šlapeta, J. R., Koudela, B., Votýpka, J., Modrý, D., Horejš, R., Lukeš, J. 2002. Coprodiagnosis of *Hammondia heydorni* in dogs by PCR based amplification of ITS 1 rRNA: Differentiation from morphologically indistinguishable oocysts of *Neospora caninum*. *Vet. J.* 163, 147–154.

1867. Slotved, H. C., Jensen, L., Lind, P. 1999. Comparison of the IFAT and Iscom-ELISA response in bovine foetuses with *Neospora caninum* infection. *Int. J. Parasitol.* 29, 1165–1174.

1868. Śmielewska-Loś, E., Pacoń, J., Jańczak, M., Ploneczka, K. 2003. Prevalence of antibodies to *Toxoplasma gondii* and *Neospora caninum* in wildlife and farmed foxes (*Vulpes vulpes*). *Electron. J. Pol. Agric. Univ.* 6, #06.

1869. Soares, H. S., Ahid, S. M. M., Bezerra, A. C. D. S., Pena, H. F. J., Dias, R. A., Gennari, S. M. 2009. Prevalence of anti-*Toxoplasma gondii* and anti-*Neospora caninum* antibodies in sheep from Mossoró, Rio Grande do Norte, Brazil. *Vet. Parasitol.* 160, 211–214.

1870. Soares, H. S., Ramos, V. N., Osava, C. F., Oliveira, S., Szabó, M. P. J., Piovezan, U., Castro, B. B., Gennari, S. M. 2016. Occurrence of antibodies against *Neospora caninum* in wild pigs (*Sus scrofa*) in the Pantanal, Mato Grosso do Sul, Brazil. *Braz. J. Vet. Res. Anim. Sci.* 53, 112–116.

1871. Soares, R. M., Lopes, E. G., Keid, L. B., Sercundes, M. K., Martins, J., Richtzenhain, L. J. 2011. Identification of *Hammondia heydorni* oocysts by a heminested-PCR (hnPCR-AP10) based on the *H. heydorni* RAPD fragment AP10. *Vet. Parasitol.* 175, 168–172.

1872. Sobrino, R., Dubey, J. P., Pabón, M., Linarez, N., Kwok, O. C., Millán, J., Arnal, M. C., Luco, D. F., López-Gatius, F., Thulliez, P., Gortázar, C., Almería, S. 2008. *Neospora caninum* antibodies in wild carnivores from Spain. *Vet. Parasitol.* 155, 190–197.

1873. Soeiro, M. N. C., Werbovetz, K., Boykin, D. W., Wilson, W. D., Wang, M. Z., Hemphill, A. 2013. Novel amidines and analogues as promising agents against intracellular parasites: A systematic review. *Parasitology* 140, 929–951.

1874. Sohn, C. S., Cheng, T. T., Drummond, M. L., Peng, E. D., Vermont, S. J., Xia, D., Cheng, S. J., Wastling, J. M., Bradley, P. J. 2011. Identification of novel proteins in *Neospora caninum* using an organelle purification and monoclonal antibody approach. *PLoS ONE* 6, e18383.

1875. Soldati, S., Kiupel, M., Wise, A., Maes, R., Botteron, C., Robert, N. 2004. Meningoencephalomyelitis caused by *Neospora caninum* in a juvenile fallow deer (*Dama dama*). *J. Vet. Med. A* 51, 280–283.

1876. Sommanustweechai, A., Vongpakorn, M., Kasantikul, T., Taewnean, J., Siriaroonrat, B., Bush, M., Pirarat, N. 2010. Systemic neosporosis in a white rhinoceros. *J. Zoo Wildl. Med.* 41, 165–168.

1877. Son, E. S., Ahn, H. J., Kim, J. H., Kim, D. Y., Nam, H. W. 2001. Determination of antigenic domain in GST fused major surface protein (Nc-p43) of *Neospora caninum*. *Korean J. Parasitol.* 39, 241–246.

1878. Sonda, S., Fuchs, N., Connolly, B., Fernandez, P., Gottstein, B., Hemphill, A. 1998. The major 36 kDa *Neospora caninum* tachyzoite surface protein is closely related to the major *Toxoplasma gondii* surface antigen. *Mol. Biochem. Parasitol.* 97, 97–108.

1879. Sonda, S., Fuchs, N., Gottstein, B., Hemphill, A. 2000. Molecular characterization of a novel microneme antigen in *Neospora caninum*. *Mol. Biochem. Parasitol.* 108, 39–51.

1880. Söndgen, P., Peters, M., Bärwald, A., Wurm, R., Holling, F., Conraths, F. J., Schares, G. 2001. Bovine neosporosis: Immunoblot improves foetal serology. *Vet. Parasitol.* 102, 279–290.

1881. Sörgel, S. C., Müller, M., Schares, G., Großmann, E., Neuss, T., Puchta, H., Kreuzer, P., Ewringmann, T., Ehrlein, J., Bogner, K. H., Schmahl, W. 2009. Beteiligung von *Neospora caninum* bei Rinderaborten in Nordbayern. *Tierärztl. Umschau* 64, 235–243.

1882. Sotiraki, S., Brozos, C., Samartzi, F., Schares, G., Kiossis, E., Conraths, F. J. 2008. *Neospora caninum* infection in Greek dairy cattle herds detected by two antibody assays in individual milk samples. *Vet. Parasitol.* 152, 79–84.

1883. Speer, C. A., Dubey, J. P. 1989. Ultrastructure of tachyzoites, bradyzoites, and tissue cysts of *Neospora caninum*. *J. Protozool.* 36, 458–463.

1884. Speer, C. A., Dubey, J. P., McAllister, M. M., Blixt, J. A. 1999. Comparative ultrastructure of tachyzoites, bradyzoites, and tissue cysts of *Neospora caninum* and *Toxoplasma gondii*. *Int. J. Parasitol.* 29, 1509–1519.

1885. Spekker, K., Czesla, M., Ince, V., Heseler, K., Schmidt, S. K., Schares, G., Däubener, W. 2009. Indoleamine 2,3-dioxygenase is involved in defense against *Neospora caninum* in human and bovine cells. *Infect. Immun.* 77, 4496–4501.

1886. Spekker, K., Leineweber, M., Degrandi, D., Ince, V., Brunder, S., Schmidt, S. K., Stuhlsatz, S., Howard, J. C., Schares, G., Degistirici, Ö., Meisel, R., Sorg, R. V., Seissler, J., Hemphill, A., Pfeffer, K., Däubener, W. 2013. Antimicrobial effects of murine mesenchymal stromal cells directed against *Toxoplasma gondii* and *Neospora caninum*: Role of immunity-related GTPases (IRGs) and guanylate-binding proteins (GBPs). *Med. Microbiol. Immunol.* 202, 197–206.

1887. Spencer, J. A., Witherow, A. K., Blagburn, B. L. 2000. A random amplified polymorphic DNA polymerase chain reaction technique that differentiates between *Neospora* species. *J. Parasitol.* 86, 1366–1368.

1888. Spencer, J. A., Higginbotham, M. J., Blagburn, B. L. 2003. Seroprevalence of *Neospora caninum* and *Toxoplasma gondii* in captive and free-ranging nondomestic felids in the United States. *J. Zoo Wildl. Med.* 34, 246–249.

1889. Spencer, J. A., Higginbotham, M. J., Young-White, R. R., Guarino, A. J., Blagburn, B. L. 2005. *Neospora caninum*: Adoptive transfer of immune lymphocytes precipitates disease in BALB/c mice. *Vet. Immunol. Immunopathol.* 106, 329–333.

1890. Špilovská, S., Reiterová, K. 2008. Seroprevalence of *Neospora caninum* in aborting sheep and goats in the eastern Slovakia. *Folia Vet.* 52, 33–35.

1891. Špilovská, S., Reiterová, K., Kováčová, D., Bobáková, M., Dubinský, P. 2009. The first finding of *Neospora caninum* and the occurrence of other abortifacient agents in sheep in Slovakia. *Vet. Parasitol.* 164, 320–323.

1891a. Špilovská, S., Moskwa, B., Reiterová, K. 2013. Kinetics of anti-*Neospora* antibodies during the period of two consecutive pregnancies in chronically infected dairy cows. *Acta Parasitol.* 58, 463–467.

1892. Špilovská, S., Reiterová, K., Antolová, D. 2015. *Neospora caninum*-associated abortions in Slovak dairy farm. *Iranian J. Parasitol.* 10, 96–101.

1893. Sreekumar, C., Hill, D. E., Fournet, V. M., Rosenthal, B. M., Lindsay, D. S., Dubey, J. P. 2003. Detection of *Hammondia heydorni*-like organisms and their differentiation from *Neospora caninum* using random-amplified polymorphic DNA-polymerase chain reaction. *J. Parasitol.* 89, 1082–1085.

1894. Sreekumar, C., Hill, D. E., Miska, K. B., Rosenthal, B. M., Vianna, M. C. B., Venturini, L., Basso, W., Gennari, S. M., Lindsay, D. S., Dubey, J. P. 2004. *Hammondia heydorni*: Evidence of genetic diversity among isolates from dogs. *Exp. Parasitol.* 107, 65–71.

1895. Sréter, T., Sebestyén, P., Dubey, J. P. 1992. Neosporosis in a dog in Hungary. *Parasitol. Hung.* 25, 5–8.

1896. Srinivasan, S., Baszler, T., Vonlaufen, N., Leepin, A., Sanderson, S. J., Wastling, J. M., Hemphill, A. 2006. Monoclonal antibody directed against *Neospora caninum* tachyzoite carbohydrate epitope reacts specifically with apical complex-associated sialylated beta tubulin. *J. Parasitol.* 92, 1235–1243.

1897. Srinivasan, S., Mueller, J., Suana, A., Hemphill, A. 2007. Vaccination with microneme protein NCMIC4 increases mortality in mice inoculated with *Neospora caninum*. *J. Parasitol.* 93, 1046–1055.

1898. Stachissini, A. V. M. 2005. Influência da infecção pelo vírus da artrite-encefalite caprina nos perfis soro-epidemiológicos em caprinos infectados pelo *Toxoplasma gondii* e *Neospora caninum*. [Influence of infection by caprine arthritis-encephalitis virus in the sero-epidemiological profiles on infected goats by *Toxoplasma gondii* and *Neospora caninum*]. *Doctor of Veterinary Medicine thesis.* Faculdade de Medicina Veterinária e Zootecnia, Universidade Estadual Paulista, Botucatu, SP, Brazil. 1–119.

1899. Ståhl, K., Björkman, C., Emanuelson, U., Rivera, H., Zelada, A., Moreno-López, J. 2006. A prospective study of the effect of *Neospora caninum* and BVDV infections on bovine abortions in a dairy herd in Arequipa, Peru. *Prev. Vet. Med.* 75, 177–188.

1900. Staska, L. M., McGuire, T. C., Davies, C. J., Lewin, H. A., Baszler, T. V. 2003. *Neospora caninum*-infected cattle develop parasite-specific CD4+ cytotoxic T lymphocytes. *Infect. Immun.* 71, 3272–3279.

1901. Staska, L. M., Davies, C. J., Brown, W. C., McGuire, T. C., Suarez, C. E., Park, J. Y., Mathison, B. A., Abbott, J. R., Baszler, T. V. 2005. Identification of vaccine candidate peptides in the NcSRS2 surface protein of *Neospora caninum* by using CD4+ cytotoxic T lymphocytes and gamma interferon-secreting T lymphocytes of infected Holstein cattle. *Infect. Immun.* 73, 1321–1329.

1902. Staubli, D., Nunez, S., Sager, H., Schares, G., Gottstein, B. 2006. *Neospora caninum* immunoblotting improves serodiagnosis of bovine neosporosis. *Parasitol. Res.* 99, 648–658.

1903. Staubli, D., Sager, H., Haerdi, C., Haessig, M., Gottstein, B. 2006. Precolostral serology in calves born from *Neospora*-seropositive mothers. *Parasitol. Res.* 99, 398–404.

1904. Staubli, D., Iten, C., Kneubühler, J., Sager, H., Müller, N., Gottstein, B. 2006. Untersuchung von bovinem Sperma auf *Neospora caninum*-DNA mittels PCR. [Search for *Neospora caninum* DNA in bull semen using PCR]. *Schweiz. Arch. Tierheilkd.* 148, 483–489.

1905. Steinman, A., Shpigel, N. Y., Mazar, S., King, R., Baneth, G., Savitsky, I., Shkap, V. 2006. Low seroprevalence of antibodies to *Neospora caninum* in wild canids in Israel. *Vet. Parasitol.* 137, 155–158.

1906. Stelmann, U. J. P., Ullmann, L. S., Langoni, H., Amorim, R. M. 2011. Equine neosporosis: Search for antibodies in cerebrospinal fluid and sera from animals with history of ataxia. *Rev. Bras. Med. Vet.* 33, 99–102.

1907. Stenlund, S., Björkman, C., Holmdahl, O. J. M., Kindahl, H., Uggla, A. 1997. Characterisation of a Swedish bovine isolate of *Neospora caninum*. *Parasitol. Res.* 83, 214–219.

1908. Stenlund, S., Kindahl, H., Magnusson, U., Uggla, A., Björkman, C. 1999. Serum antibody profile and reproductive performance during two consecutive pregnancies of cows naturally infected with *Neospora caninum*. *Vet. Parasitol.* 85, 227–234.

1909. Stenlund, S., Kindahl, H., Uggla, A., Björkman, C. 2003. A long-term study of *Neospora caninum* infection in a Swedish dairy herd. *Acta Vet. Scand.* 44, 63–71.

1910. Stieve, E., Beckmen, K., Kania, S. A., Widner, A., Patton, S. 2010. *Neospora caninum* and *Toxoplasma gondii* antibody prevalence in Alaska wildlife. *J. Wildl. Dis.* 46, 348–355.

1911. Stoessel, Z., Taylor, L. F., McGowan, M. R., Coleman, G. T., Landmann, J. K. 2003. Prevalence of antibodies to *Neospora caninum* within central Queensland beef cattle. *Aust. Vet. J.* 81, 165–166.

1912. Stokka, G. L., Lardy, G. P. 2005. Health management programs: Integrating biological and management principles in analysis, design, and implementation of programs for two-year-old beef cows. *Prof. Ani. Sci.* 21, 159–163.

1913. Straub, K. W., Cheng, S. J., Sohn, C. S., Bradley, P. J. 2009. Novel components of the Apicomplexan moving junction reveal conserved and coccidia-restricted elements. *Cell. Microbiol.* 11, 590–603.

1914. Strohbusch, M., Müller, N., Hemphill, A., Greif, G., Gottstein, B. 2008. *NcGRA2* as a molecular target to assess the parasiticidal activity of toltrazuril against *Neospora caninum*. *Parasitology* 135, 1065–1073.

1915. Strohbusch, M., Müller, N., Hemphill, A., Krebber, R., Greif, G., Gottstein, B. 2009. Toltrazuril treatment of congenitally acquired *Neospora caninum* infection in newborn mice. *Parasitol. Res.* 104, 1335–1343.

1916. Strohbusch, M., Müller, N., Hemphill, A., Margos, M., Grandgirard, D., Leib, S., Greif, G., Gottstein, B. 2009. *Neospora caninum* and bone marrow-derived dendritic cells: Parasite survival, proliferation, and induction of cytokine expression. *Parasite Immunol.* 31, 366–372.

1917. Strohmeyer, R. A., Morley, P. S., Hyatt, D. R., Dargatz, D. A., Scorza, A. V., Lappin, M. R. 2006. Evaluation of bacterial and protozoal contamination of commercially available raw meat diets for dogs. *J. Am. Vet. Med. Assoc.* 228, 537–542.

1918. Stuart, P., Zintl, A., de Waal, T., Mulcanhy, G., Hawkins, C., Lawton, C. 2013. Investigating the role of wild carnivores in the epidemiology of bovine neosporosis. *Parasitology* 140, 296–302.

1919. Sun, W. W., Meng, Q. F., Cong, W., Shan, X. F., Wang, C. F., Qian, A. D. 2015. Herd-level prevalence and associated risk factors for *Toxoplasma gondii*, *Neospora caninum*, *Chlamydia abortus* and bovine viral diarrhoea virus in commercial dairy and beef cattle in eastern, northern and northeastern China. *Parasitol. Res.* 114, 4211–4218.

1920. Sundermann, C. A., Estridge, B. H., Branton, M. S., Bridgman, C. R., Lindsay, D. S. 1997. Immunohistochemical diagnosis of *Toxoplasma gondii*: Potential for cross-reactivity with *Neospora caninum*. *J. Parasitol.* 83, 440–443.

1921. Sundermann, C. A., Estridge, B. H. 1999. Growth of and competition between *Neospora caninum* and *Toxoplasma gondii in vitro*. *Int. J. Parasitol.* 29, 1725–1732.

1922. Suteeraparp, P., Pholpark, S., Pholpark, M., Charoenchai, A., Chompoochan, T., Yamane, I., Kashiwazaki, Y. 1999. Seroprevalence of antibodies to *Neospora caninum* and associated abortion in dairy cattle from central Thailand. *Vet. Parasitol.* 86, 49–57.

1923. Suteu, O., Oltean, M., Cozma, V. 2005. First serological survey for canine neosporosis in Romania. *Bul. Univ. Stiinte Agricole Med.* 62, 591–592.

1924. Suteu, O., Titilincu, A., Modrý, D., Mihalca, A., Mircean, V., Cozma, V. 2010. First identification of *Neospora caninum* by PCR in aborted bovine foetuses in Romania. *Parasitol. Res.* 106, 719–722.

1925. Suteu, O., Mihalca, A. D., Pastiu, A. I., Györke, A., Matei, I. A., Ionica, A., Balea, A., Oltean, M., D'Amico, G., Sikó, S. B., Ionescu, D., Gherman, C. M., Cozma, V. 2014. Red foxes (*Vulpes vulpes*) in Romania are carriers of *Toxoplasma gondii* but not *Neospora caninum*. *J. Wildl. Dis.* 50, 713–716.

1926. Swift, B. L., Kennedy, P. C. 1972. Experimentally induced infection of *in utero* bovine fetuses with bovine parainfluenza 3 virus. *Am. J. Vet. Res.* 33, 57–63.

1927. Syed-Hussain, S. S., Howe, L., Pomroy, W. E., West, D. M., Smith, S. L., Williamson, N. B. 2014. Adaptation of a commercial ELISA to determine the IgG avidity in sheep experimentally and naturally infected with *Neospora caninum*. *Vet. Parasitol.* 203, 21–28.

1928. Syed-Hussain, S. S., Lowe, L., Pomroy, W. E., West, D. M., Hardcastle, M., Williamson, N. B. 2015. Vertical transmission in experimentally infected sheep despite previous inoculation with *Neospora caninum* NcNZ1 isolate. *Vet. Parasitol.* 208, 150–158.

1929. Syed-Hussain, S. S., Howe, L., Pomroy, W. E., West, D. M., Smith, S. L., Williamson, N. B. 2013. Detection of *Neospora caninum* DNA in semen of experimental infected rams with no evidence of horizontal transmission in ewes. *Vet. Parasitol.* 197, 534–542.

1930. Takashima, Y., Takasu, M., Yanagimoto, I., Hattori, N., Batanova, T., Nishikawa, Y., Kitoh, K. 2013. Prevalence and dynamics of antibodies against NcSAG1 and NcGRA7 antigens of *Neospora caninum* in cattle during the gestation period. *J. Vet. Med. Sci.* 75, 1413–1418.

1931. Talafha, A. Q., Al-Majali, A. M. 2013. Prevalence and risk factors associated with *Neospora caninum* infection in dairy herds in Jordan. *Trop. Anim. Health Prod.* 45, 479–485.

1932. Talafha, A. Q., Abutarbush, S. M., Rutley, D. L. 2015. Seroprevalence and potential risk factors associated with *Neospora* spp. infection among asymptomatic horses in Jordan. *Korean J. Parasitol.* 53, 163–167.

1933. Talevich, E., Kannan, N. 2013. Structural and evolutionary adaptation of rhoptry kinases and pseudokinases, a family of coccidian virulence factors. *BMC Evol. Biol.* 13, 117.

1934. Tampaki, Z., Mwakubambanya, R. S., Goulielmaki, E., Kaforou, S., Kim, K., Waters, A. P., Carruthers, V. B., Siden-Kiamos, I., Loukeris, T. G., Koussis, K. 2015. Ectopic expression of a *Neospora caninum* Kazal type inhibitor triggers developmental defects in *Toxoplasma* and *Plasmodium*. *PLoS ONE* 10, e0121379.

1934a. Tamponi, C., Varcasia, A., Pipia, A. P., Zidda, A., Panzalis, R., Dore, F., Dessì, G., Sanna, G., Sali, F., Bjorkman, C., Scala, A. 2015. ISCOM ELISA in milk as screening for *Neospora caninum* in dairy sheep. *Large Anim. Rev.* 21, 213–216.

1935. Tanaka, T., Nagasawa, H., Fujisaki, K., Suzuki, N., Mikami, T. 2000. Growth-inhibitory effects of interferon-γ on *Neospora caninum* in murine macrophages by a nitric oxide mechanism. *Parasitol. Res.* 86, 768–771.

1936. Tanaka, T., Hamada, T., Inoue, N., Nagasawa, H., Fujisaki, K., Suzuki, N., Mikami, T. 2000. The role of CD4+ or CD8+ T cells in the protective immune response of BALB/c mice to *Neospora caninum* infection. *Vet. Parasitol.* 90, 183–191.

1936a. Taques, I. I. G. G., Barbosa, T. R., Martini, A. C., Pitchenin, L. C., Braga, Í. A., de Melo, A. L. T., Nakazato, L., Dutra, V., de Aguiar, D. M. 2016. Molecular assessment of the transplacental transmission of *Toxoplasma gondii*, *Neospora caninum*, *Brucella canis* and *Ehrlichia canis* in dogs. *Comp. Immunol. Microbiol. Infect. Dis.* 49, 47–50.

1937. Tarantino, C., Rossi, G., Kramer, L. H., Perrucci, S., Cringoli, G., Macchioni, G. 2001. *Leishmania infantum* and *Neospora caninum* simultaneous skin infection in a young dog in Italy. *Vet. Parasitol.* 102, 77–83.

1938. Taubert, A., Zahner, H., Hermosilla, C. 2006. Dynamics of transcription of immunomodulatory genes in endothelial cells infected with different coccidian parasites. *Vet. Parasitol.* 142, 214–222.

1939. Taubert, A., Hermosilla, C., Behrendt, J., Zahner, H. 2006. Endothelzellen *in vitro* auf Kokzidieninfektionen (*Eimeria bovis*, *Toxoplasma gondii*, *Neospora caninum*) als Ausdruck einer nicht-adaptativen Immunantwort. [Responses of bovine endothelial cells in vitro to coccidian (*Eimeria bovis*, *Toxoplasma gondii*, *Neospora caninum*) infections possibly involved in innate immunity]. *Berl Munch. Tierarztl. Wochenschr.* 119, 274–281.

1940. Taubert, A., Krüll, M., Zahner, H., Hermosilla, C. 2006. *Toxoplasma gondii* and *Neospora caninum* infections of bovine endothelial cells induce endothelial adhesion molecule gene transcription and subsequent PMN adhesion. *Vet. Immunol. Immunopathol.* 112, 272–283.

1941. Tautz, D., Renz, M. 1984. Simple sequences are ubiquitous repetitive components of eukaryotic genomes. *Nucleic Acids Res.* 12, 4127–4138.

1942. Tavalla, M., Sabaghan, M., Abdizadeh, R., Khademvatan, S., Rafiei, A., Razavi Piranshahi, A. 2015. Seroprevalence of *Toxoplasma gondii* and *Neospora* spp. infections in Arab horses, southwest of Iran. *Jundishapur J. Microbiol.* 8, e14939.

1943. Tavernier, P., Sys, S. U., de Clercq, K., de Leeuw, I., Caij, A. B., de Baere, M., de Regge, N., Fretin, D., Roupie, V., Govaerts, M., Heyman, P., Vanrompay, D., Yin, L., Kalmar, I., Suin, V., Brochier, B., Dobly, A., de Craeye, S., Roelandt, S., Goossens, E., Roels, S. 2015. Serologic screening for 13 infectious agents in roe deer (*Capreolus capreolus*) in Flanders. *Infect. Ecol. Epidemiol.* 5, 29862.

1944. Teixeira, L., Marques, A., Meireles, C. S., Seabra, A. R., Rodrigues, D., Madureira, P., Faustino, A. M. R., Silva, C., Ribeiro, A., Ferreira, P., Correia da Costa, J. M., Canada, N., Vilanova, M. 2005. Characterization of the B-cell immune response elicited in BALB/c mice challenged with *Neospora caninum* tachyzoites. *Immunology* 116, 38–52.

1945. Teixeira, L., Botelho, A. S., Batista, A. R., Meireles, C. S., Ribeiro, A., Domingues, H. S., Correia da Costa, J. M., Castro, A. G., Faustino, A. M. R., Vilanova, M. 2007. Analysis of the immune response to *Neospora caninum* in a model of intragastric infection in mice. *Parasite Immunol.* 29, 23–36.

1946. Teixeira, L., Botelho, A. S., Mesquita, S. D., Correia, A., Cerca, F., Costa, R., Sampaio, P., Castro, A. G., Vilanova, M. 2010. Plasmacytoid and conventional dendritic cells are early producers of IL-12 in *Neospora caninum*-infected mice. *Immunol. Cell Biol.* 88, 79–86.

1947. Teixeira, L., Moreira, J., Melo, J., Bezerra, F., Marques, R. M., Ferreirinha, P., Correia, A., Monteiro, M. P., Ferreira, P. G., Vilanova, M. 2015. Immune response in the adipose tissue of lean mice infected with the protozoan parasite *Neospora caninum*. *Immunology* 145, 242–257.

1948. Teixeira, L., Marques, R. M., Ferreirinha, P., Bezerra, F., Melo, J., Moreira, Í., Pinto, A., Correia, A., Ferreira, P. G., Vilanova, M. 2016. Enrichment of IFN-γ producing cells in different murine adipose tissue depots upon infection with an apicomplexan parasite. *Sci. Rep.* 6, 23475.

1949. Teixeira, W. C., Silva, M. I. S., Pereira, J. G., Pinheiro, A. M., Almeida, M. Â. O., Gondim, L. F. P. 2006. Freqüência de cães reagentes para *Neospora caninum* em São Luís, Maranhão. [Frequency of *Neospora caninum* in dogs from Sao Luis, Maranhao]. *Arq. Bras. Med. Vet. Zootec.* 58, 685–687.

1949a. Teixeira, W. C., Uzêda, R. S., Gondim, L. F. P., Silva, M. I. S., Pereira, H. M., Alves, L. C., Faustino, M. A. G. 2010. Prevalência de anticorpos anti-*Neospora caninum* (Apicomplexa: Sarcocystidae) em bovinos leiteiros de propriedades rurais em três microrregiões no estado do Maranhão. [Prevalence of anti-*Neospora caninum* (Apixomplexa: Sarcocystidae) antibodies in dairy cattle in rural properties of three microrregions of Maranhao, Brazil]. *Pesq.Vet. Bras.* 30, 729–734.

1950. Tembue, A. A. S. M., Ramos, R. A. N., de Sousa, T. R., Albuquerque, A. R., da Costa, A. J., Meunier, I. M. J., Faustino, M. A. G., Alves, L. C. 2011. Serological survey of *Neospora caninum* in small ruminants from Pernambuco State, Brazil. *Rev. Bras. Parasitol. Vet.* 20, 246–248.

1951. Tennent-Brown, B. S., Pomroy, W. E., Reichel, M. P., Gray, P. L., Marshall, T. S., Moffat, P. A., Rogers, M., Driscoll, V. A., Reeve, O. F., Ridler, A. L., Ritvanen, S. 2000. Prevalence of *Neospora* antibodies in beef cattle in New Zealand. *N. Z. Vet. J.* 48, 149–150.

1952. Tenter, A. M., Barta, J. R., Beveridge, I., Duszynski, D. W., Mehlhorn, H., Morrison, D. A., Thompson, R. C. A., Conrad, P. A. 2002. The conceptual basis for a new classification of the coccidia. *Int. J. Parasitol.* 32, 595–616.

1953. Thate, F. M., Laanen, S. C. 1998. Successful treatment of neosporosis in an adult dog. *Vet. Quart.* 20, S113–S114.

1954. Thilsted, J. P., Dubey, J. P. 1989. Neosporosis-like abortions in a herd of dairy cattle. *J. Vet. Diagn. Invest.* 1, 205–209.

1955. Thobokwe, G., Heuer, C. 2004. Incidence of abortion and association with putative causes in dairy herds in New Zealand. *N. Z. Vet. J.* 52, 90–94.

1956. Thomas, W. B. 1998. Inflamatory diseases of the central nervous system in dogs. *Clin, Tech. Small Anim. Prac.* 13, 167–178.

1957. Thompson, G., Canada, N., do Carmo Topa, M., Silva, E., Vaz, F., Rocha, A. 2001. First confirmed case of *Neospora caninum*-associated abortion outbreak in Portugal. *Reprod. Dom. Anim.* 36, 309–312.

1958. Thompson, J. A., Scott, H. M. 2007. Bayesian kriging of seroprevalence to *Mycobacterium avium* subspecies *paratuberculosis* and *Neospora caninum* in Alberta beef and dairy cattle. *Can. Vet. J.* 48, 1281–1285.

1959. Thornton, R. N., Thompson, E. J., Dubey, J. P. 1991. *Neospora* abortion in New Zealand cattle. *N. Z. Vet. J.* 39, 129–133.

1960. Thornton, R. N., Gajadhar, A., Evans, J. 1994. *Neospora* abortion epdemic in a dairy herd. *N. Z. Vet. J.* 42, 190–191.

1961. Thurmond, M., Hietala, S. 1995. Strategies to control *Neospora* infection in cattle. *Bov. Pract.* 29, 60–63.

1962. Thurmond, M. C., Anderson, M. L., Blanchard, P. C. 1995. Secular and seasonal trends of *Neospora* abortion in California dairy cows. *J. Parasitol.* 81, 364–367.

1963. Thurmond, M. C., Hietala, S. K. 1996. Culling associated with *Neospora caninum* infection in dairy cows. *Am. J. Vet. Res.* 57, 1559–1562.

1964. Thurmond, M. C., Hietala, S. K. 1997. Effect of congenitally acquired *Neospora caninum* infection on risk of abortion and subsequent abortions in dairy cattle. *Am. J. Vet. Res.* 58, 1381–1385.

1965. Thurmond, M. C., Hietala, S. K. 1997. Effect of *Neospora caninum* infection on milk production in first-lactation dairy cows. *J. Am. Vet. Med. Assoc.* 210, 672–674.

1966. Thurmond, M. C., Hietala, S. K., Blanchard, P. C. 1997. Herd-based diagnosis of *Neospora caninum*-induced endemic and epidemic abortion in cows and evidence for congenital and postnatal transmission. *J. Vet. Diagn. Invest.* 9, 44–49.

1967. Thurmond, M. C., Hietala, S. K. 1999. *Neospora caninum* infection and abortion in cattle. *Curr. Vet. Ther.*, Philadelphia, W.B. Saunders Company, 4, 425–431.

1968. Thurmond, M. C., Hietala, S. K., Blanchard, P. C. 1999. Predictive values of fetal histopathology and immunoperoxidase staining in diagnosing bovine abortion caused by *Neospora caninum* in a dairy herd. *J. Vet. Diagn. Invest.* 11, 90–94.

1969. Thurmond, M. C., Johnson, W. O., Muñoz-Zanzi, C. A., Su, C. L., Hietala, S. K. 2002. A method of probability diagnostic assigment that applies Bayes theorem for use in serologic diagnostics, using an example of *Neospora caninum* infection in cattle. *Am. J. Vet. Res.* 63, 318–325.

1970. Tiemann, J. C. H., Souza, S. L. P., Rodrigues, A. A. R., Duarte, J. M. B., Gennari, S. M. 2005. Environmental effect on the occurrence of anti-*Neospora caninum* antibodies in pampas-deer (*Ozotoceros bezoarticus*). *Vet. Parasitol.* 134, 73–76.

1971. Tiemann, J. C. H., Rodrigues, A. A. R., de Souza, S. L. P., Duarte, J. M. B., Gennari, S. M. 2005. Occurrence of anti-*Neospora caninum* antibodies in Brazilian cervids kept in captivity. *Vet. Parasitol.* 129, 341–343.

1972. Tiwari, A., VanLeeuwen, J. A., Dohoo, I. R., Stryhn, H., Keefe, G. P., Haddad, J. P. 2005. Effects of seropositivity for bovine leukemia virus, bovine viral diarrhoea virus, *Mycobacterium avium* subspecies *paratuberculosis*, and *Neospora caninum* on culling in dairy cattle in four Canadian provinces. *Vet. Microbiol.* 109, 147–158.

1973. Tiwari, A., VanLeeuwen, J. A., Dohoo, I. R., Keefe, G. P., Haddad, J. P., Tremblay, R., Scott, H. M., Whiting, T. 2007. Production effects of pathogens causing bovine leukosis, bovine viral diarrhea, paratuberculosis, and neosporosis. *J. Dairy Sci.* 90, 659–669.

1974. Tomioka, Y., Sawada, M., Ochiai, K., Umemura, T. 2003. *Neospora caninum* antigens recognized by mouse IgG at different stages of infection including recrudescence. *J. Vet. Med. Sci.* 65, 745–747.

1975. Tonin, A. A., Da Silva, A. S., Thomé, G. R., Oliveira, L. S., Schetinger, M. R. C., Morsch, V. M., Flores, M. M., Fighera, R. A., Toscan, G., Vogel, F. F., Lopes, S. T. A. 2013. *Neospora caninum*: Activity of cholinesterases during the acute and chronic phases of an experimental infection in gerbils. *Exp. Parasitol.* 135, 669–674.

1976. Tonin, A. A., Da Silva, A. S., Thomê, G. R., Bochi, G. V., Schetinger, M. R. C., Moresco, R. N., Camillo, G., Toscan, G., Vogel, F. F., Lopes, S. T. A. 2014. Oxidative stress in brain tissue of gerbils experimentally infected with *Neospora caninium*. *J. Parasitol.* 100, 154–156.

1977. Tonin, A. A., Da Silva, A. S., Thomé, G. R., Schirmbeck, G. H., Cardoso, V. V., Casali, E. A., Toscan, G., Vogel, F. F., Flores, M. M., Fighera, R., Lopes, S. T. A. 2014. Changes in purine levels associated with cellular brain injury in gerbils experimentally infected with *Neospora caninum*. *Res. Vet. Sci.* 96, 507–511.

1978. Tonin, A. A., Weber, A., Ribeiro, A., Camillo, G., Vogel, F. F., Moura, A. B., Bochi, G. V., Moresco, R. N., Da Silva, A. S. 2015. Serum levels of nitric oxide and protein oxidation in goats seropositive for *Toxoplasma gondii* and *Neospora caninum*. *Comp. Immunol. Microbiol. Infect. Dis.* 41, 55–58.

1979. Tonkin, M. L., Crawford, J., Lebrun, M. L., Boulanger, M. J. 2013. *Babesia divergens* and *Neospora caninum* apical membrane antigen 1 structures reveal selectivity and plasticity in apicomplexan parasite host cell invasion. *Protein Sci.* 22, 114–127.

1980. Topazio, J. P., Weber, A., Camillo, G., Vogel, F. F., Machado, G., Ribeiro, A., Moura, A. B., Lopes, L. S., Tonin, A. A., Soldá, N. M., Bräunig, P., Da Silva, A. S. 2014. Seroprevalence and risk factors for *Neospora caninum* in goats in Santa Catarina state, Brazil. *Rev. Bras. Parasitol. Vet.* 23, 360–366.

1981. Torres, M. P., Ortega, Y. R. 2006. *Neospora caninum* antibodies in commercial fetal bovine serum. *Vet. Parasitol.* 140, 352–355.

1982. Toscan, G., Vogel, F. S. F., Cadore, G. C., Cezar, A. S., Sangioni, L. A., Pereira, R. C. F., Oliveira, L. S. S., Lopes, S. T. A. 2011. Occurrence of antibodies anti-*Neospora* spp. in cart horses and Crioula breed horses from Rio Grande do Sul State. *Arq. Bras. Med. Vet. Zootec.* 63, 258–261.

1983. Tóth, G., Gáspári, Z., Jurka, J. 2000. Microsatellites in different eukaryotic genomes: Survey and analysis. *Genome Res.* 10, 967–981.

1984. Touloudi, A., Valiakos, G., Athanasiou, L. V., Birtsas, P., Giannakopoulos, A., Papaspyropoulos, K., Kalaitzis, C., Sokos, C., Tsokana, C. N., Spyrou, V., Petrovska, L., Billinis, C. 2015. A serosurvey for selected pathogens in Greek European wild boar. *Vet. Rec. Open* 2, e000077.

1985. Tramuta, C., Lacerenza, D., Zoppi, S., Goria, M., Dondo, A., Ferroglio, E., Nebbia, P., Rosati, S. 2011. Development of a set of multiplex standard polymerase chain reaction assays for the identification of infectious agents from aborted bovine clinical samples. *J. Vet. Diagn. Invest.* 23, 657–664.

1986. Tran, J. Q., de Leon, J. C., Li, C., Huynh, M. H., Beatty, W., Morrissette, N. S. 2010. RNG1 is a late marker of the apical polar ring in *Toxoplasma gondii*. *Cytoskeleton (Hoboken.)* 67, 586–598.

1987. Tranas, J., Heinzen, R. A., Weiss, L. M., McAllister, M. M. 1999. Serological evidence of human infection with the protozoan *Neospora caninum*. *Clin. Diagn. Lab. Immunol.* 6, 765–767.

1988. Trees, A. J., Tennant, B. J., Kelly, D. F. 1991. Paresis in dogs and *Neospora caninum*. *Vet. Rec.* 129, 456.

1989. Trees, A. J. 1993. *Neospora* spp. infections in British cattle: Serological studies. *Cattle Prac.* 1, 414–418.

1990. Trees, A. J., Guy, F., Tennant, B. J., Balfour, A. H., Dubey, J. P. 1993. Prevalence of antibodies to *Neospora caninum* in a population of urban dogs in England. *Vet. Rec.* 132, 125–126.

1991. Trees, A. J., Guy, F., Low, J. C., Roberts, L., Buxton, D., Dubey, J. P. 1994. Serological evidence implicating *Neospora* species as a cause of abortion in British cattle. *Vet. Rec.* 134, 405–407.

1992. Trees, A. J., Davison, H. C., Innes, E. A., Wastling, J. M. 1999. Towards evaluating the economic impact of bovine neosporosis. *Int. J. Parasitol.* 29, 1195–1200.

1993. Trees, A. J., Williams, D. J. L. 2000. Neosporosis in the United Kingdom. *Int. J. Parasitol.* 30, 891–893.

1994. Trees, A. J., McAllister, M. M., Guy, C. S., McGarry, J. W., Smith, R. F., Williams, D. J. L. 2002. *Neospora caninum*: Oocyst challenge of pregnant cows. *Vet. Parasitol.* 109, 147–154.

1995. Trees, A. J., Williams, D. J. L. 2003. Vaccination against bovine neosporosis—The challenge is the challenge. *J. Parasitol.* 89, S198–S201.

1996. Trees, A. J., Williams, D. J. L. 2005. Endogenous and exogenous transplacental infection in *Neospora caninum* and *Toxoplasma gondii*. *Trends Parasitol.* 21, 558–561.

1997. Truppel, J. H., Montiani-Ferreira, F., Lange, R. R., Vilani, R. G. O. C., Reifur, L., Boerger, W., da Costa-Ribeiro, M. C. V., Thomaz-Soccol, V. 2010. Detection of *Neospora caninum* DNA in capybaras and phylogenetic analysis. *Parasitol. Int.* 59, 376–379.

1998. Tunev, S. S., McAllister, M. M., Anderson-Sprecher, R. C., Weiss, L. M. 2002. *Neospora caninum in vitro*: Evidence that the destiny of a parasitophorous vacuole depends on the phenotype of the progenitor zoite. *J. Parasitol.* 88, 1095–1099.

1999. Tuo, W., Davis, W. C., Fetterer, R., Jenkins, M., Boyd, P. C., Gasbarre, L. C., Dubey, J. P. 2004. Establishment of *Neospora caninum* antigen-specific T cell lines of primarily CD4+ T cells. *Parasite Immunol.* 26, 243–246.

2000. Tuo, W., Fetterer, R., Jenkins, M., Dubey, J. P. 2005. Identification and characterization of *Neospora caninum* cyclophilin that elicits gamma interferon production. *Infect. Immun.* 73, 5093–5100.

2001. Tuo, W., Fetterer, R. H., Davis, W. C., Jenkins, M. C., Dubey, J. P. 2005. *Neospora caninum* antigens defined by antigen-dependent bovine CD4+ T cells. *J. Parasitol.* 91, 564–568.

2002. Tuo, W., Zhao, Y., Zhu, D., Jenkins, M. C. 2011. Immunization of female BALB/c mice with *Neospora* cyclophilin and/or NcSRS2 elicits specific antibody response and prevents against challenge infection by *Neospora caninum*. *Vaccine* 29, 2392–2399.

2003. Uchida, M., Nagashima, K., Akatsuka, Y., Murakami, T., Ito, A., Imai, S., Ike, K. 2013. Comparative study of protective activities of *Neospora caninum* bradyzoite antigens, NcBAG1, NcBSR4, NcMAG1, and NcSAG4, in a mouse model of acute parasitic infection. *Parasitol. Res.* 112, 655–663.

2004. Uchida, Y., Ike, K., Kurotaki, T., Takeshi, M., Imai, S. 2003. Susceptibility of Djungarian hamsters (*Phodopus sungorus*) to *Neospora caninum* infection. *J. Vet. Med. Sci.* 65, 401–403.

2005. Uchida, Y., Ike, K., Kurotaki, T., Ito, A., Imai, S. 2004. Monoclonal antibodies preventing invasion of *Neospora caninum* tachyzoites into host cells. *J. Vet. Med. Sci.* 66, 1355–1358.

2006. Ueno, T. E. H., Gonçalves, V. S. P., Heinemann, M. B., Dilli, T. L. B., Akimoto, B. M., de Souza, S. L. P., Gennari, S. M., Soares, R. M. 2009. Prevalence of *Toxoplasma gondii* and *Neospora caninum* infections in sheep from Federal District, central region of Brazil. *Trop. Anim. Health Prod.* 41, 547–552.

2007. Uggla, A., Dubey, J. P., Lundmark, G., Olson, P. 1989. Encephalomyelitis and myositis in a Boxer puppy due to a *Neospora*-like infection. *Vet. Parasitol.* 32, 255–260.

2008. Uggla, A., Dubey, J. P., Funkquist, B., Segall, T. 1989. Fatal *Neospora caninum*-infektion hos riesenschnauzer. [Fatal *Neospora caninum* infection in a dog]. *Svensk Veterinärtidning* 41, 271–274.

2009. Uggla, A., Stenlund, S., Holmdahl, O. J. M., Jakubek, E. B., Thebo, P., Kindahl, H., Björkman, C. 1998. Oral *Neospora caninum* inoculation of neonatal calves. *Int. J. Parasitol.* 28, 1467–1472.

2010. Uggla, A., Mattsson, J. G., Lundén, A., Jakubek, E. B., Näslund, K., Holmdahl, O. J. M. 2000. *Neospora caninum* in Sweden. *Int. J. Parasitol.* 30, 893–896.

2011. Umemura, T., Shiraki, K., Morita, T., Shimada, A., Haritani, M., Kobayashi, M., Yamagata, S. 1992. Neosporosis in a dog: The first case report in Japan. *J. Vet. Med. Sci.* 54, 157–159.

2012. Unzaga, J. M., Moré, G., Bacigalupe, D., Rambeaud, M., Pardini, L., Dellarupe, A., De Felice, L., Gos, M. L., Venturini, M. C. 2014. *Toxoplasma gondii* and *Neospora caninum* infections in goat abortions from Argentina. *Parasitol. Int.* 63, 865–867.

2013. Uzêda, R. S., Pinheiro, A. M., Fernández, S. Y., Ayres, M. C. C., Gondim, L. F. P., Almeida, M. A. O. 2007. Seroprevalence of *Neospora caninum* in dairy goats from Bahia, Brazil. *Small Ruminant Res.* 70, 257–259.

2014. Uzeda, R. S., Costa, K. S., Santos, S. L., Pinheiro, A. M., de Almeida, M. A. O., McAllister, M. M., Gondim, L. F. P. 2007. Loss of infectivity of *Neospora caninum* oocyts maintained for a prolonged time. *Korean J. Parasitol.* 45, 295–299.

2015. Uzêda, R. S., Schares, G., Ortega-Mora, L. M., Madruga, C. R., Aguado-Martinez, A., Corbellini, L. G., Driemeier, D., Gondim, L. F. P. 2013. Combination of monoclonal antibodies improves immunohistochemical diagnosis of *Neospora caninum*. *Vet. Parasitol.* 197, 477–486.

2016. Václavek, P., Koudela, B., Modrý, D., Sedlák, K. 2003. Seroprevalence of *Neospora caninum* in aborting dairy cattle in the Czech Republic. *Vet. Parasitol.* 115, 239–245.

2017. Václavek, P., Sedlák, K., Hurková, L., Vodrážka, P., Šebesta, R., Koudela, B. 2007. Serological survey of *Neospora caninum* in dogs in the Czech Republic and a long-term study of dynamics of antibodies. *Vet. Parasitol.* 143, 35–41.

2018. Valadas, S., Gennari, S. M., Yai, L. E. O., Rosypal, A. C., Lindsay, D. S. 2010. Prevalence of antibodies to *Trypanosoma cruzi*, *Leishmania infantum*, *Encephalitozoon cuniculi*, *Sarcocystis neurona*, and *Neospora caninum* in capybara, *Hydrochoerus hydrochaeris*, from São Paulo State, Brazil. *J. Parasitol.* 96, 521–524.

2019. Valadas, S., Minervino, A. H. H., Lima, V. M. F., Soares, R. M., Ortolani, E. L., Gennari, S. M. 2010. Occurrence of antibodies anti-*Neospora caninum*, anti-*Toxoplasma gondii*, and anti-*Leishmania chagasi* in serum of dogs from Pará State, Amazon, Brazil. *Parasitol. Res.* 107, 453–457.

2020. van der Hage, M. H., Kik, M. J. L., Dorrestein, G. M. 2002. *Neospora caninum*: Myocarditis in a European pine marten (*Martes martes*). European Association of Zoo- and Wildlife Veterinarians (EAZWV) 4th Scientific Meeting, Joint with the Annual Meeting of the European Wildlife Disease Association (EWDA). May 8–12, Heidelberg, Germany, 217–220.

2021. van Ham, L. M. L., Thoonen, H., Barber, J. S., Trees, A. J., Polis, I., De Cock, H., Hoorens, J. K. 1996. *Neospora caninum* infection in the dog: Typical and atypical cases. *Vlaams Diergeneesk. Tijdsch.* 65, 326–335.

2022. van Maanen, C., Wouda, W., Schares, G., von Blumröder, D., Conraths, F. J., Norton, R., Williams, D. J. L., Esteban-Redondo, I., Innes, E. A., Mattsson, J. G., Björkman, C., Fernández-García, A., Ortega-Mora, L. M., Müller, N., Sager, H., Hemphill, A. 2004. An interlaboratory comparison of immunohistochemistry and PCR methods for detection of *Neospora caninum* in bovine foetal tissues. *Vet. Parasitol.* 126, 351–364.

2023. Vangeel, I., Méroc, E., Roelandt, S., Welby, S., Driessche, E. V., Czaplicki, G., Schoubroeck, L. V., Quinet, C., Riocreux, F., Hooyberghs, J., Houdart, P., Stede, Y. V. d. 2012. Seroprevalence of *Neospora caninum*, paratuberculosis and Q fever in cattle in Belgium in 2009–2010. *Vet. Rec.* 171, 477.

2024. VanLeeuwen, J. A., Keefe, G. P., Tremblay, R., Power, C., Wichtel, J. J. 2001. Seroprevalence of infection with *Mycobacterium avium* subspecies *paratuberculosis*, bovine leukemia virus, and bovine viral diarrhea virus in maritime Canada dairy cattle. *Can. Vet. J.* 42, 193–198.

2025. VanLeeuwen, J. A., Keefe, G. P., Tiwari, A. 2002. Seroprevalence and productivity effects of infection with bovine leukemia virus, *Mycobacterium avium* subspecies *paratuberculosis*, and *Neospora caninum* in Maritime Canadian dairy cattle. *Bov. Pract.* 36, 86–91.

2026. VanLeeuwen, J. A., Forsythe, L., Tiwari, A., Chartier, R. 2005. Seroprevalence of antibodies against bovine leukemia virus, bovine viral diarrhea virus, *Mycobacterium avium* subspecies *paratuberculosis*, and *Neospora caninum* in dairy cattle in Saskatchewan. *Can. Vet. J.* 46, 56–58.

2027. VanLeeuwen, J. A., Tiwari, A., Plaizier, J. C., Whiting, T. L. 2006. Seroprevalences of antibodies against bovine leukemia virus, bovine viral diarrhea virus, *Mycobacterium avium* subspecies *paratuberculosis*, and *Neospora caninum* in beef and dairy cattle in Manitoba. *Can. Vet. J.* 47, 783–786.

2028. VanLeeuwen, J. A., Haddad, J. P., Dohoo, I. R., Keefe, G. P., Tiwari, A., Scott, H. M. 2010. Risk factors associated with *Neospora caninum* seropositivity in randomly sampled Canadian dairy cows and herds. *Prev. Vet. Med.* 93, 129–138.

2029. VanLeeuwen, J. A., Haddad, J. P., Dohoo, I. R., Keefe, G. P., Tiwari, A., Tremblay, R. 2010. Associations between reproductive performance and seropositivity for bovine leukemia virus, bovine viral-diarrhea virus, *Mycobacterium avium* subspecies *paratuberculosis*, and *Neospora caninum* in Canadian dairy cows. *Prev. Vet. Med.* 94, 54–64.

2030. VanLeeuwen, J. A., Greenwood, S., Clark, F., Acorn, A., Markham, F., McCarron, J., O'Handley, R. 2011. Monensin use against *Neospora caninum* challenge in dairy cattle. *Vet. Parasitol.* 175, 372–376.

2030a. Van Voorhis, W. C., Adams, J. H., Adelfio, R., Ahyong, V., Akabas, M. H., Alano, P., Alday, A., Resto, Y. A., Alsibaee, A., Alzualde, A. et al. 2016. Open source drug discovery with the malaria box compound collection for neglected diseases and beyond. *PLoS Pathog.* DOI:10.1371/journal.ppat.105763,

2031. Varandas, N. P., Rached, P. A., Costa, G. H. N., de Souza, L. M., Castagnolli, K. C., da Costa, A. J. 2001. Freqüência de anticorpos anti-*Neospora caninum* e anti-*Toxolasma gondii* em cães da região nordeste do Estado de São Paulo. Correlação com neuropatias. [Frequency of antibodies for *Neospora caninum* and *Toxoplasma gondii* in dogs in northeast of São Paulo state]. *Semin. Ci. Agrár.* 22, 105–111.

2032. Varaschin, M. S., Hirsch, C., Wouters, F., Nakagaki, K. Y., Guimarães, A. M., Santos, D. S., Bezerra, P. S., Costa, R. C., Peconick, A. P., Langohr, I. M. 2012. Congenital neosporosis in goats from the State of Minas Gerais, Brazil. *Korean J. Parasitol.* 50, 63–67.

2033. Varcasia, A., Capelli, G., Ruiu, A., Ladu, M., Scala, A., Bjorkman, C. 2006. Prevalence of *Neospora caninum* infection in Sardinian dairy farms (Italy) detected by iscom ELISA on tank bulk milk. *Parasitol. Res.* 98, 264–267.

2034. Vardeleon, D., Marsh, A. E., Thorne, J. G., Loch, W., Young, R., Johnson, P. J. 2001. Prevalence of *Neospora hughesi* and *Sarcocystis neurona* antibodies in horses from various geographical locations. *Vet. Parasitol.* 95, 273–282.

2035. Vargas, J. J., Cortés, J. A. 2001. *Neospora caninum*, ¿Una zoonosis potencial? *Rev. Salud Pública* 3, 89–93.

2036. Vasileiou, N. G. C., Fthenakis, G. C., Papadopoulos, E. 2015. Dissemination of parasites by animal movements in small ruminant farms. *Vet. Parasitol.* 213, 56–60.

2037. Vega, O. L., Chávez, V. A., Falcón, P. N., Casas, A. E., Puray, Ch. N. 2010. Prevalencia de *Neospora caninum* en perros pastores de una empresa ganadera de la sierra sur del Perú. [Prevalence of *Neospora caninum* in shepherd dogs of a livestock farm in the southern highlands of Peru]. *Rev. Inv. Vet. Perú* 21, 80–86.

2038. Vemulapalli, R., Sanakkayala, N., Gulani, J., Schurig, G. G., Boyle, S. M., Lindsay, D. S., Sriranganathan, N. 2007. Reduced cerebral infection of *Neospora caninum* in BALB/c mice vaccinated with recombinant *Brucella abortus* RB51 strains expressing *N. caninum* SRS2 and GRA7 proteins. *Vet. Parasitol.* 148, 219–230.

2039. Venturini, L., Di Lorenzo, C., Venturini, C., Romero, J. 1995. Anticuerpos anti *Neospora* sp., en vacas que abortaron. [Antibodies to *Neospora* sp. in cows that had aborted]. *Vet. Arg.* 12, 167–170.

2040. Venturini, M. C., Venturini, L., Bacigalupe, D., Machuca, M., Echaide, I., Basso, W., Unzaga, J. M., Di Lorenzo, C., Guglielmone, A., Jenkins, M. C., Dubey, J. P. 1999. *Neospora caninum* infections in bovine foetuses and dairy cows with abortions in Argentina. *Int. J. Parasitol.* 29, 1705–1708.

2041. Veronesi, F., Diaferia, M., Mandara, M. T., Marenzoni, M. L., Cittadini, F., Piergili-Fioretti, D. P. 2008. *Neospora* spp. infection associated equine abortion and/or stillbirth rate. *Vet. Res. Commun.* 32(Suppl 1), S223–S226.

2042. Viana, R. B., Del Fava, C., Moura, A. C. B., Cardoso, E. C., de Araújo, C. V., Monteiro, B. M., Pituco, E. M., Vasconcellos, S. A. 2009. Ocorrência de anticorpos anti-*Neospora caninum*, *Brucella* sp. e *Leptospira* spp. em búfalos (*Bubalus bubalis*) criados na Amazônoa. [Occurrence of antibodies against *Neospora caninum*, *Brucela* sp. and *Leptospira* spp in buffaloes (*Bubalus bubalis*) raised in Amazonia, Brazil]. *Arq. Inst. Biol. (São Paulo)* 76, 453–457.

2043. Vianna, M. C. B., Sreekumar, C., Miska, K. B., Hill, D. E., Dubey, J. P. 2005. Isolation of *Neospora caninum* from naturally infected white-tailed deer (*Odocoileus virginianus*). *Vet. Parasitol.* 129, 253–257.

2044. Vidić, B., Savić, S., Boboš, S., Vidić, V., Prica, N. 2013. *Neospora caninum* in cattle: Epizootiology, diagnostics and control measures. *Proceedings of the 10th International Symposium Modern Trends in Livestock Production.* October 2–4, Belgrade, Serbia, 119–127.

2045. Villalobos, E. M. C., Ueno, H. T. E., de Souza, S. L. P., Cunha, E. M. S., Lara, M. C. C. S. H., Gennari, S. M., Soares, R. M. 2006. Association between the presence of serum antibodies against *Neospora* spp. and fetal loss in equines. *Vet. Parasitol.* 142, 372–375.

2046. Villalobos, E. M. C., Furman, K. E., Lara, M. C. C. S. H., Cunha, E. M. S., Finger, M. A., Busch, A. P. B., de Barros Filho, I. R., Deconto, I., Dornbusch, P. T., Biondo, A. W. 2012. Detection of *Neospora* sp. antibodies in cart horses from urban areas of Curitiba, Southern Brazil. *Rev. Bras. Parasitol. Vet.* 21, 68–70.

2047. Vitaliano, S. N., Silva, D. A. O., Mineo, T. W. P., Ferreira, R. A., Bevilacqua, E., Mineo, J. R. 2004. Seroprevalence of *Toxoplasma gondii* and *Neospora caninum* in captive maned wolves (*Chrysocyon brachyurus*) from southeastern and midwestern regions of Brazil. *Vet. Parasitol.* 122, 253–260.

2048. Vogel, F. S. F., Arenhart, S., Bauermann, F. V. 2006. Anticorpos anti-*Neospora caninum* em bovinos, ovinos e ubalinos no Estado do Rio Grande do Sul. *Ciência Rural* 36, 1948–1951.

2049. von Blumröder, D., Schares, G., Norton, R., Williams, D. J. L., Esteban-Redondo, I., Wright, S., Björkman, C., Frössling, J., Risco-Castillo, V., Fernández-García, A., Ortega-Mora, L. M., Sager, H., Hemphill, A., van Maanen, C., Wouda, W., Conraths, F. J. 2004. Comparison and standardisation of serological methods for the diagnosis of *Neospora caninum* infection in bovines. *Vet. Parasitol.* 120, 11–22.

2050. von Blumröder, D., Stambusch, R., Labohm, R., Klawonn, W., Dräger, K., Fasen, W., Conraths, F. J., Schares, G. 2006. Potenzielle Risikofaktoren für den serologischen Nachweis von *Neospora-caninum*-Infektionen in Rinderherden in Rheinland-Pfalz. [Potential risk factors for the serological detection of *Neospora caninum*-infections in cattle herds in Rhineland-Palatinate (Germany)]. *Tierärztl. Prax.* 34, 141–147.

2051. Vonlaufen, N., Gianinazzi, C., Müller, N., Simon, F., Björkman, C., Jungi, T. W., Leib, S. L., Hemphill, A. 2002. Infection of organotypic slice cultures from rat central nervous tissue with *Neospora caninum*: An alternative approach to study host-parasite interactions. *Int. J. Parasitol.* 32, 533–542.

2052. Vonlaufen, N., Müller, N., Keller, N., Naguleswaran, A., Bohne, W., McAllister, M. M., Björkman, C., Müller, E., Caldelari, R., Hemphill, A. 2002. Exogenous nitric oxide triggers *Neospora caninum* tachyzoite-to-bradyzoite stage conversion in murine epidermal keratinocyte cell cultures. *Int. J. Parasitol.* 32, 1253–1265.

2053. Vonlaufen, N., Guetg, N., Naguleswaran, A., Müller, N., Björkman, C., Schares, G., von Blumroeder, D., Ellis, J., Hemphill, A. 2004. *In vitro* induction of *Neospora caninum* bradyzoites in Vero cells reveals differential antigen expression, localization, and host-cell recognition of tachyzoites and bradyzoites. *Infect. Immun.* 72, 576–583.

2054. Vonlaufen, N., Naguleswaran, A., Gianinazzi, C., Hemphill, A. 2007. Characterization of the fetuin-binding fraction of *Neospora caninum* tachyzoites and its potential involvement in host-parasite interactions. *Parasitology* 134, 805–817.

2055. Vural, G., Aksoy, E., Bozkir, M., Kuçukayan, U., Erturk, A. 2006. Seroprevalence of *Neospora caninum* in dairy cattle herds in Central Anatolia, Turkey. *Veterinarski Arhiv* 76, 343–349.

2056. Waldner, C., Wildman, B. K., Hill, B. W., Fenton, R. K., Pittman, T. J., Schunicht, O. C., Jim, G. K., Guichon, P. T., Booker, C. W. 2004. Determination of the seroprevalence of *Neospora caninum* in feedlot steers in Alberta. *Can. Vet. J.* 45, 218–224.

2057. Waldner, C. L., Janzen, E. D., Ribble, C. S. 1998. Determination of the association between *Neospora caninum* infection and reproductive performance in beef herds. *J. Am. Vet. Med. Assoc.* 213, 685–690.

2058. Waldner, C. L., Janzen, E. D., Henderson, J., Haines, D. M. 1999. Outbreak of abortion associated with *Neospora caninum* infection in a beef herd. *J. Am. Vet. Med. Assoc.* 215, 1485–1490.

2059. Waldner, C. L., Henderson, J., Wu, J. T. Y., Breker, K., Chow, E. Y. W. 2001. Reproductive performance of a cow-calf herd following a *Neospora caninum*-associated abortion epidemic. *Can. Vet. J.* 42, 355–360.

2060. Waldner, C. L., Henderson, J., Wu, J. T. Y., Coupland, R., Chow, E. Y. W. 2001. Seroprevalence of *Neospora caninum* in beef cattle in northern Alberta. *Can. Vet. J.* 42, 130–132.

2061. Waldner, C. L. 2002. Pre-colostral antibodies to *Neospora caninum* in beef calves following an abortion outbreak and associated fall weaning weights. *Bovine Practitioner* 36, 81–85.

2062. Waldner, C. L., Cunningham, G., Campbell, J. R. 2004. Agreement between three serological tests for *Neospora caninum* in beef cattle. *J. Vet. Diagn. Invest.* 16, 313–315.

2063. Waldner, C. L. 2005. Serological status for *Neospora caninum*, bovine viral darrhea virus, and infectious bovine rhinotracheitis virus at pregnancy testing and reproductive performance in beef herds. *Anim. Reprod. Sci.* 90, 219–242.

2064. Wallace, R. M., Pohler, K. G., Smith, M. F., and Green, J. A. 2015. Placental PAGs: Gene origins, expression patterns, and use as markers of pregnancy. *Reproduction* 149, R115–R126.

2065. Walsh, C. P., Duncan, R. B., Zajac, A. M., Blagburn, B. L., Lindsay, D. S. 2000. *Neospora hughesi*: Experimental infections in mice, gerbils, and dogs. *Vet. Parasitol.* 92, 119–128.

2066. Walsh, C. P., Vemulapalli, R., Sriranganathan, N., Zajac, A. M., Jenkins, M. C., Lindsay, D. S. 2001. Molecular comparison of the dense granule proteins GRA6 and GRA7 of *Neospora hughesi* and *Neospora caninum*. *Int. J. Parasitol.* 31, 253–258.

2067. Walsh, R. B., Kelton, D. F., Hietala, S. K., Duffield, T. F. 2013. Evaluation of enzyme-linked immunosorbent assays performed on milk and serum samples for detection of neosporosis and leukosis in lactating dairy cows. *Can. Vet. J.* 54, 347–352.

2068. Wang, C., Wang, Y., Zou, X., Zhai, Y., Gao, J., Hou, M., Zhu, X. Q. 2010. Seroprevalence of *Neospora caninum* infection in dairy cattle in Northeastern China. *J. Parasitol.* 96, 451–452.

2069. Wang, C. R., Zhai, Y. Q., Zhao, X. C., Tan, Q. J., Chen, J., Chen, A. H., Wang, Y. 2009. [Preliminary application of PCR-based assay for the detection of *Neospora caninum* in bovine aborted fetus]. *Chin. J. Parasitol. Parasit. Dis.* 27, 140–143 (in Chinese).

2070. Wang, S., Yao, Z., Zhang, N., Wang, D., Ma, J., Liu, S., Zheng, B., Zhang, B., Liu, K., Zhang, H. 2016. Serological study of *Neospora caninum* infection in dogs in central China. *Parasite* 23, 25.

2071. Wanha, K., Edelhofer, R., Gabler-Eduardo, C., Prosl, H. 2005. Prevalence of antibodies against *Neospora caninum* and *Toxoplasma gondii* in dogs and foxes in Austria. *Vet. Parasitol.* 128, 189–193.

2072. Wapenaar, W., Jenkins, M. C., O'Handley, R. M., Barkema, H. W. 2006. *Neospora caninum*-like oocysts observed in feces of free-ranging red foxes (*Vulpes vulpes*) and coyotes (*Canis latrans*). *J. Parasitol.* 92, 1270–1274.

2073. Wapenaar, W., Barkema, H. W., O'Handley, R. M., Bartels, C. J. M. 2007. Use of an enzyme-linked immunosorbent assay in bulk milk to estimate the prevalence of *Neospora caninum* on dairy farms in Prince Edward Island, Canada. *Can. Vet. J.* 48, 493–499.

2074. Wapenaar, W., Barkema, H. W., VanLeeuwen, J. A., McClure, J. T., O'Handley, R. M., Kwok, O. C. H., Thulliez, P., Dubey, J. P., Jenkins, M. C. 2007. Comparison of serological methods for the diagnosis of *Neospora caninum* infection in cattle. *Vet. Parasitol.* 143, 166–173.

2075. Wapenaar, W., Barkema, H. W., Schares, G., Rouvinen-Watt, K., Zeijlemaker, L., Poorter, B., O'Handley, R. M., Kwok, O. C. H., Dubey, J. P. 2007. Evaluation of four serological techniques to determine the seroprevalence of *Neospora caninum* in foxes (*Vulpes vulpes*) and coyotes (*Canis latrans*) on Prince Edward Island, Canada. *Vet. Parasitol.* 145, 51–58.

2076. Watts, D. E., Benson, A. M. 2016. Prevalence of antibodies for selected canine pathogens among wolves (*Canis lupus*) from the Alaska Peninsula, USA. *J. Wildl. Dis.* 52, 506–515.

2077. Weber, A., Zetzmann, K., Ewringmann, Th. 2000. Vorkommen von Antikörpern gegen *Neospora caninum* bei Kühen in nordbayerischen Beständen mit Abortproblemen. [Prevalence of antibodies to *Neospora caninum* in aborted cowns in North Bavaria]. *Tierärztl. Umschau* 55, 28–29.

2078. Weber, F. H., Jackson, J. A., Sobecki, B., Choromanski, L., Olsen, M., Meinert, T., Frank, R., Reichel, M. P., Ellis, J. T. 2013. On the efficacy and safety of vaccination with live tachyzoites of *Neospora caninum* for prevention of *Neospora*-associated fetal loss in cattle. *Clin. Vaccine Immunol.* 20, 99–105.

2079. Wegmann, T. G., Lin, H., Guilbert, L., Mosmann, T. R. 1993. Bidirectional cytokine interactions in the maternal-fetal relationship—Is successful pregnancy a TH2 phenomenon. *Immunol. Today* 14, 353–356.

2079a. Wei, Z., Hermosilla, C., Taubert, A., He, X., Wang, X., Gong, P., Li, J., Yang, Z., Zhang, X. 2016. Canine neutrophil extracellular traps release induced by the Apicomplexan parasite *Neospora caninum in vitro*. *Front. Immunol.* 7, article 436e.

2080. Weiss, L. M., Ma, Y. F., Halonen, S., McAllister, M. M., Zhang, Y. W. 1999. The *in vitro* development of *Neospora caninum* bradyzoites. *Int. J. Parasitol.* 29, 1713–1723.

2081. Weissenböck, H., Dubey, J. P., Suchy, A., Sturm, E. 1997. Neosporose als Ursache von Encephalomalazie und Myocarditis bei Hundewelpen. [Neosporosis causing encephalomalacia and myocarditis in young dogs]. *Wien. Tierärztl. Mschr.* 84, 233–237.

2082. Wernery, U., Thomas, R., Raghavan, R., Syriac, G., Joseph, S., Georgy, N. 2008. Seroepidemiological studies for the detection of antibodies against 8 infectious diseases in dairy dromedaries of the United Arab Emirates using modern laboratory techniques—Part II. *J. Camel Prac. Res.* 15, 139–145.

2083. West, D. M., Pomroy, W. E., Collett, M. G., Hill, F. I., Ridler, A. L., Kenyon, P. R., Morris, S. T., Pattison, R. S. 2006. A possible role for *Neospora caninum* in ovine abortion in New Zealand. *Small Ruminant Res.* 62, 135–138.

2084. Weston, J. 2007. Dose-titration challenge of pregnant hoggets with *Neospora caninum* tachyzoites. *Proceedings of the Society of Sheep & Beef Cattle Veterinarians of the New Zealand Vet. Assoc.* Annual Seminar 2007. 137–144.

2085. Weston, J. F., Williamson, N. B., Pomroy, W. E. 2005. Associations between pregnancy outcome and serological response to *Neospora caninum* among a group of dairy heifers. *N. Z. Vet. J.* 53, 142–148.

2086. Weston, J. F., Howe, L., Collett, M. G., Pattison, R. S., Williamson, N. B., West, D. M., Pomroy, W. E., Syed-Hussain, S. S., Morris, S. T., Kenyon, P. R. 2009. Dose-titration challenge of young pregnant sheep with *Neospora caninum* tachyzoites. *Vet. Parasitol.* 164, 183–191.

2087. Weston, J. F., Heuer, C., Williamson, N. B. 2012. Efficacy of a *Neospora caninum* killed tachyzoite vaccine in preventing abortion and vertical transmission in dairy cattle. *Prev. Vet. Med.* 103, 136–144.

2088. Weston, J. F., Heuer, C., Parkinson, T. J., Williamson, N. B. 2012. Causes of abortion on New Zealand dairy farms with a history of abortion associated with *Neospora caninum*. *N. Z. Vet. J.* 60, 27–34.

2089. Wiengcharoen, J., Thompson, R. C. A., Nakthong, C., Rattanakorn, P., Sukthana, Y. 2011. Transplacental transmission in cattle: Is *Toxoplasma gondii* less potent than *Neospora caninum*? *Parasitol. Res.* 108, 1235–1241.

2090. Wiengcharoen, J., Nokkaew, W., Prasithpon, S., Prasomtong, P., Sukthana, Y. 2012. *Neospora caninum* and *Toxoplasma gondii* antibodies in captive elephants (*Elephanus maximus indicus*) in Kanchnaburi Province. *Thai J. Vet. Med.* 42, 235–240.

2091. Wierzchon, M., Katkiewicz, M., Marciniak, K. 2006. [Neosporosis occurrence in cattle]. *Medycyna Wet.* 62, 1041–1044 (in Polish).

2092. Wikman, M., Friedman, M., Pinitkiatisakul, S., Hemphill, A., Lövgren-Bengtsson, K., Lundén, A., Ståhl, S. 2005. Applying biotin-streptavidin binding for iscom (immunostimulating complex) association of recombinant immunogens. *Biotechnol. Appl. Biochem.* 41, 163–174.

2093. Wilkowsky, S. E., Bareiro, G. G., Mon, M. L., Moore, D. P., Caspe, G., Campero, C., Fort, M., Romano, M. I. 2011. An applied printing immunoassay with recombinant Nc-SAG1 for detection of antibodies to *Neospora caninum* in cattle. *J. Vet. Diagn. Invest.* 23, 971–976.

2094. Williams, D. J. L., McGarry, J., Guy, F., Barber, J., Trees, A. J. 1997. Novel ELISA for detection of *Neospora*-specific antibodies in cattle. *Vet. Rec.* 140, 328–331.

2095. Williams, D. J. L., Davison, H. C., Helmick, B., McGarry, J., Guy, F., Otter, A., Trees, A. J. 1999. Evaluation of a commercial ELISA for detecting serum antibody to *Neospora caninum* in cattle. *Vet. Rec.* 145, 571–575.

2096. Williams, D. J. L., Trees, A. 2000. Serological diagnosis of neosporosis in the UK. *Int. J. Parasitol.* 30, 879–880.

2097. Williams, D. J. L., Guy, C. S., McGarry, J. W., Guy, F., Tasker, L., Smith, R. F., MacEachern, K., Cripps, P. J., Kelly, D. F., Trees, A. J. 2000. *Neospora caninum*-associated abortion in cattle: The time of experimentally-induced parasitaemia during gestation determines foetal survival. *Parasitology* 121, 347–358.

2098. Williams, D. J. L., Guy, C. S., Smith, R. F., Guy, F., McGarry, J. W., McKay, J. S., Trees, A. J. 2003. First demonstration of protective immunity against foetopathy in cattle with latent *Neospora caninum* infection. *Int. J. Parasitol.* 33, 1059–1065.

2099. Williams, D. J. L., Trees, A. J. 2006. Protecting babies: Vaccine strategies to prevent foetopathy in *Neospora caninum*-infected cattle. *Parasite Immunol.* 28, 61–67.

2100. Williams, D. J. L., Guy, C. S., Smith, R. F., Ellis, J., Björkman, C., Reichel, M. P., Trees, A. J. 2007. Immunization of cattle with live tachyzoites of *Neospora caninum* confers protection against fetal death. *Infect. Immun.* 75, 1343–1348.

2101. Williams, D. J. L., Hartley, C. S., Björkman, C., Trees, A. J. 2009. Endogenous and exogenous transplacental trasmission of *Neospora caninum*—How the route of transmission impacts on epidemiology and control of disease. *Parasitology* 136, 1895–1900.

2102. Williams, J. H., Espie, I., van Wilpe, E., Matthee, A. 2002. Neosporosis in a white rhinoceros (*Ceratotherium simum*) calf. *J. S. Afr. Vet. Assoc.* 73, 38–43.

2103. Williams, J. H., Köster, L. S., Naidoo, V., Odendaal, L., Van Veenhuysen, A., De Wit, M., van Wilpe, E. 2008. Review of idiopathic eosinophilic meningitis in dogs and cats, with a detailed description of two recent cases in dogs. *J. S. Afr. Vet. Assoc.* 79, 194–204.

2104. Wilson, D. J., Orsel, K., Waddington, J., Rajeev, M., Sweeny, A. R., Joseph, T., Grigg, M. E., Raverty, S. A. 2016. *Neospora caninum* is the leading cause of bovine fetal loss in British Columbia, Canada. *Vet. Parasitol.* 218, 46–51.

2105. Winzer, P., Müller, J., Aguado-Martínez, A., Rahman, M., Balmer, V., Manser, V., Ortega-Mora, L. M., Ojo, K. K., Fan, E., Maly, D. J., Van Voorhis, W. C., Hemphill, A. 2015. *In vitro* and *in vivo* effects of the bumped kinase inhibitor 1294 in the related cyst-forming apicomplexans *Toxoplasma gondii* and *Neospora caninum*. *Antimicrob. Agents Chemother.* 59, 6361–6374.

2106. Wisniewski, M., Cabaj, W., Moskwa, B., Wedrychowicz, H. 2002. The first detection of *Neospora caninum* DNA in brains of calves in Poland. *Acta Vet.* 52, 393–400.

2107. Wobeser, B. K., Godson, D. L., Rejmanek, D., Dowling, P. 2009. Equine protozoal myeloencephalitis caused by *Neospora hughesi* in an adult horse in Saskatchewan. *Can. Vet. J.* 50, 851–853.

2108. Wolf, D., Schares, G., Cardenas, O., Huanca, W., Cordero, A., Bärwald, A., Conraths, F. J., Gauly, M., Zahner, H., Bauer, C. 2005. Detection of specific antibodies to *Neospora caninum* and *Toxoplasma gondii* in naturally infected alpacas (*Lama pacos*), llamas (*Lama glama*) and vicuñas (*Lama vicugna*) from Peru and Germany. *Vet. Parasitol.* 130, 81–87.

2109. Wolf, M., Cachin, M., Vandevelde, M., Tipold, A., Dubey, J. P. 1991. Zur klinischen Diagnostik des protozoären Myositissyndroms (*Neospora caninum*) des Welpen. [Clinical diagnosis of protozoal myositis-encephalitis syndrome (*Neospora caninum*) in puppies]. *Tierärztl. Prax.* 19, 302–306.

2110. Wolfe, A., Hogan, S., Maguire, D., Fitzpatrick, C., Vaughan, L., Wall, D., Hayden, T. J., Mulcahy, G. 2001. Red foxes (*Vulpes vulpes*) in Ireland as hosts for parasites of potential zoonotic and veterinary significance. *Vet. Rec.* 149, 759–763.

2111. Woodbine, K. A., Medley, G. F., Moore, S. J., Ramirez-Villaescusa, A., Mason, S., Green, L. E. 2008. A four year longitudinal sero-epidemiology study of *Neospora caninum* in adult cattle from 114 cattle herds in south west England: Associations with age, herd and dam-offspring pairs. *BMC Vet. Res.* 4, 35.

2112. Wooding, F. B. P., Roberts, R. M., Green, J. A. 2005. Light and electron microscope immunocytochemical studies of the distribution of pregnancy associated glycoproteins (PAGs) throughout pregnancy in the cow: Possible functional implications. *Placenta* 26, 807–827.

2113. Woods, L. W., Anderson, M. L., Swift, P. K., Sverlow, K. W. 1994. Systemic neosporosis in a California black-tailed deer (*Odocoileus hemionus columbianus*). *J. Vet. Diagn. Invest.* 6, 508–510.

2114. Wouda, W., van Knapen, F., Visser, I. J. R., Sluijter, F. J. H. 1992. Bovine abortion due to *Neospora*-like protozoa. *Med. Vet.* 9, 35–36.

2115. Wouda, W., van den Ingh, T. S. G. A. M., van Knapen, F., Sluyter, F. J. H., Koeman, J. P., Dubey, J. P. 1992. *Neospora* abortus bij het rund in Nederland. [Bovine abortion due to *Neospora* in The Netherlands]. *Tijdschr. Diergeneeskd.* 117, 599–602.

2116. Wouda, W., de Jong, J. K., van, K. F., Walvoort, H. C. 1993. [*Neospora caninum* as a cause of lameness symptoms in young dogs]. *Tijdschr. Diergeneeskd.* 118, 397–401.

2117. Duffield, T. F., Peregrine, A. S., McEwen, B. J., Hietala, S. 1999. Epidemiology of Neospora infection in Ontario Holstein dairy cows. *Proceedings of the 32 nd Annual Conference of the American Association of Bovine Practioneers.* September 23–23, Nashville, Tennessee. Bovine Practitioner No.32, 238.

2118. Wouda, W., Moen, A. R., Damsma, A., Visser, I. J. R., van Knapen, F. 1994. Bovine neosporosis: 1. Lesions and parasites in aborted fetuses. 2. Repeated transplacental transmission. *Proc. Meet. Eur. Soc. Vet. Pathol.* 12, 29.

2119. Wouda, W. 1997. *Neospora* Abortus bij het Rund. [*Neospora* abortion in cattle]. *Tijdschr. Diergeneeskd.* 122, 446–448.

2120. Wouda, W., Dubey, J. P., Jenkins, M. C. 1997. Serological diagnosis of bovine fetal neosporosis. *J. Parasitol.* 83, 545–547.

2121. Wouda, W., Moen, A. R., Visser, I. J. R., van Knapen, F. 1997. Bovine fetal neosporosis: A comparison of epizootic and sporadic abortion cases and different age classes with regard to lesion severity and immunohistochemical identification of organisms in brain, heart, and liver. *J. Vet. Diagn. Invest.* 9, 180–185.

2122. Wouda, W. 1998. *Neospora* abortion in cattle, aspects of diagnosis and epidemiology. *Doctoral thesis.* University of Utrecht, the Netherlands. 1–176.

2123. Wouda, W., Moen, A. R., Schukken, Y. H. 1998. Abortion risk in progeny of cows after a *Neospora caninum* epidemic. *Theriogenology* 49, 1311–1316.

2124. Wouda, W., Brinkhof, J., van Maanen, C., de Gee, A. L. W., Moen, A. R. 1998. Serodiagnosis of neosporosis in individual cows and dairy herds: A comparative study of three enzyme-linked immunosorbent assays. *Clin. Diagn. Lab. Immunol.* 5, 711–716.

2125. Wouda, W., Bartels, C. J. M., Moen, A. R. 1999. Characteristics of *Neospora caninum*-associated abortion storms in dairy herds in The Netherlands (1995 to 1997). *Theriogenology* 52, 233–245.

2126. Wouda, W., Dijkstra, T., Kramer, A. M. H., van Maanen, C., Brinkhof, J. M. A. 1999. Seroepidemiological evidence for a relationship between *Neospora caninum* infections in dogs and cattle. *Int. J. Parasitol.* 29, 1677–1682.

2127. Wouda, W. 2000. Diagnosis and epidemiology of bovine neosporosis: A review. *Vet. Quart.* 22, 71–74.

2128. Wouda, W., Bartels, C. J. M., Dijkstra, T. 2000. Epidemiology of bovine neosporosis with emphasis on risk factors. *Int. J. Parasitol.* 30, 884–886.

2129. Wouda, W., Bartels, C. J. M., Dijkstra, T. 2000. Some aspects of diagnosis and epidemiology of bovine neosporosis. *Proceedings of Annual ESDAR Conference 1999. Reprod. Dom. Anim.* 6(Suppl), 75–77.

2130. Wouda, W., Dijkstra, T., Kramer, A. M., Bartels, C. J. 2000. [The role of the dog in the epidemiology of neosporosis in cattle]. *Tijdschr. diegeneeskd.* 125, 614–618 (in Dutch).

2131. Wu, J. T., Dreger, S., Chow, E. Y., Bowlby, E. E. 2002. Validation of 2 commercial *Neospora caninum* antibody enzyme linked immunosorbent assays. *Can. J. Vet. Res.* 66, 264–271.

2132. Wu, X. Y., Huang, X. H. 2010. [Research progress on host cell invasion by *Neospora caninum*]. *Chin. Bull. Life Sci.* 22, 873–877 (in Chinese).

2133. Wyss, R., Sager, H., Müller, N., Inderbitzin, F., König, M., Audigé, L., Gottstein, B. 2000. Untersuchungen zum Vorkommen von *Toxoplasma gondii* und *Neospora caninum* unter fleischhygienischen Aspekten. [Distribution of *Toxoplasma gondii* and *Neospora caninum* under aspects of meat hygiene]. *Schweiz. Arch. Tierheilkd.* 142, 95–108.

2134. Xia, H. Y., Zhou, D. H., Jia, K., Zeng, X. B., Zhang, D. W., She, L. X., Lin, R. Q., Yuan, Z. G., Li, S. J., Zhu, X. Q. 2011. Seroprevalence of *Neospora caninum* infection in dairy cattle of southern China. *J. Parasitol.* 97, 172–173.

2135. Xu, M. J., Liu, Q. Y., Fu, J. H., Nisbet, A. J., Shi, D. S., He, X. H., Pan, Y., Zhou, D. H., Song, H. Q., Zhu, X. Q. 2012. Seroprevalence of *Toxoplasma gondii* and *Neospora caninum* infection in dairy cows in subtropical southern China. *Parasitology* 139, 1425–1428.

2136. Yaeger, M. J., Shawd-Wessels, S., Leslie-Steen, P. 1994. *Neospora* abortion storm in a midwestern dairy. *J. Vet. Diagn. Invest.* 6, 506–508.

2137. Yagoob, G. 2012. Seroepidemiology of *Neospora* sp. in horses in East-Azerbaijan Province of Iran. *J. Anim. Vet. Adv.* 11, 480–482.

2138. Yai, L. E. O., Cañon-Franco, W. A., Geraldi, V. C., Summa, M. E. L., Camargo, M. C. G. O., Dubey, J. P., Gennari, S. M. 2003. Seroprevalence of *Neospora caninum* and *Toxoplasma gondii* antibodies in the South American opossum (*Didelphis marsupialis*) from the city of São Paulo, Brazil. *J. Parasitol.* 89, 870–871.

2139. Yai, L. E. O., Ragozo, A. M. A., Cañón-Franco, W. A., Dubey, J. P., Gennari, S. M. 2008. Occurrence of *Neospora caninum* antibodies in capybaras (*Hydrochaeris hydrochaeris*) from São Paulo State, Brazil. *J. Parasitol.* 94, 766.

2140. Yakhchali, M., Javadi, S., Morshedi, A. 2010. Prevalence of antibodies to *Neospora caninum* in stray dogs of Urmia, Iran. *Parasitol. Res.* 106, 1455–1458.

2141. Yamage, M., Flechtner, O., Gottstein, B. 1996. *Neospora caninum*: Specific oligonucleotide primers for the detection of brain "cyst" DNA of experimentally-infected nude mice by the polymerase chain reaction (PCR). *J. Parasitol.* 82, 272–279.

2142. Yamane, I., Thomford, J. W., Gardner, I. A., Dubey, J. P., Levy, M., Conrad, P. A. 1993. Evaluation of the indirect fluorescent antibody test for diagnosis of *Babesia gibsoni* infections in dogs. *Am. J. Vet. Res.* 54, 1579–1584.

2143. Yamane, I., Kokuho, T., Shimura, K., Eto, M., Haritani, M., Ouchi, Y., Sverlow, K. W., Conrad, P. A. 1996. *In vitro* isolation of a bovine *Neospora* in Japan. *Vet. Rec.* 138, 652.

2144. Yamane, I., Kokuho, T., Shimura, K., Eto, M., Shibahara, T., Haritani, M., Ouchi, Y., Sverlow, K., Conrad, P. A. 1997. *In vitro* isolation and characterisation of a bovine *Neospora* species in Japan. *Res. Vet. Sci.* 63, 77–80.

2145. Yamane, I., Shibahara, T., Kokuho, T., Shimura, K., Hamaoka, T., Haritani, M., Conrad, P. A., Park, C. H., Sawada, M., Umemura, T. 1998. An improved isolation technique for bovine *Neospora* species. *J. Vet. Diagn. Invest.* 10, 364–368.

2146. Yamane, I., Kitani, H., Kokuho, T., Shibahara, T., Haritani, M., Hamaoka, T., Shimizu, S., Koiwai, M., Shimura, K., Yokomizo, Y. 2000. The inhibitory effect of interferon gamma and tumor necrosis factor alpha on intracellular multiplication of *Neospora caninum* in primary bovine brain cells. *J. Vet. Med. Sci.* 62, 347–351.

2147. Yang, D., Liu, J., Hao, P., Wang, J., Lei, T., Shan, D., Liu, Q. 2015. MIC3, a novel cross-protective antigen expressed in *Toxoplasma gondii* and *Neospora caninum*. *Parasitol. Res.* 114, 3791–3799.

2148. Yang, Y., Zhang, Q., Kong, Y., Ying, Y., Kwok, O. C. H., Liang, H., Dubey, J. P. 2014. Low prevalence of *Neospora caninum* and *Toxoplasma gondii* antibodies in dogs in Jilin, Henan and Anhui Provinces of the People's Republic of China. *BMC Vet. Res.* 10, 295.

2149. Yániz, J. L., López-Gatius, F., Almería, S., Carretero, T., García-Ispierto, I., Serrano, B., Smith, R. F., Dobson, H., Santolaria, P. 2009. Dynamics of heat shock protein 70 concentrations in peripheral blood lymphocyte lysates during pregnancy in lactating Holstein-Friesian cows. *Theriogenology* 72, 1041–1046.

2150. Yániz, J. L., López-Gatius, F., García-Ispierto, I., Bech-Sàbat, G., Serrano, B., Nogareda, C., Sanchez-Nadal, J. A., Almería, S., Santolaria, P. 2010. Some factors affecting the abortion rate in dairy herds with high incidence of *Neospora*-associated abortions are different in cows and heifers. *Reprod. Dom. Anim.* 45, 699–705.

2151. Yao, L., Yang, N., Liu, Q., Wang, M., Zhang, W., Qian, W. F., Hu, Y. F., Ding, J. 2009. Detection of *Neospora caninum* in aborted bovine fetuses and dam blood samples by nested PCR and ELISA and seroprevalence in Beijing and Tianjin, China. *Parasitology* 136, 1251–1256.

2152. Ybañez, R. H., Terkawi, M. A., Kameyama, K., Xuan, X., Nishikawa, Y. 2013. Identification of a highly antigenic region of subtilisin-like serine protease 1 for serodiagnosis of *Neospora caninum* infection. *Clin. Vaccine Immunol.* 20, 1617–1622.

2153. Ybañez, R. H. D., Leesombun, A., Nishimura, M., Matsubara, R., Kojima, M., Sakakibara, H., Nagamune, K., Nishikawa, Y. 2016. *In vitro* and *in vivo* effects of the phytohormone inhibitor fluridone against *Neospora caninum* infection. *Parasitol. Int.* 65, 319–322.

2154. Yeargan, M. R., Alvarado-Esquivel, C., Dubey, J. P., Howe, D. K. 2013. Prevalence of antibodies to *Sarcocystis neurona* and *Neospora hughesi* in horses from Mexico. *Parasite* 20, 29.

2155. Yildiz, K., Kul, O., Babur, C., Kilic, S., Gazyagci, A. N., Celebi, B., Gurcan, I. S. 2009. Seroprevalence of *Neospora caninum* in dairy cattle ranches with high abortion rate: Special emphasis to serologic co-existence with *Toxoplasma gondii, Brucella abortus* and *Listeria monocytogenes. Vet. Parasitol.* 164, 306–310.

2156. Yildiz, K., Yasa Duru, S., Yagci, B. B., Babur, C., Ocal, N., Gurcan, S., Karaca, S. 2009. Seroprevalence of *Neospora caninum* and coexistence with *Toxoplasma gondii* in dogs. *Türkiye Parazitol. Derg.* 33, 116–119.

2157. Yin, J., Qu, G., Cao, L., Li, Q., Fetterer, R., Feng, X., Liu, Q., Wang, G., Qi, D., Zhang, X., Miramontes, E., Jenkins, M., Zhang, N., Tuo, W. 2012. Characterization of *Neospora caninum* microneme protein 10 (NcMIC10) and its potential use as a diagnostic marker for neosporosis. *Vet. Parasitol.* 187, 28–35.

2158. Yoshimoto, M., Otsuki, T., Itagaki, K., Kato, T., Kohsaka, T., Matsumoto, Y., Ike, K., Park, E. Y. 2015. Evaluation of recombinant *Neospora caninum* antigens purified from silkworm larvae for the protection of *N. caninum* infection in mice. *J. Biosci. Bioeng.* 120, 715–719.

2159. Youn, H. J., Lakritz, J., Kim, D. Y., Rottinghaus, G. E., Marsh, A. E. 2003. Anti-protozoal efficacy of medicinal herb extracts against *Toxoplasma gondii* and *Neospora caninum. Vet. Parasitol.* 116, 7–14.

2160. Youn, H. J., Lakritz, J., Rottinghaus, G. E., Seo, H. S., Kim, D. Y., Cho, M. H., Marsh, A. E. 2004. Anti-protozoal efficacy of high performance liquid chromatography fractions of *Torilis japonica* and *Sophora flavescens* extracts on *Neospora caninum* and *Toxopasma gondii. Vet. Parasitol.* 125, 409–414.

2161. Youssefi, M. R., Arabkhazaeli, F., Hassan, A. T. M. 2009. Seroprevalence of *Neospora caninum* infection in rural and industrial cattle in northern Iran. *Iranian J. Parasitol.* 4, 20–23.

2162. Yu, J., Liu, Q., Wang, M. 2006. [Subcloning and expression of the NcSRS2 gene fragment of *Neospora caninum*]. *Chin. J. Vet. Med.* 42, 3–6 (in Chinese).

2163. Yu, J., Xia, Z., Liu, Q., Liu, J., Ding, J., Zhang, W. 2007. Seroepidemiology of *Neospora caninum* and *Toxoplasma gondii* in cattle and water buffaloes (*Bubalus bubalis*) in the People's Republic of China. *Vet. Parasitol.* 143, 79–85.

2164. Yu, J. H., Liu, Q., Xia, Z. F. 2006. Epidemiological investigation of infections with *Neospora caninum* and *Toxoplasma gondii* in cattle and water buffaloes. *Vet. Sci. China* 36, 247–251.

2165. Yu, X. L., Chen, N. H., Hu, D. M., Zhang, W., Li, X. X., Wang, B. Y., Kang, L. P., Li, X. D., Liu, Q., Tian, K. G. 2009. Detection of *Neospora caninum* from farm-bred young blue foxes (*Alopex lagopus*) in China. *J. Vet. Med. Sci.* 71, 113–115.

2166. Zanet, S., Palese, V., Trisciuoglio, A., Cantón Alonso, C., Ferroglio, E. 2013. *Encephalitozoon cuniculi, Toxoplasma gondii* and *Neospora caninum* infection in invasive eastern cottontail rabbits *Sylvilagus floridanus* in Northwestern Italy. *Vet. Parasitol.* 197, 682–684.

2167. Zanet, S., Sposimo, P., Trisciuoglio, A., Giannini, F., Strumia, F., Ferroglio, E. 2014. Epidemiology of *Leishmania infantum, Toxoplasma gondii*, and *Neospora caninum* in *Rattus rattus* in absence of domestic reservoir and definitive hosts. *Vet. Parasitol.* 199, 247–249.

2168. Zanette, M. F., Lima, V. M. F., Laurenti, M. D., Rossi, C. N., Vides, J. P., Vieira, R. F. C., Biondo, A. W., Marcondes, M. 2014. Serological cross-reactivity of *Trypanosoma cruzi*, *Ehrlichia canis*, *Toxoplasma gondii*, *Neospora caninum* and *Babesia canis* to *Leishmania infantum chagasi* tests in dogs. *Rev. Soc. Bras. Med. Trop.* 47, 105–107.

2169. Zhai, Y. Q., Zhao, J. P., Zhu, X. Q., Li, L., Wang, C. R. 2007. Research advances in the diagnosis of cattle neosporosis. *J. Anim. Vet. Adv.* 6, 1377–1387.

2170. Zhang, C. S., Liu, Q. 2006. [Serological epidemiology of *Neospora caninum* from dairy cattle]. *Chin. J. Vet. Med.* 42, 3–5 (in Chinese).

2171. Zhang, G., Huang, X., Boldbaatar, D., Battur, B., Battsetseg, B., Zhang, H., Yu, L., Li, Y., Luo, Y., Cao, S., Goo, Y. K., Yamagishi, J., Zhou, J., Zhang, S., Suzuki, H., Igarashi, I., Mikami, T., Nishikawa, Y., Xuan, X. 2010. Construction of *Neospora caninum* stably expressing TgSAG1 and evaluation of its protective effects against *Toxoplasma gondii* infection in mice. *Vaccine* 28, 7243–7247.

2172. Zhang, H., Compaore, M. K. A., Lee, E. G., Liao, M., Zhang, G., Sugimoto, C., Fujisaki, K., Nishikawa, Y., Zuan, X. 2007. Apical membrane antigen 1 is a cross-reactive antigen between *Neospora caninum* and *Toxoplasma gondii*, and the anti-NcAMA1 antibody inhibits host cell invasion by both parasites. *Mol. Biochem. Parasitol.* 151, 205–212.

2173. Zhang, H., Lee, E. G., Liao, M., Compaore, M. K. A., Zhang, G., Kawase, O., Fujisaki, K., Sugimoto, C., Nishikawa, Y., Xuan, X. 2007. Identification of ribosomal phosphoprotein P0 of *Neospora caninum* as a potential common vaccine candidate for the control of both neosporosis and toxoplasmosis. *Mol. Biochem. Parasitol.* 153, 141–148.

2174. Zhang, H., Nishikawa, Y., Yamagishi, J., Zhou, J., Ikehara, Y., Kojima, N., Yokoyama, N., Xuan, X. 2010. *Neospora caninum*: Application of apical membrane antigen 1 encapsulated in the oligomannose-coated liposomes for reduction of offspring mortality from infection in BALB/c mice. *Exp. Parasitol.* 125, 130–136.

2175. Zhang, H., Lee, E. G., Yu, L., Kawano, S., Huang, P., Liao, M., Kawase, O., Zhang, G., Zhou, J., Fujisaki, K., Nishikawa, Y., Xuan, X. 2011. Identification of the cross-reactive and species-specific antigens between *Neospora caninum* and *Toxoplasma gondii* tachyzoites by a proteomics approach. *Parasitol. Res.* 109, 899–911.

2176. Zhang, W., Deng, C., Liu, Q., Liu, J., Wang, M., Tian, K. G., Yu, X. L., Hu, D. M. 2007. First identification of *Neospora caninum* infection in aborted bovine foetuses in China. *Vet. Parasitol.* 149, 72–76.

2177. Zhao, X., Duszynski, D. W., Loker, E. S. 2001. A simple method of DNA extraction for *Eimeria* species. *J. Microbiol. Methods* 44, 131–137.

2178. Zhao, Z., Ding, J., Liu, Q., Wang, M., Yu, J., Zhang, W. 2009. Immunogenicity of a DNA vaccine expressing the *Neospora caninum* surface protein NcSRS2 in mice. *Acta Vet. Hung.* 57, 51–62.

2179. Zhao, Z. Z., Liu, Q., Wang, M. 2005. [Construction of an eukaryotic expression plasmid for the NcSRS2 gene of *Neospora caninum*]. *Acta Vet. Zootech. Sinica* 36, 819–822 (in Chinese).

2180. Zhou, M., Cao, S., Sevinc, F., Sevinc, M., Ceylan, O., Liu, M., Wang, G., Adjou Moumouni, P. F., Jirapattharasate, C., Suzuki, H., Nishikawa, Y., Xuan, X. 2016. Enzyme-linked immunosorbent assays using recombinant TgSAG2 and NcSAG1 to detect *Toxoplasma gondii* and *Neospora caninum*-specific antibodies in domestic animals in Turkey. *J. Vet. Med. Sci.*

2181. Zhu, B. Y., Hartigan, A., Reppas, G., Higgins, D. P., Canfield, P. J., Šlapeta, J. 2009. Looks can deceive: Molecular identity of an intraerythrocytic apicomplexan parasite in Australian gliders. *Vet. Parasitol.* 159, 105–111.

2182. Zimpel, C. K., Grazziotin, A. L., de Barros, I. R., Guimaraes, A. M. S., dos Santos, L. C., de Moraes, W., Cubas, Z. S., de Oliveira, M. J., Pituco, E. M., Lara, M. C. C. S. H., Villalobos, E. M. C., Silva, L. M. P., Cunha, E. M. S., Castro, V., Biondo, A. W. 2015. Occurrence of antibodies anti-*Toxoplasma gondii*, *Neospora caninum* and *Leptospira interrogans* in a captive deer herd in Southern Brazil. *Braz. J. Vet. Parasitol.* 24, 482–487.

2183. Zintl, A., Pennington, S. R., Mulcahy, G. 2006. Comparison of different methods for the solubilisation of *Neospora caninum* (Phylum *Apicomplexa*) antigen. *Vet. Parasitol.* 135, 205–213.

Index

Printed and bound by CPI Group (UK) Ltd, Croydon, CR0 4YY

24/10/2024

01778290-0013